Interactive Student Edition

Volume 1

mheducation.com/prek-12

Cover: (t to b, l to r) Kenny McCartney/Moment Open/Getty Images; YinYang/E+/Getty Images; nycshooter/Vetta/Getty Images; michaelgzc/E+/Getty Images

Send all inquiries to:
McGraw-Hill Education
8787 Orion Place
Columbus, OH 43240

ISBN: 978-0-07-662601-4 (*Interactive Student Edition*, Volume 1)
MHID: 0-07-662601-6 (*Interactive Student Edition*, Volume 1)
ISBN: 978-0-07-899749-5 (*Interactive Student Edition*, Volume 2)
MHID: 0-07-899749-6 (*Interactive Student Edition*, Volume 2)

Printed in the United States of America.

9 10 11 12 WEB 27 26 25 24 23 22

Contents in Brief

Reveal AGA® Makes Math Meaningful...

Learning on the Go!

The flexible approach of *Reveal Algebra 1, Reveal Geometry, and Reveal Algebra 2* (*Reveal AGA*) can work for you using digital only or digital and your *Interactive Student Edition* together.

Interactive Student Edition

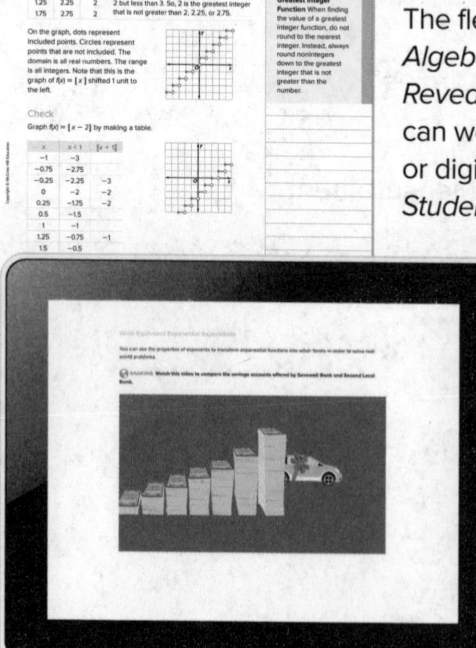

Student Digital Center

...to Reveal YOUR Full Potential!

Reveal AGA® Brings Math to Life in Every Lesson

Reveal AGA is a blended print and digital program that supports access on the go. You'll find the *Interactive Student Edition* aligns to the Student Digital Center, so you can record your digital observations in class and reference your notes later, or access just the digital center, or a combination of both! The Student Digital Center provides access to the interactive lessons, interactive content, animations, videos, and technology-enhanced practice questions.

Write down your username and password here

Username: ____________________________

Password: ____________________________

Go Online! my.mheducation.com

Web Sketchpad® Powered by The Geometer's Sketchpad®- Dynamic, exploratory, visual activities embedded at point of use within the lesson.

Animations and Videos – Learn by seeing mathematics in action.

Interactive Tools – Get involved in the content by dragging and dropping, selecting, highlighting, and completing tables.

Personal Tutors – See and hear a teacher explain how to solve problems.

eTools – Math tools are available to help you solve problems and develop concepts.

TABLE OF CONTENTS

Module 1

Tools of Geometry

Module 2

Angles and Geometric Figures

TABLE OF CONTENTS

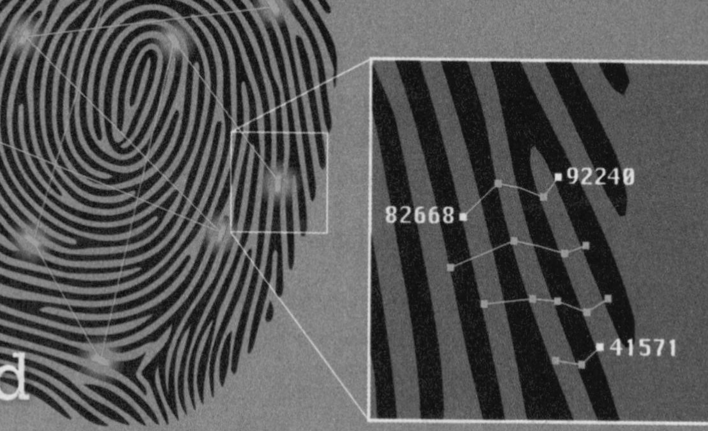

Module 3

Logical Arguments and Line Relationships

Module 4

Transformations and Symmetry

TABLE OF CONTENTS

Module 5

Triangles and Congruence

Module 6

Relationships in Triangles

TABLE OF CONTENTS

Module 7

Quadrilaterals

Module 8

Similarity

Module 9
Right Triangles and Trigonometry

Module 10

Circles

Module 11

Measurement

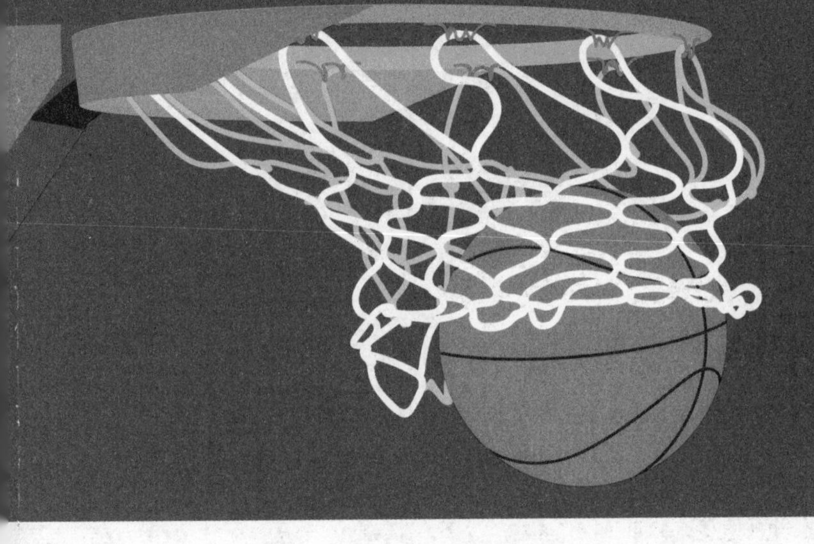

Module 12

Probability

Module 1

Tools of Geometry

Essential Question

How are points, lines, and segments used to model the real world?

What Will You Learn?

Place a check mark (✓) in each row that corresponds with how much you already know about each topic **before** starting this module.

KEY

— I don't know. — I've heard of it. — I know it!

	Before			After		
	I don't know.	I've heard of it.	I know it!	I don't know.	I've heard of it.	I know it!
analyze axiomatic systems and identify types of geometry						
analyze figures to identify points, lines, planes, and intersections of lines and planes						
find measures of line segments						
apply the Distance Formula to find lengths of line segments						
find points that partition directed line segments on number lines						
find points that partition directed line segments on the coordinate plane						
find midpoints and bisect line segments						

Foldables Make this Foldable to help you organize your notes about geometric concepts. Begin with four sheets of 11″ × 17″ paper.

1. **Fold** the four sheets of paper in half.
2. **Cut** along the top fold of the papers. Staple along the side to form a book.
3. **Cut** the right sides of each paper to create a tab for each lesson.
4. **Label** each tab with a lesson number.

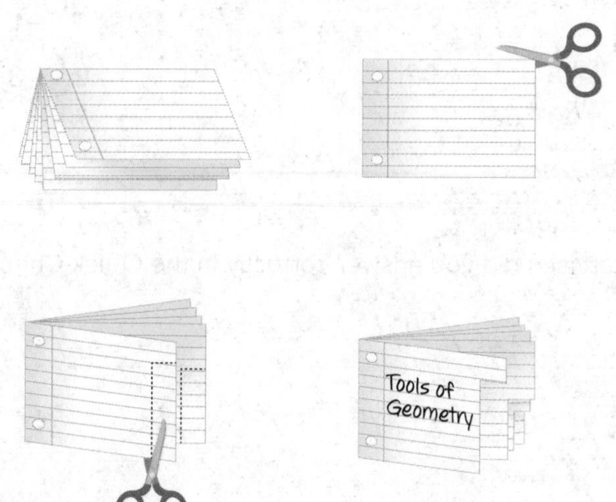

What Vocabulary Will You Learn?

Check the box next to each vocabulary term that you may already know.

- ☐ analytic geometry
- ☐ axiom
- ☐ axiomatic system
- ☐ betweenness of points
- ☐ bisect
- ☐ collinear
- ☐ congruent
- ☐ congruent segments
- ☐ coplanar
- ☐ defined term
- ☐ definition
- ☐ directed line segment
- ☐ distance
- ☐ equidistant
- ☐ fractional distance
- ☐ intersection
- ☐ line
- ☐ line segment
- ☐ midpoint
- ☐ plane
- ☐ point
- ☐ postulate
- ☐ segment bisector
- ☐ space
- ☐ synthetic geometry
- ☐ theorem
- ☐ undefined terms

Are You Ready?

Complete the Quick Review to see if you are ready to start this module.
Then complete the Quick Check.

Quick Review

Example 1

Graph and label the point Q(−3, 4) in the coordinate plane.

Start at the origin. Because the *x*-coordinate is negative, move 3 units to the left. Then move 4 units up because the *y*-coordinate is positive. Draw a dot and label it *Q*.

Example 2

Evaluate the expression $[-2-(-7)]^2+(1-8)^2$.

Follow the order of operations.

$[-2-(-7)]^2+(1-8)^2$

$= 5^2+(-7)^2$ Subtract in parentheses.

$= 25+49$ Evaluate exponents.

$= 74$ Add.

Quick Check

Graph and label each point on the coordinate plane.

1. $W(-5, 2)$
2. $X(0, 4)$
3. $Y(-3, -1)$
4. $Z(4, -2)$

Evaluate each expression.

5. $(4-2)^2+(7-3)^2$
6. $(-5-3)^2+(3-4)^2$
7. $[-1-(-9)]^2+(5-3)^2$
8. $[-3-(-4)]^2+[-1-(-6)]^2$

How did you do?

Which exercises did you answer correctly in the Quick Check? Shade those exercise numbers below.

① ② ③ ④ ⑤ ⑥ ⑦ ⑧

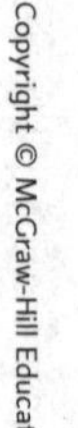

The Geometric System

Today's Goals
- Apply axioms to draw conclusions.
- Identify examples of synthetic and analytic geometry.

Today's Vocabulary
axiomatic system
undefined terms
definition
defined term
axiom
postulate
theorem
synthetic geometry
analytic geometry

Explore Using a Game to Explore Axiomatic Systems

Online Activity Use a real-world situation to complete the Explore.

INQUIRY What are the characteristics of a good set of rules?

Learn The Axiomatic System of Geometry

Geometry is an axiomatic system based on logical reasoning and axioms.

The Axiomatic System of Geometry	
An **axiomatic system** has a set of axioms from which theorems can be derived.	
undefined terms	words, usually readily understood, that are not formally explained by means of more basic words and concepts
definition	assigns properties to a mathematical object
defined term	a term that has a definition and can be explained using undefined terms and/or defined terms
axiom or **postulate**	statement that is accepted as true without proof
theorem	statement or conjecture that can be proven true using undefined terms, definitions, and axioms

Undefined terms are used to write definitions. Undefined terms and definitions are used to create axioms. Undefined terms, definitions, and axioms are used to prove theorems.

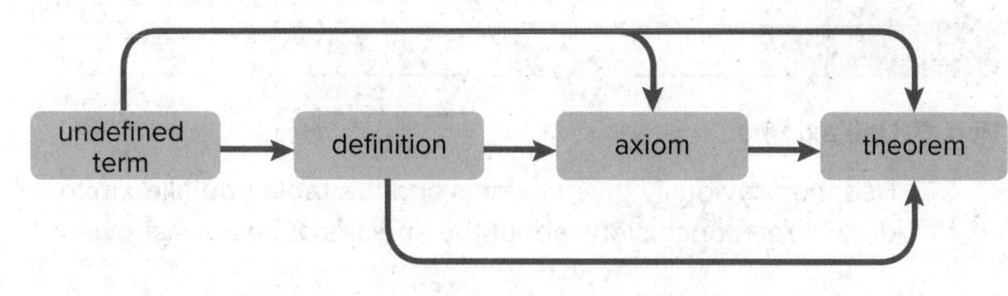

One real-world axiomatic system that is probably familiar to you is the set of rules to a game. The rules are the axioms, and they are used to evaluate the legality of each play.

Math History Minute

Thales (c. 624–546 B.C.) was a Greek mathematician, philosopher, and astronomer, and is the first known individual attributed with a mathematical discovery. He inspired Euclid, Plato, and Aristotle, who considered him to be the first philosopher in the Greek tradition.

Your Notes

Example 1 Apply an Axiomatic System

ANIMALS **In the fictional country of Rythoth, blue animals are from the mountains, and red animals are from the valleys. These animals are categorized into three distinct classes: mammals, birds, and reptiles. Mammals are covered by hair or fur, birds are covered by feathers, and reptiles are covered by scales.**

Talk About It!
What conclusion cannot be made from the provided axioms?

Part A Categorize the animals.

Write the name of each animal in the corresponding categories in the table.

Birthplace	Mammal	Bird	Reptile
Mountains			
Valleys			

Part B Use axioms.

Use the previously given axioms and the table you filled in to draw three conclusions about the species of animals shown.

- The ____ is a mammal from the mountains of Rythoth.
- The ____ is a reptile from the valleys of Rythoth.
- The ____ is a bird from the valleys of Rythoth.

Study Tip

Theorems Theorems, or conclusions, made from a set of axioms must be true in every situation. It takes only one example that contradicts the conjecture to show that a theorem or conclusion is not true.

Check

PLANETS The fictional galaxy of Yogul contains at least 20 planets including Mothera, Sothera, and Kothera. An animal can live on any planet in the Yogul galaxy that contains its biome. Lizards and scorpions live in the desert. Frogs and monkeys live in tropical forests. Bears and foxes can be found in the tundra. The biomes of each planet are permanent and will not change over time.

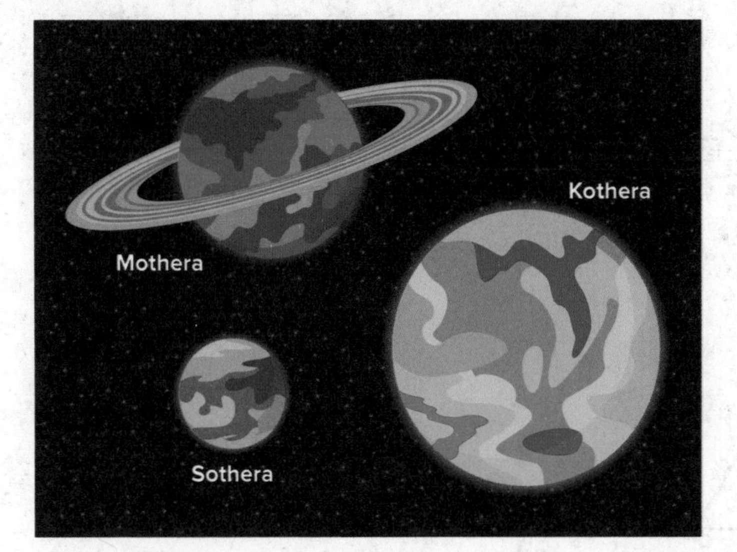

Color	Biome
	desert
	tropical forest
	tundra

Use the axioms given to determine what conclusions can be made about the planets of Yogul. Select all that apply. ____

A. Bears and foxes can live on Sothera.

B. Lizards and scorpions can only live on Mothera.

C. Only frogs and monkeys can survive on Kothera.

D. Bears and foxes can survive on Sothera at temperatures as low as −20°F.

E. All animals can live on Kothera.

F. Scorpions and lizards can live on Mothera.

Learn Types of Geometry

There are several types of geometry that are built upon different sets of postulates including synthetic geometry and analytic geometry.

Synthetic geometry is the study of geometric figures without the use of coordinates. Synthetic geometry is sometimes called *pure geometry* or *Euclidean geometry*.	**Analytic geometry** is the study of geometry using a coordinate system. Analytic geometry is sometimes called *coordinate geometry* or *Cartesian geometry*.

Think About It!

What is an advantage of using analytic geometry instead of synthetic geometry?

Go Online You can complete an Extra Example online.

Example 2 Identify Types of Geometry

Classify each figure as illustrating *synthetic geometry* or *analytic geometry*.

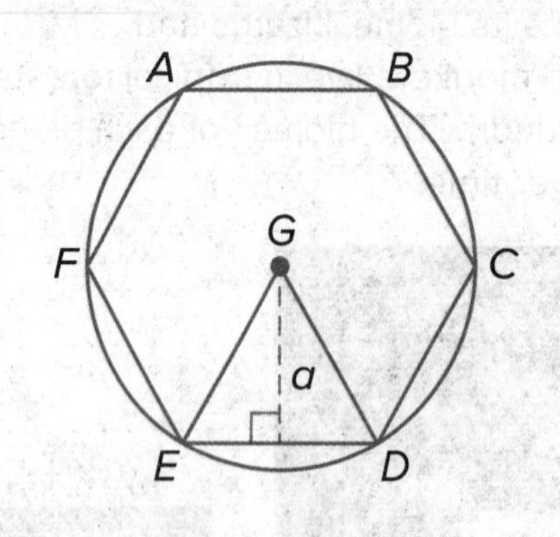

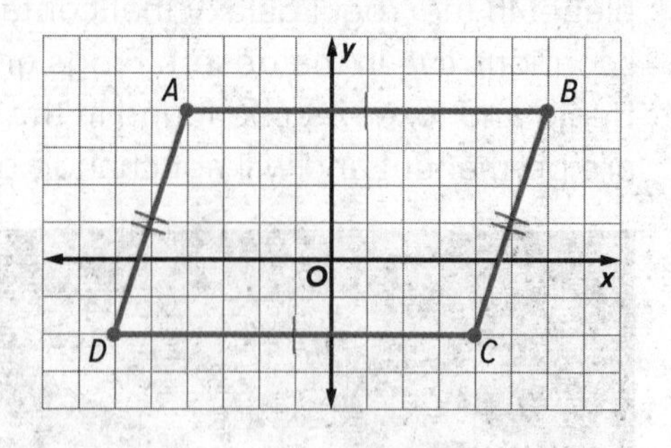

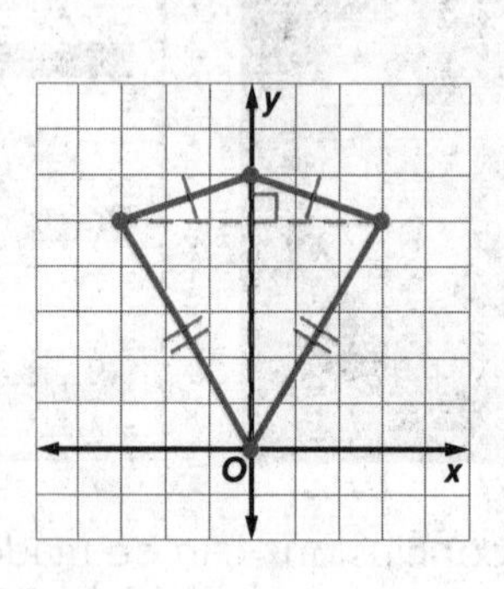

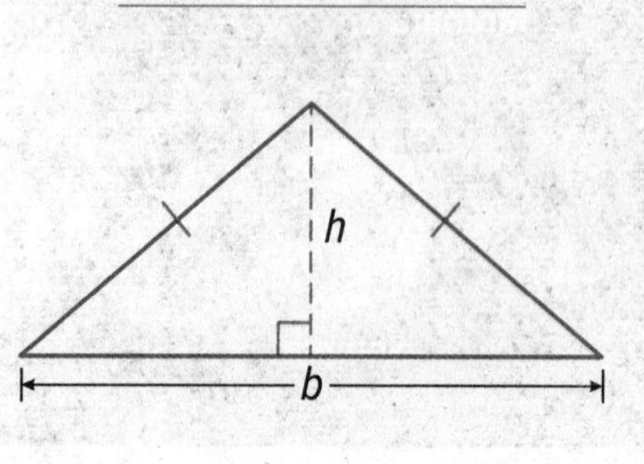

Check

Classify each figure as illustrating *synthetic geometry* or *analytic geometry*.

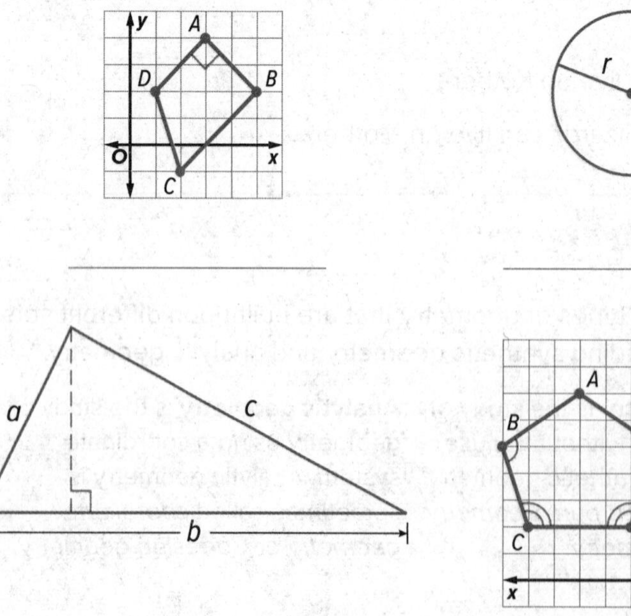

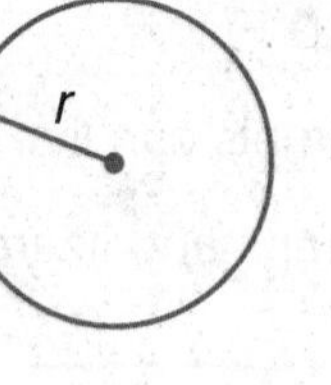

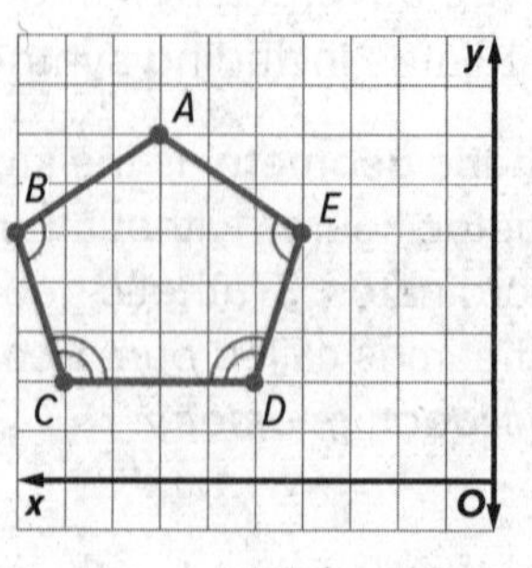

Practice

Go Online You can complete your homework online.

Example 1

1. **BASKETBALL** The Badgers' basketball team has 10 players. During practice, half of the players wear red jerseys numbered 1-5, and the other half wear yellow jerseys numbered 6-10. The yellow team wins the practice game 32-26.
   - Kylie wears number 5 and scores 9 points.
   - Kelsey's team wins the game.
   - Marie and Kylie are on opposing teams.

   Use the axioms to make three conclusions about the game played.

2. **PRINTING** Rico's T-shirt Company sells customized short sleeve T-shirts, long sleeve T-shirts, and sweatshirts. Each type of shirt sells in multiples of 5. It costs \$25.00 for 5 short sleeve T-shirts, \$30.00 for 5 long sleeve T-shirts, and \$40.00 for 5 sweatshirts. Short sleeve and long sleeve T-shirts can be made in any color except navy or black. Sweatshirts are only made in navy and black.
   - Mercedes bought green shirts for \$55.00.
   - Quinn bought 10 navy sweatshirts.
   - Rachel paid \$30.00 for several red shirts.
   - Hector bought black and yellow shirts for \$65.00.

   Use the axioms to make four conclusions about the shirts sold.

3. **LANDSCAPING** Tom owns a landscaping business. He charges \$40 for a yard cleanup, \$50 to mow a lawn, and \$75 to mulch a yard. On average, it takes Tom 25 minutes for a yard cleanup, 40 minutes to mow a lawn, and 2 hours to mulch a yard. Tom's clients are Mr. Hansen, Ms. Martinez, and Mrs. Johnson.
   - Mr. Hansen paid \$125 for lawn services this week.
   - Tom spent more than an hour at Ms. Martinez' house this week.
   - Mrs. Johnson wrote Tom a check for \$165 for the week.
   - Tom made \$405 from his three clients this week.

   Use the axioms to make four conclusions about the landscaping that Tom did.

4. **CUPCAKES** Olivia's Cupcake Shoppe sells small and large cupcakes in three flavors.
   - Niamh paid \$3 for a cupcake with buttercream icing.
   - Bethany bought a small vanilla cupcake.
   - Mateo paid \$3.50 for a cupcake with strawberry icing and a chocolate cupcake.

   Use the axioms to make two conclusions about the cupcakes that were purchased.

Olivia's
cupcake shoppe
Flavors
• Chocolate with vanilla icing •
• Vanilla with strawberry icing •
• Strawberry with buttercream icing •
Sizes
Small........$1.75 Large........$3.00

Example 2

Classify each figure as illustrating *synthetic geometry* or *analytic geometry*.

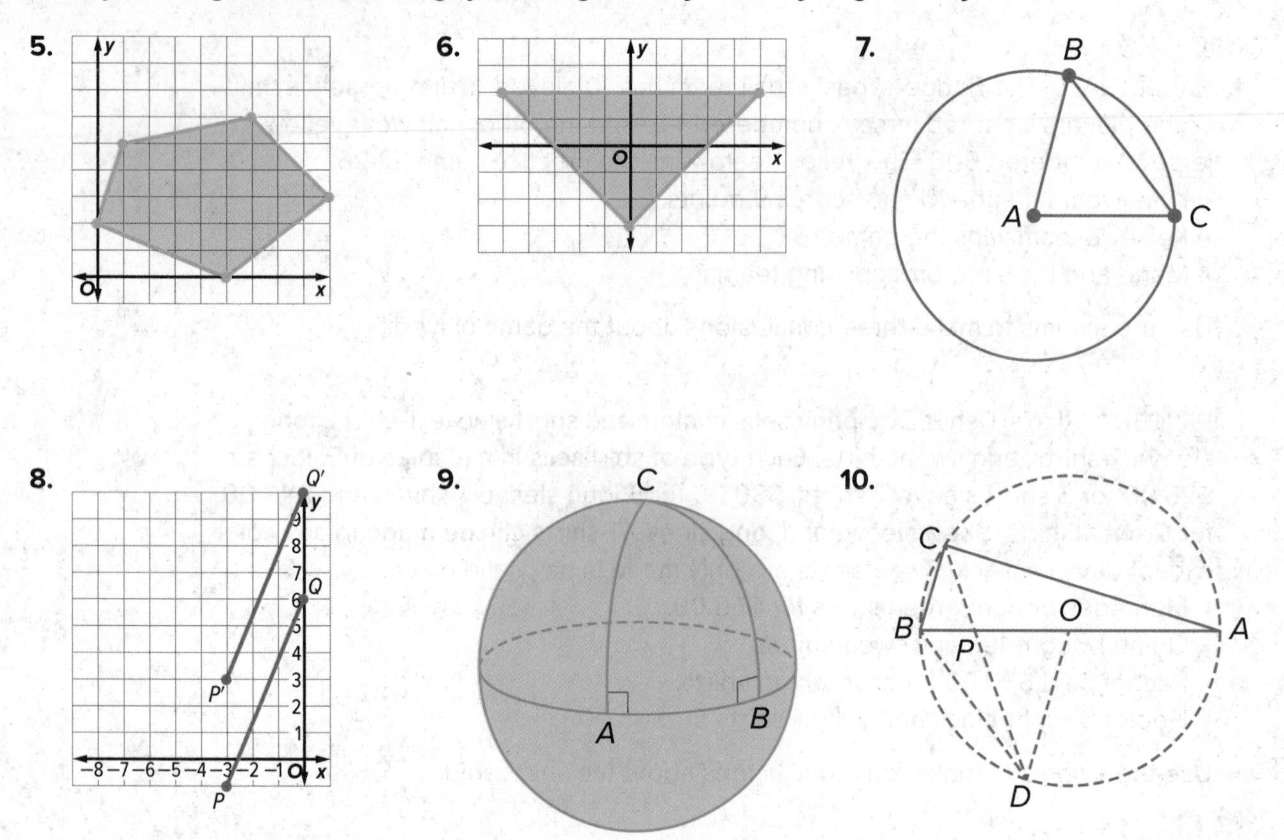

Mixed Exercises

11. RESTAURANT Damon sells three types of salads at his restaurant: cobb, wedge, and spinach. Each salad is served with 2 dinner rolls. The price of the cobb salad is $7.99, the price of the wedge salad is $8.99, and the price of the spinach salad is $5.99. Grilled chicken can be added to any salad for an additional $2.00.

- Malik spent $7.99 on a salad.
- Pedro and Deandra each spent $8.99 on their salads.
- Rafael ate a wedge salad.
- Drake did not add chicken to his salad.

Use the axioms to make a conclusion about the salads that are eaten.

12. CLASSROOM Mrs. Fields teaches high school geometry. Her classroom tools include a compass, straightedge, pencil, and protractor. Does Mrs. Fields likely teach *analytic geometry* or *synthetic geometry*? Explain your reasoning.

13. REASONING Theo is stuck on a problem on a test. The problem is asking him to use a given formula to find the distance between two points on a graph. Is Theo using analytic *geometry* or *synthetic geometry*? Explain your reasoning.

14. USE A SOURCE Survey a group of students in your classroom about favorite colors. Write three axioms about the data you collected. Then use your axioms to write a conclusion. Explain your reasoning.

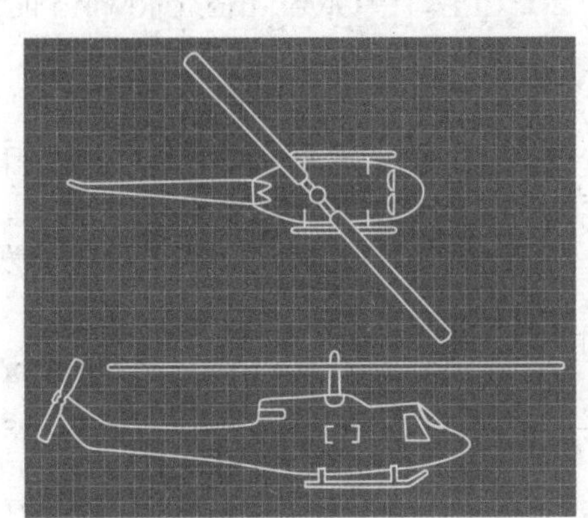

15. STATE YOUR ASSUMPTION Sydney is an engineer. She is using a blueprint for a project that is drawn on a grid, as shown. Is Sydney likely using *analytic geometry* or *synthetic geometry*? Explain any assumptions that you make.

16. Mr. Sail assigns a project where students identify shapes that represent real-world objects. Is this an example of *analytic geometry* or *synthetic geometry*? Explain your reasoning.

17. CONSTRUCT ARGUMENTS Consider the following axiomatic system for bus routes.

- Each bus route lists the stops in the order at which they are visited by the bus.
- Each route visits at least four distinct stops.
- No route visits the same stop twice, except for the first stop, which is always the same as the last stop.
- There is a stop called Downtown, which is visited by each route.
- Every stop other than Downtown is visited by at most two routes.

The city has stops at Downtown, King St, Maxwell Ave, Stadium District, State St, Grace Blvd, and Charlotte Ave. Are the following three routes a model for the axiomatic system? Justify your argument.
ROUTE 1: Downtown, King St, Stadium District, State St, Downtown
ROUTE 2: Stadium District, State St, Grace Blvd, Maxwell Ave, Downtown, Stadium District
ROUTE 3: King St, Stadium District, Downtown, Maxwell Ave, Stadium District, King St

18. SHOPPING The Clothing Shop is having a sale. All clothes are 20% off, and all accessories are 30% off.

- Jaisa bought two necklaces.
- Sheree bought a shirt and a purse.

Use the axioms to make one conclusion about Jaisa or Sheree's purchases.

19. WRITE Write a comparison of the rules and plays of a game and the elements of an axiomatic system. Then choose a game or sport for which you know the rules. Explain a rule from the game or sport and a play from the game. Does the play violate or fall within the rule? Explain.

20. **CREATE** Given the following list of axioms, draw a model to properly represent the information.

- There exist five points.
- Each line contains only these five points.
- There exist two lines.
- Each line contains at least two points.

21. **WHICH ONE DOESN'T BELONG?** Three-point geometry is a finite subset of geometry with the following four axioms:

- There exists exactly three distinct points.
- Each pair of distinct points are on exactly one line.
- Not all the points are on the same line.
- Each pair of distinct lines intersect in at least one point.

Which of the following does not satisfy all the axioms of three-point geometry? Justify your conclusion.

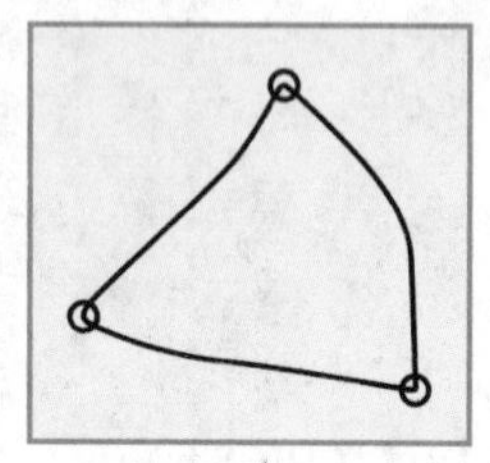

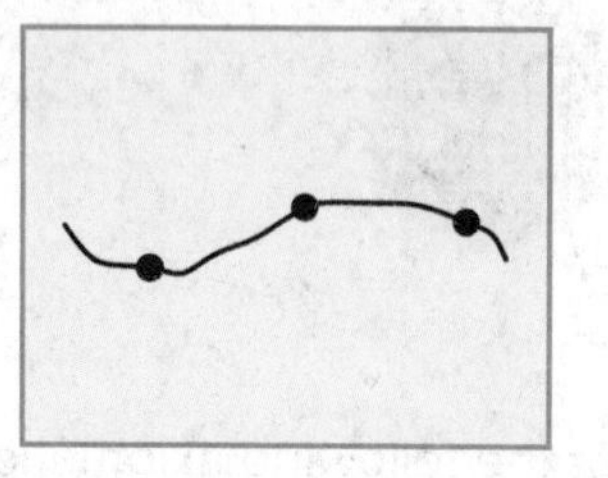

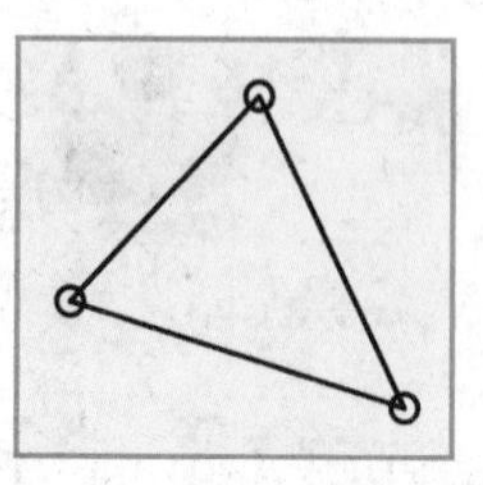

22. **FIND THE ERROR** Grant read the following axioms for a video game he is playing.

- There are four keys hidden on each level.
- Each level ends when the player collects the third key.
- The game has 10 levels.

From these axioms, Grant concluded:

- to complete the game, he will need to find 30 keys.
- there are 40 keys in the game.
- he can collect all 40 keys in the game.

Are Grant's conclusions correct? Explain your reasoning.

23. **WHICH ONE DOESN'T BELONG?** Using your understanding of analytic and synthetic geometry, which of the following figures does not belong? Justify your conclusion.

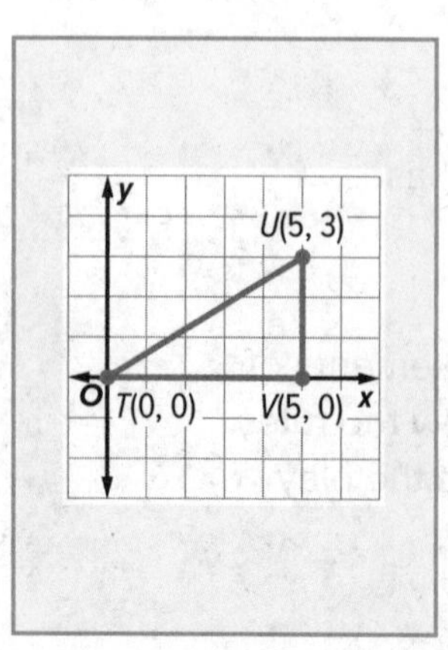

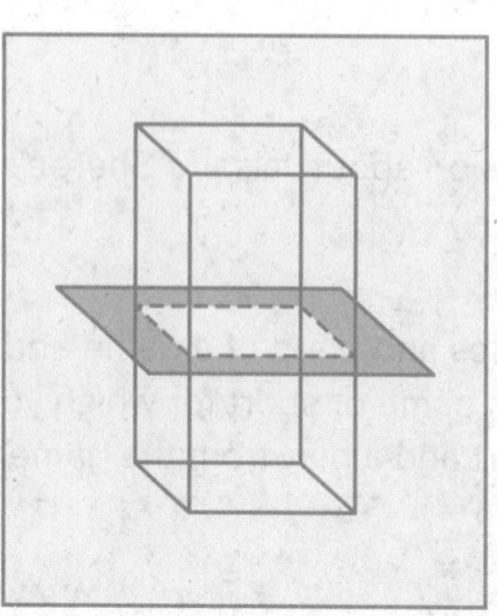

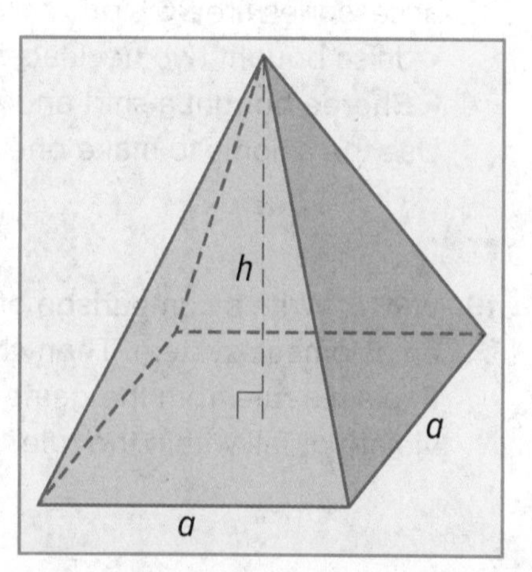

Points, Lines, and Planes

Learn Points, Lines, and Planes

In geometry, *point*, *line*, and *plane* are considered undefined terms because they are usually readily understood and are not formally explained by means of more basic words and concepts.

You are already familiar with the terms point, line, and plane from algebra. You graphed on a coordinate *plane* and found ordered pairs that represented *points* on *lines*. In geometry, these terms have a similar meaning.

Undefined Terms	
A **point** is a location. It has neither shape nor size. Named by a capital letter Example point *A*	A
A **line** is made up of points and has no thickness or width. There is exactly one line through any two points. Named by the letters representing two points on the line or a lowercase script letter Example line *m*, line *PQ* or $\overleftrightarrow{PQ}$, line *QP* or $\overleftrightarrow{QP}$	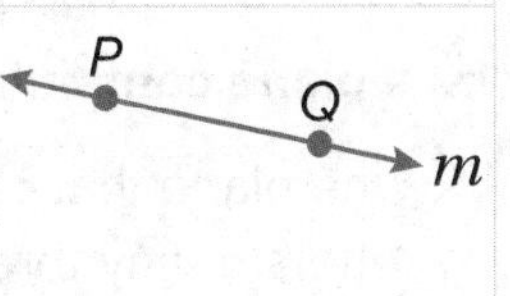
A **plane** is a flat surface made up of points that extends infinitely in all directions. There is exactly one plane through any three points not on the same line. Named by a capital script letter or by the letters naming three points that are not all on the same line Example plane *K*, plane *BCD*, plane *CDB*, plane *DCB*, plane *DBC*, plane *CBD*, plane *BDC*	D, B, C, K

Space is defined as a boundless three-dimensional set of all points. Space can contain lines and planes.

Collinear points are points that lie on the same line. *Noncollinear* points do not lie on the same line.

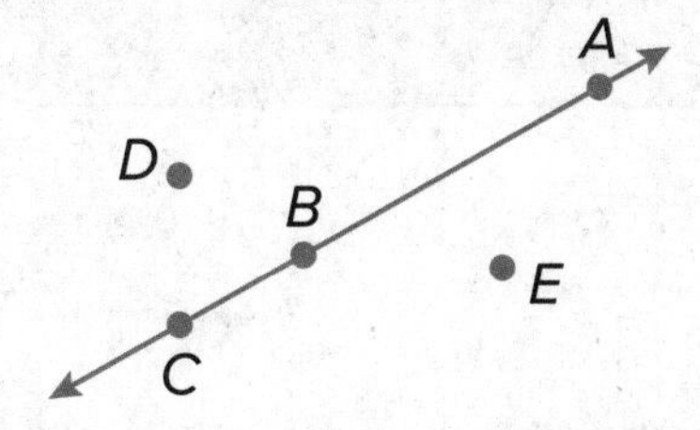

Points *A*, *B*, and *C* are collinear.

Coplanar points are points that lie in the same plane. *Noncoplanar* points do not lie in the same plane.

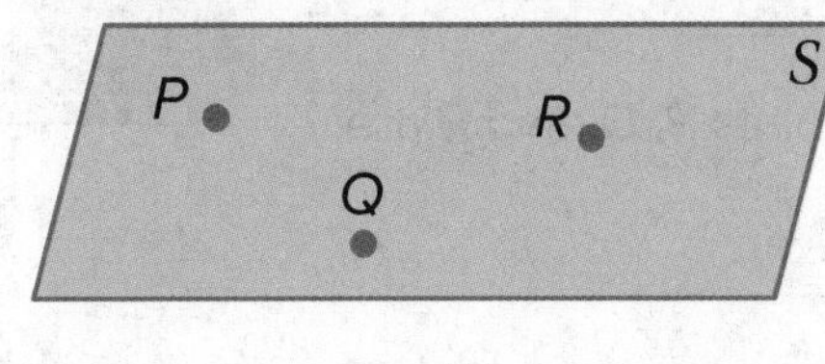

Points *P*, *Q*, and *R* are coplanar.

Today's Goals
- Identify points, lines, and planes.
- Identify intersections of lines and planes.

Today's Vocabulary
point
line
plane
space
collinear
coplanar
intersection

Talk About It!
Can three points be both noncollinear and noncoplanar? Justify your argument.

Your Notes

Example 1 Name Lines and Planes

Use the figure to name each of the following.

a. a line containing point *Q*

The line can be named as line c, or any two of the three points on the line can be used to name the line.

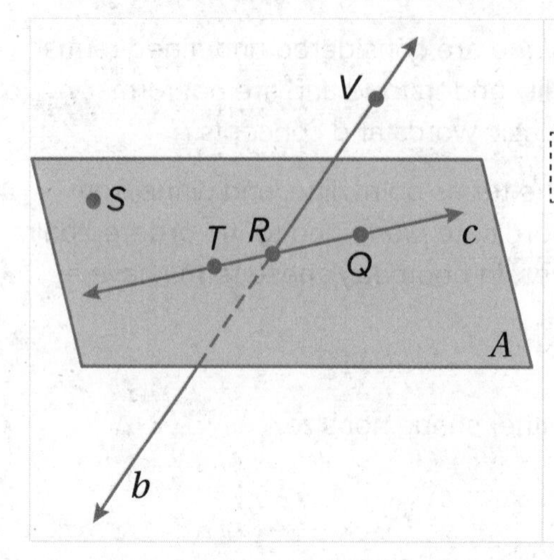

Write the additional names for line c below.

b. a plane containing point *S* and point *T*

One plane that can be named is plane A. You can also use the letters of any three *noncollinear* points to name this plane. Plane *TRS* and plane *TQS* can be used to name this plane.

Circle another correct name for plane A.

plane *QST* plane *STV* plane *QVS* plane *VST*

Study Tip

Additional Planes Although not drawn, there is another plane that contains point *S* and point *T*. Because points *S*, *T*, and *V* are noncollinear, points *S* and *T* are in plane *STV*.

Example 2 Model Points, Lines, and Planes

STUDENT DESK **Name the geometric terms modeled by the objects in the picture.**

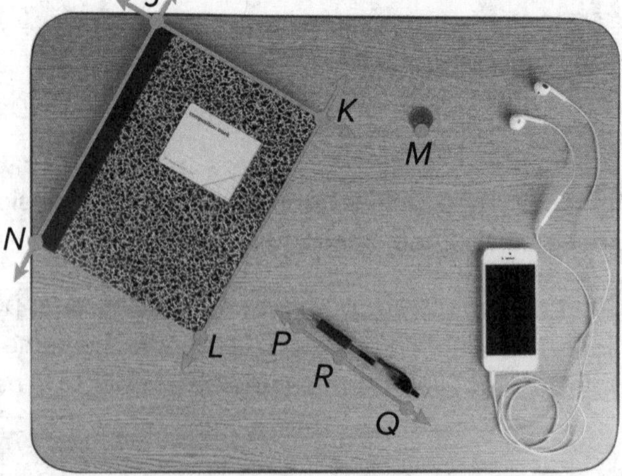

The notebook models ________.

The edges of the notebook model lines *JK*, *KL*, and ____.

The quarter models point ____ in space.

Points *N*, *L*, and *K* are ________.

Points *P*, *Q*, and *R* are ________.

Go Online You can complete an Extra Example online.

Explore Intersections of Three Planes

Online Activity Use a concrete model to complete the Explore.

> **INQUIRY** What figures can be formed by the intersection of three planes?

Learn Intersections of Lines and Planes

The **intersection** of two or more geometric figures is the set of points they have in common. Two lines intersect in a point. Lines can intersect planes, and planes can intersect each other.

Example 3 Draw Geometric Figures

Draw and label a figure to represent the relationship.

$\overleftrightarrow{QR}$ and $\overleftrightarrow{ST}$ intersect at U for $Q(-3, -2)$, $R(4, 1)$, $S(2, 3)$, and $T(-1, -5)$ on the coordinate plane. Point V is coplanar with these points but not collinear with $\overleftrightarrow{QR}$ and $\overleftrightarrow{ST}$.

Graph each point and draw $\overleftrightarrow{QR}$ and $\overleftrightarrow{ST}$.

Label the intersection point as U.

An infinite number of points are coplanar with Q, R, S, T, and U but are not collinear with $\overleftrightarrow{QR}$ and $\overleftrightarrow{TS}$. In the graph, one such point is $V(-2, 3)$.

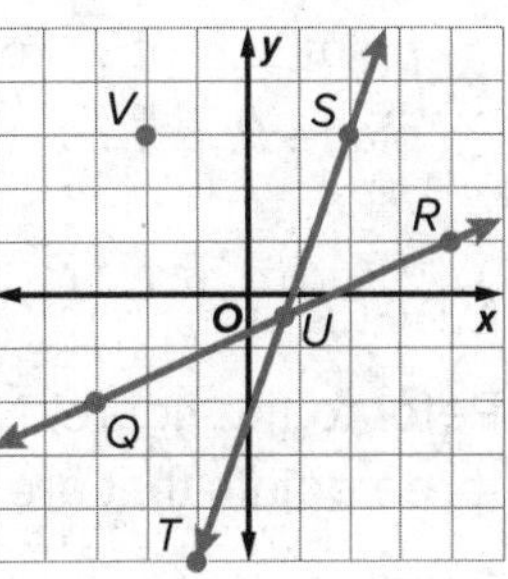

Check

Draw and label a figure to represent the relationship.

$\overleftrightarrow{JK}$ and $\overleftrightarrow{LM}$ intersect at P for $J(-4, 3)$, $K(6, -3)$, $L(-4, -5)$, and $M(3, 3)$ on the coordinate plane. Point Q is coplanar with these points, but not collinear with $\overleftrightarrow{JK}$ and $\overleftrightarrow{LM}$.

Go Online You can complete an Extra Example online.

Study Tip

Dimensions A point has no dimension. A line exists in one dimension. However, a circle is two-dimensional, and a pyramid is three-dimensional.

Study Tip

Three-Dimensional Drawings Because it is impossible to show an entire plane in a figure, edged shapes with different shades of color are used to represent planes.

Example 4 Interpret Drawings

Refer to the figure.

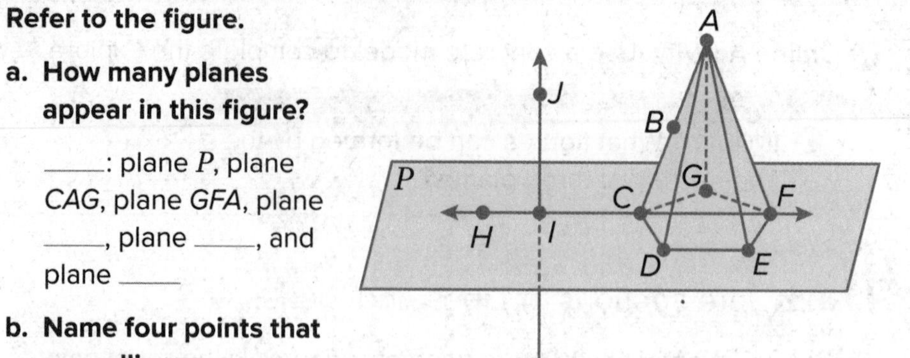

a. How many planes appear in this figure?

____: plane P, plane CAG, plane GFA, plane ____, plane ____, and plane ____

b. Name four points that are collinear.

Points H, I, ____, and ____ are collinear.

c. Name the intersection of plane GAC and plane P.

Plane GAC intersects plane P in ____.

d. At what point do $\overleftrightarrow{JI}$ and $\overleftrightarrow{DC}$ intersect? Explain.

It does ____ appear that these lines intersect. $\overleftrightarrow{DC}$ lies in plane P, but only point ____ of $\overleftrightarrow{JI}$ lies in plane P.

Check

Refer to the figure. Name three points that are collinear.

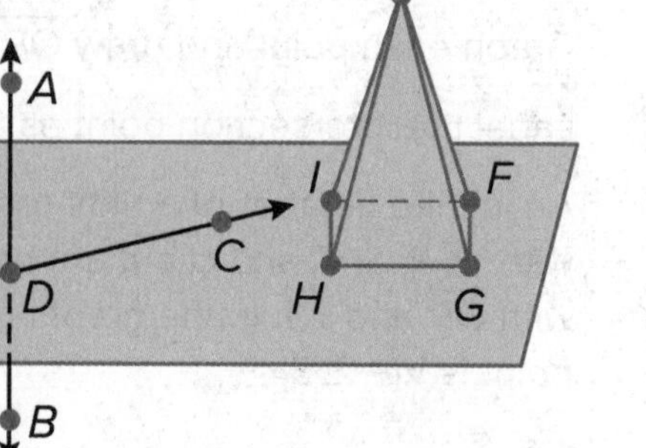

Points ____, ____, and ____ are collinear.

Example 5 Model Intersections

AVIATION A biplane has two main wings that are stacked one above the other. Struts connect the wings and are used for support. Flying wires run diagonally from the main body of the plane to the wings and between the stacked wings.

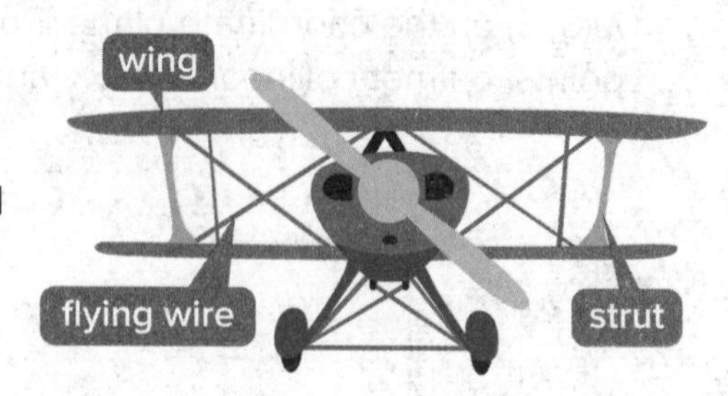

Complete the statements regarding the geometric terms modeled by the biplane.

Each wing models a ____.

The intersection of a strut and a wing models a ____.

The crossing of two flying wires models a ____.

Go Online You can complete an Extra Example online.

Name Isabella Metivier Period 4 Date 11-7-22

Practice

Go Online You can complete your homework online.

Example 1

Refer to the figure for Exercises 1-7 .

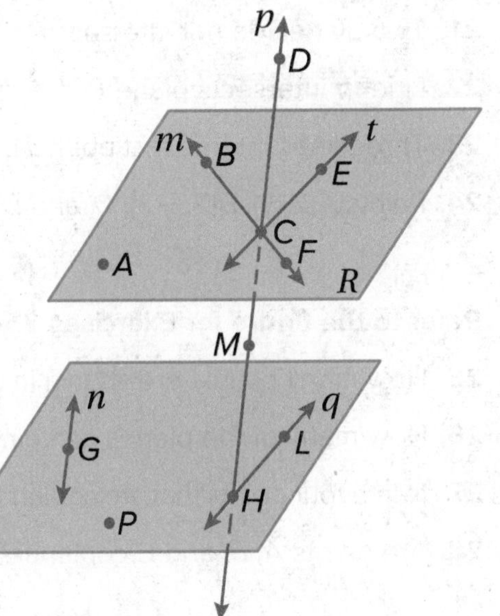

1. Name the lines that are only in plane Q. Ng, HLq

2. How many planes are labeled in the figure? 2

3. Name the plane containing the lines m and t. A

4. Name the intersection of lines m and t. MTC

5. Name a point that is *not* coplanar with points A, B, and C.

6. Are points F, M, G, and P coplanar? Explain.

7. Does line n intersect line q? Explain.

Example 2

Name the geometric terms modeled by each object or phrase.

8. roof of a house

9. a tabletop

10. bridge support beam

11. a chessboard

12.

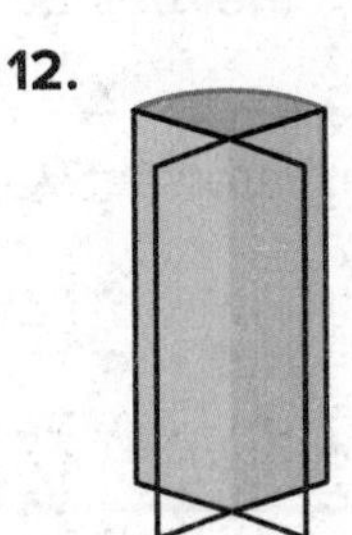

13.

14. a wall and the floor

15. the edge of a table

16. two connected walls

17. a blanket

18. a telephone pole

19. a tablet computer

Example 3

USE TOOLS **Draw and label a figure for each relationship.**

20. Points X and Y lie on $\overleftrightarrow{CD}$.

21. Two planes do not intersect.

22. Line m intersects plane R at a single point.

23. Three lines intersect at point J but do not all lie in the same plane.

24. Points $A(2, 3)$, $B(2, -3)$, C, and D are collinear, but A, B, C, D, and F are not.

Example 4

Refer to the figure for Exercises 25–28.

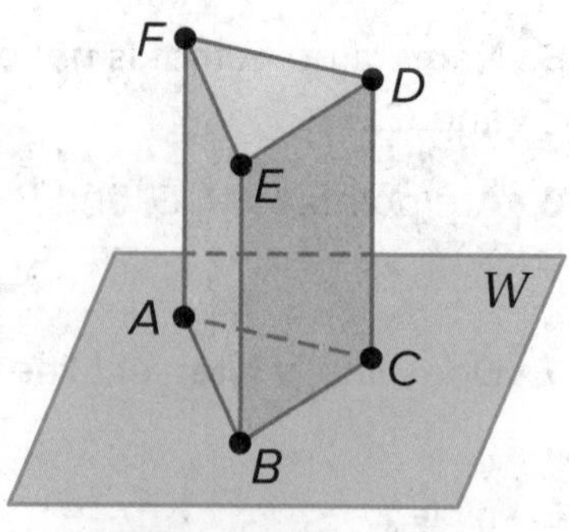

25. How many planes are shown in the figure?

26. How many of the planes contain points F and E?

27. Name four points that are coplanar.

28. Are points A, B, and C coplanar? Explain.

Example 5

29. BUILDING The roof and exterior walls of a house represent intersecting planes. Using the image, name all the lines that are formed by the intersecting planes.

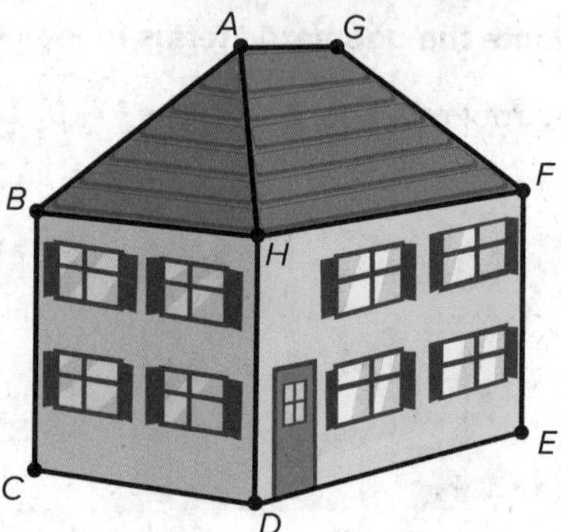

30. If the surface of a lake represents a plane, what geometric term is represented by the intersection of a fishing line and the lake's surface?

31. ART Perspective drawing is a method that artists use to create paintings and drawings of three-dimensional objects. The artist first draws the horizon line and two vanishing points along the horizon. Buildings or other objects are created by drawing receding lines and vertical lines.

a. Where do the receding lines and horizon lines intersect?

b. Identify examples of planes within this picture.

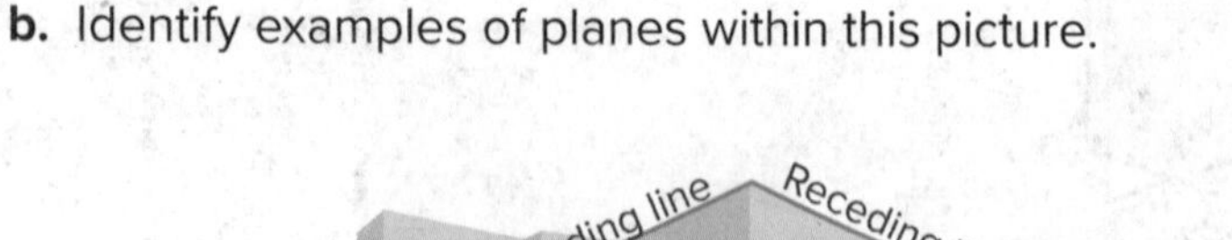

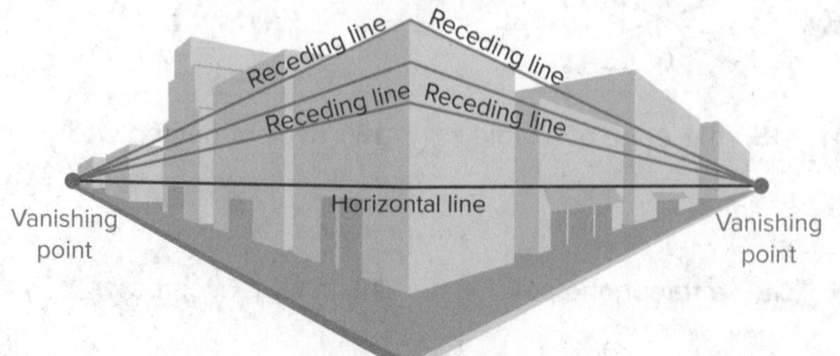

Name ____________________ Period ____________ Date ____________

Mixed Exercises

USE TOOLS **Draw and label a figure for each relationship.**

32. $\overleftrightarrow{LM}$ and $\overleftrightarrow{NP}$ are coplanar but do not intersect.

33. $\overleftrightarrow{FG}$ and $\overleftrightarrow{JK}$ intersect at $P(4, 3)$, where point F is at $(-2, 5)$ and point J is at $(7, 9)$.

34. Lines s and t intersect, and line v does not intersect either one.

Refer to the figure for Exercises 35-38.

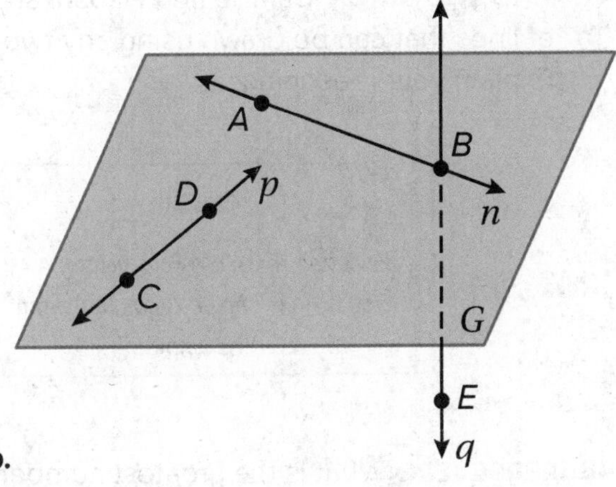

35. Name a line that contains point E.

36. Name a point contained in line n.

37. What is another name for line p?

38. Name the plane containing lines n and p.

USE TOOLS **Draw and label a figure for each relationship.**

39. Point K lies on $\overleftrightarrow{RT}$.

40. Plane J contains line s.

41. $\overleftrightarrow{YP}$ lies in plane B and contains point C, but does not contain point H.

42. Lines q and f intersect at point Z in plane U.

43. Name the geometric term modeled by the object.

44. Name the geometric term modeled by a partially-opened folder.

45. **CREATE** Sketch three planes that intersect in a point.

46. **ANALYZE** Is it possible for two points on the surface of a prism to be neither collinear nor coplanar? Justify your argument.

47. **FIND THE ERROR** Camille and Hiroshi are trying to determine the greatest number of lines that can be drawn using any two of four random points. Is either correct? Explain your reasoning.

Camille	Hiroshi
Because there are four points, $4 \cdot 3$ or 12 lines can be drawn between the points.	You can draw $3 + 2 + 1$ or 6 lines between the points.

48. **PERSEVERE** What is the greatest number of planes determined using any three of the points *A*, *B*, *C*, and *D* if no three points are collinear?

49. **WRITE** A *finite plane* is a plane that has boundaries or does not extend indefinitely. The sides of the cereal box shown are finite planes. Give a real-life example of a finite plane. Is it possible to have a real-life object that is an infinite plane? Explain your reasoning.

50. **CREATE** Sketch three planes that intersect in a line.

Line Segments

Explore Using Tools to Determine Betweenness of Points

Online Activity Use a pencil and straightedge to complete the Explore.

INQUIRY How can a line segment be divided into any number of line segments?

Today's Goals
- Calculate measures of line segments.
- Apply the definition of congruent line segments to find missing values.

Today's Vocabulary
line segment
betweenness of points
congruent
congruent segments

Learn Betweenness of Points

A **line segment** is a measurable part of a line that consists of two points, called endpoints, and all of the points between them. The two endpoints are used to name the segments.

You know that for any two real numbers a and b, there is a real number n between a and b such that $a < n < b$. This relationship also applies to points on a line and is called **betweenness of points.**

Key Concept • Betweenness of Points

Point C is between A and B if and only if A, B, and C are collinear and $AC + CB = AB$.

Talk About It!
What is an example of how the betweenness of points can be applied to the real world?

Example 1 Find Measurements by Adding

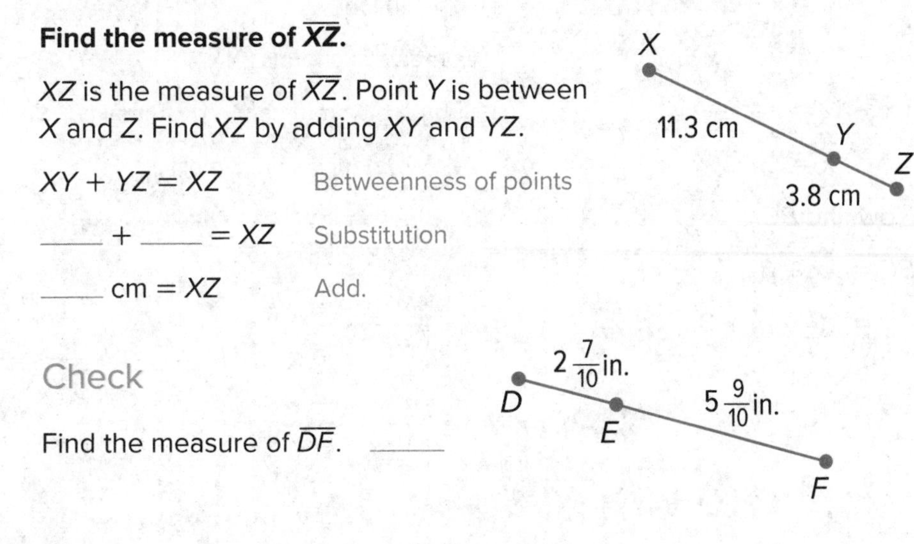

Find the measure of $\overline{XZ}$.

XZ is the measure of $\overline{XZ}$. Point Y is between X and Z. Find XZ by adding XY and YZ.

$XY + YZ = XZ$	Betweenness of points
____ + ____ $= XZ$	Substitution
____ cm $= XZ$	Add.

Check

Find the measure of $\overline{DF}$. ____

Go Online You can complete an Extra Example online.

Your Notes

Problem-Solving Tip

Draw a Diagram Draw a diagram to help you see and correctly interpret a situation that has been described in words.

Think About It!

How can you check your solution for x?

Think About It!

Once you find BC, how could you find AC without evaluating $AC = 4x - 12$?

Example 2 Find Measurements by Subtracting

Find the measure of $\overline{QR}$.

Point Q is between points P and R.

$13\frac{3}{4}$ ft, $6\frac{5}{8}$ ft, P, Q, R

$PQ + QR = PR$	Betweenness of points
____ $+ QR =$ ____	Substitution
$QR =$ ____ ft	Subtract $6\frac{5}{8}$ from each side and simplify.

Check

Find the measure of $\overline{PQ}$. Round your answer to the nearest tenth, if necessary.

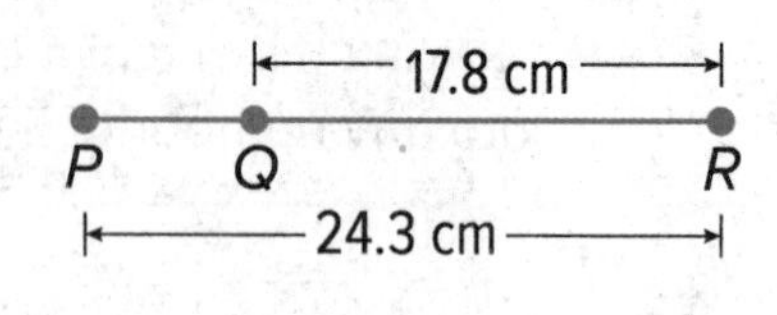

____ cm

Example 3 Write and Solve Equations to Find Measurements

Find the value of x and BC if B is between A and C, $AC = 4x - 12$, $AB = x$, and $BC = 2x + 3$.

Step 1 Plot two points and label them A and C. Connect the points.

Step 2 Plot point B between points A and C.

$4x - 12$, x, $2x + 3$, A, B, C

Step 3 Label segments AB, BC, and AC with their given measures.

Step 4 Use betweenness of points to write an equation and solve for x.

$AC = AB + BC$	Betweenness of points
$4x - 12 = x + 2x + 3$	Substitution
$4x - 12 =$ ____ $x + 3$	Combine like terms.
$x - 12 =$ ____	Subtract $3x$ from each side. Simplify.
$x =$ ____	Add 12 to each side. Simplify.

Now find BC.

$BC = 2x + 3$	Given
$= 2($____$) + 3$	$x = 15$
$=$ ____	

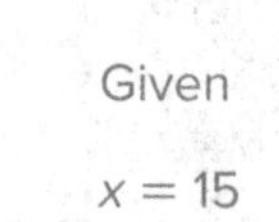

Go Online You can complete an Extra Example online.

Apply Example 4 Use Betweenness of Points

SPACE NEEDLE **Darrell is visiting the Space Needle in Seattle, Washington. He knows that the total height of the Space Needle is 605 feet. The distance from the ground to the observation deck is 10 feet more than six times the distance from the observation deck to the top of the Space Needle. Help Darrell find the distance from the ground to the observation deck.**

1 What is the task?

Describe the task in your own words. Then list any questions that you may have. How can you find answers to your questions?

____

____

____

2 How will you approach the task? What have you learned that you can use to help you complete the task?

____

____

____

3 What is your solution?

Use your strategy to solve the problem.

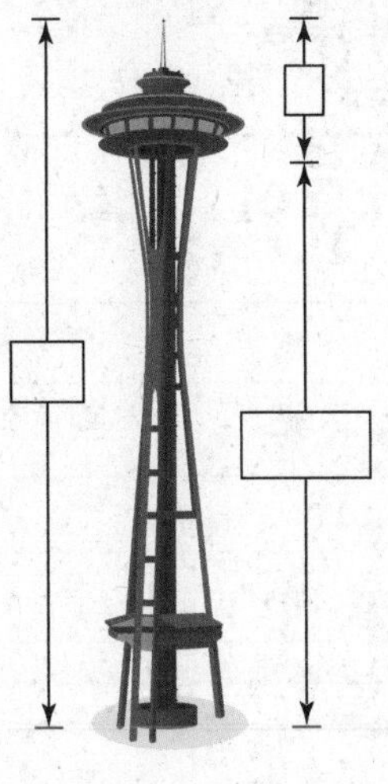

What equation represents the distance from the ground to the top of the Space Needle?

____

What is the distance from the ground to the observation deck?

____ ft

4 How can you know that your solution is reasonable?

Write About It! Write an argument that can be used to defend your solution.

____

____

____

Study Tip

Congruent Segments Use a consecutive number of tick marks for each new pair of congruent segments in a figure. The segments with two tick marks are congruent, and the segments with three tick marks are congruent.

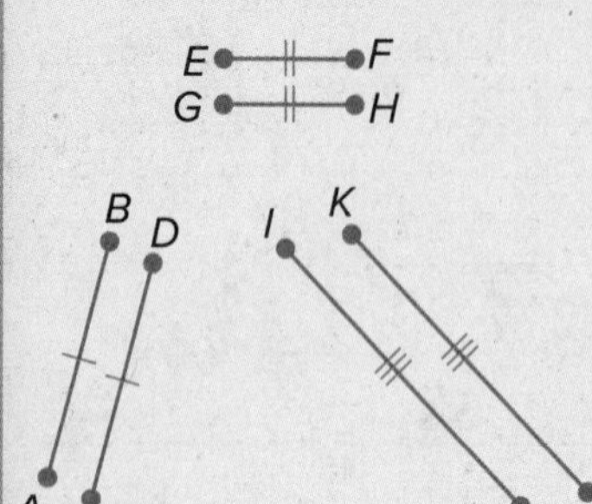

Learn Line Segment Congruence

If two geometric figures have exactly the same shape and size, then they are **congruent**. Two segments that have the same measure are **congruent segments**.

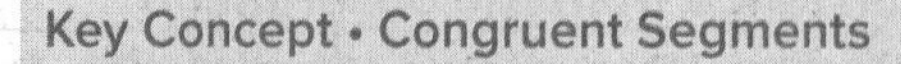

Key Concept • Congruent Segments

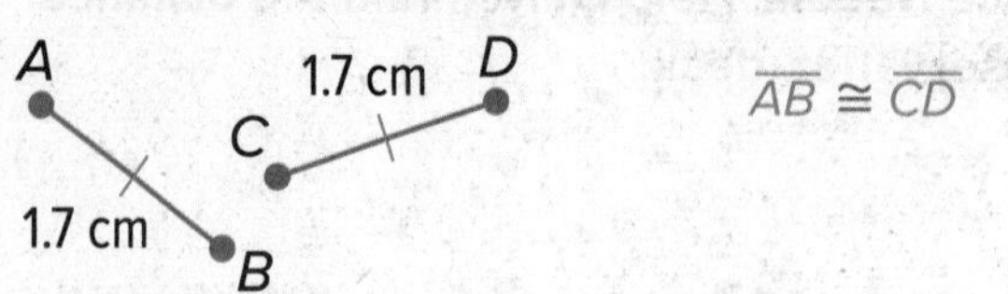

$\cong$ is read *is congruent to*. Red slashes on the figure also indicate congruence.

Segment AB is congruent to segment CD.

Congruent segments have the same measure.

Example 5 Write and Solve Equations by Using Congruence

Find the value of x if Q is between P and R, $PQ = 6x + 20$, $QR = 2(x + 6)$, and $\overline{PQ} \cong \overline{QR}$.

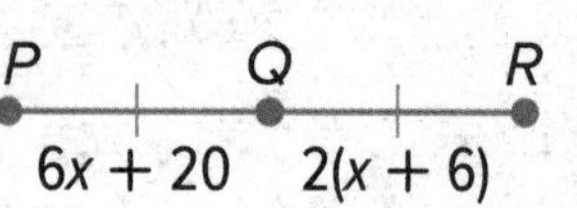

Write the justifications in the correct order. You may use a justification more than once.

Definition of congruence Distributive Property
Divide each side by 4. Simplify. Substitution
Subtract $2x$ from each side. Subtract 20 from each side.

$PQ = QR$ __________

$6x + 20 = 2(x + 6)$ __________

$6x + 20 = 2x + 12$ __________

$6x + 20 - 2x = 2x + 12 - 2x$ __________

$4x + 20 = 12$ __________

$4x + 20 - 20 = 12 - 20$ __________

$4x = -8$ __________

$\frac{4x}{4} = \frac{-8}{4}$ __________

$x = -2$ __________

Study Tip

Equal vs. Congruent Lengths are *equal*, and segments are *congruent*. It is correct to say that $AB = CD$ and $\overline{AB} \cong \overline{CD}$. However, it is not correct to say that $\overline{AB} = \overline{CD}$ or that $AB \cong CD$.

Watch Out!

Check Your Answer Sometimes solutions will result in negative segment lengths. If this occurs, review your work carefully. Either an error was made, or there is no solution.

Check

Find the value of x if U is between T and V, $TU = 7x + 35$, $UV = 4(x + 7)$, and $\overline{TU} \cong \overline{UV}$.

$x =$ ____

Go Online You may want to complete the construction activities for this lesson.

Mixed Exercises

34. Find the length of $\overline{UW}$ if W is between U and V, $UV = 16.8$ centimeters, and $VW = 7.9$ centimeters.

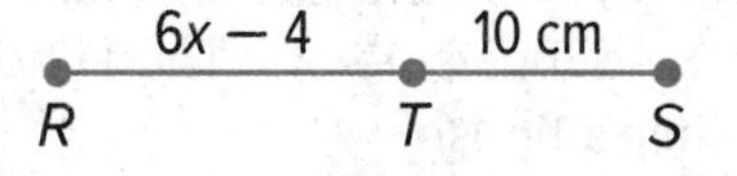

35. Find the value of x if $RS = 24$ centimeters.

36. Find the length of $\overline{LO}$ if M is between L and O, $LM = 7x - 9$, $MO = 14$ inches, and $LO = 10x - 7$.

37. Find the value of x if $\overline{PQ} \cong \overline{RS}$, $PQ = 9x - 7$, and $RS = 29$.

38. Find the measure of $\overline{NL}$.

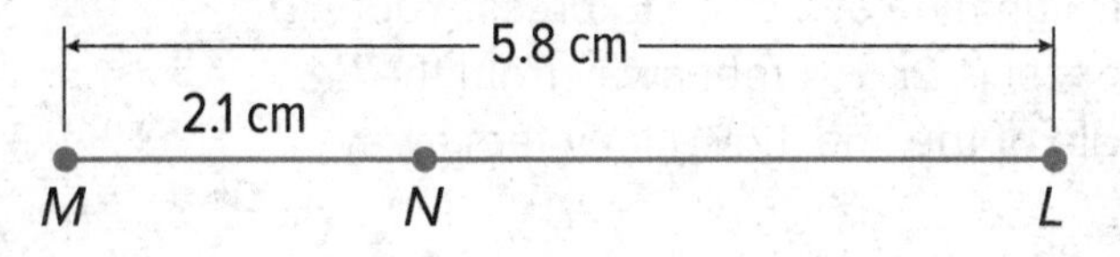

39. PRECISION If point P is between A and M, write a true statement.

40. HIKING A hiking trail is 20 kilometers long. Park organizers want to build 5 rest stops for hikers with one on each end of the trail and the other 3 spaced evenly between. How much distance will separate successive rest stops?

41. RACE The map shows the route of a race. You are at Y, 6000 feet from the first checkpoint A. The second checkpoint B is located at the midpoint between A and the end of the race Z. The total race is 3.1 miles. How far apart are the two checkpoints?

42. FIELD TRIP The marching band at Jefferson High School is taking a field trip from Lansing, Michigan, to Detroit, Michigan. The bus driver was told to stop 53 miles into the trip. If the rest of the trip is 41 miles and the entire journey can be represented by the expression $3x + 16$, find the value of x.

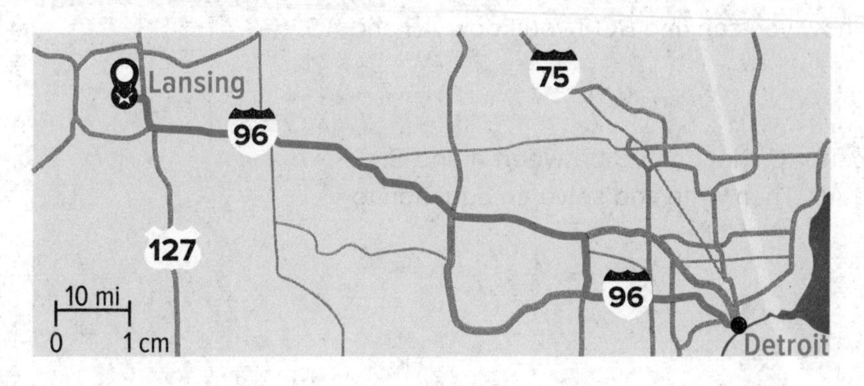

43. DISTANCE Madison lives between Anoa and Jamie as depicted on the line segment. The distance between Anoa's house and Madison's house is represented by $3x + 2$ miles, the distance between Madison's house and Jamie's house is represented by $3x + 4$ miles, and the distance between Anoa's house and Jamie's house is represented by $9x - 3$ miles. Find the value of x. Then find the distance between Madison's house and Jamie's house.

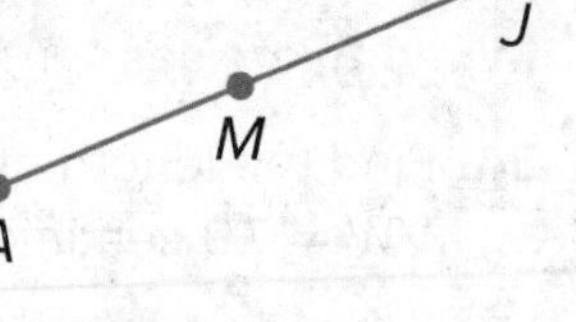

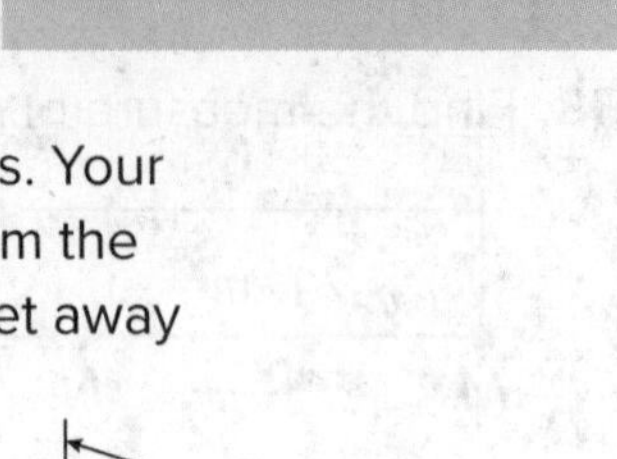

44. FIREFIGHTING A firefighter training course is taking place in a high-rise building. The high-rise building where they practice is 48 stories high. If the emergency happens on the top floor and the firefighters have already gone 29 stories, how many stories do they still need to go?

45. CAFE You are waiting at the end of a long straight line at Coffee Express. Your friend Denzel is $r + 12$ feet in front of you. Denzel is $2r + 4$ feet away from the front of the line. If Denzel is in the exact middle of the line, how many feet away are you from the front of the line?

46. REASONING For $\overline{AC}$, write and solve an equation to find AB.

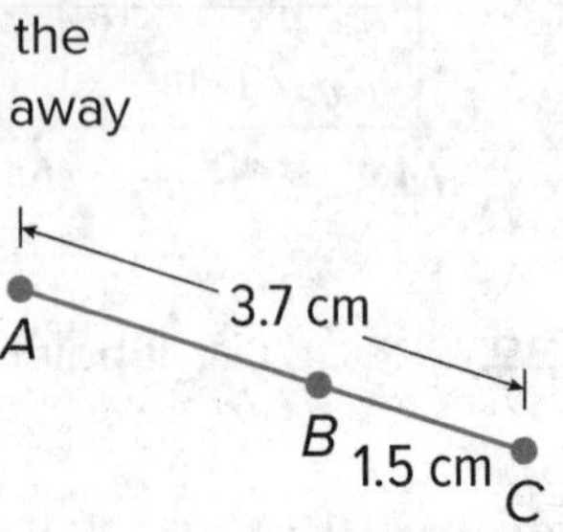

47. PERSEVERE Point K is between points J and L. If $JK = x^2 - 4x$, $KL = 3x - 2$, and $JL = 28$, find JK and KL.

48. ANALYZE Determine whether the statement *If point M is between points C and D, then CD is greater than either CM or MD* is *sometimes, always,* or *never* true. Justify your argument.

49. PERSEVERE Point C is located between points B and D. Also, $BC = 5x + 7$, $CD = 3y + 4$, $BD = 38$, and $BD = 2x + 8y$. Find the values of x and y.

50. WRITE If point B is between points A and C, explain how you can find AC if you know AB and BC. Explain how you can find BC if you know AB and AC.

51. CREATE Sketch line segment AC. Plot point B between A and C. Use a ruler to find AC and AB. Then write and solve an equation to find BC.

Distance

Learn Distance on a Number Line

The **distance** between two points is the length of the segment between the points. The coordinates of the points can be used to find the length of the segment.

Key Concept • Distance Formula on Number Line

P Q

x_1 x_2

If P has coordinate x_1 and Q has coordinate x_2, then
$PQ = |x_2 - x_1|$ or $|x_1 - x_2|$.

Because $\overline{PQ}$ is the same as $\overline{QP}$, the order in which you name the endpoints is not important when calculating distance.

Example 1 Find Distance on a Number Line

Use the number line.

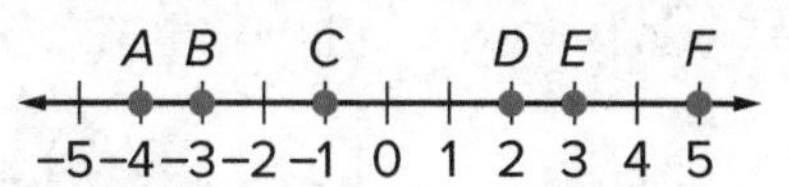

Find *CF*.

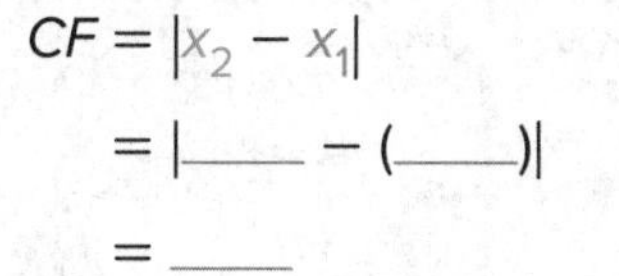

$CF = \|x_2 - x_1\|$	Distance Formula
$= \|\_\_\_\_ - (\_\_\_\_)\|$	$x_1 = -1$ and $x_2 = 5$
$= \_\_\_\_$	Simplify.

Check

Use the number line.

A B C D E
−6 −5 −4 −3 −2 −1 0 1 2 3 4 5 6 7 8

Find *AE*.

A. −12 B. 2 C. 12 D. 13

Today's Goals
- Find the length of a line segment on a number line.
- Find the distance between two points on the coordinate plane.

Today's Vocabulary
distance

Think About It!
Why do you think the Distance Formula uses absolute value?

Think About It!
Compare and contrast the length of $\overline{CF}$ and the length of $\overline{FC}$.

Go Online You can complete an Extra Example online.

Your Notes

Example 2 Determine Segment Congruence

Determine whether $\overline{CB}$ and $\overline{DF}$ are congruent.

A B C D E F

−5 −4 −3 −2 −1 0 1 2 3 4 5

The coordinates of C and B are −1 and −3. The coordinates of D and F are 2 and 5. Find the length of each segment.

$CB = \lvert x_2 - x_1 \rvert$	Distance Formula
$= \lvert \_\_\_\_ - (-1) \rvert$	Substitute.
$= \lvert \_\_\_\_ \rvert$	Subtract.
$= \_\_\_\_$	Simplify.

The length of $\overline{CB}$ is 2 units.

$DF = \lvert x_2 - x_1 \rvert$	Distance Formula
$= \lvert 5 - \_\_\_\_ \rvert$	Substitute.
$= \lvert \_\_\_\_ \rvert$	Subtract.
$= \_\_\_\_$	Simplify.

The length of $\overline{DF}$ is 3 units.

Because $CB \neq DF$, the segments are not congruent.

Check

Determine whether $\overline{AC}$ and $\overline{BD}$ are congruent.

A B C D E

−6 −5 −4 −3 −2 −1 0 1 2 3 4 5 6 7 8

The segments ______ congruent.

Watch Out!

Subtraction with Negatives Remember that subtracting a negative number is like adding a positive number.

Explore Use the Pythagorean Theorem to Find Distances

Online Activity Use dynamic geometry software to complete the Explore.

INQUIRY How can you find the distance between two points on the coordinate plane?

Go Online A derivation of the Distance Formula is available.

Learn Distance on the Coordinate Plane

The endpoints of a segment on the coordinate plane can be used to find the length of that segment by using the Distance Formula.

Key Concept • Distance Formula on the Coordinate Plane

If P has coordinates (x_1, y_1) and Q has coordinates (x_2, y_2), then
$PQ = \sqrt{(x_2 - x_1)^2 + (y_2 - y_1)^2}$.

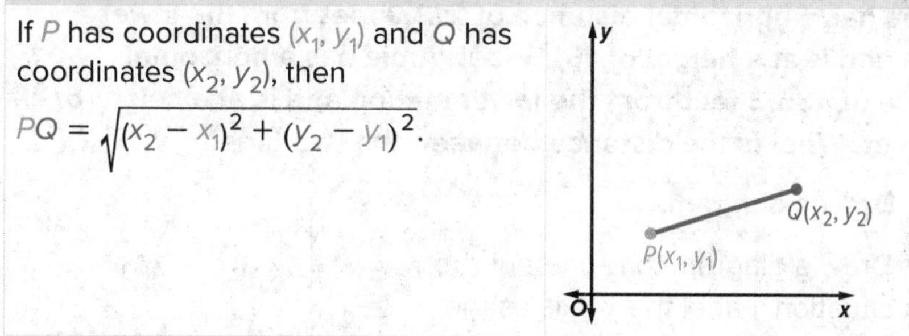

Example 3 Find Distance on the Coordinate Plane

Find the distance between $J(4, 3)$ and $K(-3, -7)$.

Let $J(4, 3)$ be (x_1, y_1) and $K(-3, -7)$ be (x_2, y_2).

$JK = \sqrt{(x_2 - x_1)^2 + (y_2 - y_1)^2}$ Distance Formula

$= \sqrt{(\_\_\_ - x_1)^2 + (\_\_\_ - y_1)^2}$ Substitute x_2 and y_2.

$= \sqrt{(-3 - \_\_\_)^2 + (-7 - \_\_\_)^2}$ Substitute x_1 and y_1.

$= \sqrt{(\_\_\_)^2 + (\_\_\_)^2}$ Subtract.

$= \sqrt{49 + \_\_\_}$ Simplify.

$= \sqrt{149}$ Simplify.

The distance between J and K is $\sqrt{149}$ or approximately _____ units.

Go Online An alternate method is available for this example.

Check

Find the distance between A and B. _____

y
B(5, 9)
A(−6, −4)
x

Go Online You can complete an Extra Example online.

Think About It!

Compare and contrast the Distance Formula on a number line with the Distance Formula on the coordinate plane.

Watch Out!

Simplify Radicals Do not forget to leave your answer in simplest radical form when using the Distance Formula or the Pythagorean Theorem.

Example 4 Calculate Distance

INCLINE **Chelsea and Amie are sitting in separate cars on the Monongahela Incline. Chelsea is traveling up Mount Washington and Amie is traveling down. When the two girls notice each other, Chelsea has a horizontal distance of 212.0 feet from the lower station and is at a height of 151.6 feet. Amie has a horizontal distance of 435.3 feet from the lower station and is at a height of 311.3 feet. What is the distance between the two girls?**

Step 1 Draw a diagram.

Draw a diagram to represent the situation. Label the x-axis as the "Horizontal Distance from Lower Station (in feet)." Label the y-axis as the "Height (in feet)." Use a scale of 50 on the x-axis and the y-axis.

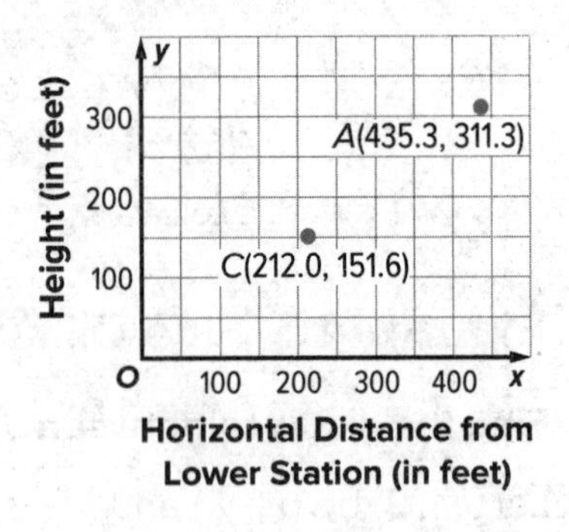

Step 2 Use the Distance Formula.

$(x_1, y_1) = (212.0, 151.6)$ and $(x_2, y_2) = (435.3, 311.3)$

$D = \sqrt{(x_2 - x_1)^2 + (y_2 - y_1)^2}$ Distance Formula

$= \sqrt{(\_\_\_\_ - 212.0)^2 + (311.3 - \_\_\_\_)^2}$ Substitute.

$= \sqrt{\_\_\_\_^2 + 159.7^2}$ Subtract.

$= \sqrt{49{,}862.89 + \_\_\_\_}$ Square each term.

$= \sqrt{\_\_\_\_}$ Add.

$\approx \_\_\_\_$ Take the positive square root.

Chelsea and Amie are approximately ______ feet apart.

Think About It!

Does your answer seem reasonable? Why or why not?

Check

SNOWBOARDING Manuel wants to go snowboarding with his friend. The closest ski and snowboard resort is approximately 20 miles west and 50 miles north of his house. Manuel picks up his friend who lives 15 miles south and 10 miles east of Manuel's house. How far away are the two boys from the resort?

______ mi

Go Online You can complete an Extra Example online.

Name _______________ Period _______ Date _______

Practice

Go Online You can complete your homework online.

Example 1

Use the number line to find each measure.

J K L M N P
−7 −6 −5 −4 −3 −2 −1 0 1 2 3 4 5 6

1. *JL*

2. *JK*

3. *KP*

4. *NP*

5. *JP*

6. *LN*

Use the number line to find each measure.

E F G H J K L
−6 −4 −2 0 2 4 6 8 10

7. *JK*

8. *LK*

9. *FG*

10. *JG*

11. *EH*

12. *LF*

Use the number line to find each measure.

J K L M N
−6 −4 −2 0 2 4 6 8 10

13. *LN*

14. *JL*

Example 2

Determine whether the given segments are congruent. Write *yes* or *no*.

A B C D E F
−10 −9 −8 −7 −6 −5 −4 −3 −2 −1 0 1 2 3 4 5 6 7 8 9 10

15. $\overline{AB}$ and $\overline{EF}$

16. $\overline{BD}$ and $\overline{DF}$

17. $\overline{AC}$ and $\overline{CD}$

18. $\overline{AC}$ and $\overline{DE}$

19. $\overline{BE}$ and $\overline{CF}$

20. $\overline{CD}$ and $\overline{DF}$

Example 3

Find the distance between each pair of points.

21. **22.** **23.**

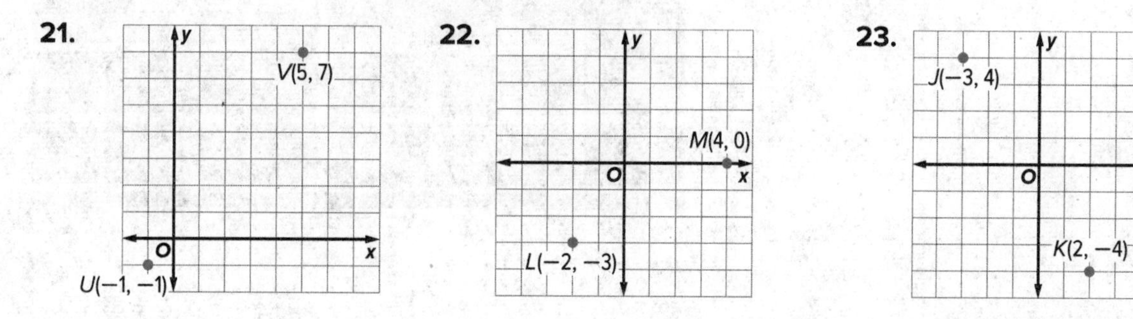

24. $A(2, 6)$, $N(5, 10)$

25. $R(3, 4)$, $T(7, 2)$

26. $X(-3, 8)$, $Z(-5, 1)$

Example 4

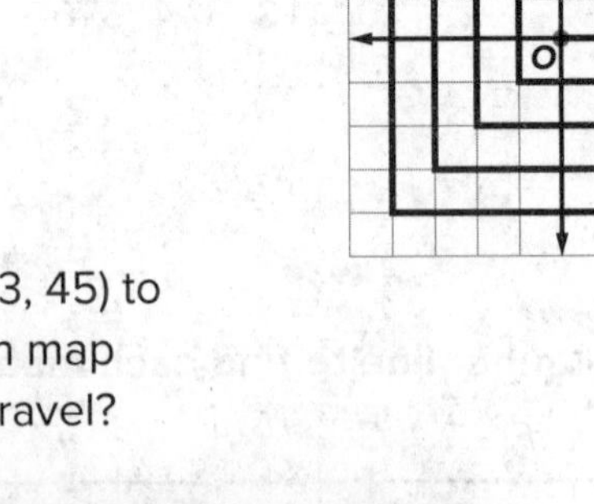

27. SPIRALS Denise traces the spiral shown in the figure. The spiral begins at the origin. What is the shortest distance between Denise's starting point and her ending point?

28. ZOOLOGY A tiny songbird called the blackpoll warbler migrates each fall from North America. A tracking study showed one bird flew from Vermont at map coordinates (63, 45) to Venezuela at map coordinates (67, 10) in three days. If each map coordinate represents 75 kilometers, how far did the bird travel?

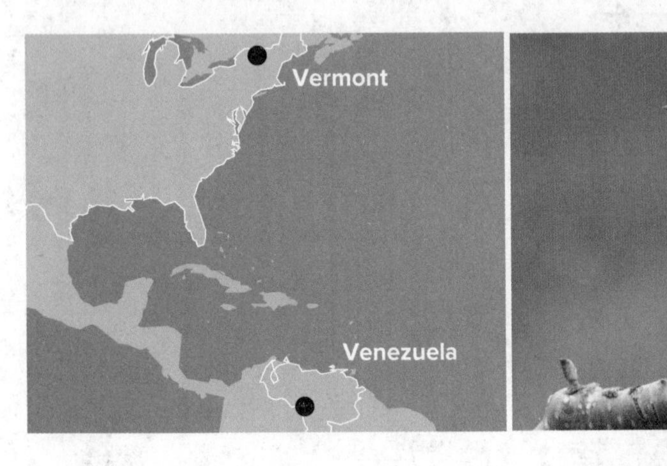

29. CONSTRUCT ARGUMENTS Mariah is training for a sprint-distance triathlon. She plans on cycling from her house to the library, shown on the grid with a scale in miles. If the cycling portion of the triathlon is 12 miles, will Mariah have cycled at least $\frac{2}{3}$ of that distance during her bike ride? Justify your argument.

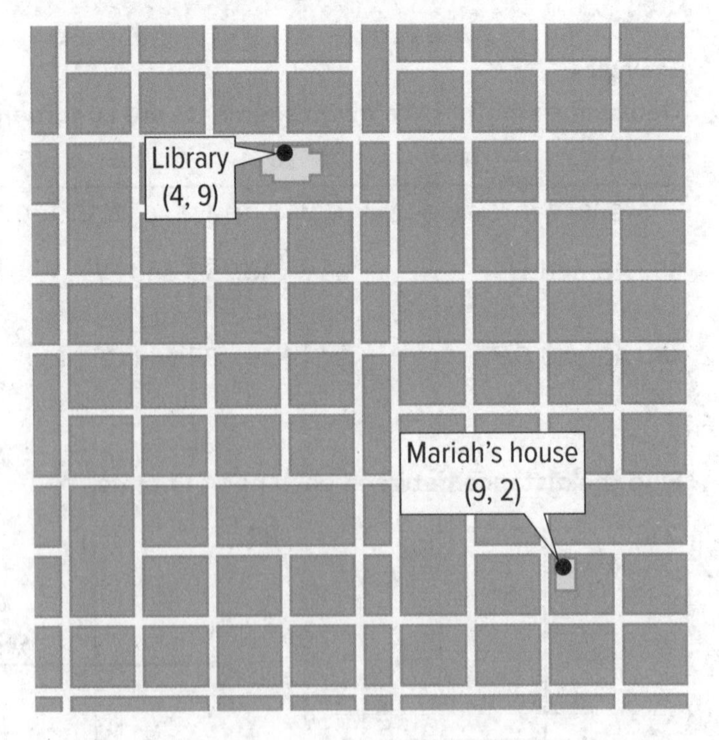

Name _______________ Period _______________ Date _______________

30. SPORTS The distance between each base on a baseball infield is 90 feet. The third baseman throws a ball from third base to point P. To the nearest foot, how far did the player throw the ball?

2nd Base
90 ft
P
Infield
30 ft
3rd Base
1st Base
Home Plate

Mixed Exercises

Find the distance between each pair of points. Round to the nearest tenth, if necessary.

31. $M(-4, 9)$, $N(-5, 3)$

32. $C(2, 4)$, $D(5, 7)$

33. $A(5, 1)$, $B(3, 6)$

34. $V(4, 4)$, $X(5, 8)$

35. $S(6, 4)$, $T(3, 2)$

36. $M(-1, 8)$, $N(-3, 3)$

37. $W(-8, 1)$, $Y(0, 6)$

38. $B(3, -4)$, $C(5, -5)$

39. $R(6, 11)$, $T(3, -7)$

40. $A(-3, 8)$ and $B(-1, 4)$

41. $M(4, -3)$ and $N(-2, 1)$

42. $X(-3, 5)$ and $Y(4, 2)$

43. Use the number line to determine whether $\overline{SV}$ and $\overline{UX}$ are congruent. Write *yes* or *no*.

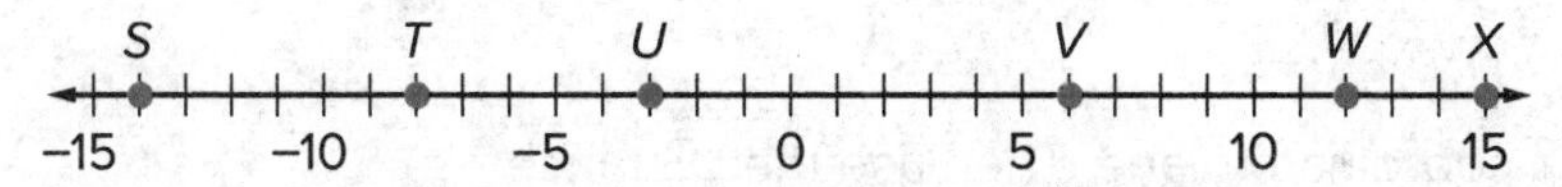

Name the point(s) that satisfy the given condition.

44. two points on the x-axis that are 10 units from (1, 8)

45. two points on the y-axis that are 25 units from (–24, 3)

46. Refer to the figure. Are $\overline{VT}$ and $\overline{SU}$ congruent?

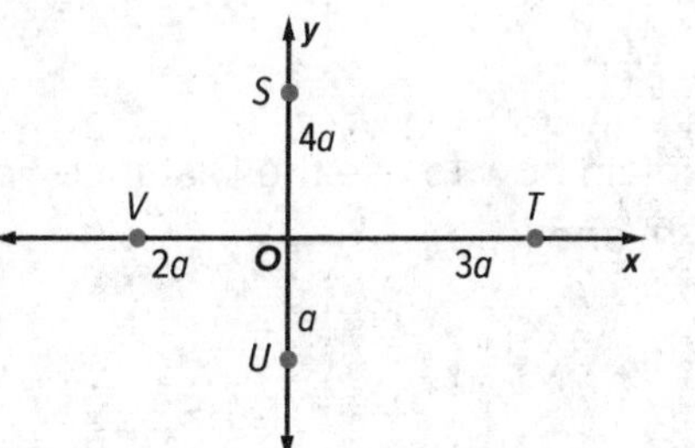

47. KNITTING Mei is knitting a scarf with diagonal stripes. Before she began, she laid out the pattern on a coordinate grid where each unit represented 2 inches. On the grid, the first stripe began at (2, 0) and ended at (5, 4). All the stripes are the same length. How many inches long is each stripe on the scarf?

48. ART A terracotta bowl artifact has a triangular pattern around the top, as shown. All the triangles are about the same size and can be represented on a coordinate plane with vertices at points (0, 6.8), (4.5, 6.8), and (2.25, 0). If each unit represents 1 centimeter, what is the approximate perimeter of each triangle, to the nearest tenth of a centimeter?

49. ANALYZE Consider rectangle $QRST$ with $QR = ST = 4$ centimeters and $RS = QT = 2$ centimeters. If point U is on $\overline{QR}$ such that $QU = UR$ and point V is on $\overline{RS}$ such that $RV = VS$, then is $\overline{QU}$ congruent to $\overline{RV}$? Justify your argument.

50. WRITE Explain how the Pythagorean Theorem and the Distance Formula are related.

51. PERSEVERE Point P is located on the segment between point $A(1, 4)$ and point $D(7, 13)$. The distance from A to P is twice the distance from P to D. What are the coordinates of point P?

52. CREATE Plot points Y and Z on a coordinate plane. Then use the Distance Formula to find YZ.

53. PERSEVERE Suppose point A is located at (1, 3) on a coordinate plane. If AB is 10 and the x-coordinate of point B is 9, explain how to use the Distance Formula to find the y-coordinate of point B.

54. WRITE Explain how to use the Distance Formula to find the distance between points (a, b) and (c, d).

Locating Points on a Number Line

Explore Locating Points on a Number Line with Fractional Distance

Online Activity Use dynamic geometry software to complete the Explore.

INQUIRY What general method can you use to locate a point some fraction of the distance from one point to another point on a number line?

Today's Goals
- Find a point on a directed line segment on a number line that is a given fractional distance from the initial point.
- Find a point that partitions a directed line segment on a number line in a given ratio.

Today's Vocabulary
directed line segment

fractional distance

Learn Locating Points on a Number Line with Fractional Distance

While a line segment has two endpoints, a **directed line segment** has an initial endpoint and a terminal endpoint.

Using a directed line segment enables you to calculate the coordinate of an intermediary point some fraction of the length of the segment, or **fractional distance,** from the initial endpoint.

Key Concept • Locating a Point at Fractional Distances on a Number Line

Find the coordinate of a point that is $\frac{a}{b}$ of the distance from point C to point D.

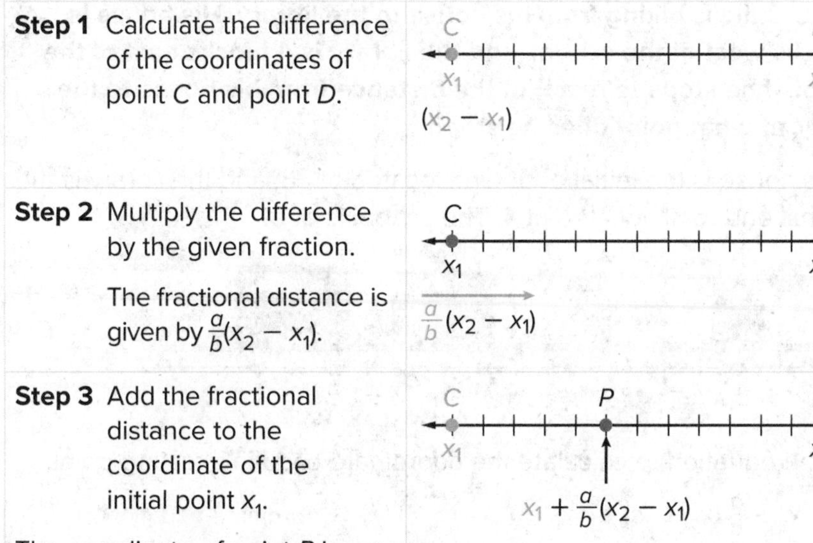

Step 1 Calculate the difference of the coordinates of point C and point D.

Step 2 Multiply the difference by the given fraction.

The fractional distance is given by $\frac{a}{b}(x_2 - x_1)$.

Step 3 Add the fractional distance to the coordinate of the initial point x_1.

The coordinate of point P is given by $x_1 + \frac{a}{b}(x_2 - x_1)$.

The coordinate of a point on a line segment with endpoints x_1 and x_2 is given by $x_1 + \frac{a}{b}(x_2 - x_1)$, where $\frac{a}{b}$ is the fraction of the distance.

Watch Out!

Don't Use Absolute Value When finding the distance from an initial endpoint to a terminal endpoint on a directed line segment, don't use absolute value. The difference created by $(x_2 - x_1)$ can be positive or negative. The sign of the difference will indicate the direction of the directed line segment.

Talk About It!

In the Key Concept, what phrase helped you identify the initial endpoint? What phrase helped you identify the terminal endpoint?

Your Notes

Example 1 Locate a Point at a Fractional Distance

Find B on $\overline{AC}$ that is $\frac{1}{4}$ of the distance from A to C.

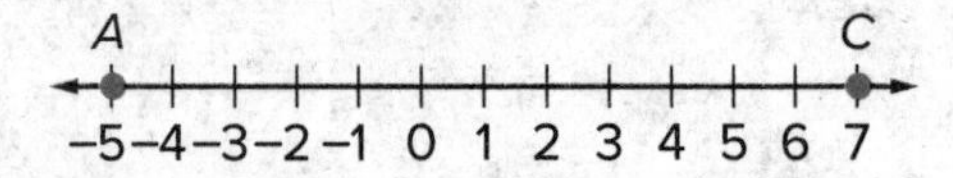

Point A is the initial endpoint, and point C is the terminal endpoint.

A C
−5 −4 −3 −2 −1 0 1 2 3 4 5 6 7

Use the equation to calculate the coordinate of point B.

$B = x_1 + \frac{a}{b}(x_2 - x_1)$ Coordinate equation

$= \_\_\_\_ + \frac{1}{4}(7 - \_\_\_\_)$ $x_1 = -5$, $x_2 = 7$, and $\frac{a}{b} = \frac{1}{4}$

$= \_\_\_\_$ Simplify.

Point B is located at -2 on the number line.

A B C
−5 −4 −3 −2 −1 0 1 2 3 4 5 6 7

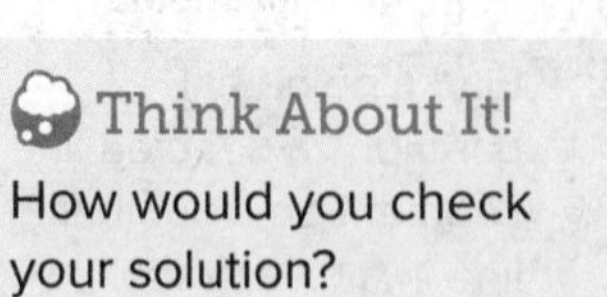

Think About It!

How would you check your solution?

Check

Find X on $\overline{BE}$ that is $\frac{3}{5}$ of the distance from B to E.

B E
−5 −4 −3 −2 −1 0 1 2 3 4 5 6 7 8 9 10

A. 2 **B.** 3 **C.** 5 **D.** 6

Example 2 Locate a Point at a Fractional Distance in the Real World

BIKING Julio is biking from his house to the library. His house is 8 blocks west of the school, and the library is 4 blocks east of the school. If he stops to rest $\frac{1}{3}$ of the distance from his house to the library, at what point does he stop?

Julio's house is the initial endpoint, located at −8, and the library is the terminal endpoint, located at 4. The school is at 0.

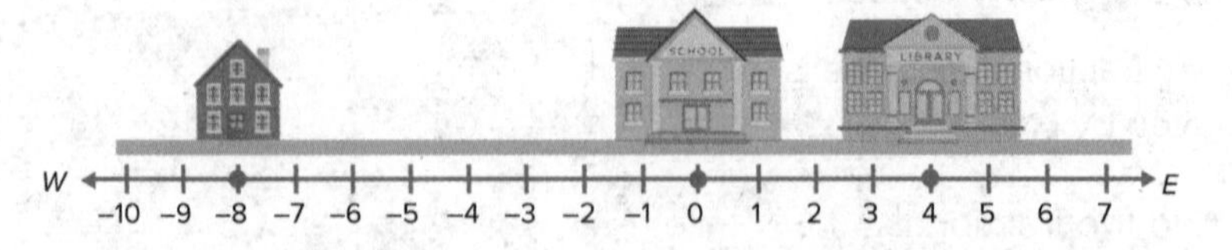

Use the equation to calculate the coordinate of Julio's resting point.

$B = x_1 + \frac{a}{b}(x_2 - x_1)$ Coordinate equation

$= \_\_\_\_ + \_\_\_\_ [4 - (-8)]$ $x_1 = -8$, $x_2 = 4$, and $\frac{a}{b} = \frac{1}{3}$

$= \_\_\_\_$ Simplify.

Think About It!

What would the coordinate be if Julio wanted to rest $\frac{1}{3}$ of the distance if he is going from the library to his house?

Go Online You can complete an Extra Example online.

Check

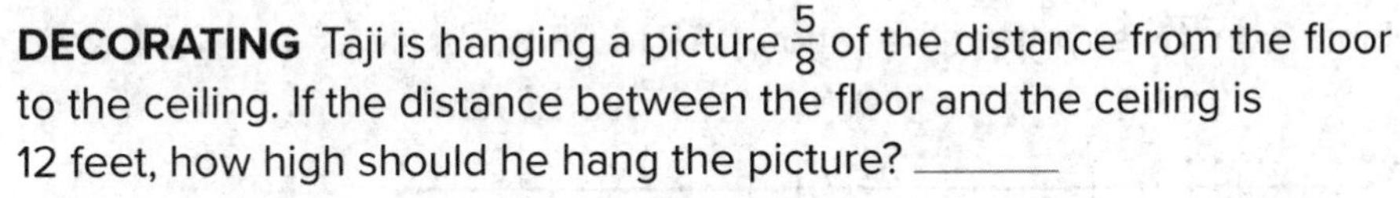

DECORATING Taji is hanging a picture $\frac{5}{8}$ of the distance from the floor to the ceiling. If the distance between the floor and the ceiling is 12 feet, how high should he hang the picture? _____

Learn Locating Points on a Number Line with a Given Ratio

You can calculate the coordinate of an intermediary point that partitions the directed line segment into a given ratio.

Key Concept • Section Formula on a Number Line

If C has coordinate x_1 and D has coordinate x_2, then a point P that partitions the line segment in a ratio of $m{:}n$ is located at coordinate $\frac{nx_1 + mx_2}{m + n}$, where $m \neq -n$.

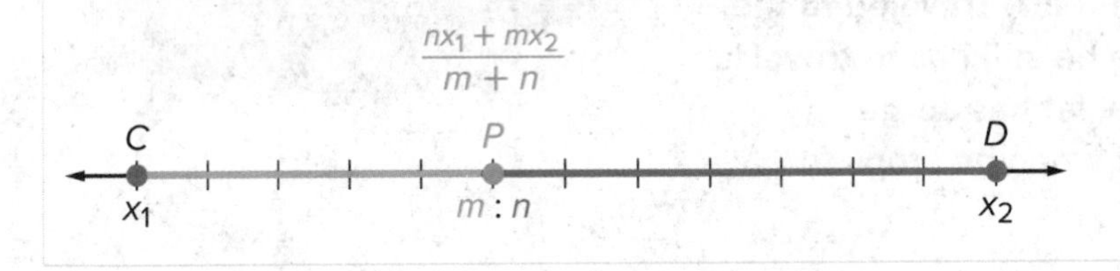

Example 3 Locate a Point on a Number Line When Given a Ratio

Find B on $\overline{AC}$ such that the ratio of AB to BC is 3:4.

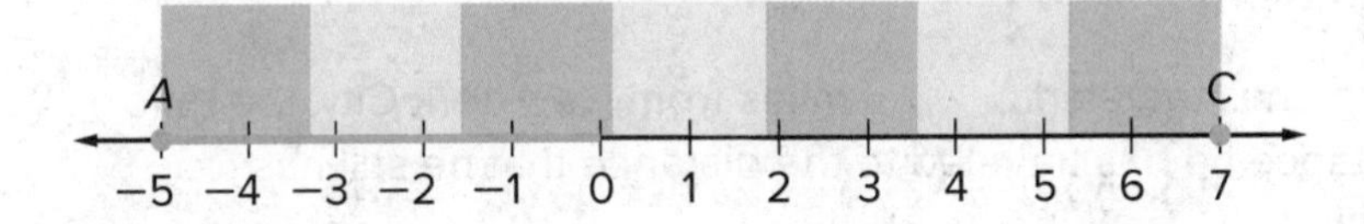

Use the Section Formula to determine the coordinate of point B.

$B = \frac{nx_1 + mx_2}{m + n}$ Section Formula

$= \frac{}{} = \frac{1}{7}$ $m = 3$, $n = 4$, $x_1 = -5$, and $x_2 = 7$

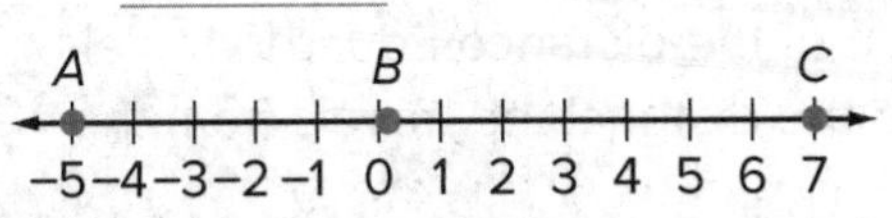

So, B is located at _____ on the number line.

(continued on the next page)

Go Online You may want to complete the Concept Check to check your understanding.

Study Tip

Checking Solutions When using the Section Formula, you can check your solution by converting the given ratio into a fraction. Use this fraction and the coordinate equation to find the fractional distance from your initial endpoint to your terminal endpoint. If you don't calculate the same coordinate, you have made an error.

Check

Find P on $\overline{AF}$ such that the ratio of AP to PF is 1:3.

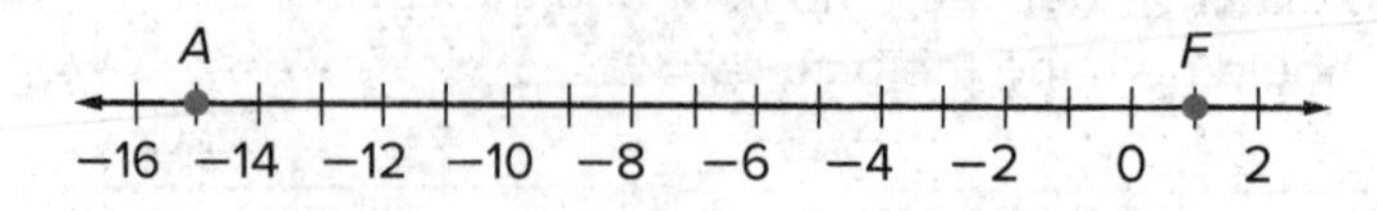

P is located at ______ on the number line.

Example 4 Partition a Directed Line Segment

Think About It!
How can you use estimation to check your answer?

ROAD TRIP **Jorge is traveling 2563 miles from New York City to San Francisco by car. He plans on stopping for gas when the ratio of the distance he has already traveled to the distance he still has to travel is 2:5. How far has Jorge traveled when he stops for gas?**

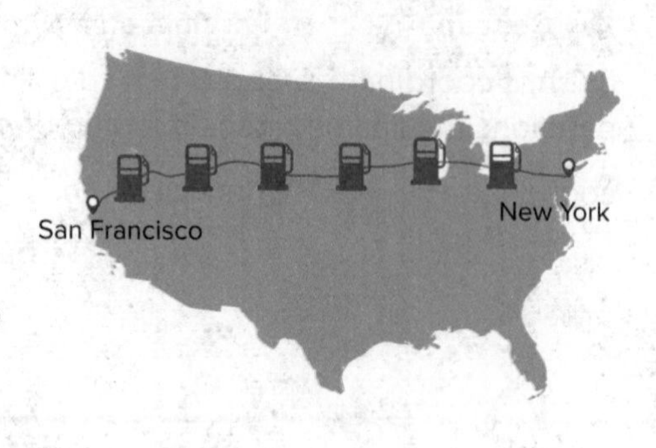

Use the Section Formula to determine how far Jorge will travel before he stops for gas.

$B = \frac{nx_1 + mx_2}{m + n}$ Section Formula

$= \underline{\qquad\qquad} = 732.3$ $m = 2, n = 5, x_1 = 0,$ and $x_2 = 2563$

When Jorge has traveled ______ miles from New York City, the ratio of the distance he has traveled to the distance that he still has to travel is 2:5.

Check

ERRANDS Eduardo travels 30 miles from his house to the bike shop. When Eduardo goes to the bike shop, he always stops at a local pizza place that is along the way. The ratio of the distance Eduardo travels from his house to the pizza place to the distance he travels from the pizza place to the bike shop is 2:3.

How far is the pizza place from Eduardo's house?

______ mi

Go Online You can complete an Extra Example online.

Name ______________________ Period ____________ Date ____________

Practice

Go Online You can complete your homework online.

Examples 1 and 3

Refer to the number line.

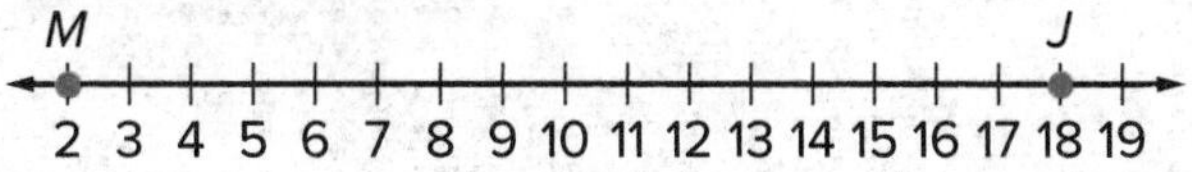

1. Find the coordinate of point *B* that is $\frac{1}{4}$ of the distance from *M* to *J*.

2. Find the coordinate of point *C* that is $\frac{7}{8}$ of the distance from *M* to *J*.

3. Find the coordinate of point *D* that is $\frac{7}{16}$ of the distance from *M* to *J*.

4. Find the coordinate of point *X* such that the ratio of *MX* to *XJ* is 3:1.

5. Find the coordinate of point *X* such that the ratio of *MX* to *XJ* is 2:3.

6. Find the coordinate of point *X* such that the ratio of *MX* to *XJ* is 1:1.

Refer to the number line.

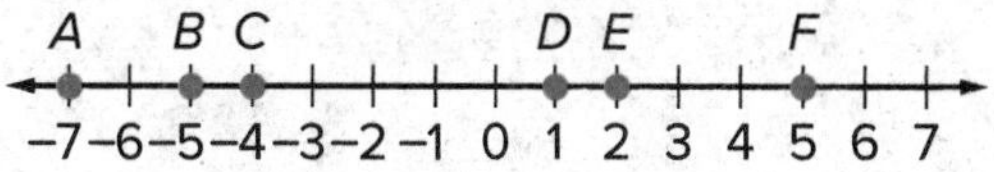

7. Find the coordinate of point *G* that is $\frac{2}{3}$ of the distance from *B* to *D*.

8. Find the coordinate of point *H* that is $\frac{1}{5}$ of the distance from *C* to *F*.

9. Find the coordinate of point *J* that is $\frac{1}{6}$ of the distance from *A* to *E*.

10. Find the coordinate of point *K* that is $\frac{4}{5}$ of the distance from *A* to *F*.

11. Find the coordinate of point *X* such that the ratio of *AX* to *XF* is 1:3.

12. Find the coordinate of point *X* such that the ratio of *BX* to *XF* is 3:2.

13. Find the coordinate of point *X* such that the ratio of *CX* to *XE* is 1:1.

14. Find the coordinate of point *X* such that the ratio of *FX* to *XD* is 5:3.

Refer to the number line.

A B C D F

−5 −4 −3 −2 −1 0 1 2 3 4 5

15. Find the coordinate of point X on $\overline{AF}$ that is $\frac{1}{3}$ of the distance from A to F.

16. Find the coordinate of point Y on $\overline{AC}$ that is $\frac{1}{4}$ of the distance from A to C.

Refer to the number line.

A W X Y Z E

−10 −9 −8 −7 −6 −5 −4 −3 −2 −1 0 1 2 3 4 5 6 7 8 9 10

17. Which point on $\overline{AE}$ is $\frac{2}{3}$ of the distance from A to E?

18. Point X is what fractional distance from E to A?

19. Find the coordinate of point M on $\overline{AE}$ that is $\frac{1}{5}$ of the distance from A to E.

Refer to the number line.

F G H J K L

−15 −10 −5 0 5

20. The ratio of FX to XK is 1:1. Which point is located at X?

21. Find the coordinate of Q on $\overline{FL}$ such that the ratio of FQ to QL is 12:7.

Examples 2 and 4

22. TRAVEL Caroline is taking a road trip on I-70 in Kansas. She stops for gas at mile marker 36. Her destination is at mile marker 353 in Topeka, but she decides to stop at an attraction $\frac{3}{4}$ of the way after stopping for gas. At about which mile marker did Caroline stop to visit the attraction?

Name ______________________ Period ____________ Date ____________

23. HIKING A hiking trail is 24 miles from start to finish. There are two rest areas located along the trail.

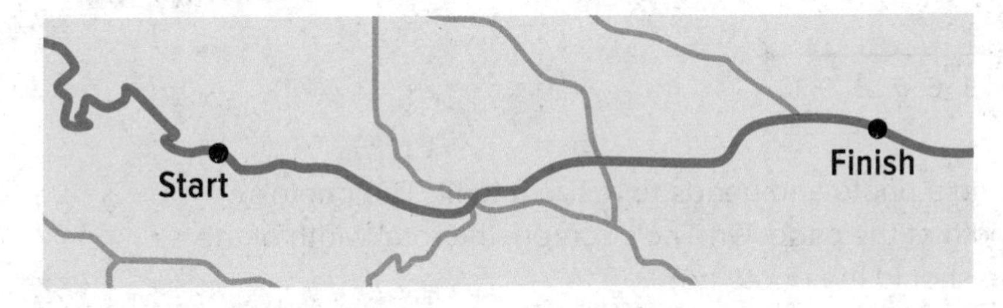

a. The first rest area is located such that the ratio of the distance from the start of the trail to the rest area and the distance from the rest area to the end of the trail is 2:9. To the nearest hundredth of a mile, how far is the first rest area from the starting point of the trail?

b. Kadisha claims that the distance she has walked and that the distance she has left to walk has a ratio of 5:7. How many miles has Kadisha walked?

24. Melany wants to hang a canvas, which is 8 feet wide, on his wall. Where on the canvas should Melany mark the location of the hangers if the canvas requires a hanger every $\frac{1}{5}$ of its length, excluding the edges? Justify your answer.

25. MIGRATION Many American White Pelicans migrate each year, with hundreds of them stopping to rest in various locations along the way. The ratio of the distance some flocks travel from their summer home to one stopover to the distance from the stopover to the winter home is 3:4. If the total distance that the pelicans migrate is 1680 miles, how long is the distance from the summer home to the stopover?

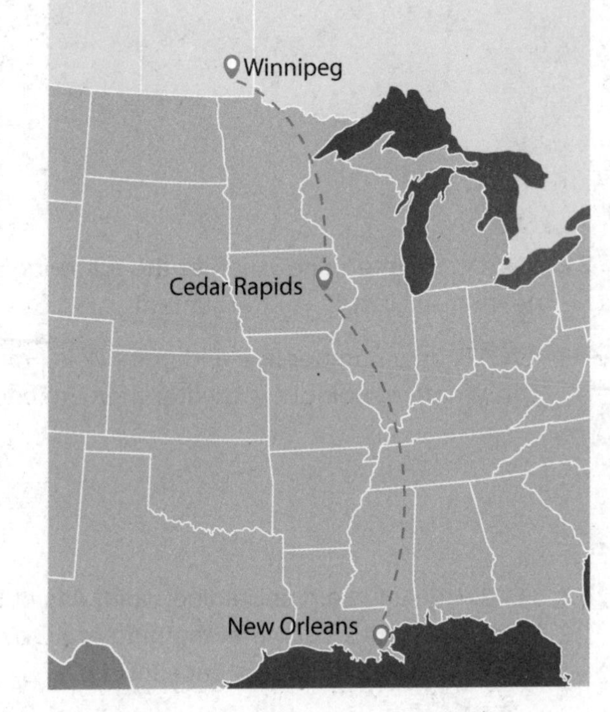

Mixed Exercises

26. Write an equation that can be used to find the coordinate of point K that is $\frac{2}{5}$ of the distance from Q to R.

Q R

−5 −4 −3 −2 −1 0 1 2 3 4 5 6 7 8 9 10

27. SOCIAL MEDIA Tito is posting a photo and needs to resize it to fit. The photo's width should fill $\frac{4}{5}$ of the width of the page. On Tito's screen, the total width of the page is 3 inches. How wide should the photo be?

28. NEONATAL At birth, the ratio of a baby's head length to the length of the rest of its body is 1:3. If a baby's total body length is 22 inches, how long is the baby's head?

29. CREATE Draw a segment and label it $\overline{AB}$. Using only a compass and a straightedge, construct a segment $\overline{CD}$ such that $CD = 5\frac{1}{4}AB$. Explain and then justify your construction.

30. WRITE Naoki wants to center a canvas, which is 8 feet wide, on his bedroom wall, which is 17 feet wide. Where on the wall should Naoki mark the location of the nails, if the canvas requires nails every $\frac{1}{5}$ of its length, excluding the edges? Explain your solution process.

31. ANALYZE Determine whether the following statement is *sometimes, always,* or *never* true. Justify your argument.

If $\overline{XY}$ is on a number line and point W is $\frac{2}{5}$ of the distance from X to Y, then the coordinate of point W is greater than the coordinate of point X.

32. PERSEVERE On a number line, point A is at 5, and point B is at −10. Point C is on $\overline{AB}$ such that the ratio of AC to CB is 1:3. Find D on $\overline{BC}$ that is $\frac{3}{8}$ of the distance from B to C.

Locating Points on a Coordinate Plane

Today's Goals
- Find a point on a directed line segment on the coordinate plane that is a given fractional distance from the initial point.
- Find a point that partitions a directed line segment on the coordinate plane in a given ratio.

Explore Applying Fractional Distance

Online Activity Use a real-world situation to complete the Explore.

INQUIRY How do we use fractional distances in the real world?

Learn Locating Points on the Coordinate Plane with Fractional Distance

You can find a point on a directed line segment that is a fractional distance from an endpoint on the coordinate plane.

Key Concept • Locating a Point at a Fractional Distance on the Coordinate Plane

The coordinates of a point on a line segment that is $\frac{a}{b}$ of the distance from initial endpoint $A(x_1, y_1)$ to terminal endpoint $C(x_2, y_2)$ are given by $\left(x_1 + \frac{a}{b}(x_2 - x_1), y_1 + \frac{a}{b}(y_2 - y_1)\right)$, where $\frac{a}{b}$ is the fraction of the distance if $b \neq 0$.

Watch Out!

Determine the Initial Endpoint Direction is important when determining a point that is a fractional distance on a directed line segment. Identify the initial endpoint you move from and the terminal endpoint you move toward.

Example 1 Fractional Distances on the Coordinate Plane

Find C on $\overline{AB}$ that is $\frac{3}{4}$ of the distance from A to B.

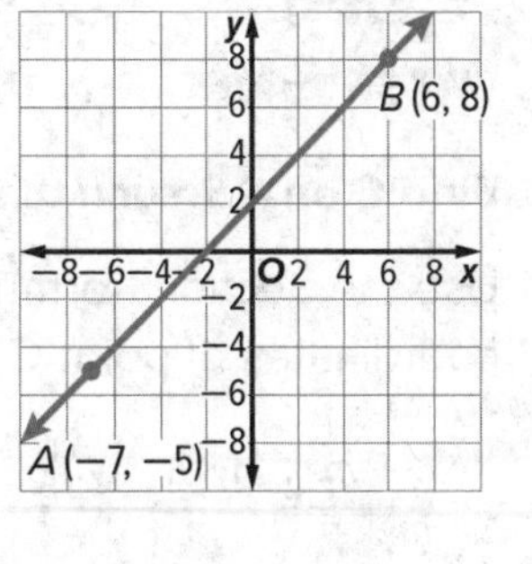

Step 1 Identify the endpoints.

Identify the initial and terminal endpoints.

$(x_1, y_1) = (\_\_\_\_, \_\_\_\_)$ and $(x_2, y_2) = (\_\_\_\_, \_\_\_\_)$

Step 2 Find the x- and y-coordinates.

Find the coordinates of C using the formula for fractional distance.

$\left(x_1 + \frac{a}{b}(x_2 - x_1), y_1 + \frac{a}{b}(y_2 - y_1)\right)$ Fractional Distance Formula

$\left(\_\_\_ + \frac{3}{4}[\_\_\_ - \_\_\_], \_\_\_ + \frac{3}{4}[\_\_\_ - \_\_\_]\right)$ Substitution

Point C is located at (2.75, 4.75).

Study Tip

Checking Coordinates You can check that you have computed the coordinates of C correctly by finding the lengths of $\overline{AC}$ and $\overline{AB}$. If $\frac{AC}{AB}$ is not equal to $\frac{3}{4}$, then you have made an error.

Think About It!

What are the coordinates of a point that is $\frac{3}{4}$ of the distance from B to A?

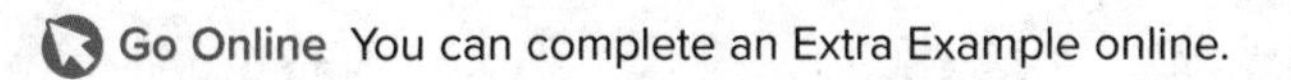
Go Online You can complete an Extra Example online.

Your Notes

Check

Find P on $\overline{QR}$ that is $\frac{1}{6}$ of the distance from Q to R.

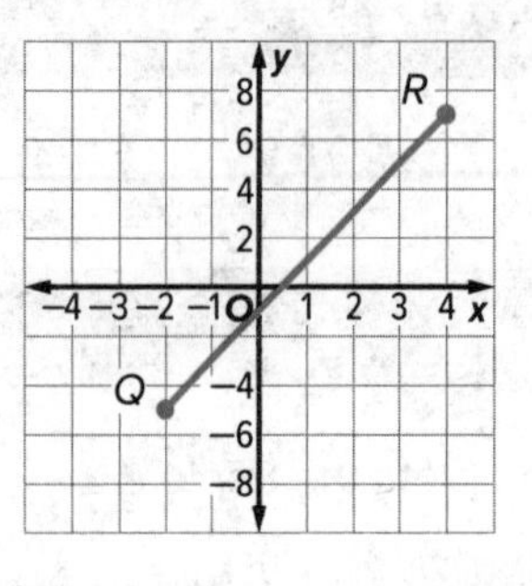

Coordinates of point P ______

Learn Locating Points on the Coordinate Plane with a Given Ratio

The Section Formula can be used to locate a point that partitions a directed line segment on the coordinate plane.

Key Concept • Section Formula on the Coordinate Plane

If A has coordinates (x_1, y_1) and C has coordinates (x_2, y_2), then a point B that partitions the line segment in a ratio of $m{:}n$ has coordinates

$B\left(\frac{nx_1 + mx_2}{m+n}, \frac{ny_1 + my_2}{m+n}\right)$, where $m \neq n$.

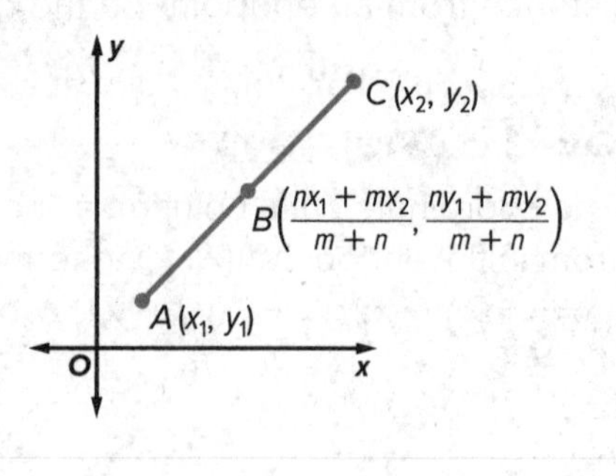

Example 2 Locate a Point on the Coordinate Plane When Given a Ratio

Find C on $\overline{AB}$ such that the ratio of AC to CB is 1:2.

Use the Section Formula to determine the coordinates of point C.

B(6, 2)
A (−5, −2)

$\left(\frac{nx_1 + mx_2}{m+n}, \frac{ny_1 + my_2}{m+n}\right)$ Section Formula

$= \left(\frac{2(-5) + 1(6)}{1+2}, \frac{2(-2) + 1(2)}{1+2}\right)$ Substitute.

$= \left(-\frac{4}{3}, -\frac{2}{3}\right)$ Simplify.

Point C is located at $\left(-\frac{4}{3}, -\frac{2}{3}\right)$.

Talk About It!

How could you check the coordinates of point C?

Check

Find S on $\overline{QR}$ such that the ratio of QS to SR is 2:1.

A. (4, 8)

B. (2, 3)

C. (1, 1)

D. (0, −1)

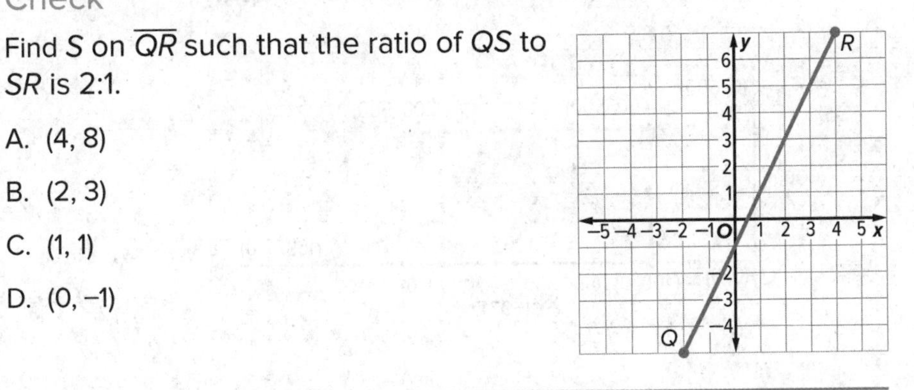

Example 3 Partition a Directed Line Segment on the Coordinate Plane

ZIP LINES **Kendrick is riding a zip line. The zip line is 1800 meters long and starts at a platform 600 meters above the ground. After he jumps, someone takes a picture of his descent. When the picture is taken, the ratio of the distance Kendrick has traveled to the distance he has remaining is 1:2. The picture will show the horizontal distance from 400 meters to 1200 meters from the base of the platform and the vertical distance from ground level to a height of 500 meters. Will Kendrick be in the frame of the picture?**

To determine whether Kendrick is in the frame of the picture, first, determine the horizontal distance x of the zip line. Then, use this information to determine Kendrick's location using the Section Formula.

Step 1 Determine the horizontal distance x of the zip line.

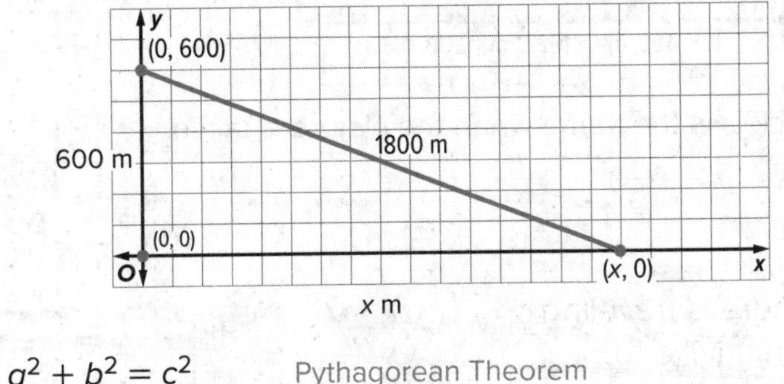

$a^2 + b^2 = c^2$ Pythagorean Theorem

$600^2 + x^2 = 1800^2$ Substitute.

$x \approx 1697.1$ Solve.

The horizontal distance of the zip line is about 1697.1 meters.

(continued on the next page)

Go Online You can complete an Extra Example online.

Step 2 Model the area captured by the photograph.

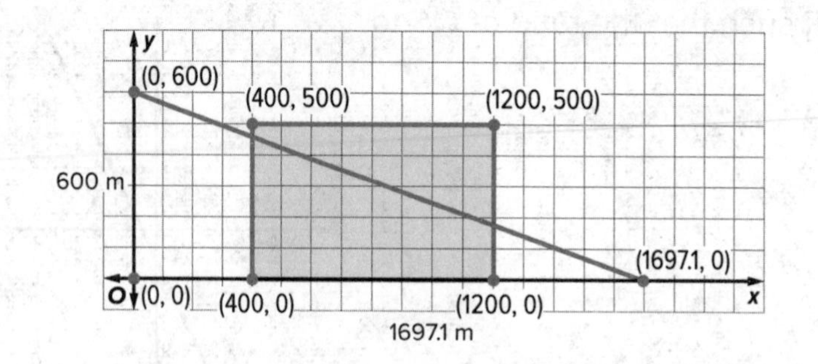

Step 3 Determine Kendrick's location on the zip line.

Use the Section Formula to calculate Kendrick's coordinates.

$\left(\frac{nx_1 + mx_2}{m+n}, \frac{ny_1 + my_2}{m+n}\right)$ Section Formula

$= \left(\frac{2(0) + 1(1697.1)}{1+2}, \frac{2(600) + 1(0)}{1+2}\right)$ Substitute.

$= (565.7, \_\_\_\_)$ Simplify.

Kendrick is at (565.7, 400) when the picture is taken.

Step 4 Graph Kendrick's location to determine whether he is in the frame.

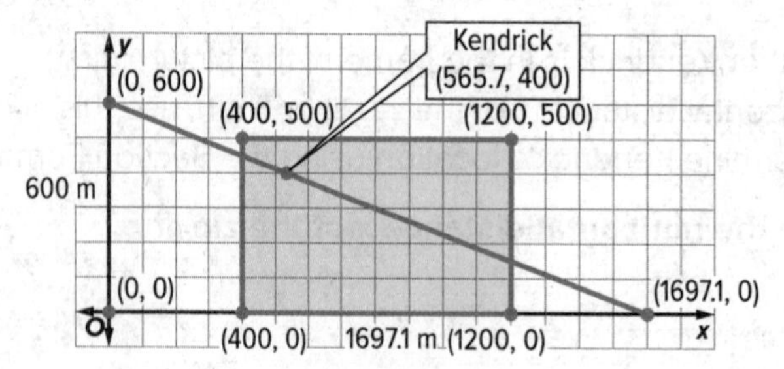

Yes. Kendrick is in the frame when the picture is taken.

Check

TRAVEL Andre is traveling from Jeffersonville to Springfield. He plans to stop for a break when the distance he has traveled and the distance he has left to travel have a ratio of 3:7. Where should Andre stop for his break?

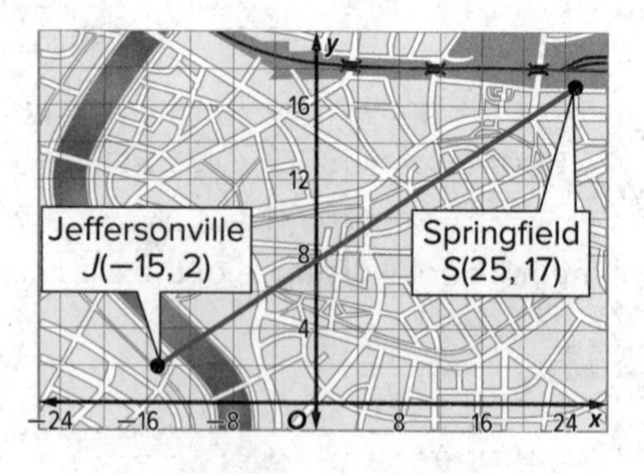

A. (13, 12.5) B. (22, 12.5) C. (−3, 6.5) D. (−12, 6.5)

Go Online You can complete an Extra Example online.

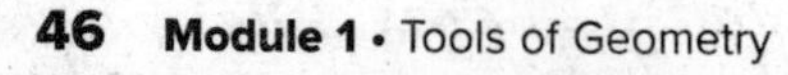

Name ______________________ Period ________ Date ________

Practice

Go Online You can complete your homework online.

Example 1

Find the coordinates of point X on the coordinate plane for each situation.

1. Point X on $\overline{AB}$ is $\frac{1}{5}$ of the distance from A to B.

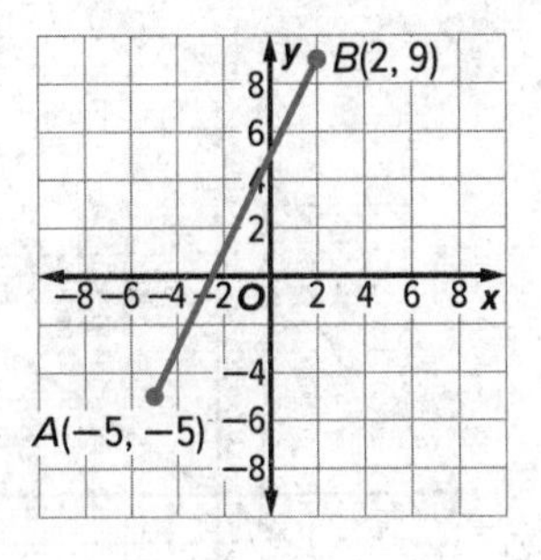

2. Point X on $\overline{RS}$ is $\frac{1}{6}$ of the distance from R to S.

R(−4, 2)
S(2, −2)

3. Point X on $\overline{JK}$ is $\frac{1}{3}$ of the distance from J to K.

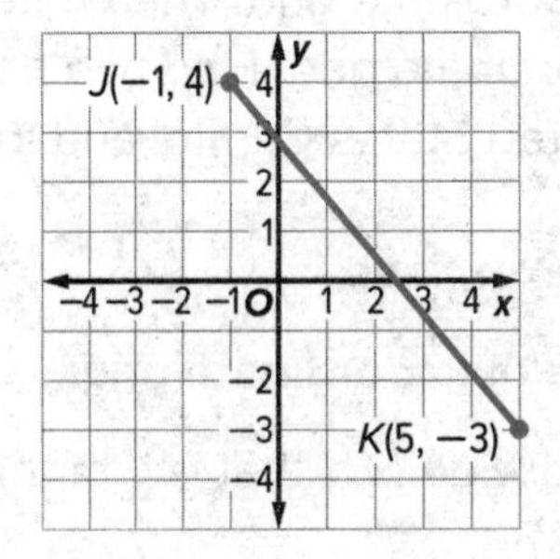

Example 2

Refer to the coordinate grid.

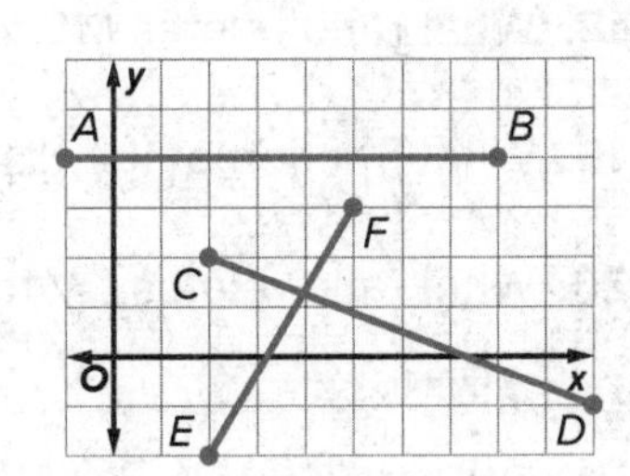

4. Find point X on $\overline{AB}$ such that the ratio of AX to XB is 1:3.

5. Find point Y on $\overline{CD}$ such that the ratio of DY to YC is 2:1.

6. Find point Z on $\overline{EF}$ such that the ratio of EZ to ZF is 2:3.

Examples 1 and 2

Refer to the coordinate grid.

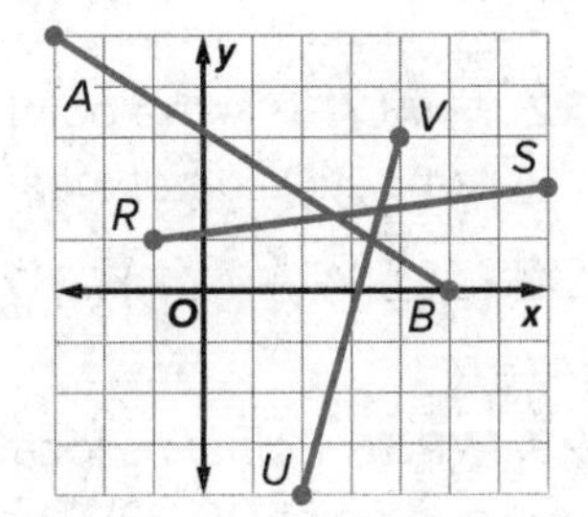

7. Find point C on $\overline{AB}$ that is $\frac{1}{5}$ of the distance from A to B.

8. Find point Q on $\overline{RS}$ that is $\frac{5}{8}$ of the distance from R to S.

9. Find point W on $\overline{UV}$ that is $\frac{1}{7}$ of the distance from U to V.

10. Find point D on $\overline{AB}$ that is $\frac{3}{4}$ of the distance from A to B.

11. Find point Z on $\overline{RS}$ such that the ratio of RZ to ZS is 1:3.

12. Find point G on $\overline{AB}$ such that the ratio of AG to GB is 3:2.

13. Find point E on $\overline{UV}$ such that the ratio of UE to EV is 3:4.

Example 3

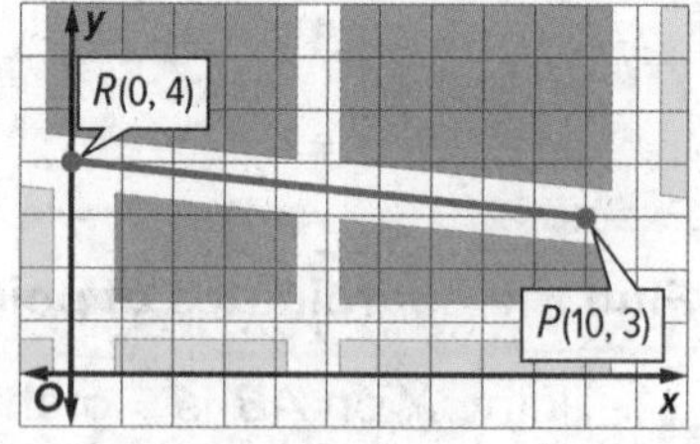

14. MAPS Leila is walking from the park at point P to a restaurant at point R. She wants to stop for a break when the distance she has traveled and the distance she has left to travel has a ratio of 3:5. At which point should Leila stop for her break?

15. CITY PLANNING The United States Capitol is located at (2, −4) on a coordinate grid. The White House is located at (−10, 16) on the same coordinate grid. Find two points on the straight line between the United States Capitol and the White House such that the ratio is 1:3.

Mixed Exercises

Refer to the coordinate grid.

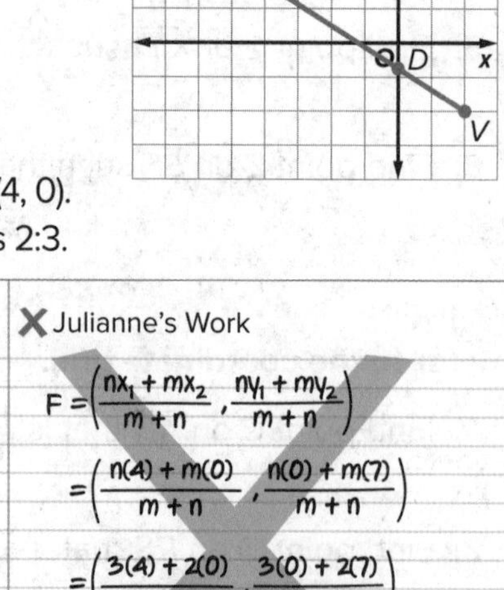

16. Find X on $\overline{MN}$ that is $\frac{3}{4}$ of the distance from M to N.

17. Find Y on $\overline{MN}$ such that the ratio of MY to YN is 1:3.

Point D is located on $\overline{MV}$. The coordinates of D are $\left(0, -\frac{3}{4}\right)$.

18. What ratio relates MD to DV?

19. What fraction of the distance from M to V is MD?

20. What ratio relates DV to MD?

21. FIND THE ERROR Point W is located at (0, 7), and point X is located at (4, 0). Julianne wants to find point F on $\overline{WX}$ such that the ratio of WF to FX is 2:3.

a. What error did Julianne make when solving this problem?

b. What are the correct coordinates of point F?

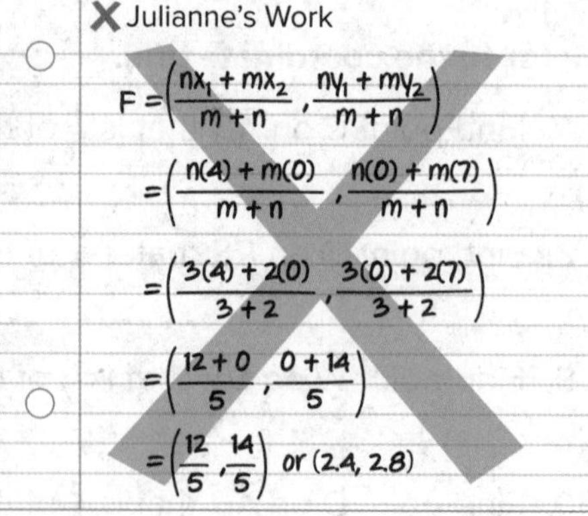

Julianne's Work
$F = \left(\frac{nx_1 + mx_2}{m+n}, \frac{ny_1 + my_2}{m+n}\right)$
$= \left(\frac{n(4) + m(0)}{m+n}, \frac{n(0) + m(7)}{m+n}\right)$
$= \left(\frac{3(4) + 2(0)}{3+2}, \frac{3(0) + 2(7)}{3+2}\right)$
$= \left(\frac{12+0}{5}, \frac{0+14}{5}\right)$
$= \left(\frac{12}{5}, \frac{14}{5}\right)$ or (2.4, 2.8)

22. ANALYZE Is the point one-third of the distance from (x_1, y_1) to (x_2, y_2) *sometimes*, *always*, or *never* the point $\left(\frac{x_1 + x_2}{3}, \frac{y_1 + y_2}{3}\right)$? Justify your argument.

23. WRITE Point P is located on the segment between point $A(1, 4)$ and point $D(7, 13)$. The distance from A to P is twice the distance from P to D. Explain how to find the fractional distance that P is from A to D. What are the coordinates of point P?

24. PERSEVERE Point $C(6, 9)$ is located on the segment between point $A(4, 8)$ and point B. Point C is $\frac{1}{4}$ of the distance from A to B. What are the coordinates of point B?

25. CREATE Draw a line on a coordinate plane. Label two points on the line F and G. Locate a third point on the line between points F and G and label this point H. The point H on $\overline{FG}$ is what fractional distance from F to G?

Midpoints and Bisectors

Explore Midpoints

Online Activity Use paper folding to complete the Explore.

INQUIRY What general formula can you use to find the midpoint of a line segment?

Learn Midpoints on a Number Line

The **midpoint** of a segment is the point halfway between the endpoints of the segment. A point is **equidistant** from other points if it is the same distance from them. The midpoint separates the segment into two segments with a ratio of 1:1. So, you can use the Section Formula to derive the Midpoint Formula.

Key Concept • Midpoint on a Number Line

If $\overline{AB}$ has endpoints at x_1 and x_2 on a number line, then the midpoint M of $\overline{AB}$ has coordinate $M = \frac{x_1 + x_2}{2}$.

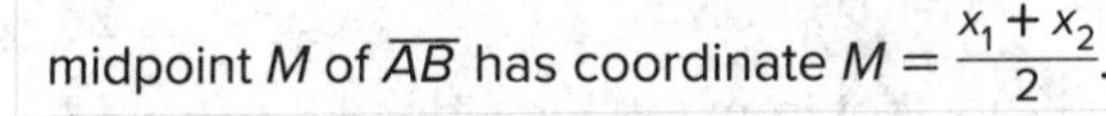

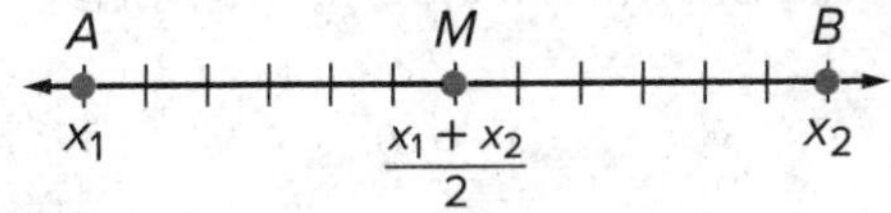

Example 1 Find the Midpoint on a Number Line

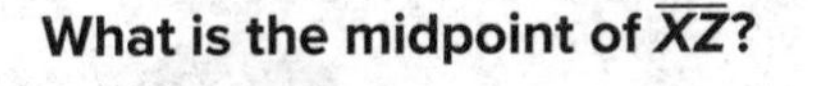

What is the midpoint of $\overline{XZ}$?

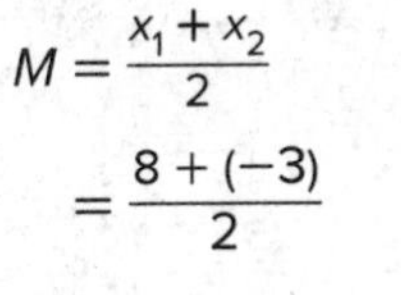

$M = \frac{x_1 + x_2}{2}$ Midpoint Formula

$= \frac{8 + (-3)}{2}$ Substitution

$=$ ____ or ____ Simplify.

The midpoint of $\overline{XZ}$ is ____.

Go Online You can complete an Extra Example online.

Today's Goals

- Find the coordinate of a midpoint on a number line.
- Find the coordinates of the midpoint or endpoint of a line segment on the coordinate plane.
- Find missing values using the definition of a segment bisector.

Today's Vocabulary

midpoint
equidistant
bisect
segment bisector

Watch Out!

Ratios Remember that 1:1 refers to the ratio of the distances, not to the measures of the segments.

Think About It!

Would your answer be different if you reversed the order of x_1 and x_2?

Your Notes

Check

What is the midpoint of $\overline{AF}$?

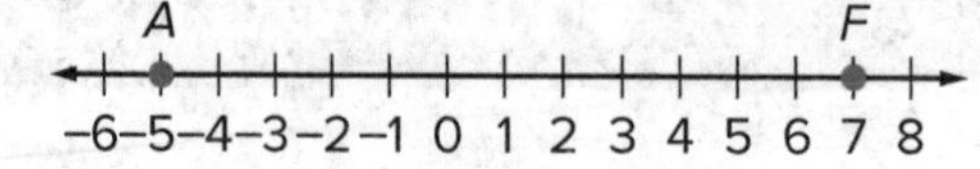

____

Think About It!

How else could Aponi have located the midpoint?

Example 2 Midpoints in the Real World

SIGNS **Aponi works at a vintage clothing store. She wants to hang a new sign so it is centered above the dressing-room doors. Given that the dressing-room doors have the same width, find the point along the wall that Aponi should hang the new sign.**

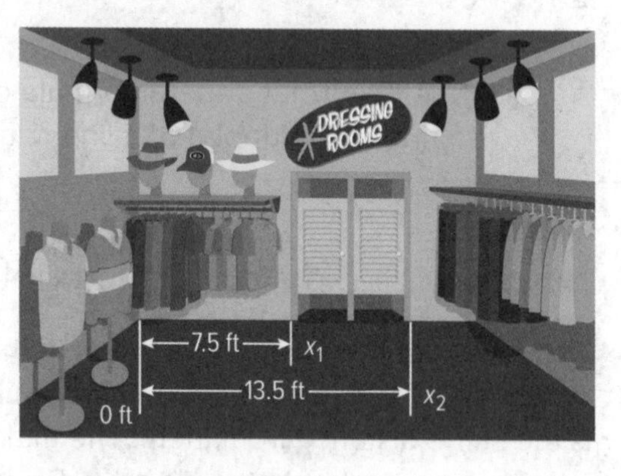

$M = \frac{x_1 + x_2}{2}$ Midpoint Formula

$= \frac{7.5 + (13.5)}{2}$ Substitution

$=$ ____ or ____ Simplify.

Aponi should hang the sign ____ feet from the left side of the wall.

Check

DISTANCE Jorge travels from his school on 38th Street to the library on 62nd Street. He stops halfway there to take a break. Where does Jorge stop to rest?

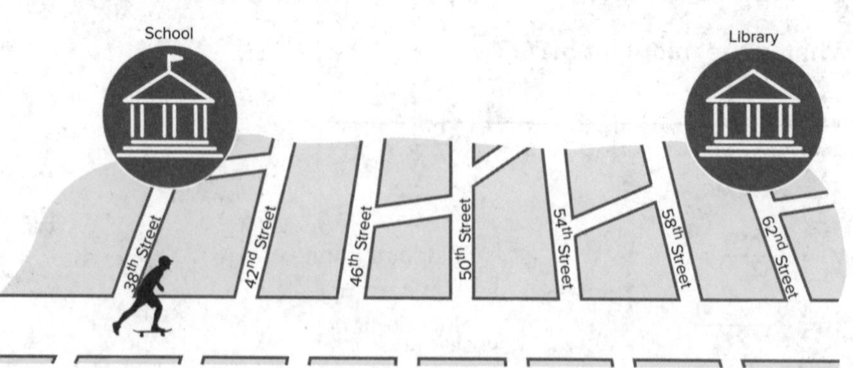

Jorge stops at ____________.

Go Online You can complete an Extra Example online.

Learn Midpoints on the Coordinate Plane

The Section Formula can be used to derive the Midpoint Formula for a segment on the coordinate plane.

Because the midpoint separates the line segment into a ratio of 1:1, substitute 1 for m and n into the formula.

$M = \left(\frac{nx_1 + mx_2}{m+n}, \frac{ny_1 + my_2}{m+n}\right)$ Section Formula

$= \left(\frac{(1)x_1 + (1)x_2}{1+1}, \frac{(1)y_1 + (1)y_2}{1+1}\right)$ Substitution

$= \left(\frac{x_1 + x_2}{2}, \frac{y_1 + y_2}{2}\right)$ Midpoint Formula

Key Concept • Midpoint Formula on the Coordinate Plane

If $\overline{PQ}$ has endpoints at $P(x_1, y_1)$ and $Q(x_2, y_2)$ on the coordinate plane, then the midpoint M of $\overline{PQ}$ has coordinates $M\left(\frac{x_1 + x_2}{2}, \frac{y_1 + y_2}{2}\right)$.

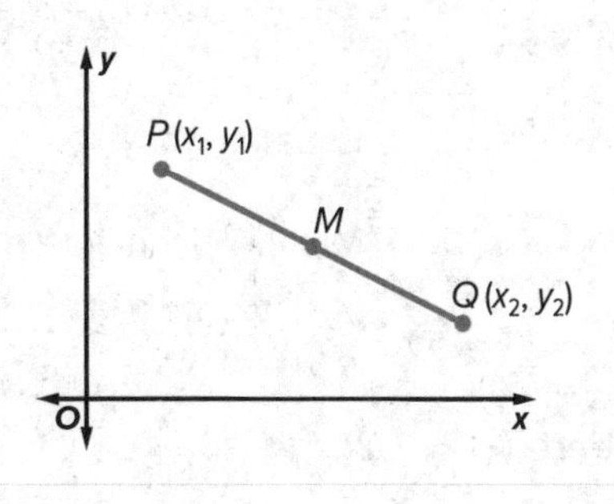

Example 3 Find the Midpoint on the Coordinate Plane

Find the coordinates of M, the midpoint of $\overline{AB}$, for $A(-2, 1)$ and $B(8, 3)$.

$M = \left(\frac{x_1 + x_2}{2}, \frac{y_1 + y_2}{2}\right)$ Midpoint Formula

$= \left(\frac{-2 + 8}{2}, \frac{1 + 3}{2}\right)$ Substitution

$= \left(\frac{6}{2}, \frac{4}{2}\right)$ or $(3, 2)$ Simplify.

Check

Find the coordinates of B, the midpoint of $\overline{AC}$, for $A(-3, -2)$ and $C(5, 10)$.

(____, ____)

Talk About It!

Would the coordinates of the midpoint be different if you use point A as (x_2, y_2) and point B as (x_1, y_1)? Explain.

Go Online You can complete an Extra Example online.

Watch Out!

Midpoint Formula The Midpoint Formula only uses addition and division. Think of the midpoint as the average of the *x*- and *y*-coordinates of the given endpoints.

Study Tip

Check for Reasonableness Always graph the given information and the calculated coordinates of the midpoint to check the reasonableness of your answer.

Think About It!

How can you use the graph to determine whether your answer is reasonable?

Example 4 Find Missing Coordinates

Find the coordinates of *A* if $P\left(3, \frac{1}{2}\right)$ is the midpoint of $\overline{AB}$ and *B* has coordinates (8, 3).

First, substitute the known information into the Midpoint Formula. Let *A* be (x_1, y_1) and *B* be (x_2, y_2).

$$M = \left(\frac{x_1 + x_2}{2}, \frac{y_1 + y_2}{2}\right) \quad \text{Midpoint Formula}$$

$$\left(3, \frac{1}{2}\right) = \left(\frac{x_1 + 8}{2}, \frac{y_1 + 3}{2}\right) \quad \text{Substitution}$$

Next, write two equations to solve for x_1 and y_1.

______ Equation for x_1

____ $= x_1 + 8$ Multiply each side by 2.

____ $= x_1$ Solve.

______ Equation for y_1

____ $= y_1 + 3$ Multiply each side by 2.

____ $= y_1$ Solve.

The coordinates of *A* are (−2, −2).

Plot the points on a coordinate plane to check your answer for reasonableness.

y

O *x*

Check

Find the coordinates of *Q* if *R*(6, −1) is the midpoint of $\overline{QS}$ and *S* has coordinates (12, 4). ______

Go Online You can complete an Extra Example online.

Learn Bisectors

Because the midpoint separates the segment into two congruent segments, we can say that the midpoint **bisects** the segment. Any segment, line, plane, or point that bisects a segment is called a **segment bisector**.

Example 5 Find Missing Measures

Find the measure of $\overline{RT}$ if T is the midpoint of $\overline{RQ}$.

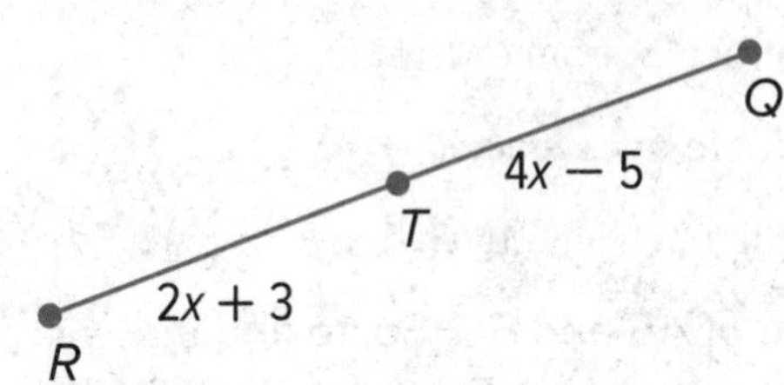

Because T is the midpoint, $RT = TQ$. Use this equation to solve for x.

$RT = TQ$	Definition of midpoint
$2x + 3 = 4x - 5$	Substitution
$3 = 2x - 5$	Subtract $2x$ from each side.
$8 = 2x$	Add 5 to each side.
$4 = x$	Divide each side by 2.

Substitute 4 for x in the equation for RT.

$RT = 2x + 3$	Equation for RT
$= 2(4) + 3$	Substitution
$= 11$	Simplify.

Think About It!

Is there a way to find the length of $\overline{TQ}$ without calculating when you know the length of $\overline{RT}$? Why or why not?

Check

Find the measure of $\overline{RS}$ if S is the midpoint of $\overline{RT}$.

R $7x - 5$ S $6x + 4$ T

A. 56

B. 58

C. 112

D. 116

Go Online You can complete an Extra Example online.

Example 6 Find the Total Length

Find the measure of $\overline{AC}$ if B is the midpoint of $\overline{AC}$.

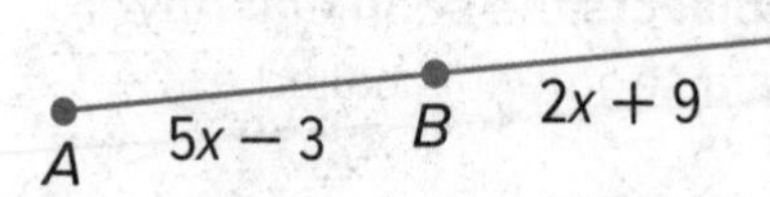

Because B is the midpoint, $AB = BC$. Use this equation to solve for x.

$AB = BC$	Definition of midpoint
$5x - 3 =$ ____	Substitution
____$x - 3 =$ ____	Subtract $2x$ from each side.
$3x =$ ____	Add 3 to each side.
$x =$ ____	Divide each side by 3.

The length of $\overline{AC}$ is equal to the sum of AB and BC. So, to find the length of $\overline{AC}$, substitute 4 for x in the expression $5x - 3 + 2x + 9$.

$AC = 5x - 3 + 2x + 9$	Length of $\overline{AC}$
$= 5(\_\_\_) - 3 + 2(\_\_\_) + 9$	$x = 4$
$=$ ____ $- 3 +$ ____ $+ 9$	Multiply.
$=$ ____	Simplify.

The measure of $\overline{AC}$ is 34.

Think About It!
What concept are we using when we say that $AC = AB + BC$?

Check

Find the measure of $\overline{AC}$ if B is the midpoint of $\overline{AC}$. Round your answer to the nearest tenth, if necessary.

A
3m + 4
B
8m + 7
C

____

Pause and Reflect

Did you struggle with anything in this lesson? If so, how did you deal with it?

Record your observations here.

Go Online You may want to complete the construction activities for this lesson.

Go Online You can complete an Extra Example online.

Name ______________________ Period ____________ Date ____________

Practice

Go Online You can complete your homework online.

Example 1

Use the number line to find the coordinate of the midpoint of each segment.

J K L M N P

−7 −6 −5 −4 −3 −2 −1 0 1 2 3 4 5 6

1. $\overline{KM}$ **2.** $\overline{JP}$ **3.** $\overline{LN}$

4. $\overline{MP}$ **5.** $\overline{LP}$ **6.** $\overline{JN}$

Use the number line to find the coordinate of the midpoint of each segment.

E F G H J K L

−6 −4 −2 0 2 4 6 8 10

7. $\overline{FK}$ **8.** $\overline{HK}$ **9.** $\overline{EF}$

10. $\overline{FG}$ **11.** $\overline{JL}$ **12.** $\overline{EL}$

USE TOOLS **Use the number line to find the coordinate of the midpoint of each segment.**

A B C D E

−6 −4 −2 0 2 4 6 8 10 12

13. $\overline{DE}$ **14.** $\overline{BC}$

15. $\overline{BD}$ **16.** $\overline{AD}$

Example 2

17. HOME IMPROVEMENT Callie wants to build a fence halfway between her house and her neighbor's house. How far away from Callie's house should the fence be built?

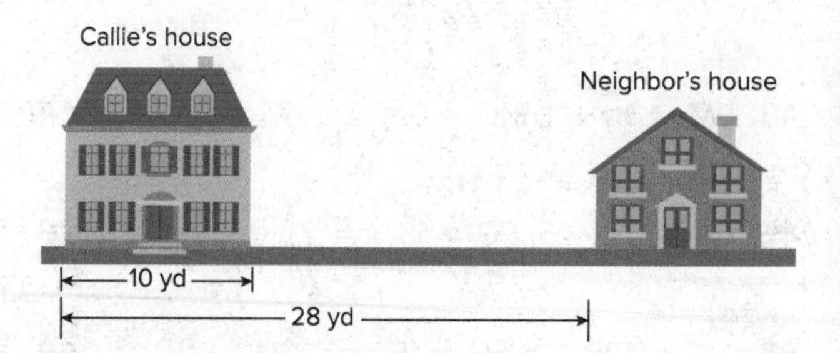

18. DINING Calvino's home is located at the midpoint between Fast Pizza and Pizza Now. Fast Pizza is a quarter mile away from Calvino's home. How far away is Pizza Now from Calvino's home? How far apart are the two pizzerias?

Example 3

Find the coordinates of the midpoint of a segment with the given endpoints.

19. (5, 11), (3, 1)

20. (7, −5), (3, 3)

21. (−8, −11), (2, 5)

22. (7, 0), (2, 4)

23. (−5, 1), (2, 6)

24. (−4, −7), (12, −6)

25. (2, 8), (8, 0)

26. (9, −3), (5, 1)

27. (22, 4), (15, 7)

28. (12, 2), (7, 9)

29. (−15, 4), (2, −10)

30. (−2, 5), (3, −17)

31. (2.4, 14), (6, 6.8)

32. (−11.2, −3.4), (−5.6, −7.8)

Example 4

Find the coordinates of the missing endpoint if *B* is the midpoint of $\overline{AC}$.

33. $C(-5, 4)$, $B(-2, 5)$

34. $A(1, 7)$, $B(-3, 1)$

35. $A(-4, 2)$, $B(6, -1)$

36. $C(-6, -2)$, $B(-3, -5)$

37. $A(4, -0.25)$, $B(-4, 6.5)$

38. $C\left(\frac{5}{3}, -6\right)$, $B\left(\frac{8}{3}, 4\right)$

Examples 5 and 6

Suppose *M* is the midpoint of $\overline{FG}$. Find each missing measure.

39. $FM = 5y + 13$, $MG = 5 - 3y$, $FG = ?$

40. $FM = 3x - 4$, $MG = 5x - 26$, $FG = ?$

41. $FM = 8a + 1$, $FG = 42$, $a = ?$

42. $MG = 7x - 15$, $FG = 33$, $x = ?$

43. $FM = 3n + 1$, $MG = 6 - 2n$, $FG = ?$

44. $FM = 12x - 4$, $MG = 5x + 10$, $FG = ?$

45. $FM = 2k - 5$, $FG = 18$, $k = ?$

46. $FG = 14a + 1$, $FM = 14.5$, $a = ?$

47. $MG = 13x + 1$, $FG = 15$, $x = ?$

48. $FG = 11x - 15.6$, $MG = 10.9$, $x = ?$

Mixed Exercises

Find the coordinates of the missing endpoint if *P* is the midpoint of $\overline{NQ}$.

49. $N(2, 0)$, $P(5, 2)$

50. $N(5, 4)$, $P(6, 3)$

51. $Q(3, 9)$, $P(-1, 5)$

52. Find the value of y if M is the midpoint of $\overline{LN}$.

$9y - 4$ $\quad$ $6y + 5$

L $\quad$ M $\quad$ N

53. CAMPING Troop 175 is designing a new campground by first mapping everything on a coordinate grid. They found locations for the mess hall and their cabins. They want the bathrooms to be halfway between these two places. What are the coordinates of the location of the bathrooms?

Cabins

Mess Hall

54. GAME DESIGN A computer software designer is creating a new video game. The designer wants to create a secret passage that is halfway between the castle and the bridge. Where should the secret passage be located?

Bridge

Castle

55. SCAVENGER HUNT Pablo is going to ask Bianca to prom by sending her on a scavenger hunt. At the end of the scavenger hunt, Pablo will be standing halfway between the gazebo and the ice cream shop in town. Where should Pablo stand?

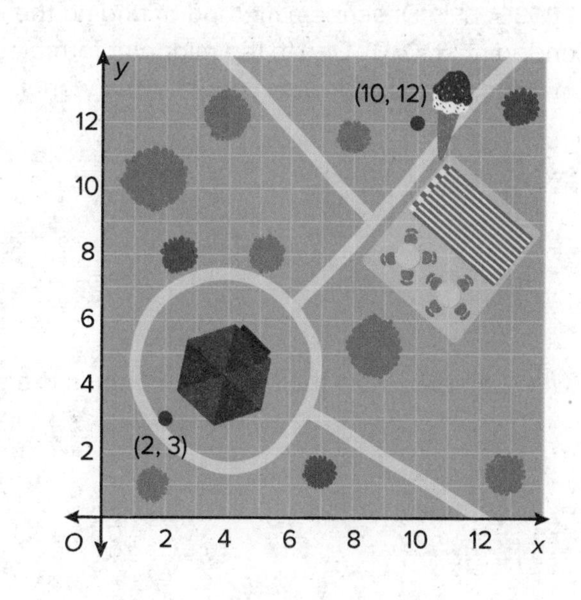

56. WALKING Javier walks from his home at point K to the Internet café at point O. If the school at point W is exactly halfway between Javier's house and the Internet café, how far does Javier walk?

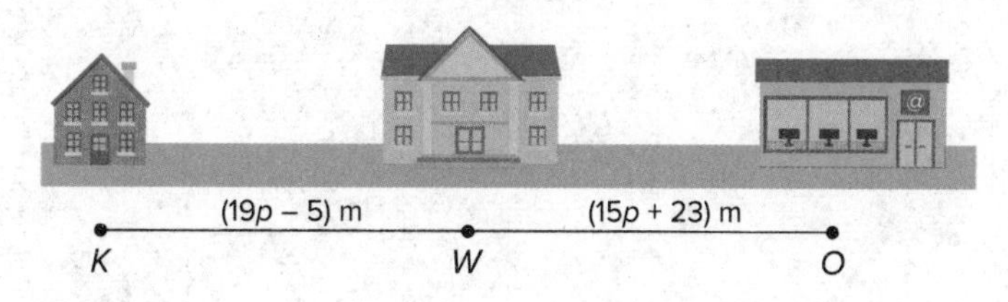

57. **SCHOOL LIFE** Bryan is at the library doing a research paper. He leaves the library at point A and walks to the soccer field for a game at point C. The supermarket at point B is exactly halfway between the library and the soccer field. After Bryan's first soccer game, he walks to the supermarket to buy a snack, and then he walks back to the soccer field for his second game. Not including the time spent at the soccer game, how far does Bryan walk?

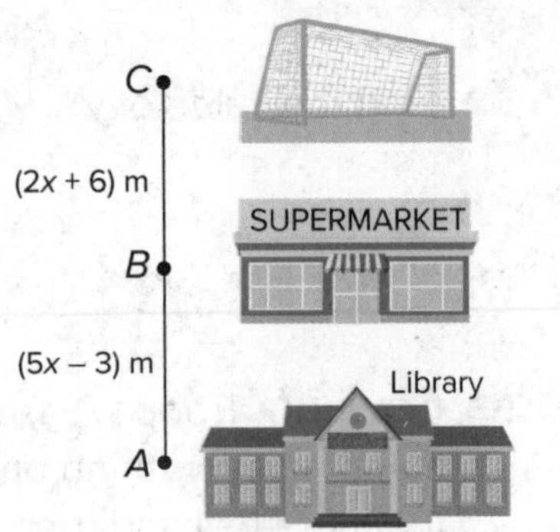

58. **REASONING** A drone flying over a field of corn identifies a dry area. The coordinates of the vertices of the area are shown. To what coordinates should the portable irrigation system be sent to water the dry area? Explain your reasoning.

(56, 1012)
(888, 1055)
(144, 289)
(895, 332)

59. **PERSEVERE** Describe a method of finding the midpoint of a segment that has one endpoint at (0, 0). Derive the midpoint formula, give an example using your method, and explain why your method works.

60. **WRITE** Explain how the Midpoint Formula is a special case of the Section Formula.

61. **CREATE** Construct $\overline{AC}$ given $\overline{AB}$ if B is the midpoint of $\overline{AC}$.

Review

Essential Question

How are points, lines, and segments used to model the real world?

Module Summary

Lesson 1-1

The Geometric System

- An axiomatic system has a set of axioms from which theorems can be derived.
- Synthetic geometry is the study of geometric figures without the use of coordinates.
- Analytic geometry is the study of geometry using a coordinate system.

Lessons 1-2 through 1-4

Points, Lines, Line Segments, and Planes

- The terms *point, line,* and *plane* are undefined terms because they are readily understood and are not formally explained by means of more basic words and concepts.
- Collinear points are points that lie on the same line. Coplanar points are points that lie in the same plane.
- The intersection of two or more geometric figures is the set of points they have in common.
- Point C is between A and B if and only if A, B, and C are collinear and $AC + CB = AB$.
- Two segments that have the same measure are congruent segments.
- The distance between two points on a number line is the absolute value of their difference.
- The distance between two points on a coordinate plane, (x_1, y_1) and (x_2, y_2), is $\sqrt{(x_2 - x_1)^2 + (y_2 - y_1)^2}$.

Lessons 1-5 and 1-6

Locating Points

- If C has coordinate x_1 and D has coordinate x_2, then a point P that partitions the line segment in a ratio of $m{:}n$ is located at coordinate $\frac{nx_1 + mx_2}{m + n}$.
- The coordinates of point B that is $\frac{a}{b}$ of the distance from point $A(x_1, y_1)$ to point $C(x_2, y_2)$ are $\left(x_1 + \frac{a}{b}(x_2 - x_1), y_1 + \frac{a}{b}(y_2 - y_1)\right)$.

Lesson 1-7

Midpoints and Bisectors

- If $\overline{AB}$ has endpoints at x_1 and x_2 on a number line, then the midpoint M of $\overline{AB}$ has coordinate $M = \frac{x_1 + x_2}{2}$
- A midpoint separates a segment into two congruent parts, so it bisects the segment.

Study Organizer

Foldables

Use your Foldable to review this module. Working with a partner can be helpful. Ask for clarification of concepts as needed.

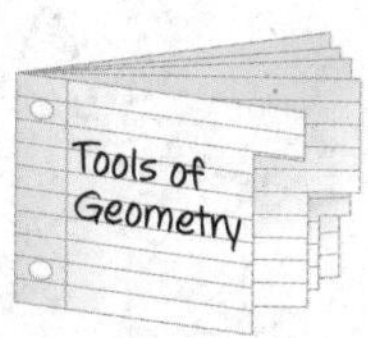

Name ______________________ Period __________ Date __________

Test Practice

1. **TABLE ITEM** Select the geometric terms modeled by the real-world objects. (Lesson 1-2)

Real-World Object	Point	Line	Plane
Electronic Tablet			
Pool Stick			
Scoop of Ice Cream			
Light Pole			
Emoji			

2. **MULTI-SELECT** Use the figure to name all planes containing point *W*. (Lesson 1-2)

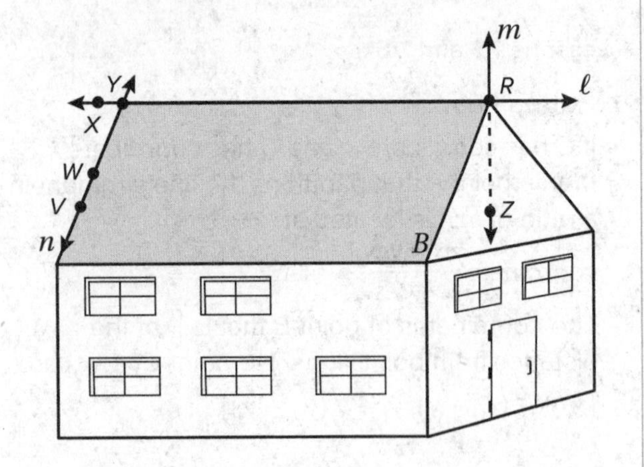

Ⓐ plane *VWY*

Ⓑ plane *VWX*

Ⓒ plane *RYV*

Ⓓ plane *VWZ*

Ⓔ plane *RYX*

3. **OPEN RESPONSE** What geometric figures do the pages of the book represent? (Lesson 1-2)

4. **MULTIPLE CHOICE** Which sequence identifies the correct order for completing the construction to copy a line segment using a compass and straightedge? (Lesson 1-3)

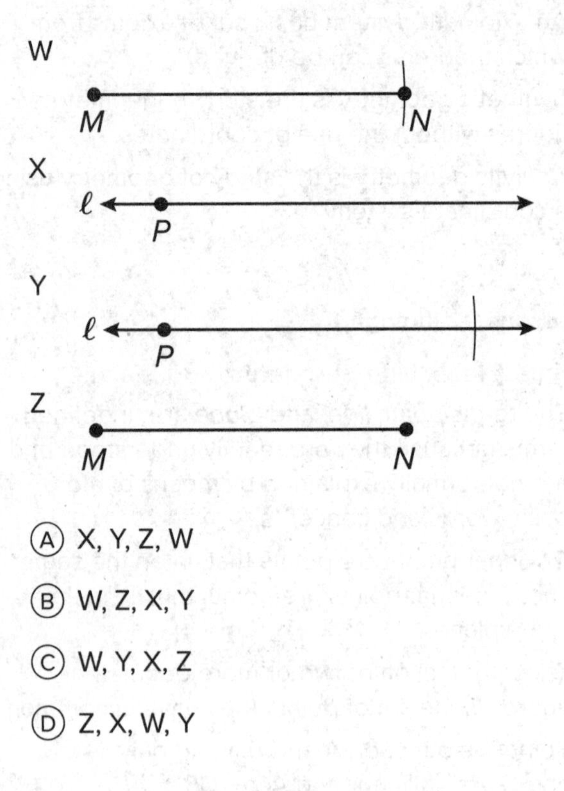

Ⓐ X, Y, Z, W

Ⓑ W, Z, X, Y

Ⓒ W, Y, X, Z

Ⓓ Z, X, W, Y

Name _______________ Period _______ Date _______

5. **OPEN RESPONSE** Find the value of x if Q is between P and R, $PQ = 5x - 10$, $QR = 3(x + 4)$, and $\overline{PQ} \cong \overline{QR}$. (Lesson 1-3)

P Q R

$5x - 10$ $3(x + 4)$

6. **OPEN RESPONSE** On a straight highway, the distance from Loretta's house to a park is 43 miles. Her friend Jamal lives along this same highway between Loretta's house and the park. The distance from Loretta's house to Jamal's house is 31 miles. How many miles is it from Jamal's house to the park? (Lesson 1-3)

7. **MULTIPLE CHOICE** Find the distance between the two points on a coordinate plane. (Lesson 1-4)

$A(5, 1)$ and $B(-3, -3)$

Ⓐ $4\sqrt{5}$

Ⓑ $4\sqrt{3}$

Ⓒ $2\sqrt{2}$

Ⓓ $2\sqrt{3}$

8. **OPEN RESPONSE** True or false: $\overline{XY} \cong \overline{WZ}$ (Lesson 1-4)

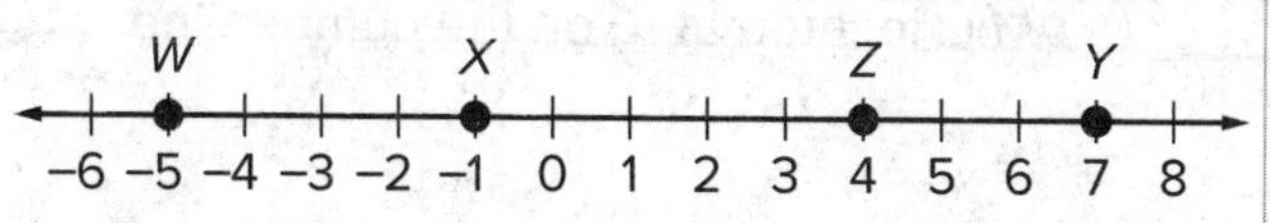

9. **MULTIPLE CHOICE** The coordinates of A and B on a number line are -7 and 9. The coordinates of C and D on a number line are -4 and 12. Are $\overline{AB}$ and $\overline{CD}$ congruent? If yes, what is the length of each segment? (Lesson 1-4)

Ⓐ no

Ⓑ yes; 16

Ⓒ yes; −16

Ⓓ yes; 8

10. **OPEN RESPONSE** The coordinate of point X on $\overline{PQ}$ that is $\frac{3}{4}$ of the distance from P to Q is ___. (Lesson 1-5)

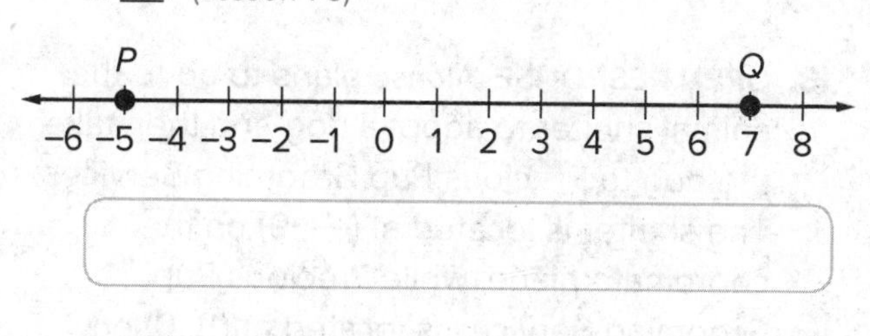

11. **MULTIPLE CHOICE** On a number line, point S is located at -3 and point T is located at 9. Where is point R located on $\overline{ST}$ if the ratio of SR to RT is 3:4? (Lesson 1-5)

Ⓐ $\frac{27}{7}$

Ⓑ $2\frac{1}{4}$

Ⓒ $1\frac{1}{4}$

Ⓓ $\frac{15}{7}$

Name ______________________ Period ____________ Date ____________

12. MULTIPLE CHOICE Find point R on $\overline{ST}$ such that the ratio of SR to RT is 1:2. (Lesson 1-6)

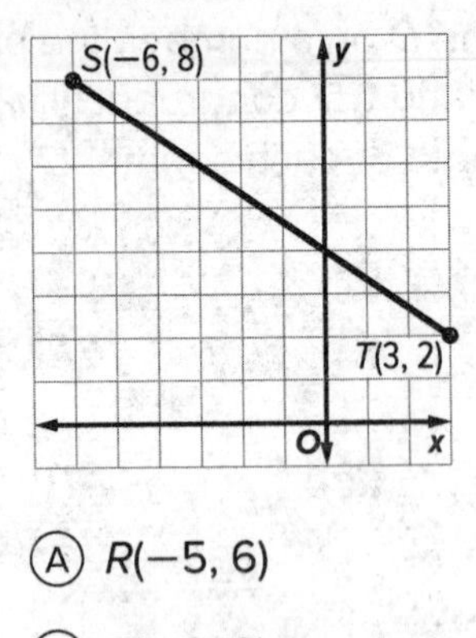

Ⓐ $R(-5, 6)$

Ⓑ $R(-3, 6)$

Ⓒ $R(-1.5, 5)$

Ⓓ $R(0, 4)$

13. OPEN RESPONSE Alonso plans to go to the animal shelter to adopt a dog and then take the dog to Precious Pup Grooming Services. The shelter is located at (−1, 9) on the coordinate plane, while Precious Pup Grooming Services is located at (11, 0) on the coordinate plane. Find the location of Alonso's home if it is $\frac{1}{3}$ of the distance from the shelter to Precious Pup Grooming Services. (Lesson 1-6)

14. OPEN RESPONSE Find the coordinates of A if $M(6, -1)$ is the midpoint of $\overline{AB}$, and B has the coordinates (8, −7). (Lesson 1-7)

15. MULTIPLE CHOICE Find the measure of $\overline{YZ}$ if Y is the midpoint of $\overline{XZ}$. (Lesson 1-7)

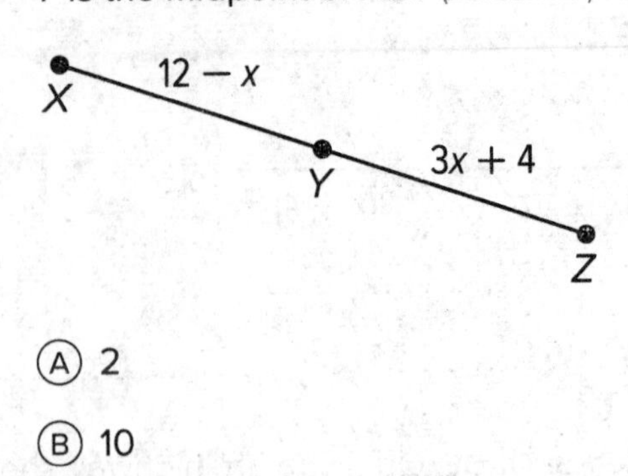

Ⓐ 2

Ⓑ 10

Ⓒ 16

Ⓓ 20

16. MULTIPLE CHOICE Find the y-coordinate of the point M, the midpoint of $\overline{AB}$, for $A(-3, 3)$ and $B(5, 7)$. (Lesson 1-7)

Ⓐ −1

Ⓑ 1

Ⓒ 2

Ⓓ 5

17. MULTIPLE CHOICE Points A and B are plotted on a number line. What is the location of M, the midpoint of $\overline{AB}$, for A at −9 and B at 28? (Lesson 1-7)

Ⓐ M is located at 18.5 on the number line.

Ⓑ M is located at 14 on the number line.

Ⓒ M is located at 9.5 on the number line.

Ⓓ M is located at $\frac{10}{3}$ on the number line.

Module 2

Angles and Geometric Figures

Essential Question

How are angles and two-dimensional figures used to model the real world?

What Will You Learn?

Place a check mark (✓) in each row that corresponds with how much you already know about each topic **before** starting this module.

KEY

— I don't know. — I've heard of it. — I know it!

	Before			After		
	I don't know	I've heard of it	I know it!	I don't know	I've heard of it	I know it!
apply the definitions of angles, parts of angles, congruent angles, and angle bisectors to calculate angle measures						
apply the characteristics of complementary and supplementary angles and parallel and perpendicular lines to calculate angle measures						
apply the characteristics of perpendicular lines to calculate angle measures						
find perimeters, circumferences, and areas of two-dimensional geometric shapes						
reflect, translate, and rotate figures						
solve for unknown measures of three-dimensional figures by calculating surface areas and volumes						
model three-dimensional geometric figures with orthographic drawings						
determine levels of precision and accuracy						
determine the correct numbers of significant figures in recorded measurements						

Foldables Make this Foldable to help you organize your notes about angles and geometric figures. Begin with two sheets of grid paper.

1. **Fold** in half along the width.
2. **On** the first sheet, cut 5 centimeters along the fold at the ends.
3. **On** the second sheet, cut in the center, stopping 5 centimeters at the ends.
4. **Insert** the first sheet through the second sheet and align the folds. Label with lesson numbers.

First Sheet

Second Sheet

What Vocabulary Will You Learn?

Check the box next to each vocabulary term that you may already know.

- ☐ accuracy
- ☐ adjacent angles
- ☐ angle
- ☐ angle bisector
- ☐ angle of rotation
- ☐ approximate error
- ☐ area
- ☐ base of a pyramid or cone
- ☐ bases of a prism or cylinder
- ☐ center of rotation
- ☐ circumference
- ☐ complementary angles
- ☐ component form
- ☐ concave
- ☐ cone
- ☐ congruent angles
- ☐ convex
- ☐ cylinder
- ☐ edge of a polyhedron
- ☐ equiangular polygon
- ☐ equilateral polygon
- ☐ exterior
- ☐ face of a polyhedron
- ☐ geometric model
- ☐ image
- ☐ interior
- ☐ line of reflection
- ☐ linear pair
- ☐ net
- ☐ opposite rays
- ☐ orthographic drawing
- ☐ perimeter
- ☐ perpendicular
- ☐ Platonic solid
- ☐ polygon
- ☐ polyhedron
- ☐ precision
- ☐ preimage
- ☐ prism
- ☐ pyramid
- ☐ ray
- ☐ reflection
- ☐ regular polygon
- ☐ regular polyhedron
- ☐ rigid motion
- ☐ rotation
- ☐ sides
- ☐ significant figures
- ☐ sphere
- ☐ straight angle
- ☐ supplementary angles
- ☐ surface area
- ☐ transformation
- ☐ translation
- ☐ translation vector
- ☐ vertex
- ☐ vertex of a polyhedron
- ☐ vertical angles
- ☐ volume

Are You Ready?

Complete the Quick Review to see if you are ready to start this module. Then complete the Quick Check.

Quick Review

Example 1

Solve $5x + 2 = 90$.

$5x + 2 = 90$	Original equation.
$5x = 88$	Subtract 2 from each side.
$x = 17.6$	Divide each side by 5.

Example 2

Evaluate $2(3)(4) + 2(3)(5) + 2(4)(5)$.

$2(3)(4) + 2(3)(5) + 2(4)(5)$	Original expression
$= 24 + 30 + 40$	Multiply.
$= 94$	Add.

Quick Check

Solve each equation.

1. $3x - 9 = 180$

2. $2x + 10x - 9 = 90$

3. $15x + 42 = 12x + 51$

4. $9x + 1 = 17x - 31$

Evaluate each expression.

5. $6(15)(22)$

6. $0.5(8)(9)$

7. $2(6)(7) + 2(6)(10) + 2(7)(10)$

8. $0.5(5)(12) + 0.5(5)(12) + 5(14) + 12(14) + 13(14)$

How Did You Do?

Which exercises did you answer correctly in the Quick Check? Shade those exercise numbers below.

① ② ③ ④ ⑤ ⑥ ⑦ ⑧

Angles and Congruence

Today's Goals

- Analyze figures using the definitions of angles and parts of angles.
- Calculate angle measures using the definitions of congruent angles and angle bisectors.
- Analyze figures using the characteristics of adjacent angles, linear pairs of angles, and vertical angles.

Today's Vocabulary

ray
opposite rays
straight angle
angle
sides
vertex
interior
exterior
congruent angles
angle bisector
adjacent angles
linear pair
vertical angles

Explore Angles Formed by Intersecting Lines

Online Activity Use dynamic geometry software to complete the Explore.

INQUIRY What angle relationships are formed by two intersecting lines?

Learn Angles

Lines and portions of lines intersect to form angles.

A **ray** is the part of a line consisting of a point on the line, called the *endpoint of the ray,* together with all of the collinear points on one side of the endpoint.

Two collinear rays with a common endpoint are **opposite rays**. Opposite rays form a **straight angle**, which has a measure of 180°.

An **angle** is a pair of rays that have a common endpoint.

The rays are called **sides** of the angle. The common endpoint is the **vertex**.

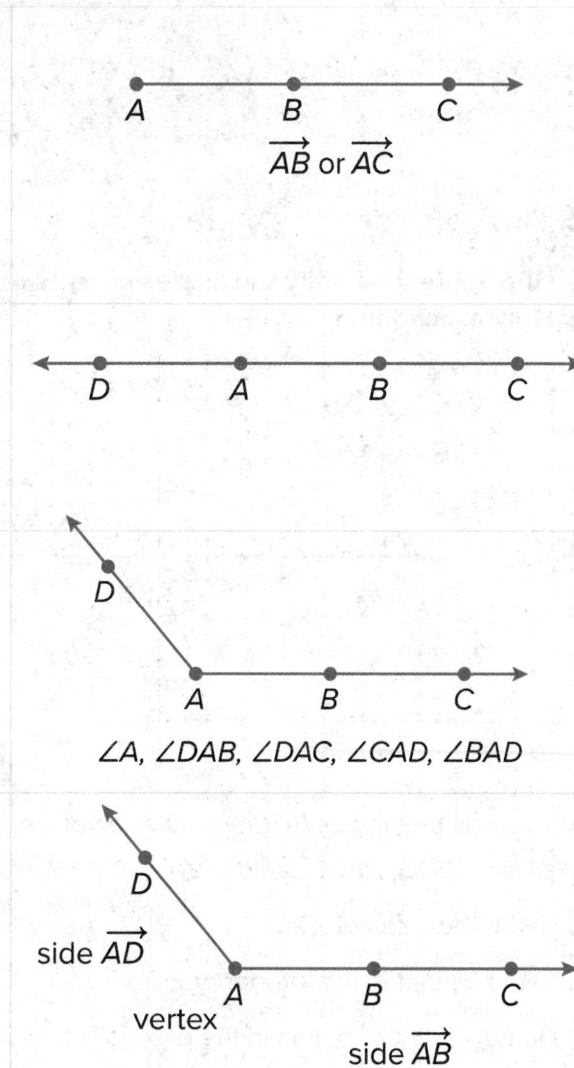

(*continued on the next page*)

Study Tip

Naming Angles When naming an angle using three letters, the first letter represents a point on one side of the angle, the second letter must always represent the vertex, and the third letter represents a point on the other side of the angle. Name an angle using a single letter only when there is exactly one angle located at that vertex.

Your Notes

Go Online You can watch a video to see how to use a protractor to measure and draw angles.

Think About It!

Can a point be in the interior of one angle and the exterior of another angle? If so, give an example.

An angle divides a plane into three distinct parts.

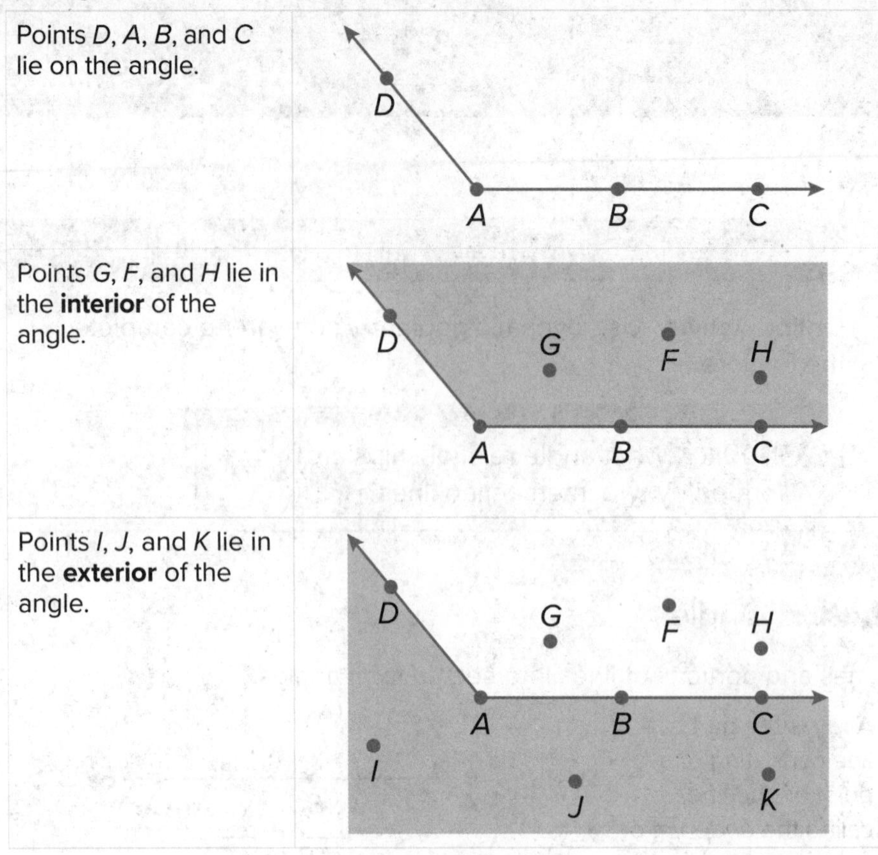

Points D, A, B, and C lie on the angle.
Points G, F, and H lie in the **interior** of the angle.
Points I, J, and K lie in the **exterior** of the angle.
D
A
B
C
G
F
H
I
J
K

Example 1 Identify Angles

Use the figure to identify the angles or parts of angles that satisfy each given condition.

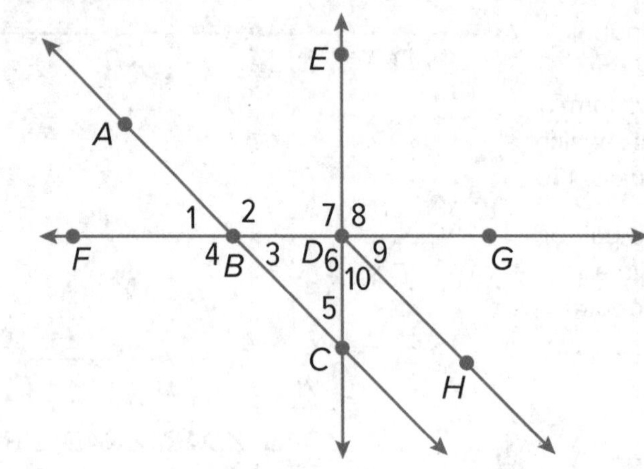

a. Name all the angles that have D as a vertex.

$\angle EDF, \angle EDG, \angle FDC, \angle GDC, \angle$ ____, $\angle$ ____, $\angle FDH$

b. Name the sides of $\angle 2$. ____

c. Name a point in the interior of $\angle FDE$. ____

d. Name a point or points in the exterior of $\angle FDE$. ____

Go Online You can complete an Extra Example online.

Check

Use the figure to identify the angles or parts of angles that satisfy the given condition. Which angle has sides $\overrightarrow{DB}$ and $\overrightarrow{DC}$? Select all that apply.

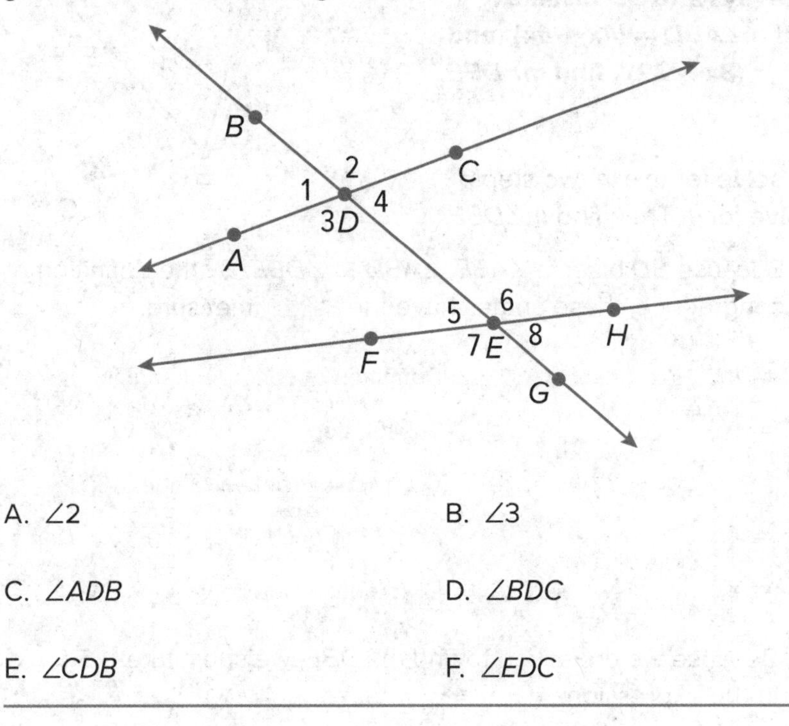

A. ∠2

B. ∠3

C. ∠ADB

D. ∠BDC

E. ∠CDB

F. ∠EDC

Learn Congruent Angles

The measure of an angle is the measure in degrees of the space between the sides of an angle. Angles that have the same measure are **congruent angles**. Congruent angles are indicated on the figure by matching numbers of arcs.

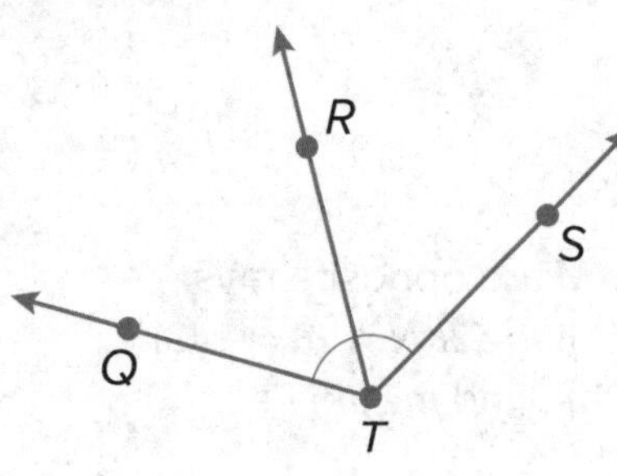

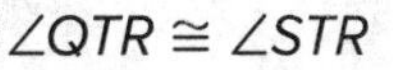
$\angle QTR \cong \angle STR$

A ray or segment that divides an angle into two congruent parts is an **angle bisector**. In the figure, $\overrightarrow{TR}$ bisects ∠QTS.

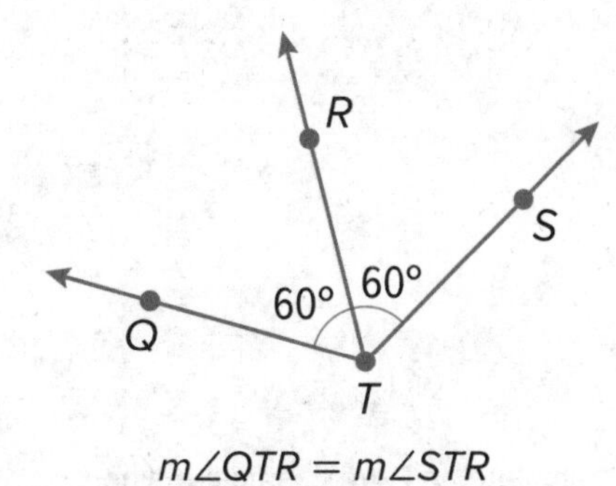

$m\angle QTR = m\angle STR$

Example 2 Congruent Angles and Angle Bisectors

In the figure, $\overrightarrow{BA}$ and $\overrightarrow{BC}$ are opposite rays and $\overrightarrow{BD}$ bisects $\angle ABE$. If $m\angle ABD = (4x + 14)°$ and $m\angle DBE = (8x - 32)°$, find $m\angle DBE$.

We can solve for this in two steps. First, solve for x. Then find $m\angle DBE$.

Step 1: Because $\overrightarrow{BD}$ bisects $\angle ABE$, $\angle ABD \cong \angle DBE$. By the definition of congruence, these angles have the ______ measure.

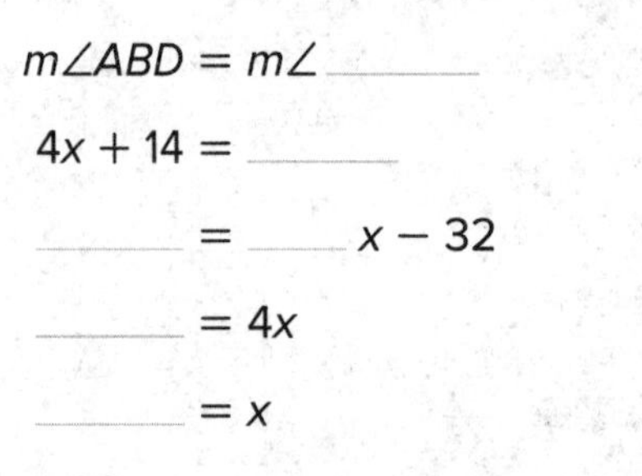

$m\angle ABD = m\angle$ ______	Definition of congruent angles
$4x + 14 =$ ______	Substitution
______ $=$ ______ $x - 32$	Subtract $4x$ from each side.
______ $= 4x$	Add 32 to each side.
______ $= x$	Divide each side by 4.

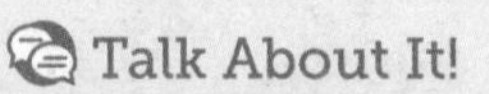

Talk About It!

Suppose $\overrightarrow{BE}$ is an angle bisector of $\angle DBC$. What is $m\angle EBC$? Explain your solution process.

Step 2: Because we are asked to find $m\angle DBE$, we substitute 11.5 for x in the expression.

$m\angle DBE = 8$ ______ $- 32$	Given
$= 8$ ______ $- 32$	Substitute.
$=$ ______ $- 32$	Multiply.
$=$ ______	Subtract.

$m\angle DBE =$ ______ °

Check

In the figure, $\overrightarrow{KJ}$ and $\overrightarrow{KM}$ are opposite rays, and $\overrightarrow{KN}$ bisects $\angle JKL$. If $m\angle JKN = (8x - 13)°$ and $m\angle NKL = (6x + 11)°$, find $m\angle JKN$.

$m\angle JKN =$ ______ °

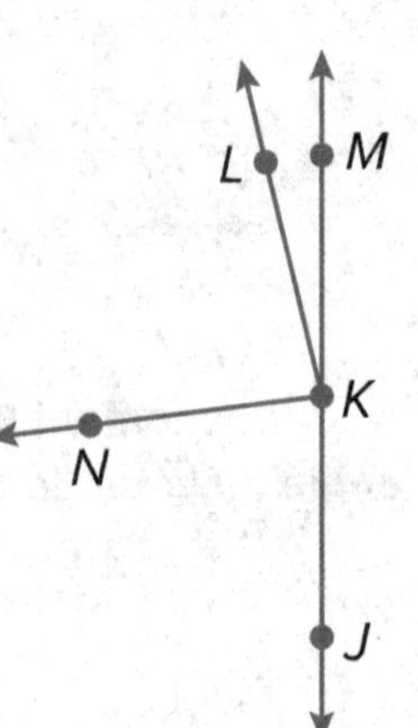

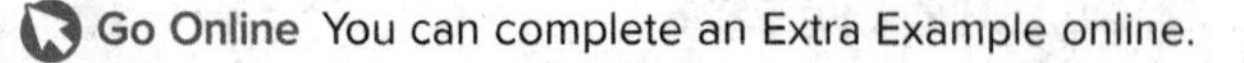
Go Online You can complete an Extra Example online.

Learn Special Angle Pairs

There are three special angle pairs.

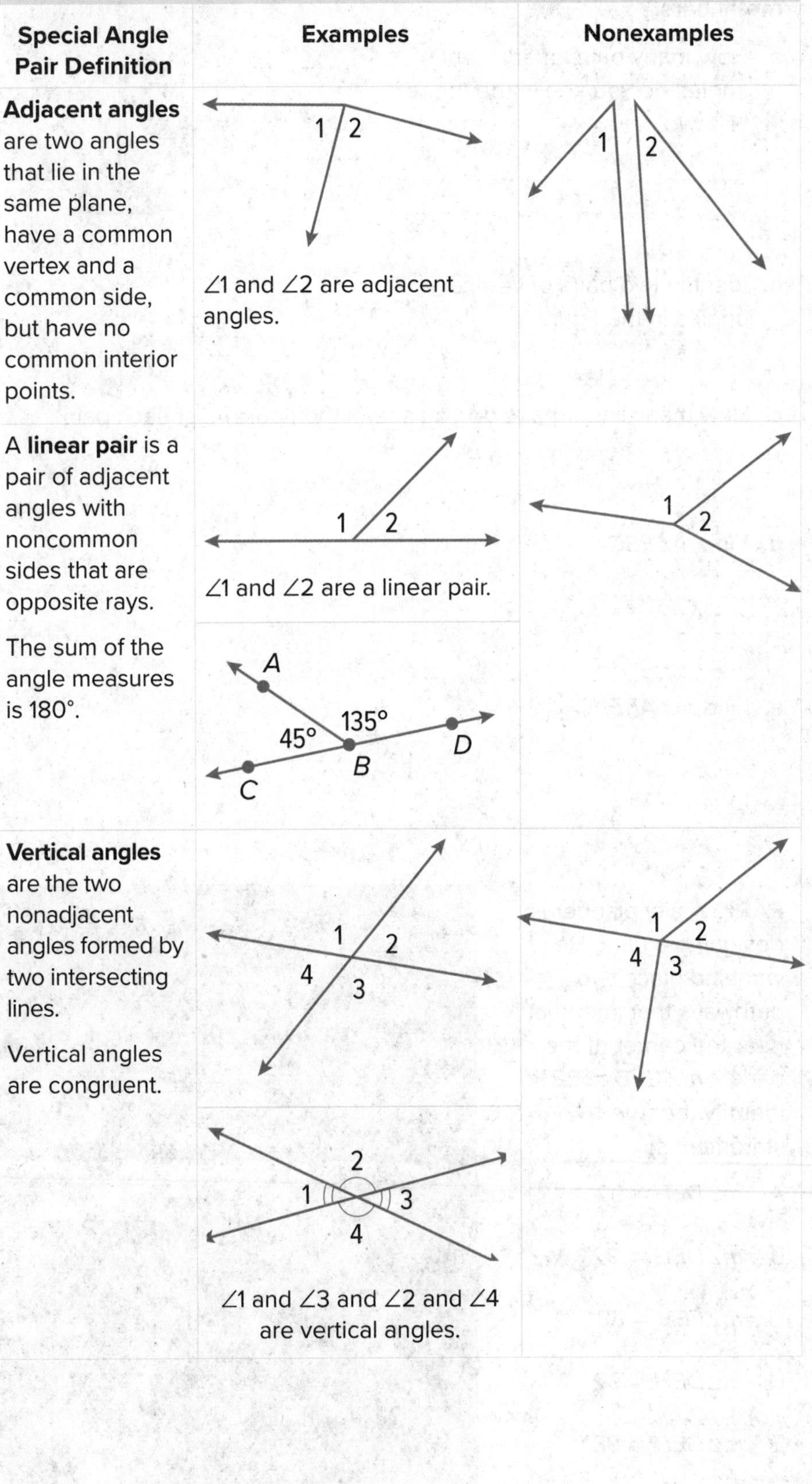

Key Concept • Special Angle Pairs

Special Angle Pair Definition	Examples	Nonexamples
Adjacent angles are two angles that lie in the same plane, have a common vertex and a common side, but have no common interior points.	∠1 and ∠2 are adjacent angles.	
A **linear pair** is a pair of adjacent angles with noncommon sides that are opposite rays. The sum of the angle measures is 180°.	∠1 and ∠2 are a linear pair.	
Vertical angles are the two nonadjacent angles formed by two intersecting lines. Vertical angles are congruent.	∠1 and ∠3 and ∠2 and ∠4 are vertical angles.	

Example 3 Vertical Angles and Angle Pairs

HOME DECOR **The office lamp is made using two intersecting metal bars.**

a. How many pairs of adjacent angles do you see in the figure? List two pairs.

b. Identify two pairs of vertical angles in the figure.

c. How many linear pairs do you see in the figure? List each pair.

d. Find $m\angle EBC$.

e. Find $m\angle ABE$.

Think About It!

Can vertical angles also be adjacent angles? Explain.

Check

PARK A city planner is designing a park. He wants to place two pathways that intersect near the center of the park. If $m\angle GED = 88°$, identify the true statement(s).

A. $m\angle DEF = 92°$

B. $m\angle DEG = 92°$

C. $m\angle FEH = 88°$

D. $m\angle DEH = 92°$

E. $m\angle GEH = 88°$

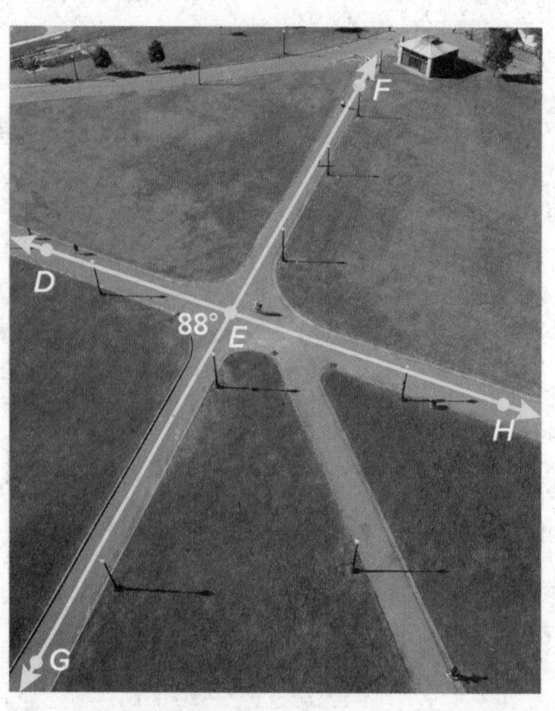

Go Online You can complete an Extra Example online.

Go Online

You may want to complete the construction activities for this lesson.

Name ____________ Period ________ Date ________

Find the value of each variable.

34. **35.** **36.**

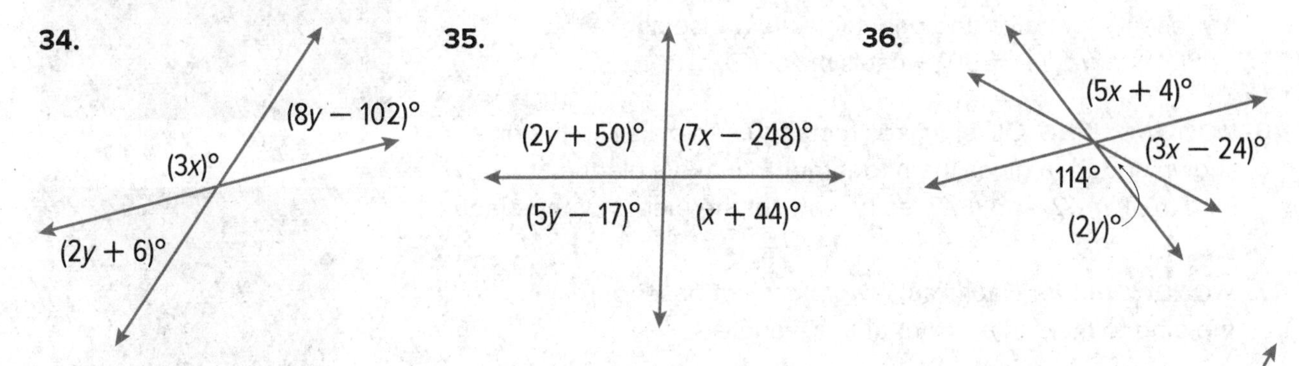

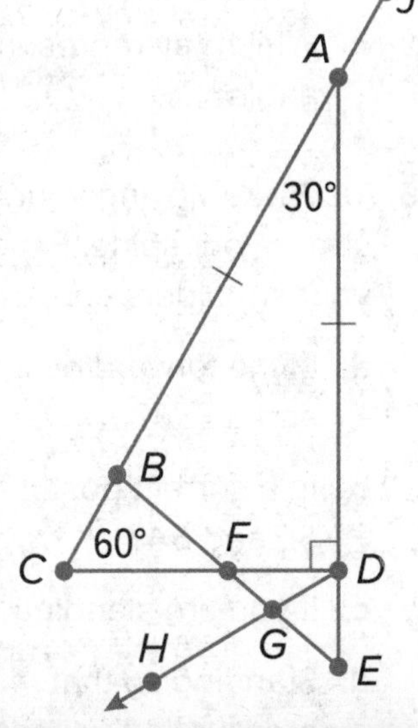

Name an angle or angle pair that satisfies each condition.

37. two adjacent angles

38. two vertical angles

39. a linear pair that has vertex F

Use the picture at the right.

40. Name four rays.

41. Name three angles.

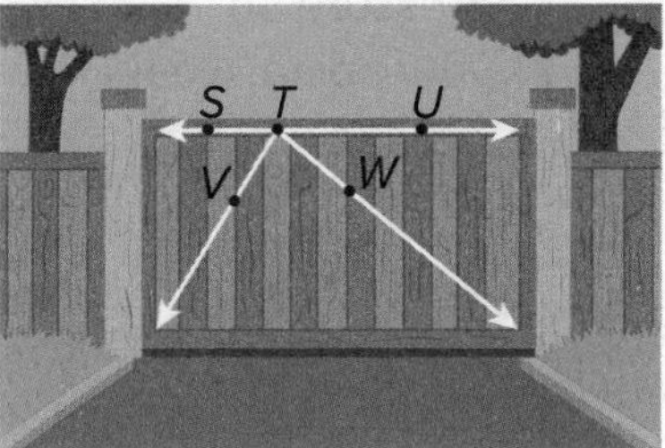

TRAFFIC **In the traffic circle around the Arc de Triomphe in Paris, France, there are eight lanes of traffic. Tell whether each angle pair satisfies the given condition.**

42. vertical angles

a. $\angle ZCY$ and $\angle TCU$ **b.** $\angle XCW$ and $\angle SCT$

c. $\angle QCR$ and $\angle WCV$ **d.** $\angle TCU$ and $\angle UCT$

43. linear pair

a. $\angle RCU$ and $\angle WCU$ **b.** $\angle QCR$ and $\angle SCR$

c. $\angle VCX$ and $\angle WCY$ **d.** $\angle ZCR$ and $\angle UCW$

44. adjacent angles

a. $\angle WCU$ and $\angle RCU$ **b.** $\angle QCS$ and $\angle SCR$

c. $\angle VCW$ and $\angle QCR$ **d.** $\angle VCX$ and $\angle VCU$

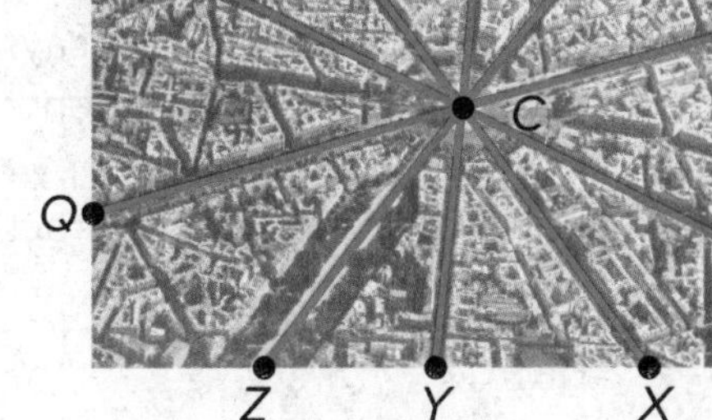

45. **POOL** Felipe uses a computer program to model the paths of pool balls. $\angle GFH$ is a straight angle that represents the rail of the pool table. If $\overrightarrow{FK}$ bisects $\angle JFL$, and $m\angle JFL = 90°$, what is $m\angle LFK$?

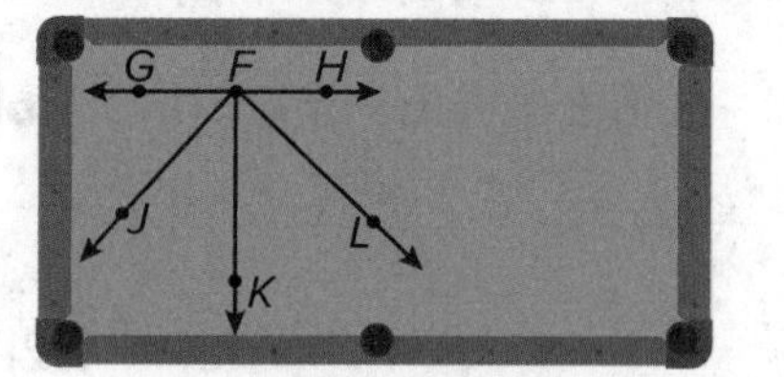

46. **WOODWORKING** Oliver makes rectangular blocks like the one shown and then glues them together to make a plaque. Find $m\angle 1$, $m\angle 2$, and $m\angle 3$, so he can cut the pieces of the plaque.

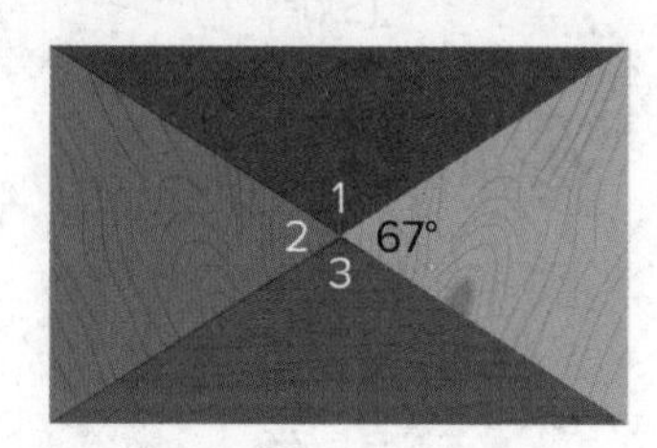

47. **WOODWORKING** Naomi cuts two pieces of baseboard molding to meet in a corner at a 90° angle.

a. To what degree should she set her table saw for the cut?

b. Which ray represents the angle bisector of the molding angle?

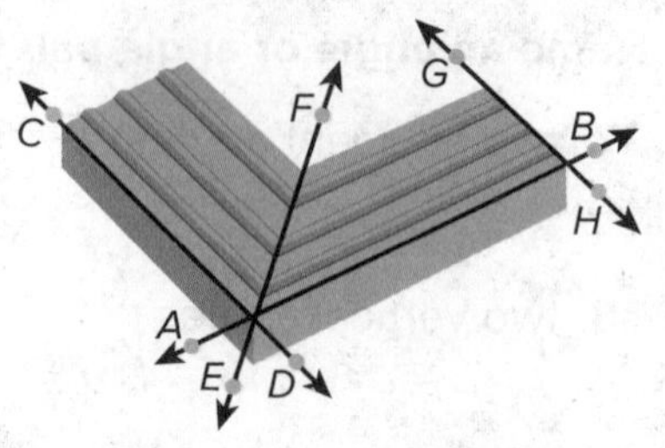

48. **TEXTING** Moving your head forward to look at a screen can stress your spine. Experts recommend aligning your ears with your shoulders and arms.

a. In the forward posture, what is the relationship between $\angle CSE$ and $\angle ESA$?

b. In a correct posture, what is the relationship between $\overrightarrow{SE}$ and $\overrightarrow{SA}$?

c. If you are standing so that $m\angle CSE = 26°$, what is $m\angle ESA$?

d. Standing so that $m\angle CSE \geq 15°$ puts more than 27 pounds of pressure on your spine. If there is 34 pounds of pressure on your spine, what inequality describes $m\angle ESA$?

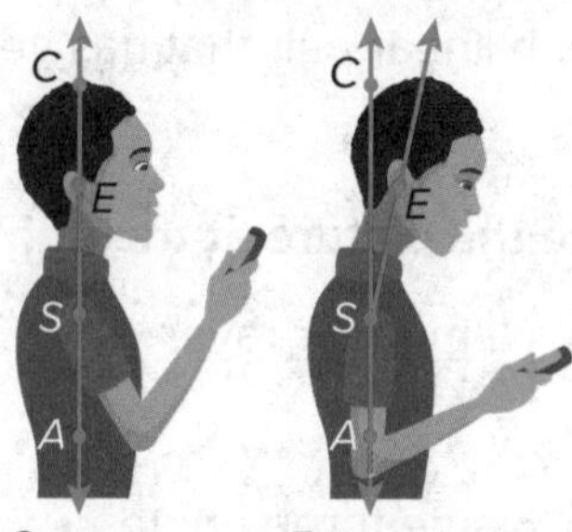

Correct posture

Forward posture

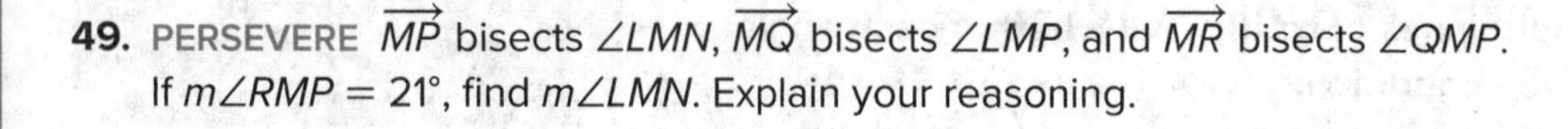

49. **PERSEVERE** $\overrightarrow{MP}$ bisects $\angle LMN$, $\overrightarrow{MQ}$ bisects $\angle LMP$, and $\overrightarrow{MR}$ bisects $\angle QMP$. If $m\angle RMP = 21°$, find $m\angle LMN$. Explain your reasoning.

50. **ANALYZE** Maria constructed a copy of $\angle PVQ$ and labeled it $\angle FGH$.

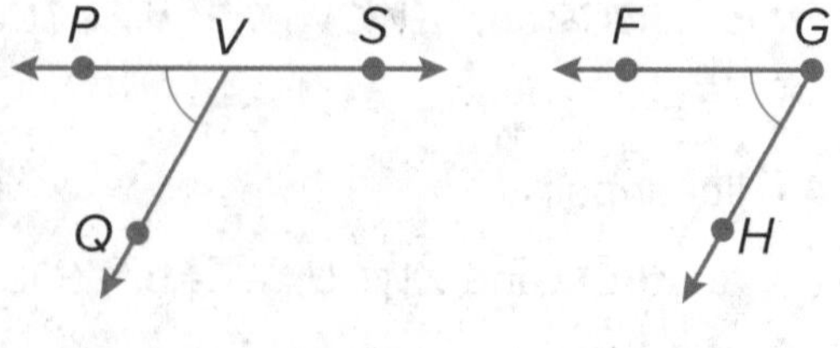

a. Are $\angle FGH$ and $\angle QVS$ a linear pair? Explain.

b. Maria must also copy $\angle QVS$. Sal says she can create a copy of $\angle QVS$ if she extends $\overrightarrow{GH}$ past G. Mona says Maria can create a copy of $\angle QVS$ by extending $\overrightarrow{GF}$ past G. Who is correct? Justify your argument.

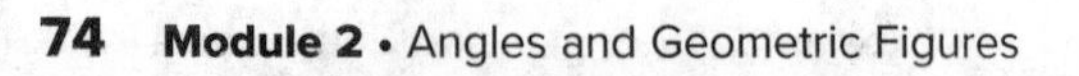

Angle Relationships

Today's Goals

- Calculate angle measures using the characteristics of complementary and supplementary angles.
- Calculate angle measures using the characteristics of perpendicular lines.
- Demonstrate understanding of what can and cannot be assumed from a diagram.

Today's Vocabulary

complementary angles

supplementary angles

perpendicular

Explore Complementary and Supplementary Angles

Online Activity Use dynamic geometry software to complete the Explore.

INQUIRY How do complementary angles compare to supplementary angles?

Learn Complementary and Supplementary Angles

Complementary and Supplementary Angles

Complementary Angles	Supplementary Angles
Definition	
two angles with measures that have a sum of 90°	two angles with measures that have a sum of 180°
Examples	

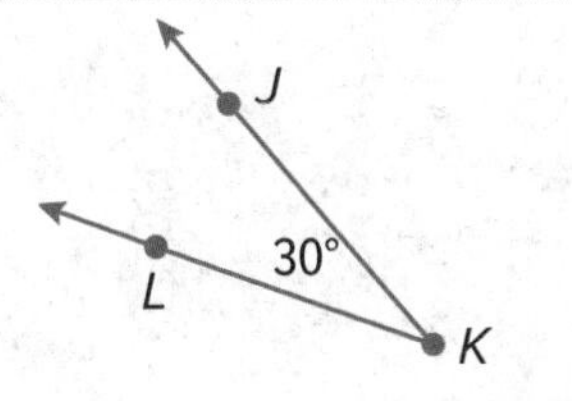

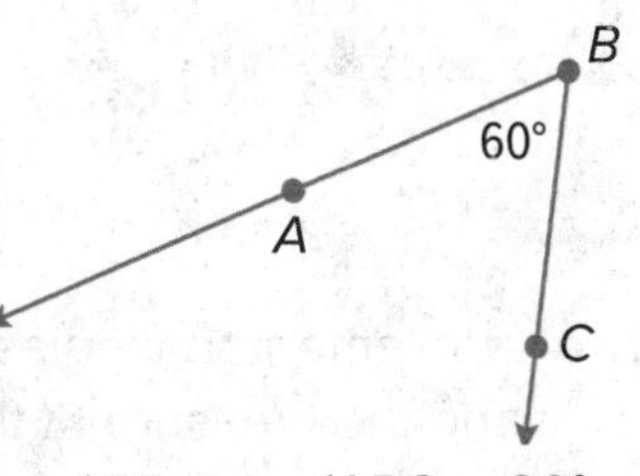

$m\angle JKL + m\angle ABC = 90°$
$30° + 60° = 90°$

F
110°
E
D
G
70°
H
J

$m\angle DEF + m\angle GHJ = 180°$
$110° + 70° = 180°$

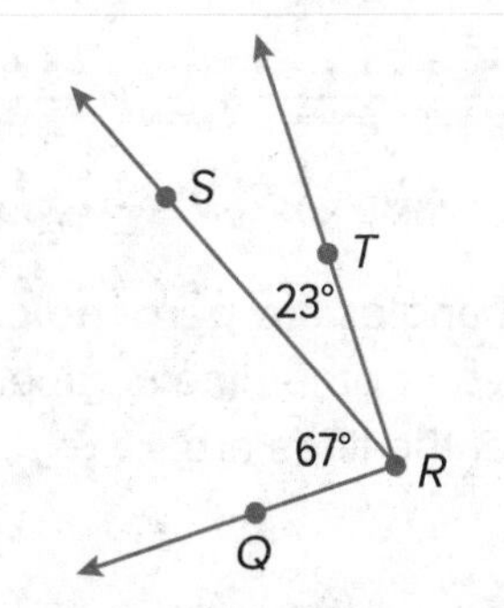

$m\angle QRS + m\angle SRT = 90°$
$67° + 23° = 90°$

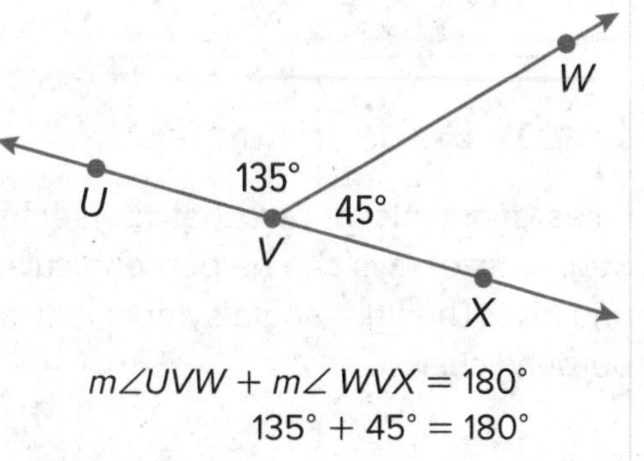

$m\angle UVW + m\angle WVX = 180°$
$135° + 45° = 180°$

Think About It!

A linear pair is ______, supplementary while two supplementary angles are ______ a linear pair.

Study Tip

Complementary and Supplementary Angles Pairs of angles that are complementary or supplementary do not have to be adjacent angles.

Your Notes

Example 1 Complementary and Supplementary Angles

Find the measures of two complementary angles if the measure of the larger angle is five more than four times the measure of the smaller angle.

If two angles are complementary, then the sum of the angle measures is 90°. To find the measures of each angle, first write an equation. Let x = the measure of the smaller angle. Then the measure of the larger angle is $4x + 5$.

Step 1

First, solve for x.

$x + 4x + 5 =$ ____ Complementary angle measures add to 90°.

____ $+ 5 = 90$ Combine like terms.

$5x =$ ____ Subtract 5 from each side.

$x =$ ____ Divide each side by 5.

So, the measure of the smaller angle is ____°.

Step 2

Next, find the measure of the larger angle.

$4x + 5 = 4($____$) + 5$ Substitute 17 for x.

$=$ ____ $+ 5$ Multiply.

$=$ ____ Solve.

The measures of the angles are 17° and 73°.

CHECK

Does your answer seem reasonable?

Talk About It!

Adrian claims that if two complementary angles are both acute, then a pair of supplementary angles must both be obtuse. Do you agree? Explain why or why not.

Check

The difference between the measures of two supplementary angles is 18°. The measure of the smaller angle is ____°, and the measure of the larger angle is ____°.

Learn Perpendicularity

Lines, segments, or rays that intersect at right angles are **perpendicular**. Segments or rays can be perpendicular to lines or other line segments and rays. The right angle symbol indicates that the lines are perpendicular.

Go Online You can complete an Extra Example online.

Perpendicular lines intersect to form four right angles.	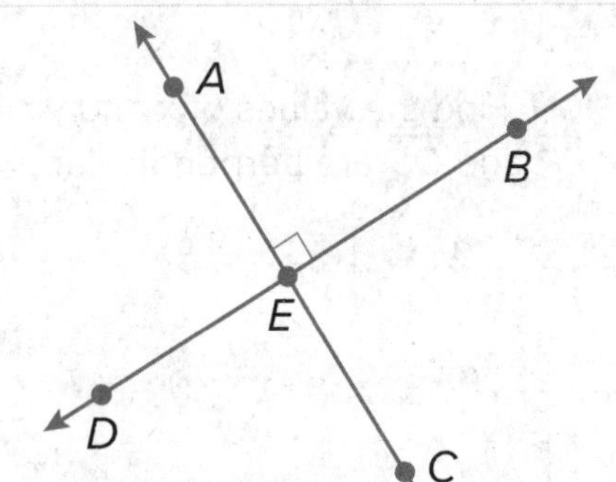 $\angle AEB$, $\angle BEC$, $\angle CED$, and $\angle DEA$ are right angles.
Perpendicular lines intersect to form congruent adjacent angles.	A B E D C $\angle AEB \cong \angle BEC$ $\angle AEB$, $\angle BEC$, $\angle CED$, and $\angle DEA$ are right angles.

Example 2 Perpendicular Lines

TANGRAMS **The tangram is a puzzle consisting of seven flat shapes called *tans* which are put together to form shapes. Find the values of *x* and *y* such that $\overleftrightarrow{AD}$ and $\overleftrightarrow{EC}$ in the tangram are perpendicular.**

F, $(7x + 10)°$, $(8x + 5)°$, D, E, B, $(3y + 15)°$, A, C

If $\overleftrightarrow{AD}$ and $\overleftrightarrow{EC}$ are perpendicular, then $m\angle ABC = 90°$ and $m\angle EBD = 90°$.

Step 1 Solve for *y*.

$3y + 15 =$ ____ $\quad m\angle ABC = 90°$

$y =$ ____ $\quad$ Solve for *y*.

Step 2 Solve for *x*.

$m\angle EBF + m\angle FBD =$ ________ $\quad$ sum of parts = whole

$7x + 10 + 8x + 5 = 90$ $\quad$ Substitution

$x =$ ____ $\quad$ Solve for *x*.

Study Tip

Symbols $\perp$ is read *is perpendicular to.*
Example: $\overline{AC} \perp \overline{DB}$

Think About It!

Besides right angles, how else can you describe $\angle ABC$ and $\angle EBD$?

Check

DESIGN Find the values of x and y such that $\overleftrightarrow{PR}$ and $\overrightarrow{QS}$ are perpendicular.

$x =$ ____

$y =$ ____

Explore Interpreting Diagrams

Online Activity Use dynamic geometry software to complete the Explore.

INQUIRY What information can be assumed from a diagram, and what information cannot be assumed?

Learn Interpreting Diagrams

In geometry, figures are sketches that are used to depict a situation. They are not drawn to reflect total accuracy. Certain relationships can be assumed from a figure, but most cannot.

Interpreting Diagrams

Can Be Assumed

All points and lines shown are coplanar.

G, H, and J are collinear.

$\overrightarrow{HM}$, $\overrightarrow{HL}$, $\overrightarrow{HK}$, and $\overleftrightarrow{GJ}$ intersect at H.

H is between G and J.

L is in the interior of $\angle MHK$.

$\angle GHM$ and $\angle MHL$ are adjacent angles.

$\angle GHL$ and $\angle LHJ$ are a linear pair.

$\angle JHK$ and $\angle KHG$ are supplementary.

Cannot Be Assumed

Lines that appear perpendicular may not be perpendicular.

Angles that appear congruent may not be congruent.

Segments that appear congruent may not be congruent.

The list of statements that can be assumed is not a complete list. There are more special pairs of angles than those listed.

Go Online You can complete an Extra Example online.

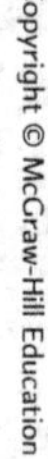

Because points of intersection can be assumed, you can identify vertical angles from the figure.

Because linear pairs can be assumed from the figure, you can apply known characteristics of a linear pair, such as supplementary angles.

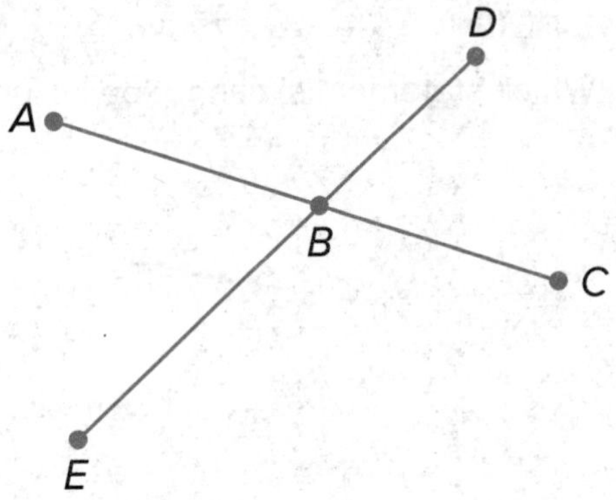

$\angle ABD$ and $\angle CBE$ are vertical angles.

$\angle EBA$ and $\angle ABD$ form a linear pair, so $m\angle EBA + m\angle ABD = 180°$.

Example 3 Interpreting Diagrams

Determine whether each statement can be assumed from the figure. Explain.

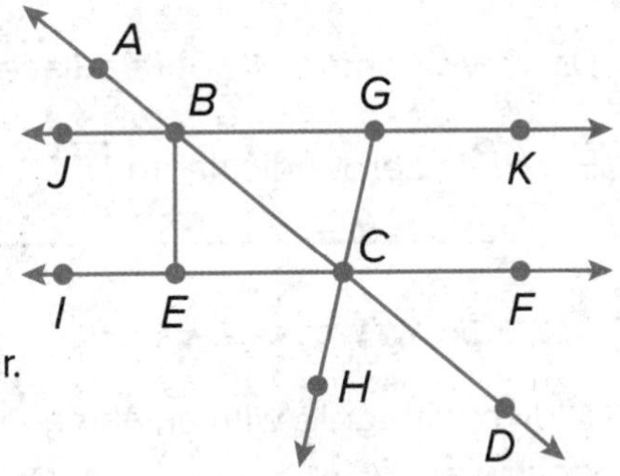

a. $\overrightarrow{CE}$ and $\overrightarrow{CF}$ are opposite rays.

Yes; C is a common endpoint.

b. $\angle BGC$ and $\angle KGC$ form a linear pair.

_____; their noncommon sides are opposite rays.

c. $\angle ABJ$ and $\angle CBG$ are vertical angles.

Yes; these angles are nonadjacent and are formed by _____ _____.

d. $\angle BCG$ and $\angle DCF$ are congruent.

_____; these angles are not vertical angles. There isn't enough information given to determine this.

e. $\overline{BE}$ and $\overleftrightarrow{IF}$ are perpendicular.

_____; there isn't enough information given to determine this.

f. $\angle EBC$ and $\angle GBC$ are complementary angles.

_____; there isn't any information about perpendicularity or angle measure so this cannot be determined.

g. $\angle ICH$ and $\angle HCD$ are adjacent angles.

_____; these angles share a common side.

h. $\overline{BC}$ is an angle bisector of $\angle ECG$.

_____; there isn't any information about congruent angles so this cannot be determined.

Think About It!

If you are given that $\overline{BE} \perp \overline{IC}$, can you determine whether $\angle BEI \cong \angle BEC$? Explain your solution process.

Watch Out!

Congruence and Perpendicularity Remember that congruent angles or segments and perpendicular or parallel lines cannot be assumed from a figure.

Check

Which statement(s) cannot be assumed from the figure?

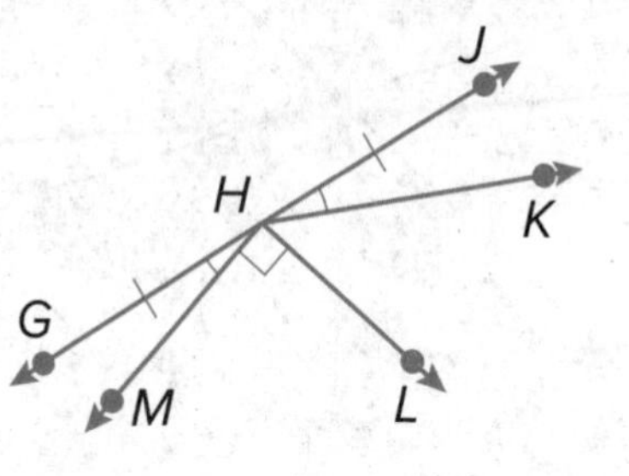

A. $\angle KHJ$ and $\angle GHM$ are complementary.

B. $\angle GHK$ and $\angle JHK$ are a linear pair.

C. $\overrightarrow{HL}$ is perpendicular to $\overrightarrow{HJ}$.

D. $\angle GHM$ and $\angle MHK$ are adjacent angles.

E. $\overrightarrow{HL}$ is perpendicular to $\overrightarrow{HM}$.

Pause and Reflect

Did you struggle with anything in this lesson? If so, how did you deal with it?

Record your observations here.

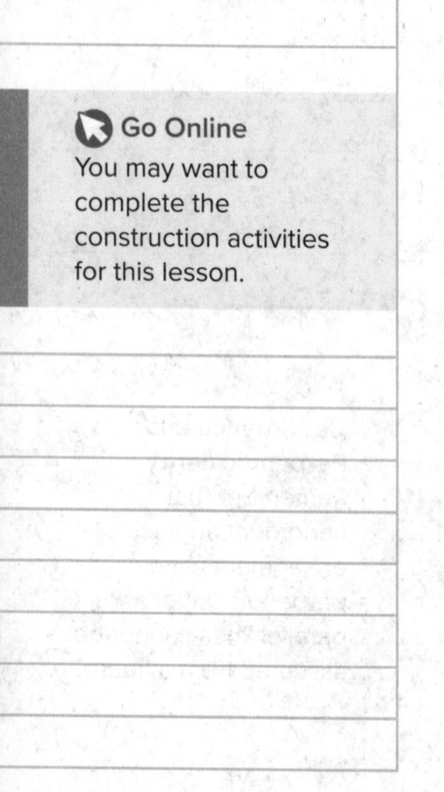

Go Online
You may want to complete the construction activities for this lesson.

Go Online You can complete an Extra Example online.

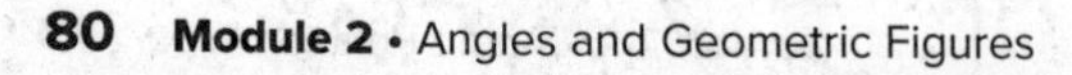

Name Isabella Metivier Period 4 Date 11-21-22

Practice

Go Online You can complete your homework online.

Example 1

1. Find the measures of two supplementary angles if the difference between the measures of the two angles is 35°.

2. $\angle E$ and $\angle F$ are complementary. The measure of $\angle E$ is 54° more than the measure of $\angle F$. Find the measure of each angle.

3. The measure of an angle's supplement is 76° less than the measure of the angle. Find the measures of the angle and its supplement.

4. $\angle Q$ and $\angle R$ are complementary. The measure of $\angle Q$ is 26° less than the measure of $\angle R$. Find the measure of each angle.

5. The measure of the supplement of an angle is three times the measure of the angle. Find the measures of the angle and its supplement.

6. The bascule bridge shown is opening from its horizontal position to its fully vertical position. So far, the bridge has lifted 35° in 21 seconds. At this rate, how much longer will it take for the bridge to reach its vertical position?

sleepy

xtiltx

Example 2

7. Rays BA and BC are perpendicular. Point D lies in the interior of $\angle ABC$. If $m\angle ABD = (3r + 5)°$ and $m\angle DBC = (5r - 27)°$, find $m\angle ABD$ and $m\angle DBC$.

8. $\overleftrightarrow{WX}$ and $\overleftrightarrow{YZ}$ intersect at point V. If $m\angle WVY = (4a + 58)°$ and $m\angle XVY = (2b - 18)°$, find the values of a and b such that $\overleftrightarrow{WX}$ is perpendicular to $\overleftrightarrow{YZ}$.

9. Refer to the figure at the right. If $m\angle 2 = (a + 15)°$ and $m\angle 3 = (a + 35)°$, find the value of a such that $\overrightarrow{HL} \perp \overrightarrow{HJ}$. 38

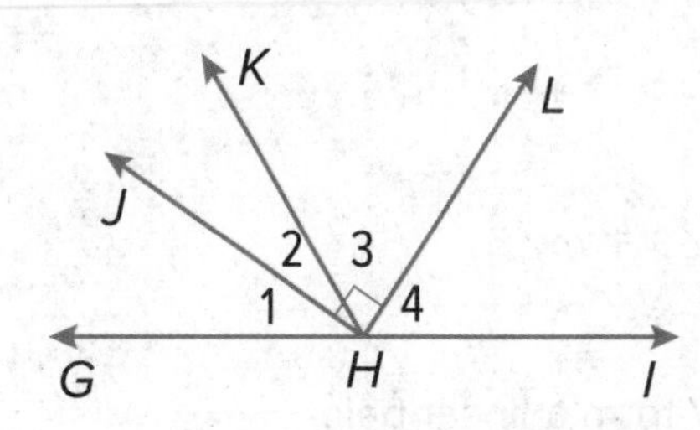

10. Rays DA and DC are perpendicular. Point B lies in the interior of $\angle ADC$. If $m\angle ADB = (3a + 10)°$ and $m\angle BDC = 13a°$, find a, $m\angle ADB$, and $m\angle BDC$.

75

Example 3

Determine whether each statement can be assumed from the given figure. Explain.

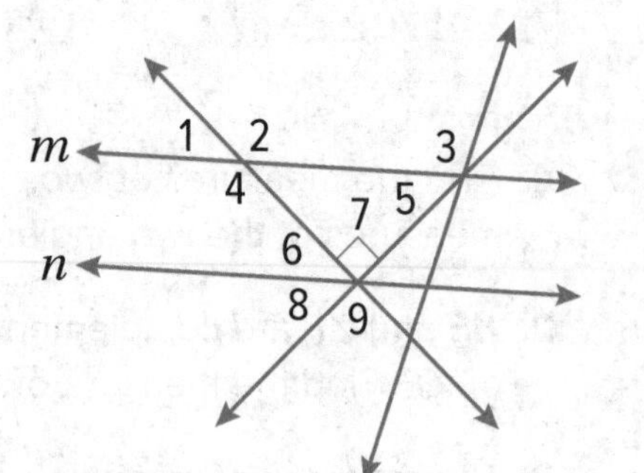

11. $\angle 6$ and $\angle 8$ are complementary.

12. $\angle 7$ and $\angle 8$ form a linear pair.

13. $\angle 2$ and $\angle 4$ are vertical angles.

14. $m\angle 9 = m\angle 6 + m\angle 8$

Mixed Exercises

15. The measure of the supplement of an angle is 60° less than four times the measure of the complement of the angle. Find the measure of the angle.

16. $\angle 6$ and $\angle 7$ form a linear pair. Twice the measure of $\angle 6$ is twelve more than four times the measure of $\angle 7$. Find the measure of each angle.

Refer to the figure at the right.

17. If $m\angle ADB = (6x - 4)°$ and $m\angle BDC = (4x + 24)°$, find the value of x such that $\angle ADC$ is a right angle.

18. If $m\angle FDE = (3x - 15)°$ and $m\angle FDB = (5x + 59)°$, find the value of x such that $\angle FDE$ and $\angle FDB$ are supplementary.

19. If $m\angle BDC = (8x + 12)°$ and $m\angle FDB = (12x - 32)°$, find $m\angle FDE$.

Determine whether each statement can be assumed from the given figure. Explain.

20. $\angle 4$ and $\angle 7$ are vertical angles.

21. $\angle 3 \cong \angle 6$

22. $m\angle 5 = m\angle 3 + m\angle 6$

23. $\angle 5$ and $\angle 7$ form a linear pair.

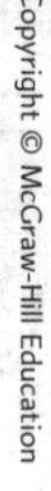

For Exercises 24 and 25, lines *p* and *q* intersect to form adjacent angles 1 and 2.

24. If $m\angle 1 = (7x + 6)°$ and $m\angle 2 = (8x - 6)°$, find the value of x such that p is perpendicular to q.

25. If $m\angle 1 = (4x - 3)°$ and $m\angle 2 = (3x + 8)°$, find the value of x such that $\angle 1$ is supplementary to $\angle 2$.

26. COLOR GUARD Shannon is designing a new rectangular flag for the school's color guard and is determining the angles at which to cut the fabric. She wants the measure of $\angle 2$ to be three times as great as the measure of $\angle 1$. She thinks the measures of $\angle 3$ and $\angle 4$ should be equal. Finally, she wants the measure of $\angle 6$ to be half that of $\angle 5$. Determine the measures of the angles.

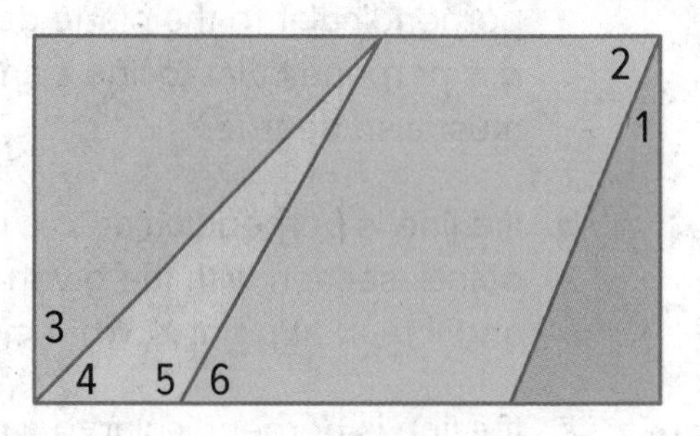

27. STRING ART String art is created by wrapping string around nails or wires to form patterns. Use the string art pattern to find the values of x, y, and z.

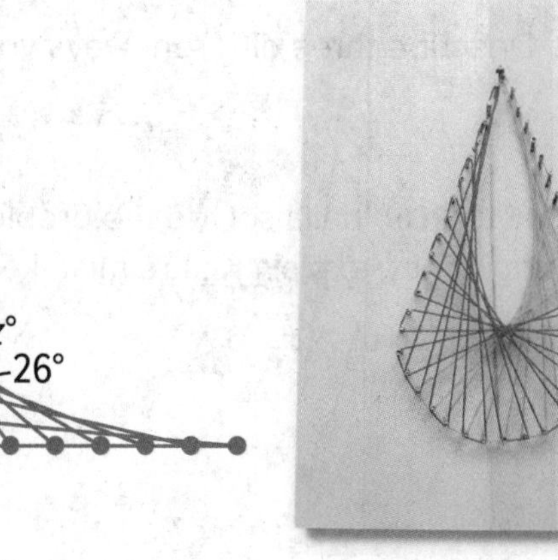

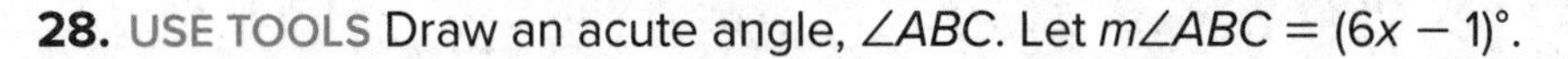

28. USE TOOLS Draw an acute angle, $\angle ABC$. Let $m\angle ABC = (6x - 1)°$.

a. Use a protractor to determine the measure of $\angle ABC$. Use this measure to determine the value of x.

b. Explain how you would determine the measure of an angle that is complementary to $\angle ABC$.

c. Explain how you would determine the measure of an angle that is supplementary to $\angle ABC$.

29. ANALYZE Are there angles that do not have a complement? Justify your argument.

30. PERSEVERE If a line, line segment, or ray is perpendicular to a plane, then it is perpendicular to every line, line segment, or ray in the plane that intersects it.

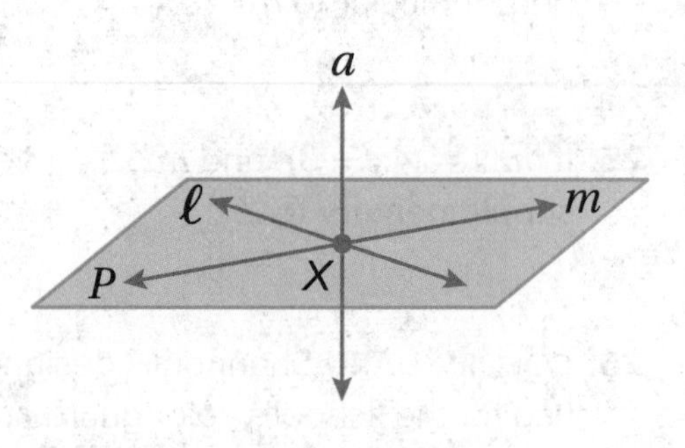

a. If a line is perpendicular to each of two intersecting lines at their point of intersection, then the line is perpendicular to the plane determined by them. If line a is perpendicular to line ℓ and line m at point X, what must also be true?

b. If a line is perpendicular to a plane, then any line perpendicular to the given line at the point of intersection with the given plane is in the given plane. If line a is perpendicular to plane P and line m at point X, what must also be true?

c. If a line is perpendicular to a plane, then every plane containing the line is perpendicular to the given plane. If line a is perpendicular to plane P, what must also be true?

31. WRITE Describe three different ways you can determine that an angle is a right angle.

32. FIND THE ERROR Kaila solved the problem, as shown. Is her solution correct? If it is, explain your reasoning. If not, explain Kaila's mistake and correct the work.

If $m\angle F = (6x - 9)°$ and $m\angle G = (2x + 13)°$, find the value of x such that $\angle F$ and $\angle G$ are supplementary.

$$(6x - 9)° + (2x + 13)° = 90°$$
$$8x° - 4° = 90°$$
$$8x° = 86°$$
$$x = 10.75$$

33. CREATE Create $\angle 1$ along with its complement and supplement by drawing only a line and two rays.

34. WHICH ONE DOESN'T BELONG Three students used the figure to write a statement. Is each statement correct? Justify your conclusion.

Samar: $\angle WZU$ is a right angle.

Jana: $\angle YZU$ and $\angle UZV$ are supplementary.

Antonio: $\angle VZU$ is adjacent to $\angle YZX$.

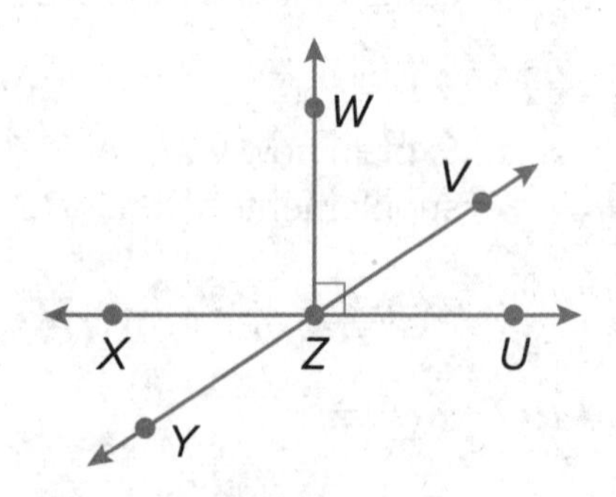

35. ANALYZE Do all angles have a supplement? Explain.

Two-Dimensional Figures

Learn Perimeter, Circumference, and Area

A **polygon** is a closed plane figure with at least three straight sides.

The **perimeter** of a polygon is the sum of the lengths of the sides of the polygon. Some shapes have special formulas for perimeter, but all are derived from the basic definition of perimeter.

The **circumference** of a circle is the distance around the circle.

Area is the number of square units needed to cover a surface.

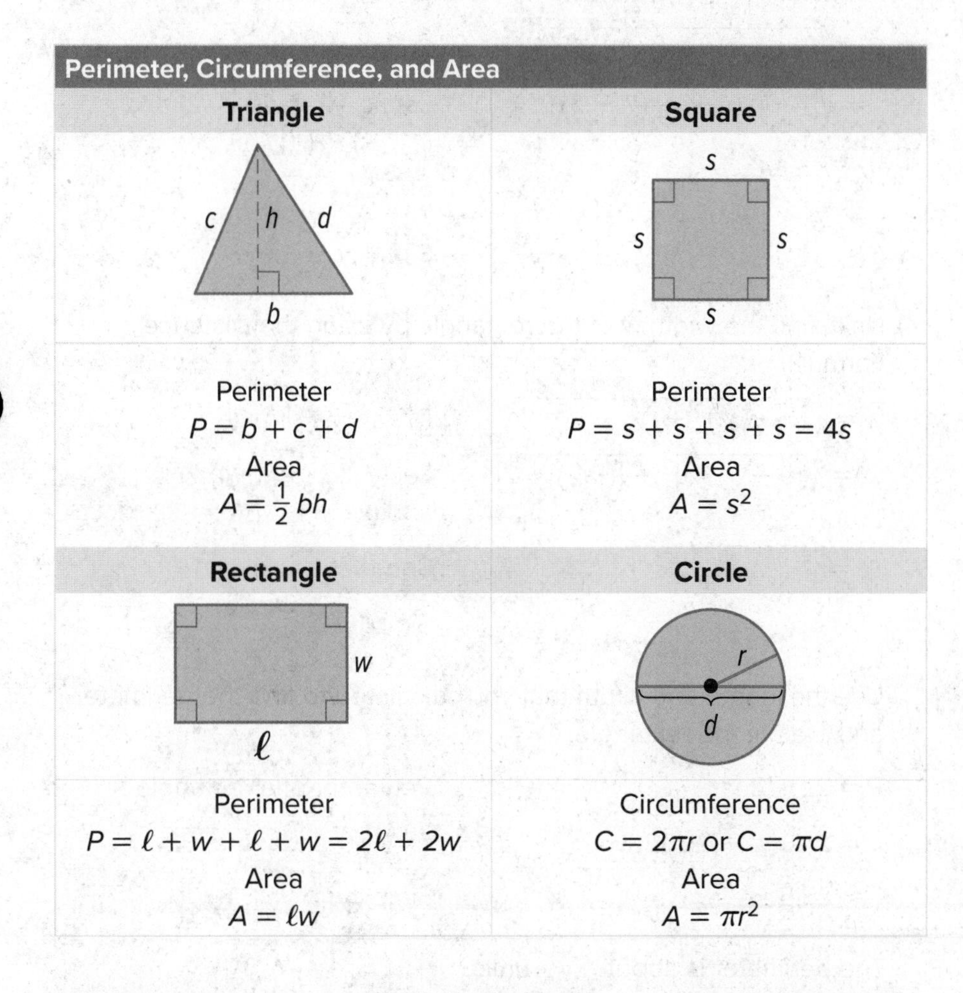

Perimeter, Circumference, and Area

Triangle	Square
Perimeter $P = b + c + d$ Area $A = \frac{1}{2}bh$	Perimeter $P = s + s + s + s = 4s$ Area $A = s^2$
Rectangle	**Circle**
Perimeter $P = \ell + w + \ell + w = 2\ell + 2w$ Area $A = \ell w$	Circumference $C = 2\pi r$ or $C = \pi d$ Area $A = \pi r^2$

You can use the Distance Formula to find the perimeter and area of a polygon graphed on a coordinate plane. You can also use the Distance Formula to calculate the radius of a circle and then use the appropriate equations for circumference and area.

An **equilateral polygon** has all sides congruent. An **equiangular polygon** has all angles congruent. A **regular polygon** is a convex polygon that is both equilateral and equiangular.

Today's Goals

- Find perimeters, circumferences, and areas of two-dimensional geometric shapes.
- Calculate the measures of real-world objects.

Today's Vocabulary

polygon
perimeter
circumference
area
equilateral polygon
equiangular polygon
regular polygon
concave
convex
geometric model

Go Online You can watch a video to see how to find the perimeter and area of a figure on the coordinate plane.

Your Notes

Study Tip

Concave and Convex

Polygons can be **concave** or **convex**. Suppose the line containing each side is drawn. If any of the lines contain any point in the interior of the polygon, then it is concave. Otherwise, it is convex.

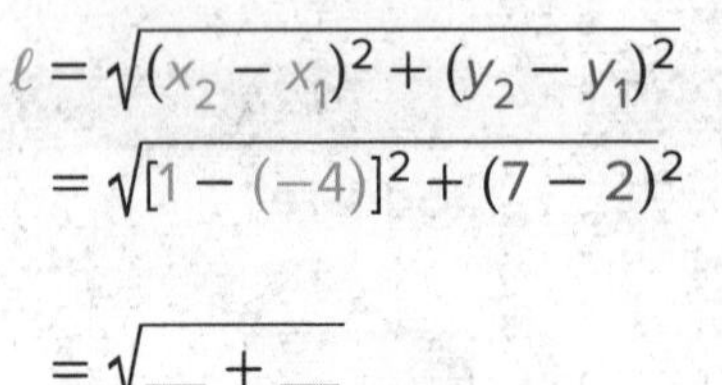

Study Tip

Perimeter vs. Area

Because calculating the area of a figure involves multiplying two dimensions, *square units* are used. There is only one dimension used when finding the perimeter, thus, it is given simply in *units*.

Think About It!

How can you use the coordinate grid to check your answer?

Example 1 Find Perimeter, Circumference, and Area

Find the perimeter or circumference and area of each figure.

a. Rectangle *ABCD*

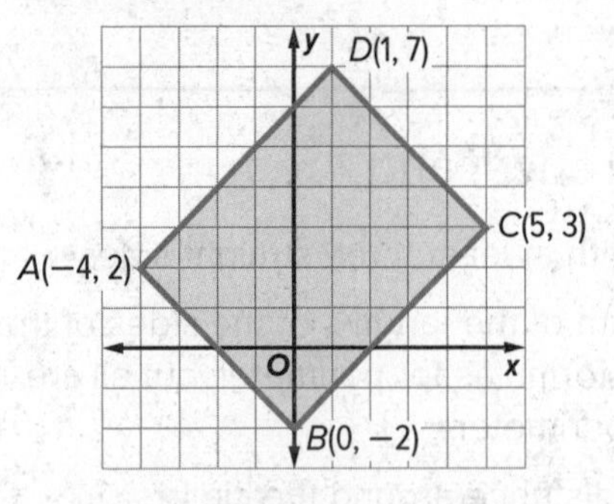

First, find the length ℓ of the rectangle by using the Distance Formula.

$\ell = \sqrt{(x_2 - x_1)^2 + (y_2 - y_1)^2}$	Distance Formula
$= \sqrt{[1 - (-4)]^2 + (7 - 2)^2}$	Let $(x_1, y_1) = A(-4, 2)$ and $(x_2, y_2) = D(1, 7)$.
$= \sqrt{\_\_\_ + \_\_\_}$	Subtract.
$= \_\_\_$	Simplify.

Next, find the width w of the rectangle by using the Distance Formula.

$w = \sqrt{(x_2 - x_1)^2 + (y_2 - y_1)^2}$	Distance Formula
$= \sqrt{[0 - (-4)]^2 + [(-2) - 2]^2}$	Let $(x_1, y_1) = A(-4, 2)$ and $(x_2, y_2) = B(0, -2)$.
$= \sqrt{\_\_\_ + \_\_\_}$	Subtract.
$= \_\_\_$	Simplify.

Use the length and width that you calculated to find the perimeter and area of the rectangle.

$P = 2\ell + 2w$	Perimeter of a rectangle
$= 2\,\_\_\_ + 2\sqrt{32}$	$\ell = \sqrt{50}$ and $w = \sqrt{32}$
≈ 25.5	Simplify.

The perimeter is about ___ units.

$A = \ell w$	Area of a rectangle
$= \_\_\_ \times \sqrt{32}$	$\ell = \sqrt{50}$ and $w = \sqrt{32}$
$= 40$	Simplify.

The area is ___ square units.

b. Circle C

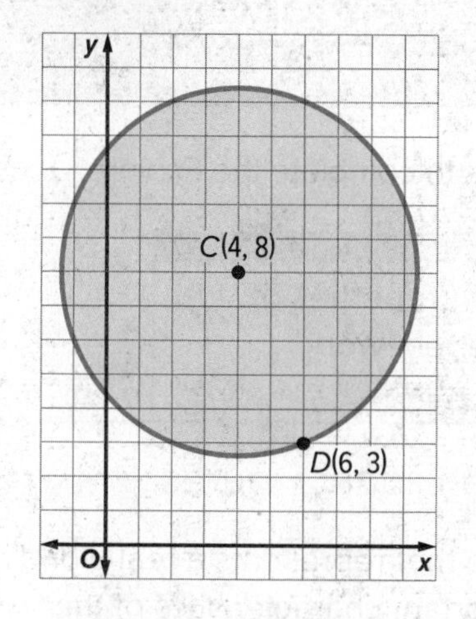

Use the Distance Formula to calculate the length of the radius of the circle.

$r = \sqrt{(x_2 - x_1)^2 + (y_2 - y_1)^2}$ Distance Formula

$= \sqrt{(6 - 4)^2 + (3 - 8)^2}$ $C(4, 8)$ and $D(6, 3)$

$= \sqrt{2^2 + (-5)^2}$ Subtract.

$= \sqrt{29}$ Simplify.

Use the value of *r* to find the circumference and area of the circle.

$C = 2\pi r$ Circumference

$= 2\pi\sqrt{29}$ or about 33.8 $r = \sqrt{29}$

The circumference of the circle is about 33.8 units.

$A = \pi r^2$ Area of a circle

$= \pi(\sqrt{29})^2$ $r = \sqrt{29}$

$= 29\pi$ or about 91.1 Simplify.

The area of the circle is about 91.1 square units.

Check

Find the circumference and area of the circle. Round to the nearest tenth if necessary.

$C \approx$ ______ units

$A \approx$ ______ units^2

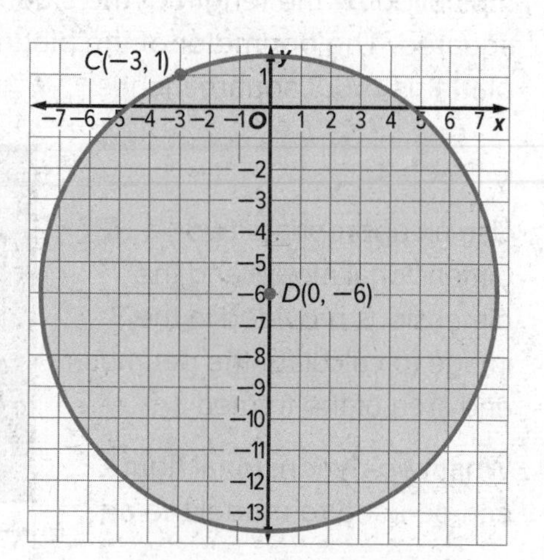

Explore Modeling Objects by Using Two-Dimensional Figures

Online Activity Use real-world objects to complete the Explore.

> **INQUIRY** How can you apply the properties of two-dimensional figures to solve real-world problems?

Learn Modeling with Two-Dimensional Figures

A **geometric model** is a geometric figure that represents a real-world object. A good model shows all the important characteristics of the object it represents, although some of the detail may be lost.

Drafters use two-dimensional geometric models to create technical drawings that communicate an object's function or construction. Scientists may use two-dimensional models to record an object's general shape or mechanics in a field notebook. You can use two-dimensional models to estimate the perimeter, circumference, and area of objects.

Example 2 Modeling with Two-Dimensional Figures

Study Tip

Approximations Because the corners of the serving platter are rounded, we can only approximate the perimeter and area of the serving platter using the formulas for a square.

Use an appropriate two-dimensional model and the dimensions provided in the image to calculate the perimeter and area of the plate.

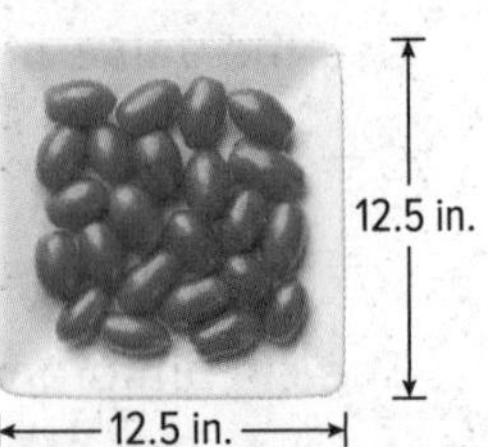

What two-dimensional figure can be used to model the serving platter? ________

What are the perimeter and area of the serving platter? Round to the nearest tenth, if necessary.

Perimeter $= 4s = 4($______$) =$ ____ in.

Area $= s^2 = ($______$)^2 = 156.3$ in^2

Because the platter is a square, the perimeter of the platter is 4 multiplied by the length of the side. The area is the length of the side squared. The perimeter of the platter is 50 inches, and the area of the platter is 156.3 square inches.

Check

Use an appropriate two-dimensional model and the dimensions provided in the image to calculate the perimeter and area of the framed art.

What two-dimensional figure can be used to model the art?

____

$P =$ ______ cm; $A =$ ______ cm^2

Example 3 Using a Two-Dimensional Model

BUSINESS **Isaiah owns a small café.**

Part A A new fire code states that there must be 15 square feet of free space for every customer in the café. How many people can be in the café?

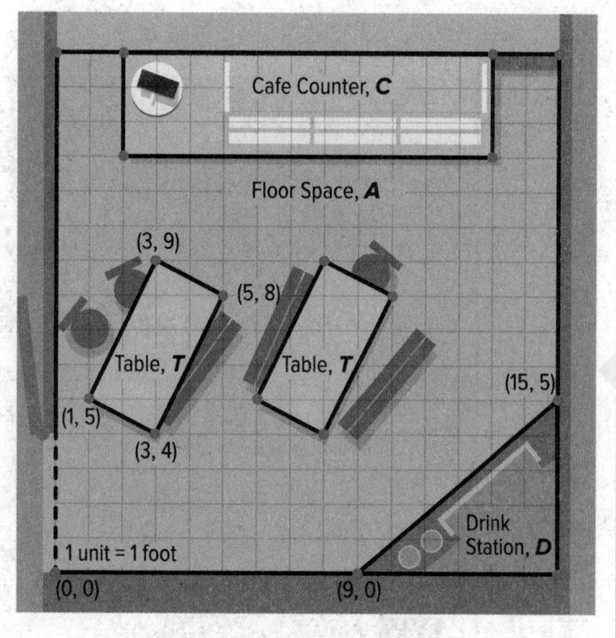

Step 1 Find the amount of free space available.

Find the total area of the café.

Area of the café =
15×15 or ____ ft^2

Then, find the area of the counter and the drink station.

$C = 3 \times 11$ or ____ ft^2

$D = \frac{1}{2}(5 \cdot 6)$ or ____ ft^2

Find the areas of the tables by using the Distance Formula.

$\ell = \sqrt{(3-1)^2 + (9-5)^2}$ or $\sqrt{20}$ and $w = \sqrt{(3-1)^2 + (4-5)^2}$ or $\sqrt{5}$

$T = \ell \cdot w = \sqrt{20} \cdot \sqrt{5}$ or ____ ft^2

Find the amount of free space available for Isaiah's customers.

$A =$ area of the café $- C - D - 2T$

$= 225 - 33 - 15 - (2 \times 10)$ or ____ ft^2

Step 2 Find the number of people that can be in the café.

$157 \text{ ft}^2 \cdot \frac{1 \text{ person}}{15 \text{ ft}^2} \approx 10.5$ or 10 people

The café can hold 10 people.

Part B Isaiah wants to hang garland around the tables and the drink station. How much garland does Isaiah need?

Find the sum of the perimeters of the tables and drink station.

length of garland = 2 • perimeter of table + perimeter of drink station

$= 2(2\sqrt{20} + 2\sqrt{5}) + (6 + 5 + \sqrt{(15-9)^2 + (5-0)^2})$

$\approx$ ____ feet

Isaiah would need at least 45.6 feet of garland.

Problem-Solving Tip

Evaluate Your Answer It can be tempting to complete the final calculation in a multi-step exercise and conclude that you have arrived at the answer. However, always remember to define appropriate quantities when solving a real-world problem. In this example, it does not make sense to have 10.5 people. You can determine that a correct answer for this exercise must be a whole number.

Study Tip

Radical Form Leave answers in radical form until the last calculation. This will prevent compounding errors caused by rounding throughout steps within a problem.

Check

LANDSCAPING Monica is redesigning her backyard. She has created the following blueprint to model her design.

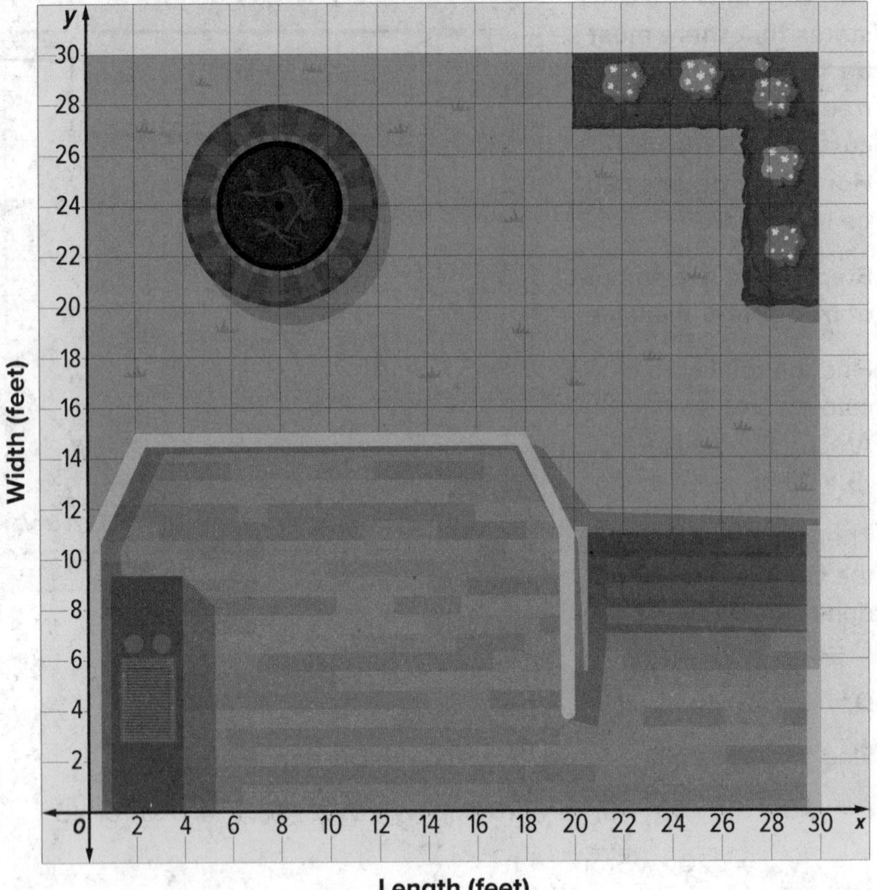

Part A

Monica wants to have at least 300 square feet of grass available in the backyard for her dog. Is there enough space for her dog? If there is, then how much area is available?

A. no

B. yes; 387.7 ft^2

C. yes; 396.7 ft^2

D. yes; 472.7 ft^2

Part B

Monica wants to build a fence in the backyard. She does not want to enclose the edge of the deck that extends from (0, 0) to (30, 0). If Monica wants to enclose the rest of the backyard, including the side edges of the deck and the side edge of the stairs, then how many feet of material are needed to complete the project?

____ feet

Go Online You can complete an Extra Example online.

Name ______________________ Period ____________ Date ____________

Practice

Go Online You can complete your homework online.

Example 1

Find the perimeter or circumference and area of each figure if each unit on the graph measures 1 centimeter. Round answers to the nearest tenth, if necessary.

1. **2.** **3.**

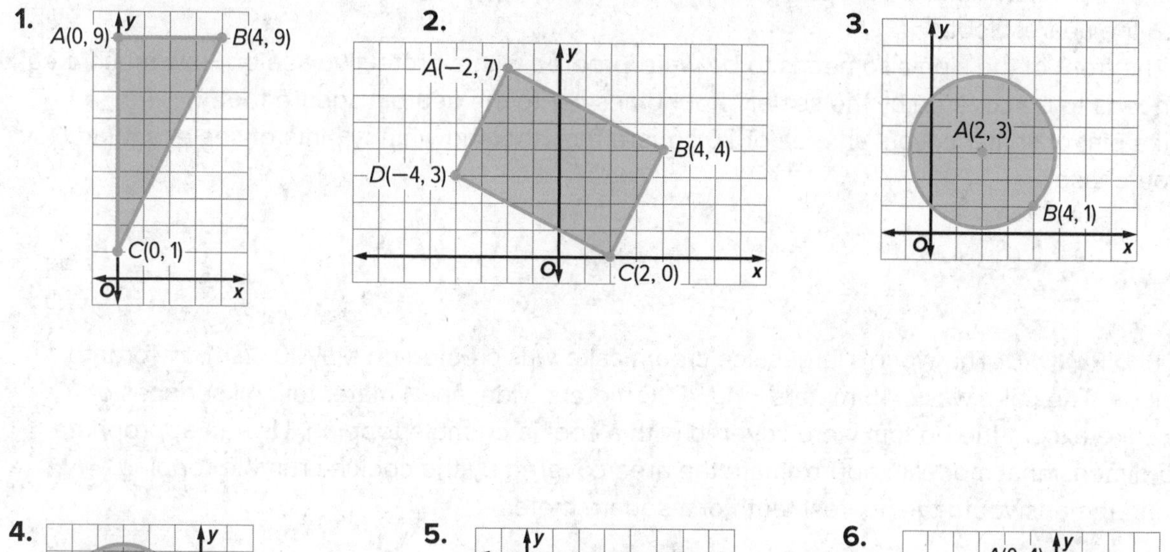

4.

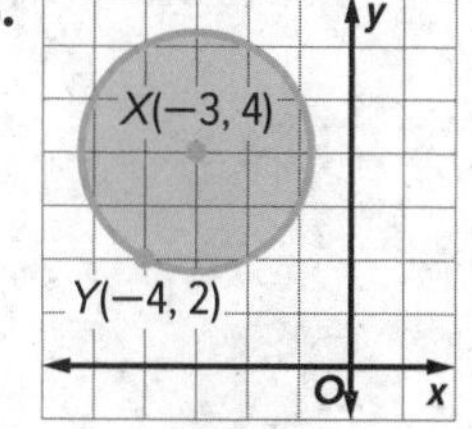

5.

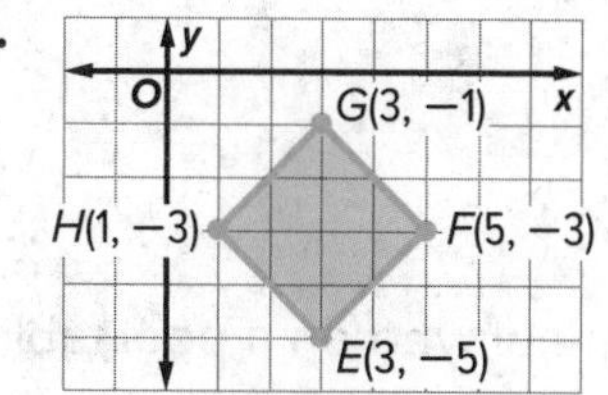

6.

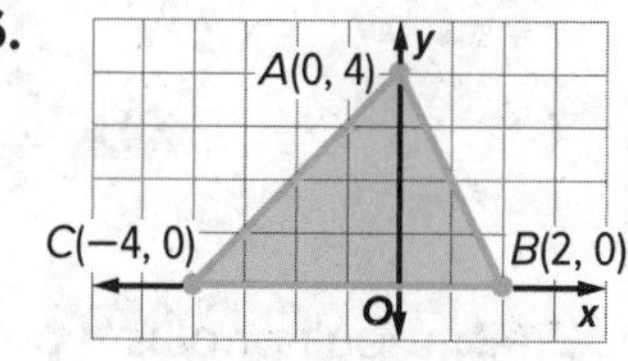

Example 2

Use a two-dimensional model and the dimensions provided to calculate the perimeter or circumference and area of each object. Round to the nearest tenth, if necessary.

7.

8.

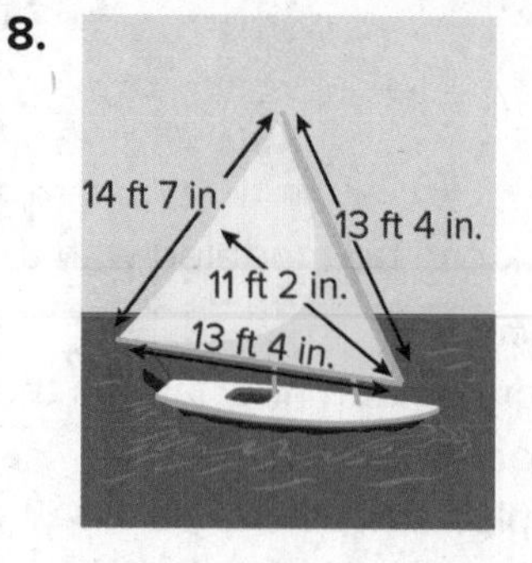

9.

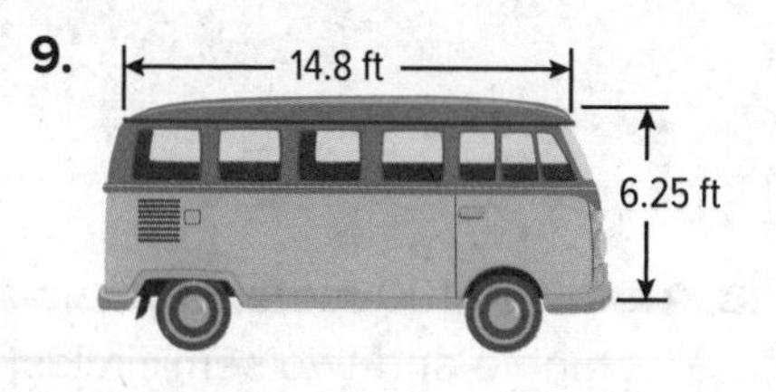

Example 3

10. DESIGN Dev is designing a new sign for his art studio. However, he needs to make several improvements to the sign before it is ready to be hung.

a. Dev wants to add a metal trim around the perimeter of the sign. How much trim should Dev purchase? Round answer to the nearest foot.

b. The front of the sign also needs to be waterproofed with a protective sealer. How much area needs to be covered by the sealer? Round answer to the nearest square foot.

c. If a pint of sealer covers an area of 20 square feet, then how many pints of sealer should Dev purchase?

11. WORLD RECORD The world's largest ice cream cake was created on May 10, 2011, in Toronto, Canada. The cake was 4.45 meters long, 4.06 meters wide, and 1 meter tall. All surfaces of the cake except the bottom were covered with a cookie crumble topping. Use an appropriate two-dimensional model to approximate the area covered by the cookie crumble topping. Round the answer to the nearest tenth of a square meter.

12. POOL Eight-ball pool is a popular game played on a pool table that has six pockets. In eight-ball pool, there are 7 striped balls, 7 solid-colored balls, and a black eight ball. At the beginning of each game, players position the 15 balls in a rack in preparation for the first shot.

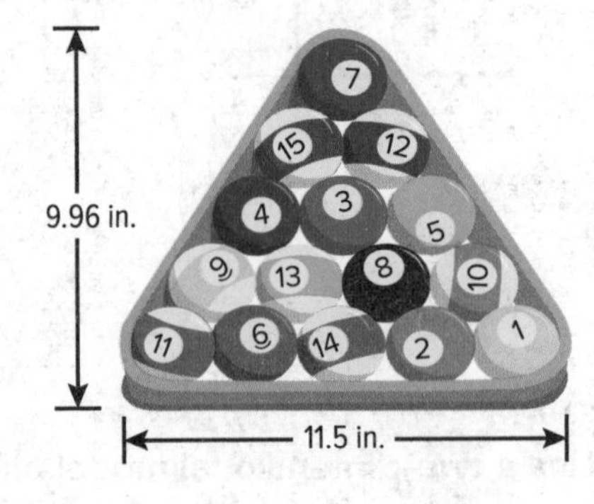

a. Find the area contained by the rack using an appropriate two-dimensional model. Round the answer to the nearest tenth of a square inch.

b. Approximate the area covered by a single ball to the nearest tenth of a square inch.

13. TRACK A 400-meter Olympic-size track can be modeled with a rectangle and two semicircles.

a. If an athlete runs around the track once, then how far has the athlete traveled to the nearest meter?

b. What assumption can be used to explain the difference between your answer in **part a** and the actual length around the track?

c. Each lane is 1.22 meters wide. If the athlete runs in the center of the inside lane, then how far has she traveled after a single lap to the nearest meter?

d. How far inside the track should the athlete be positioned to run exactly 400 meters? Round the answer to the nearest centimeter.

Name ______ Period ______ Date ______

Mixed Exercises

Identify the figure with the given vertices. Find the perimeter and area of the figure.

14. $A(3, 5)$, $B(3, 1)$, $C(0, 1)$

15. $Q(-3, 2)$, $R(1, 2)$, $S(1, -4)$, $T(-3, -4)$

16. $G(-4, 1)$, $H(4, 1)$, $I(0, -2)$

17. $K(-1, 1)$, $L(3, 4)$, $M(6, 0)$, $N(2, -3)$

18. Rectangle *WXYZ* has a length that is 5 more than three times its width.

a. Draw and label a figure for rectangle *WXYZ*.

b. Write an algebraic expression for the perimeter of the rectangle.

c. Find the width if the perimeter is 58 millimeters. Explain how you can check that your answer is correct.

d. Use a ruler to draw and label $\overline{PQ}$, which is congruent to the segment representing the length of rectangle *WXYZ*. What is the measure of $\overline{PQ}$?

19. FENCING The figure shows Derek's house and his backyard on a coordinate grid. Derek is planning to fence in the play area in his backyard. Part of the play area is enclosed by the house and does not need to be fenced. Each unit on the coordinate grid represents 5 feet. The cost for the fencing materials and installation is \$10 per foot. How much will it cost Derek to install the fence? Explain.

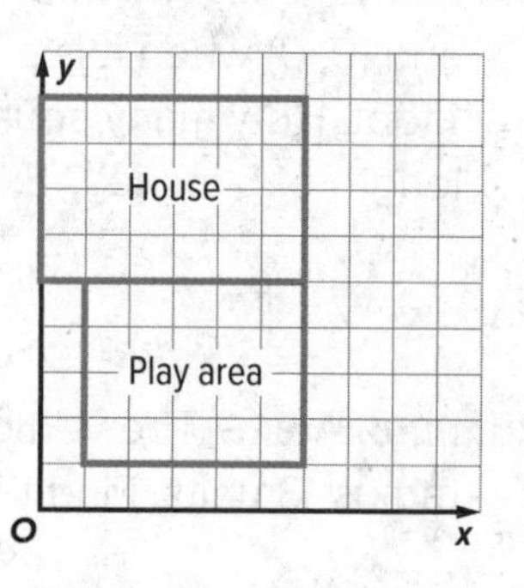

20. Explain a method to find the area of $\triangle QRS$ given that $\overline{RT} \perp \overline{QS}$. Then find the area. Show your work.

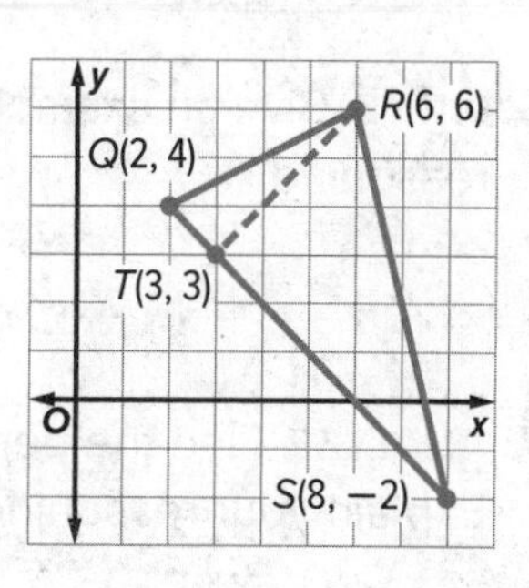

21. **SONAR** Sonar is used by oceanographers to locate marine animals and to map the contours of the ocean floor. Sonar sends out sound pulses, called pings, and receives the returning sound echo. Sonar uses the returning sound echo to detect the location of animals or the distance from a rock formation. If each unit on the coordinate grid measures 1 mile, then what area does the sonar system cover? Round to the nearest tenth.

22. Two vertices of square *ABCD* are *C*(5, 8) and *D*(2, 4).

a. Do you need to find the coordinates for the other two vertices to find the perimeter and area of the square? Justify your argument.

b. Find the perimeter and area of square *ABCD*. Show your work.

23. The coordinate grid shows an equilateral triangle that fits inside a square.

a. Find the area of the square. Show your work.

b. Find the area of the triangle. Show your work.

c. Find the area of the square that is not covered by the triangle. Write an exact value and then round to the nearest tenth. Justify your reasoning.

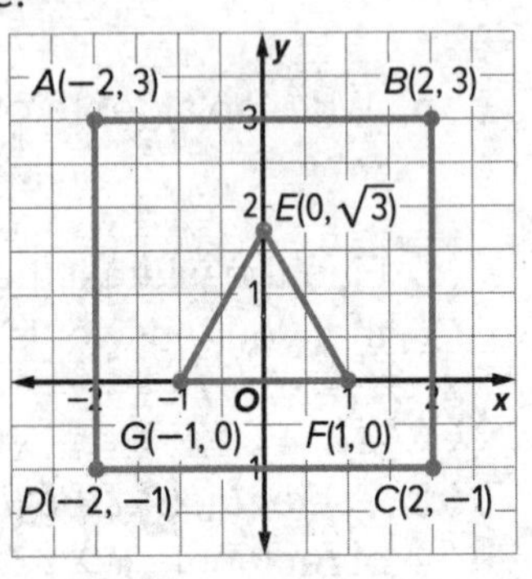

24. **PERSEVERE** The floor plan of a rectangular room has the coordinates (0, 12.5), (20, 12.5), (20, 0), and (0, 0) when it is placed on the coordinate plane. Each unit on the coordinate plane measures 1 foot. How many square tiles will it take to cover the floor of the room if the tiles have a side length of 5 inches? Explain.

25. **PERSEVERE** The vertices of a rectangle with side lengths of 10 and 24 units are on a circle of radius 13 units. Find the area between the figures.

26. **WRITE** Give an example of a polygon that is equiangular but not a regular polygon. Explain your reasoning.

27. **ANALYZE** Find the perimeter of equilateral triangle *KLM* given the vertices *K*(−2, 1) and *M*(10, 6). Explain your reasoning.

Transformations in the Plane

Explore Introducing Transformations

Online Activity Use graphing technology to complete the Explore.

INQUIRY How are reflections, translations, and rotations similar?

Today's Goals

- Analyze figures to identify the types of rigid motions represented.
- Calculate the coordinates of the vertices of images given the coordinates of the preimages.

Today's Vocabulary

transformation
preimage
image
rigid motion
reflection
translation
rotation
line of reflection
center of rotation
translation vector
component form
angle of rotation

Learn Identifying Transformations

A **transformation** is a function that takes points in the plane as inputs and gives other points as outputs. In a transformation, the **preimage** is mapped onto the **image**. A **rigid motion**, also called a *congruence transformation* or an *isometry*, is a transformation that preserves distance and angle measure.

The three main types of rigid motions are shown below. The preimage is shown in blue, and the image is shown in green. Prime notation is used to indicate transformations. If A is the preimage, then A' is the image after one transformation.

Key Concept • Reflections, Translations, and Rotations

A **reflection** or *flip* is a transformation in a line called the **line of reflection**. Each point of the preimage and its image are the same distance from the line of reflection.	A **translation** or *slide* is a transformation that moves all points of the original figure the same distance in the same direction.	A **rotation** or *turn* is a transformation about a fixed point (called the **center of rotation**), through a specific angle, and in a specific direction. Each point of the original figure and its image are the same distance from the center.
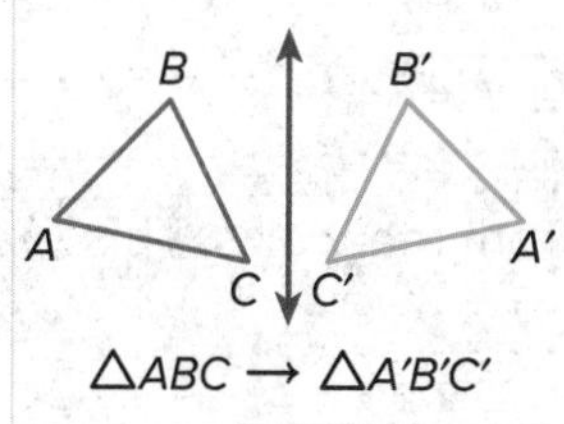 $\triangle ABC \rightarrow \triangle A'B'C'$	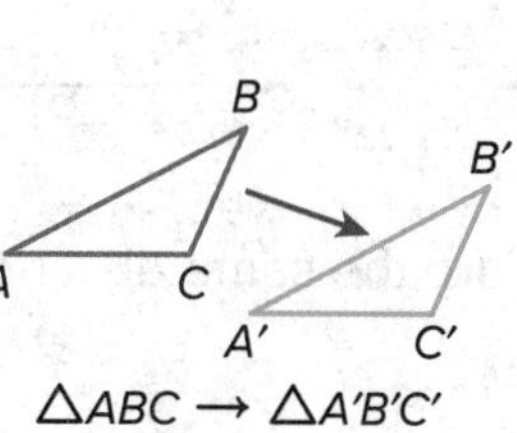 $\triangle ABC \rightarrow \triangle A'B'C'$	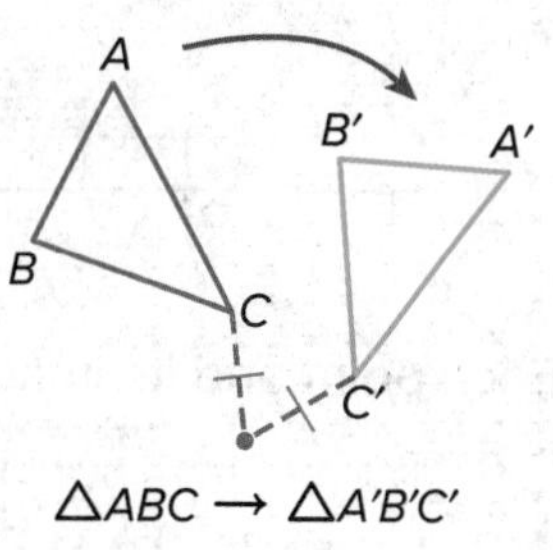 $\triangle ABC \rightarrow \triangle A'B'C'$

Study Tip

Rigid Motion A rigid motion is also called a *rigid transformation*. The two terms can be used interchangeably.

Your Notes

Study Tip

Identifying Transformations Look for lines of reflection or centers of rotation when identifying transformations. In a reflection, each point of the preimage and its corresponding point of the image are the same distance from the line of reflection. In a rotation, each point of the preimage and its corresponding point of the image are the same distance from the center of rotation.

Example 1 Identify Transformations in the Real World

HOBBIES **Identify the type of rigid motion shown in the photo as a *reflection*, *translation*, or *rotation*.**

The landscape is mirrored in the water. This is an example of a ________.

Check

CHECKERS In the game of checkers, players move their pieces on the diagonal. Identify the type of rigid motion shown as a *reflection*, *translation*, or *rotation*.

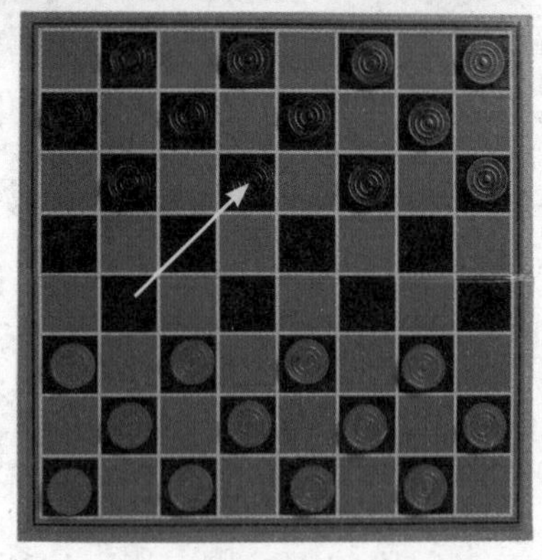

The type of rigid motion is a ________.

Example 2 Identify Transformations on the Coordinate Plane

Identify the type of rigid motion shown as a *reflection*, *translation*, or *rotation*.

a.

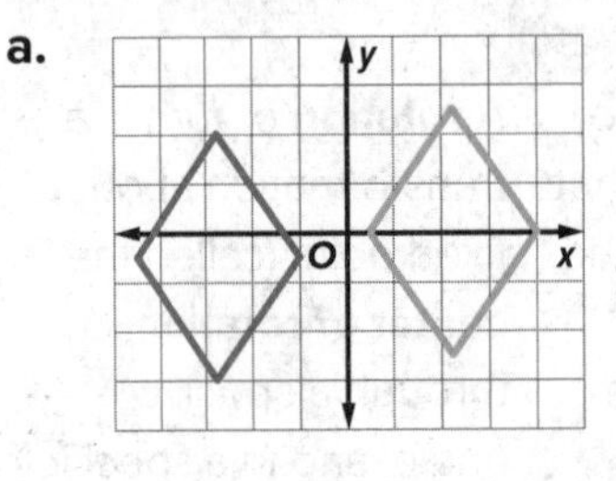

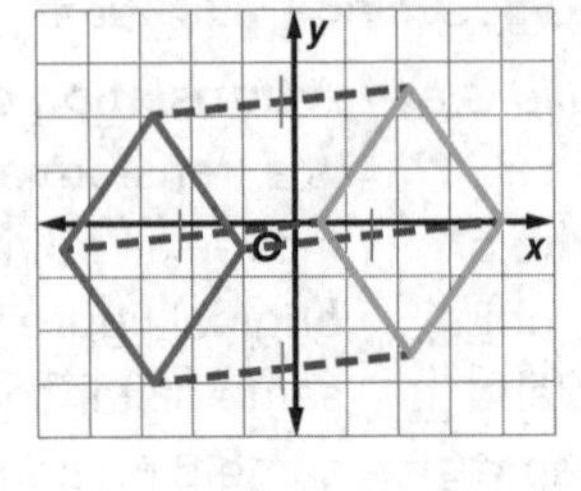

Each vertex and its image can be connected by lines with the same length and slope. This is a ________.

b.

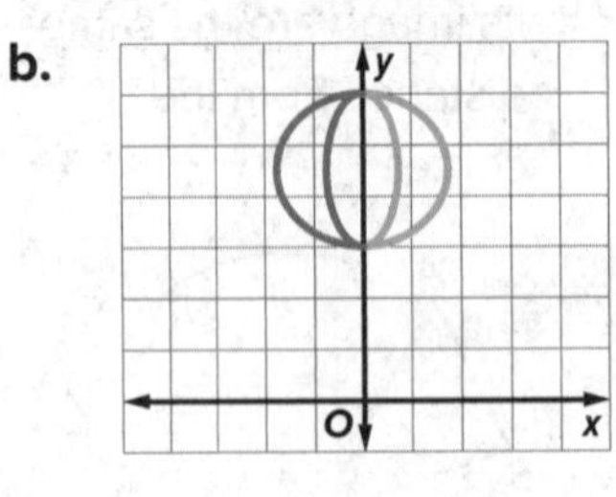

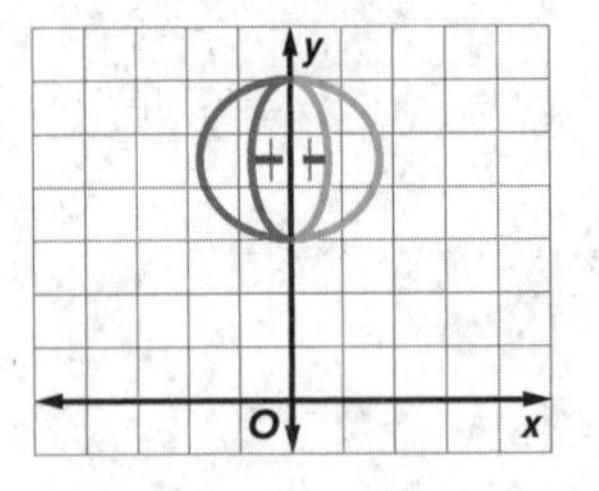

Each point and its image are the same distance from the y-axis. This is a ________.

Go Online You can complete an Extra Example online.

c.

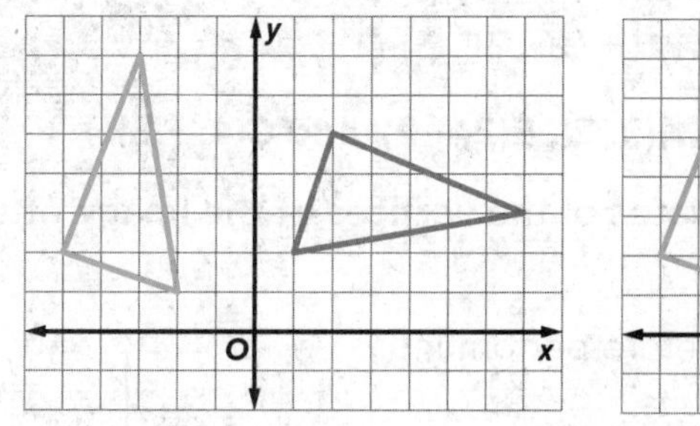

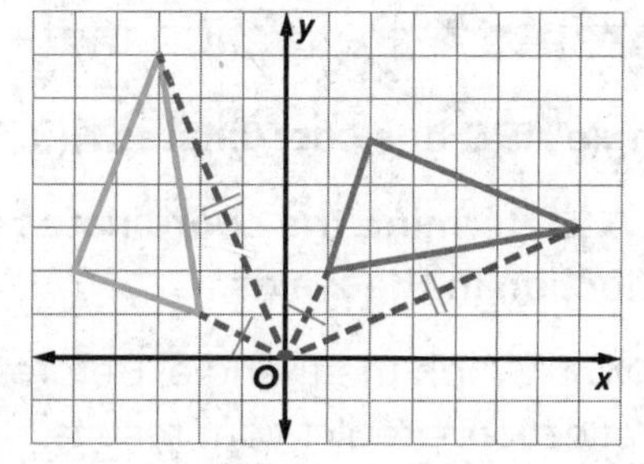

Each vertex and its image are the same distance from the origin. The angles formed by each pair of corresponding points and the origin are congruent. This is a ________________.

Check

The type of rigid motion shown is a ________________.

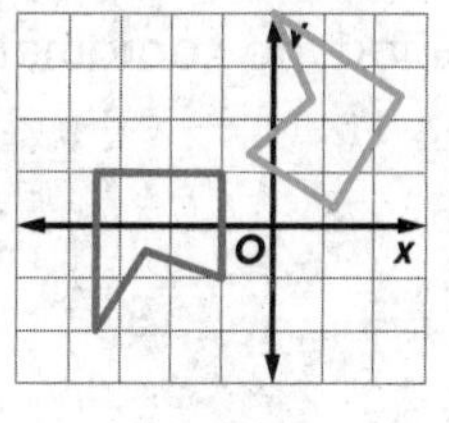

Learn Representing Reflections

In a reflection, each point of the preimage and its corresponding point on the image are the same distance from the line of reflection.

A reflection can be described as a function in which the preimage is reflected in the line of reflection. The points of the preimage are the input, and the corresponding points on the image are the output.

Key Concept • Reflections in the *x*- or *y*-axis

	Reflections in the *x*-axis	**Reflections in the *y*-axis**
Words	To reflect a point in the *x*-axis, multiply its *y*-coordinate by −1.	To reflect a point in the *y*-axis, multiply its *x*-coordinate by −1.
Symbols	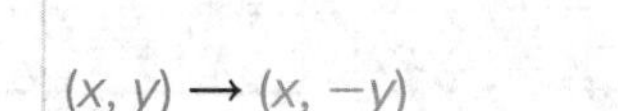 $(x, y) \rightarrow (x, -y)$	$(x, y) \rightarrow (-x, y)$
Example	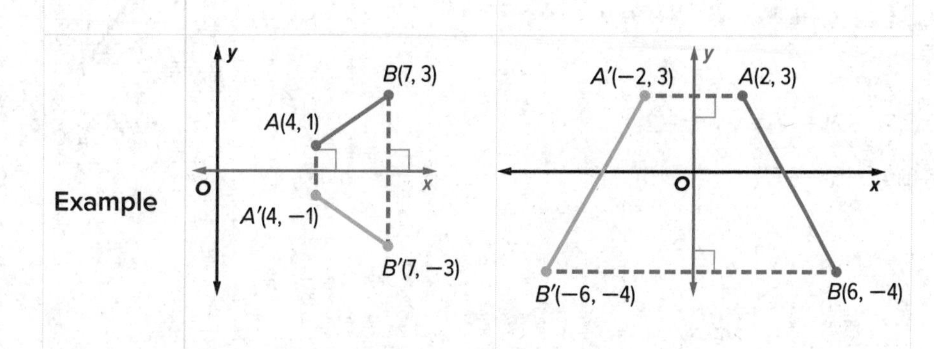	

Study Tip

What Is Preserved? Because it is a rigid motion, all lengths and angle measures are preserved in a reflection.

Example 3 Reflection in the x- or y-Axis

Triangle *ABC* has coordinates *A*(3, 2), *B*(2, −2), and *C*(4, −5).

Part A Determine the coordinates of the vertices of the image after a reflection in the *x*-axis.

> **Think About It!**
> Suppose the coordinates of *A* are (5, −2) and the coordinates of *A*′ are (5, 2). Describe the transformation of *A*.

PREDICT Graph the triangle. Before performing the reflection, predict your results.

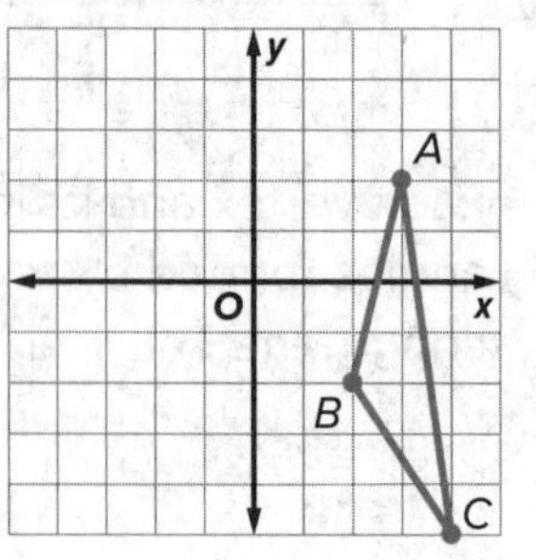

____

____

Multiply the y-coordinate of each vertex by −1.

Find the coordinates of the vertices of the image.

$(x, y) \rightarrow (x, -y)$

$A(3, 2) \rightarrow A'(3, -2)$

$B(2, -2) \rightarrow B'($_____, _____$)$

$C(4, -5) \rightarrow C'($_____, _____$)$

CHECK The image matches the prediction.

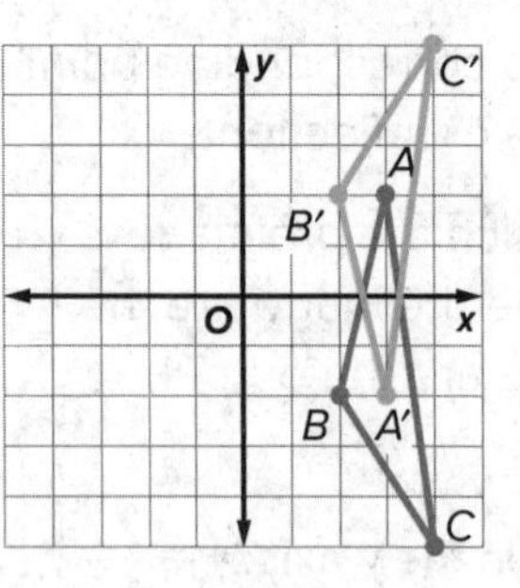

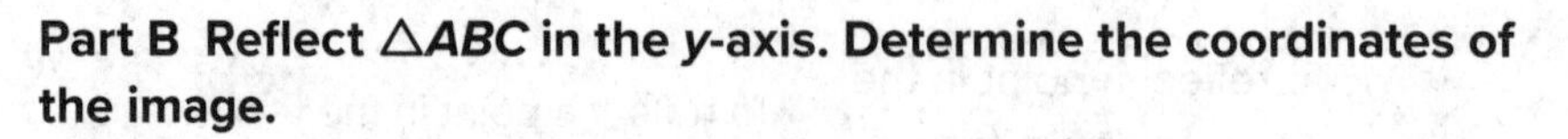

Part B Reflect △*ABC* in the *y*-axis. Determine the coordinates of the image.

PREDICT Before performing the reflection, predict your results.

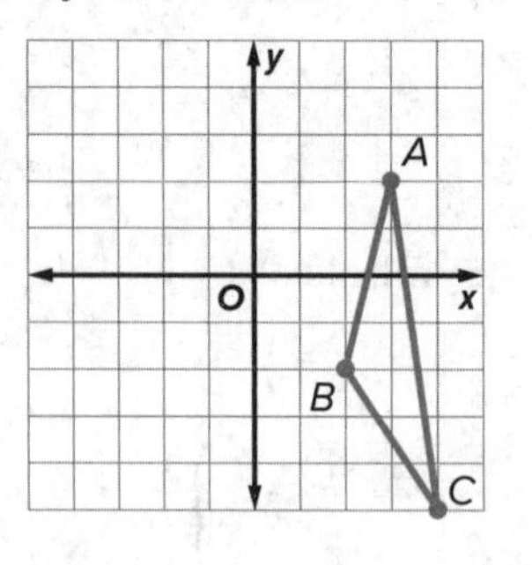

____

____

Multiply the x-coordinate of each vertex by −1.

Find the coordinates of the vertices of the image.

$(x, y) \rightarrow (-x, y)$

$A(3, 2) \rightarrow A'($_____$, 2)$

$B(2, -2) \rightarrow B'($_____$, -2)$

$C(4, -5) \rightarrow C'($_____$, -5)$

Go Online You can complete an Extra Example online.

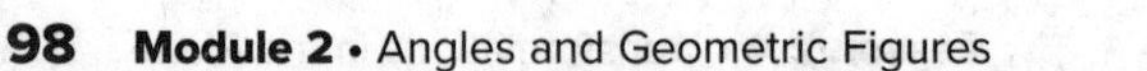

c.

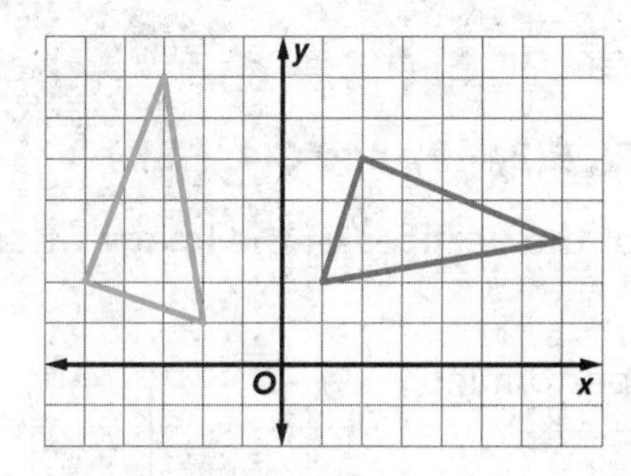

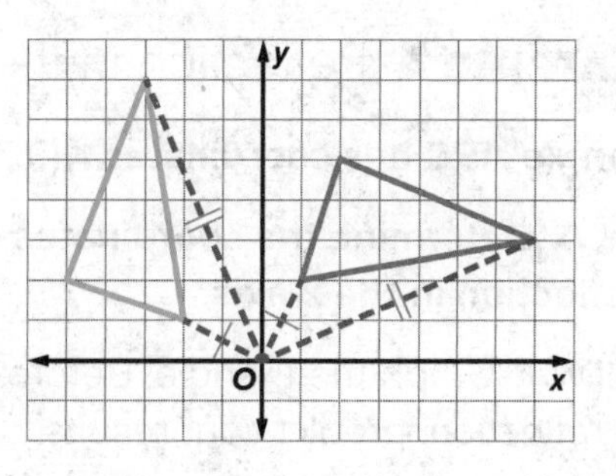

Each vertex and its image are the same distance from the origin. The angles formed by each pair of corresponding points and the origin are congruent. This is a ______________.

Check

The type of rigid motion shown is a ______________.

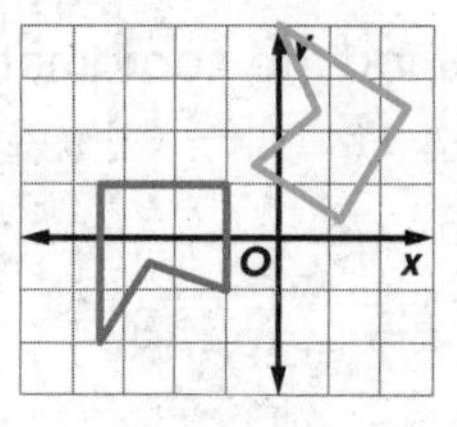

Learn Representing Reflections

In a reflection, each point of the preimage and its corresponding point on the image are the same distance from the line of reflection.

A reflection can be described as a function in which the preimage is reflected in the line of reflection. The points of the preimage are the input, and the corresponding points on the image are the output.

Key Concept • Reflections in the *x*- or *y*-axis

	Reflections in the *x*-axis	Reflections in the *y*-axis
Words	To reflect a point in the *x*-axis, multiply its *y*-coordinate by −1.	To reflect a point in the *y*-axis, multiply its *x*-coordinate by −1.
Symbols	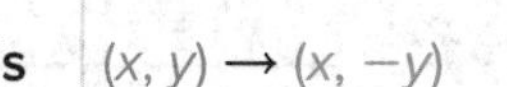 $(x, y) \rightarrow (x, -y)$	$(x, y) \rightarrow (-x, y)$
Example	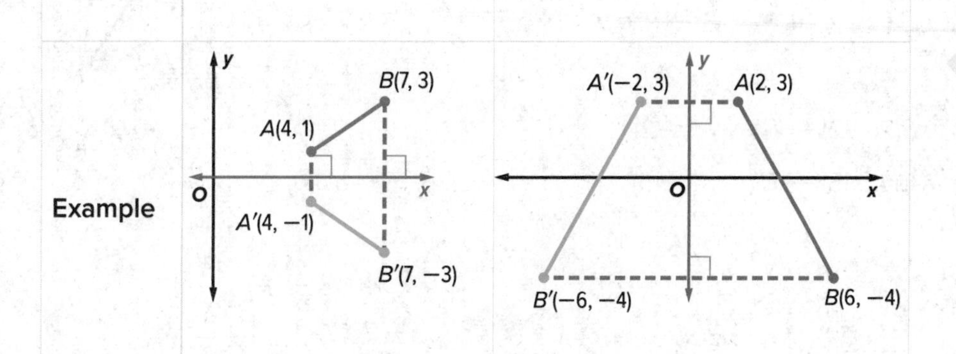	

Study Tip

What Is Preserved? Because it is a rigid motion, all lengths and angle measures are preserved in a reflection.

Example 3 Reflection in the x- or y-Axis

Triangle *ABC* has coordinates *A*(3, 2), *B*(2, −2), and *C*(4, −5).

Part A Determine the coordinates of the vertices of the image after a reflection in the *x*-axis.

> **Think About It!**
> Suppose the coordinates of *A* are (5, −2) and the coordinates of *A′* are (5, 2). Describe the transformation of *A*.

PREDICT Graph the triangle. Before performing the reflection, predict your results.

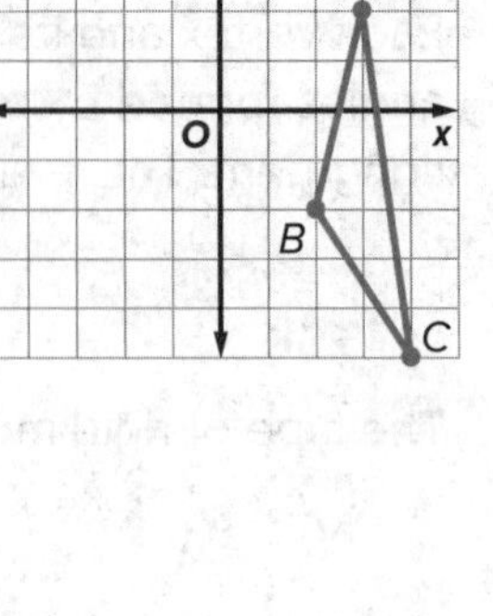

Multiply the *y*-coordinate of each vertex by −1.

Find the coordinates of the vertices of the image.

$(x, y) \rightarrow (x, -y)$

$A(3, 2) \rightarrow A'(3, -2)$

$B(2, -2) \rightarrow B'($_____, _____$)$

$C(4, -5) \rightarrow C'($_____, _____$)$

CHECK The image matches the prediction.

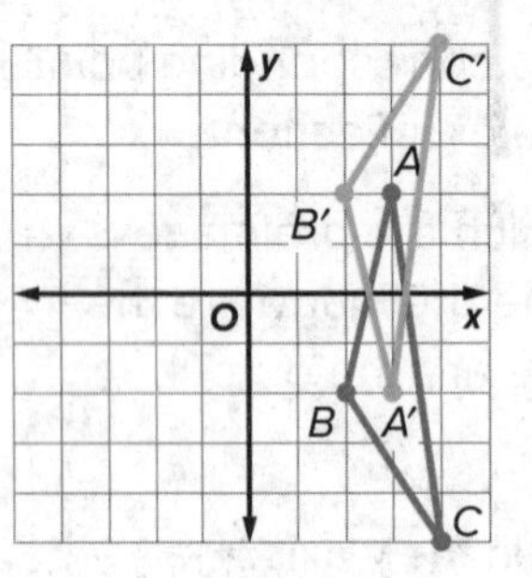

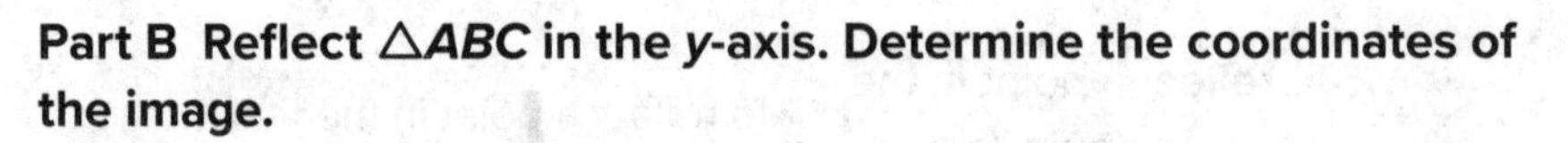

Part B Reflect △*ABC* in the *y*-axis. Determine the coordinates of the image.

PREDICT Before performing the reflection, predict your results.

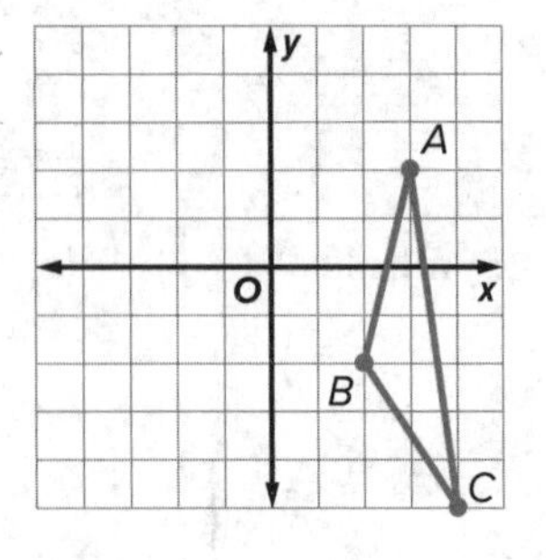

Multiply the *x*-coordinate of each vertex by −1.

Find the coordinates of the vertices of the image.

$(x, y) \rightarrow (-x, y)$

$A(3, 2) \rightarrow A'($_____$, 2)$

$B(2, -2) \rightarrow B'($_____$, -2)$

$C(4, -5) \rightarrow C'($_____$, -5)$

Go Online You can complete an Extra Example online.

CHECK The image matches the prediction.

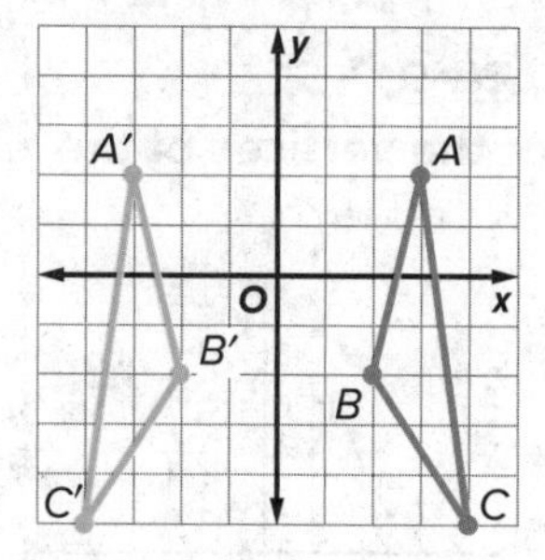

Check

Triangle JKL has coordinates $J(2, -8)$, $K(6, -7)$, and $L(4, -2)$. Determine the coordinates of the vertices of the image after a reflection in the x-axis.

A. $J'(2, 8)$, $K'(6, 7)$, $L'(4, 2)$

B. $J'(-2, -8)$, $K'(-6, -7)$, $L'(-4, -2)$

C. $J'(-2, 8)$, $K'(-6, 7)$, $L'(-4, 2)$

D. $J'(2, -8)$, $K'(6, -7)$, $L'(4, -2)$

Learn Representing Translations

A translation is a function in which all of the points of a figure move the same distance in the same direction.

A preimage is translated along a **translation vector**. The translation vector describes the magnitude and direction of the slide if the magnitude is the length of the vector from its initial point to its terminal point.

To describe a translation in the coordinate plane, it is helpful to write the vector in component form. A vector in **component form** is written as $\langle x, y\rangle$, which describes the vector in terms of its horizontal component x and vertical component y.

Key Concept • Translations	
Words	To translate a point along vector $\langle a, b\rangle$, add a to the x-coordinate and add b to the y-coordinate.
Symbols	$(x, y) \rightarrow (x + a, y + b)$
Example	$P(-2, 3)$ translated along vector $\langle 7, 4\rangle$ is $P'(-2 + 7, 3 + 4)$ or $P'(5, 7)$.

Study Tip

What Is Preserved? Because it is a rigid motion, all lengths and angle measures are preserved in a translation.

Example 4 Translations

For quadrilateral *QRST* with vertices *Q*(−8, −2), *R*(−9, −5), *S*(−4, −7), and *T*(−4, −2), find the coordinates of the vertices of the image after a translation along the vector ⟨7, 1⟩.

PREDICT Graph the quadrilateral. Before performing the translation, predict your results.

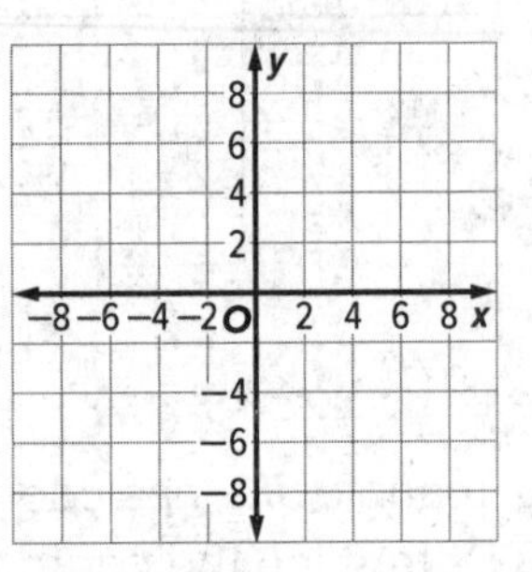

A translation along ⟨7, 1⟩ will move the figure 7 units to the right and 1 unit up.

Find the coordinates of the vertices of the image.

$(x, y) \rightarrow (x + 7, y + 1)$

$Q(-8, -2) \rightarrow Q'(-8 + 7, -2 + 1)$ or $Q'(-1, -1)$

$R(-9, -5) \rightarrow R'(-9 + 7, -5 + 1)$ or $R'(-2, -4)$

$S(-4, -7) \rightarrow S'(-4 + 7, -7 + 1)$ or $S'(3, -6)$

$T(-4, -2) \rightarrow T'(-4 + 7, -2 + 1)$ or $T'(3, -1)$

CHECK The image matches the prediction.

Check

Quadrilateral *ABCD* has vertices *A*(−3, 1), *B*(−5, 3), *C*(−2, 5), and *D*(−1, 3). What are the coordinates of the vertices of the image after a translation along vector ⟨5, −3⟩?

A. *A*′(2, −2), *B*′(0, 0), *C*′(3, 2), and *D*′(4, 0)

B. *A*′(−8, −2), *B*′(−10, 0), *C*′(−7, 2), and *D*′(−6, 0)

C. *A*′(2, 4), *B*′(0, 6), *C*′(3, 8), and *D*′(4, 6)

D. *A*′(−8, 4), *B*′(−10, 6), *C*′(−7, 8), and *D*′(−6, 6)

Pause and Reflect

Did you struggle with anything in this lesson? If so, how did you deal with it?

Record your observations here.

Go Online You can complete an Extra Example online.

Learn Representing Rotations

A rotation is a function that moves every point of a preimage through a specified angle and direction about a fixed point, called the center of rotation. Under a rotation, each point and its image are at the same distance from the center of rotation. In this lesson, you can assume that the origin is the center of rotation. The specified angle is called the **angle of rotation**.

The direction of a rotation can be clockwise or counterclockwise. In this course, you can assume that all rotations are counterclockwise unless stated otherwise.

clockwise   counterclockwise

When a point is rotated 90°, 180°, or 270° counterclockwise about the origin, you can use the following rules. A rotation of 360° will map the image onto the preimage.

Key Concept • Rotations in the Coordinate Plane

90° Rotation

To rotate a point 90° counterclockwise about the origin, multiply the y-coordinate by −1 and then interchange the x- and y-coordinates.

Symbols $(x, y) \rightarrow (-y, x)$

Example

$P'(-2, 5)$, $90°$, $P(5, 2)$, O, x, y

180° Rotation

To rotate a point 180° counterclockwise about the origin, multiply the x- and y-coordinates by −1.

Symbols $(x, y) \rightarrow (-x, -y)$

Example

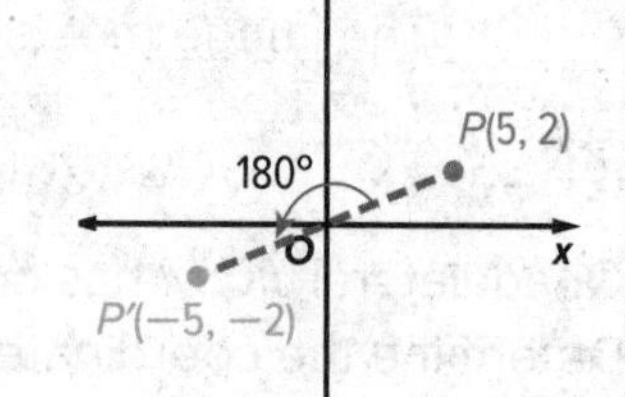

270° Rotation

To rotate a point 270° counterclockwise about the origin, multiply the x-coordinate by −1 and then interchange the x- and y-coordinates.

Symbols $(x, y) \rightarrow (y, -x)$

Example

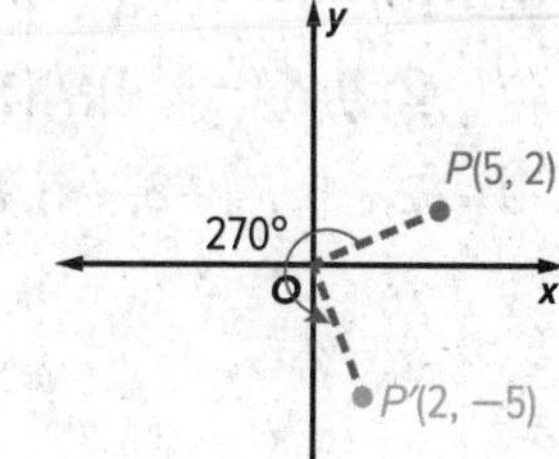

Study Tip

What Is Preserved? Because it is a rigid motion, all lengths and angle measures are preserved in a rotation.

Talk About It!

Would two successive 90° rotations counterclockwise about the origin result in the same image as a 180° rotation clockwise about the origin? Explain.

Think About It!

Describe the figure after the rotation.

Example 5 Rotations

Parallelogram *FGHJ* has vertices *F*(2, 1), *G*(7, 1), *H*(6, −3), and *J*(1, −3). What are the coordinates of the vertices of its image after a rotation of 180° about the origin?

PREDICT Graph parallelogram *FGHJ*.

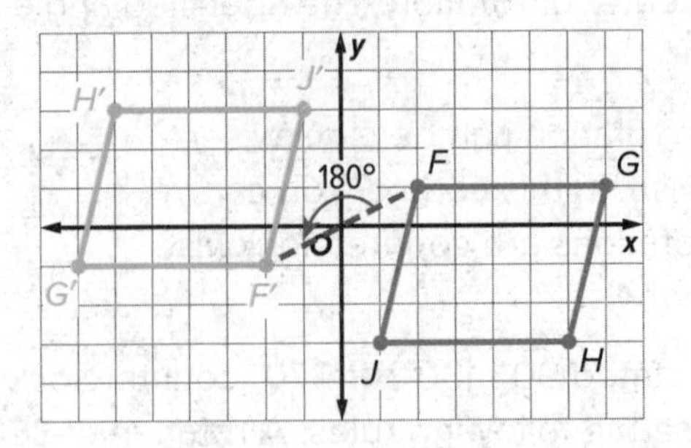

Before performing the rotation, predict your results.

To rotate a point 180° counterclockwise about the origin, multiply the *x*- and *y*-coordinates by −1. Find the coordinates of the vertices of the image.

$(x, y) \rightarrow (-x, -y)$

$F(2, 1) \rightarrow F'(-2, -1)$

$G(7, 1) \rightarrow G'(-7, -1)$

$H(6, -3) \rightarrow H'($____, ____$)$

$J(1, -3) \rightarrow J'($____, ____$)$

CHECK The image meets the prediction.

Study Tip

Rotations

Verify the coordinates of the image by graphing the preimage and image on the coordinate grid.

Check

Quadrilateral *JKLM* has coordinates *J*(1, 2), *K*(4, 3), *L*(6, 1), and *M*(3, 1). Determine the coordinates of the vertices of the image after a 270° rotation about the origin.

A. *J'*(2, −1), *K'*(3, −4), *L'*(1, −6), and *M'*(1, −3)

B. *J'*(2, 1), *K'*(3, 4), *L'*(1, 6), and *M'*(1, 3)

C. *J'*(−2, 1), *K'*(−3, 4), *L'*(−1, 6), and *M'*(−1, 3)

D. *J'*(−2,−1), *K'*(−3,−4), *L'*(−1,−6), and *M'*(−1,−3)

Go Online You can complete an Extra Example online.

Name Period Date

24. COMBINATION LOCKS Benicio locks his safe by setting each of the three dials to 8. To unlock the safe, he turns the left dial 90° counterclockwise, the middle dial 270° clockwise, and the right dial 180° counterclockwise. Which three numbers, in order, unlock the safe?

25. BEEKEEPING A beekeeper uses a frame of partial honeycomb cells that bees fill with honey and complete with wax. When the honey is ready for harvest, the beekeeper turns the tap allowing the honey to flow out of the hive without disturbing the bees. By what transformation are the sides of the partial honeycomb cells related when the tap is closed? when the tap is open?

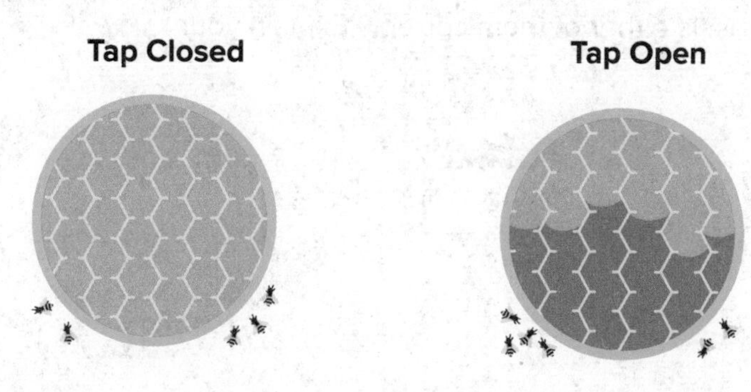

Find the coordinates of the figure with the given coordinates after the transformation on the plane. Then graph the preimage and image.

26. preimage: $J(-3, 0)$, $K(-2, 4)$, $L(-1, 0)$, image: triangle QRS, translation of JKL along vector $\langle 5, -4 \rangle$

27. preimage: $A(1, 3)$, $B(1, 1)$, $C(4, 1)$, image: triangle DEF, rotation of ABC 270° counterclockwise about the origin

28. FIND THE ERROR Saurabh and Elena visit a craft fair and notice a quilt with a pattern. Saurabh claims the pattern is made using translations. Elena believes that the pattern is made using rotations. Who is correct? Justify your argument.

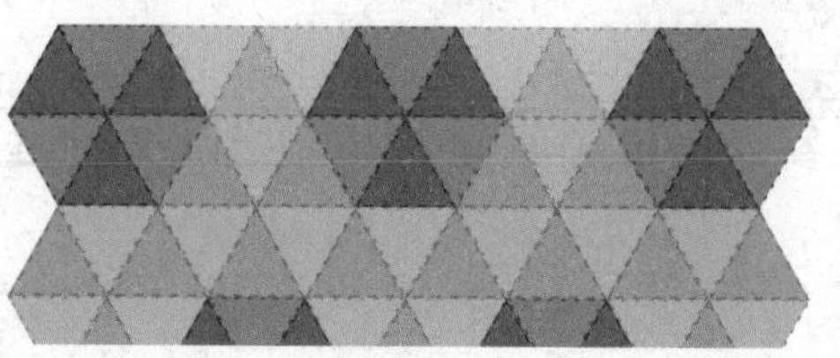

29. The vertices of $\triangle ABC$ are $A(-1, 1)$, $B(4, 2)$, and $C(1, 5)$. The vertices of $\triangle DEF$ are $D(-1, -1)$, $E(4, -2)$, and $F(1, -5)$ such that $\triangle ABC \cong \triangle DEF$. Identify the congruence transformation.

30. STRUCTURE $\overline{XY}$ has endpoints $X(-5, 6)$ and $Y(0, 4)$, the image of $\overline{XY}$ has the endpoints $X'(6, 5)$ and $Y'(4, 0)$, and $\overline{XY} \cong \overline{X'Y'}$. Identify the transformation.

31. STRUCTURE The vertices of quadrilateral *FGHJ* are $F(2, -3)$, $G(-2, -5)$, $H(-3, 6)$, and $J(3, 5)$. The vertices of quadrilateral *KLMN* are $K(5, -5)$, $L(1, -7)$, $M(0, 4)$, and $N(6, 3)$ such that $FGHJ \cong KLMN$. If quadrilateral *FGHJ* is the preimage and quadrilateral *KLMN* is the image, identify the transformation.

32. ANALYZE The image of $\triangle ABC$ reflected in the *y*-axis is $\triangle A'B'C'$.

a. Describe the result of reflecting $\triangle A'B'C'$ in the *y*-axis. Explain.

b. Describe the result of reflecting $\triangle A'B'C'$ in the *x*-axis. Explain.

33. FIND THE ERROR Antwan and Diamond are finding the coordinates of the image of $P(2, 3)$ after a reflection in the *x*-axis. Is either of them correct? Explain your reasoning.

34. WRITE In the diagram, $\triangle DEF$ is called a *glide reflection* of $\triangle ABC$. Based on the diagram, define a glide reflection. Explain your reasoning.

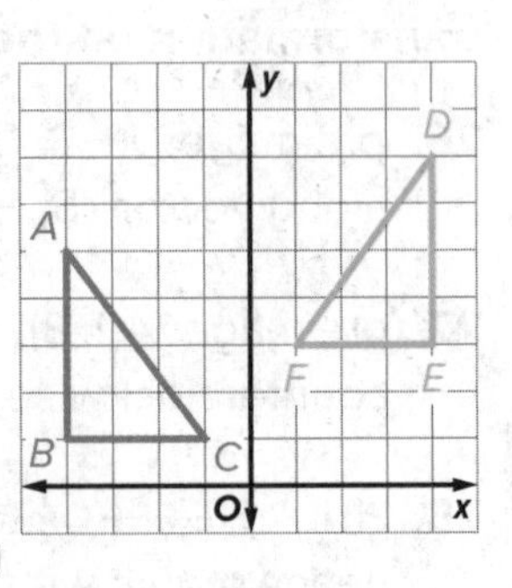

35. CREATE Draw a polygon on the coordinate plane that when reflected in the *y*-axis looks exactly like the original figure.

36. ANALYZE Is the reflection of a figure in the *x*-axis equivalent to the rotation of that same figure 180° about the origin? Explain.

Three-Dimensional Figures

Learn Identifying Three-Dimensional Figures

A **polyhedron** is a closed three-dimensional figure made up of flat polygonal regions. A **face of a polyhedron** is a flat surface on the polyhedron. An **edge of a polyhedron** is a line segment where the faces of the polyhedron intersect. The **vertex of a polyhedron** is the intersection of three edges of the polyhedron. The **bases of a prism or cylinder** are the two parallel congruent faces of the solid. The **base of a pyramid or cone** is the face of the solid opposite the vertex of the solid.

Types of Solids

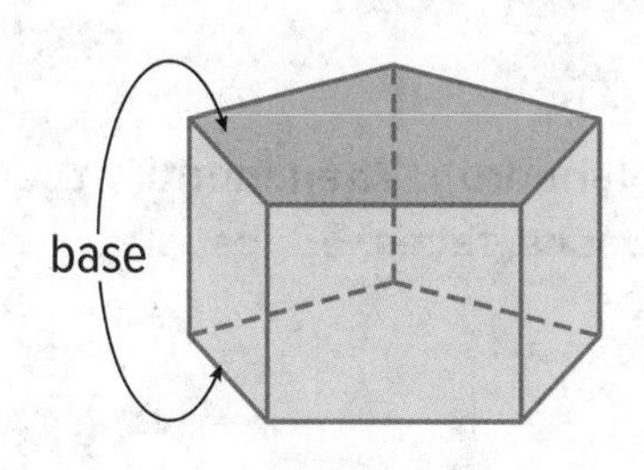

A **prism** is a polyhedron that has two parallel congruent bases, connected by parallelogram faces.

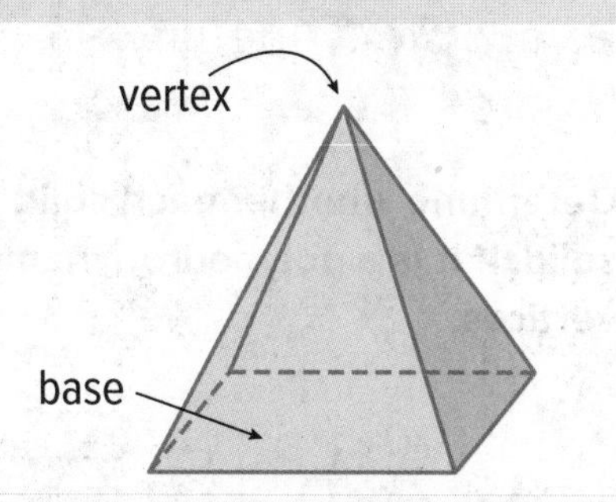

A **pyramid** is a polyhedron that has a polygonal base and three or more triangular faces that meet at a common vertex.

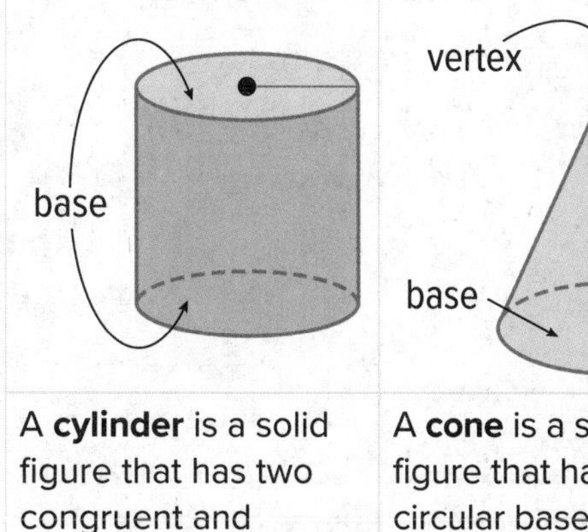

A **cylinder** is a solid figure that has two congruent and parallel circular bases connected by a curved surface.

A **cone** is a solid figure that has a circular base connected by a curved surface to a single vertex.

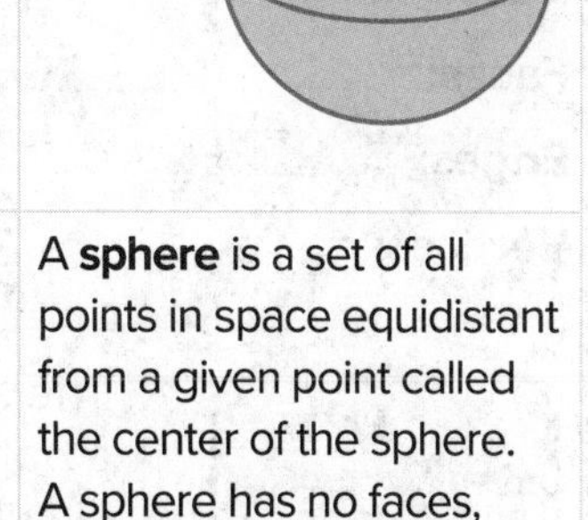

A **sphere** is a set of all points in space equidistant from a given point called the center of the sphere. A sphere has no faces, edges, or vertices.

Polyhedra, or *polyhedrons,* are named by the shapes of their bases.

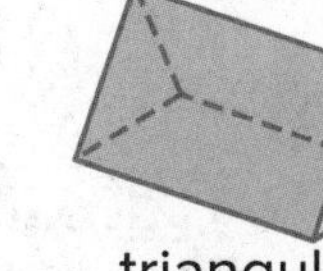
triangular prism

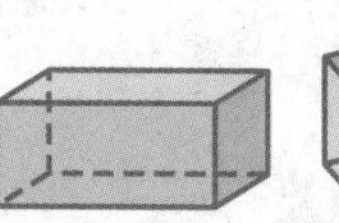
rectangular prism

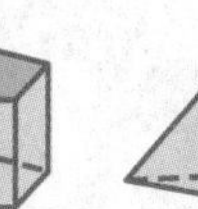
pentagonal prism

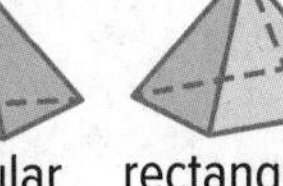
triangular pyramid

rectangular pyramid

pentagonal pyramid

Today's Goals

- Identify and determine characteristics of three-dimensional figures.
- Calculate surface areas and volumes.

Today's Vocabulary

polyhedron
face of a polyhedron
edge of a polyhedron
vertex of a polyhedron
bases of a prism or cylinder
base of a pyramid or cone
prism
pyramid
cylinder
cone
sphere
regular polyhedron
Platonic solids
surface area
volume

Study Tip

Right vs. Oblique In right prisms, the bases are connected to each other by rectangular faces. However, in oblique prisms, at least one face is not a rectangle.

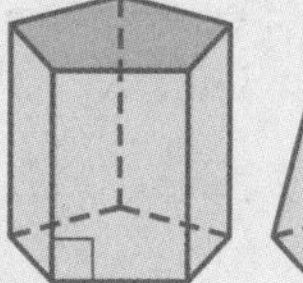
right prism

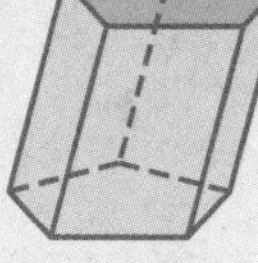
oblique prism

Your Notes

Math History Minute

The Platonic solids are named for the Greek philosopher **Plato (c. 428 BCE–c. 348 BCE)**, who theorized in his dialogue, the *Timaeus*, that the classical elements were made of them. Of all the Platonic solids, the Greeks believed that the dodecahedron represented the entire universe.

A polyhedron is a **regular polyhedron** if all of its faces are regular congruent polygons and all of the edges are congruent. There are exactly five types of regular polyhedra, called **Platonic solids** because Plato used them extensively.

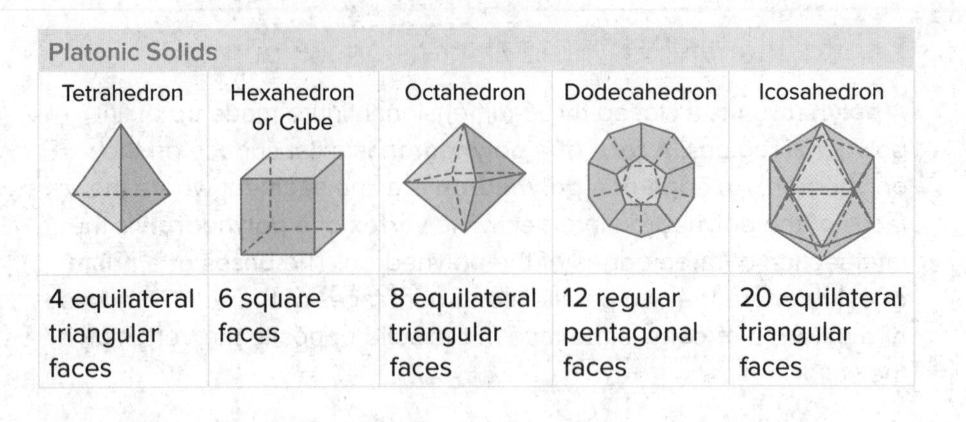

Platonic Solids

Tetrahedron	Hexahedron or Cube	Octahedron	Dodecahedron	Icosahedron
4 equilateral triangular faces	6 square faces	8 equilateral triangular faces	12 regular pentagonal faces	20 equilateral triangular faces

Example 1 Identify Properties of Three-Dimensional Figures

Determine whether each solid is a polyhedron. Then identify the solid. If it is a polyhedron, name the bases, faces, edges, and vertices.

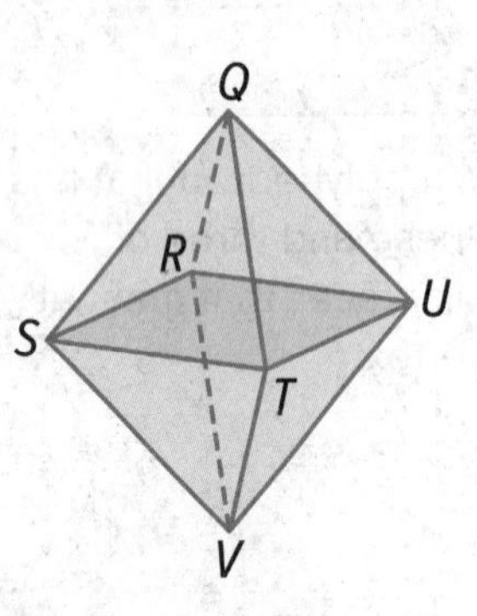

The solid _____ formed by polygonal faces, so it is a ___________. There is no base, but the solid has _____ equilateral triangular faces, so it is an ___________.

Bases: none

Faces: ___________

Edges: ___________

Vertices: ___________

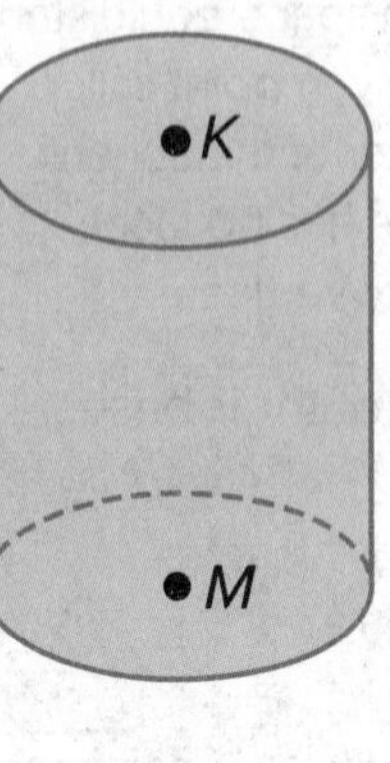

The solid has a curved surface, so it _______ a polyhedron. It has two congruent and parallel circular bases, so it is a _________.

Go Online You can complete an Extra Example online.

Example 2 Model Three-Dimensional Figures

Identify the three-dimensional figure that can model the beverage container. State whether the model is a polyhedron.

The beverage container can be modeled by a __________.
Because the model has a curved surface it ________ a polyhedron.

Check

Identify the three-dimensional figure that can model the top of the camping lodge. State whether the model is a polyhedron.

The top of the camping lodge can be modeled by a

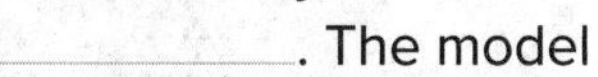

______________. The model ______ a polyhedron.

> **Study Tip**
>
> **Approximations** When modeling a real-world object, often the object cannot be perfectly modeled by a three-dimensional figure. Thus, three-dimensional figures provide only approximate measures for an object.

Explore Measuring Real-World Objects

Online Activity Use dynamic geometry software to complete the Explore.

> **INQUIRY** How can you apply the properties of three-dimensional figures to solve real-world problems?

> **Talk About It!**
>
> What is the relationship between the volume of a prism and the volume of a regular pyramid that have the same base and height? How does this compare to the relationship between the volume of a cylinder and the volume of a cone that have the same height and congruent bases?

Learn Measuring Three-Dimensional Figures

Often a geometric figure is used to model a real-world object to estimate a measurement. **Surface area** is the sum of the areas of all faces and side surfaces of a three-dimensional figure. **Volume** is the measure of the amount of space enclosed by a three-dimensional figure.

Prism	Right Pyramid	Cylinder	Cone	Sphere
P, h	ℓ	r, h	r, h, ℓ	r
$S = Ph + 2B$	$S = \frac{1}{2}P\ell + B$	$S = 2\pi rh + 2\pi r^2$	$S = \pi r\ell + \pi r^2$	$S = 4\pi r^2$
$V = Bh$	$V = \frac{1}{3}Bh$	$V = \pi r^2 h$	$V = \frac{1}{3}\pi r^2 h$	$V = \frac{4}{3}\pi r^3$

S = total surface area V = volume h = height of a solid
P = perimeter of the base B = area of base ℓ = slant height, r = radius

Watch Out!

Height vs. Slant Height The *height* of a pyramid or cone is not the same as its *slant height*. The height of a pyramid or cone is the perpendicular distance from the center of the base to the vertex. The slant height is the height of the face or curved surface of a pyramid or cone, respectively, that joins the base to the vertex.

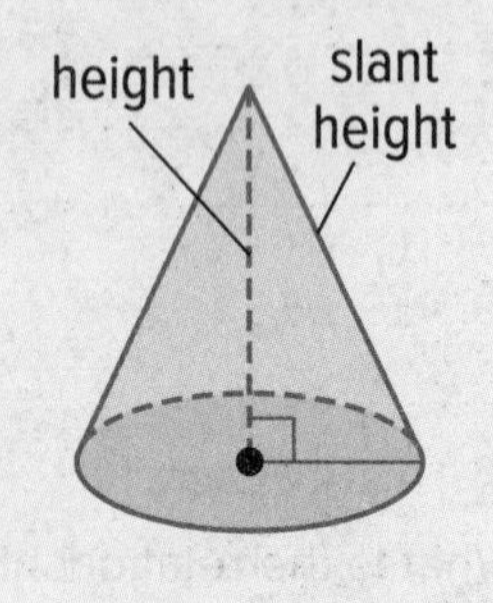

Watch Out!

Square and Cubic Units Remember to use square units when measuring surface area and cubic units when measuring volume.

Think About It!

How could you write the surface area of the cone in terms of pi?

Example 3 Find Measurements of Three-Dimensional Figures

Find the surface area and volume of the cone. Round each measure to the nearest tenth, if necessary.

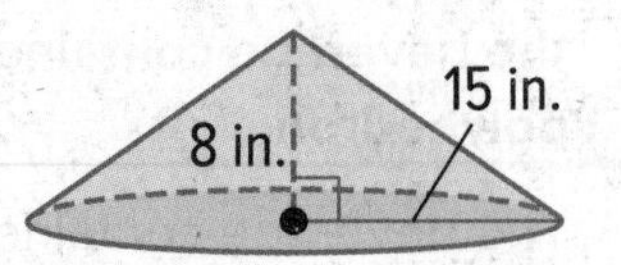

Surface Area

Because the radius of the base is 15 inches and the height of the cone is 8 inches, you can use the Pythagorean Theorem to find the slant height.

$c^2 = a^2 + b^2$ Pythagorean Theorem

$c^2 = 8^2 + 15^2$ $a = 8$ and $b = 15$

$c^2 = 289$ Simplify.

$c = \sqrt{289}$ or 17 Simplify.

The slant height is 17 inches, and the radius is 15 inches. Use the formula for the surface area of a cone.

$S = \pi r \ell + \pi r^2$ Surface area of cone

$= \pi$ ______ $(17) + \pi$ ______ ^{2} $r = 15$ in. and $\ell = 17$ in.

≈ 1508.0 Use a calculator.

The surface area of the cone is about 1508.0 square inches.

Volume

Use the formula for the volume of a cone.

$V = \frac{1}{3}\pi r^2 h$ Volume of cone

$= \frac{1}{3}\pi$ ______ (8) or about 1885.0 $r = 15$ in. and $h = 8$ in.

The volume of the cone is about 1885.0 cubic inches.

Check

Find the surface area and volume of the rectangular prism. Round each measure to the nearest tenth, if necessary.

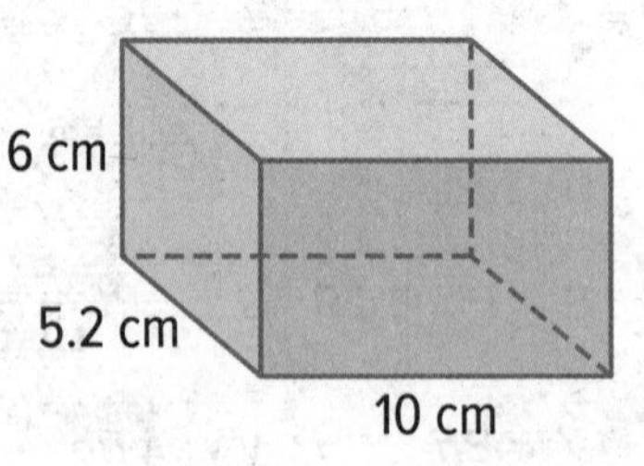

$S =$ ______ cm^2

$V =$ ______ cm^3

Example 4 Calculate Measurements by Using Three-Dimensional Models

NEW YEAR'S EVE **The New Year's Eve ball is a geodesic sphere that is 12 feet in diameter. It weighs 11,875 pounds, is lit by 32,256 LED lights, and is covered with 2688 crystal triangles.**

Part A How many lights are contained on the ball's surface within an area of 4 square feet?

Step 1 Find the surface area of the ball.

Because the diameter is 12 feet, the radius of the sphere is 6 feet.

$S = 4\pi\_\_\_\_^2$	Surface area of a sphere
$= 4\pi\_\_\_\_^2$	$r = 6$
$= 144\pi$ or about 452.4	Use a calculator.

The surface area of the ball is about _____ square feet.

Step 2 Determine the number of lights within an area of 4 square feet.

$$4 \text{ ft}^2 \times \frac{32{,}256 \text{ lights}}{452.4 \text{ ft}^2} = 285.2 \text{ or } \_\_\_\_ \text{ lights}$$

There are _____ lights within an area of 4 square feet.

Part B Tony is repairing a section of the ball that has a volume of 8 cubic feet. How much does the section weigh?

Step 1 Find the volume of the ball.

$V = \frac{4}{3}\pi\_\_\_\_^3$	Volume of a sphere
$= \frac{4}{3}\pi\_\_\_\_^3$	$r = 6$
$= \_\_\_\_\pi$ or about 904.8	Use a calculator.

Step 2 Determine the weight of the section.

$$8 \text{ ft}^3 \times \frac{11{,}875 \text{ lb}}{904.8 \text{ ft}^3} = 105.0$$

The section of the ball weighs about _____ pounds.

Check

POOLS Mateo is building a new pool. A cross section of the pool is shown.

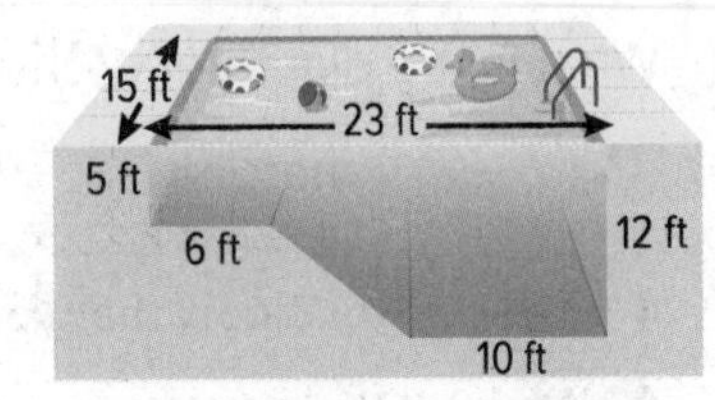

Part A What is the volume of the pool to the nearest tenth?

$V =$ _____ ft^3

Part B Mateo needs to install a protective liner to cover the walls and flat base of the deep end of the pool. How much liner is required to cover the deep end of the pool in square feet?

A. 570 ft^2 B. 750 ft^2 C. 900 ft^2 D. 1800 ft^2

Think About It!

What assumption did you make about the New Year's Eve ball to solve the problem?

Apply Example 5 Solve for Unknown Values

WATER PARK Destiny visits a water park where a new cyclone water ride has opened. On the new ride, a tunnel flows into a large funnel. The slant height of the funnel is 29 meters, and the surface area of the funnel is 910 square meters. When the funnel is at its widest, what is the diameter? Round your answer to the nearest tenth.

Study Tip

Assumptions When modeling, you assume that a three-dimensional object is perfectly modeled by a geometric solid. Because this is not always true, you should assume that your calculations for surface area and volume will be approximations.

1 What is the task?

Describe the task in your own words. Then list any questions that you may have. How can you find answers to your questions?

2 How will you approach the task? What have you learned that you can use to help you complete the task?

3 What is your solution?

Use your strategy to solve the problem.

What three-dimensional solid can you use to model the funnel?

What equation will you use?

What is the diameter of the funnel?

_______ meters

4 How can you know that your solution is reasonable?

Write About It! Write an argument that can be used to defend your solution.

Go Online You can complete an Extra Example online.

Name ______________________ Period __________ Date __________

Practice

Go Online You can complete your homework online.

Example 1

Determine whether each solid is a polyhedron. Then identify the solid. If it is a polyhedron, name the bases, faces, edges, and vertices.

1. **2.** **3.**

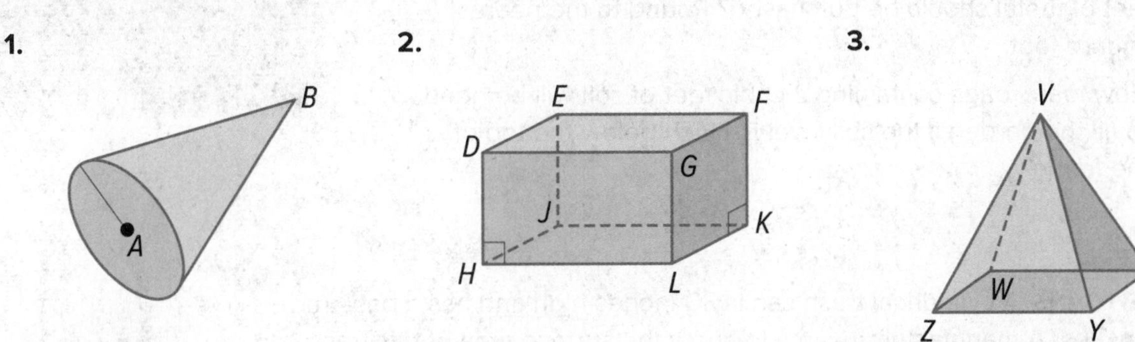

Example 2

Identify the three-dimensional figure that can model each object. State whether the model is or is not a polyhedron.

4. **5.** **6.**

Example 3

Find the surface area and volume of each solid. Round each measure to the nearest tenth, if necessary.

7. **8.** **9.**

10. **11.** **12.**

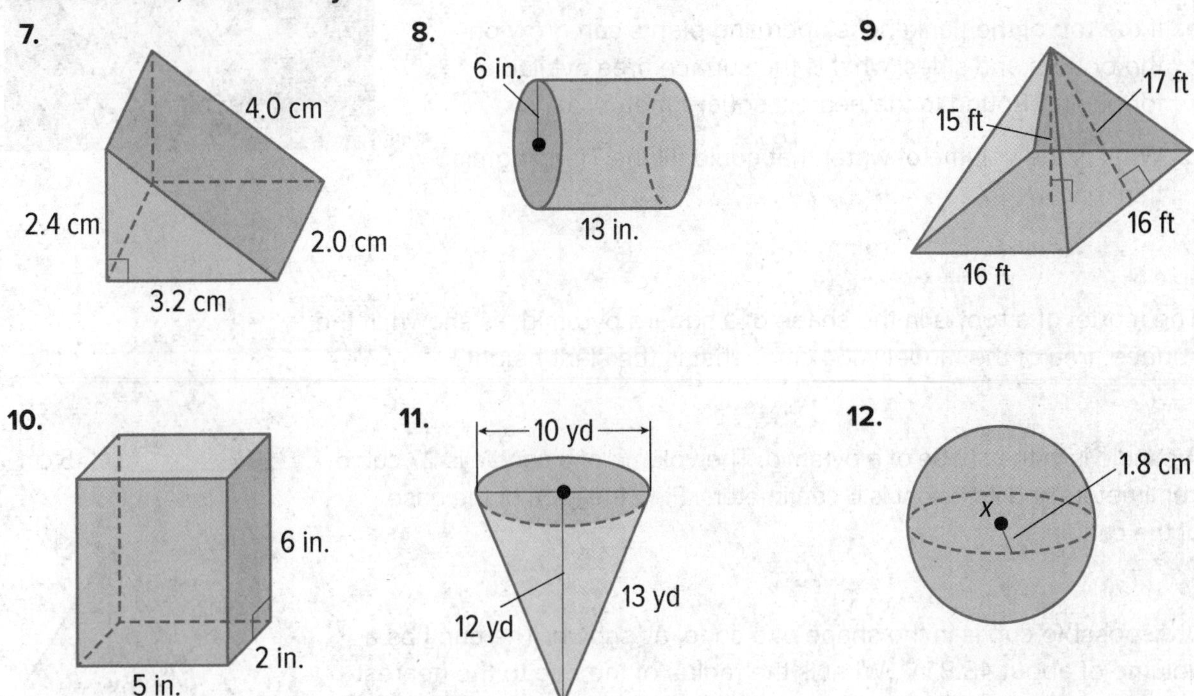

Example 4

13. GARDENING The plans for constructing a raised vegetable garden use corrugated metal in a wooden frame. The finished garden is 4 feet long, 30 inches wide, and 32 inches tall.

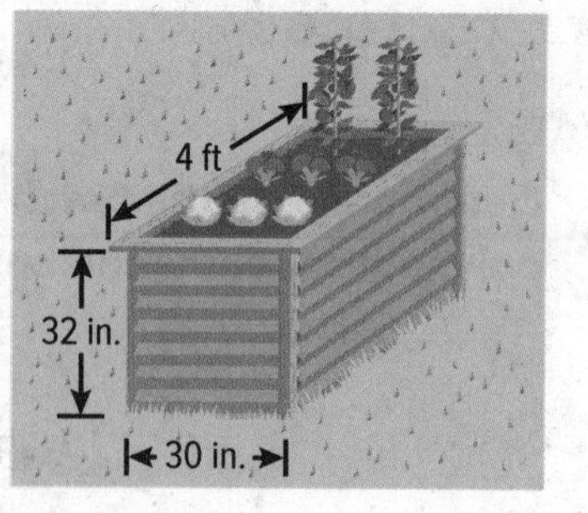

a. The metal is only used on the lateral faces, so how many square feet of metal should be purchased? Round to the nearest square foot.

b. How many bags containing 2 cubic feet of soil will be needed to fill the garden if the soil level is 1 inch below the top of the frame?

14. TRASH CANS A cylindrical trash can is 30 inches high and has a base radius of 7 inches. A manufacturer wants to know the surface area of this trash can, including the top of the lid. What is the surface area? Round to the nearest square inch.

30 in.

7 in.

15. ALGAE A scientist has a fish tank in the shape of a rectangular prism. The tank is 18 inches high, 14 inches wide, and 30 inches long. After one month, the scientist found that the sides and bottom of his fish tank were covered with algae. The scientist wants to run tests on the algae to help determine why it started to grow. How much algae is there for the scientist to test?

16. GEOLOGY A *tiankeng* is a sinkhole with nearly vertical walls. The Tianpingmiao tiankeng is approximately cylindrical with a diameter of 180 meters and a depth of 420 meters.

a. If the top of the tiankeng is open and plants can grow on the bottom and sides, what is the surface area available for plants? Round to the nearest square meter.

b. What is the volume of water that could fill the Tianpingmiao tiankeng?

Example 5

17. The model of a roof is in the shape of a square pyramid, as shown. If the surface area of the model is 64 cm^2, what is the slant height?

? cm

5 cm

18. A candle is in the shape of a pyramid. The volume of a candle is 27 cubic centimeters and its height is 6 centimeters. Find the area of the base of the candle.

19. A disposable cup is in the shape of a cone, as shown. The cup has a volume of about 48.8 in^3. What is the radius of the cup to the nearest inch?

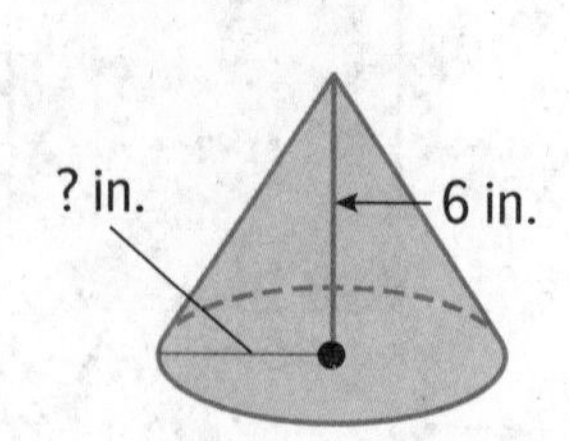

Name ______________________ Period ____________ Date ______________

Mixed Exercises

20. PLANETS For a time, Johannes Kepler thought that the Platonic solids were related to the orbits of the planets. He made models of each of the Platonic solids. He made a frame of each of the Platonic solids by fashioning together wooden edges. How many edges did Kepler have to make for the cube?

21. SILO A silo used for storing grain is shaped like a cylinder with a cone on top. The radius of the base of the cylinder and cone is 8 feet. The height of the cylindrical part is 25 feet, and the height of the cone is 6 feet.

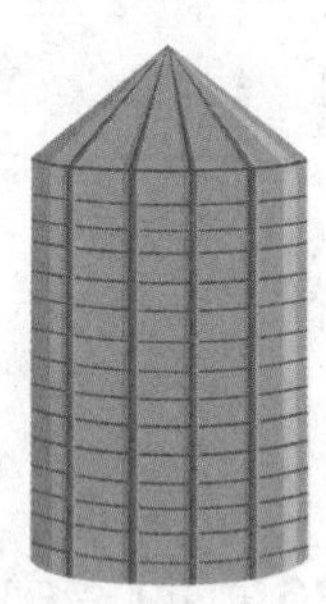

a. What is the volume of the cylindrical part of the silo? Round to the nearest cubic foot.

b. What is the volume of the conical part of the silo? Round to the nearest cubic foot.

c. What is the volume of the entire silo? Round to the nearest cubic foot.

22. USE A SOURCE Find a real object that can be modeled with one or more three-dimensional figures. Identify the best three-dimensional model and calculate the surface area and volume of the object.

23. A garden shop sells pyramid-shaped lawn ornaments that each have a base area of 900 square centimeters and a height of 40 centimeters. The lawn ornaments are made of concrete, granite, or marble.

Material	Density (kg/m^3)
Concrete	2371
Granite	2691
Marble	2711

a. What is the volume of one lawn ornament in cubic meters? Explain.

b. Find the weights of three of these ornaments that are each made from a different material. Round to the nearest tenth of a kilogram.

c. What generalization can you make about the relationships among the volume of an ornament, the weight of the lawn ornament, and the density of the material used to make it?

24. REASONING The volume of a new extra large toy tennis ball for pets is about 221 cubic centimeters. If 3 extra large toy tennis balls are packaged and sold in cylindrical package as shown, what is the approximate volume of the cylindrical package? Explain.

25. **FIND THE ERROR** Alex and Sia are calculating the surface area of the rectangular prism shown. Is either of them correct? Explain your reasoning.

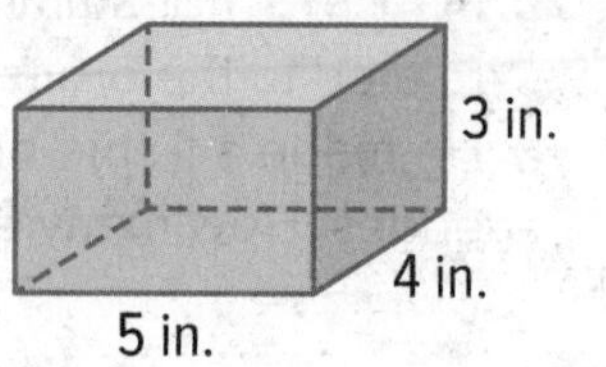

Alex
$(5 \cdot 3) \cdot 6$ faces
$= 90$ in^2

Sia
$2(5 \cdot 4 \cdot 3)$
$= 120$ in^2

26. **ANALYZE** When a polygon is *inscribed* in a circle, all of the vertices of the polygon lie on the circle. Consider a pyramid and a prism that have bases that are regular polygons inscribed in a circle. What solid results if the number of sides of the bases is increased infinitely?

27. **WRITE** Which solid has a greater volume: cone with a base radius of 7 centimeters and a height of 28 centimeters or a pyramid with base area of 154 square centimeters and height of 28 centimeters? Explain your reasoning.

28. **CREATE** Draw an irregular 14-sided polyhedron that has two congruent bases.

29. **PERSEVERE** Find the volume of a cube that has a total surface area of 54 square millimeters.

30. **ANALYZE** Is a cube a regular polyhedron? Justify your argument.

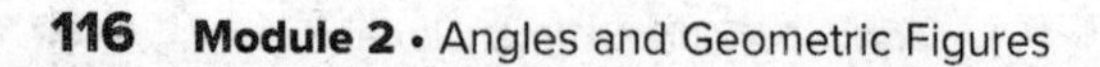

Lesson 2-6

Two-Dimensional Representations of Three-Dimensional Figures

Explore Representing Three-Dimensional Figures

Online Activity Use two-dimensional drawings to complete the Explore.

INQUIRY How can you accurately represent a three-dimensional figure with two-dimensional drawings?

Today's Goals

- Identify the orthographic drawings that best model selected three-dimensional figures.
- Calculate surface areas of three-dimensional figures represented by nets, and determine the correct nets for three-dimensional geometric figures.

Today's Vocabulary

orthographic drawing
net

Learn Representing Three-Dimensional Figures with Orthographic Drawings

The two-dimensional views of the top, left, front, and right sides of an object are called an **orthographic drawing**.

Example 1 Make a Model from an Orthographic Drawing

Make a model of a figure from the orthographic drawing shown.

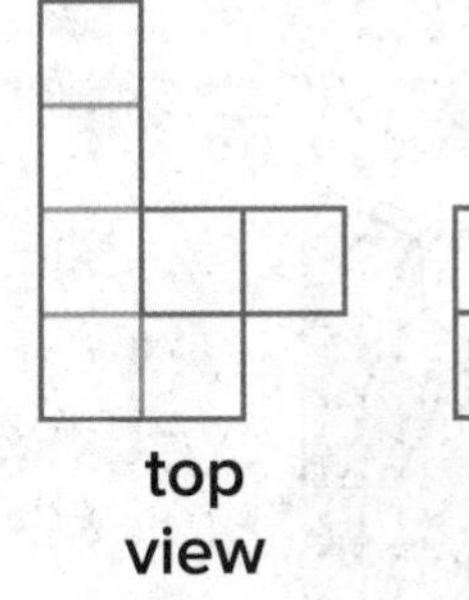
top view

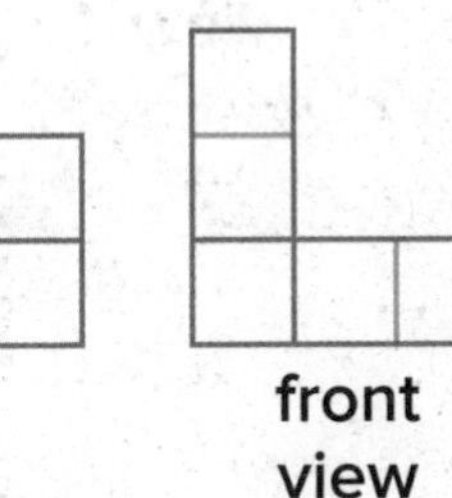
left view

front view

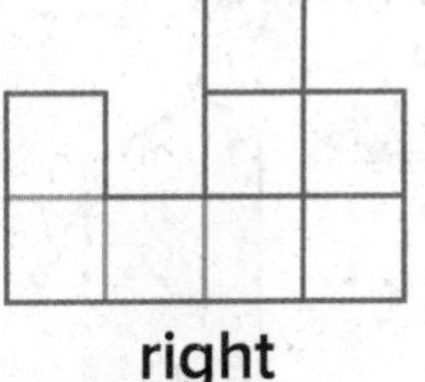
right view

Step 1 **Create the base of the model.** Start with a base that matches the top view.

front right

Step 2 **Use the front view.**

- The front left side is 3 blocks high.
- The front middle and right sides are 1 block high.
- Highlighted segments indicate breaks where columns or rows of blocks appear at different depths.
- The highest block in the front left column is farther back than the 2 blocks below it.
- The third block on the bottom row is farther back than the first 2 blocks in the row.

front right

(continued on the next page)

Your Notes

Step 3 Use the left view. Use the left view to find where the breaks in the front view occur.

- The first column is 2 blocks high.
- The second column is 1 block high.
- The third column is 3 blocks high.
- The fourth column is 2 blocks high.
- Remove any unnecessary blocks.

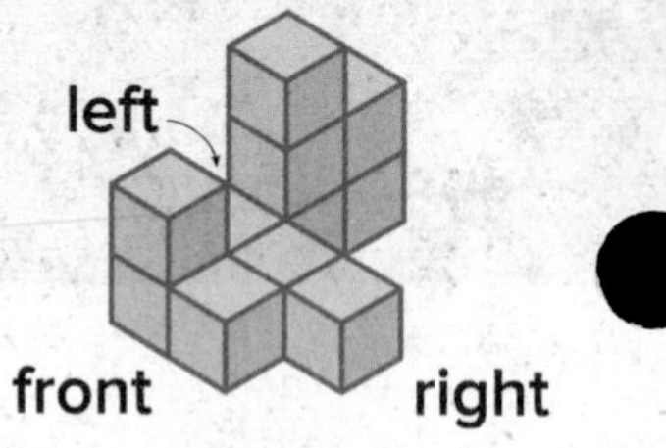

Step 4 Check your model. Use the right view to confirm that you have made the correct model.

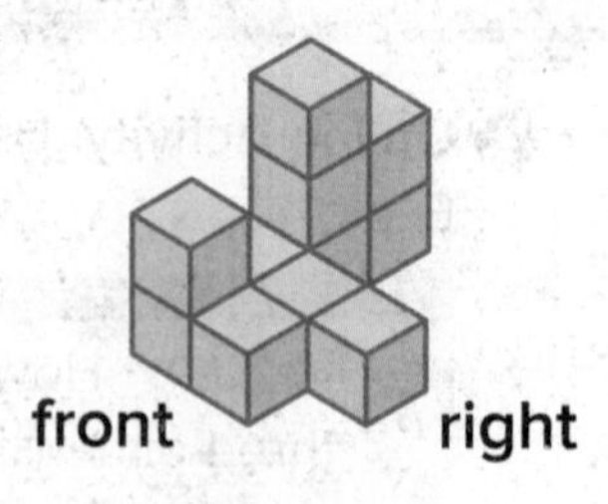

Study Tip

Left View It is often difficult to accurately interpret the left view of an orthographic drawing. Imagine holding a model and rotating it counterclockwise to see the left side. The outline after this rotation will be the left view of the model.

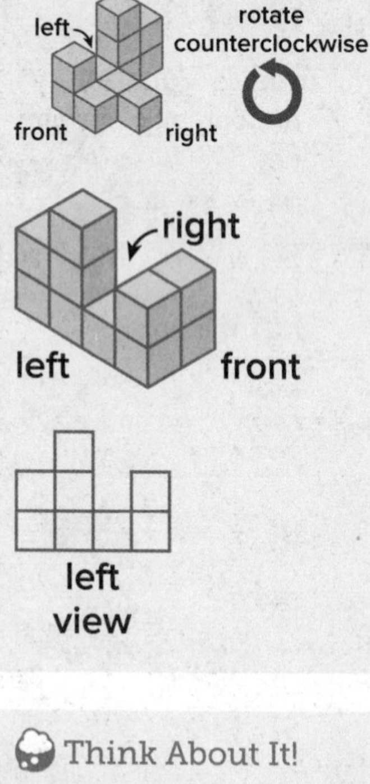

Think About It!

Why is a back view not used in an orthographic drawing?

Check

Which model corresponds to the orthographic drawing?

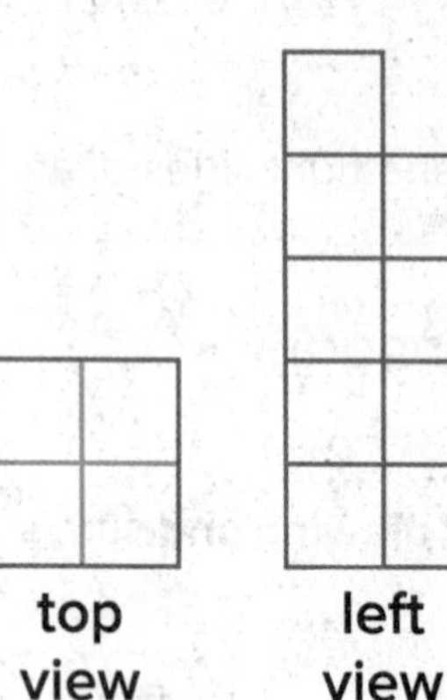

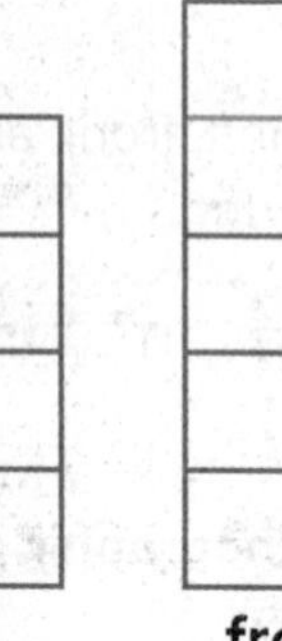

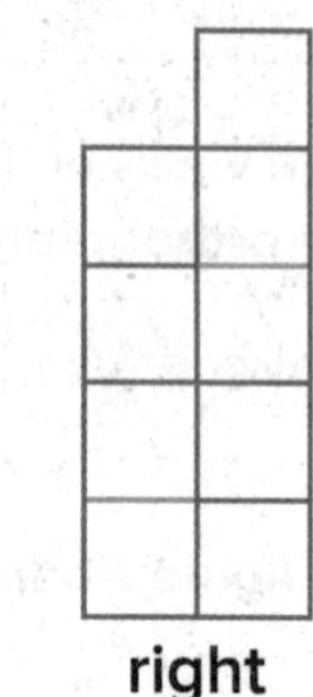

A. front right

B.

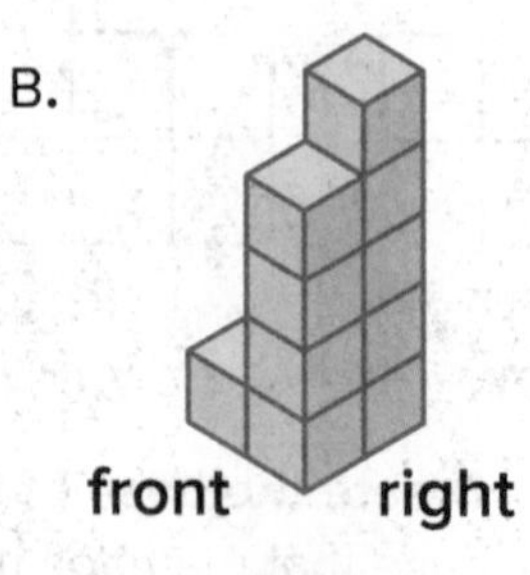

C. front right

D.

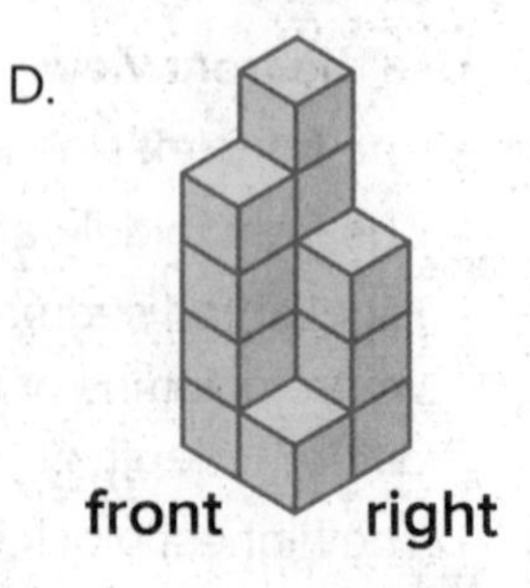

Go Online You can complete an Extra Example online.

Example 2 Make an Orthographic Drawing

Make an orthographic drawing of the figure shown.

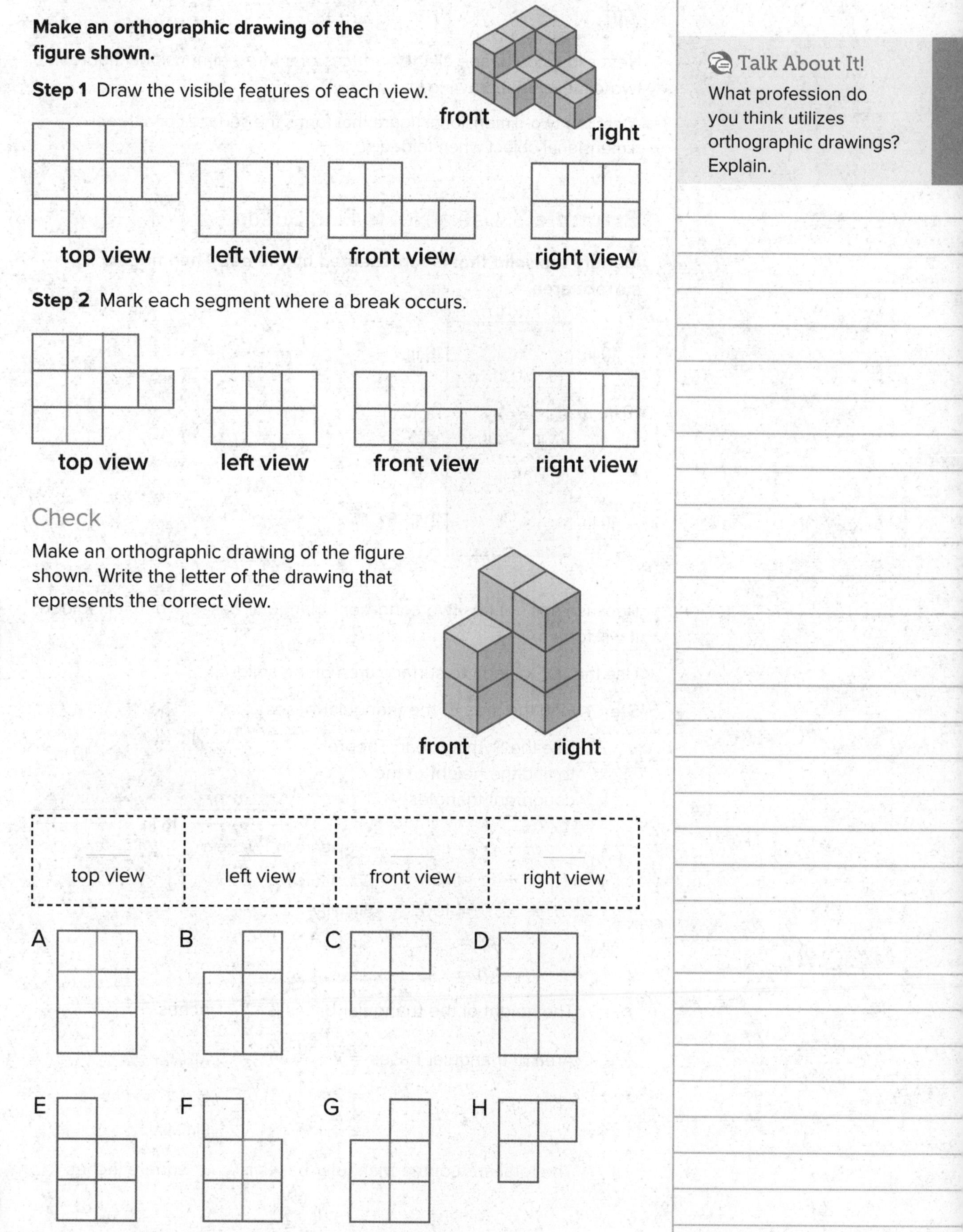

Step 1 Draw the visible features of each view.

top view | left view | front view | right view

Step 2 Mark each segment where a break occurs.

top view | left view | front view | right view

Check

Make an orthographic drawing of the figure shown. Write the letter of the drawing that represents the correct view.

front
right

____ top view | ____ left view | ____ front view | ____ right view

A B C D

E F G H

Go Online You can complete an Extra Example online.

Talk About It!

What profession do you think utilizes orthographic drawings? Explain.

Learn Representing Three-Dimensional Figures with Nets

Nets allow you to see all the surfaces of a three-dimensional figure in a two-dimensional drawing.

A **net** is a two-dimensional figure that forms the surfaces of a three-dimensional object when folded.

Example 3 Use a Net to Find Surface Area

Identify the solid that is represented by the net. Then find its surface area.

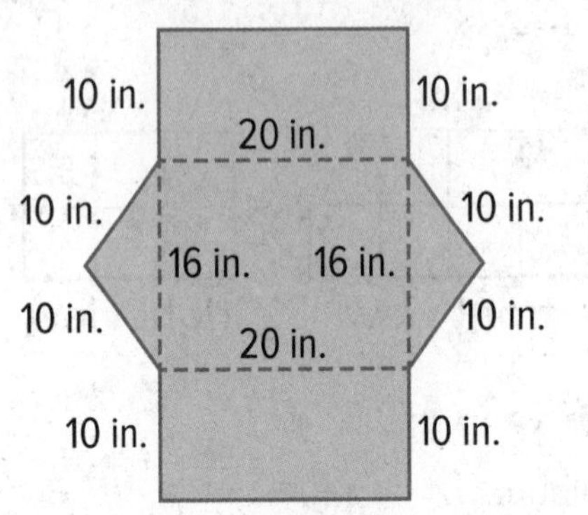

Because this net has two congruent triangular bases, when it is folded, it will form a ______.

Use the net to find the surface area of the solid.

Step 1 Find the area of the triangular bases.

Use the Pythagorean Theorem to find the height of the congruent triangles.

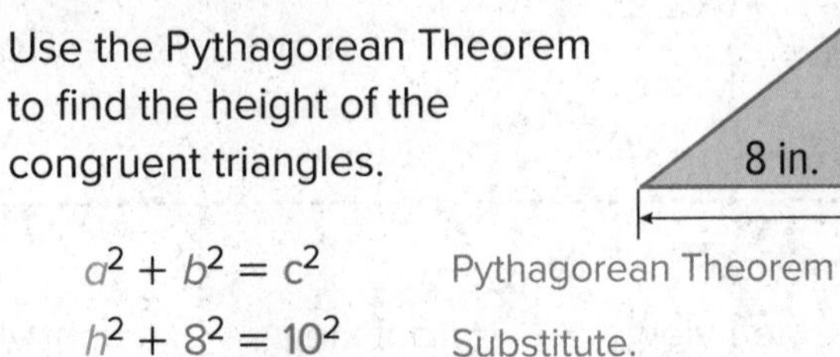

$a^2 + b^2 = c^2$	Pythagorean Theorem
$h^2 + 8^2 = 10^2$	Substitute.
$h^2 +$ ____ $= 100$	Simplify.
$h^2 =$ ____	Subtract.
$h =$ ____	Solve.

The height of the triangular bases is ____ inches.

Area of triangular bases $= 2 \times \frac{1}{2}bh$	Area of 2 $\cong$ $\triangle$s
$= 2 \times \frac{1}{2}(16)(6)$	$b = 16$ and $h = 6$
$=$ ____	Simplify.

The total area of the triangular bases is ____ square inches.

Step 2 Find the total surface area of the triangular prism.

$S = 96 + 2(\_\_\_\_)(20) + 16(\_\_\_\_)$ Area of triangular bases plus area of three rectangles

$= 96 + 400 + 320$ or 816 in^2 Simplify.

The surface area of the triangular prism is ______ square inches.

Check

Identify the solid that is represented by the net. Then find its surface area.

A. square pyramid; 104 cm^2

B. tetrahedron; $64 + 64\sqrt{21} \text{ cm}^2$

C. tetrahedron; 88 cm^2

D. square pyramid; $64 + 32\sqrt{21} \text{ cm}^2$

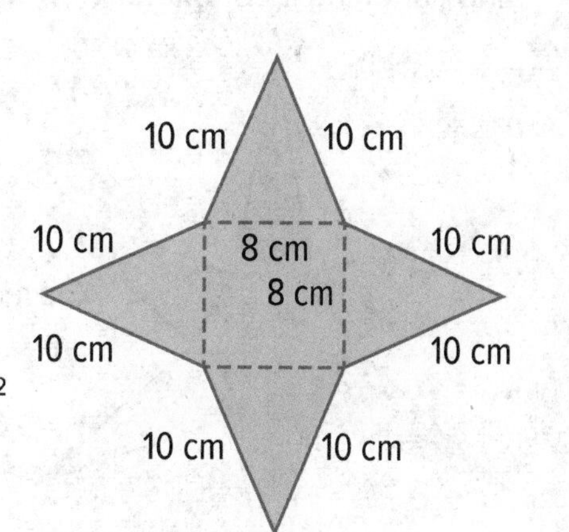

Think About It!

If a solid can be represented by more than one net, will the surface area of the solid change? Explain.

Example 4 Identify Platonic Solids

Identify the Platonic solid that is represented by the net.

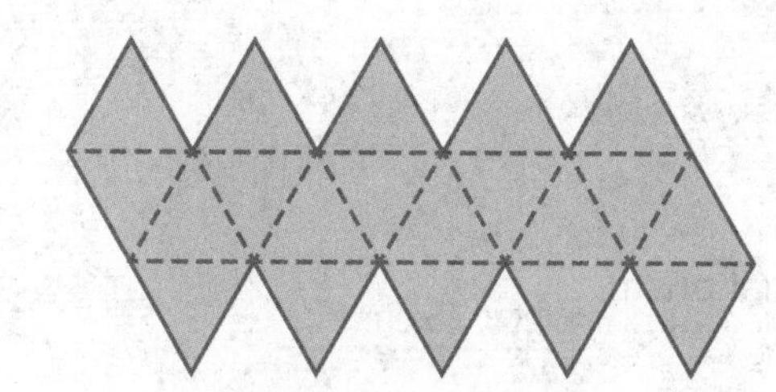

Because this net has 20 equilateral triangles, it represents a net of a(n) ______________.

Check

Identify the Platonic solid that is represented by the net.

A. decahedron

B. pentagonal prism

C. dodecahedron

D. icosahedron

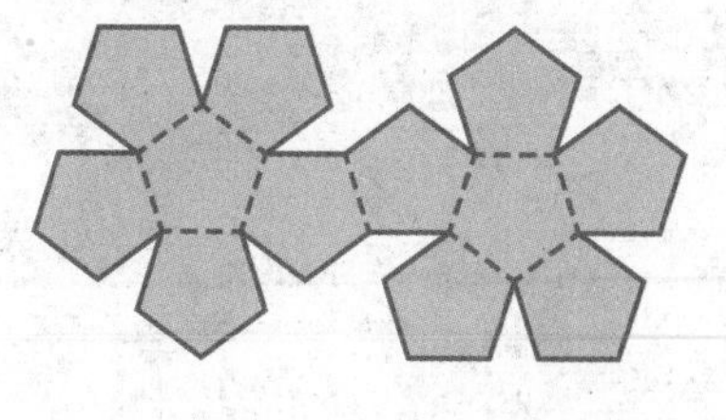

Go Online You can complete an Extra Example online.

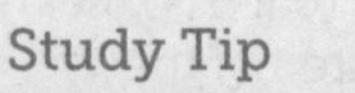

Study Tip

Approximations A net for a sphere can be created using several adjoining pointed ellipses or by creating a polyhedron with a large number of sides. However, because paper can only curve in one direction, it is impossible to make a perfect sphere. Thus, the spheres made from nets will be approximations.

net sphere

Example 5 Draw Nets for Three-Dimensional Figures

Draw a net for the hexagonal pyramid.

To draw the net of a three-dimensional solid, visualize cutting the solid along one or more of its edges, opening up the solid, and flattening it completely.

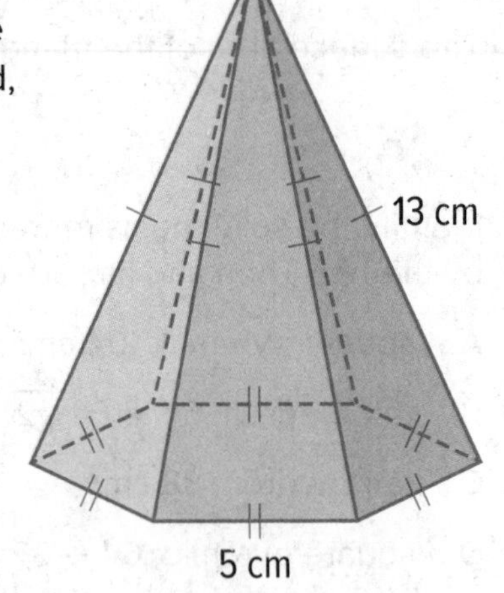

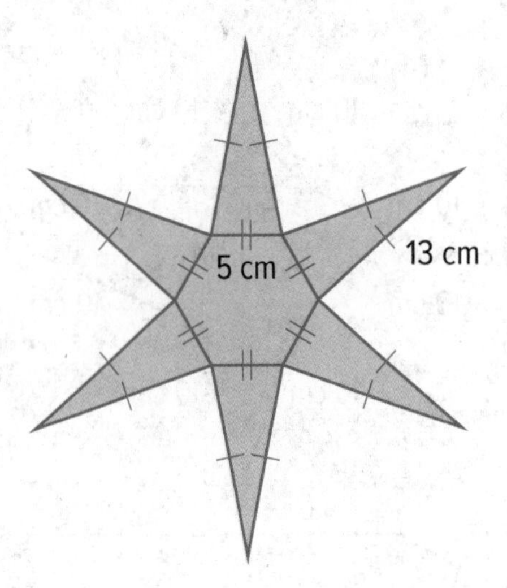

Check

Draw a net for the regular pentagonal prism.

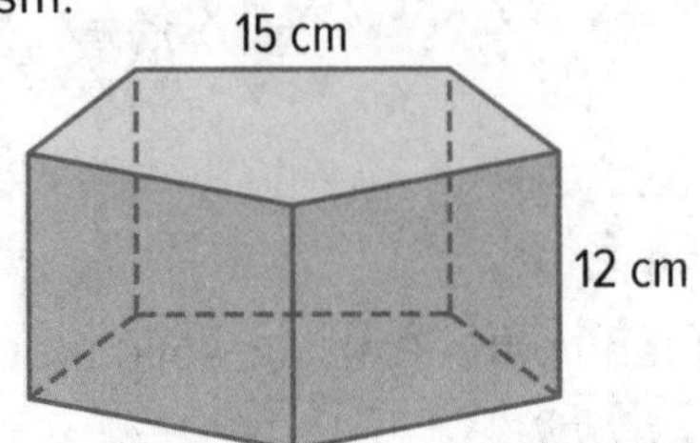

Go Online You can complete an Extra Example online.

Example 6 Represent a Real-World Object with a Net

TENTS **Draw a net to represent the three-dimensional figure that can be used to model the tent.**

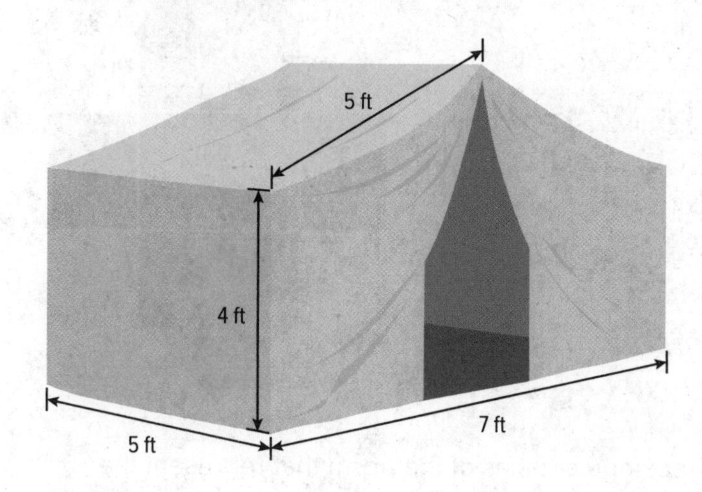

The tent can be modeled by a(n) ____________.

Step 1 Start by drawing the bottom of the tent.

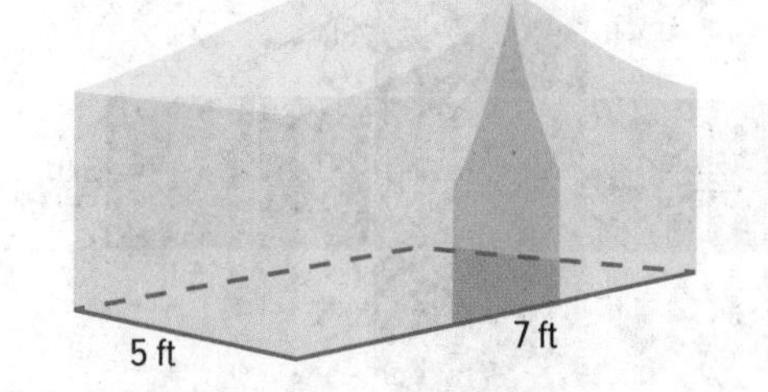

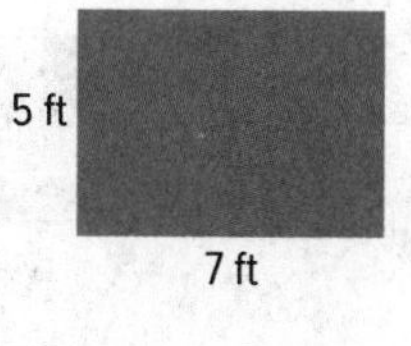

Step 2 Next, draw the pentagonal bases of the prism in the net. The pentagonal faces will attach to the rectangle at the 7-foot edges.

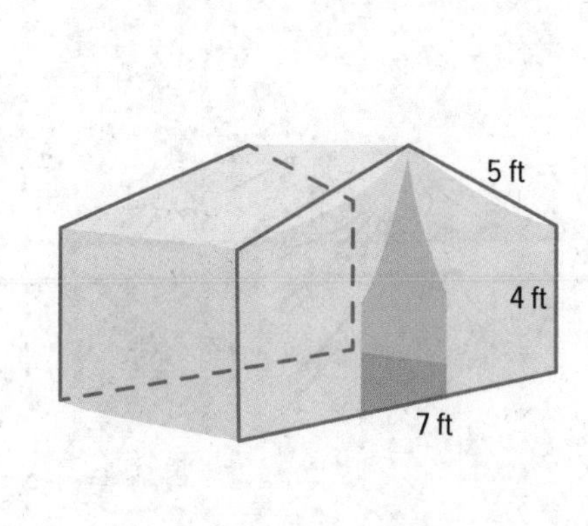

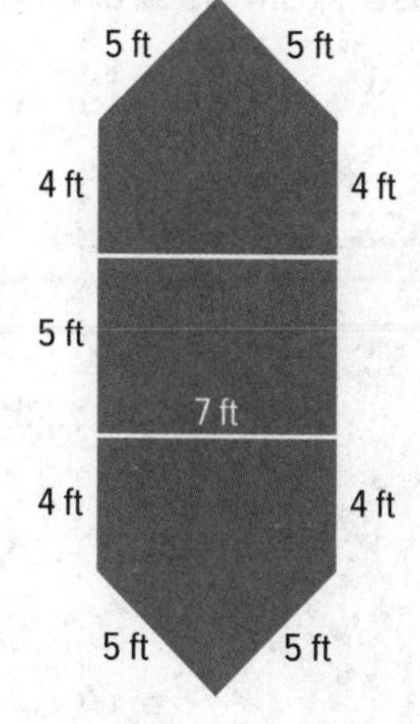

(*continued on the next page*)

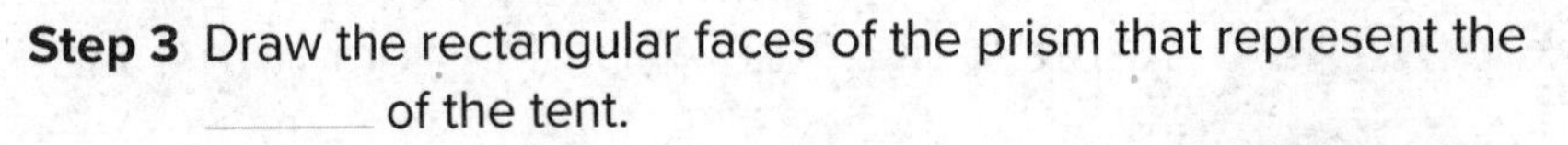

Step 3 Draw the rectangular faces of the prism that represent the _______ of the tent.

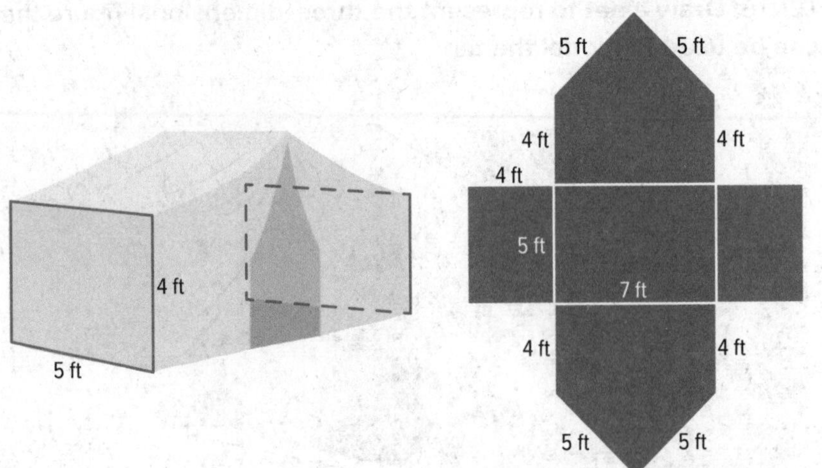

Think About It!

Is there more than one way to correctly draw a net? Explain using the net from Example 6.

Step 4 Draw the rectangular faces of the prism that represent the roof of the tent. Compare your net to the original figure to ensure that the dimensions are correct.

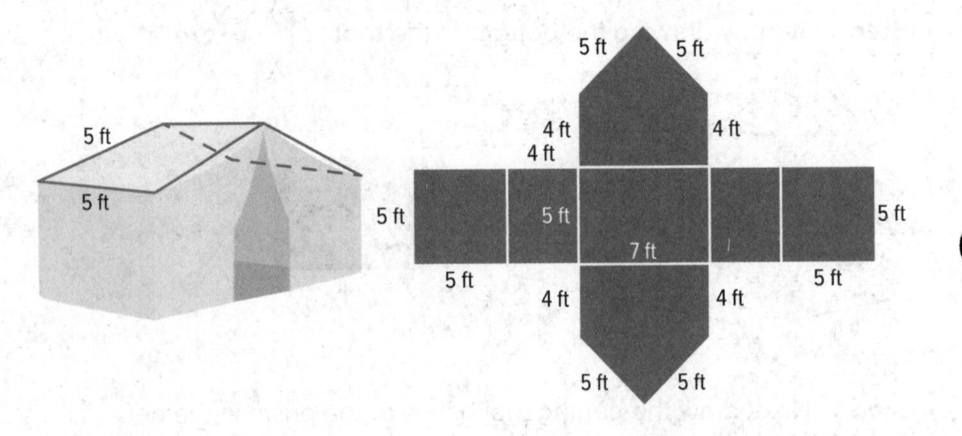

Check

GIFT WRAPPING Draw a net to represent the three-dimensional figure that can be used to model the gift box.

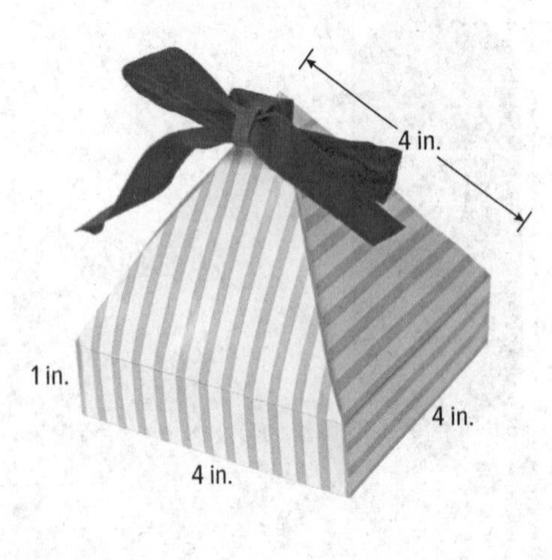

Go Online You can complete an Extra Example online.

Name ______________________ Period ____________ Date ____________

Practice

Go Online You can complete your homework online.

Example 1

Make a model of a figure for each orthographic drawing.

1.

top view left view front view right view

2.

top view left view front view right view

Example 2

Make an orthographic drawing of each figure.

3.

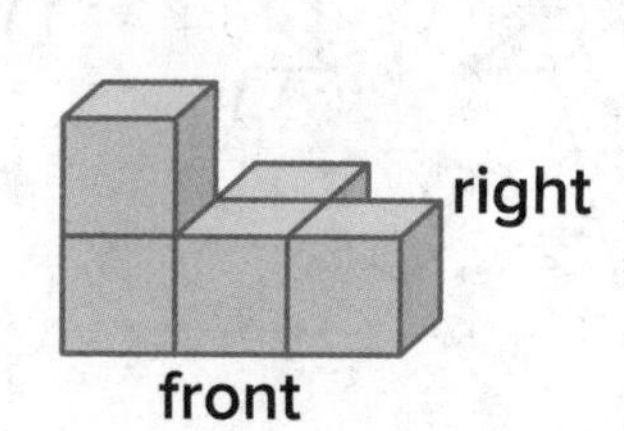

4.

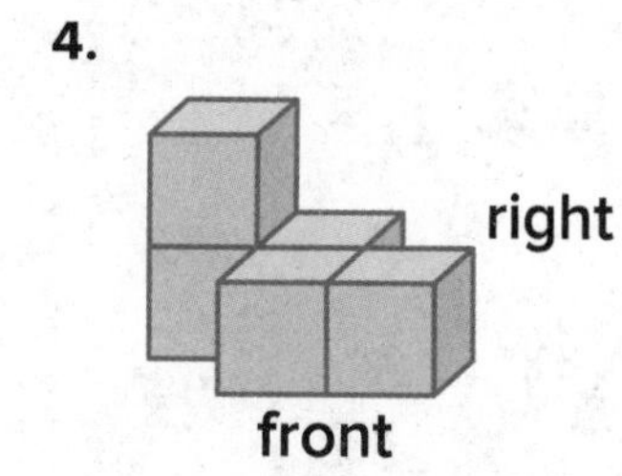

5.

right

front

6.

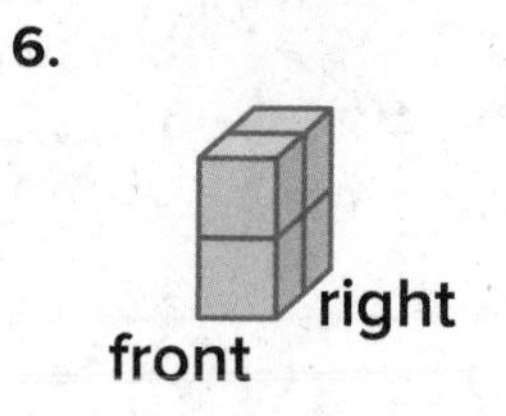

7.

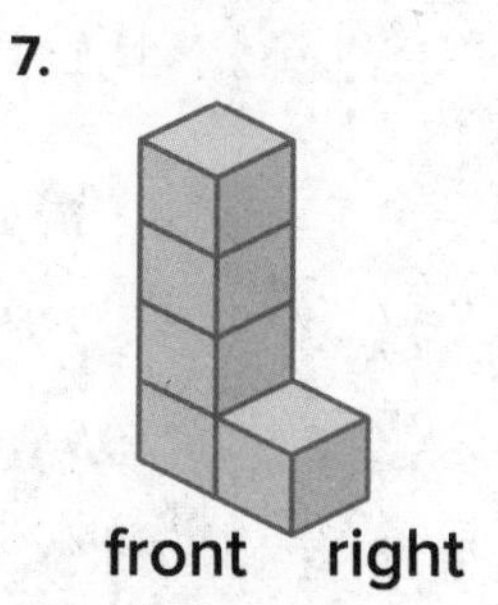

8.

right

front

Example 3

Make a model of the solid that is represented by each net. Then identify the solid and find its surface area.

9.

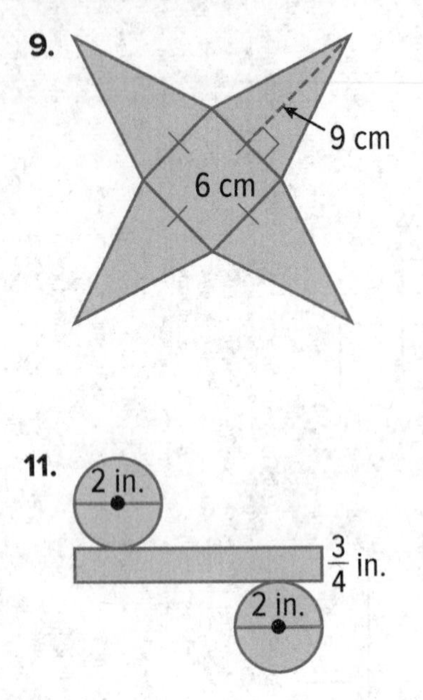

10.

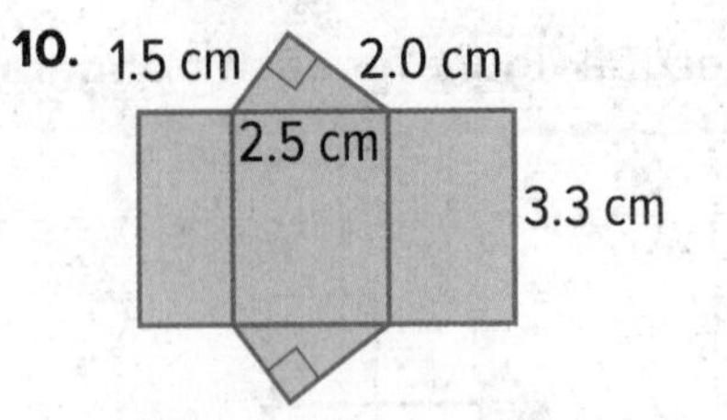

11.

12.

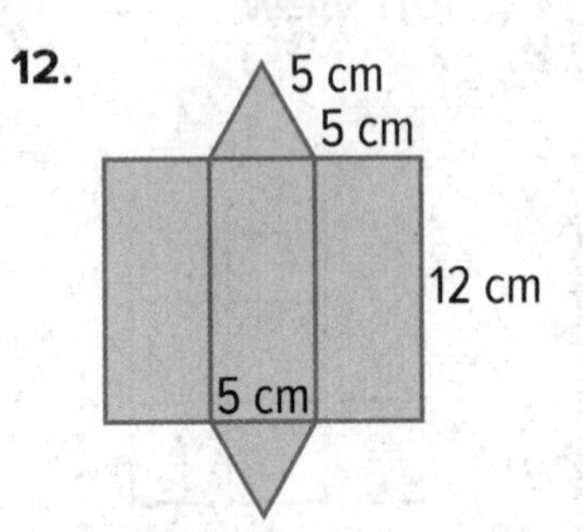

Example 4

Identify the Platonic solid that is represented by the net.

13.

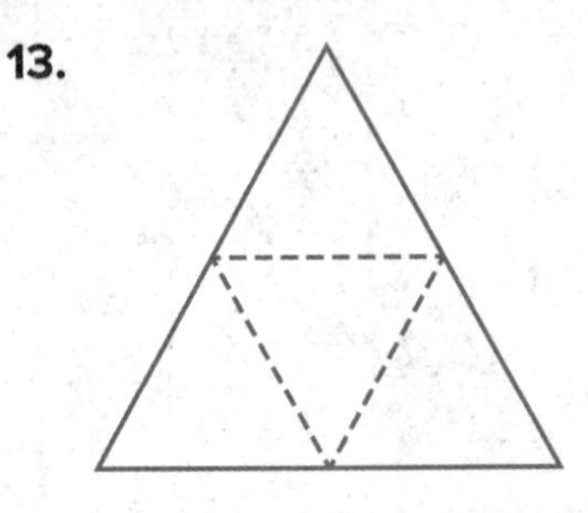

14.

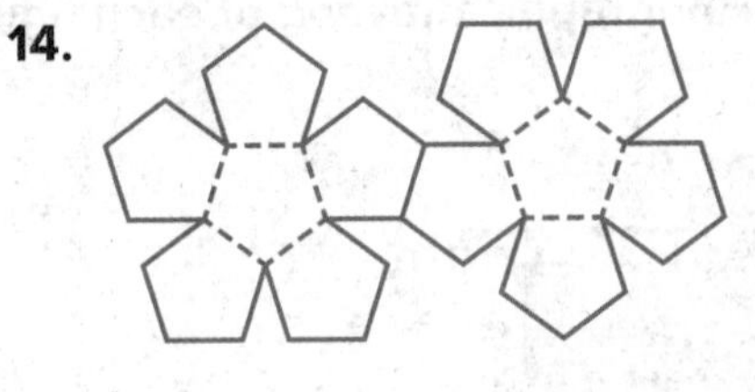

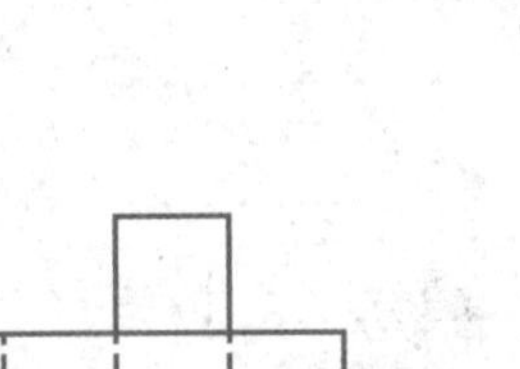

15.

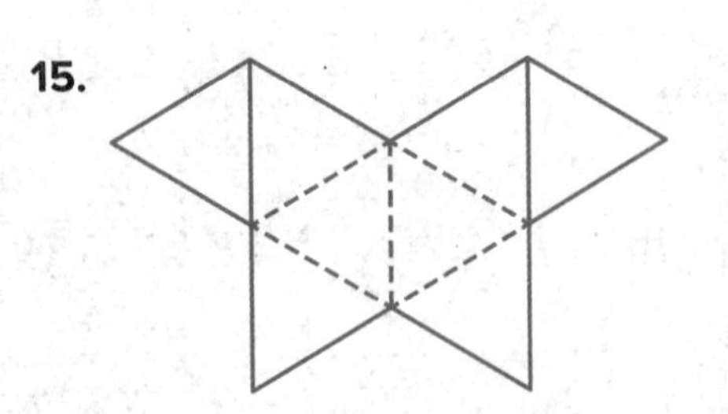

16.

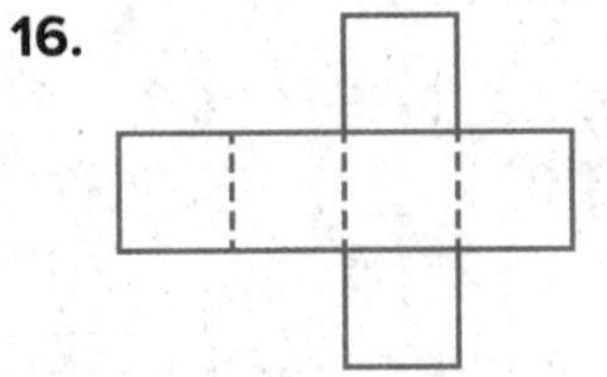

Examples 5 and 6

Draw a net for each solid or object.

17.

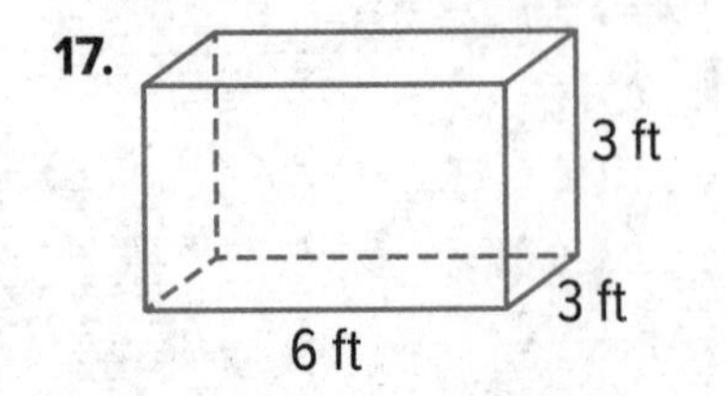

18.

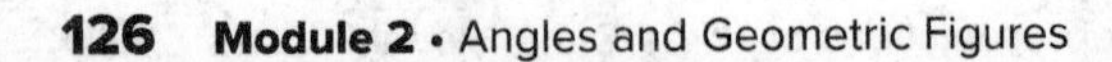

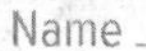

Name ______________________ Period ____________ Date ____________

Draw a net for each solid or object.

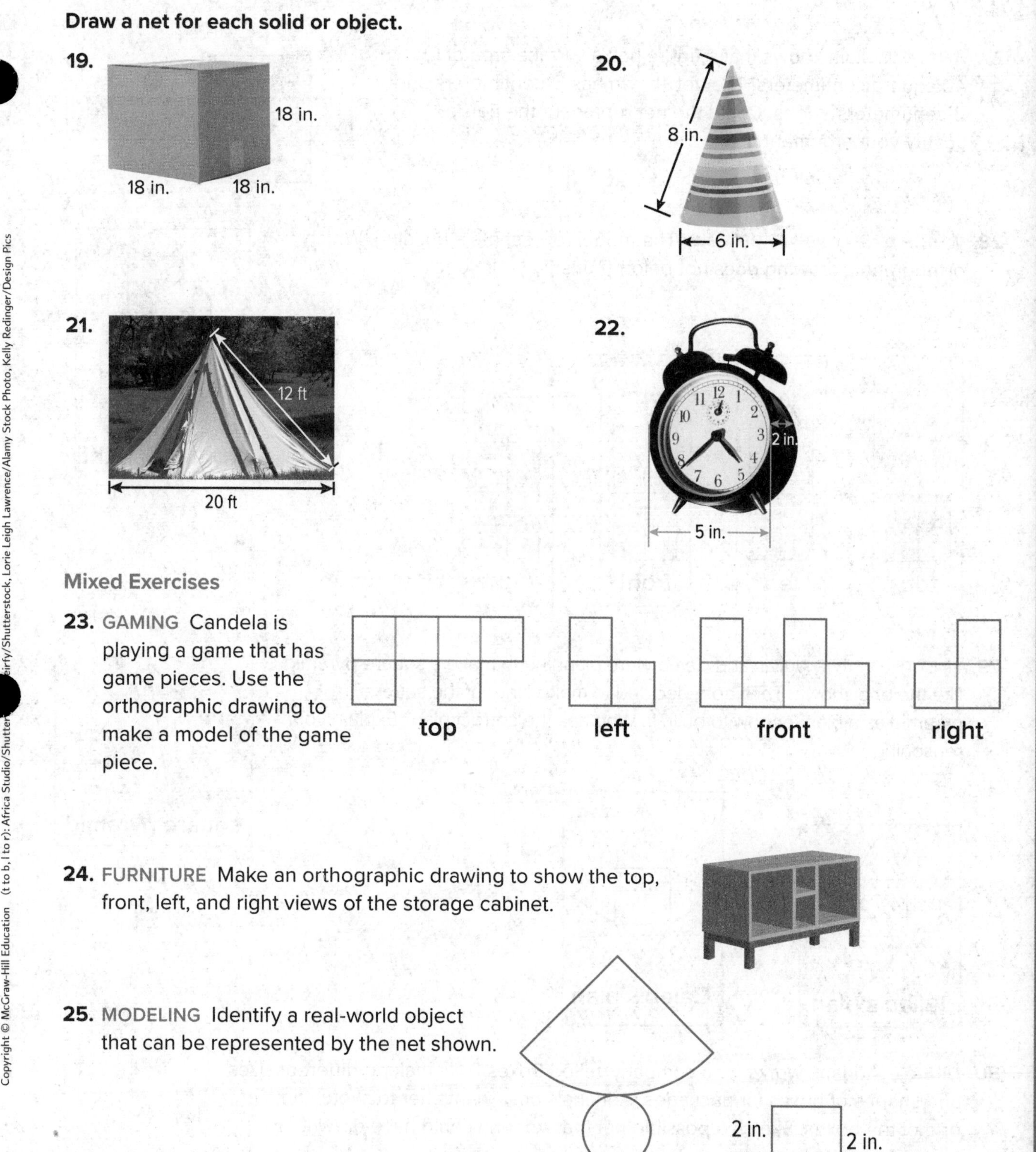

19.

20.

21.

22.

Mixed Exercises

23. GAMING Candela is playing a game that has game pieces. Use the orthographic drawing to make a model of the game piece.

24. FURNITURE Make an orthographic drawing to show the top, front, left, and right views of the storage cabinet.

25. MODELING Identify a real-world object that can be represented by the net shown.

26. GIFT WRAP Olive is wrapping a gift for her sister in a box that has the net shown here. How many square inches of wrapping paper will it take to cover the box?

2 in.
2 in.
6 in.
3 in.
3 in.

27. **ANALYZE** Julia knows that a figure has a surface area of 40 square centimeters. The net shown has 5-centimeters and 2-centimeters edges. Could the net represent the figure? Justify your argument.

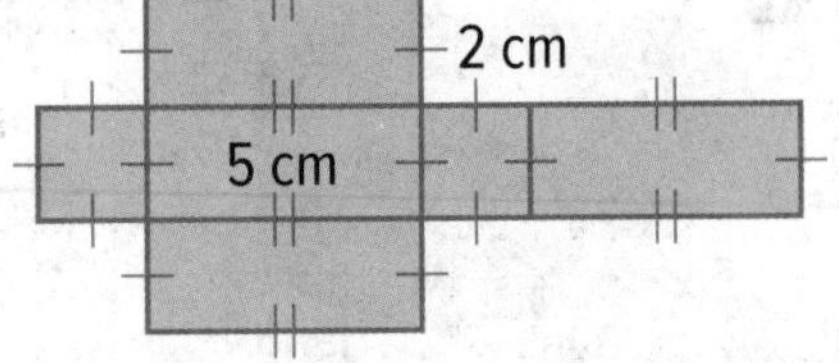

28. **WHICH ONE DOESN'T BELONG** The model represents a building. Which orthographic drawing does not belong? Justify your conclusion.

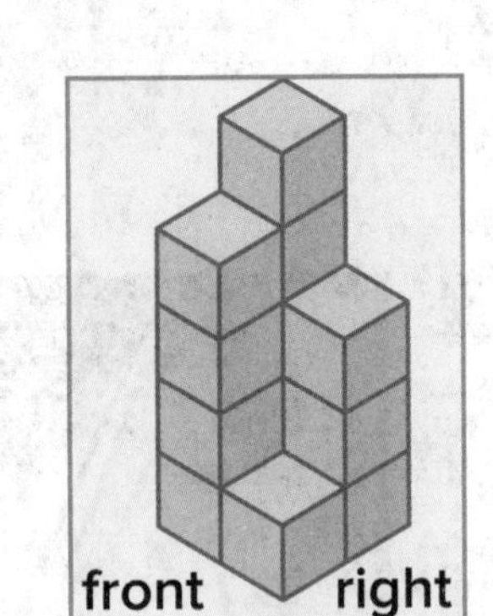

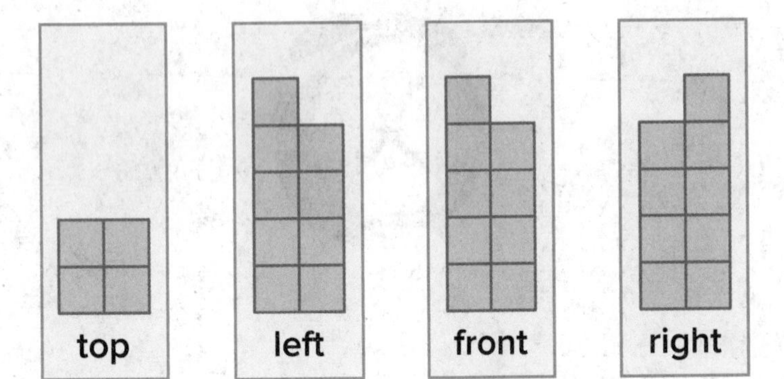

29. **FIND THE ERROR** Julian and Caleb were planning to make a square pyramid like the one shown. They both decided to make a net of the square pyramid as a plan for how to build it. Who has the correct plan? Explain your reasoning.

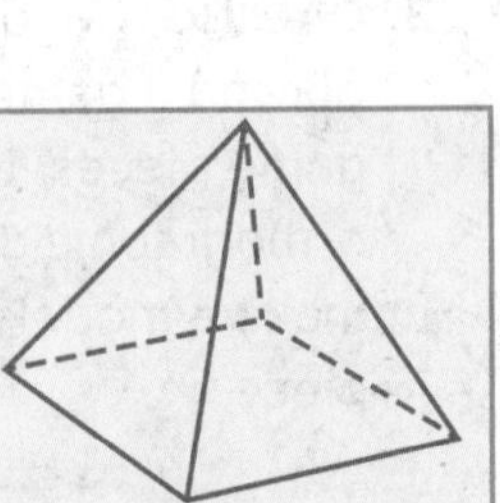
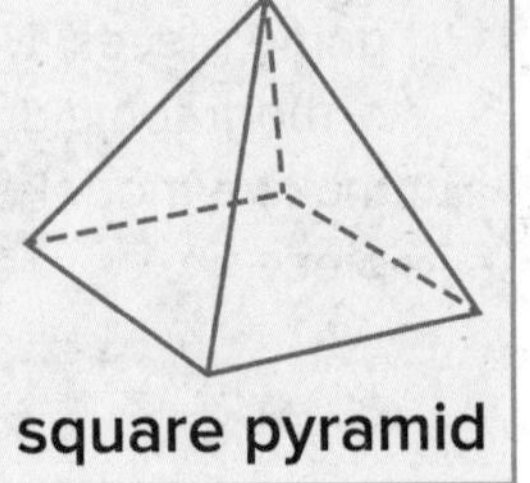

Julian's plan

30. **CREATE** Adriana works for a company called Boxes R Us making different sizes and shapes of boxes for packages. Adriana's boss wants her to sketch nets of one of the boxes. Sketch a possible net that Adriana could have drawn.

31. **WRITE** Describe the similarities and differences in orthographic drawings and nets.

32. **PERSEVERE** How many Platonic solids are there? Give a description of each solid that includes the number of two-dimensional shapes that meet at each vertex, number of faces, number of vertices, and number of edges.

Precision and Accuracy

Explore Precision and Accuracy in Basketball

Online Activity Use a real-world situation to complete the Explore.

> **INQUIRY** How are the concepts of precision and accuracy similar, and how are they different?

Learn Precision and Accuracy

Precision is the repeatability, or reproducibility, of a measurement. It depends only on the smallest unit of measure available on a measuring tool. Suppose you are told that a segment measures 8 centimeters. The length, to the nearest centimeter, of each segment shown below is 8 centimeters.

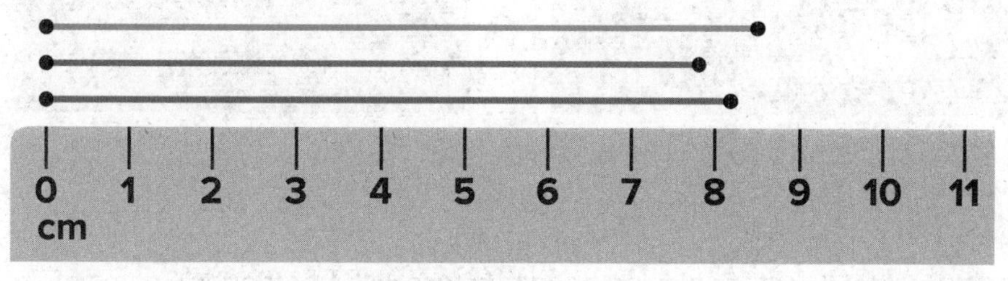

Accuracy is the nearness of a measurement to the true value of the measure. Consider the target practice results shown below.

The targets demonstrate the various levels of precision and accuracy.

accurate and precise

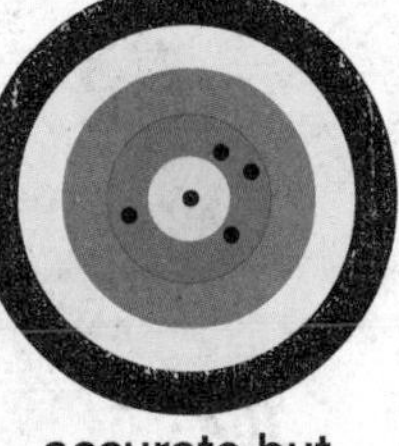
accurate but not precise

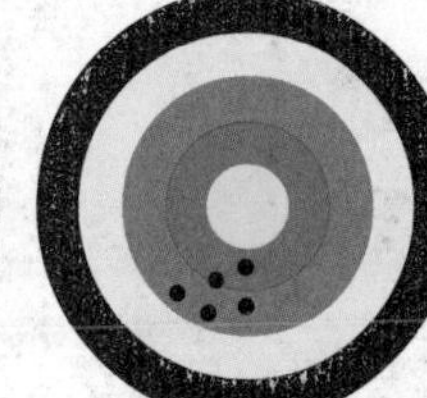
precise but not accurate

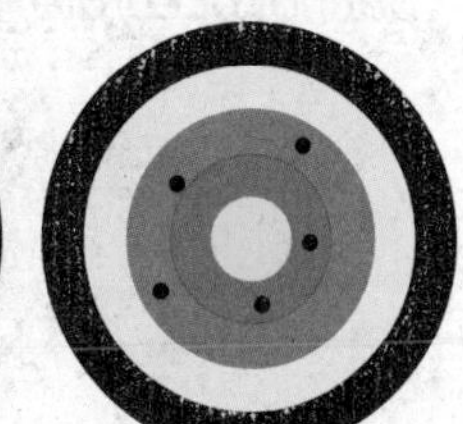
not accurate and not precise

Today's Goals
- Determine the levels of precision and accuracy in real-world scenarios.
- Calculate the approximate error of measurements.
- Choose the appropriate level of accuracy of measurements when reporting quantities.

Today's Vocabulary
precision
accuracy
approximate error

Think About It!
How do you determine how to round when measuring a line segment?

Think About It!
Why are only the first and second targets accurate?

Your Notes

Study Tip

Accuracy vs. Precision Remember that precision refers to the clustering of a group of measurements and is dependent on reproducing a certain measurement. Accuracy only refers to how close a measurement is to the true value of the measure. A measurement can be accurate without being precise or vice versa.

Example 1 Identify Precision and Accuracy

HORSESHOES **Ally and Isha are playing horseshoes. In this game, each player has two horseshoes that are thrown as close as possible to the stake in the middle of the pit. The results for four innings are shown below.**

Label each horseshoe pit as *accurate but not precise, precise but not accurate, accurate and precise,* or *not accurate and not precise.*

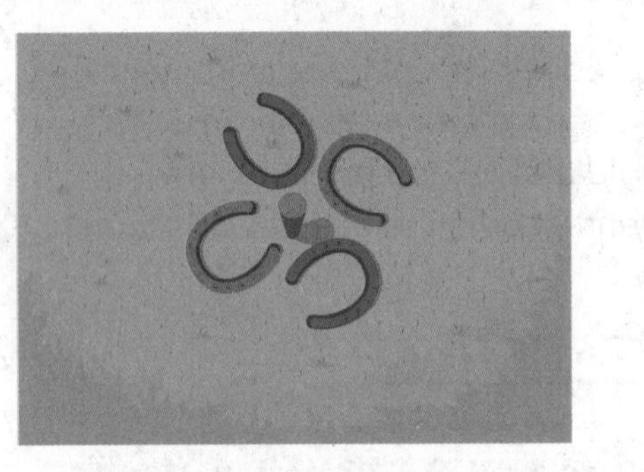

Check

LAWN GAMES Kate is playing bean bag toss with her friends. Teams of two take turns tossing bean bags at a raised platform with a hole at the far end. When a team throws, or knocks, their own bag into the hole, they receive 3 points. Label each board as *accurate but not precise, precise but not accurate, accurate and precise,* or *not accurate and not precise.*

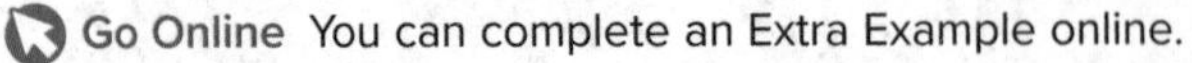

Go Online You can complete an Extra Example online.

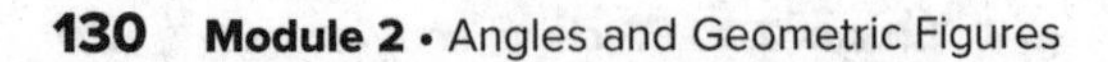

Learn Approximate Error

In the physical world, measurements are always approximate. The **approximate error** of a measurement can help you determine how accurate your calculations can be using the measurement.

Key Concept • Approximate Error

The positive difference between an actual measurement and an approximate or estimated measurement is its approximate error E_a.

$E_a = |\text{actual measurement} - \text{estimated measurement}|$

Example 2 Find Approximate Error

A student weighs a 10-gram precision mass on three different scales. Find the approximate error for each measurement.

a. spring scale: 9.86 grams

$E_a = |\text{actual measurement} - \text{estimated measurement}|$
$= |10 - 9.86|$ or ____ g

b. lab scale: 9.92 grams

$E_a = |\text{actual measurement} - \text{estimated measurement}|$
$= |10 -$ ____$|$ or ____ g

c. food scale: 10.3 grams

$E_a = |\text{actual measurement} - \text{estimated measurement}|$
$= |10 -$ ____$|$ or ____ g

Check

The temperature in Portland, Oregon, is 35° F. Declan measures the temperature outside his house. The thermometer measures 34.2° F. What is the approximate error of the temperature?

____ °F

Learn Calculating with Rounded Measurements

When rounding to a place value, look at the value immediately to the right of that position. If the value is 5 or greater, then round up.

42.64 rounds to 42.6. Because $4 < 5$, do not round to the next tenth.

42.57 rounds to 42.6. Because $7 \geq 5$, round to the next tenth.

Given a measurement of 42.6 centimeters rounded to the nearest tenth, the actual measurement could be any value in a range of values that round to 42.6.

$42.55 \leq \text{actual measurement} < 42.65$

Think About It!

In what real-world situation would it be helpful to find an approximate error?

Think About It!

Why is it important to calculate approximate errors when using scales?

Example 3 Calculate with Rounded Measurements

CARPETING **Alejandro wants to carpet his bedroom. He measures the dimensions of his bedroom and rounds to the nearest foot. The carpet he chose costs $2.63 per square foot.**

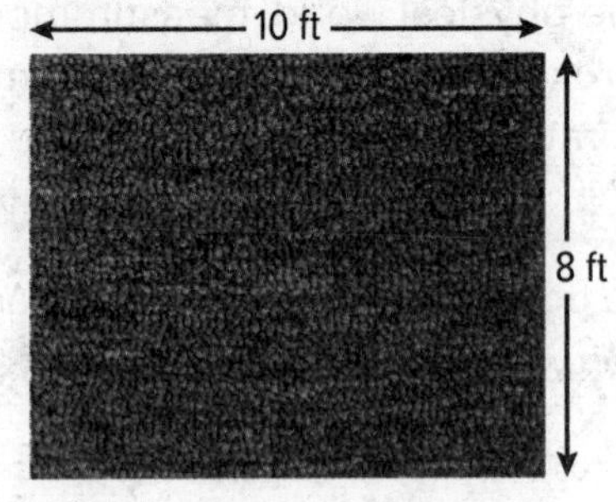

Part A What is the possible range for how much it will cost to carpet Alejandro's bedroom?

Step 1 **Find the possible range for the area of the room.**

9.5 feet $\leq$ actual length < 10.5 feet

7.5 feet $\leq$ actual width < 8.5 feet

least possible area = $9.5 \cdot 7.5$ or 71.25 ft^2
greatest possible area = $10.5 \cdot 8.5$ or 89.25 ft^2

The area is at least 71.25 square feet but less than 89.25 square feet.

Step 2 **Find the cost to buy carpet for the room.**

cost for least possible area: 71.25 • $ ______ = $187.39
cost for greatest possible area: 89.25 • $ ______ = $ ______

The cost would be at least $187.39 but less than $234.73.

Part B Alejandro checks the dimensions of the room, and measures it to be 9.8 feet by 8.2 feet to the nearest tenth of a foot. How does this change the range for the cost of the carpeting?

Step 1 **Find the possible range for the area of the room.**

9.75 feet $\leq$ actual length < 9.85 feet

8.15 feet $\leq$ actual width < 8.25 feet

least possible area = 9.75 • ______ or 79.4625 ft^2
greatest possible area = ______ • 8.25 or ______ ft^2

Step 2 **Find the cost to buy carpet for the room.**

cost for least possible area: ______ • $2.63 = $208.99
cost for greatest possible area: 81.2625 • $ ______ = $ ______

The cost would be at least $208.99 but less than $213.72.

When the measurements were rounded to the nearest foot, the range of costs was more than $45. With the measurements rounded to the nearest tenth, the range of costs is about $4.75. Rounding to the nearest tenth creates a more accurate range for the cost of the carpeting.

Talk About It!

How would a recorded time be affected in a stopwatch rounded to the nearest second versus to the nearest millisecond?

Name ______________________ Period ____________ Date ____________

Practice

Go Online You can complete your homework online.

Example 1

1. **PRECISION** A manufacturer claims that its rice cakes are packaged with 20 in each package. A sample of 12 packages is counted for accuracy. The sample yields a count of {18, 17, 17, 17, 18, 18, 18, 17, 18, 17, 18, 17} rice cakes. How accurate and precise is the manufacturer's claim? Explain your reasoning.

Example 2

2. **SCALES** A 10-pound weight is weighed on two different scales. Find the approximate error of each weight.
   a. digital bathroom scale: 9.59 pounds
   b. food scale: 10.09 pounds

3. **PHYSICS** A circuit has amperage of 0.01 milliamp. A multimeter measures the amperage of the circuit at 0.06 milliamp. What is the approximate error?

4. **CARPENTRY** A door frame is 2.13 meters high. Chandra measures the height of the door frame with a carpenter's rule. She measures 2.22 meters. What is the approximate error of the height?

5. **COOKING** Water boils at 212.0°F. Jeremiah uses a kitchen thermometer to measure the temperature of a pot of boiling water. The thermometer measures 213.1°F. What is the approximate error of the temperature?

Example 3

6. **CONSTRUCTION** The public works department is repaving some of the roads in the city. The materials needed to repave this section of the road cost $2.50 per square foot.

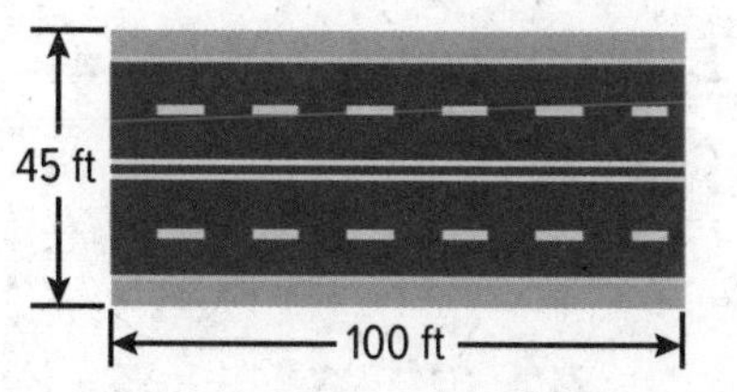

   a. What is the possible range for the area of the road?

   b. What is the possible range for the cost, c, of the materials needed to repave this section of the road?

7. **GRASS SEED** George is buying grass seed for his lawn. Grass seed is sold at $0.40 per square yard.

17 yd

12 yd

a. What is the least value for the length of the lawn?

b. What is the greatest value for the width of the lawn?

c. What is the possible range for the area of the lawn?

d. What is the possible range for the cost of the grass seed?

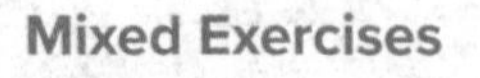

Mixed Exercises

8. **THERMOMETER** The thermostat on a heated pool is set at 76.5°F. A thermometer in the pool is shown. What is the approximate error of the temperature?

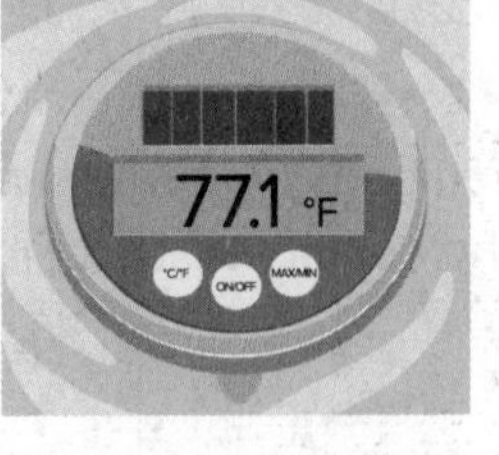

9. **SANDWICHES** A sandwich shops claims to sell foot-long sandwiches. Xiao uses a ruler to measure her sandwich. The ruler measures 11 inches. What is the approximate error of the length?

10. **SPEED** A police officer uses a radar detector to measure the speed of Roya's car. Roya's speedometer reads 55 miles per hour. The radar detector measures her speed at 56.71 miles per hour. What is the approximate error of the speed?

11. **BICYCLES** An assembly line supervisor weighs three 25-pound bicycle frames on a scale. Find the approximate error of each weight.

a. Bicycle A: 25.11 pounds

b. Bicycle B: 24.99 pounds

c. Bicycle C: 24.36 pounds

12. **DELI** Josephina works at a deli. She is testing the scales at the deli to make sure they are accurate. She uses a weight that is exactly 1 pound and gets the following results shown in the table. Which scale is the most accurate?

Scale	Weight (lb)
1	1.013
2	1.01
3	0.97

Name ______________________ Period __________ Date __________

13. GARDEN Mr. Granger wants to spread fertilizer on his vegetable garden that has dimensions 41.5 feet by 30.8 feet. The fertilizer he chose costs $0.75 per square foot for adequate coverage. What is the possible range for the cost of the fertilizer that is needed to cover the vegetable garden?

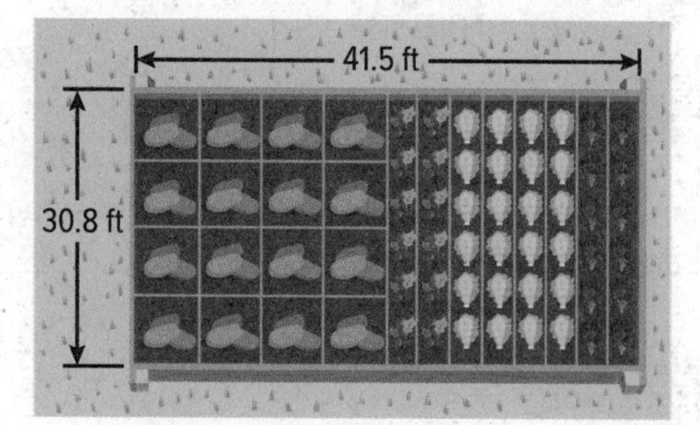

14. HEIGHT Lucas was proud of how much he had grown over the last six months since his grandma had seen him last. He told her that he was 6 feet 3 inches. His grandma didn't believe him, so she measured him again, and he was 6 feet 1 inch. What is the approximate error of Lucas' height?

15. PAINT You measure a wall of your room as 8 feet high and 12 feet wide. You want to apply wallpaper to only this wall. The wallpaper is expensive and will cost $1.25 per square foot. What is the possible range for the cost of the wallpaper?

16. Four measurements were taken three different times. The correct measurement is 52.4 cm. Determine whether the set of measurements is *accurate, precise, both,* or *neither*. Explain your reasoning.

a. 56.1 cm, 48.9 cm, 24.2 cm, 5 cm

b. 73.1 cm, 74.0 cm, 73.5 cm, 73.7 cm

c. 52.6 cm, 52.5 cm, 52.2 cm, 52.3 cm

17. **WRITE** Many people confuse the definitions of accuracy and precision. What is the difference between accuracy and precision? Give an example of a set of four numbers that represents accurate and precise measurements for a cut of meat at a steakhouse that advertises a 16-ounce ribeye steak special on Tuesday nights.

18. **WRITE** Isabel says that if a set of measurements is accurate, then it is also precise. If you agree, explain your reasoning. If you disagree, provide a counterexample.

19. **PERSEVERE** Jayden measures and labels the dimensions of a box.

4.92 in.

7.28 in.

15.3 in.

a. Calculate the areas of the faces of the box.

b. Determine the surface area of the box.

c. Determine the range of values that should contain the actual (true) measure of the surface area of the box. Explain your reasoning.

d. Suppose that Jayden had incorrectly measured the first dimension as 15.1 inches. Find the surface area of the box using this measure.

20. **CREATE** A manufacturer claims that its bags of sweetener contain 9.7 ounces in each bag. Create a sample of weights of 10 bags of sweetener such that the sample is precise and accurate. Explain your reasoning.

Representing Measurements

Explore Significant Figures

Online Activity Use the guiding exercises to complete the Explore.

INQUIRY How can you determine the number of significant figures in a measurement?

Learn Determining Significant Figures

Using significant figures allows you to maintain the correct level of precision when you are working with measurements. The **significant figures**, or *significant digits*, of a number are the digits that are used to express a measure to the appropriate degree of accuracy.

Key Concept • Significant Figures

Rules	Examples
Nonzero digits are always significant.	2.14 **3 significant figures: 2, 1, and 4**
In whole numbers, zeros are significant if they fall between nonzero digits.	5078 **4 significant figures: 5, 0, 7, and 8**
In decimal numbers greater than or equal to 1, every digit is significant.	7.60 **3 significant figures: 7, 6, and 0**
In decimal numbers less than 1, the first nonzero digit and every digit to the right are significant.	0.029 **2 significant figures: 2 and 9**

Example 1 Determine Significant Figures

Determine the number of significant figures in each measurement.

0.0320 inches

This is a decimal number ____ than 1. The first nonzero digit is ____, and there are two digits to the right of 3; 2 and ____. So, this measurement has ____ significant figures.

107,000 centimeters

Because this is a whole number, zeros are only significant if they fall between nonzero digits. There is one zero that falls between 1 and ____. So, this measurement has ____ significant figures.

Today's Goals

- Determine the correct number of significant figures in recorded measurements.
- Round measurements to the correct number of significant figures.

Today's Vocabulary

significant figures

Watch Out!

Look for the Decimal

If a number has no decimal place, then zeros are only significant if they fall between nonzero digits. For example, 165,000 has only 3 significant figures because all three zeros fall after the 5. If a number does have a decimal place, then follow the rules listed.

Talk About It!

What is the purpose of using significant figures?

Think About It!

Create three different measurements that each have four significant figures.

Your Notes

Check

Determine the number of significant figures in each measurement.

a. 0.03927 milliliter has _____ significant figures

b. 5,134,180 pounds has _____ significant figures

Example 2 Find Significant Figures by Using Tools

Find the possible range for the length of the segment using the correct number of significant figures.

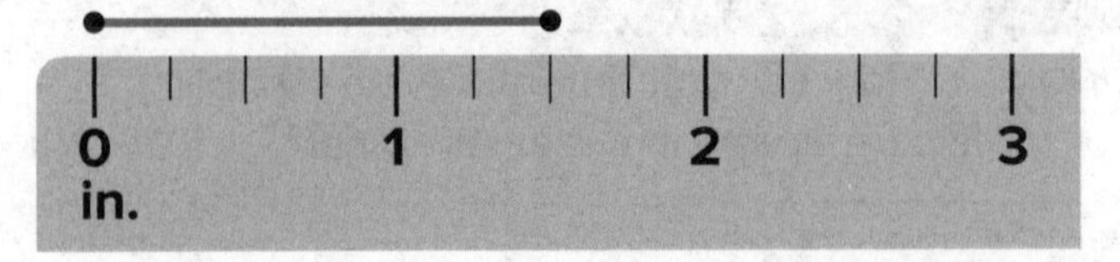

The length of the segment is approximately $1\frac{1}{2}$ inches.

This measurement was given to the nearest $\frac{1}{4}$ inch, so the possible range of this measurement is within $\frac{1}{2}$ (_____) or _____ inch of the measured length.

The exact measurement is between $1\frac{3}{8}$ and _____ inches or _____ and 1.625 inches.

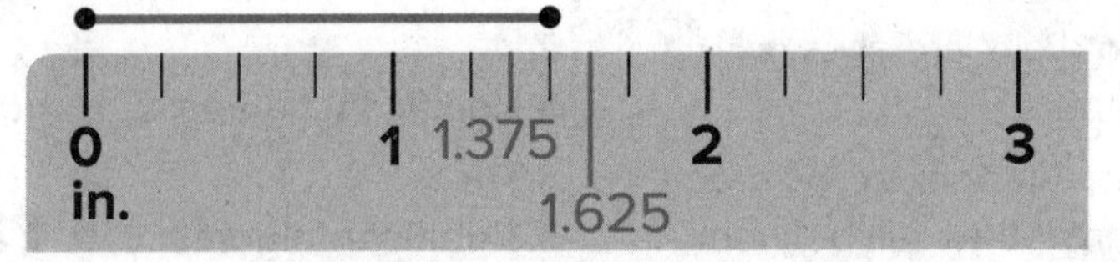

Due to the precision of the ruler, the length of the segment has _____ significant figures.

Check

Find the possible range for the length of the segment.

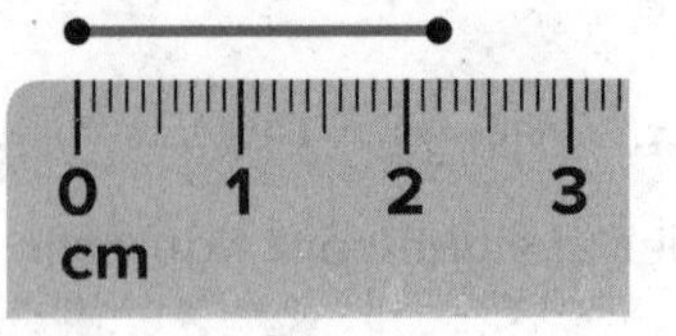

A. 2.0 cm to 2.2 cm

B. 2.00 cm to 3.00 cm

C. 2.15 cm to 2.25 cm

D. 2.08 cm to 2.12 cm

Go Online You can complete an Extra Example online.

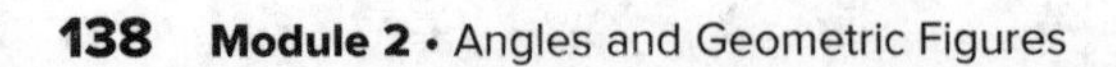

Learn Calculating with Significant Figures

When you are calculating with significant figures, the accuracy of the result is limited by the least accurate measurement.

Key Concept • Calculations with Significant Figures	
Addition and Subtraction	**Multiplication and Division**
When using addition and subtraction, a calculation cannot have more digits to the right of the decimal point than either of the original numbers.	When using multiplication and division, the number of significant figures in the final product or quotient is determined by the original number that has the *fewest* number of figures.

Numbers that are not measured are not considered when determining significant figures. For example, if you have 5 cereal boxes that weigh 14 ounces each, then the significant figures used in a calculation would be determined from the measurement, 14 ounces, not the quantity.

Significant figures are also not affected by conversion factors. For example, when using the conversion 12 inches = 1 foot, the significant figures are determined by the original measurement being converted.

Talk About It!

Why is it important to have a standard method for calculations with significant figures?

Example 3 Calculate with Significant Figures

Find each measurement rounded to the correct number of significant figures.

a. volume of an 837.24-mL sample after 276.516 mL is removed

837.24 has 2 digits after the decimal, and 276.516 has ____. So the result should have ____ digits after the decimal.

Find the difference. Then round to the hundredths place.

837.24 − ________ = 560.724 or ________ mL.

b. area of the rectangle

$A = (4.25)(2.5)$

$=$ ________

Using significant figures, the area is ____ square inches.

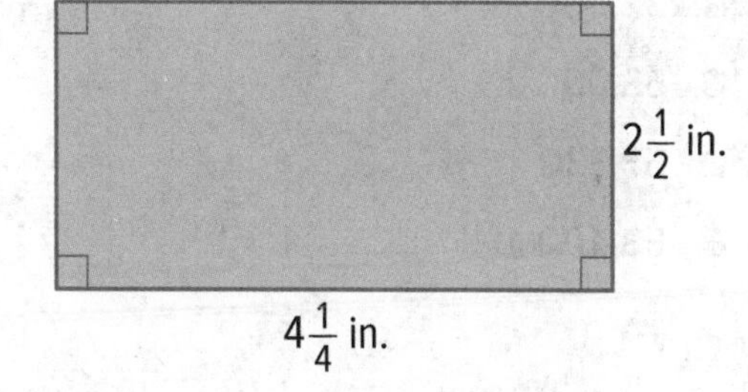

Check

A mixing bowl contains 8.5 fluid ounces of water. If 4.25 fluid ounces are removed from the bowl, how many fluid ounces of water remain? Round to the correct number of significant figures.

____ fl oz

Go Online You can complete an Extra Example online.

Example 4 Use Significant Figures in the Real World

Think About It!

What assumption did you make while solving this problem?

FLOWER GARDEN **The world's second largest flower garden is Keukenhof Park in Lisse, the Netherlands, which covers 79 acres and contains 7 million flower bulbs. About 30 gardeners work together each year to design the flower formations. How many bulbs do they use in each square yard of the garden? Round to the correct number of significant figures.**

$$\frac{\text{total bulbs}}{\text{total area}} = \frac{7{,}000{,}000 \text{ bulbs}}{79 \text{ acres}}$$

$$= \frac{7{,}000{,}000 \text{ bulbs}}{79 \text{ acres}} \times \frac{1 \text{ acre}}{4840 \text{ yd}^2} \quad 1 \text{ acre} = 4840\text{yd}^2$$

$$= \underline{\qquad\qquad} \text{ bulbs/yd}^2 \quad \text{Simplify.}$$

Because the number of bulbs is a quantity, not a measurement, the number of significant figures is determined by the given measurement, which is 79 acres. Because 79 has _____ significant figures, the final product should also have _____ significant figures. Thus, they use _____ bulbs in each square yard of the garden.

Check

SNOW REMOVAL Mark's snow plow truck can clear 1600 tons of snow in an hour. How many pounds can Mark's truck clear in a minute? Round to the correct number of significant figures. (Hint: 1 T = 2000 lb)

A. 20,000 lb

B. 32,000 lb

C. 53,000 lb

D. 53,330 lb

E. 53,400 lb

Go Online You can complete an Extra Example online.

Example 5 Use Tools to Calculate Measurements

The radius of a circle has the measurement shown. What is the possible range for the area of the circle? Round to the correct number of significant figures.

Step 1 Find the possible range for the length of the radius.

The approximate length of the segment is 7.6 centimeters. This measurement is given to the nearest _____ centimeter, so the approximate error is $\frac{1}{2}(0.1)$ or _____ centimeter. Therefore, the exact length is between 7.55 and _____ centimeters.

Step 2 Determine the number of significant figures.

Because the range of the length is between 7.55 and 7.65 centimeters, the length has _____ significant figures.

Step 3 Calculate the area of the circle.

The area of a circle is equal to _____, where r is the length of the radius. Complete the expressions to calculate the least and greatest possible areas of the circle.

least possible area: $\pi(\_\_\_\_\_)^2 \approx 179.0786352$ cm^2

greatest possible area: $\pi(\_\_\_\_\_)^2 \approx 183.8538561$ cm^2

Using significant figures, the area of the circle is between 179 and 184 square centimeters.

Check

The radius of a circle has the measurement shown. What is the possible range for the area of the circle? Round to the correct number of significant figures.

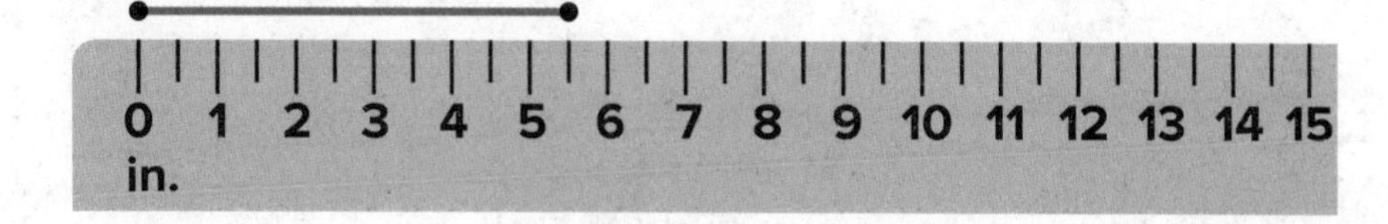

The area of the circle is between _____ and _____ square inches.

Pause and Reflect

Did you struggle with anything in this lesson? If so, how did you deal with it?

Record your observations here.

Name ______________________ Period __________ Date __________

Practice

Go Online You can complete your homework online.

Example 1

Determine the number of significant digits in each measurement.

1. 54.023

2. 0.923

3. 0.30

4. 100.58

5. 0.0002

6. 101.01

7. ACADEMICS The students in Miss Li's class are measuring the height of a chalkboard. Miss Li asked the students to write the measurement with 4 significant digits. Which student correctly followed her instructions?

Student	Measure
Sasha	48.5 centimeters
Michelle	48.53 centimeters
Alwan	49 centimeters
Remmie	48.530 centimeters

Example 2

Find the possible range for each length of the segment using the correct number of significant figures.

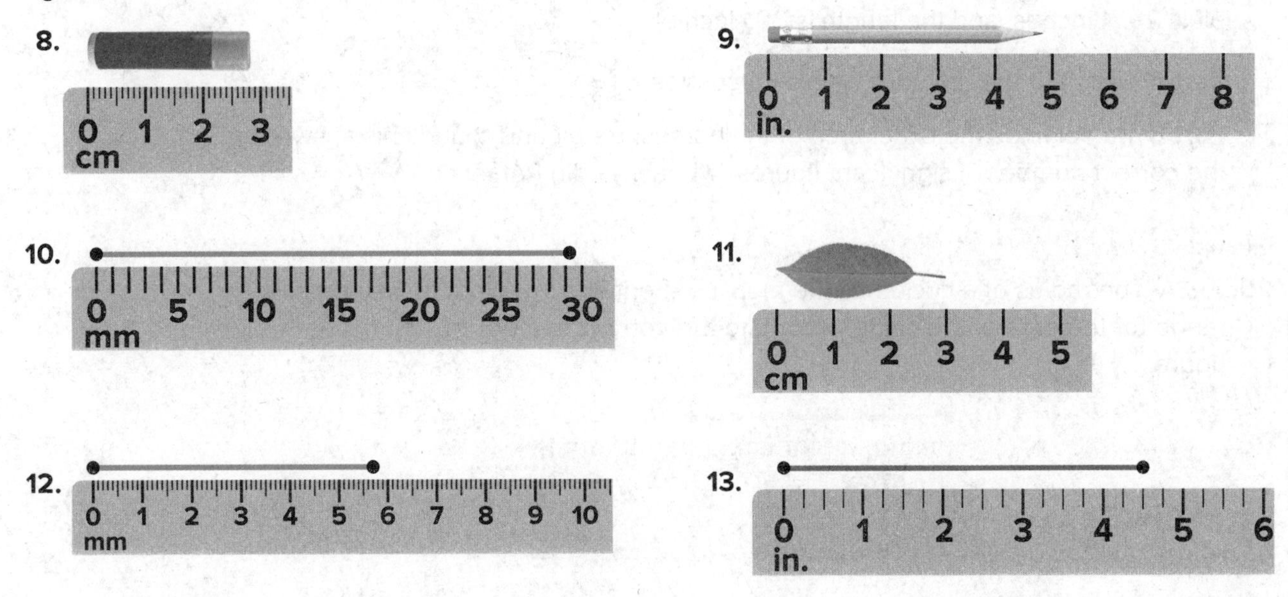

Example 3

The base of a triangle is fixed at 2.218 millimeters. Determine the number of significant figures of the area of the triangle with each given height.

14. 1.86 mm

15. 0.099 mm

16. 0.1279 mm

17. 2.109 mm

18. 11.0 mm

19. 1.7 mm

20. Using significant figures, which of the following students wrote a calculation that could have a sum or difference of 51.9?

Student	Calculation
Juliana	48.222 + 3.769
Lori	48.22 + 3.76
Nobu	48.222 + 3.7
Jerome	48.2 + 3.769

Example 4

21. CHEMISTRY Angel has 8.341 mL of saline. She pours 1.1 mL of saline into another solution. How much saline does Angel have left? Round your measurement to the correct number of significant figures.

22. A parallelogram with base b and height h has area, A, given by the formula $A = bh$. Find the area of the given parallelogram. Round your measurement to the correct number of significant figures.

3.91 cm

1.2 cm

23. AREA Find the area of a triangle with a height of 4.90 centimeters and a base length of 6.174 centimeters. Round your measurement to the correct number of significant figures.

24. DIMENSIONS Rafael is building a horseshoe pit in his backyard. The width of the pit is 29.71 inches, and the length is 30.1 inches.

Part A Estimate the area of Rafael's horseshoe pit.

Part B Rafael finds the exact area of the horseshoe pit and rounds his answer to the correct number of significant figures. What area did Rafael find?

Example 5

25. AREA The radius of a circle has the measurement shown. What is the possible range for the area of the circle? Round to the correct number of significant figures.

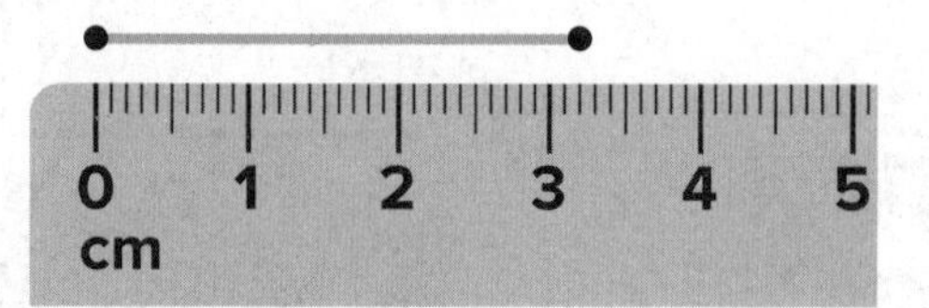

26. CIRCUMFERENCE The radius of a circle has the measurement shown. What is the possible range for the circumference of the circle? Round to the correct number of significant figures.

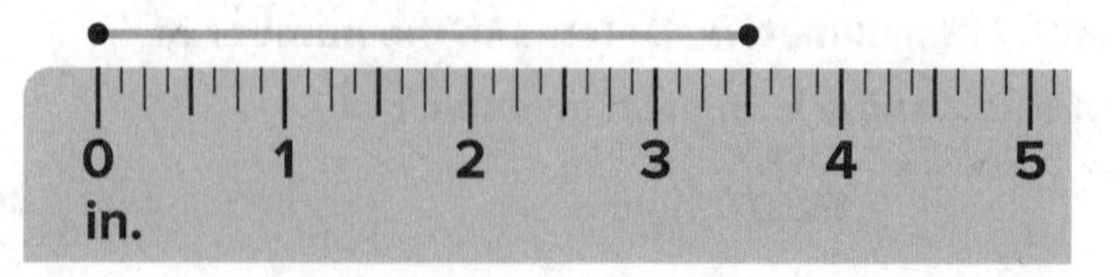

Mixed Exercises

Determine the number of significant digits in each measurement.

27. 53.74

28. 0.03298

29. 10.500

30. 6,102.0

31. 7,109,100

32. 0.110

Name _______________ Period _______________ Date _______________

33. CHEMISTRY A beaker contains a sample of NaCl weighing 49.8767 grams. If the empty beaker weighs 49.214 grams, what is the weight of the NaCl? Round to the correct number of significant figures.

34. COOKING Jordan makes a sandwich on a paper plate weighing 32.47 grams. The bread weighs 60.13 grams. Jordan adds 12.3 grams of turkey, 2.4 grams of mayonnaise, and 3.0 grams of lettuce. What is the final weight of the plate and sandwich?

35. CHEMISTRY Three chemists weigh an item using different scales. The values they report are shown on the scales. How many significant figures should be used for each measurement?

36. SWIMMING POOL A rectangular swimming pool measures 24.2 feet by 76 feet.

a. Find the perimeter of the pool. Round to the correct number of significant figures.

b. Find the area of the pool. Round to the correct number of significant figures.

37. AREA Find the area of the given triangle. Round your measure to the correct number of significant figures.

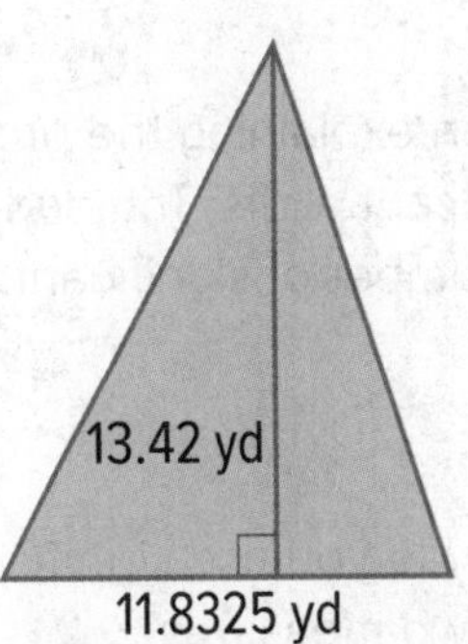

38. MURAL Krista is painting a rectangular wall that has an area of 247 square feet. If she can paint 5.25 square feet in an hour, about how long will it take for Krista to finish the mural on her own? Round to the correct number of significant figures.

39. MASS Suppose that you measured the volume of a rock to be 2.3 cm^3 and you know the density to be 3.6 g/cm^3. What is the mass of the rock? Round your measure to the correct number of significant figures.

40. DRIVING Keandra is taking a trip to visit her extended family and makes a stop somewhere during her trip. The distance between Cincinnati, where Keandra started, and Dayton, where Keandra made the stop is 54 miles. The distance between Dayton and Toledo, where Keandra's family lives is 150.2 miles. How far did Keandra travel on her trip? Round to the correct number of significant figures.

41. VOLUME A rectangular box has a length of 10.876 inches, a width of 4.34 inches, and a height of 13.22 inches. What is the volume of the rectangular prism? Round to the correct number of significant figures.

42. TRAVEL You estimate that your car gets 28 miles per gallon. The cost of gas per gallon is shown. How much does it cost you to travel 455 miles? Round to the correct number of significant figures.

43. FIND THE ERROR A student found that the dimensions of a rectangle were 1.40 centimeters and 1.60 centimeters. She was asked to report the area using the correct number of significant figures. She reported the area as 2.2 square centimeters. What error did the student make? Explain your reasoning.

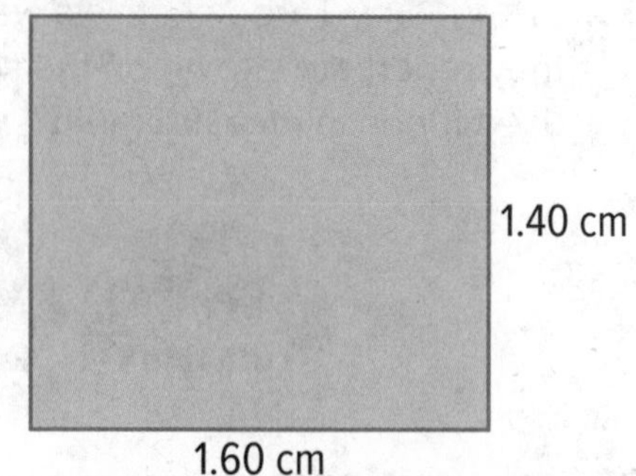

44. PERSEVERE The Sun is an excellent source of electrical energy. A field of solar panels yields 19.23 Watts per square foot. Determine the amount of electricity produced by a field of solar panels that is 410 feet by $201\frac{1}{3}$ yards.

45. WRITE When explaining the process of finding the perimeter of a triangle using significant digits, Trinidad claimed that 0.045 inch and 0.0045 inch have the same number of significant figures. Is she correct? Explain your answer.

46. WRITE How do you use significant figures to determine how to report a sum or product of two measures?

47. ANALYZE Determine whether the following statement is *sometimes*, *always*, or *never* true. Justify your argument.
Zeros are significant figures.

48. CREATE The swim team measures time to the hundredth of a second. Amanda's time was slower than Jocelyn's time in the 100-meter freestyle. What are possible times for Amanda and Jocelyn if each has times with 4 significant digits?

Review

Essential Question

How are angles and two-dimensional figures used to model the real world?

Module Summary

Lessons 2-1 and 2-2

Angles

- Angles that have the same measure are congruent angles.
- A ray or segment that divides an angle into two congruent parts is an angle bisector.
- Relationships between special angle pairs can be used to find missing measures.
- Complementary angles are two angles with measures that have a sum of 90°. Supplementary angles are two angles that have measures that have a sum of 180°.
- Certain relationships can be assumed from a figure, but most cannot.

Lessons 2-3 and 2-4

Two-Dimensional Figures

- The perimeter of a polygon is the sum of the lengths of the sides of the polygon.
- The circumference of a circle is the distance around the circle.
- Area is the number of square units needed to cover a surface.
- A transformation is a function that takes points in the plane as inputs and gives other points as outputs.
- A rigid motion is a transformation that preserves distance and angle measure.
- The three main types of rigid motions are reflection, translation, and rotation.

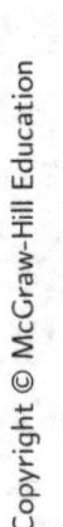

Lessons 2-5 and 2-6

Three-Dimensional Figures

- Surface area is the sum of the areas of all faces and side surfaces of a three-dimensional figure.
- Volume is the measure of the amount of space enclosed by a three-dimensional figure.
- A three-dimensional figure can be modeled by an orthographic drawing, which shows its top, left, front, and right views.
- A net is a two-dimensional figure that forms the surfaces of a three-dimensional object when folded.

Lessons 2-7 and 2-8

Measurements

- Precision is the repeatability, or reproducibility, of a measurement.
- Accuracy is the nearness of a measurement to the true value of the measure.
- The significant figures, or significant digits, of a number are the digits that contribute to its precision in a measurement.

Study Organizer

Foldables

Use your Foldable to review this module. Working with a partner can be helpful. Ask for clarification of concepts as needed.

Name ________________ Period ________ Date ________

Test Practice

1. MULTI-SELECT Select all the angles for which $\overrightarrow{HA}$ and $\overrightarrow{HE}$ are the sides. (Lesson 2-1)

Ⓐ $\angle AHE$

Ⓑ $\angle AGE$

Ⓒ $\angle EHA$

Ⓓ $\angle EGA$

Ⓔ $\angle AHB$

2. OPEN RESPONSE In the figure, $\overrightarrow{CD}$ and $\overrightarrow{CB}$ are opposite rays, and $\overrightarrow{CA}$ bisects $\angle BCE$.

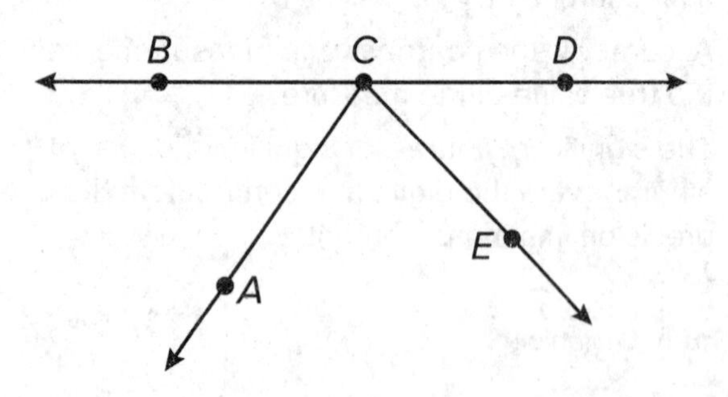

Suppose $m\angle ECA = 14x - 2$ and $m\angle ACB = 12x + 8$. What is $m\angle ECA$? (Lesson 2-1)

3. OPEN RESPONSE Describe how you would construct an angle bisector using paper-folding. (Lesson 2-1)

4. MULTIPLE CHOICE Two angles are supplementary. The measure of the larger angle is 12 less than 3 times the measure of the smaller angle. Find the measure of the larger angle. (Lesson 2-2)

Ⓐ 25.5°

Ⓑ 48°

Ⓒ 64.5°

Ⓓ 132°

5. MULTIPLE CHOICE Which value of x will make $\overleftrightarrow{AB}$ perpendicular to $\overleftrightarrow{CD}$? (Lesson 2-2)

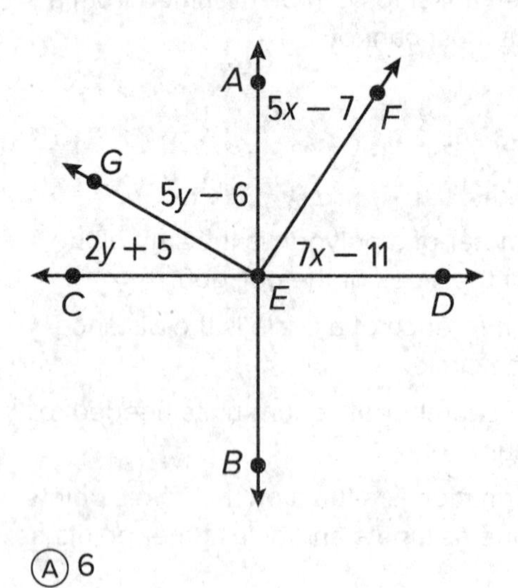

Ⓐ 6

Ⓑ 9

Ⓒ 11

Ⓓ 13

Name ______________________ Period __________ Date __________

6. **MULTIPLE CHOICE** What is the **best** estimate for the area of the triangle, in square units? (Lesson 2-3)

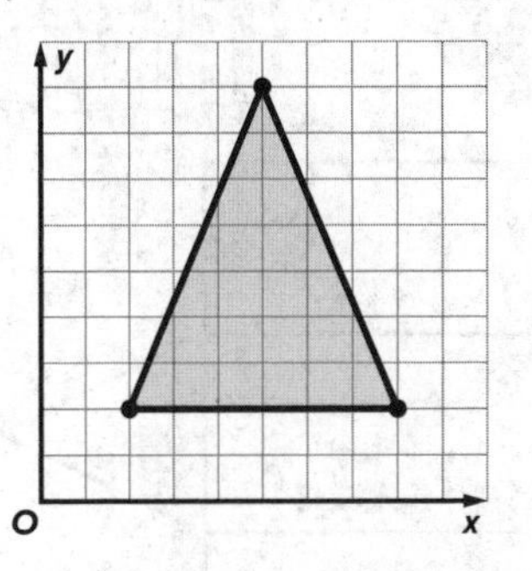

Ⓐ 5.25 square units

Ⓑ 10.5 square units

Ⓒ 21 square units

Ⓓ 42 square units

7. **OPEN RESPONSE** Find the perimeter of the rectangle. Then, find the area of the rectangle. (Lesson 2-3)

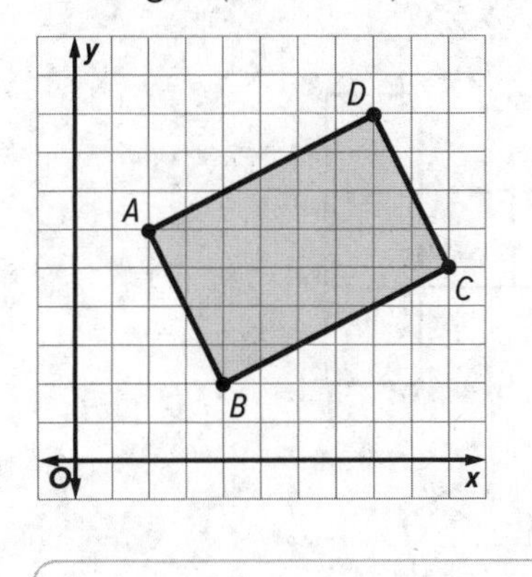

8. **OPEN RESPONSE** Find the perimeter of the triangle. Round your answer to the nearest hundredth. (Lesson 2-3)

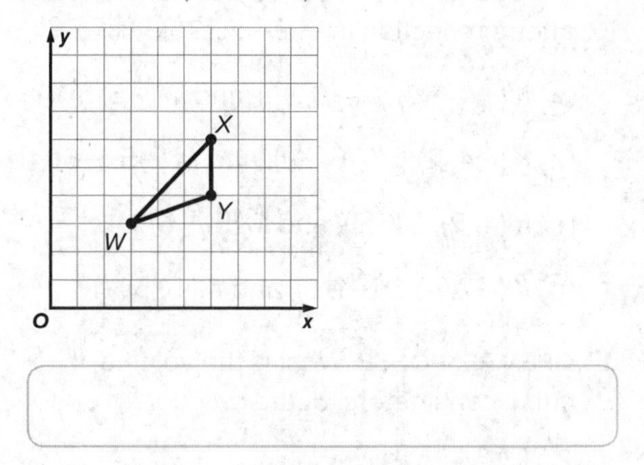

9. **MULTIPLE CHOICE** What are the coordinates of the image of $\triangle JKL$ after a 180° clockwise rotation about the origin? (Lesson 2-4)

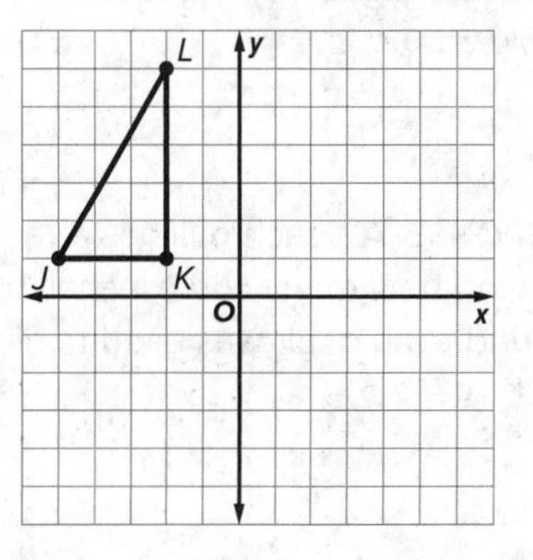

Ⓐ $J'(-5, 1)$, $K'(-2, 1)$, and $L'(-2, 6)$

Ⓑ $J'(5, -1)$, $K'(2, -1)$, and $L'(2, -6)$

Ⓒ $J'(-1, -5)$, $K'(-1, -2)$, and $L'(-6, -2)$

Ⓓ $J'(1, 5)$, $K'(1, 2)$, and $L'(6, 2)$

10. **OPEN RESPONSE** Kyle is creating a video game. Every time the main character jumps in the game, the image follows a translation using the function mapping $(x, y) \rightarrow (x + 4, y + 9)$. If the main character is located at $(-5, 0)$, what will the new location be after a jump? (Lesson 2-4)

11. MULTIPLE CHOICE $\triangle KLM$ has coordinates $K(4, -2)$, $L(6, -1)$, and $M(5, 5)$. What would be the coordinates of the vertices of the image after a reflection in the x-axis? (Lesson 2-4)

Ⓐ $K'(-4, -2)$, $L'(-6, -1)$, and $M'(-5, 5)$

Ⓑ $K'(-4, 2)$, $L'(-6, -1)$, and $M'(-5, -5)$

Ⓒ $K'(4, 2)$, $L'(6, 1)$, and $M'(5, -5)$

Ⓓ $K'(-2, 4)$, $L'(-1, 6)$, and $M'(5, 5)$

12. OPEN RESPONSE What is the volume, in cubic centimeters, of the cylinder? (Lesson 2-5)

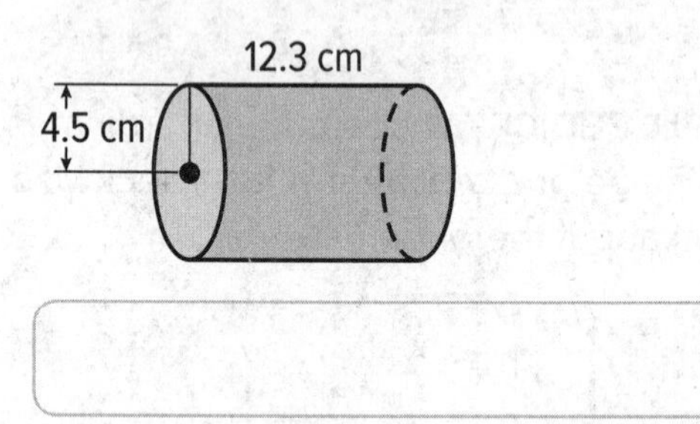

13. OPEN RESPONSE A beach ball has a radius of 8 inches. Find how many cubic inches to the nearest hundredth of air was used to fill the beach ball. (Lesson 2-5)

14. OPEN RESPONSE Identify two three-dimensional shapes that are represented by the grain silo. (Lesson 2-6)

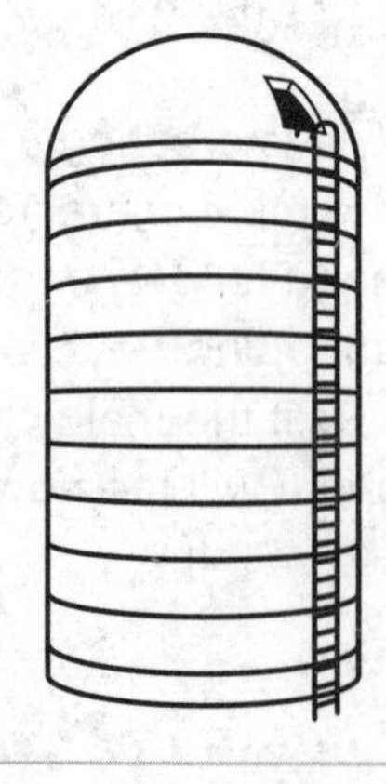

15. MULTIPLE CHOICE Which net could be used to represent the storage shed? (Lesson 2-6)

Ⓐ

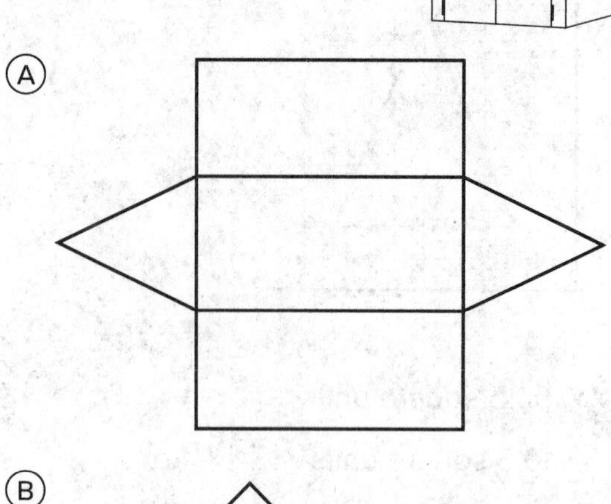

Ⓑ

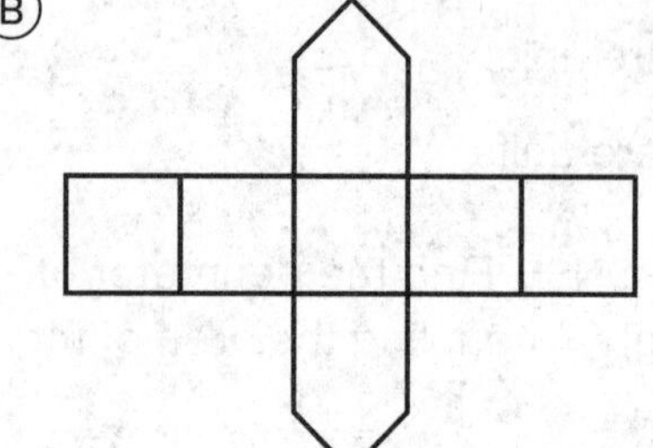

Ⓒ

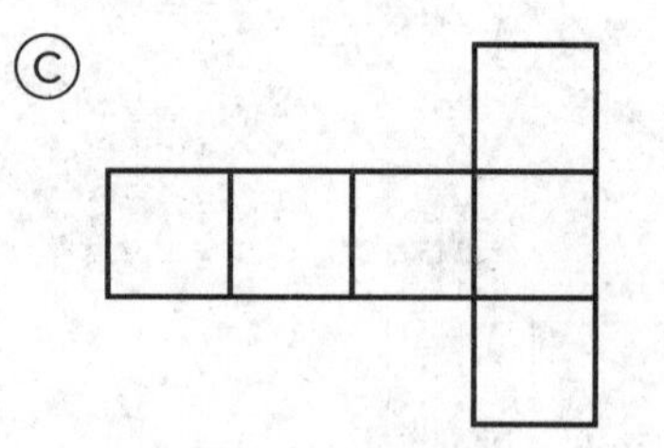

Ⓓ

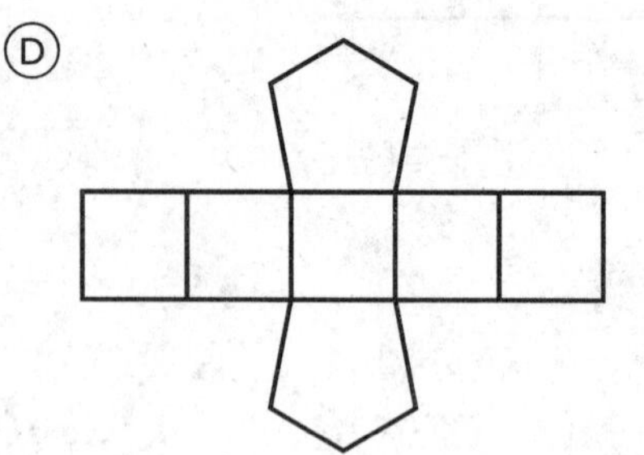

Module 3

Logical Arguments and Line Relationships

Essential Question

What makes a logical argument, and how are logical arguments used in geometry?

What Will You Learn?

Place a check mark (✓) in each row that corresponds with how much you already know about each topic **before** starting this module.

KEY

— I don't know. — I've heard of it. — I know it!

	Before			After		
	I don't know.	I've heard of it.	I know it!	I don't know.	I've heard of it.	I know it!
make and analyze conjectures based on inductive reasoning						
disprove conjectures by using counterexamples						
determine truth values of statements, negations, conjunctions, and disjunctions						
write and analyze conditionals and biconditionals using logic						
distinguish correct logic or reasoning from that which is flawed using the Laws of Detachment and Syllogism						
construct viable arguments by writing paragraph proofs						
construct viable arguments by writing flow proofs						
prove statements about segments and angles by writing two-column proofs						
identify and use relationships between pairs of angles						
identify and use parallel and perpendicular lines using the slope criteria						
solve problems using distances and parallel and perpendicular lines						

Foldables Make this Foldable to help you organize your notes about logic, reasoning, and proof. Begin with one sheet of notebook paper.

1. **Fold** lengthwise to the holes.
2. **Cut** five tabs in the top sheet.
3. **Label** the tabs as shown.

Reasoning and proof

Vocabulary

Reasoning

Proof

Parallels

Perpendiculars and Distance

What Vocabulary Will You Learn?

Check the box next to each vocabulary term that you may already know.

☐ alternate exterior angles	☐ deductive argument	☐ paragraph proof
☐ alternate interior angles	☐ deductive reasoning	☐ parallel lines
☐ biconditional statement	☐ disjunction	☐ parallel planes
☐ compound statement	☐ equidistant	☐ proof
☐ conclusion	☐ exterior angles	☐ skew lines
☐ conditional statement	☐ flow proof	☐ slope
☐ conjecture	☐ hypothesis	☐ slope criteria
☐ conjunction	☐ if-then statement	☐ statement
☐ consecutive interior angles	☐ inductive reasoning	☐ transversal
☐ contrapositive	☐ interior angles	☐ truth value
☐ converse	☐ inverse	☐ two-column proof
☐ corresponding angles	☐ logically equivalent	☐ valid argument
☐ counterexample	☐ negation	

Are You Ready?

Complete the Quick Review to see if you are ready to start this module. Then complete the Quick Check.

Quick Review

Example 1

Solve $36x - 14 = 16x + 58$.

$36x - 14 = 16x + 58$	Original equation
$20x - 14 = 58$	Subtract $16x$ from each side.
$20x = 72$	Add 14 to each side.
$x = 3.6$	Divide each side by 20.

Example 2

If $m\angle BXA = 3x + 5$ and $m\angle DXE = 56$, find x.

B X E C D A

$m\angle BXA = m\angle DXE$

Vertical $\angle$s are $\cong$.

$3x + 5 = 56$	Substitution.
$3x = 51$	Subtract 5 from each side.
$x = 17$	Divide each side by 3.

Quick Check

Solve each equation.

1. $8x - 10 = 6x$
2. $18 + 7x = 10x + 39$
3. $3(11x - 7) = 13x + 25$
4. $3x + 8 = 0.5x + 35$

Refer to the figure above.

5. Identify a pair of vertical angles that appear to be obtuse.
6. If $m\angle DXB = 116$ and $m\angle EXA = 3x + 2$, find x.
7. If $m\angle BXC = 90$, $m\angle CXD = 6x - 13$, and $m\angle DXE = 10x + 7$, find x.

How did you do?

Which exercises did you answer correctly in the Quick Check? Shade those exercise numbers below.

① ② ③ ④ ⑤ ⑥ ⑦

Conjectures and Counterexamples

Explore Using Inductive Reasoning to Make Conjectures

Online Activity Use dynamic geometry software to complete the Explore.

INQUIRY How can you use observations and patterns to make predictions?

Today's Goals
- Write and analyze conjectures by using inductive reasoning.
- Disprove conjectures by using counterexamples.

Today's Vocabulary
inductive reasoning
conjecture
counterexample

Learn Inductive Reasoning and Conjecture

Inductive reasoning is the process of reaching a conclusion based on a pattern of examples. When you assume that an observed pattern will continue, you are applying inductive reasoning. You can use inductive reasoning to make an educated guess based on known information and specific examples. This educated guess is also known as a **conjecture**.

Example 1 Patterns and Conjectures

Write a conjecture that describes the pattern in the sequence. Then use your conjecture to find the next term in the sequence.

Appointment times: 8:30 A.M., 9:15 A.M., 10:00 A.M., 10:45 A.M., . . .

Step 1 Look for a pattern.

8:30 A.M. 9:15 A.M. 10:00 A.M. 10:45 A.M.

+45 min +45 min +45 min

Step 2 Make a conjecture.

Each appointment time is _____ minutes after the previous appointment time. The next appointment time will be 10:45 A.M. + 0:45 or _____ A.M.

Check

Write a conjecture that describes the pattern in the sequence. Then use your conjecture to find the next term in the sequence.

$\frac{1}{2}$, 1, 2, 4, . . .

The next number in the sequence is _______ the preceding number.

The next number in the sequence is _____.

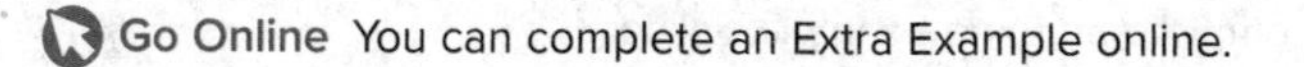
Go Online You can complete an Extra Example online.

Your Notes

Think About It!
It is possible for more than one conjecture to correctly describe a pattern. Each number in the sequence 2, 4, 8, 16, 32, . . . is 2^n where $n \geq 1$. What other conjecture can be made about the values?

Example 2 Algebraic Conjectures

Make a conjecture about the sum of the squares of two consecutive natural numbers. List or draw some examples that support your conjecture.

Step 1 List examples.

$1^2 + 2^2 =$ ____ $\qquad 6^2 + 7^2 =$ ____

$2^2 + 3^2 =$ ____ $\qquad 10^2 + 11^2 =$ ____

Step 2 Look for a pattern.

Notice that all the sums are odd numbers.

Step 3 Make a conjecture.

The sum of the squares of two consecutive natural numbers is an ____ number.

Check

Make a conjecture about the sum of two odd numbers.

The sum of two odd numbers is always a(n) ______ number.

Example 3 Geometric Conjectures

Make a conjecture about the relationship between the segments joining opposite vertices of isosceles trapezoids.

Step 1 Draw several examples.

An isosceles trapezoid is a trapezoid with two opposite congruent legs.

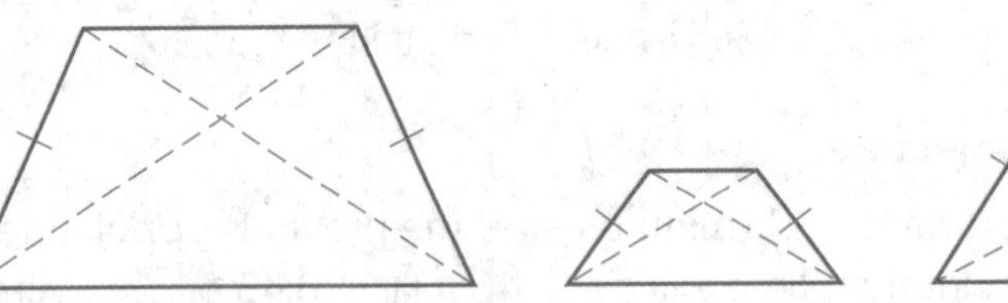

Step 2 Look for a pattern.

Notice that the segments joining opposite vertices of each isosceles trapezoid appear to have the same measure. Use a ruler or compass to confirm this.

Step 3 Make a conjecture.

The segments joining opposite vertices of an isosceles trapezoid are congruent.

Go Online You can complete an Extra Example online.

Check

Make a conjecture about the relationships between AD and AB, if C is the midpoint of $\overline{AB}$ and D is the midpoint of $\overline{AC}$.

AD is ___________ of AB.

> **Think About It!**
>
> Could the pattern of the data change over time? Explain your reasoning.

Example 4 Make Conjectures from Data

GAS PRICES **The table shows the average price of gasoline in the United States for the years 2010 through 2018. Make a conjecture about the price of gas in 2019. Explain how this conjecture is supported by the data given.**

Year	Price (dollars per gallon)
2010	2.84
2011	3.58
2012	3.68
2013	3.58
2014	3.44
2015	2.43
2016	2.14
2017	2.42
2018	2.84

Look for patterns in the data.

The price of gasoline increased from 2010 to 2012. From 2012 to 2016, the price of gas decreased, at first at a steady rate, and then more dramatically. Beginning in 2017, the price of gas began to increase at a steady rate.

> **Use a Source**
>
> Find data about the digital music revenue in the United States in recent years. Make a conjecture about the future trends in digital music revenue.

The data shows that the price of gas follows an oscillating pattern, increasing in price for several years before decreasing in price for several years.

Conjecture: In 2019, the price of gas will continue to increase.

Check

HEARING LOSS Almost 50% of young adults between the ages of 12 and 35 years old are exposed to damaging levels of sound from the use of personal electronic devices. The intensity of a sound and the time spent listening to a sound highly affects the amount of damage that can be done to someone's hearing. The intensity of a sound to the human ear is measured in A-weighted decibels, or dBA. For every 3 decibels over 85 decibels, the exposure time it takes to cause hearing damage is cut in half. How long does it take to cause hearing damage at 106 decibels? Write your answer as a decimal.

Decibel Level (dBA)	Exposure Time (hours)
85	8
88	4
91	2
94	1
97	$\frac{1}{2}$
100	$\frac{1}{4}$

______ minutes

Go Online You can complete an Extra Example online.

Learn Counterexamples

To show that a conjecture is true for all cases, you must prove it. It only takes one example that contradicts the conjecture, however, to show that a conjecture is not always true. This example is called a **counterexample**, and it can be a number, a drawing, or a statement.

Think About It!

What is a counterexample for the conjecture *All silvery coins are nickels*?

Example 5 Find Counterexamples

Find a counterexample to show that each conjecture is false.

a. If n is a real number, then $-n$ is a negative number.

When n is -4, $-n$ is $-(-4)$ or 4, which is a positive number. Because $-n$ is not negative, this is a counterexample.

What does the prefix *counter-* mean? How does this meaning relate to a counterexample?

b. If $\angle ABC \cong \angle DBE$, then $\angle ABC$ and $\angle DBE$ are vertical angles.

When points A, B, and D are noncollinear and points E, B, and C are noncollinear, the conjecture is ______.

In the figure, $\angle ABC \cong \angle DBE$, but $\angle ABC$ and $\angle DBE$ ______ vertical angles.

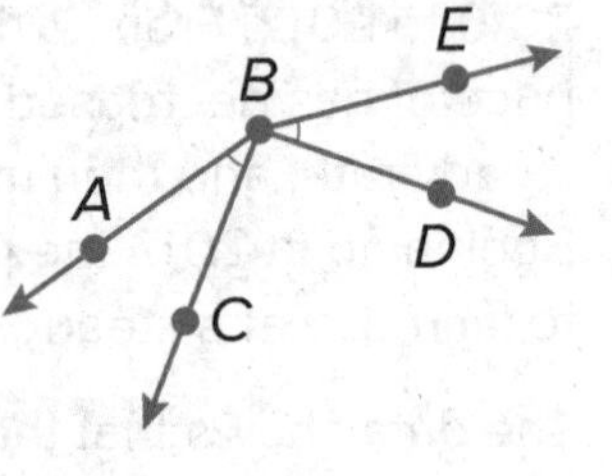

Check

Find a counterexample to show that each conjecture is false.

a. If n is a real number, then $\frac{1}{n} < n$. Select all that apply.

A. $n = -3$　B. $n = \frac{1}{4}$　C. $n = 1$　D. $n = 5$　E. $n = 100$

b. If a line intersects a segment at its midpoint, then the line is perpendicular to the segment. Draw a diagram to represent the counterexample.

Problem-Solving Tip

Draw a Diagram
Remember that a counterexample can be a number, a drawing, or a statement that proves a conjecture to be false. If you are struggling to find a counterexample, try drawing a diagram. This will allow you to analyze the situation and determine the validity of the conjecture.

Go Online You can complete an Extra Example online.

Name ______________________ Period __________ Date __________

Practice

Go Online You can complete your homework online.

Example 1

Write a conjecture that describes the pattern in each sequence. Then use your conjecture to find the next term in the sequence.

1. 4, 8, 12, 16, 20

2. 2, 22, 222, 2222

3. $1, \frac{1}{2}, \frac{1}{4}, \frac{1}{8}$

4. $6, \frac{11}{2}, 5, \frac{9}{2}, 4$

5. Arrival times: 3:00 P.M., 12:30 P.M., 10:00 A.M., . . .

6. Percent humidity: 100%, 93%, 86%, . . .

7.

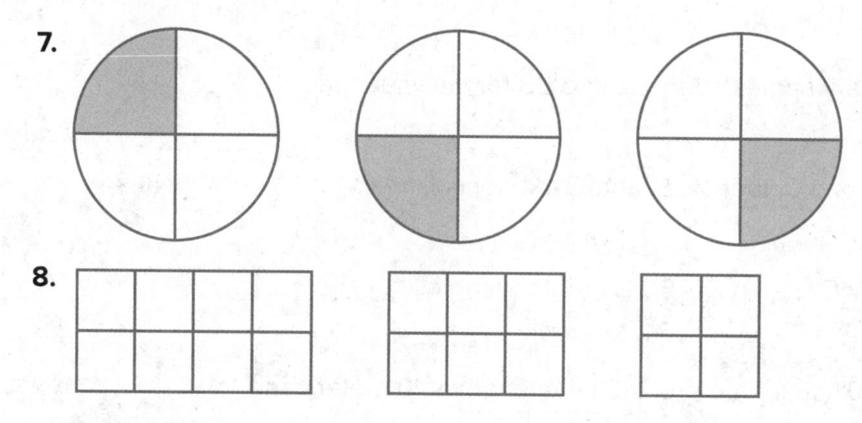

8.

Examples 2 and 3

Make a conjecture about each value or geometric relationship.

9. the product of two odd numbers

10. the product of two and a number, plus one

11. the relationship between a and c if $ab = bc$, $b \neq 0$

12. the relationship between a and b if $ab = 1$

13. the relationship between two intersecting lines that form four congruent angles

14. the relationship between the angles of a triangle with all sides congruent

15. the relationship between NP and PQ if point P is the midpoint of $\overline{NQ}$

16. the relationship between the volume of a prism and a pyramid with the same base and equal heights

Example 4

17. RAMPS Xio is rolling marbles down a ramp. Every second that passes, she measures how far the marbles travel. She records the information in the table shown below.

Second	1st	2nd	3rd	4th
Distance (cm)	20	60	100	140

Make a conjecture about how far the marble will roll in the fifth second.

Example 5

Determine whether each conjecture is *true* or *false*. Find a counterexample for any false conjecture.

18. If n is a prime number, then $n + 1$ is not prime.

19. If x is an integer, then $-x$ is positive.

20. If $\angle 2$ and $\angle 3$ are supplementary angles, then $\angle 2$ and $\angle 3$ form a linear pair.

21. If you have three points A, B, and C, then A, B, and C are noncollinear.

22. If in $\triangle ABC$, $(AB)^2 + (BC)^2 = (AC)^2$, then $\triangle ABC$ is a right triangle.

23. If the area of a rectangle is 20 square meters, then the length is 10 meters and the width is 2 meters.

Mixed Exercises

24. REASONING Given: $2a^2 = 72$. Conjecture: $a = 6$. Write a counterexample.

25. CONSTRUCT ARGUMENTS Barbara is in charge of the award medals for a sporting event. She has 31 medals to present to various individuals on 6 competing teams. She asserts that at least one team will end up with more than 5 medals. Do you believe her assertion? Justify your argument.

26. USE TOOLS Miranda is developing a chart that shows her ancestry. She makes the three sketches shown. The first dot represents herself. The second sketch represents herself and her parents. The third sketch represents herself, her parents, and her grandparents. Sketch what you think would be the next figure in the sequence.

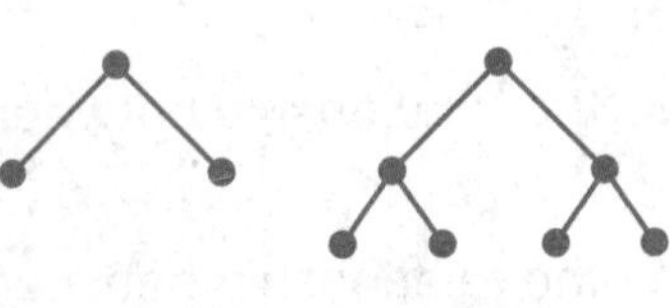

27. **REGULARITY** The figure shows a sequence of squares each made out of identical square tiles.

a. Starting from zero tiles, how many tiles do you need to make the first square? How many tiles do you have to add to the first square to get the second square? How many tiles do you have to add to the second square to get the third square?

b. Make a conjecture about the list of numbers that you started writing in your answer to **part a**.

c. Make a conjecture about the sum of the first n odd numbers.

28. **STRUCTURE** Adric made the following pattern by connecting points with line segments.

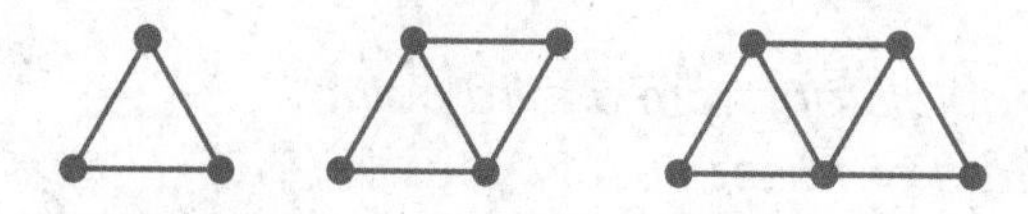

a. Suppose Adric continues the pattern. How many line segments will he need to make 4 triangles? 5 triangles?

b. Suppose Adric makes n triangles. Make a conjecture about the number of line segments he will need to make the triangles.

c. Compare the number of line segments to the number of points in each step of the pattern. How many more line segments than points will there be if Adric continues the pattern to 4 triangles? 5 triangles? Extend the pattern to make a conjecture stating how many more line segments than points are needed to draw n triangles.

29. A prime number is a number, other than 1, that is divisible by only itself and 1. Lucille read that prime numbers are very important in cryptography, so she decided to find a systematic way of producing prime numbers. After some experimenting, she conjectured that $2^n - 1$ is a prime for all whole numbers $n > 1$. Find a counterexample to this conjecture.

30. A line segment of length 1 is repeatedly shortened by removing one third of its remaining length, as shown.

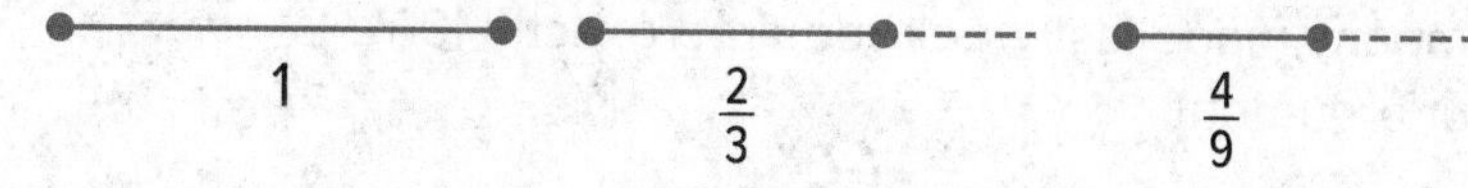

Find and use a pattern to make a conjecture about the length of the line segment after being shortened n times.

31. **PERSEVERE** If you draw points on a circle and connect every pair of points, then the circle is divided into regions. For example, two points form two regions, three points form four regions, and four points form eight regions.

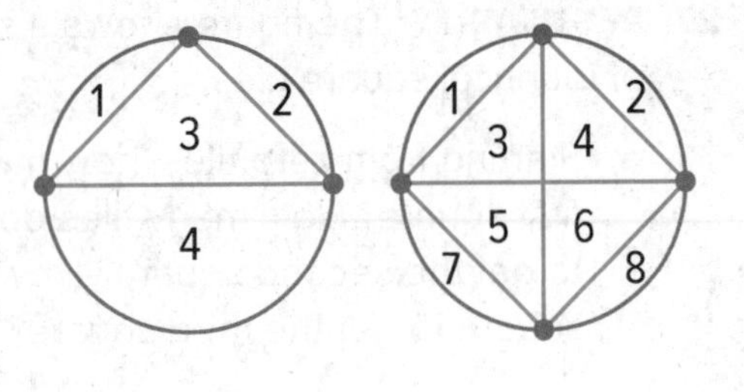

a. Make a conjecture about the relationship between the number of points on a circle and the number of regions formed in the circle.

b. Does your conjecture hold true when there are six points? Support your answer with a diagram.

32. **CREATE** Write a number sequence that can be generated by two different patterns. Explain your patterns.

33. **ANALYZE** Consider the conjecture *If two points are equidistant from a third point, then the three points are collinear.* Is the conjecture *true* or *false*? Justify your argument. If false, give a counterexample.

In Exercises 34 and 35, use undefined terms, definitions, and/or postulates to explain why each conjecture is true.

34. **WRITE** Wei drew the figure at the right. Then she stated the following conjecture: A plane contains at least two lines.

R

B

A

C

35. **WRITE** Andre drew the figure at the right. Then he stated the following conjecture: Every line contains at least one line segment.

ℓ

Q

P

36. **ANALYZE** Kayla owns a company that makes patios and garden paths out of square tiles. The figures show the patterns used to make paths of different lengths.

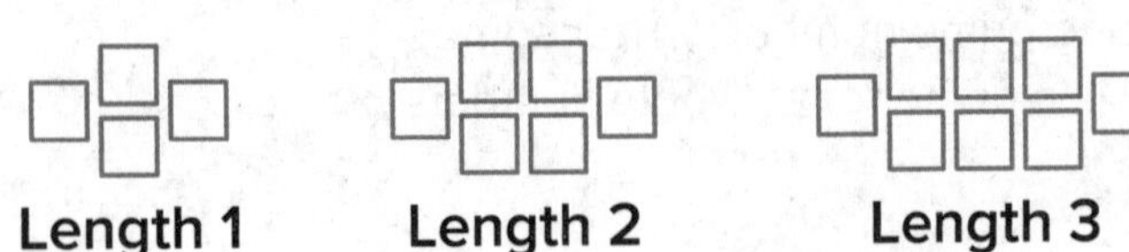

a. Kayla would like to find the number of tiles needed to make a path of any length. Look for a pattern and make a conjecture about the number of tiles needed to make a path of length n.

b. Kayla said that one path her company made last week required exactly 103 tiles. Is this possible? Justify your argument.

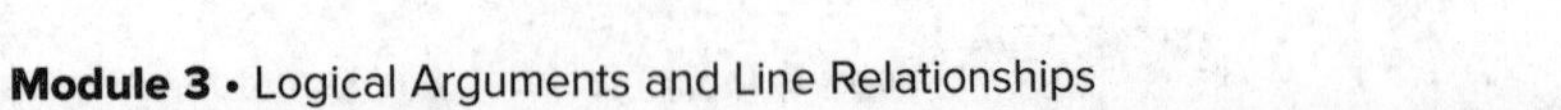

Statements, Conditionals, and Biconditionals

Explore Truth Values

Online Activity Use the video to complete the Explore.

INQUIRY How can you determine the truth value of a statement?

Learn Using Logic

A **statement** is any sentence that is either true T or false F, but not both. **Truth value** is the truth or falsity of a statement. Statements are often represented using a letter such as p or q.

If a statement is represented by p, then *not p* is the **negation** of the statement. The negation of a statement has the opposite meaning, as well as the opposite truth value, of the original statement. The negation of a statement p is *not p* or $\sim p$.

Two or more statements joined by the word *and* or *or* form a **compound statement**. A compound statement using the word *and* is called a **conjunction**. A conjunction is true only when both statements that form it are true. A conjunction is written as p and q or $p \wedge q$.

A compound statement using the word *or* is called a **disjunction**. A disjunction is true if at least one of the statements is true. A disjunction is written as *p or q* or $p \vee q$.

Today's Goals

- Write compound statements for conjunctions and disjunctions and determine truth values of statements.
- Identify hypotheses and conclusions of conditional statements and write related conditionals.
- Write and analyze biconditional statements and determine truth values of biconditional statements.

Today's Vocabulary

statement
truth value
negation
compound statement
conjunction
disjunction
conditional statement
if-then statement
hypothesis
conclusion
converse
inverse
contrapositive
logically equivalent
biconditional statement

Example 1 Truth Values of Conjunctions

Use the statements to write a compound statement for each conjunction. Then find the truth values. Explain your reasoning.

p: The figure is a trapezoid.

q: The figure has four congruent sides.

r: The figure has four right angles.

a. p and r

p and r: The figure is a trapezoid, and the figure has four right angles. Although r is true, p is ______. So, *p and r* is ______.

b. $\sim p \wedge q$

$\sim p \wedge q$: The figure is not a trapezoid, and the figure has four congruent sides. Both $\sim p$ and q are true, so $\sim p \wedge q$ is ______.

Go Online You can complete an Extra Example online.

Think About It!

Give an example of a true conjunction.

Your Notes

Watch Out!

Negation Just as the opposite of an integer is not always negative, the negation of a statement is not always false. The negation of a statement has the opposite truth value of the original statement.

Example 2 Truth Values of Disjunctions

Use the statements to write a compound statement for the disjunction $p \vee \sim r$. Then find its truth value. Explain your reasoning.

p: $\angle ABC$ and $\angle CBD$ are complementary.

q: $\angle ABC$ and $\angle CBD$ are vertical angles.

r: $\overline{AB} \cong \overline{BD}$

$p \vee \sim r$: $\angle ABC$ and $\angle CBD$ are complementary, or $\overline{AB}$ and $\overline{BD}$ are not congruent.

$p \vee \sim r$ is ______, because p is ______ and $\sim r$ is ______.

Study Tip

If and Then The word *if* is not part of the hypothesis, and the word *then* is not part of the conclusion. However, these words can indicate where the hypothesis and conclusion begin. Consider the conditional below.

If Felipe has band practice, then he will come home after dinner.

Felipe has band practice is the hypothesis, and *Felipe will come home after dinner* is the conclusion.

Learn Conditionals

A **conditional statement** is a compound statement that consists of a premise, or *hypothesis*, and a *conclusion*, which is false only when its premise is true and its conclusion is false.

Conditional Statements and Related Conditionals

Words	Examples
An **if-then statement** is a compound statement of the form "if p, then q," where p and q are statements. Symbols: $p \rightarrow q$; read *if p, then q*, or *p implies q*	
The **hypothesis** of a conditional statement is the phrase immediately following the word *if*. Symbols: $p \rightarrow q$; read *if p, then q*, or *p implies q*	If it rains, then the parade will be canceled.
The **conclusion** of a conditional statement is the phrase immediately following the word *then*. Symbols: $p \rightarrow q$; read *if p, then q*, or *p implies q*	
The **converse** is formed by exchanging the hypothesis and conclusion of the conditional. Symbols: $q \rightarrow p$, read *if q*, then *p*, or *q implies p*	If the parade is canceled, then it has rained.
The **inverse** is formed by negating both the hypothesis and conclusion of the conditional. Symbols: $\sim p \rightarrow \sim q$, read *if not p, then not q*	If it does not rain, then the parade will not be canceled.
The **contrapositive** is formed by negating both the hypothesis and the conclusion of the converse of the conditional. Symbols: $\sim q \rightarrow \sim p$, read *if not q, then not p*	If the parade is not canceled, then it does not rain.

Study Tip

Logically Equivalent A conditional and its contrapositive are either both true or both false. Similarly, the converse and inverse of a conditional are either both true or both false. Statements with the same truth value are said to be **logically equivalent**.

Go Online You can complete an Extra Example online.

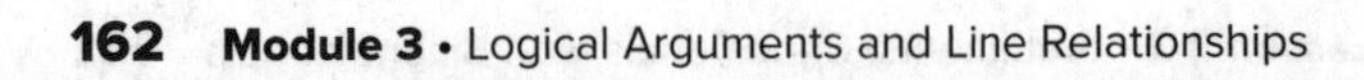

Example 3 Identify the Hypothesis and Conclusion

Identify the hypothesis and conclusion of each conditional statement.

a. If a polygon has six sides, then it is a hexagon.

Hypothesis: A polygon has six sides.

Conclusion: The polygon is a hexagon.

b. Another performance will be scheduled if the first one is sold out.

Notice that the word *if* appears in the second portion of the sentence.

Hypothesis: ______________________

Conclusion: ______________________

Check

Identify the hypothesis and conclusion of each conditional statement.

a. If the forecast is rain, then I will take an umbrella.

Hypothesis: ______________________

Conclusion: ______________________

b. A number is divisible by 10 if its last digit is a 0.

Hypothesis: ______________________

Conclusion: ______________________

Example 4 Write a Conditional in If-Then Form

Identify the hypothesis and conclusion for each conditional statement. Then write the statement in if-then form.

a. Four quarters can be exchanged for a $1 bill.

Hypothesis: You have four quarters.

Conclusion: You can exchange them for a $1 bill.

If-then: If you have four quarters, then you can exchange them for a $1 bill.

b. The sum of the measures of two supplementary angles is 180°

Hypothesis: ______________________

Conclusion: ______________________

If-then: ______________________

Go Online You can complete an Extra Example online.

Think About It!

If a conditional is true, are the converse and inverse *sometimes*, *always*, or *never* true? Support your answer with an example.

Think About It!

How do you identify the hypothesis and conclusion of a conditional statement when the statement is not in if-then form?

Check

Identify the hypothesis and conclusion of the conditional statement *A polygon with two sets of parallel sides is a parallelogram.* Then write the statement in if-then form.

Hypothesis: ____________________

Conclusion: ____________________

____

____

Example 5 Related Conditionals

NATURE **The tang is a saltwater fish that inhabits shallow coral reefs in tropical areas. Tangs are a part of the Acanthuridae family along with surgeonfish and unicornfish. All members of the Acanthuridae family are saltwater fish. Write the converse, inverse, and contrapositive of the true conditional statement *Tangs are fish that live in salt water.* Determine whether each related conditional is *true* or *false*. If a statement is false, then find a counterexample.**

Conditional: *If a fish is a tang, then it lives in salt water.*

Converse: *If a fish lives in salt water, then* ____________________.

Counterexample: A surgeonfish lives in salt water, but it is not a tang.

Therefore, the converse is __________.

Inverse: *If a fish is not a tang, then it* __________ *live in salt water.*

Counterexample: A surgeonfish is not a tang, but it ______ live in salt water. Therefore, the inverse is ______.

Contrapositive: *If a fish does not live in* ____________________, *then it is not a tang.*

Based on the information above, this statement is ________.

Think About It!

Write a conditional statement in which the converse, inverse, and contrapositive are true.

Check

MUSIC Symphony orchestras contain instruments from 4 musical families: strings, woodwinds, brass, and percussion. However, string orchestras only contain string instruments. String instruments include the violin, viola, cello, bass, and harp. Write the converse, inverse, and contrapositive of the true conditional statement *If an orchestra is a string orchestra, then it contains string instruments*. Determine whether each related conditional is *true* or *false*. If the statement is false, find a counterexample.

Converse: ____________________

Inverse: ____________________

Contrapositive: ____________________

Go Online You can complete an Extra Example online.

Learn Biconditionals

You can use logic and *biconditional statements* to indicate exclusivity in situations. For example, Aarón is applying for admission into culinary school. He must earn a 3.5 GPA or higher this semester to be accepted. You can express this as two if-then statements.

- If he earns a 3.5 GPA or higher this semester, then he will be accepted.
- If Aarón is accepted into culinary school, then he has earned a 3.5 GPA or higher for the semester.

Biconditional Statement	
Words	A **biconditional statement** is the conjunction of a conditional and its converse.
Symbols	$(p \rightarrow q) \wedge (q \rightarrow p) \rightarrow (p \leftrightarrow q)$, read *p if and only if q*

So, the biconditional statement for the example above is *Aarón will be accepted into culinary school if and only if he earns a 3.5 GPA or higher this semester.*

Example 6 Write Biconditionals

Write the conditional and converse for each statement. Determine the truth values of the conditionals and converses. If false, find a counterexample. Write a biconditional statement if possible.

a. Rasha listens to music when she is in study hall.

Conditional: ______________________

Is the conditional statement *true* or *false*? If false, provide a counterexample. ______________________

Converse: ______________________

Is the converse *true* or *false*? If false, provide a counterexample.

____

Because the converse is false, a biconditional statement cannot be written.

b. If two lines are parallel, then they have the same slope.

Conditional: ______________________

Converse: ______________________

The conditional and the converse are ______. So, a biconditional ______ be written.

Biconditional: Two lines are parallel if and only if ______________________

____

Go Online You can complete an Extra Example online.

Study Tip

If and Only If
The phrase *if and only if* can be abbreviated with *iff*.

Think About It!

Compare the mathematical meanings of the symbols $\rightarrow$ and $\leftrightarrow$ in $p \rightarrow q$ and $p \leftrightarrow q$.

Think About It!

If a biconditional is true, what do you know about the conditional and converse? If a biconditional is false, what do you know about the conditional and converse?

Check

Write the conditional and converse for the statement. Determine the truth values of the conditional and converse. If false, find a counterexample. Write a biconditional statement if possible.

Isosceles triangles have at least two congruent sides.

Conditional: ____________________

Converse: ____________________

The conditional is ________, and the converse is ________.

Biconditional: ____________________

____

Example 7 Determine Truth Values of Biconditionals

Write each biconditional as a conditional and its converse. Then determine whether the biconditional is *true* or *false*. If it is false, give a counterexample.

Two angles are complements if and only if their measures have a sum of 90°.

Write the biconditional statement as a conditional.

____

____

Write the converse of your conditional statement.

____

____

The conditional and the converse are true, so the biconditional is true.

Go Online
An alternate method is available for this example.

Check

Write the biconditional as a conditional and its converse. Then, determine whether the biconditional is *true* or *false*. If false, give a counterexample.

$x > -2$, if and only if x is positive.

Conditional: ____________________

Converse: ____________________

The biconditional is ________, because ____________________.

Go Online You can complete an Extra Example online.

Name ______________________ Period __________ Date __________

Practice

Go Online You can complete your homework online.

Examples 1 and 2

Use the statements to write a compound statement for each conjunction or disjunction. Then find the truth values. Explain your reasoning.

p: $-3 - 2 = -5$

q: Vertical angles are congruent.

r: $2 + 8 > 10$

1. p and q
2. $p \wedge r$
3. $q \vee \sim r$
4. $r \vee q$
5. $\sim p \wedge \sim q$
6. $\sim r \vee \sim p$

Example 3

Identify the hypothesis and conclusion of each conditional statement.

7. "If there is no struggle, there is no progress." (Frederick Douglass).
8. If two angles are adjacent, then they have a common side.
9. If you lead, then I will follow.
10. If $3x - 4 = 11$, then $x = 5$.
11. If two angles are vertical, then they are congruent.

Example 4

Identify the hypothesis and conclusion for each conditional statement. Then write each statement in if-then form.

12. Get a free water bottle with a one-year membership.
13. Everybody at the party received a gift.
14. The intersection of two planes is a line.
15. The area of a circle is πr^2.
16. Collinear points lie on the same line.
17. A right angle measures 90 degrees.

Example 5

Write the converse, inverse, and contrapositive of each true conditional statement. Determine whether each related conditional is *true* or *false*. If a statement is false, then find a counterexample.

18. AIR TRAVEL Ulma is waiting to board an airplane. Over the speakers she hears a flight attendant say "If you are seated in rows 10 to 20, you may now board."
19. RAFFLE If you have five dollars, then you can buy five raffle tickets.
20. GEOMETRY If two angles are complementary, then the angles are acute.
21. MEDICATION A medicine bottle says "If you will be driving, then you should not take this medicine."

Example 6

Write the conditional and converse for each statement. Determine the truth values of the conditionals and converses. If false, find a counterexample. Write a biconditional statement if possible.

22. 89 is an even number if it is divisible by 2.

23. The game will be cancelled if it is raining.

24. Laura's soccer team plays on Saturdays.

Example 7

Write each biconditional as a conditional and its converse. Then determine whether the biconditional is *true* or *false*. If it is false, give a counterexample.

25. A polygon is a quadrilateral if and only if it has four sides.

26. An angle is acute if and only if it has a measure less than 90°.

Mixed Exercises

27. Find the truth value of $(p \wedge q) \vee r$.

p: $(-4)^2 > 0$

q: An isosceles triangle has at least two congruent sides.

r: Two angles, whose measure have a sum of 90, are supplements.

28. Suppose p and q are both false. What is the truth value of $(p \wedge {\sim}q) \vee {\sim}p$?

29. What is the truth value of $({\sim}p \vee q) \wedge r$ if p is true, q is false, and r is true?

30. What is the truth value of $({\sim}p \wedge q) \vee r$ if p is true, q is false, and r is true?

31. CHOCOLATE Luca has a bag of miniature chocolate bars that come in two distinct types: dark and milk. Luca picks a chocolate out of the bag. Use the following statements to determine whether the statement ${\sim}({\sim}p \vee {\sim}q)$ is true.

p: the chocolate bar is dark chocolate

q: the chocolate bar is milk chocolate

32. Clark says that a parallelogram is a quadrilateral with equal opposite angles. Write his statement in if-then form.

33. REASONING Kalia asked Elijah whether his hockey team won the game last night and whether he scored a goal. Elijah said "yes." Kalia then asked Goldi whether she or Elijah scored a goal at the game. Goldi said "yes." What can you conclude about whether or not Goldi scored?

34. PRECISION If I roll two 6-sided dice and the sum of the numbers is 11, then one die must be a 5. Write the converse, inverse, and contrapositive of the true conditional statement. Determine whether each related conditional is *true* or *false*. If a statement is false, then find a counterexample.

For Exercises 35 and 36, use the following statement.

If a ray bisects an angle, then it divides the angle into two congruent angles.

35. Write the inverse of the given statement.

36. Write the contrapositive of the given statement.

37. Write the statement *All right angles are congruent* in if-then form.

38. Use the segment to write a statement that has the same truth value as $3 = 5$.

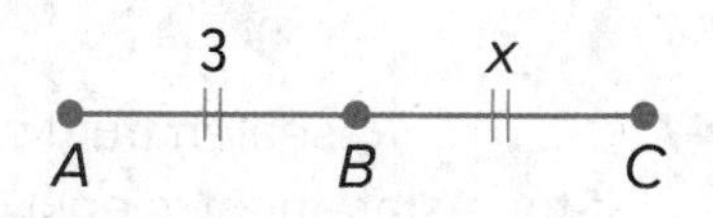

39. CREATE Consider a situation that can be represented with an if-then statement.

a. Write a true if-then statement for which the converse is false.

b. Write the converse, inverse, and contrapositive of your sentence.

c. Give the truth value of each statement you wrote for part **b.**

40. ANALYZE You are evaluating a conditional statement in which the hypothesis is true, but the conclusion is false. Is the inverse of the statement true or false? Justify your argument.

PERSEVERE To negate a statement containing the words *all* or *for every*, you can use the phrase *at least one* or *there exists*. To negate a statement containing the phrase *there exists*, use the phrase *for all* or *for every*.

p: All polygons are convex. $\sim p$: *At least one* polygon is *not* convex.

q: *There exists* a problem that has no solution. $\sim q$: *For every* problem, there is a solution.

Sometimes there are phrases that may be implied. For example, *The square of a real number is nonnegative* implies the following conditional and its negation.

p: *For every* real number x, $x^2 \geq 0$.

$\sim p$: *There exists* a real number x, such that $x^2 < 0$.

Use the information above to write the negation of each statement.

41. Every student at Hammond High School has a locker.

42. All squares are rectangles.

43. There exists a real number x, such that $x^2 = x$.

44. There exists a student who has at least one class in the C-Wing.

45. Every real number has a real square root.

46. There exists a segment that has no midpoint.

47. CREATE Research truth tables online. Then make a truth table to prove that an if-then statement is equivalent to its contrapositive and its inverse is equivalent to its converse.

48. WRITE Describe the relationship among a conditional, its converse, its inverse, and its contrapositive.

49. FIND THE ERROR Nicole and Kiri are evaluating the conditional *If 15 is prime, then 20 is divisible by 4.* Both think that the conditional is true, but their reasoning differs. Is either of them correct? Explain your reasoning.

Nicole	Kiri
The conclusion is true because 20 is divisible by 4. So, the conditional is true.	The hypothesis is false because 15 is not prime. So, the conditional is true.

50. CREATE Write a conditional statement for which the converse, inverse, and contrapositive are all true. Explain your reasoning.

Deductive Reasoning

Explore Applying Laws of Deductive Reasoning by Using Venn Diagrams

Online Activity Use dynamic geometry software to complete the Explore.

INQUIRY How can you use Venn diagrams to determine the truth value of a statement?

Today's Goals
- Apply the Law of Detachment to determine the validity of conclusions.
- Apply the Law of Syllogism to make valid conclusions from given statements.

Today's Vocabulary
deductive reasoning
valid argument

Learn The Law of Detachment

Unlike inductive reasoning, which uses a specific pattern of examples or observations to make a general conclusion, **deductive reasoning** uses general facts, rules, definitions, or properties to reach specific *valid* conclusions from given statements. An argument is **valid** if it is impossible for all the premises, or supporting statements, of the argument to be true and for its conclusion to be false. One law related to deductive reasoning is the Law of Detachment.

Key Concept • Law of Detachment	
Words	If $p \rightarrow q$ is a true statement and p is true, then q is true.
Example	*Given*: If a car is out of gas, then it will not start. Sarah's car is out of gas. *Valid Conclusion*: Sarah's car will not start.

Example 1 Inductive and Deductive Reasoning

Determine whether each conclusion is based on *inductive* or *deductive* reasoning.

a. If a student is late returning a library book, then he or she will be charged a $2 late fee. Chang returned a library book late, so he concludes that he will be charged a $2 late fee.

Chang is basing his conclusion on the library's policies, so he is using __________ reasoning.

b. Every time Tamika has worn her favorite jersey to a football game, her school's team has won the game. Tamika is wearing her favorite jersey to the football game tonight, so she concludes that her school's team will win the game.

Tamika is basing her conclusion on a specific pattern of observations, so she is using __________ reasoning.

Go Online You can complete an Extra Example online.

Your Notes

Study Tip

Given Information All information provided after the *Given* can be assumed to be true. The *Conclusion* cannot be assumed to be true.

Study Tip

Arguments An argument consists of reasons, proof, or evidence to support a position. A *logical argument*, such as the one shown in part **a**, is supported by the rules of logic. These rules include the Law of Detachment. This is different from a *statistical argument*, which is supported by examples or data.

Check

Determine whether each conclusion is based on *inductive* or *deductive* reasoning.

a. Newton's first law of motion states that an object at rest will remain at rest unless acted on by an unbalanced force. Elisa watches a soccer ball roll across the field. She concludes that an unbalanced force has acted upon the soccer ball. __________

b. Mrs. Jackson notices that her family's data usage is increasing by approximately 2500 megabytes of data every month. So, she concludes that her family's data usage next month will be 2500 megabytes greater than this month's data usage. __________

Example 2 The Law of Detachment

Determine whether each conclusion is valid based on the given information. Write *valid* or *invalid*. Explain your reasoning.

a. Given: To go on the field trip, a student must turn in a permission slip. Mariana turned in her permission slip.

Conclusion: Mariana can go on the field trip.

Step 1 Identify the hypothesis and conclusion.

Because a student must turn in a permission slip to go on the field trip, the phrase *a student must turn in a permission slip* is the __________ of the conditional statement.

p: A student turns in a permission slip.

q: The student can go on the field trip.

Step 2 Analyze the conclusion.

The given statement *Mariana turned in her permission slip* satisfies the hypothesis, so p is _____. By the Law of Detachment, *Mariana can go on the field trip,* which matches q, is a true or valid conclusion.

b. Given: If a figure is a square, then it is a polygon.

Figure *A* is a polygon.

Conclusion: Figure *A* is a square.

Step 1 Identify the hypothesis and conclusion.

p: A figure is __________

q: It is a __________

Step 2 Analyze the conclusion.

The given statement *Figure A is a polygon* satisfies the conclusion q of a true conditional. However, knowing that a conditional statement and its conclusion are true __________ make the hypothesis true. Figure *A* _______ be a triangle. The conclusion is *invalid*.

Go Online You can complete an Extra Example online.

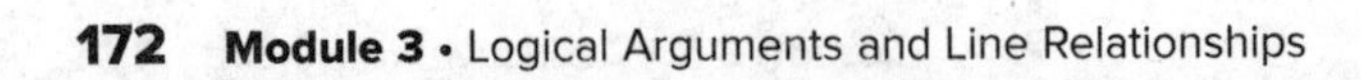

Check

Determine whether the conclusion is valid based on the given information. Select the correct answer and justification.

a. Given: If three points are noncollinear, then they determine a plane. Points A, B, and C lie in plane G.

Conclusion: Points A, B, and C are noncollinear.

A. Valid; points A, B, and C determine plane G. Therefore, they are noncollinear.

B. Valid; because points A, B, and C are noncollinear, they determine plane G.

C. Invalid; points A, B, and C determine plane G. Therefore, they are noncollinear.

D. Invalid; points A, B, and C can be collinear and lie in plane G.

b. Given: If Dakota goes to the video game store, then he will buy a new game. Dakota went to the video game store this afternoon.

Conclusion: Dakota bought a new game.

A. Invalid; because the statement *Dakota bought a new game* does not satisfy the hypothesis of the conditional statement, the conclusion is not true.

B. Valid; because the statement *Dakota went to the video game store this afternoon* satisfies the conclusion of the conditional statement, the hypothesis of the conditional is true.

C. Valid; because the statement *Dakota went to the video game store this afternoon* satisfies the hypothesis of the conditional statement, the conclusion is true.

D. Invalid; because the statement *Dakota went to the video game store this afternoon* satisfies only the hypothesis, the conclusion is not true.

Learn The Law of Syllogism

One law that is related to deductive reasoning is the Law of Syllogism. This law allows you to draw conclusions from two true conditional statements when the conclusion of one statement is the hypothesis of the other.

Key Concept • Law of Syllogism	
Words	If $p \rightarrow q$ and $q \rightarrow r$ are true statements, then $p \rightarrow r$ is a true statement.
Example	Given: If you get a job, then you will earn money. If you earn money, then you will buy a car. Valid Conclusion: If you get a job, then you will buy a car.

Talk About It!

Do you think that the order of the given statements is important when applying the Law of Syllogism? Justify your argument.

Example 3 The Law of Syllogism

SLEEP **Scientists have found that the quality and amount of sleep greatly impact learning and memory. Lack of sleep causes students to have trouble focusing and receiving new information. Sleep deprivation also makes it difficult to retrieve previously-learned information. Draw a valid conclusion from the given statements, if possible.**

Given: If you are tired, then you will not do well on your test.
If you do not get enough sleep, then you will be tired.

Step 1 Identify the hypothesis and conclusion that are the same.

Determine whether the conclusion of one statement is the hypothesis of the other statement.

Given: If you are tired, then you will not do well on your test.

If you do not get enough sleep, then you will be tired.

Reorder the given statements so the conclusion of the first statement is the hypothesis of the second statement. This will allow you to make a valid conclusion using the Law of Syllogism.

Given: If you do not get enough sleep,	p: You do not get enough sleep.
then you will be tired.	q: You will be tired.
If you are tired, then you will not do well on your test.	r: You will not do well on your test.

Step 2 Represent the statements with symbols.

Let p, q, and r represent the parts of the given conditional statements. Analyze the logic of the given conditional statements using symbols.

Statement 1: $p \rightarrow q$

Statement 2: $q \rightarrow r$

Because both statements are true and the conclusion of the first statement is the hypothesis of the second statement, $p \rightarrow r$ by the Law of Syllogism. A valid conclusion is *If you do not get enough sleep, then you will not do well on your test.*

Check

GRAND CANYON The Grand Canyon covers an area of 1900 square miles and contains 277 miles of the Colorado River. Since the Grand Canyon became a national park in 1919, over 193 million people have visited.

Draw a valid conclusion from the given statements, if possible.

Given: If Ebony takes a vacation, then she will go to the Grand Canyon. If Ebony goes to the Grand Canyon, then she will hike to the Colorado River.

____

Go Online You can complete an Extra Example online.

Study Tip

True vs. Valid Conclusions A true conclusion is not the same as a valid conclusion. True conclusions reached using invalid reasoning are still invalid.

Think About It!

Can the Law of Syllogism be applied if the two given statements have the same conclusion? Justify your argument.

Go Online to practice what you've learned about deductive reasoning in the Put It All Together over Lessons 3-1 through 3-3.

Name _______________ Period _______________ Date _______________

Practice

Go Online You can complete your homework online.

Example 1

Determine whether each conclusion is based on *inductive* or *deductive* reasoning.

1. At Fumio's school, if a student is late five times, then the student will receive a detention. Fumio has been late to school five times. Therefore, he will receive a detention.

2. A dental assistant notices that a patient has never been on time for an appointment. She concludes that the patient will be late for her next appointment.

3. A person must have a membership to work out at a gym. Jessie is working out at that gym. Jessie has a membership to that gym.

4. If Emilio decides to go to a concert tonight, then he will miss football practice. Tonight, Emilio went to a concert. Emilio missed football practice.

5. Every Wednesday, Jacy's mother calls. Today is Wednesday, so Jacy concludes that her mother will call.

6. Whenever Juanita has attended a tutoring session, she notices that her grades have improved. Juanita attends a tutoring session, and she concludes her grades will improve.

Example 2

Determine whether each conclusion is valid based on the given information. Write *valid* or *invalid*. Explain your reasoning.

7. **Given:** Right angles are congruent. $\angle 1$ and $\angle 2$ are right angles.
   Conclusion: $\angle 1 \cong \angle 2$

8. **Given:** If a figure is a square, then it has four right angles. Figure *ABCD* has four right angles.
   Conclusion: Figure *ABCD* is a square.

9. **Given:** If you leave your lights on while your car is off, then your battery will die. Your battery is dead.
   Conclusion: You left your lights on while your car was off.

10. **Given:** If Dennis gets a part-time job, then he can afford a car payment. Dennis can afford a car payment.
    Conclusion: Dennis got a part-time job.

11. **Given:** If 75% of the prom tickets are sold, then the prom will be held at the country club. 75% of the prom tickets were sold.
    Conclusion: The prom will be held at the country club.

Example 3

Use the Law of Syllogism to draw a valid conclusion from each set of given statements, if possible. If no valid conclusion can be drawn, write *no valid conclusion* and explain your reasoning.

12. If you interview for a job, then you wear a suit.
If you interview for a job, then you will update your resume.

13. If Tina has a grade point average of 3.0 or greater, she will be on the honor role.
If Tina is on the honor role, then she will have her name in the school paper.

14. If two lines are perpendicular, then they intersect to form right angles.
Lines *s* and *r* form right angles.

15. If the measure of an angle is between 90° and 180°, then it is obtuse.
If an angle is obtuse, then it is not acute.

16. If two lines in a plane are not parallel, then they intersect.
If two lines intersect, then they intersect in a point.

17. If a number ends in 0, then it is divisible by 2.
If a number ends in 4, then it is divisible by 2.

Mixed Exercises

CONSTRUCT ARGUMENTS **Draw a valid conclusion from the given statements, if possible. Then state whether your conclusion was drawn using the Law of Detachment or the Law of Syllogism. If no valid conclusion can be drawn, write *no valid conclusion*. Justify your argument.**

18. Given: If a figure is a square, then all the sides are congruent. Figure *ABCD* is a square.

19. Given: If two angles are complementary, the sum of the measures of the angles is 90°. $\angle 1$ and $\angle 2$ are complementary angles.

20. Given: Ballet dancers like classical music. If you like classical music, then you enjoy the opera.

21. Given: If you are an athlete, then you enjoy sports. If you are competitive, then you enjoy sports.

22. Given: If a polygon is regular, then all of its sides are congruent. All of the sides of polygon *WXYZ* are congruent.

23. Given: If Terryl completes a course with a grade of C, then he will not receive credit. If Terryl does not receive credit, he will have to take the course again.

Name ______________________ Period ____________ Date ____________

USE TOOLS **Determine whether each conclusion is valid based on the given information. Write *valid* or *invalid*. Explain your reasoning using a Venn diagram.**

24. Given: If the temperature drops below 32°F, it may snow. The temperature did not drop below 32°F on Monday.

Conclusion: It did not snow on Monday.

25. Given: All vegetarians do not eat meat. Theo is a vegetarian.

Conclusion: Theo does not eat meat.

26. TUTORING Marla sometimes stays after school to tutor classmates. If it is Tuesday, then Marla tutors chemistry. If Marla tutors chemistry, then she arrives home at 4 P.M. Today Marla arrived home at 4 P.M. Can it be concluded that today is Tuesday? Explain your reasoning.

27. MUSIC Composer Ludwig van Beethoven wrote 9 symphonies and 5 piano concertos. If you lived in Vienna in the early 1800s, then you could attend a concert conducted by Beethoven himself. Write a valid conclusion to the hypothesis: *If Mozart could not attend a concert conducted by Beethoven, . . .*

28. DIRECTIONS Paolo has an appointment to see a financial advisor on the fifteenth floor of an office building. When he gets to the building, the people at the front desk tell him that if he wants to go to the fifteenth floor, then he must take the red elevator. While looking for the red elevator, a guard informs him that if he wants to find the red elevator, then he must find the replica of Michelangelo's David. When he finally got to the fifteenth floor, his financial advisor greeted him asking, "What did you think of the Michelangelo?" How did Paolo's financial advisor conclude that Paolo must have seen the Michelangelo statue?

29. SIGNS Two signs are posted outside a trampoline park. Inside the trampoline park, you see a child with a parent. Write a valid conclusion based on the given information about the age of the child.

30. LOGIC As Maite's mother left for work, she quickly gave Maite some instructions. "If you need me, call my cell phone. If I do not answer, then it means I'm in a meeting. The meeting will not last more than 30 minutes, and I will call you back when the meeting is over." Later that day, Maite tried to call her mother's cell phone, but her mother was in a meeting and could not answer the phone. Maite concludes that she will have to wait no more than 30 minutes before she gets a call back from her mother. What law of logic did Maite use to draw this conclusion?

31. ENERGY Use deductive reasoning to draw a valid conclusion from the following statements: If a heat wave occurs, then air conditioning will be used more frequently; if air conditioning is used more frequently, then energy costs will be higher; there is a heat wave in Florida. If no valid conclusion can be drawn, then write *no valid conclusion* and explain your reasoning.

32. WRITE Explain why the Law of Syllogism cannot be used to draw a conclusion from these conditionals.

If you wear winter gloves, then you will have warm hands.

If you do not have warm hands, then your gloves are too thin.

33. PERSEVERE Use symbols for *conjunction*, *disjunction*, and *implies* to represent the Law of Detachment and the Law of Syllogism symbolically. Let p represent the hypothesis, and let q represent the conclusion.

34. CREATE Write a pair of statements in which the Law of Syllogism can be used to reach a valid conclusion. Specify the conclusion that can be reached.

35. ANALYZE Students in Mr. Kendrick's class are divided into two groups for an activity. Students in Group A must always tell the truth. Students in Group B must always lie. Jonah and Janeka are in Mr. Kendrick's class. When asked whether he and Janeka are in group A or B, Jonah says, "We are both in Group B." To which group does each student belong? Justify your argument.

36. WRITE Compare and contrast inductive and deductive reasoning when making conclusions and proving conjectures.

37. CREATE Write three statements that illustrate the Law of Syllogism.

38. CREATE Write three statements that illustrate the Law of Detachment.

39. WHICH ONE DOESN'T BELONG? Use statements (1) and (2). Determine which statement does not belong. Justify your conclusion.

(1) *If a triangle is equilateral, then it has three congruent sides.*

(2) *If all the sides of a triangle are congruent, then each angle measures 60°.*

A If a triangle is not equilateral, then it cannot have congruent angles.

B A figure with three congruent sides is always an equilateral triangle.

C If a triangle is not equilateral, then none of the angles measures 60°.

D If a triangle is equilateral, then each of its angles measures 60°.

Writing Proofs

Explore Algebraic Proof

Online Activity Use guiding exercises to complete the Explore.

INQUIRY How can you write an algebraic proof?

Today's Goals
- Analyze figures to identify and use postulates about points, lines, and planes.
- Analyze and construct viable arguments in a two-column format.
- Analyze and construct viable arguments in a flow proof format.
- Analyze and construct viable arguments in a paragraph proof format.

Today's Vocabulary
proof
two-column proof
deductive argument
flow proof
paragraph proof

Learn Postulates About Points, Lines, and Planes

Recall that a postulate or axiom is a statement accepted as true without proof. The postulates listed below about points, lines, and planes cannot be proven, but they can be used as reasons in proofs.

Postulates: Points, Lines, and Planes	
3.1	Through any two points, there is exactly one line.
3.2	Through any three noncollinear points, there is exactly one plane.
3.3	A line contains at least two points.
3.4	A plane contains at least three noncollinear points.
3.5	If two points lie in a plane, then the entire line containing those points lies in that plane.
3.6	If two lines intersect, then their intersection is exactly one point.
3.7	If two planes intersect, then their intersection is a line.

Example 1 Identify Postulates

ARCHITECTURE **Explain how the photo illustrates that each statement is true. Then state the postulate that can be used to show that the statement is true.**

a. Lines n and p intersect at point *D*.

The top edges of the building are represented by lines n and p. The lines intersect at the corner point, *D*. Postulate 3.6 states that if two lines intersect, then their intersection is exactly one point.

(continued on the next page)

Go Online You can complete an Extra Example online.

Your Notes

b. Points *A*, *B*, and *D* determine a plane.

Points *A*, *B*, and *D* are three noncollinear points on the front face of the building. By Postulate 3.2, through any three noncollinear points, there is exactly one plane.

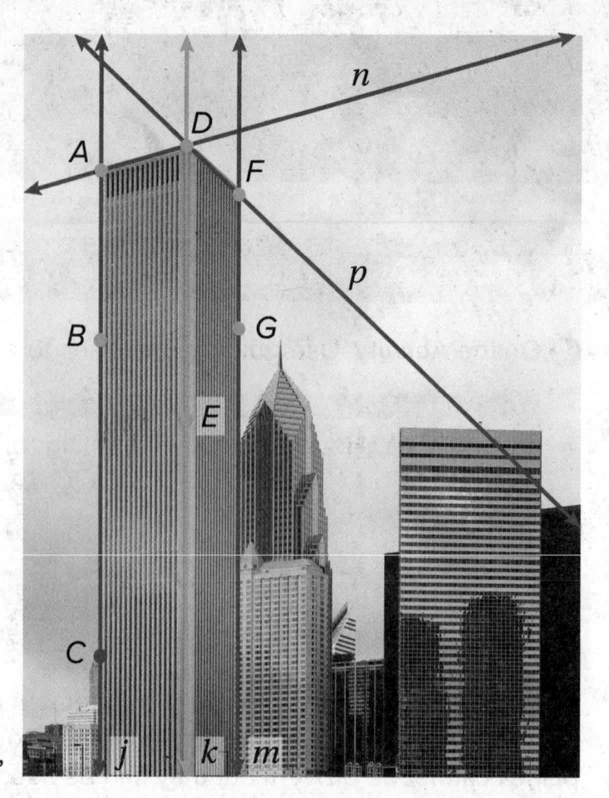

c. The plane that contains *A*, *D*, and *E* intersects the plane that contains *F*, *G*, and *E* in line k.

The front face of the building can be represented by the plane that contains *A*, *D*, and *E*. The side face of the building can be represented by the plane that contains *F*, *G*, and *E*. These planes intersect at the corner of the building represented by line k. Postulate 3.7 states that if two planes intersect, then their intersection is a line.

Check

ANCIENT MONUMENTS The image illustrates the statement $\overleftrightarrow{AB}$ *is the only line through A and B*. Which postulate proves that this statement is true?

Example 2 Use Postulates

Determine whether each statement is *always, sometimes*, or *never* true. Justify your argument.

a. The intersection of three planes is a line.

_________; if three planes intersect, then their intersection could be a line or a point.

b. Line *r* contains only point *P*.

_________; Postulate 3.3 states that a line contains at least _____ points.

c. Through points *H* and *K*, there is exactly one line.

_________; Postulate _______ states that through any two points, there is exactly _______ line.

Check

Determine whether the statement is *always*, *sometimes*, or *never* true. Justify your argument.

Two intersecting lines determine a plane. ________________________

____

____

____

____

Learn Two-Column Proofs

A **proof** is a logical argument in which each statement is supported by a statement that is accepted as true. These supporting statements can include definitions, postulates, and theorems. A **two-column proof** is a proof that contains statements and reasons that are organized in a two-column format. You can develop a **deductive argument** to prove a statement by building a logical chain of statements and reasons.

Key Concept • How to Write a Proof

Step 1 List the given information. Draw a diagram if needed.

Step 2 Create a deductive argument that links the given information to the statement that you are proving.

Step 3 Justify each statement with a reason. Reasons include definitions, postulates, theorems, and algebraic properties.

Step 4 State what it is that you have proven.

Go Online You can complete an Extra Example online.

Think About It!

Martin claims that this is a true statement. *Through any three points, there is exactly one plane*. Do you agree? Explain.

Example 3 Two-Column Proof

Complete the two-column proof by selecting the correct statements and reasons.

Given: Q is the midpoint of $\overline{PR}$.

Prove: $\overline{PQ} \cong \overline{QR}$

Statements	Reasons
1. Q is the midpoint of $\overline{PR}$.	1. Given
2. ______	2. ______
3. $\overline{PQ} \cong \overline{QR}$	3. ______

STATEMENTS/REASONS:
Definition of midpoint
Definition of congruence
$PQ = QR$
Betweenness of points
Q is between P and R.

Once a conjecture has been proven true, it can be used as a reason in other proofs. The conjecture proven above is known as the Midpoint Theorem.

Theorem 3.1: Midpoint Theorem

If M is the midpoint of $\overline{AB}$, then $\overline{AM} \cong \overline{MB}$.

Study Tip

Midpoint Theorem and Definition The definition of *midpoint* is in terms of equality, and the Midpoint Theorem is in terms of congruence.

Check

Complete the two-column proof by selecting the correct statements and reasons.

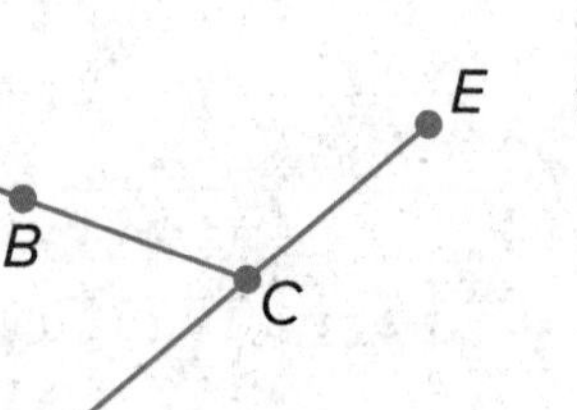

Given: B is the midpoint of $\overline{AC}$. C is the midpoint of $\overline{DE}$. $AB = CE$.

Prove: $BC = DC$

Statements	Reasons
1. B is the midpoint of $\overline{AC}$. C is the midpoint of $\overline{DE}$.	1. Given
2. ______	2. Definition of midpoint
3. $AB = CE$	3. Given
4. ______	4. ______
5. $BC = DC$	5. Substitution

STATEMENTS/REASONS:
$AB = BC$, $DC = CE$
Substitution
$BC = CE$
Transitive Property
$AC = DE$
$AB = DC$, $BC = CE$

Learn Flow Proofs

A **flow proof** uses boxes and arrows to show the logical progression of an argument. The statement is in the box, and the reason is below it. Arrows indicate the order of the steps.

Go Online You can complete an Extra Example online.

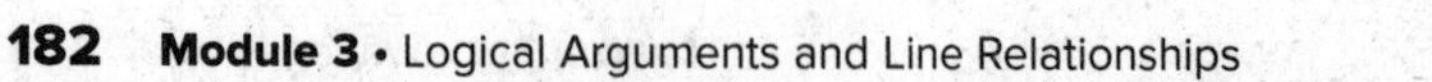

Example 4 Flow Proofs

Write each statement and reason in the correct box to complete the flow proof.

Given: P is the midpoint of $\overline{JK}$.

Prove: $x = 2$

Proof:

P is the midpoint of $\overline{JK}$.
Given

$JP = PK$
Definition of congruence

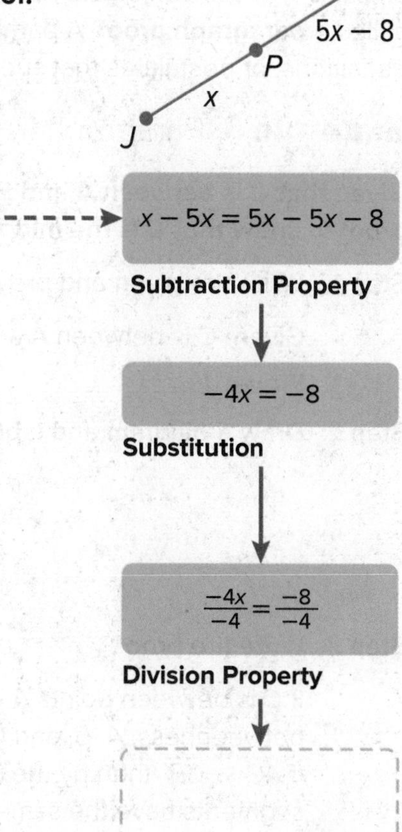

STATEMENTS/REASONS:

$x = 2$

Substitution

$x = 5x - 8$

Substitution

$\overline{JP} \cong \overline{PK}$

Midpoint Theorem

Think About It!

Can you eliminate a step from the proof? Explain.

Check

Write each statement and reason in the correct box to complete the flow proof.

Given: $FG = HK$

Prove: $x = 7$

Proof:

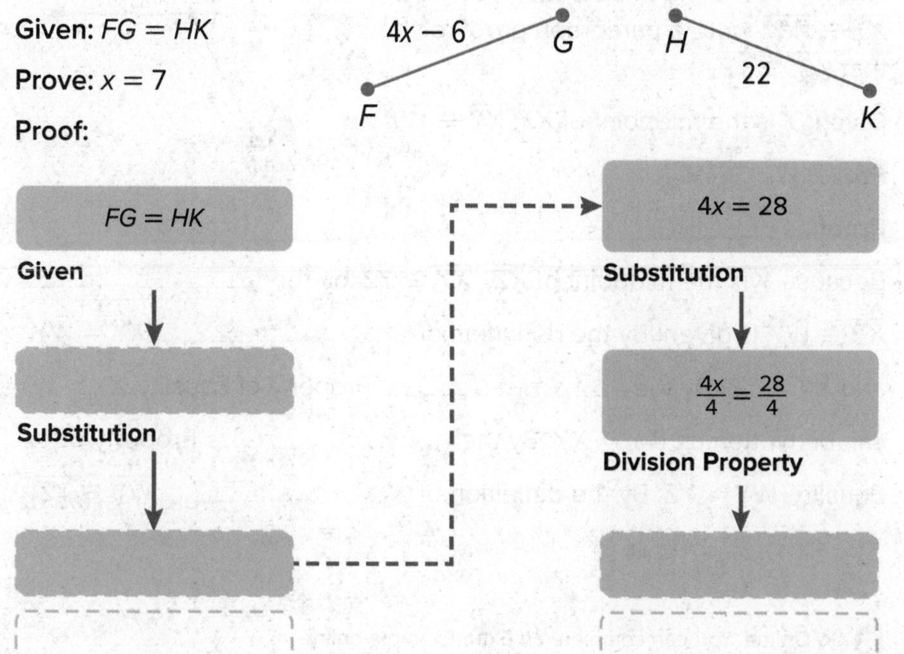

STATEMENTS/REASONS:

Addition Property

$x = 7$

Substitution

$4x - 6 = 22$

$4x - 6 + 6 = 22 + 6$

Learn Paragraph Proofs

Another way to prove a conjecture is to write a paragraph that explains why the conjecture for a given situation is true. This is called a **paragraph proof**. A paragraph proof includes the theorems, definitions, or postulates that support each statement.

Example 5 Paragraph Proof

Given that C is between A and B and $\overline{AC} \cong \overline{CB}$, write a paragraph proof to show that C is the midpoint of $\overline{AB}$.

Step 1: Write the given and prove statements.

Given: C is between A and B and $\overline{AC} \cong \overline{CB}$.

Prove: ____________________

Step 2: Draw a diagram and label any given information.

Step 3: Write the proof.

If C is between points A and B, then by the definition of betweenness, A, B, and C are collinear and $AC + CB = AB$. If $\overline{AC} \cong \overline{CB}$, then by the definition of congruence, the segments have the same measure, which means that $AC = CB$. From the definition of midpoint of a segment, if C is between points A and B and $AC = CB$, then C is the midpoint of $\overline{AB}$.

Check

Given that Y is the midpoint of $\overline{XZ}$ and $\overline{XY} \cong \overline{WY}$, write a paragraph proof to show that $\overline{WY} \cong \overline{YZ}$.

X Y Z W

Given: Y is the midpoint of $\overline{XZ}$; $\overline{XY} \cong \overline{WY}$

Prove: $\overline{WY} \cong \overline{YZ}$

Proof:

Because Y is the midpoint of $\overline{XZ}$, $\overline{XY} \cong \overline{YZ}$ by the ____________________.

$\overline{XY} \cong \overline{WY}$ is given. By the definition of ____________________, $XY = WY$ and $XY = YZ$. By the ____________________ Property of Equality, $XY = WY$ can be written as $WY = XY$. By the ____________________ Property of Equality, $WY = YZ$. By the definition of ____________________ $\overline{WY} \cong \overline{YZ}$.

Go Online You can complete an Extra Example online.

Avoid a Common Error

A common error in writing proofs is to skip steps in the logical progression of the argument.

Study Tip

Properties of Equality

The following properties are true for any real numbers a, b, and c.

Reflexive Property of Equality	$a = a$
Symmetric Property of Equality	If $a = b$, then $b = a$.
Transitive Property of Equality	If $a = b$ and $b = c$, then $a = c$.

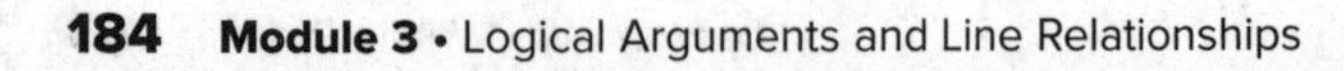

Name ______________________ Period __________ Date __________

Practice

Go Online You can complete your homework online.

Example 1

MUSIC **Explain how the figure illustrates that each statement is true. Then state the postulate that can be used to show that each statement is true.**

1. Planes O and M intersect in line t.

2. Line p lies in plane N.

SIGNS **In the figure, $\overleftrightarrow{DG}$ and $\overrightarrow{DP}$ are in plane *J* and *H* lies on $\overleftrightarrow{DG}$. State the postulate that can be used to show that each statement is true.**

3. Points *G* and *P* are collinear.

4. Points *D*, *H*, and *P* are coplanar.

Example 2

CONSTRUCT ARGUMENTS **Determine whether each statement is *always*, *sometimes*, or *never* true. Justify your argument.**

5. There is exactly one plane that contains noncollinear points *A*, *B*, and *C*.

6. There are at least three lines through points *J* and *K*.

7. If points *M*, *N*, and *P* lie in plane *X*, then they are collinear.

8. Points *X* and *Y* are in plane *Z*. Any point collinear with points *X* and *Y* is in plane *Z*.

9. The intersection of two planes can be a point.

10. Points *A*, *B*, and *C* determine a plane.

Example 3

11. PROOF Point *Y* is the midpoint of $\overline{XZ}$. Point *W* is collinear with *X*, *Y*, and *Z*. *Z* is the midpoint of $\overline{YW}$. Write a two-column proof to prove that $\overline{XY} \cong \overline{ZW}$.

12. PROOF Write a two-column proof to prove that $w = 3.5$.

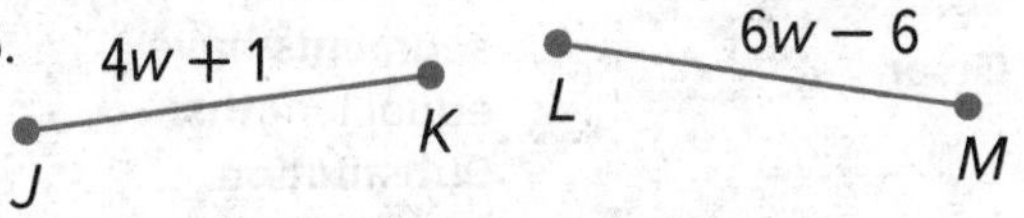

Given: $\overline{JK} \cong \overline{LM}$

Prove: $w = 3.5$

$4w + 1$ (J to K) $6w - 6$ (L to M)

13. PROOF Complete the two-column proof.

Given: $SR = RT$, $SR = UR$, and $RT = RV$

Prove: R is the midpoint of $\overline{ST}$. R is the midpoint of $\overline{UV}$.

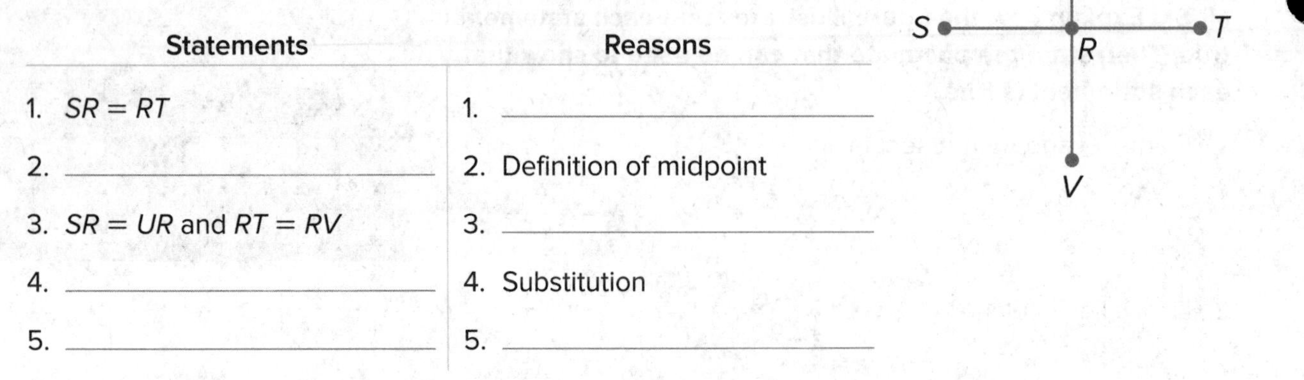

Statements	Reasons
1. $SR = RT$	1. ________
2. ________	2. Definition of midpoint
3. $SR = UR$ and $RT = RV$	3. ________
4. ________	4. Substitution
5. ________	5. ________

14. PROOF Complete the two-column proof to prove that $x = 1.25$.

Given: H is the midpoint of $\overline{FG}$.

Prove: $x = 1.25$

Statements	Reasons
1. H is the midpoint of $\overline{FG}$.	1. ________
2. ________	2. Definition of midpoint
3. ________	3. Congruent segments have equal lengths.
4. $2x + 7 = 12x - 5.5$	4. ________
5. ________	5. Addition Property of Equality
6. ________	6. ________
7. ________	7. Division Property of Equality
8. $x = 1.25$	8. ________

Example 4

15. PROOF Point L is the midpoint of $\overline{JK}$. $\overline{JK}$ intersects $\overline{MK}$ at K. If $\overline{MK} \cong \overline{JL}$, write a flow proof to prove that $\overline{LK} \cong \overline{MK}$.

16. PROOF Complete the flow proof to prove that if $\overline{MN} \cong \overline{PQ}$, $MN = 5x - 10$, and $PQ = 4x + 10$, then $MN = 90$.

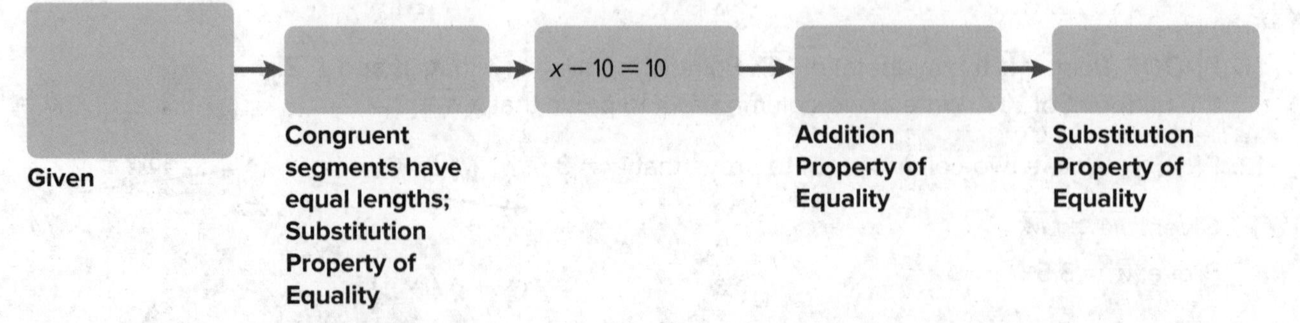

Name ______________________ Period __________ Date __________

Example 5

17. PROOF In the figure at the right, point B is the midpoint of $\overline{AC}$ and point C is the midpoint of $\overline{BD}$. Write a paragraph proof to prove that $AB = CD$.

A B C D

18. PROOF Write a paragraph proof to prove that if $PQ = 4(x - 3) + 1$, $QR = x + 10$, and $x = 7$, then $\overline{PQ} \cong \overline{QR}$.

$4(x-3)+1$ $x+10$
P Q R

Mixed Exercises

19. What postulate can be used to show the following statement is true?
Line m contains points A and F.

20. ROOFING Fai and Max are building a new roof. They wanted a roof with two sloping planes that intersect in a curved arch. Is this possible?

21. Carson claims that a line will always intersect a plane at only one point, and he draws this picture to show his reasoning. Iza thinks it is possible for a line to intersect a plane at more than one point. Who is correct? Explain.

P

22. REASONING The figure shows a straight portion of the course for a city marathon. The water station W is located at the midpoint of $\overline{AB}$.

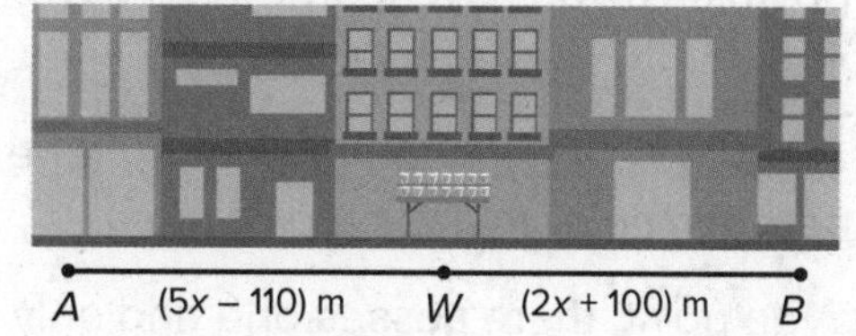

A $(5x - 110)$ m W $(2x + 100)$ m B

a. What is the length of the course from point A to point W?

b. Write a paragraph proof for your answer to part **a.**

c. Explain how you used a definition in your paragraph proof.

23. AIRLINES An airline company wants to provide service to San Francisco, Los Angeles, Chicago, Dallas, Washington D.C., and New York City. The company's president draws lines between each pair of cities in the list on a map. No three of the cities are collinear. How many lines did the president draw?

24. SMALL BUSINESSES A small company has 16 employees. The owner placed 16 points on a sheet of paper in such a way that no 3 were collinear. Each point represented a different employee. He then connected two points with a line segment if they represented coworkers in the same department.

a. What is the maximum number of line segments that can be drawn between pairs among the 16 points?

b. When the owner finished the diagram, he found that his company was split into two groups, one with 10 people and the other with 6. All the people within a group were in the same department, but nobody from one group was from the other group. How many line segments were there?

25. FIND THE ERROR Omair and Ana were working on a paragraph proof to prove that if $\overline{AB}$ is congruent to $\overline{BD}$ and A, B, and D are collinear, then B is the midpoint of $\overline{AD}$. Each student started his or her proof in a different way. Is either of them correct? Explain your reasoning.

Omair	Ana
If B is the midpoint of $\overline{AD}$, then B divides $\overline{AD}$ into two congruent segments.	$\overline{AB}$ is congruent to $\overline{BD}$, and A, B, and D are collinear.

26. CREATE Draw a figure that satisfies five of the seven postulates you have learned. Explain which postulates you chose and how your figure satisfies each postulate.

27. PERSEVERE Use the following true statements and the definitions and postulates you have learned to answer each question.

Two planes are perpendicular if and only if one plane contains a line perpendicular to the second plane.

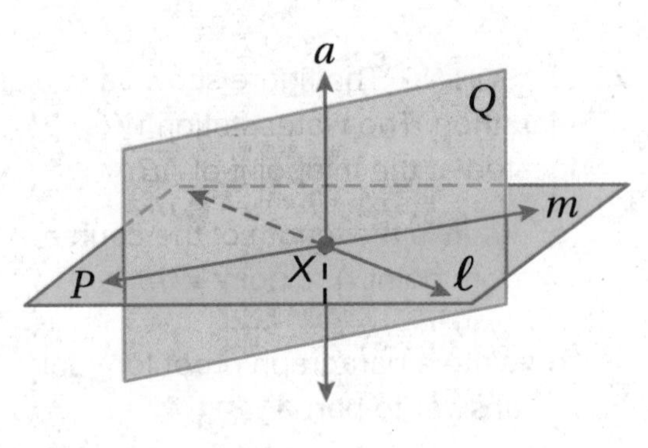

a. Through a given point, there passes one and only one plane perpendicular to a given line. If plane Q is perpendicular to line ℓ at point X and line ℓ lies in plane P, what must also be true?

b. Through a given point, there passes one and only one line perpendicular to a given plane. If plane Q is perpendicular to plane P at point X and line a lies in plane Q, what must also be true?

28. WRITE How does writing a proof require logical thinking?

ANALYZE **Determine whether each statement is *sometimes*, *always*, or *never* true. Justify your argument.**

29. Through any three points, there is exactly one plane.

30. A plane contains at least two distinct lines.

Proving Segment Relationships

Explore Segment Relationships

Online Activity Use dynamic geometry software to complete the Explore.

INQUIRY How can you use what you have already learned to prove segment relationships?

Today's Goals

- Prove theorems about line segments by using the Segment Addition Postulate.
- Prove theorems about line segments by using properties of segment congruence.

Learn Segment Addition

When you use a ruler to measure the length of an object, you match the mark for zero at one endpoint of the object. Then you look for the ruler mark that corresponds to the other endpoint. This illustrates the Ruler Postulate.

Postulate 3.8: Ruler Postulate	
Words	The points on any line or line segment can be put into one-to-one correspondence with real numbers.
Example	Given any two points A and B on a line, if A corresponds to zero, then B corresponds to a positive real number. 

In this figure, point B is said to be between points A and C. You can also say that $AB + BC = AC$ by the Segment Addition Postulate.

A B C

Postulate 3.9: Segment Addition Postulate	
Words	If A, B, and C are collinear, then point B is between A and C if and only if $AB + BC = AC$.
Example	AB, BC; A, B, C; AC

Think About It!

Determine whether the statement is true or false. If it is false, provide a counterexample.

If A, B, C, D, and E are collinear with $AC = 10$, B between A and C, C between B and D, D between C and E, and $AC = BD = CE$, then $AB = BC = DE$.

Your Notes

STATEMENTS/REASONS:

Segment Addition Postulate

$QT = RV$

Substitution Property of Equality

$QR + RT = RT + TV$

Example 1 Segment Addition Postulate

Write the correct statements and reasons to complete the two-column proof.

Given: $\overline{QT} \cong \overline{RV}$

Prove: $\overline{QR} \cong \overline{TV}$

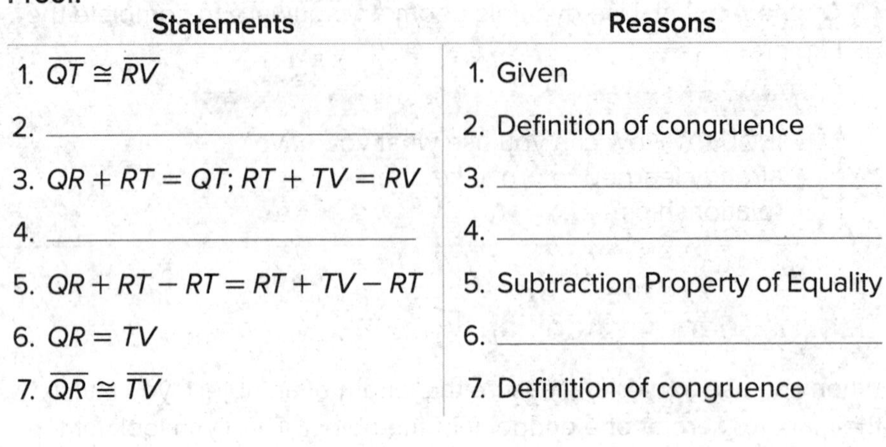

Proof:

Statements	Reasons
1. $\overline{QT} \cong \overline{RV}$	1. Given
2. ______	2. Definition of congruence
3. $QR + RT = QT$; $RT + TV = RV$	3. ______
4. ______	4. ______
5. $QR + RT - RT = RT + TV - RT$	5. Subtraction Property of Equality
6. $QR = TV$	6. ______
7. $\overline{QR} \cong \overline{TV}$	7. Definition of congruence

Check

Write the correct statement and reason to complete the two-column proof.

REASONS:

Definition of congruence

Given

Segment Addition Postulate

Substitution Property of Equality

Transitive Property of Equality

Given: $\overline{CE} \cong \overline{FE}$; $\overline{ED} \cong \overline{EG}$

Prove: $\overline{CD} \cong \overline{FG}$

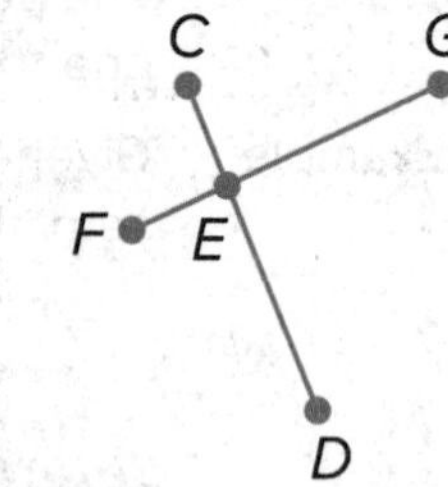

Proof:

Statements	Reasons
1. $\overline{CE} \cong \overline{FE}$; $\overline{ED} \cong \overline{EG}$	1. ______
2. $CE = FE$; $ED = EG$	2. ______
3. $CE + ED = CD$	3. ______
4. $FE + EG = CD$	4. ______
5. $FE + EG = FG$	5. ______
6. $CD = FG$	6. ______
7. $\overline{CD} \cong \overline{FG}$	7. ______

Go Online You can complete an Extra Example online.

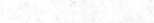

Learn Segment Congruence

You learned that segment measures are reflexive, symmetric, and transitive. Because segments with the same measure are congruent, these properties apply to segment congruence.

Go Online A proof of Theorem 3.2 is available.

Theorem 3.2: Properties of Segment Congruence	
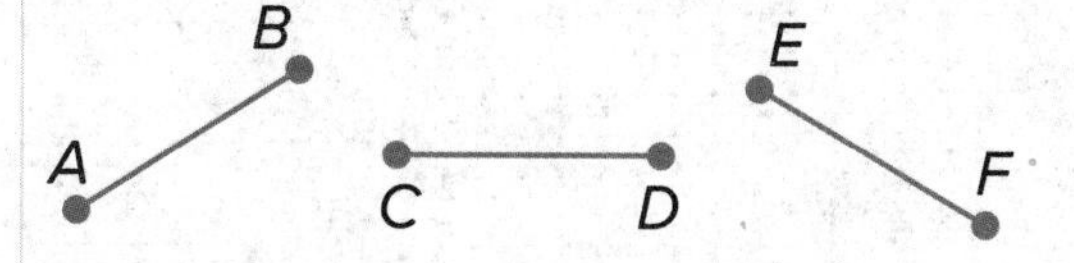	
Reflexive Property of Congruence	$\overline{AB} \cong \overline{AB}$
Symmetric Property of Congruence	If $\overline{AB} \cong \overline{CD}$, then $\overline{CD} \cong \overline{AB}$.
Transitive Property of Congruence	If $\overline{AB} \cong \overline{CD}$, and $\overline{CD} \cong \overline{EF}$, then $\overline{AB} \cong \overline{EF}$.

You will prove the Reflexive and Symmetric Properties of Congruence in Exercise 12.

Talk About It!

Is there an Addition Property of Congruence? Explain.

Example 2 Prove Segment Congruence

Write the correct statement and reasons to complete the two-column proof.

Given: R is the midpoint of $\overline{QS}$.
T is the midpoint of $\overline{VS}$.
$\overline{QR} \cong \overline{VT}$

Prove: $\overline{RS} \cong \overline{TS}$

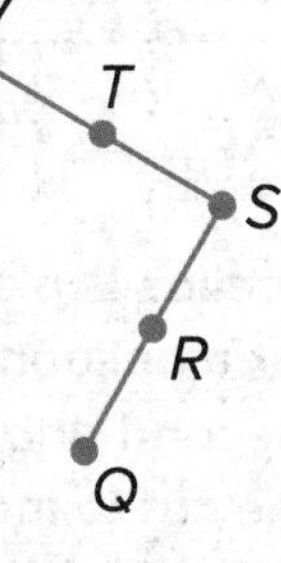

Proof:

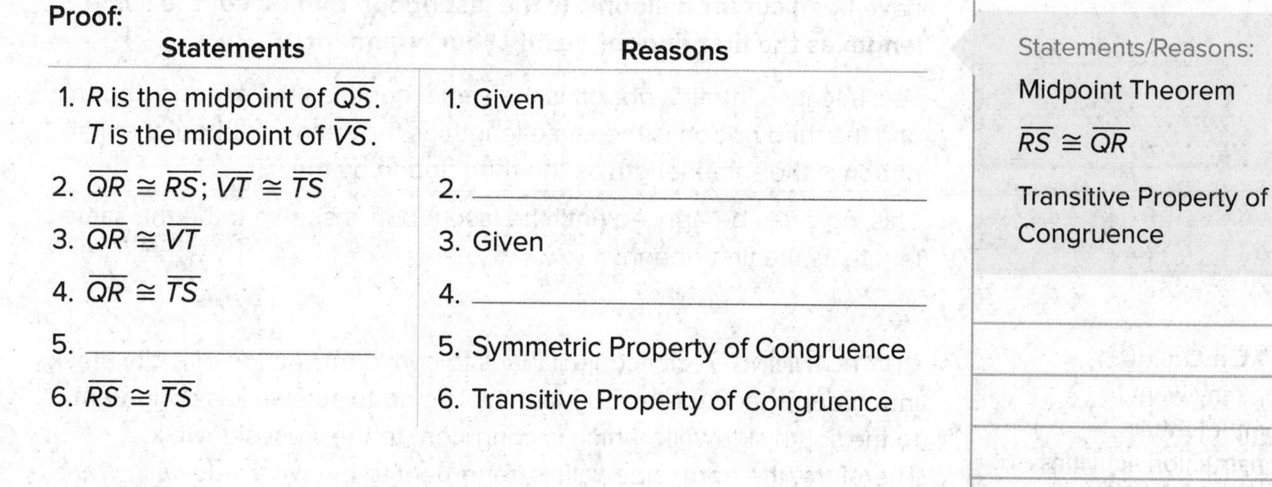

Statements	Reasons
1. R is the midpoint of $\overline{QS}$. T is the midpoint of $\overline{VS}$.	1. Given
2. $\overline{QR} \cong \overline{RS}$; $\overline{VT} \cong \overline{TS}$	2. __________
3. $\overline{QR} \cong \overline{VT}$	3. Given
4. $\overline{QR} \cong \overline{TS}$	4. __________
5. __________	5. Symmetric Property of Congruence
6. $\overline{RS} \cong \overline{TS}$	6. Transitive Property of Congruence

Statements/Reasons:

Midpoint Theorem

$\overline{RS} \cong \overline{QR}$

Transitive Property of Congruence

Go Online You can complete an Extra Example online.

REASONS:

Add.

Definition of congruence

Divide by 2.

Given

Midpoint Theorem

Segment Addition Postulate

Substitution Property of Equality

Check

Write the correct reasons to complete the two column proof.

Given: $\overline{GJ} \cong \overline{GI}$

K is the midpoint of $\overline{GJ}$.

H is the midpoint of $\overline{GI}$.

Prove: $\overline{GK} \cong \overline{GH}$

G, K, H, J, I

Proof:

Statements	Reasons
1. K is the midpoint of $\overline{GJ}$; H is the midpoint of $\overline{GI}$	1. ______
2. $\overline{GK} \cong \overline{KJ}$; $\overline{GH} \cong \overline{HI}$	2. ______
3. $GK = KJ$; $GH = HI$	3. ______
4. $\overline{GJ} \cong \overline{GI}$	4. ______
5. $GJ = GI$	5. ______
6. $GJ = GK + KJ$; $GI = GH + HI$	6. ______
7. $GK + KJ = GH + HI$	7. ______
8. $GK + GK = GH + GH$	8. ______
9. $2GK = 2GH$	9. ______
10. $GK = GH$	10. ______
11. $\overline{GK} \cong \overline{GH}$	11. ______

Example 3 Determine Congruence

CITY PLANNING **Marcellus is planning a birthday party. He measures a length of ribbon for a balloon, and then uses this ribbon to measure and cut a second ribbon. He continues this pattern of using the last ribbon that he cut to measure the next ribbon until 10 ribbons have been cut for balloons. Is the last ribbon that he cut the same length as the first ribbon? Justify your argument.**

Yes; because the first ribbon is the same length as the second ribbon and the third ribbon is the same length as the second ribbon, the first ribbon is the same length as the third ribbon by the ______.

This logic can be applied until the last ribbon is shown to be the same length as the first ribbon.

Check

CITY PLANNING A city council plans to convert a section of a city block into green space for the community. The north sidewalk is congruent to the south sidewalk, which is congruent to the west sidewalk. Therefore, the north sidewalk is congruent to the west sidewalk. What theorem, postulate, or property justifies this statement?

______

Go Online You may want to complete the construction activities for this lesson.

Go Online You can complete an Extra Example online.

Name ______ Period ______ Date ______

Practice

Go Online You can complete your homework online.

Example 1

1. **PROOF** Write the correct statements and reasons to complete the two-column proof.

Given: C is the midpoint of $\overline{AE}$.
C is the midpoint of $\overline{BD}$.
$\overline{AE} \cong \overline{BD}$

Prove: $\overline{AC} \cong \overline{CD}$

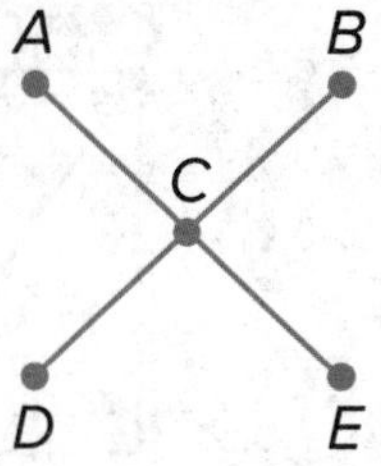

Proof:

Statements	Reasons
1. ___?___ ___?___ ___?___	1. Given
2. $AC = CE$ $BC = CD$	2. ___?___
3. $AE = BD$	3. ___?___
4. ___?___ ___?___	4. Segment Addition Property
5. $AC + CE = BC + CD$	5. ___?___
6. $AC + AC = CD + CD$	6. ___?___
7. ___?___	7. Simplify.
8. ___?___	8. Division Property
9. $\overline{AC} \cong \overline{CD}$	9. ___?___

2. **PROOF** Write the correct statements and reasons to complete the two-column proof.

Given: $\overline{SU} \cong \overline{LR}$
$\overline{TU} \cong \overline{LN}$

Prove: $\overline{ST} \cong \overline{NR}$

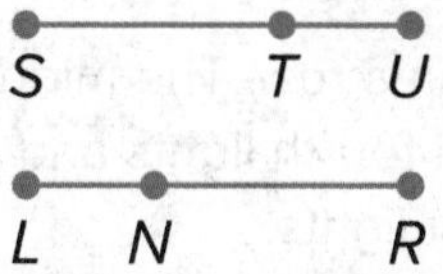

Proof:

Statements	Reasons
1. $\overline{SU} \cong \overline{LR}$, $\overline{TU} \cong \overline{LN}$	1. ___?___
2. ___?___	2. Definition of $\cong$ segments
3. $SU = ST + TU$ $LR = LN + NR$	3. ___?___
4. $ST + TU = LN + NR$	4. ___?___
5. $ST + LN = LN + NR$	5. ___?___
6. $ST + LN - LN = LN + NR - LN$	6. ___?___
7. ___?___	7. Substitution Property
8. $\overline{ST} \cong \overline{NR}$	8. ___?___

Example 2

PROOF Write a two-column proof to prove each geometric relationship.

3. If $\overline{VZ} \cong \overline{VY}$ and $\overline{WY} \cong \overline{XZ}$, then $\overline{VW} \cong \overline{VX}$.

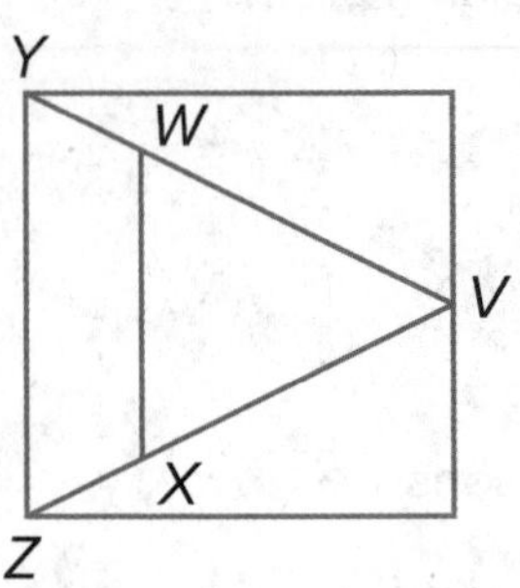

4. If E is the midpoint of $\overline{DF}$ and $\overline{CD} \cong \overline{FG}$, then $\overline{CE} \cong \overline{EG}$.

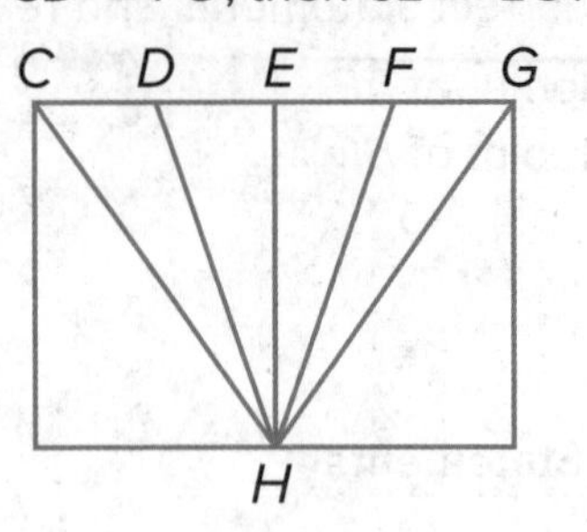

Example 3

5. **FAMILY** Maria is 11 inches shorter than her sister Clara. Luna is 11 inches shorter than her brother Chad. If Maria is shorter than Luna, how do the heights of Clara and Chad compare? What else can be concluded if Maria and Luna are the same height?

6. **LUMBER** Byron works in a lumberyard. His boss just cut a dozen planks and asked Byron to double check that they are all the same length. The planks were numbered 1 through 12. Byron took out plank number 1 and checked that the other planks are all the same length as plank 1. He concluded that they must all be the same length. Explain how you know that plank 7 and plank 10 are the same length even though they were never directly compared to each other.

7. **NEIGHBORHOODS** Karla, Lola, and Mandy live in three houses that are on the same line. Lola lives between Karla and Mandy. Karla and Mandy live a mile apart. Is it possible for Lola's house to be a mile from both Karla's and Mandy's houses?

Mixed Exercises

8. **PROOF** Five lights, A, B, C, D, and E, are aligned in a row. The middle light is the midpoint of the segment between the second and fourth lights and also the midpoint of the segment between the first and last lights.

a. Draw a figure to illustrate the situation.

b. Complete this proof.

Given: C is the midpoint of $\overline{BD}$ and $\overline{AE}$.

Prove: $AB = DE$

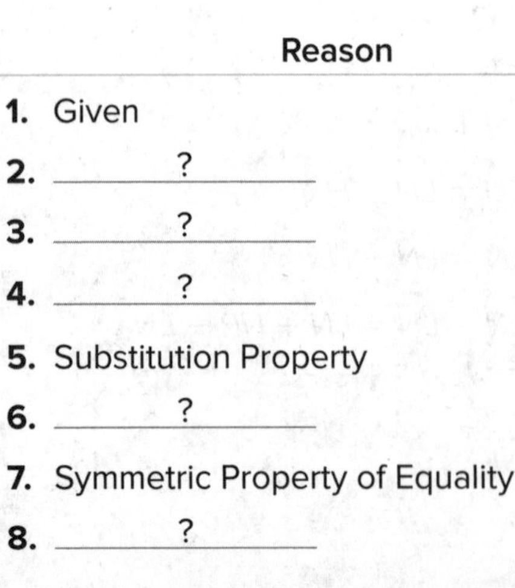

Statement	Reason
1. C is the midpoint of $\overline{BD}$ and $\overline{AE}$.	1. Given
2. $BC = CD$ and ___?___	2. ___?___
3. $AC = AB + BC$, $CE = CD + DE$	3. ___?___
4. $AC - BC = AB$	4. ___?___
5. ___?___	5. Substitution Property
6. $CE - CD = DE$	6. ___?___
7. $AB = CE - CD$	7. Symmetric Property of Equality
8. ___?___	8. ___?___

Name ______________________ Period __________ Date __________

9. **PROOF** $\overline{AC} \cong \overline{GI}$, $\overline{FE} \cong \overline{LK}$, and $AC + CF + FE = GI + IL + LK$. Prove that $\overline{CF} \cong \overline{IL}$.

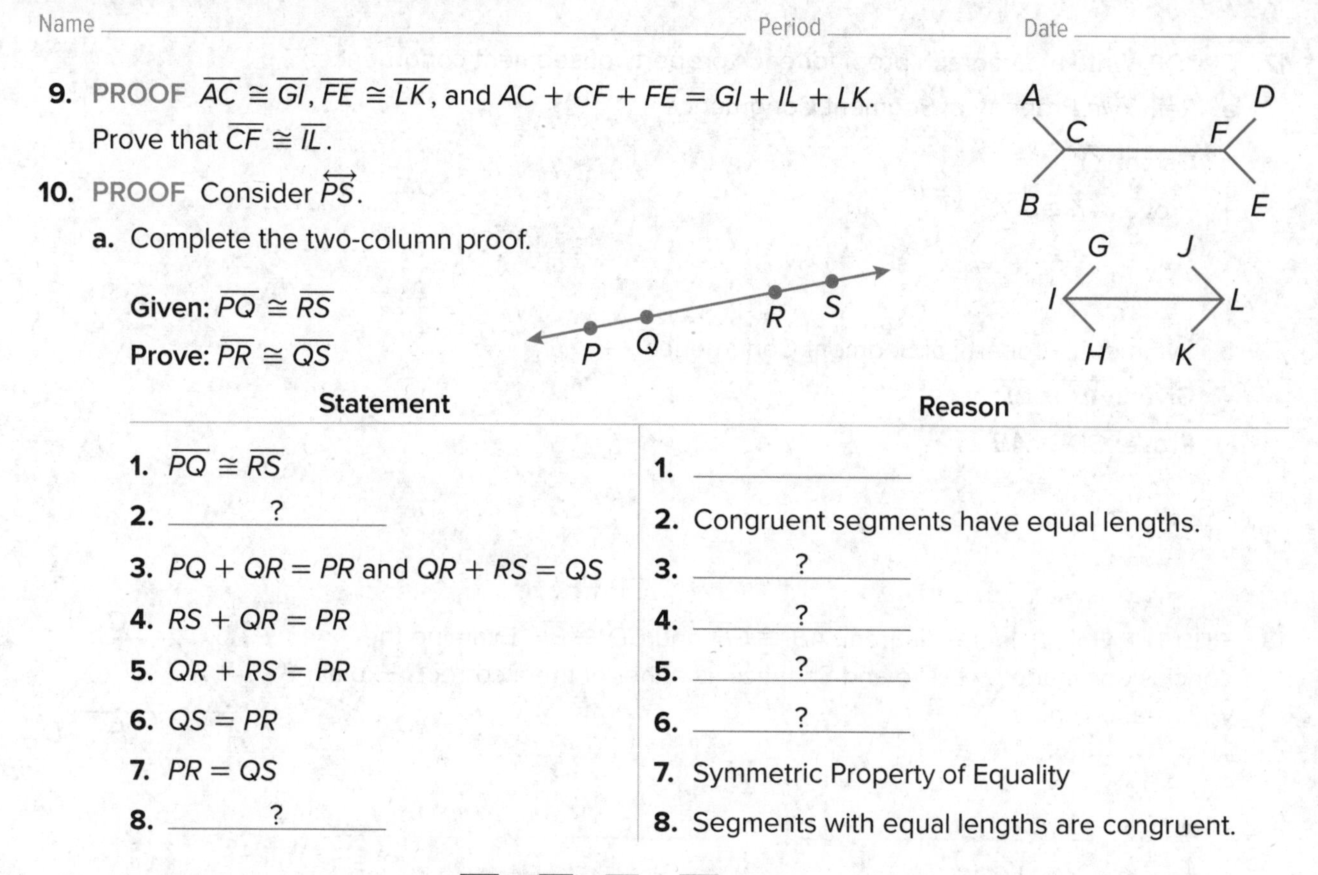

10. **PROOF** Consider $\overleftrightarrow{PS}$.

a. Complete the two-column proof.

Given: $\overline{PQ} \cong \overline{RS}$

Prove: $\overline{PR} \cong \overline{QS}$

Statement	Reason
1. $\overline{PQ} \cong \overline{RS}$	1. ______
2. ___?___	2. Congruent segments have equal lengths.
3. $PQ + QR = PR$ and $QR + RS = QS$	3. ___?___
4. $RS + QR = PR$	4. ___?___
5. $QR + RS = PR$	5. ___?___
6. $QS = PR$	6. ___?___
7. $PR = QS$	7. Symmetric Property of Equality
8. ___?___	8. Segments with equal lengths are congruent.

b. Can it also be proved that $\overline{PQ} \cong \overline{RS}$ if $\overline{PR} \cong \overline{QS}$? Explain.

11. **PROOF** A city planner is designing a new park. The park has two straight paths, $\overline{AB}$ and $\overline{CD}$, which are the same length. A monument, M, is located at the midpoint of both paths.

a. The city planner thinks that the length of $\overline{AM}$ will be the same as the length of $\overline{CM}$. Explain why this makes sense.

b. Complete the two-column proof.

Given: $\overline{AB} \cong \overline{CD}$; M is the midpoint of $\overline{AB}$ and $\overline{CD}$.

Prove: $\overline{AM} \cong \overline{CM}$

Statement	Reason
1. ___?___	1. Given
2. $AB = CD$	2. ___?___
3. $\overline{AM} \cong \overline{MB}$; $\overline{CM} \cong \overline{MD}$	3. ___?___
4. $AM = MB$; $CM = MD$	4. Congruent segments have equal lengths.
5. $AM + MB = AB$; $CM + MD = CD$	5. ___?___
6. $AM + MB = CM + MD$	6. ___?___
7. $AM + AM = CM + CM$	7. Substitution Property of Equality
8. $2AM = 2CM$	8. ___?___
9. ___?___	9. Division Property of Equality
10. ___?___	10. Segments with equal lengths are congruent.

12. PROOF Write a paragraph proof for each property of segment congruence.

a. Reflexive Property of Segment Congruence

Given: $\overline{XY}$

Prove: $\overline{XY} \cong \overline{XY}$

b. Symmetric Property of Segment Congruence

Given: $\overline{AB} \cong \overline{CD}$

Prove: $\overline{CD} \cong \overline{AB}$

13. FIND THE ERROR In the diagram, $\overline{AB} \cong \overline{CD}$ and $\overline{CD} \cong \overline{BF}$. Examine the conclusions made by Leslie and Shantice. Is either of them correct? Explain your reasoning.

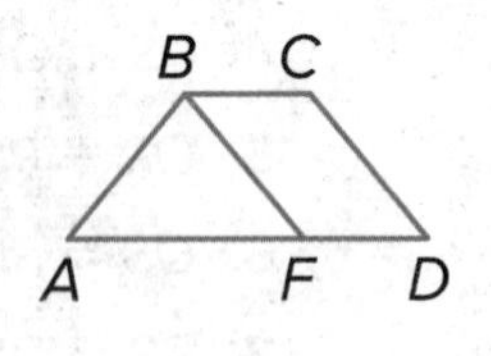

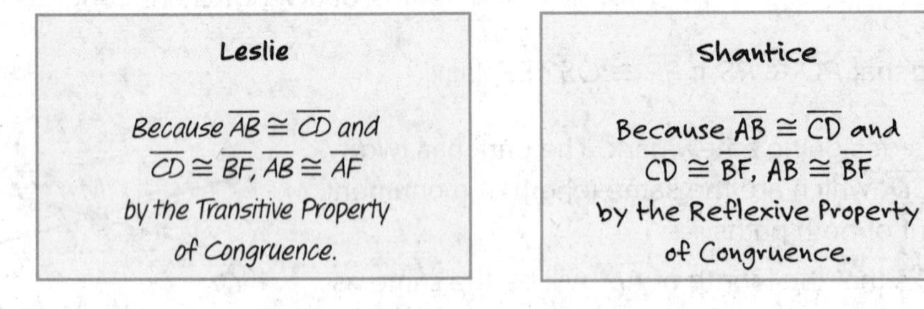

14. PROOF *ABCD* is a square. Prove that $\overline{AC} \cong \overline{BD}$.

15. CREATE Draw a representation of the Segment Addition Postulate in which the segment is two inches long, contains four collinear points, and contains no congruent segments.

16. CREATE Write an example of the Transitive Property and the Substitution Property that illustrates the difference between them.

17. FIND THE ERROR Justin knows that point *R* is the midpoint of $\overline{QS}$, and he knows that this means that $QR = RS$. He says that $PR = PQ + QR$ by the Segment Addition Postulate. So, $PR = PQ + RS$ by substitution. Do you agree with Justin's reasoning? Explain your reasoning.

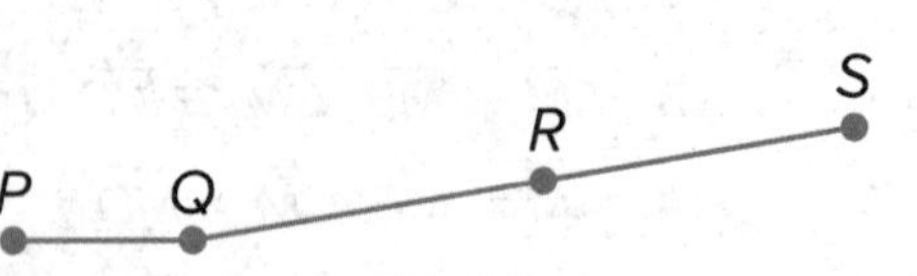

18. WRITE Compare and contrast paragraph proofs and two-column proofs.

19. PROOF Write a paragraph proof to prove that if *P*, *Q*, *R*, and *S* are collinear, $\overline{PQ} \cong \overline{RS}$, and *Q* is the midpoint of $\overline{PR}$, then *R* is the midpoint of $\overline{QS}$.

Proving Angle Relationships

Explore Angle Relationships

Online Activity Use dynamic geometry software to complete the Explore.

INQUIRY How is the complement of a given $\angle A$ related to an angle congruent to $\angle A$?

Today's Goals

- Prove theorems about angles by using the Angle Addition Postulate.
- Prove theorems about angles by using properties and theorems of angle congruence.
- Prove theorems about right angles.

Learn Angle Addition

A protractor is used to measure angles. The Protractor Postulate illustrates the relationship between angle measures and real numbers. You will use these theorems and postulates to find angle measures.

Postulate 3.10: Protractor Postulate

Every angle has a measure that is between 0 and 180.

Postulate 3.11: Angle Addition Postulate

D is in the interior of $\angle ABC$ if and only if $m\angle ABD + m\angle DBC = m\angle ABC$.

Theorem 3.3: Supplement Theorem

If two angles form a linear pair, then they are supplementary angles.

Theorem 3.4: Complement Theorem

If the noncommon sides of two adjacent angles form a right angle, then the angles are complementary angles.

You will prove Theorems 3.3 and 3.4 in Exercises 19–20.

Example 1 Angle Addition Postulate

What is $m\angle 3$ if $m\angle 1 = 23°$ and $m\angle ABC = 131°$?

Choose from the reasons provided to justify each step.

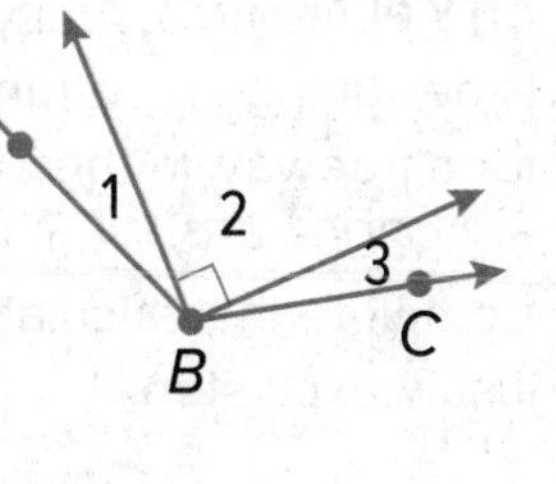

$m\angle 1 + m\angle 2 + m\angle 3 = m\angle ABC$ ____________

$23° + 90° + m\angle 3 = 131°$ ____________

$113° + m\angle 3 = 131°$ ____________

$113° + m\angle 3 - 113° = 131° - 113°$ ____________

$m\angle 3 = 18°$ ____________

Reasons:

Angle Addition Postulate

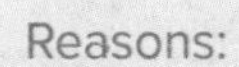

Betweenness of points

Subtraction Property

Substitution Property

Your Notes

Check

What is $m\angle 3$ if $m\angle 2 = 26°$? ______°

What is $m\angle 4$ if $m\angle 5 = (2x°)$ and $m\angle 4 = (x + 9)°$? ______°

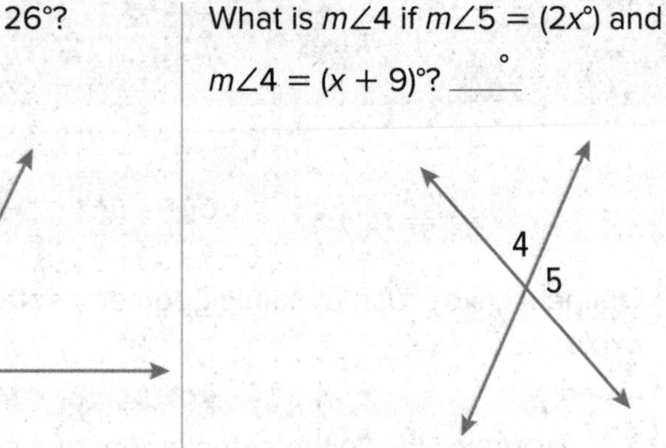

Example 2 Complement and Supplement Theorems

SHELVING **Mae Lin is installing shelves in her room. One of the brackets she chose for her shelves is shown. If $m\angle 3 = 55°$, what is $m\angle 4$?**

Choose from the reasons provided to justify each step.

REASONS:

Complement Theorem

Substitution Property

Subtraction Property

Supplement Theorem

$m\angle 3 + m\angle 4 = 180°$ ______

$55° + m\angle 4 = 180°$ ______

$m\angle 4 = 125°$ ______

Check

CITY PLANNING A city planner is designing an entrance ramp for a freeway. In the diagram, $m\angle ACD = 45°$. What is $m\angle BCA$? Complete the calculations and justify each step.

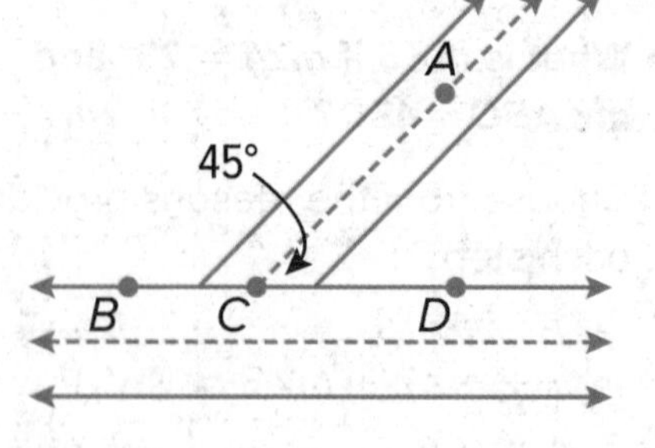

$m\angle BCA + m\angle ACD =$ ______ ______

$m\angle BCA +$ ______ $=$ ______ ______

$m\angle BCA =$ ______ ______

Go Online You can complete an Extra Example online.

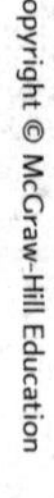

Learn Congruent Angles

The properties of algebra that apply to the congruence of segments and the equality of their measures also hold true for the congruence of angles and the equality of their measures.

Theorem 3.5: Properties of Angle Congruence
Reflexive Property of Congruence $\angle 1 \cong \angle 1$
Symmetric Property of Congruence If $\angle 1 \cong \angle 2$, then $\angle 2 \cong \angle 1$.
Transitive Property of Congruence If $\angle 1 \cong \angle 2$ and $\angle 2 \cong \angle 3$, then $\angle 1 \cong \angle 3$.

> **Go Online** Proofs of the Reflexive Property of Congruence and the Transitive Property of Congruence are available.

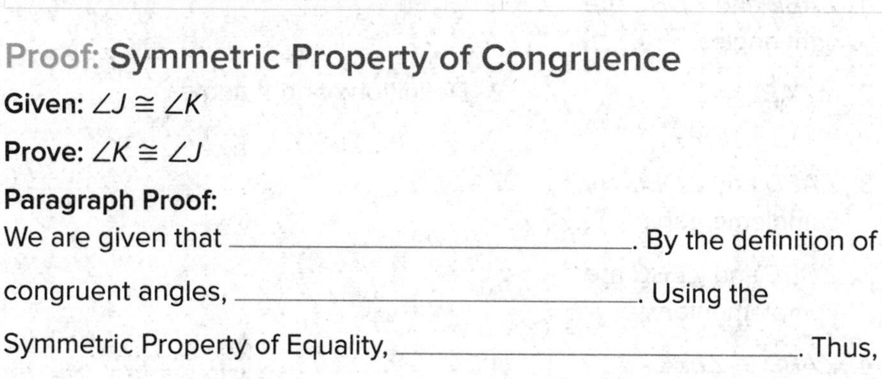

Proof: Symmetric Property of Congruence

Given: $\angle J \cong \angle K$

Prove: $\angle K \cong \angle J$

Paragraph Proof:

We are given that ________________. By the definition of congruent angles, ________________. Using the Symmetric Property of Equality, ________________. Thus, ________________ by the definition of congruent angles.

> **Talk About It!** Explain the difference between the Complement Theorem and the Congruent Complements Theorem.

Theorems	
Theorem 3.6: Congruent Supplements Theorem Angles supplementary to the same angle or to congruent angles are congruent. **Abbreviation** $\angle$s suppl. to same $\angle$ or $\cong$ $\angle$s are $\cong$.	If $m\angle 1 + m\angle 2 = 180°$ and $m\angle 2 + m\angle 3 = 180°$, then $\angle 1 \cong \angle 3$.
Theorem 3.7: Congruent Complements Theorem Angles complementary to the same angle or to congruent angles are congruent. **Abbreviation** $\angle$s compl. to same $\angle$ or $\cong$ $\angle$s are $\cong$.	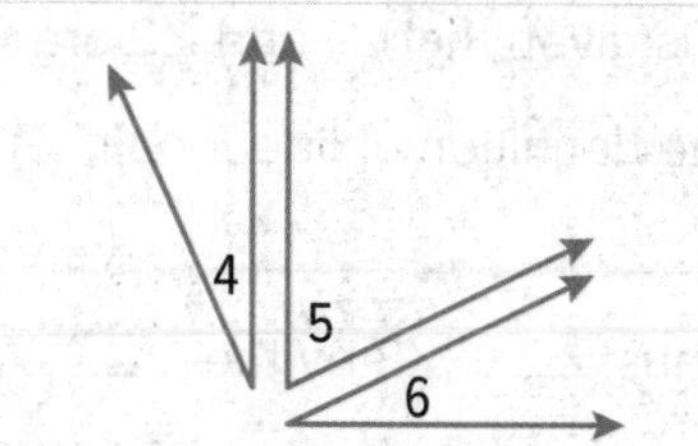 If $m\angle 4 + m\angle 5 = 90°$ and $m\angle 5 + m\angle 6 = 90°$, then $\angle 4 \cong \angle 6$.
Theorem 3.8: Vertical Angles Theorem If two angles are vertical angles, then they are congruent.	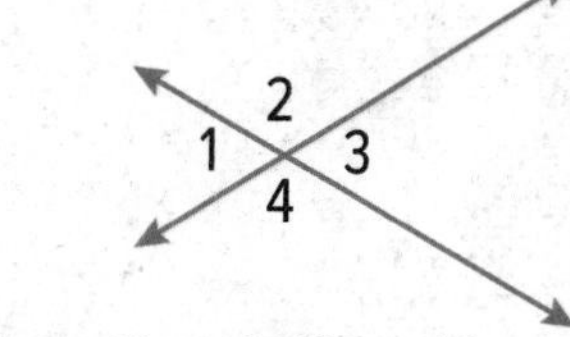 $\angle 1 \cong \angle 3$ and $\angle 2 \cong \angle 4$

You will prove one case of Theorems 3.6 and 3.7 in Exercises 21-22. You will prove the second case of each theorem in Exercise 31.

Example 3 Congruent Supplements and Complements

In the figure, $\angle ABE$ and $\angle DBC$ are right angles.

Select from the reasons provided to complete the proof.

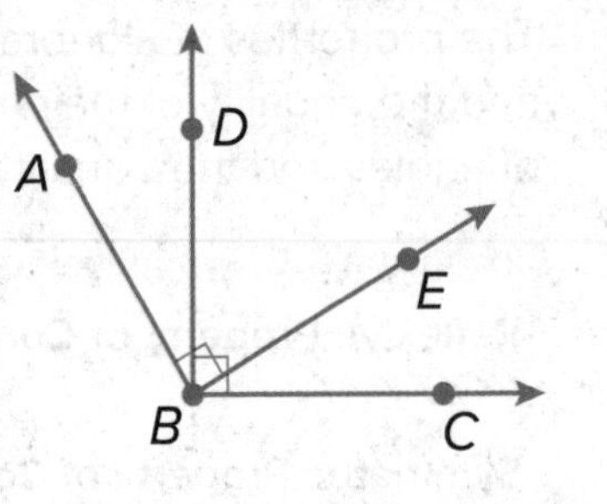

Given: $\angle ABE$ and $\angle DBC$ are right angles.

Prove: $\angle ABD \cong \angle EBC$

REASONS:

Complement Theorem

Congruent Complements Theorem

Definition of complementary angles

Definition of congruence

Proof:

Statements	Reasons
1. $\angle ABE$ and $\angle DBC$ are right angles.	1. Given
2. $m\angle ABE = 90°$; $m\angle DBC = 90°$	2. Definition of right angle
3. $\angle ABD$ and $\angle DBE$ are complementary; $\angle DBE$ and $\angle EBC$ are complementary.	3. ________
4. $\angle ABD \cong \angle EBC$	4. ________

Check

Complete the proof.

Given: $\angle 1$ and $\angle 3$ are supplementary.

Prove: $\angle 2 \cong \angle 3$

Proof:

It is given that $\angle 1$ and $\angle 3$ are ________. By the definition of linear pair, $\angle 1$ and $\angle 2$ are a linear pair. So, by the ________, $\angle 1$ and $\angle$__ are supplementary.

Thus, $\angle 2 \cong \angle 3$ by the ________.

Go Online You can complete an Extra Example online.

Example 4 Vertical Angles

Complete the proof.

Choose from the statements and reasons provided.

Given: $\angle 1 \cong \angle 2$

Prove: $\angle 3 \cong \angle 4$

Proof:

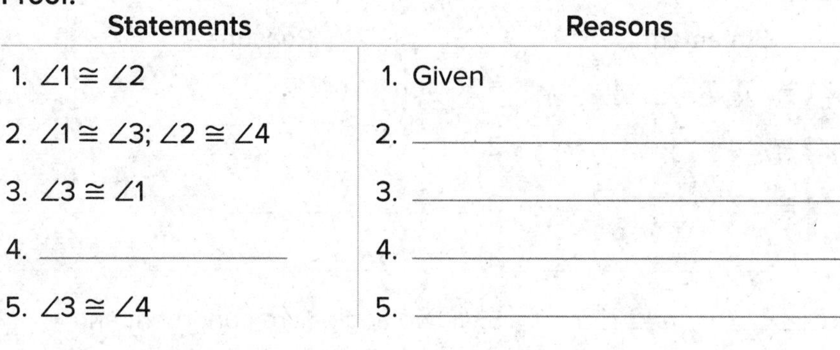

Statements	Reasons
1. $\angle 1 \cong \angle 2$	1. Given
2. $\angle 1 \cong \angle 3$; $\angle 2 \cong \angle 4$	2. ________
3. $\angle 3 \cong \angle 1$	3. ________
4. ________	4. ________
5. $\angle 3 \cong \angle 4$	5. ________

STATEMENTS/REASONS:

- Symmetric Property of Congruence
- Transitive Property of Congruence
- Vertical Angles Theorem
- $\angle 3 \cong \angle 2$
- $\angle 4 \cong \angle 2$
- $\angle 4 \cong \angle 3$

Check

Complete the proof. Choose from the reasons provided.

Given: $\angle 4 \cong \angle 7$

Prove: $\angle 5 \cong \angle 6$

Proof:

Statements	Reasons
1. $\angle 4 \cong \angle 7$	1. Given
2. $\angle 5 \cong \angle 4$ and $\angle 7 \cong \angle 6$	2. ________
3. $\angle 5 \cong \angle 7$	3. ________
4. $\angle 5 \cong \angle 6$	4. ________

REASONS:

- Vertical Angles Theorem
- Definition of vertical angles
- Transitive Property of Congruence
- Symmetric Property of Congruence
- Supplement Theorem
- Definition of linear pair

Go Online You can complete an Extra Example online.

Learn Right Angle Theorems

You can prove the following theorems about right angles using what you already know about angle measures.

Theorem 3.9	Perpendicular lines intersect to form four right angles.
Theorem 3.10	All right angles are congruent.
Theorem 3.11	Perpendicular lines form congruent adjacent angles.
Theorem 3.12	If two angles are congruent and supplementary, then each angle is a right angle.
Theorem 3.13	If two congruent angles form a linear pair, then they are right angles.

You will prove Theorem 3.9 and Theorems 3.11 through 3.13 in Exercises 23–26.

Go Online
A proof of Theorem 3.10 is available.

STATEMENTS/REASONS:

- Reflexive Property of Congruence
- Symmetric Property of Congruence
- Transitive Property of Congruence
- Vertical Angles Theorem
- ∠1 and ∠2 are right angles.
- ∠1 ≅ ∠2

REASONS:

- Perpendicular lines intersect to form four right angles.
- All right angles are congruent.
- Perpendicular lines form congruent adjacent angles.
- If two angles are congruent and supplementary, then each angle is a right angle.
- If two congruent angles form a linear pair, then they are right angles.

Example 5 Right Angle Theorems in Proofs

Complete the proof.

Choose from the statements and reasons provided.

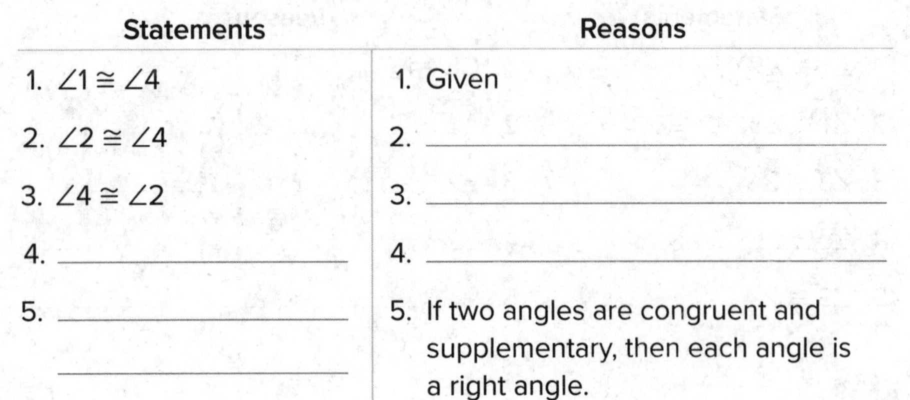

Given: ∠1 ≅ ∠4

Prove: ∠1 and ∠2 are right angles.

Proof:

Statements	Reasons
1. ∠1 ≅ ∠4	1. Given
2. ∠2 ≅ ∠4	2. ________
3. ∠4 ≅ ∠2	3. ________
4. ________	4. ________
5. ________ ________	5. If two angles are congruent and supplementary, then each angle is a right angle.

Check

Complete the proof. Choose from the reasons provided.

Given: ∠1 ≅ ∠4
Lines j and k are perpendicular.

Prove: ∠2 ≅ ∠4

Proof:

Statements	Reasons
1. Lines j and k are perpendicular.	1. Given
2. ∠2 ≅ ∠1	2. ________
3. ∠1 ≅ ∠4	3. Given
4. ∠2 ≅ ∠4	4. Transitive Property of Congruence

Go Online You can complete an Extra Example online.

Name ______________________ Period ____________ Date ____________

Practice

Go Online You can complete your homework online.

Example 1

Find the measure of each angle.

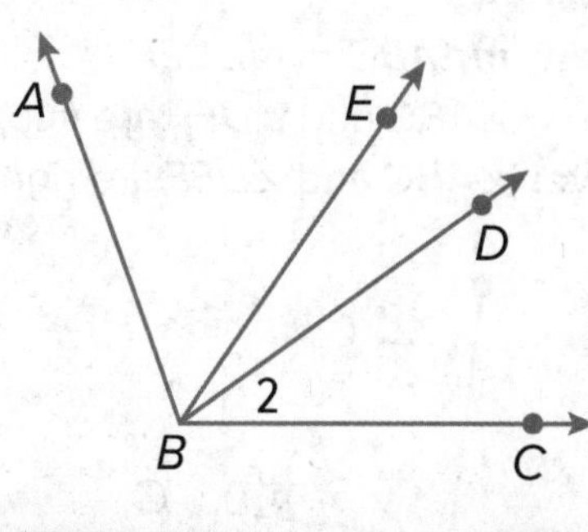

1. Find $m\angle ABC$ if $m\angle ABD = 70°$ and $m\angle DBC = 43°$.
2. If $m\angle EBC = 55°$ and $m\angle EBD = 20°$, find $m\angle 2$.
3. Find $m\angle ABD$ if $m\angle ABC = 110°$ and $m\angle 2 = 36°$.

Example 2

4. FLAGS The Alabama state flag is white and has two diagonal red stripes. If the $m\angle 1 = 112°$, what is $m\angle 2$?

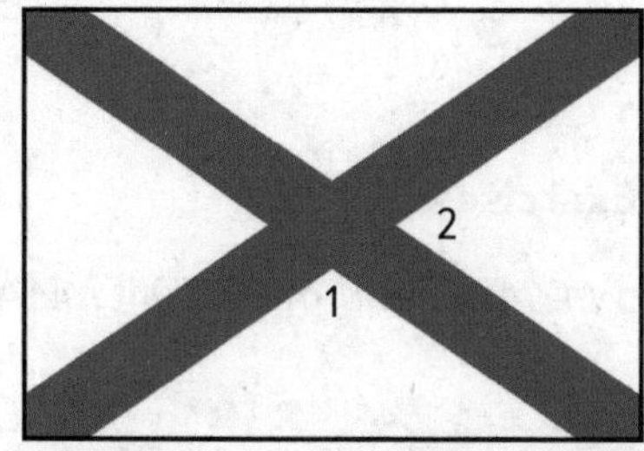

5. CONSTRUCTION Alan has installed a new window above the entrance of an office building. If $m\angle 2 = 44°$, what is $m\angle 1$?

Example 3

PROOF **Write a two-column proof.**

6. **Given:** $\angle 2 \cong \angle 4$
   Prove: $\angle 1 \cong \angle 3$

7. **Given:** $\angle 1 \cong \angle 3$
   Prove: $\angle 2 \cong \angle 4$

Example 4

PROOF **Write a two-column proof.**

8. **Given:** $\angle 5 \cong \angle 7$
   Prove: $\angle 5 \cong \angle 8$

Example 5

PROOF **Write a two-column proof.**

9. **Given:** $m\angle ABC = m\angle DEF$

$\angle ABC$ and $\angle DEF$ are supplementary.

Prove: $\angle ABC$ and $\angle DEF$ are right angles.

10. **Given:** $\angle 1 \cong \angle 2$; $m \perp p$

Prove: $\angle 2 \cong \angle 3$

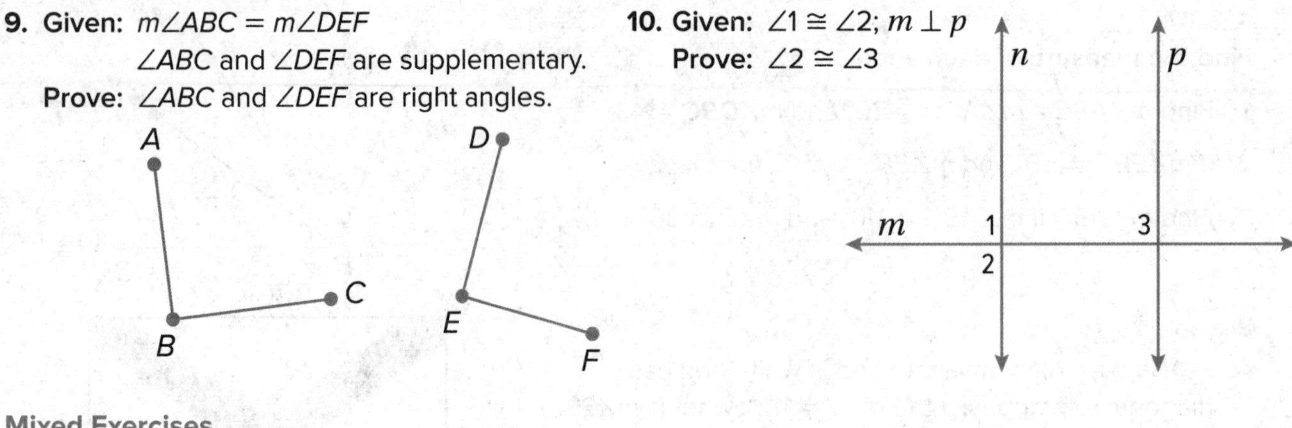

Mixed Exercises

11. Find $m\angle ABC$ and $m\angle CBD$ if $m\angle ABD = 120°$.

12. Find $m\angle JKL$ and $m\angle LKM$ if $m\angle JKM = 140°$.

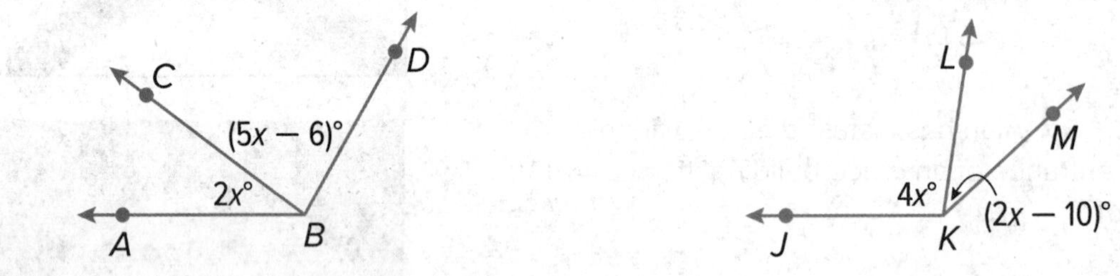

Find the measure of each numbered angle and name the theorems that you used to justify your work.

13. $m\angle 6 = (2x - 21)°$
$m\angle 7 = (3x - 34)°$

14. $m\angle 5 = m\angle 6$

15. $\angle 2$ and $\angle 3$ are complementary.
$\angle 1 \cong \angle 4$ and $m\angle 2 = 28°$.

16. $\angle 2$ and $\angle 4$ and $\angle 4$ and $\angle 5$ are supplementary.
$m\angle 4 = 105°$.

17. $m\angle 9 = (3x + 12)°$
$m\angle 10 = (x - 24)°$

18. $m\angle 3 = (2x + 23)°$
$m\angle 4 = (5x - 112)°$

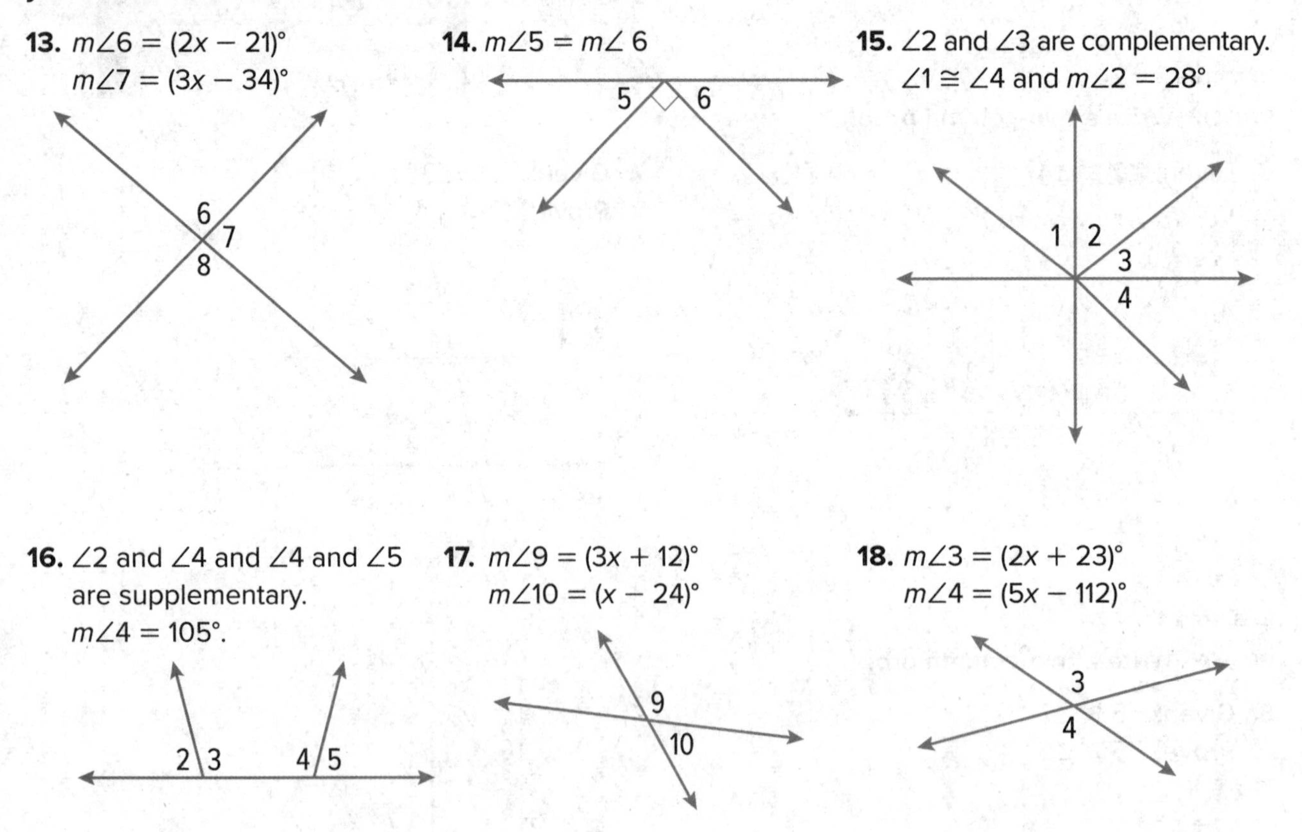

Name _______________________________ Period _____________ Date _____________

PROOF **Write a two-column proof for each theorem.**

19. Supplement Theorem

Given: $\angle PQT$ and $\angle TQR$ form a linear pair.
Prove: $\angle PQT$ and $\angle TQR$ are supplementary.

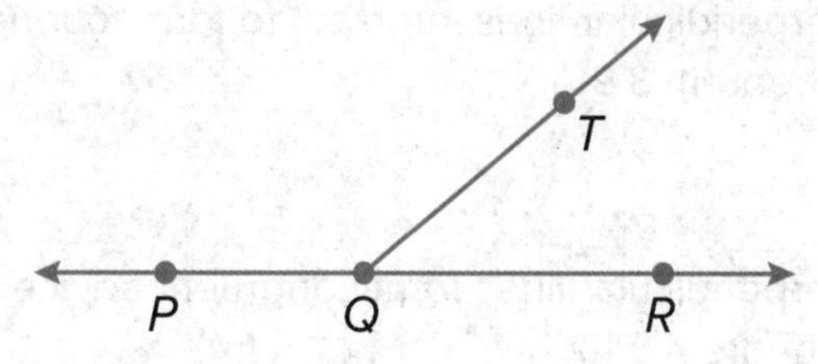

20. Complement Theorem

Given: $\angle ABC$ is a right angle.
Prove: $\angle ABD$ and $\angle CBD$ are complementary.

21. Congruent Supplements Theorem (Case 1)

Given: $\angle 1$ and $\angle 2$ are supplementary.
$\angle 2$ and $\angle 3$ are supplementary.
Prove: $\angle 1 \cong \angle 3$

22. Congruent Complements Theorem (Case 1)

Given: $\angle 4$ and $\angle 5$ are complementary.
$\angle 5$ and $\angle 6$ are complementary.
Prove: $\angle 4 \cong \angle 6$

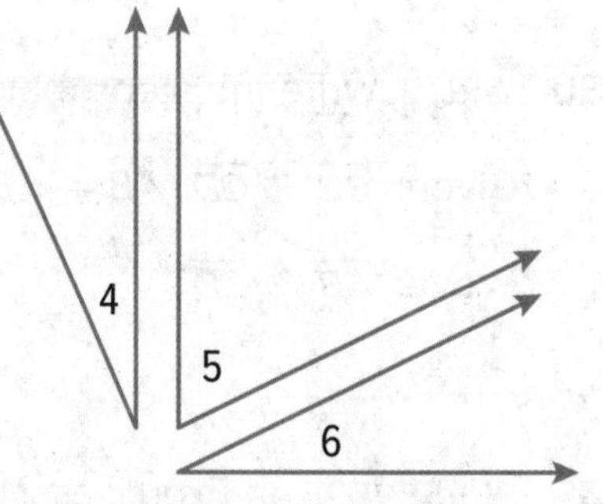

PROOF Use the figure to write a proof of each theorem.

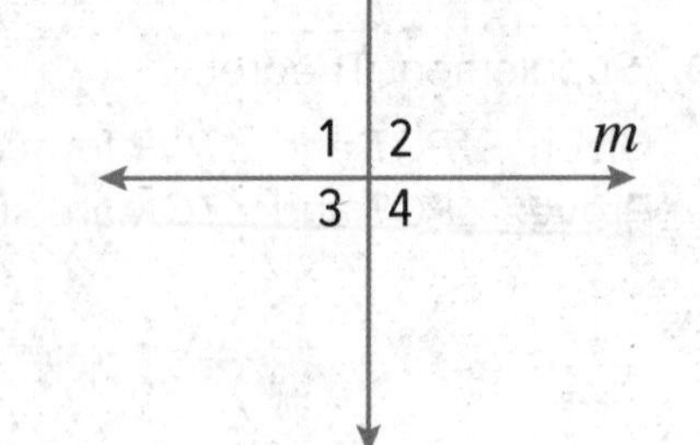

23. Perpendicular lines intersect to form four right angles. (Theorem 3.9)

24. Perpendicular lines form congruent adjacent angles. (Theorem 3.11)

25. If two angles are congruent and supplementary, then each angle is a right angle. (Theorem 3.12)

26. If two congruent angles form a linear pair, then they are right angles. (Theorem 3.13)

27. CONSTRUCT ARGUMENTS For a school project, students are making a giant icosahedron, which is a large solid with twenty identical triangular faces. John is in charge of quality control. He must make sure that the measures of all the angles in all the triangles are the same. He does this by using a precut template and comparing the corner angles of every triangle to the template. How does this assure that the angles in all the triangles will be congruent to each other?

28. ANALYZE Find $m\angle C$ if $\angle C \cong \angle A$, $m\angle A = 3x°$, $m\angle B = (x + 20)°$, and $\angle A$ and $\angle B$ are supplementary. Justify your argument.

29. CREATE Draw $\angle WXZ$ such that $m\angle WXZ = 45°$. Construct $\angle YXZ \cong \angle WXZ$. Make a conjecture about the measure of $\angle WXY$, and then prove your conjecture.

30. WRITE Write the steps that you would use to complete the proof.

Given: $\overline{BC} \cong \overline{CD}$, $AB = \frac{1}{2}BD$

Prove: $\overline{AB} \cong \overline{CD}$

A B C D

31. PERSEVERE In Exercises 21 and 22, you proved one case of the Congruent Supplements Theorem and one case of the Congruent Complements Theorem. Explain why there is another case for each of these theorems. Then write a proof of this second case for each theorem.

Parallel Lines and Transversals

Learn Parallel Lines and Transversals

If two lines do not intersect, then they are either parallel or skew.

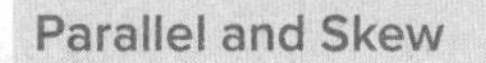

Parallel and Skew

Parallel Lines

Parallel lines are coplanar lines that do not intersect.

Example $\overleftrightarrow{JK} \parallel \overleftrightarrow{LM}$

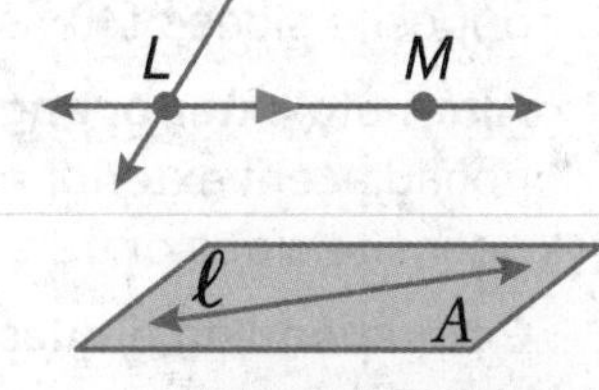

Skew Lines

Skew lines are lines that do not intersect and are not coplanar.

Example Lines ℓ and m are skew.

Parallel Planes

Parallel planes are planes that do not intersect.

Example Planes A and B are parallel.

If segments or rays are contained within lines that are parallel or skew, then the segments or rays are parallel or skew.

Example 1 Identify Parallel and Skew Relationships

Identify each of the following using the cube shown. Assume lines and planes that appear to be parallel or perpendicular are parallel or perpendicular, respectively.

a. all lines skew to $\overleftrightarrow{BC}$

$\overleftrightarrow{AF}$, $\overleftrightarrow{DE}$, $\overleftrightarrow{FG}$, and ______

b. all lines parallel to $\overleftrightarrow{EH}$

$\overleftrightarrow{AB}$, $\overleftrightarrow{CD}$, or ______

c. all planes parallel to plane *DCH*

Plane ______ is the only plane parallel to plane *DCH*.

A line that intersects two or more lines in a plane at different points is called a **transversal**. In the diagram, line t is a transversal of lines q and r. Notice that line t forms a total of eight angles with lines q and r. These angles and specific pairings of these angles are given special names.

Today's Goals

- Identify special angle pairs, parallel and skew lines, and transversals.
- Find values by applying theorems about parallel lines and transversals.

Today's Vocabulary

parallel lines
skew lines
parallel planes
transversal
interior angles
exterior angles
consecutive interior angles
alternate interior angles
alternate exterior angles
corresponding angles

Study Tip

Parallel Lines The statement $\overleftrightarrow{JK} \parallel \overleftrightarrow{LM}$ is read as *line JK is parallel to line LM*. In a figure, arrowheads are used to indicate that lines are parallel.

Talk About It!

Can a two-dimensional figure contain skew lines? Justify your argument.

Your Notes

Study Tip

Same-Side Interior Angles Consecutive interior angles are also called *same-side interior angles*.

Transversal Angle Pair Relationships	
Four **interior angles** lie in the region between lines q and r.	∠3, ∠4, ∠5, ∠6
Four **exterior angles** lie in the two regions that are not between lines q and r.	∠1, ∠2, ∠7, ∠8
Consecutive interior angles are interior angles that lie on the same side of transversal t.	∠4 and ∠5, ∠3 and ∠6
Alternate interior angles are nonadjacent interior angles that lie on opposite sides of transversal t.	∠3 and ∠5, ∠4 and ∠6
Alternate exterior angles are nonadjacent exterior angles that lie on opposite sides of transversal t.	∠1 and ∠7, ∠2 and ∠8
Corresponding angles lie on the same side of transversal t and on the same side of lines q and r.	∠1 and ∠5, ∠2 and ∠6, ∠3 and ∠7, ∠4 and ∠8

Example 2 Classify Angle Pair Relationships

Classify the relationship between each pair of angles as *alternate interior, alternate exterior, corresponding,* or *consecutive interior angles*.

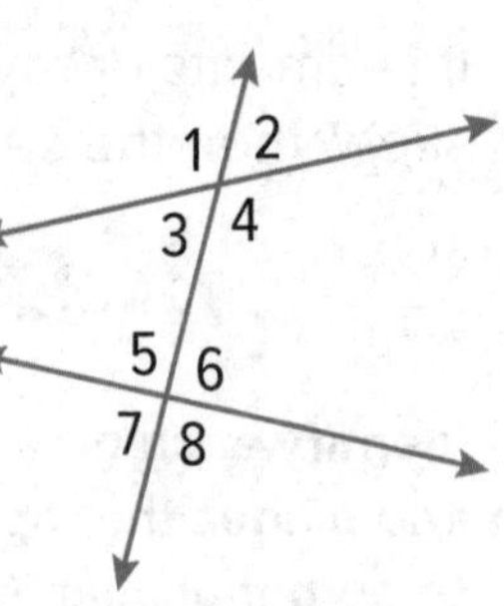

∠4 and ∠5 are

____

∠3 and ∠7 are

____

∠3 and ∠5 are

____

∠1 and ∠8 are

____

Example 3 Identify Transversals and Classify Angle Pairs

Identify the transversal connecting each pair of angles in the photo. Then classify the relationship between each pair of angles.

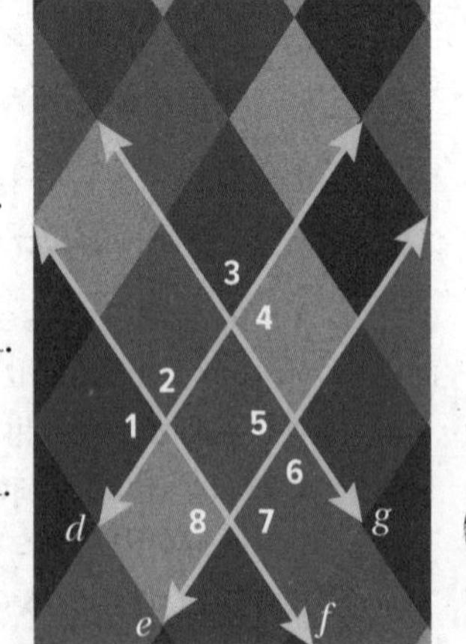

The transversal connecting ∠1 and ∠8 is line ____.

These are ____________ angles.

The transversal connecting ∠3 and ∠6 is line ____.

These are ______________ angles.

The transversal connecting ∠6 and ∠7 is line ____.

These are ________________ angles.

Go Online You can complete an Extra Example online.

Explore Relationships Between Angles and Parallel Lines

Online Activity Use dynamic geometry software to complete the Explore.

INQUIRY How do parallel lines affect the relationships between special angle pairs?

Learn Angles and Parallel Lines

If two lines are parallel and cut by a transversal, then there are special relationships in the angle pairs formed by the lines.

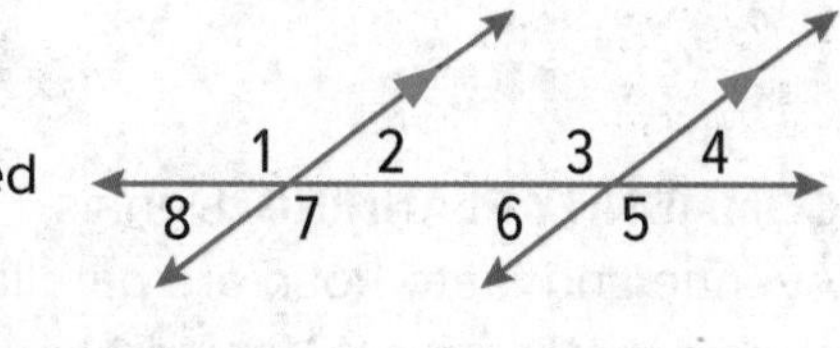

Theorem 3.14: Corresponding Angles Theorem	
If two parallel lines are cut by a transversal, then each pair of corresponding angles is congruent.	$\angle 1 \cong \angle 3$, $\angle 2 \cong \angle 4$, $\angle 5 \cong \angle 7$, $\angle 6 \cong \angle 8$
Theorem 3.15: Alternate Interior Angles Theorem	
If two parallel lines are cut by a transversal, then each pair of alternate interior angles is congruent.	$\angle 2 \cong \angle 6$, $\angle 3 \cong \angle 7$
Theorem 3.16: Consecutive Interior Angles Theorem	
If two parallel lines are cut by a transversal, then each pair of consecutive interior angles is supplementary.	$\angle 2$ and $\angle 3$, $\angle 6$ and $\angle 7$
Theorem 3.17: Alternate Exterior Angles Theorem	
If two parallel lines are cut by a transversal, then each pair of alternate exterior angles is congruent.	$\angle 1 \cong \angle 5$, $\angle 4 \cong \angle 8$

You will prove Theorems 3.16 and 3.17 in Exercises 47 and 48.

A special relationship also exists when the transversal of two parallel lines is a perpendicular line.

Theorem 3.18: Perpendicular Transversal Theorem
In a plane, if a line is perpendicular to one of two parallel lines, then it is perpendicular to the other.

You will prove Theorem 3.18 in Exercise 48.

Study Tip

Angle Relationships Theorems 3.15–3.17 generalize the relationships between specific pairs of angles. If you get confused about the relationships, you can verify them using only corresponding angles, vertical angles, and linear pairs.

Go Online Proofs of Theorems 3.14 and 3.15 are available.

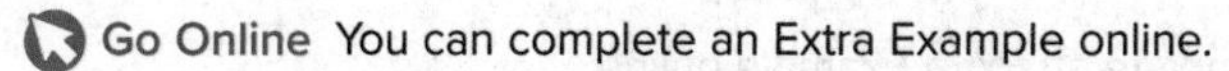
Go Online You can complete an Extra Example online.

Example 4 Use Theorems About Parallel Lines

Go Online An alternate method is available for this example.

RAILROADS **Crossties i and k are parallel. Both crossties are intersected by crosstie h. If $m\angle 1 = 42°$, find $m\angle 7$.**

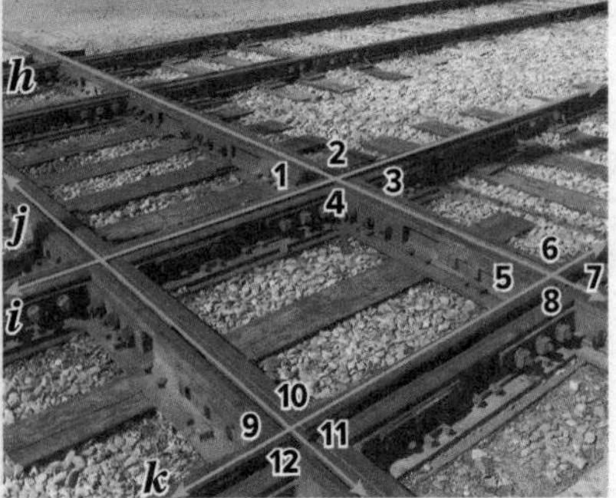

$\angle 7 \cong \angle 1$	Alternate Exterior Angles Theorem
________	Definition of congruent angles
$m\angle 7 =$ ___°	Substitution

The measure of $\angle 7$ is 42°.

Check

COMMUNITY PLANNING Dennis Avenue and State Road are parallel streets that intersect Newport Lane along the south side of Oak Creek Park. If $m\angle 3 = 62°$, find $m\angle 4$.

___°

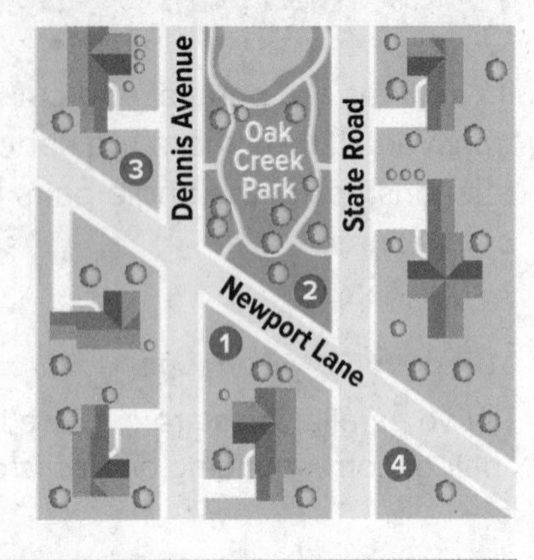

Example 5 Find Values of Variables

Study Tip

Precision Theorems 3.14–3.17 only apply to *parallel* lines cut by a transversal. You should assume that lines are parallel only if the information is given or the lines are marked with parallel arrows.

Use the figure to find the value of the indicated variable. Justify your reasoning.

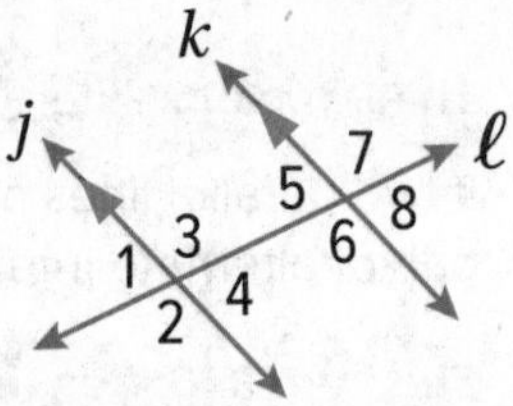

a. If $m\angle 3 = (4x + 7)°$ and $m\angle 6 = (5x - 13)°$, find x.

$\angle 3 \cong \angle 6$	________ Angles Theorem
$m\angle 3 = m\angle 6$	Definition of ________ angles
$4x + 7 = 5x - 13$	Substitution
$x = 20$	Simplify.

b. Find y if $m\angle 8 = 68°$ and $m\angle 3 = (3y - 2)°$.

$\angle 5 \cong \angle 8$	________ Angles Theorem
$m\angle 5 = m\angle$___	Definition of ________ angles
$m\angle 5 =$ ___°	Substitution

Because lines j and k are parallel, $\angle 5$ and $\angle 3$ are supplementary by the ____________ Angles Theorem.

$m\angle 3 + m\angle 5 = 180°$	Definition of supplementary angles
______ + ___ = 180	Substitution
$3y + 66 = 180$	Simplify.
$y =$ ___	Simplify.

Go Online You can complete an Extra Example online.

Name ______________________ Period __________ Date __________

Practice

Go Online You can complete your homework online.

Example 1

Identify each of the following using the figure shown. Assume lines and planes that appear to be parallel or perpendicular are parallel or perpendicular, respectively.

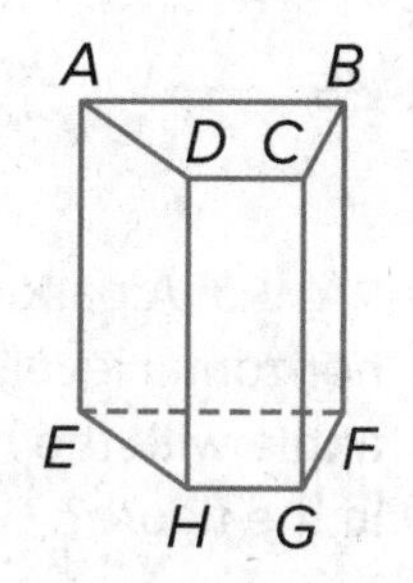

1. three segments parallel to $\overline{AE}$

2. a segment skew to $\overline{AB}$

3. a pair of parallel planes

4. a segment parallel to $\overline{AD}$

5. three segments parallel to $\overline{HG}$

6. five segments skew to $\overline{BC}$

7. How could you characterize the relationship between faces *ABCD* and *DCGH*? Explain.

Examples 2 and 3

Identify the transversal connecting each pair of angles. Then classify the relationship between each pair of angles as *alternate interior*, *alternate exterior*, *corresponding*, or *consecutive interior* angles.

8. $\angle 4$ and $\angle 5$

9. $\angle 5$ and $\angle 15$

10. $\angle 12$ and $\angle 14$

11. $\angle 7$ and $\angle 15$

12. $\angle 2$ and $\angle 12$

13. $\angle 3$ and $\angle 6$

14. $\angle 1$ and $\angle 9$

15. $\angle 3$ and $\angle 9$

16. $\angle 10$ and $\angle 16$

17. $\angle 5$ and $\angle 13$

For Exercises 18 and 19, use the figure.

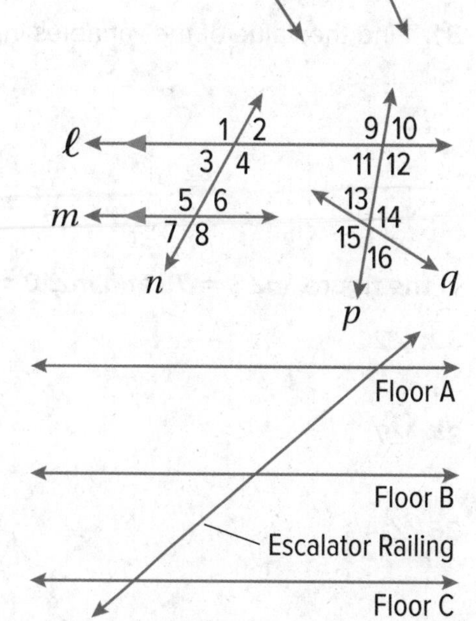

18. What type of angles are $\angle 3$ and $\angle 10$?

19. State the transversal that connects $\angle 11$ and $\angle 13$.

20. ESCALATORS An escalator at a shopping mall runs up several levels. The escalator railing can be modeled by a straight line running past horizontal lines that represent the floors. Describe the relationships of these lines.

Example 4

In the figure, $m\angle 7 = 100°$. Find the measure of each angle.

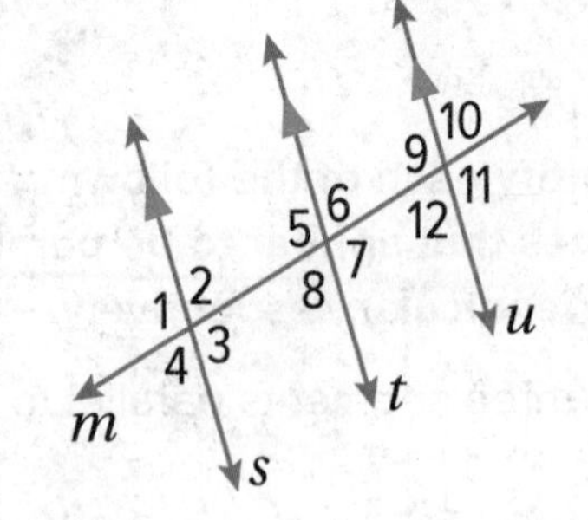

21. $\angle 9$

22. $\angle 6$

23. $\angle 8$

24. $\angle 2$

25. $\angle 5$

26. $\angle 11$

27. RAMPS A parking garage ramp rises to connect two horizontal levels of a parking lot. The ramp makes a 10° angle with the horizontal. What is the measure of angle 1 in the figure?

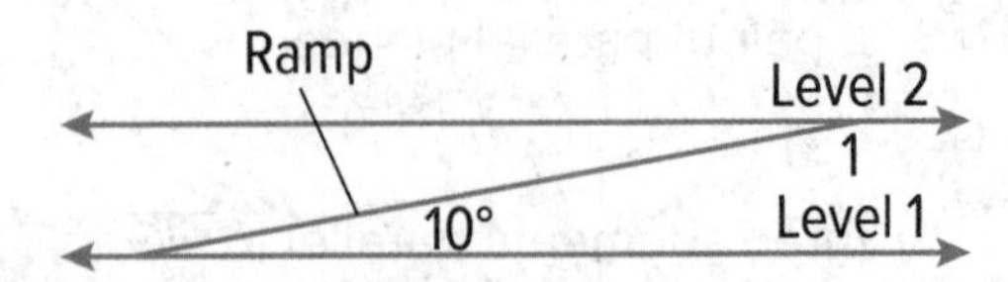

28. CITY ENGINEERING Seventh Avenue runs perpendicular to 1st and 2nd Streets, which are parallel. However, Maple Avenue makes a 115° angle with 2nd Street. What is the measure of angle 1?

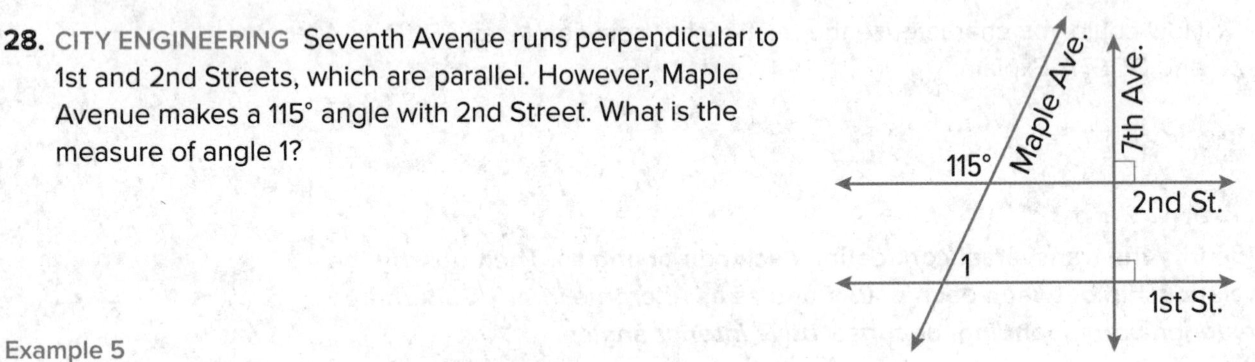

Example 5

Find the value of the variables in each figure. Explain your reasoning.

29.

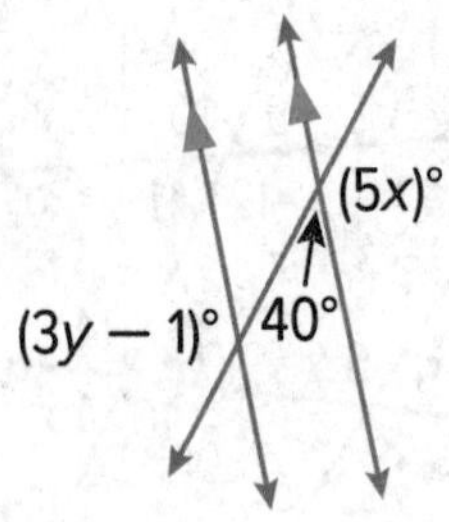

30.

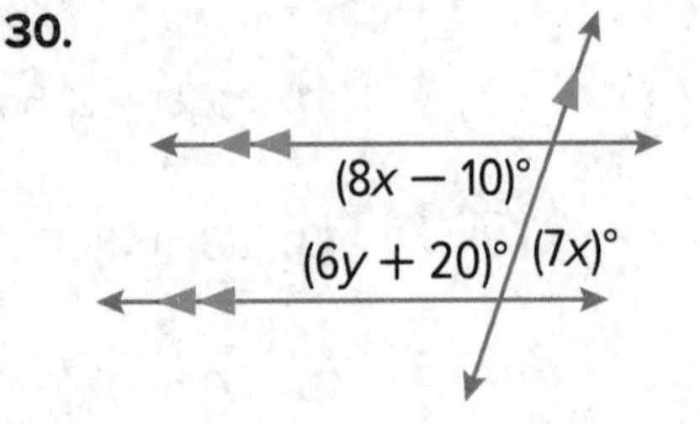

31. Find the value of the variables in the figure.

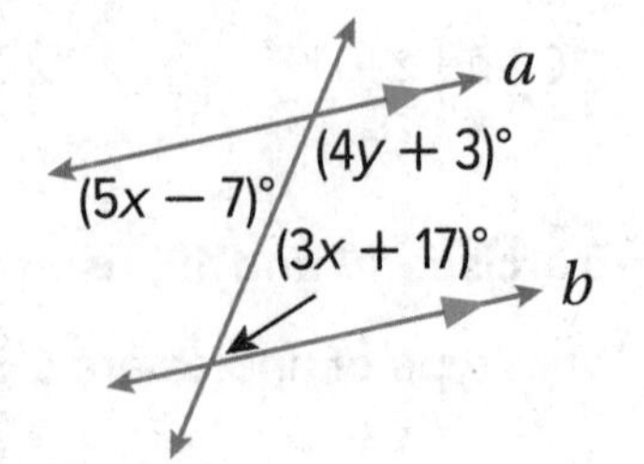

Mixed Exercises

In the figure, $m\angle 3 = 75$ and $m\angle 10 = 105°$. Find the measure of each angle.

32. $\angle 2$

33. $\angle 5$

34. $\angle 7$

35. $\angle 15$

36. $\angle 14$

37. $\angle 9$

Name ______________________ Period __________ Date __________

USE A MODEL **Lines a and b are parallel and are cut by transversal t to form interior angles $\angle 7$, $\angle 8$, $\angle 9$, and $\angle 10$. $\angle 7$ and $\angle 8$ are consecutive interior angles, and $m\angle 7 = 94°$. $\angle 8$ and $\angle 10$ are alternate interior angles. Find the measure of each angle.**

38. $\angle 10$

39. $\angle 9$

40. $\angle 8$

41. CARPENTRY A carpenter is building a podium. The side panel of the podium is cut from a rectangular piece of wood. The rectangle must be sawed along the dashed line in the figure. What is the measure of $\angle 1$? Explain your reasoning.

1

116°

42. MAPPING Use the figure.

a. Connor lives at the angle that forms an alternate interior angle with Georgia's residence. Label the location of Connor's home on the map.

b. Quincy lives at the angle that forms a consecutive interior angle with Connor's residence. Label the location of Quincy's home on the map.

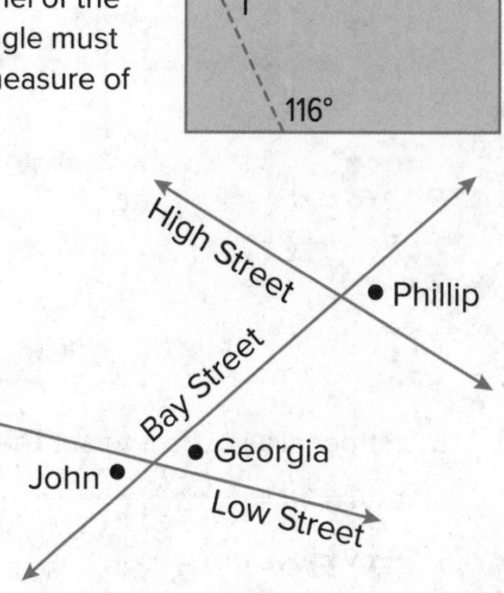

43. USE A SOURCE Research the flag for the Solomon Islands. Sketch the flag. Label angles formed by the yellow stripe, or transversal. Describe the relationship between the angles you labeled on the flag.

44. PRECISION Find the values of x and y in the trapezoid. Justify your answer.

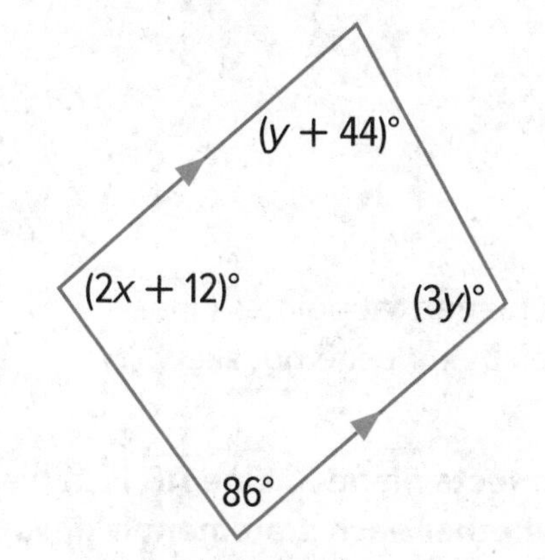

45. PROOF In the figure, lines m and n are parallel and lines p and q are parallel. Write a paragraph proof to prove that if $m\angle 1 - m\angle 4 = 25°$, then $m\angle 9 - m\angle 12 = 25°$.

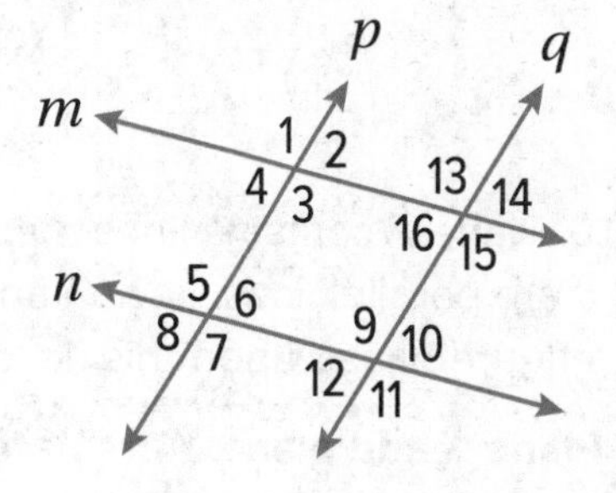

46. In the figure, $m\angle 4 = 118°$. Find each angle measure. Justify each step.

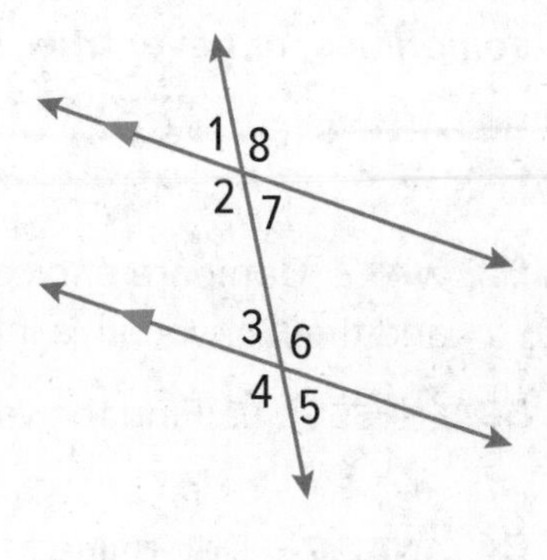

a. $m\angle 8$

b. $m\angle 7$

47. PROOF Write a paragraph proof of the Alternate Exterior Angles Theorem. Given: $q \parallel r$; Prove: $\angle 1 \cong \angle 7$.

ℓ
4 1 3 2
q
8 5 7 6
r

48. PROOF **Write a two-column proof to prove each theorem.**

a. Consecutive Interior Angles Theorem

Given: $q \parallel r$

Prove: $\angle 2$ and $\angle 5$ are supplementary

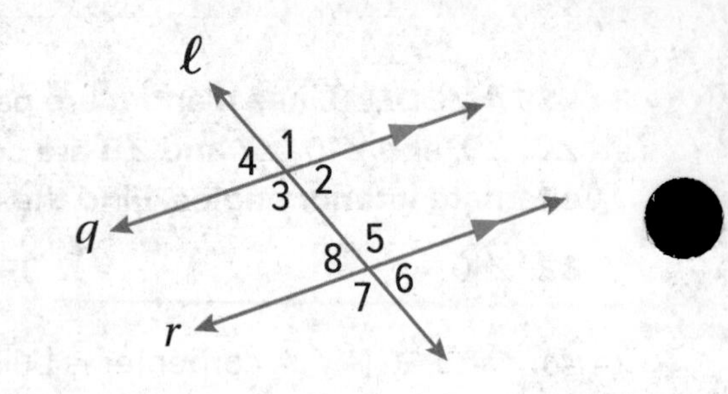

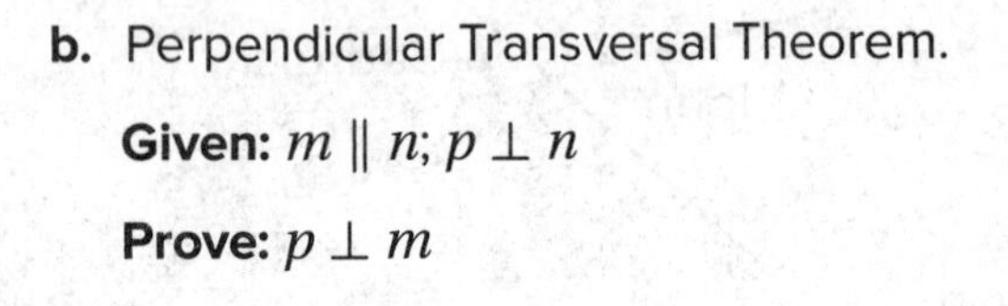

b. Perpendicular Transversal Theorem.

Given: $m \parallel n$; $p \perp n$

Prove: $p \perp m$

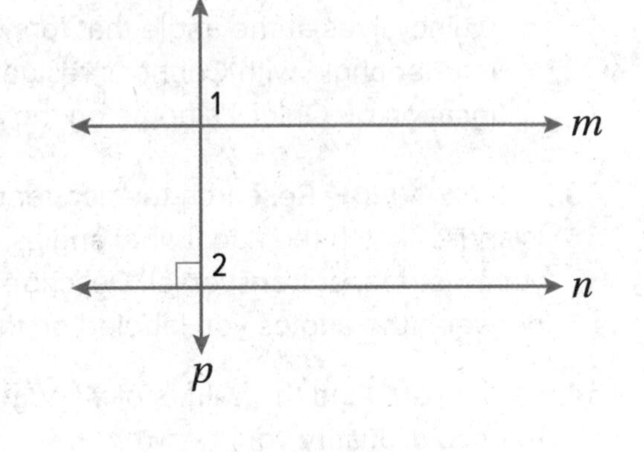

49. CREATE Plane P contains lines a and b. Line c intersects plane P at point J. Lines a and b are parallel lines, lines a and c are skew, and lines b and c are not skew. Draw a figure based upon this description.

ANALYZE **Plane X and plane Y are parallel and plane Z intersects plane X. Line $\overleftrightarrow{AB}$ is in plane X, line $\overleftrightarrow{CD}$ is in plane Y, and line $\overleftrightarrow{EF}$ is in plane Z. Determine whether each statement is *always*, *sometimes*, or *never* true. Justify your argument.**

50. $\overleftrightarrow{AB}$ is skew to $\overleftrightarrow{CD}$.

51. $\overleftrightarrow{AB}$ intersects $\overleftrightarrow{EF}$.

52. WRITE Compare and contrast the Alternate Interior Angles Theorem and the Consecutive Interior Angles Theorem.

53. PERSEVERE Find the values of x and y.

$(y^2)°$

$x°$ $(8y - 15)°$

54. ANALYZE Determine the minimum number of angle measures you would have to know to find the measures of all the angles formed by two parallel lines cut by a transversal. Justify your argument.

55. WRITE Can a pair of planes be described as skew? Explain.

Slope and Equations of Lines

Learn Slope Criteria for Parallel and Perpendicular Lines

Slope is the ratio of the change in the *y*-coordinate (rise) to the corresponding change in the *x*-coordinate (run) as you move from one point to another along a line. The **slope criteria** outlines a method for proving the relationship between lines based on a comparison of the slopes of the lines. You can use the slopes of two lines to determine whether the lines are parallel, perpendicular, or neither.

Postulate 3.12: Slope Criteria for Parallel and Perpendicular Lines

Slopes of Parallel Lines
Two distinct nonvertical lines have the same slope if and only if they are parallel. All vertical lines are parallel.

Slopes of Perpendicular Lines
Two nonvertical lines are perpendicular if and only if the product of their slopes is −1. Vertical and horizontal lines are perpendicular.

Example 1 Determine Line Relationships When Given Points

Determine whether $\overleftrightarrow{AB}$ and $\overleftrightarrow{CD}$ are *parallel, perpendicular,* or *neither* for $A(3, 6)$, $B(-9, 2)$, $C(5, 4)$, and $D(2, 3)$. Graph each line to verify your answer.

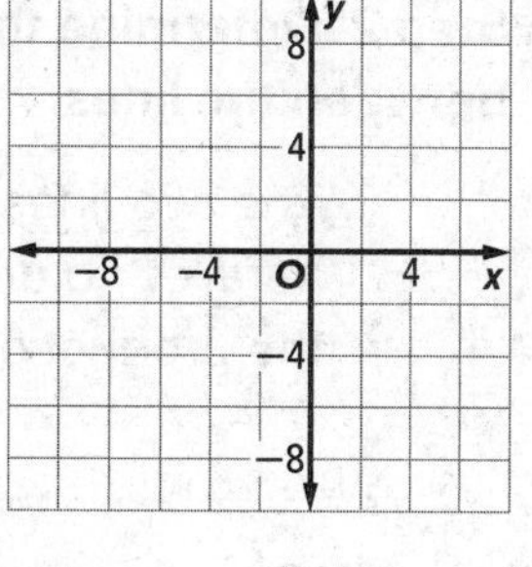

Step 1 Find the slope of each line.

slope $= \frac{y_2 - y_1}{x_2 - x_1}$, where $x_1 \neq x_2$

slope of $\overleftrightarrow{AB} = \frac{6 - 2}{3 - (-9)} = \frac{4}{12}$ or ____

slope of $\overleftrightarrow{CD} = \frac{4 - 3}{5 - 2}$ or ____

Step 2 Determine the relationship.

The two lines have the same slope, so they are ________.

Check

Determine whether $\overleftrightarrow{AB}$ and $\overleftrightarrow{CD}$ are *parallel, perpendicular,* or *neither* for $A(14, 13)$, $B(-11, 0)$, $C(-3, 7)$, and $D(-4, -5)$. Graph each line to verify your answer. ________

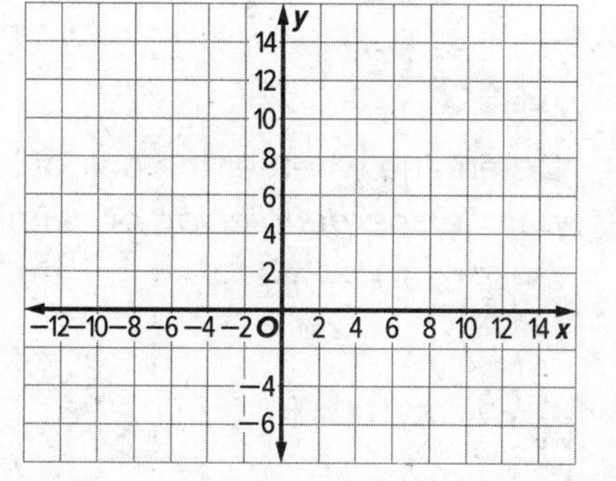

Go Online You can complete an Extra Example online.

Today's Goals
- Classify lines as parallel, perpendicular, or neither by comparing the slopes of the lines.
- Classify lines as parallel, perpendicular, or neither by comparing the equations of the lines.

Today's Vocabulary
slope
slope criteria

Go Online
You may want to complete the Concept Check to check your understanding.

Talk About It!
Feng argues that you could have graphed the points and determined whether the lines were parallel, perpendicular, or neither just by looking at the graph. Do you agree? What useful question would you ask Feng to determine whether his argument is reasonable?

Your Notes

Example 2 Determine Line Relationships When Given Graphs

Determine whether each pair of lines is *parallel, perpendicular,* or *neither.*

a. $\overleftrightarrow{RS}$ and $\overleftrightarrow{TU}$

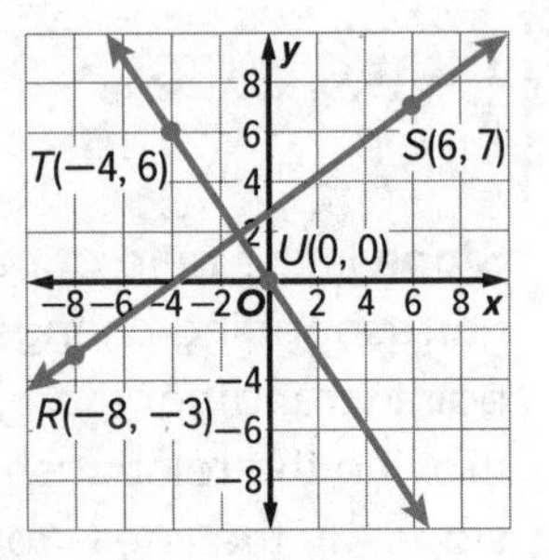

Step 1 Find the slope of each line.

$$\text{slope} = \frac{y_2 - y_1}{x_2 - x_1}, \text{ where } x_1 \neq x_2$$

$$\text{slope of } \overleftrightarrow{RS} = \frac{7 - (-3)}{6 - (-8)} = \frac{10}{14} \text{ or } \frac{5}{7}$$

$$\text{slope of } \overleftrightarrow{TU} = \frac{0 - 6}{0 - (-4)} = -\frac{6}{4} \text{ or } -\frac{3}{2}$$

Step 2 Determine the relationship, if any, between the lines.

The two lines do not have the same slope, so they are not parallel. The product of the slopes of the lines is $\left(\frac{5}{7}\right)\left(-\frac{3}{2}\right)$ or $-\frac{15}{14}$. Because the product of the slopes is not −1, the two lines are not perpendicular. So, $\overleftrightarrow{RS}$ and $\overleftrightarrow{TU}$ are neither parallel nor perpendicular.

Study Tip

Slopes of Perpendicular Lines If a line ℓ has a slope of $\frac{a}{b}$, then the slope of a line perpendicular to line ℓ is the negative reciprocal, $-\frac{b}{a}$, because $\frac{a}{b}\left(-\frac{b}{a}\right) = -1$.

b. $\overleftrightarrow{EF}$ and $\overleftrightarrow{DG}$

Step 1 Find the slope of each line.

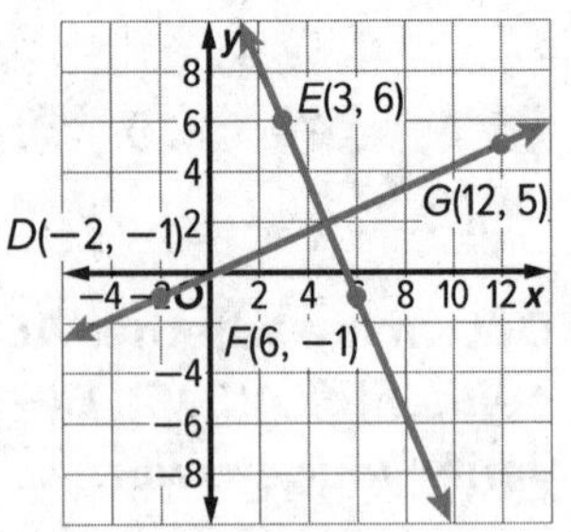

$$\text{slope} = \frac{y_2 - y_1}{x_2 - x_1}, \text{ where } x_1 \neq x_2$$

$$\text{slope of } \overleftrightarrow{EF} = \frac{-1 - 6}{6 - 3} = \_\_\_\_$$

$$\text{slope of } \overleftrightarrow{DG} = \frac{5 - (-1)}{12 - (-2)} = \frac{6}{14} \text{ or } \_\_\_\_$$

Step 2 Determine the relationship, if any, between the lines.

The two lines ______ have the same slope, so they ______ parallel. To determine whether the lines are perpendicular, find the product of their slopes.

______________ 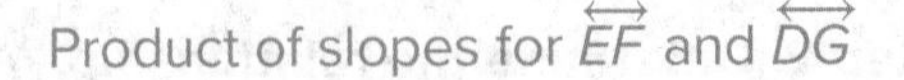
Product of slopes for $\overleftrightarrow{EF}$ and $\overleftrightarrow{DG}$

Because the product of their slopes is −1, the two lines are ______________.

Check

Determine whether the pair of lines is *parallel, perpendicular*, or *neither*.

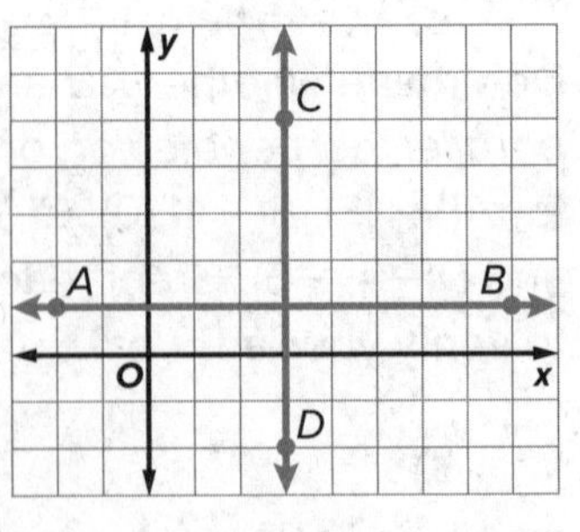

Go Online You can complete an Extra Example online.

Explore Equations of Lines

Online Activity Use dynamic geometry software to complete the Explore.

> **INQUIRY** How do the equations of parallel lines compare to the equations of perpendicular lines?

Learn Equations of Lines

An equation of a nonvertical line can be written in different but equivalent forms.

Key Concept • Nonvertical Line Equations

The slope-intercept form of a linear equation is $y = mx + b$, where m is the slope of the line and b is the y-intercept.	slope: $y = mx + b$ $y = 3x + 8$; y-intercept
The point-slope form of a linear equation is $y - y_1 = m(x - x_1)$, where (x_1, y_1) is any point on the line and m is the slope of the line.	point (3, 5): $y - 5 = -2(x - 3)$; slope

The equations of horizontal and vertical lines involve only one variable.

Key Concept • Horizontal and Vertical Line Equations

The equation of a horizontal line is $y = b$, where b is the y-intercept of the line.	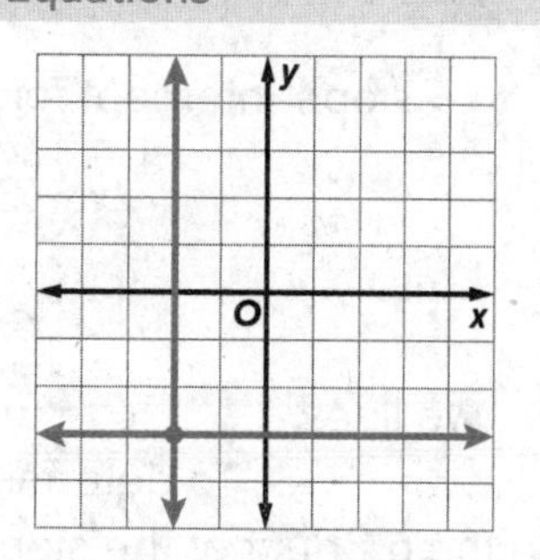
The equation of a vertical line is $x = a$, where a is the x-intercept of the line.	

When given the equations of two lines, you can compare the equations to determine the relationship between the lines.

Math History Minute

French mathematician **Gaspard Monge (1746–1818)** is known as the father of the point-slope form of the linear equation. He is also credited with first stating in print the relationship between the slopes of perpendicular lines as $aa' + 1 = 0$. For his work in mathematics, his name is one of 72 names inscribed on the base of the Eiffel Tower.

Example 3 Determine Line Relationships When Given Equations

Determine whether each pair of lines is *parallel, perpendicular,* or *neither.*

a. $y = 3x - 2$; $y - 0 = -\frac{1}{3}(x - 2)$

slope-intercept form   point-slope form

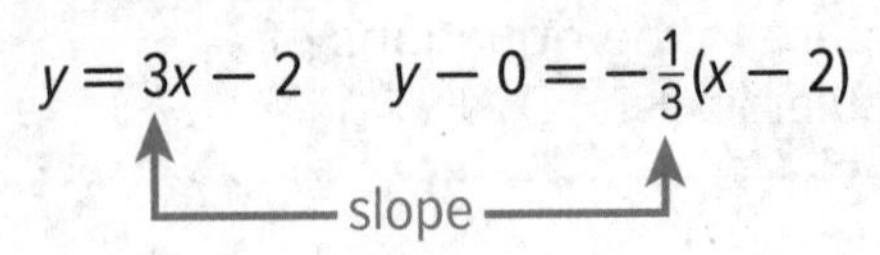

The two lines ________ the same slope, so the lines are ________. To determine whether the lines are perpendicular, find the product of the slopes.

________ Product of slopes

Because the product of their slopes is ____, the two lines are ________.

b. $y = 3$; $x = 1$

$y = 3$ — horizontal line — slope of 0

$x = 1$ — vertical line — undefined slope

Vertical and horizontal lines are always perpendicular.

c. $y - 5 = -\frac{3}{4}(x + 2)$; $y = -\frac{3}{4}x + 2$

point-slope form   slope-intercept form

$y - 5 = -\frac{3}{4}(x + 2)$ $y = -\frac{3}{4}x + 2$

slope

Because the slopes of both lines are ________, the lines are parallel.

d. $y = 2x + 3$; $y - 1 = \frac{1}{2}(x + 2)$

slope-intercept form   point-slope form

$y = 2x + 3$ $y - 1 = \frac{1}{2}(x + 2)$

slope

The two lines ________ the same slope, so the lines are ________. To determine whether the lines are perpendicular, find the product of the slopes.

________ Product of slopes

Because the product of the slopes ________ −1, the two lines are not perpendicular. So, the two lines are ________ parallel ________ perpendicular.

e. $x = -2$; $x = 4$

Both lines are vertical with ________ slope. Vertical lines are always ________.

Go Online You can complete an Extra Example online.

Watch Out!

Vertical Lines If you calculate the slope of the line $x = 1$ using the slope formula, you get $m = \frac{-1 - 4}{1 - 1} = -\frac{5}{0}$ or an undefined slope. You cannot find the product of the slope of $x = 1$ and $y = 3$. However, vertical and horizontal lines are always perpendicular.

Study Tip

Zero and Undefined Slope If the change in y values is 0, then the line is horizontal. If the change in x values is 0, then the line is vertical.

Check

Determine whether each pair of lines is *parallel*, *perpendicular*, or *neither*.

a. $y = 3x - 9$; $y = -\frac{1}{3}x + 2$ ________

b. $y = \frac{9}{7}x - \frac{19}{7}$; $y - 1 = \frac{9}{7}(x + 3)$ ________

c. $x = -3$; $x = 4$ ________

Example 4 Use Slope to Graph a Line

DESIGN **Valentina is designing a park using grid paper. She wants to build a sidewalk that connects with the fountain at $P(0, 1)$ and is perpendicular to the existing sidewalk that passes through points $Q(-6, -2)$ and $R(0, -6)$. Graph the line that represents the new sidewalk.**

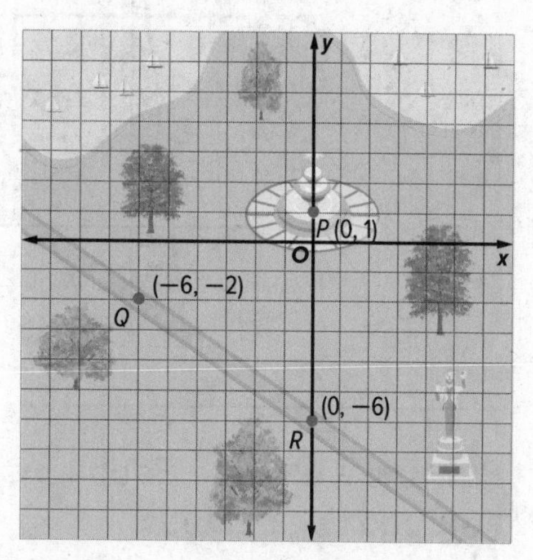

The slope of the existing sidewalk, $\overleftrightarrow{QR}$ is $\frac{-6 - (-2)}{0 - (-6)} = -\frac{4}{6}$ or ____.

Because $-\frac{2}{3}\left(\frac{3}{2}\right) =$ ____, the slope of the line perpendicular to $\overleftrightarrow{QR}$ through P is ____.

Graph the line that represents the new sidewalk.

Step 1 Plot a point at $P(0, 1)$.

Step 2 Move up 3 units and then right 2 units. Plot a second point at this location.

Step 3 Graph the line connecting these two points.

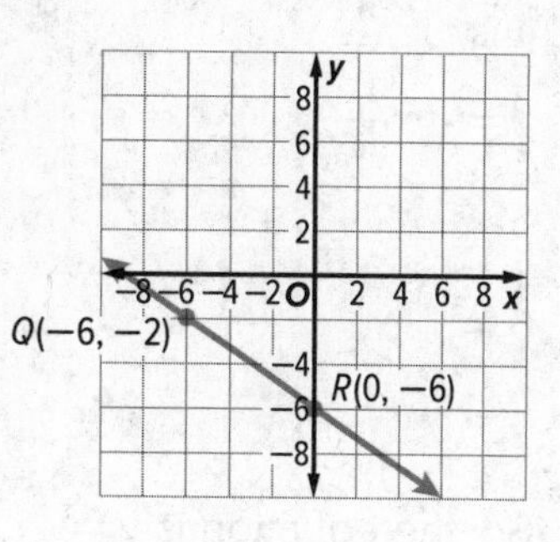

Go Online You can complete an Extra Example online.

Check

MAPS Isabella is creating a map of her town's metro lines. She knows that the A Line and the E Line are parallel. On her map, the equation that represents the A Line is $y = 8x + 11$ and the E Line passes through (9, 5). Write the equation in slope-intercept form that represents the E Line. ________

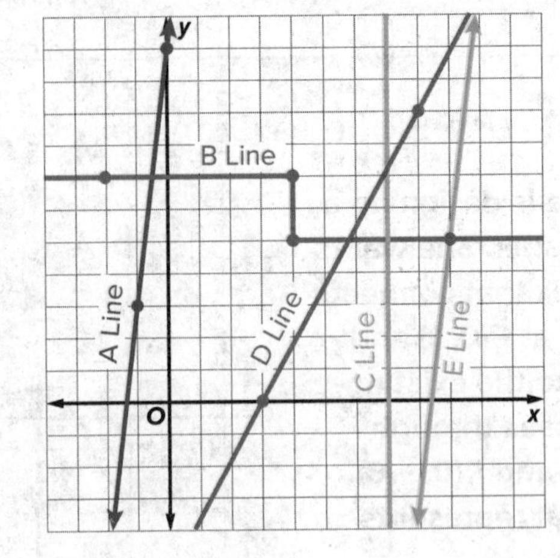

> **Think About It!**
> Kennedy suggests that there is another line parallel to $y = -\frac{3}{4}x + 3$ that contains the point (−3, 6). She says that the equation of the line is $y - 6 = -\frac{3}{4}(x + 3)$. Do you agree? Explain your reasoning.

Example 5 Write Equations of Parallel and Perpendicular Lines

Write an equation in slope-intercept form for the line parallel to $y = -\frac{3}{4}x + 3$ containing (− 3, 6).

The slope of $y = -\frac{3}{4}x + 3$ is $-\frac{3}{4}$, so the slope of the line parallel to it is ________.

$y = mx + b$	Slope-intercept form
$6 = -\frac{3}{4}(-3) + b$	$m = -\frac{3}{4}$ and $(x, y) = (-3, 6)$
$6 =$ ________ $+ b$	Simplify.
________ $= b$	Subtract $\frac{9}{4}$ from each side.

So, the equation is $y =$ ________.

> **Go Online**
> An alternate method is available for this example.

Check

Write an equation in slope-intercept form for the line parallel to $y = \frac{1}{2}x + \frac{5}{2}$ containing $\left(\frac{3}{2}, 1\right)$.

____

Go Online You can complete an Extra Example online.

Name ________________ Period ________ Date ________

Practice

Go Online You can complete your homework online.

Example 1

Determine whether $\overleftrightarrow{AB}$ and $\overleftrightarrow{CD}$ are *parallel*, *perpendicular*, or *neither*. Graph each line to verify your answer.

1. $A(1, 5)$, $B(4, 4)$, $C(9, -10)$, $D(-6, -5)$
2. $A(-6, -9)$, $B(8, 19)$, $C(0, -4)$, $D(2, 0)$
3. $A(4, 2)$, $B(-3, 1)$, $C(6, 0)$, $D(-10, 8)$
4. $A(8, -2)$, $B(4, -1)$, $C(3, 11)$, $D(-2, -9)$
5. $A(8, 4)$, $B(4, 3)$, $C(4, -9)$, $D(2, -1)$
6. $A(4, -2)$, $B(-2, -8)$, $C(4, 6)$, $D(8, 5)$

Example 2

Determine whether each pair of lines is *parallel*, *perpendicular*, or *neither*.

7. 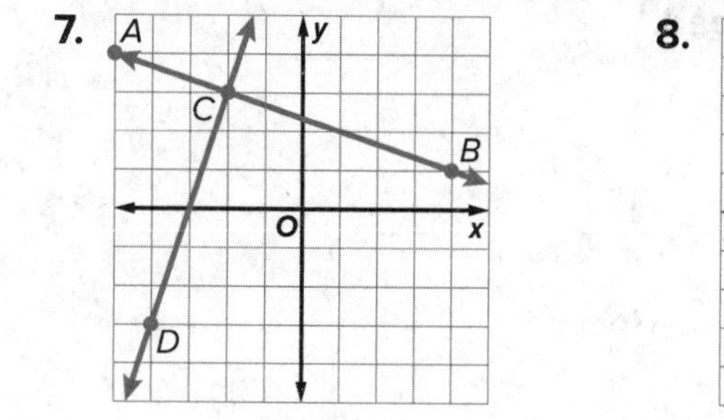

8. 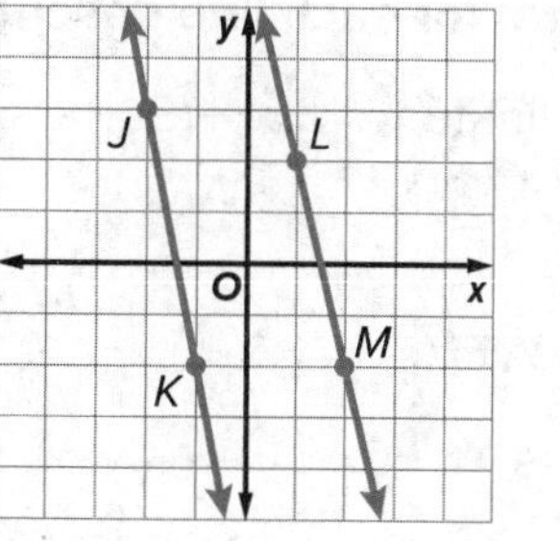

9. 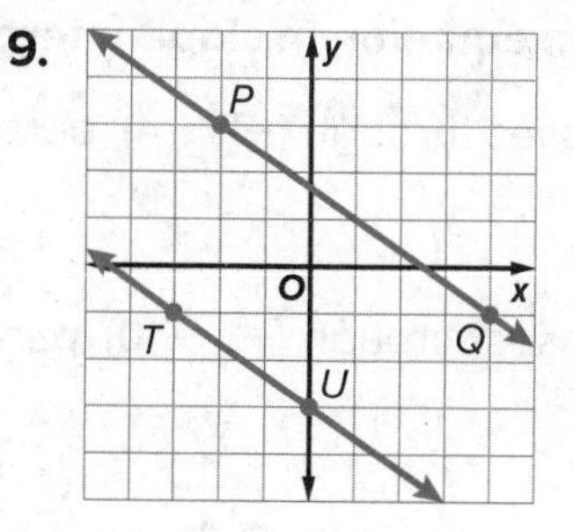

Example 3

Determine whether each pair of lines is *parallel*, *perpendicular*, or *neither*.

10. $y = 2x + 4$, $y = 2x - 10$
11. $y = -\frac{1}{2}x - 12$, $y - 3 = 2(x + 2)$
12. $y - 4 = 3(x + 5)$, $y + 3 = -\frac{1}{3}(x + 1)$
13. $y - 3 = 6(x + 2)$, $y + 3 = -\frac{1}{3}(x - 4)$
14. $x = -2$, $y = 10$
15. $y = 5$, $y = -3$

Example 4

Graph the line that satisfies each condition.

16. passes through $A(2, -5)$, parallel to $\overleftrightarrow{BC}$ with $B(1, 3)$ and $C(4, 5)$
17. passes through $X(1, -4)$, parallel to $\overleftrightarrow{YZ}$ with $Y(5, 2)$ and $Z(-3, -5)$
18. passes through $K(3, 7)$, perpendicular to $\overleftrightarrow{LM}$ with $L(-1, -2)$ and $M(-4, 8)$
19. passes through $D(-5, -6)$, perpendicular to $\overleftrightarrow{FG}$ with $F(-2, -9)$ and $G(1, -5)$
20. SKIING Gavin is working on an animated film about skiing. The figure shows a ski slope, represented by $\overleftrightarrow{AB}$, and one of the chairs on the chair lift, represented by point C.

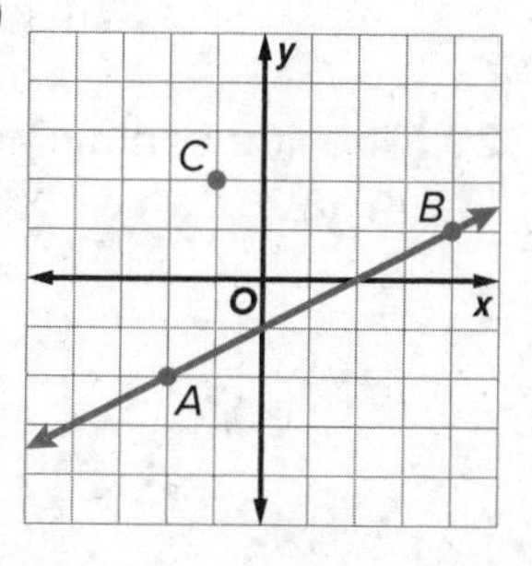

   a. The chair needs to move along a straight line that is parallel to $\overleftrightarrow{AB}$. What is the equation of this line?

   b. The top of the chair lift occurs at $y = 20$. Explain how Gavin can find the coordinates of the chair when it reaches the top of the chair lift.

21. REASONING The director of a marching band uses a coordinate plane to design the band's formations. During one formation, a drummer marches from point A to point B and then turns 90° to her right and marches until she reaches the x-axis.

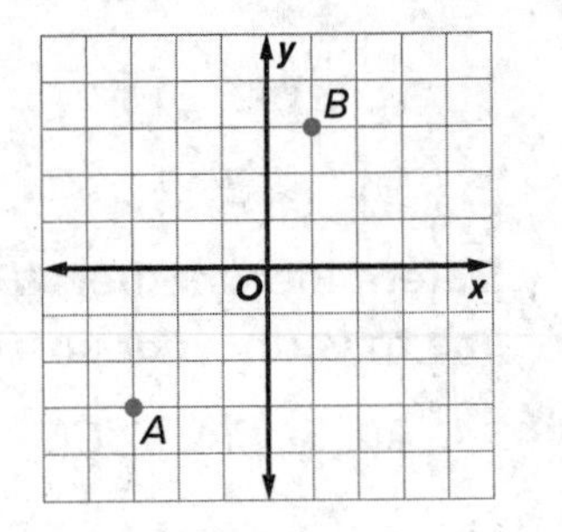

a. When the drummer marches from point B to the x-axis, what is the equation of the line that she marches along?

b. The director wants to know whether the drummer will cross the x-axis at a point where the x-coordinate is greater than or less than 5. Explain how the director can answer this question.

Example 5

Write an equation in slope-intercept form for each line described.

22. passes through (−7, −4), perpendicular to $y = \frac{1}{2}x + 9$

23. passes through (−1, −10), parallel to $y = 7$

24. passes through (6, 2), parallel to $y = -\frac{2}{3}x + 1$

25. passes through (−2, 2), perpendicular to $y = -5x - 8$

Mixed Exercises

Find the value of x or y that satisfies the given conditions. Then graph the line.

26. The line containing (4, −2) and (x, −6) is perpendicular to the line containing (−2, −9) and (3, −4).

27. The line containing (−4, 9) and (4, 3) is parallel to the line containing (−8, 1) and (4, y).

28. The line containing (8, 7) and (7, −6) is perpendicular to the line containing (2, 4) and (x, 3).

29. The line containing (1, −3) and (3, y) is parallel to the line containing (5, −6) and (9, y).

Name ______________________ Period ____________ Date ____________

Write equations in slope-intercept form for a line that is *parallel* and a line that is *perpendicular* to the given line and that passes through the given point.

30. passes through $B(6, 3)$

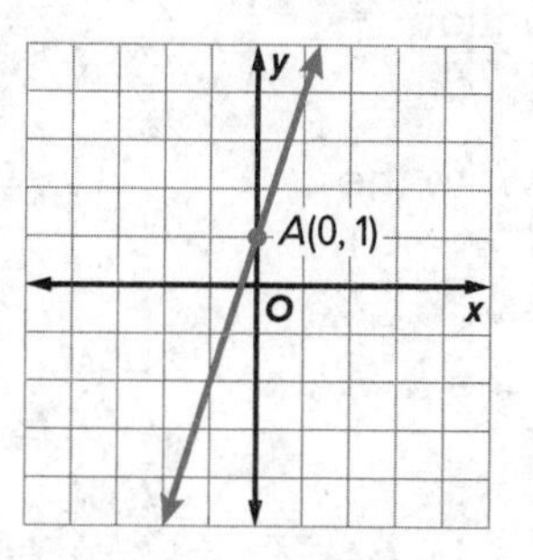

31. passes through $S(-2, -4)$

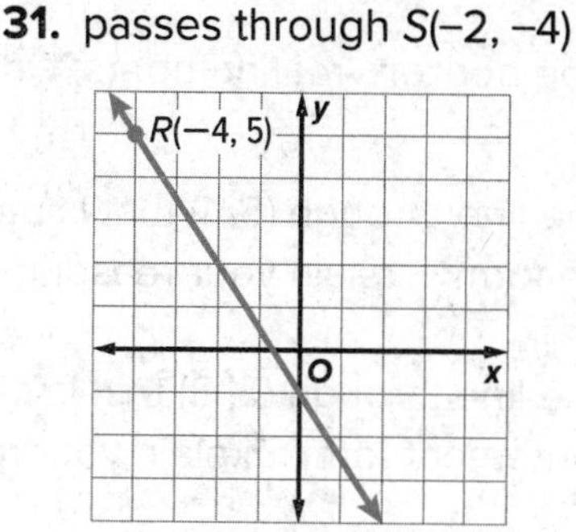

PRECISION **Determine whether any of the lines in each figure are parallel or perpendicular. Justify your answers.**

32.

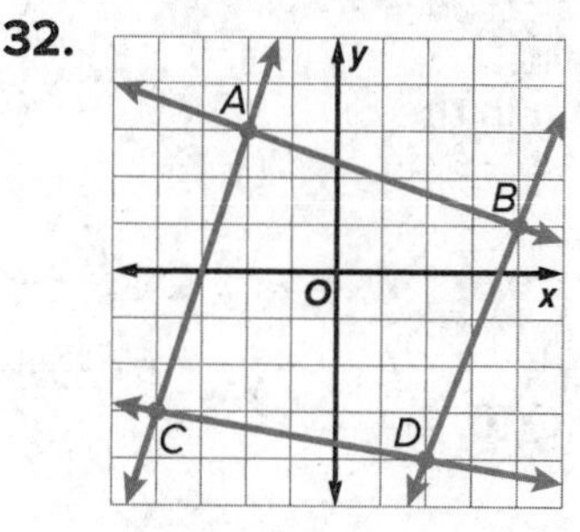

33.

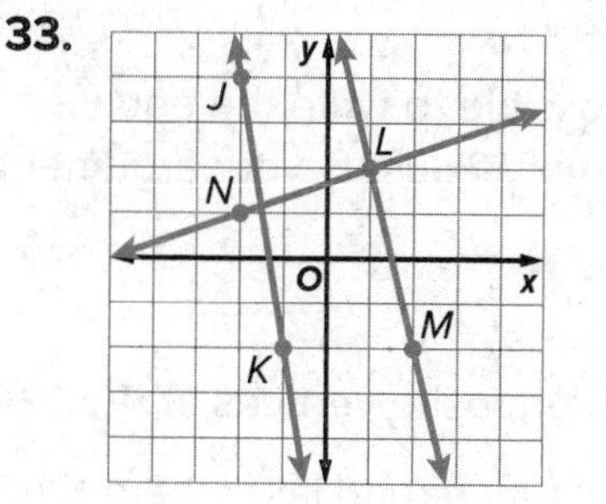

34.

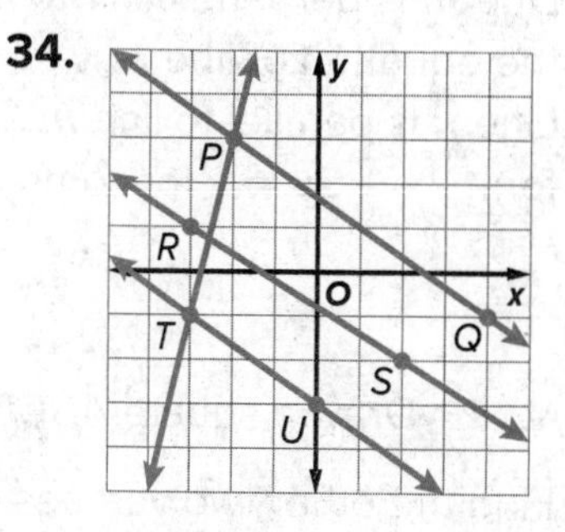

35. CITY BLOCKS The figure shows a map of part of a city consisting of two pairs of parallel roads. If a coordinate grid is applied to this map, Ford Street would have a slope of −3.

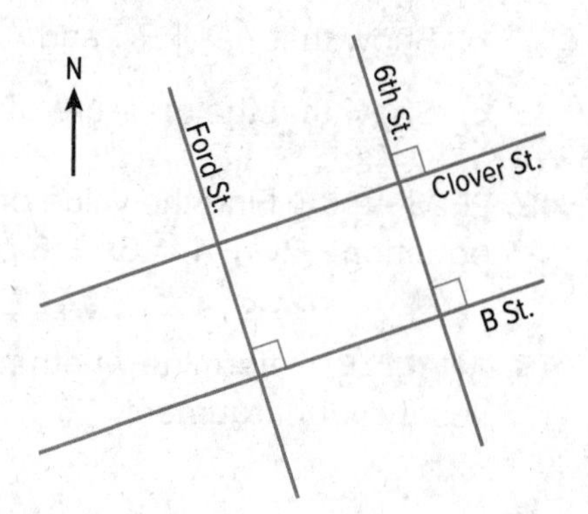

a. The intersection of B Street and Ford Street is 150 yards east of the intersection of Ford Street and Clover Street. How many yards south is it?

b. What is the slope of 6th Street? Explain.

c. What are the slopes of Clover and B Streets? Explain.

d. The intersection of B Street and 6th Street is 600 yards east of the intersection of B Street and Ford Street. How many yards north is it?

36. REASONING $\overleftrightarrow{AB}$ is parallel to $\overleftrightarrow{CD}$. The coordinates of A, B, and C are $A(-3, 1)$, $B(6, 4)$, and $C(1, -1)$. What is a possible set of coordinates for point D? Describe the reasoning you used to find the coordinates.

37. USE A MODEL A video game designer is using a coordinate plane to plan the path of a helicopter. She has already determined that the helicopter will move along straight segments from P to Q to R. The designer wants the next part of the path, $\overline{RS}$, to be perpendicular to $\overline{QR}$, and she wants point S to lie on the y-axis. What should the coordinates of point S be? Justify your answer.

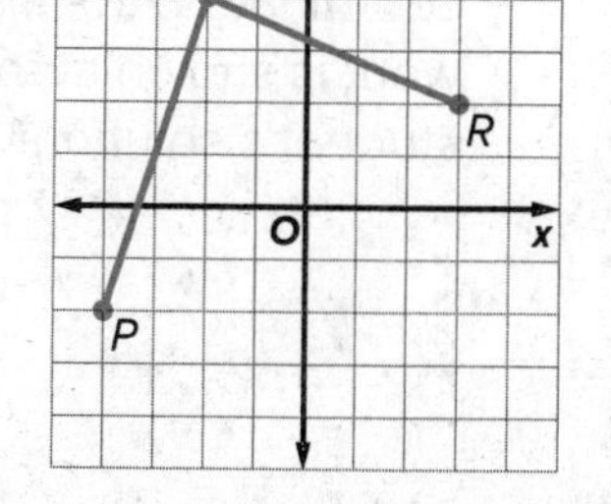

38. Line p passes through (1, 3) and (4, 7), and line q passes through (0, −2) and (a, b).

a. Find the slopes of lines p and q.

b. Find possible values of a and b if $p \parallel q$.

39. STRUCTURE Let a and b be nonzero real numbers. Line p has the equation $y = ax + b$.

a. Find the equation of the line through (5, 1) that is parallel to line p. Write the equation in point-slope form. Explain your reasoning.

b. Find the equation of the line through (2, 3) that is perpendicular to line p. Write the equation in slope-intercept form. Explain your reasoning.

40. CONSTRUCT ARGUMENTS The equation of line ℓ is $3y - 2x = 6$.

a. Line m is perpendicular to line ℓ and passes through the point $P(6, -2)$. Find the equation of line m.

b. Line n is parallel to line m. Is it possible to write the equation of line n in the form $2x + 3y = k$ for some constant k? Justify your argument.

41. ANALYZE Draw a square $ABCD$ with opposite vertices at $A(2, -4)$, and $C(10, 4)$.

a. Find the other two vertices of the square and label them B and D.

b. Show that $\overline{AD} \parallel \overline{BC}$ and $\overline{AB} \parallel \overline{DC}$.

c. Show that the measure of each angle inside the square is equal to 90°.

42. PERSEVERE Find the value of n so that the line perpendicular to the line with the equation $-2y + 4 = 6x + 8$ passes through the points $(n, -4)$ and $(2, -8)$.

43. ANALYZE Determine whether the points at (−2, 2), (2, 5), and (6, 8) are collinear. Justify your argument.

44. CREATE Write equations for a pair of perpendicular lines that intersect at the point at (−3, −7).

45. WRITE Write biconditionals to determine whether lines are parallel or perpendicular using slopes.

46. FIND THE ERROR A student was asked to find the equation of the line perpendicular to $\overleftrightarrow{AB}$ that passes through point P, given that A, B, and P have coordinates $A(0, 3)$, $B(2, 2)$, and $P(1, 4)$. The student's work is shown at the right. Do you agree with the student's solution? Explain your reasoning.

Slope of $\overleftrightarrow{AB} = \frac{2-3}{2-0} = -\frac{1}{2}$.

So, the slope of the required line is 2. The equation of this line is $y = 2x + b$. The line passes through $P(1,4)$.

To find b: $1 = 2(4) + b$

$1 = 8 + b$

$-7 = b$

So, the equation is $y = 2x - 7$.

Proving Lines Parallel

Explore Intersecting Lines

Online Activity Use dynamic geometry software to complete the Explore.

INQUIRY If a pair of alternate exterior or alternate interior angles is congruent, what relationship is formed?

Today's Goals

- Apply angle relationship theorems to identify parallel lines and find missing values.

Learn Identifying Parallel Lines

Corresponding angles are congruent when the lines cut by the transversal are parallel. The converse of this relationship is also true.

Theorem 3.19: Converse of Corresponding Angles Theorem
If two lines are cut by a transversal so that corresponding angles are congruent, then the lines are parallel.

Postulate 3.13: Parallel Postulate
If given a line and a point not on the line, then there exists exactly one line through the point that is parallel to the given line.

Study Tip

Euclid's Postulates The father of modern geometry, Euclid (c. 300 B.C.), realized that only a few postulates were needed to prove the theorems in his day. The Parallel Postulate is one of Euclid's five original postulates.

Parallel lines that are cut by a transversal create several pairs of congruent angles. These special angle pairs can be used to prove that a pair of lines is parallel.

Theorem 3.20: Alternate Exterior Angles Converse	
If two lines in a plane are cut by a transversal so that a pair of alternate exterior angles is congruent, then the lines are parallel.	If $\angle 1 \cong \angle 5$, then $a \parallel b$.

Theorem 3.21: Consecutive Interior Angles Converse	
If two lines in a plane are cut by a transversal so that a pair of consecutive interior angles is supplementary, then the lines are parallel.	If $m\angle 7 + m\angle 6 = 180°$, then $a \parallel b$.

Theorem 3.22: Alternate Interior Angles Converse	
If two lines in a plane are cut by a transversal so that a pair of alternate interior angles is congruent, then the lines are parallel.	If $\angle 7 \cong \angle 3$, then $a \parallel b$.

Theorem 3.23: Perpendicular Transversal Converse
If two lines in a plane are perpendicular to the same line, then the lines are parallel.

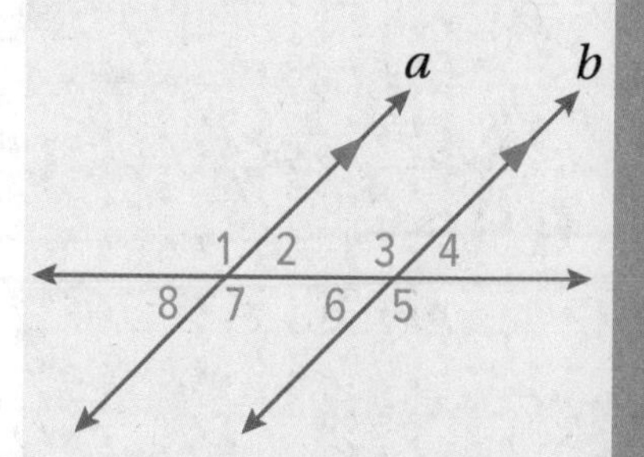

You will prove Theorems 3.20, 3.22, and 3.23 in Exercises 20, 19, and 18, respectively.

Go Online Proofs of Theorems 3.19 and 3.21 are available.

Your Notes

Watch Out!

Error Analysis Students should recognize that line ℓ is the transversal in part **a**, line a is the transversal in part **b**, and line m is the transversal in part **c**.

Example 1 Identify Parallel Lines

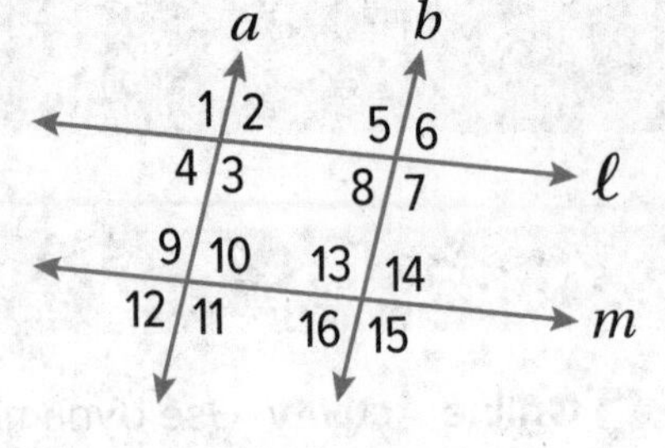

Use the given information to determine which lines, if any, are parallel. State the postulate or theorem that justifies your answer.

a. $\angle 2 \cong \angle 8$

$\angle 2$ and $\angle 8$ are ________________ ________ of lines a and b.

Because $\angle 2 \cong \angle 8$, $a \parallel b$ by the ____________________.

b $\angle 3 \cong \angle 11$

$\angle 3$ and $\angle 11$ are ________________ of lines ℓ and m. Because $\angle 3 \cong \angle 11$, $\ell \parallel m$ by the ____________________.

c. $\angle 12 \cong \angle 14$

$\angle 12$ and $\angle 14$ are 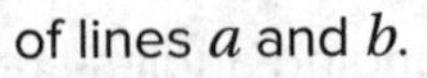 of lines a and b.
Because $\angle 12 \cong \angle 14$, $a \parallel b$ by the ____________________.

Check

Use the given information to determine which lines, if any, are parallel. State the postulate or theorem that justifies your answer.

a. $\angle 1 \cong \angle 15$

A. $\ell \parallel m$; Alternate Exterior Angles Converse

B. $n \parallel k$; Alternate Exterior Angles Converse

C. $\ell \parallel m$; Converse of Corresponding Angles Theorem

D. It is not possible to determine whether the lines are parallel.

b. $m\angle 3 + m\angle 10 = 180$

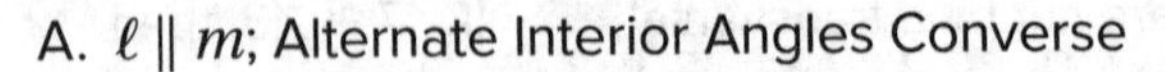

c. $\angle 3 \cong \angle 5$

A. $\ell \parallel m$; Alternate Interior Angles Converse

B. $\ell \parallel m$; Consecutive Interior Angles Converse

C. $n \parallel k$; Alternate Interior Angles Converse

D. It is not possible to determine whether the lines are parallel.

Go Online You can complete an Extra Example online.

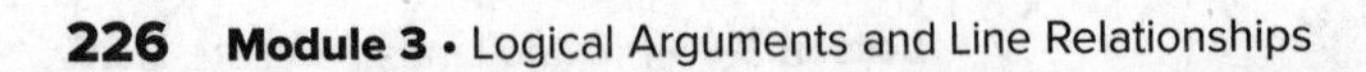

Example 2 Use Angle Relationships

Find the value of y so that $e \parallel f$.

From the figure, you know that line d is perpendicular to line e. For lines e and f to be parallel, line f must also be perpendicular to line d. If line f is perpendicular to line d, then $(4y + 10)° = 90°$. Solve for y.

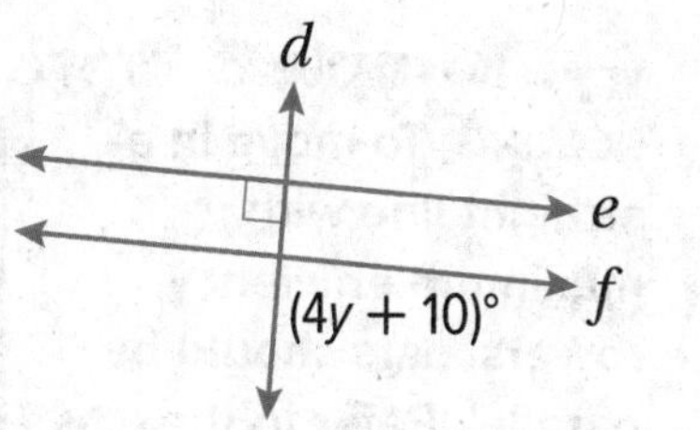

_____ $= 4y + 10$ Definition of perpendicular

_____ $= 4y$ Subtract 10 from each side.

_____ $= y$ Divide each side by 4.

Check

a. Find the value of x so that $m \parallel n$. Identify the postulate or theorem you used.

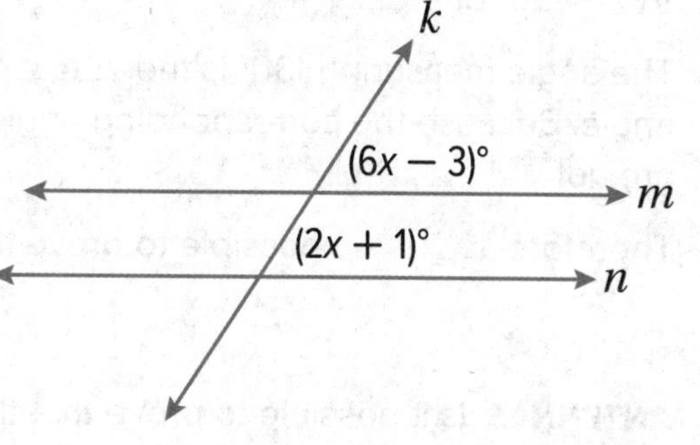

$x =$ _____; ____________________

b. Find $m\angle LMN$ so that $a \parallel b$.

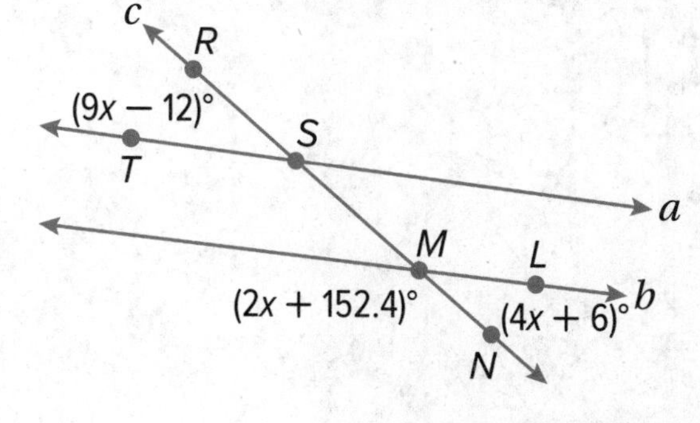

A. 3.6° B. 20.4° C. 100° D. 159.6°

Think About It!

Lakeisha argues that we do not have enough information to determine the correct value of y. She says that the two angles are not corresponding angles; therefore, it does not help to assume that the two angles are congruent. What theorems can you use to prove that line e is parallel to line f?

Go Online You can complete an Extra Example online.

Study Tip

Assumptions When applying geometric theorems to real-world objects, we often make assumptions about the relationships between the objects being represented. In Example 3, the boat is the transversal and we are trying to determine whether the oars are parallel lines.

Talk About It!

What other information can be used to show that the oars are not parallel?

Study Tip

Proving Lines Parallel When two parallel lines are cut by a transversal, the angle pairs formed are either congruent or supplementary. When a pair of lines forms angles that do not meet this criterion, the lines cannot be parallel.

Go Online

You may want to complete the construction activities for this lesson.

The angle pair relationships formed by a transversal can be used to prove that two lines are parallel.

Example 3 Prove Lines Parallel

ROWING **To move in a straight line with maximum efficiency, rowers' oars should be parallel. Refer to the photo at the right. Is it possible to prove that any of the oars are parallel? Justify your answer.**

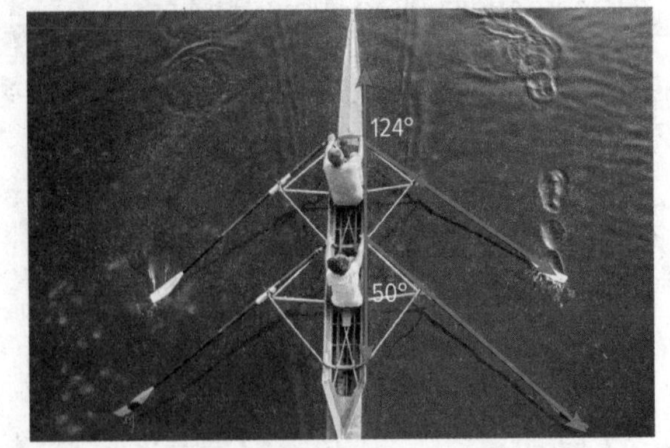

The angle that forms a linear pair with the 50° angle has a measure of 180° − 50° or ______.

The angle measuring 130° is the ____________ angle to the ______ angle. Because the corresponding angles ______ congruent, the lines are not ______.

Therefore, it ______ possible to prove that the oars are parallel.

Check

ANTENNAS Is it possible to prove that the support poles of the antenna complex are parallel? Justify your answer.

A. No; because the consecutive interior angles are not supplementary, the support poles cannot be parallel.

B. No; because the alternate interior angles are not supplementary, the support poles cannot be parallel.

C. Yes; because the alternate interior angles are supplementary, the support poles are parallel.

D. Yes; because the consecutive interior angles are congruent, the support poles are parallel.

Go Online You can complete an Extra Example online.

Name ______________________ Period __________ Date __________

Practice

Go Online You can complete your homework online.

Example 1

Use the given information to determine which lines, if any, are parallel. State the postulate or theorem that justifies your answer.

1. $\angle 3 \cong \angle 7$

2. $\angle 9 \cong \angle 11$

3. $\angle 2 \cong \angle 16$

4. $m\angle 5 + m\angle 12 = 180°$

Given the following information, determine which lines, if any, are parallel. State the theorem that justifies your answer.

5. $\angle 1 \cong \angle 6$

6. $m\angle 7 + m\angle 6 = 180°$

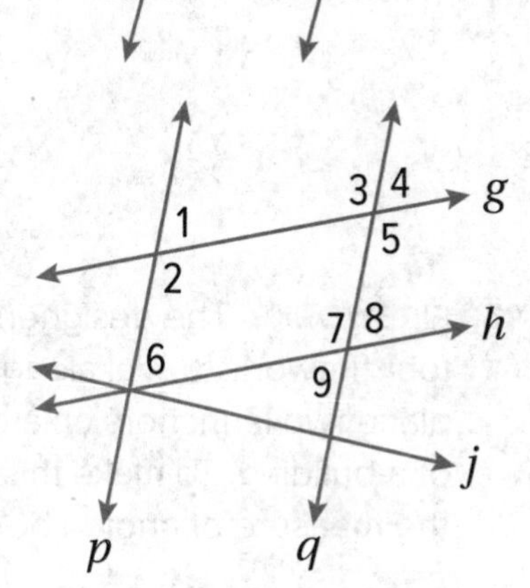

Example 2

Find the value of x so that $\ell \parallel m$.

7.

8.

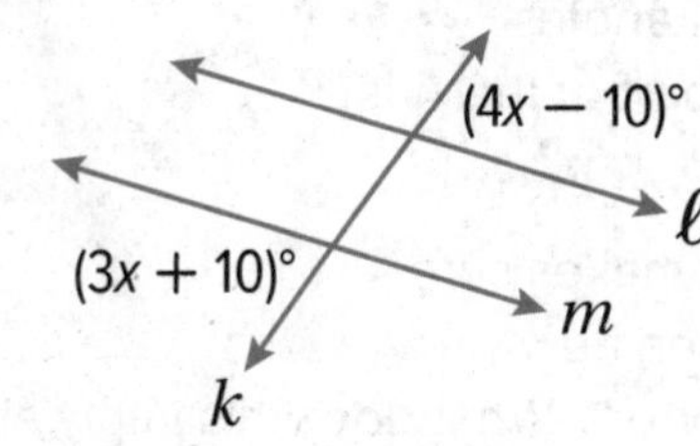

9.

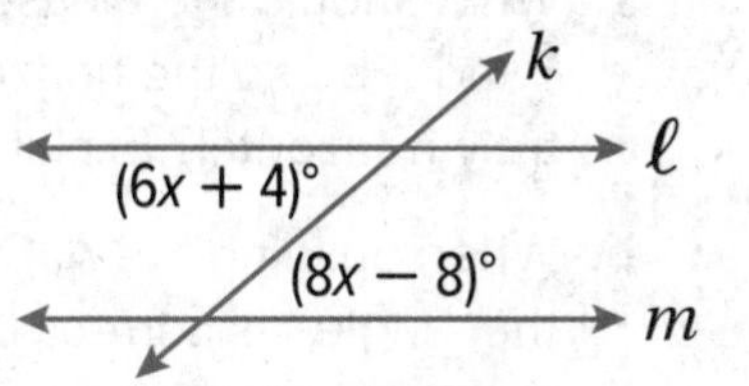

10.

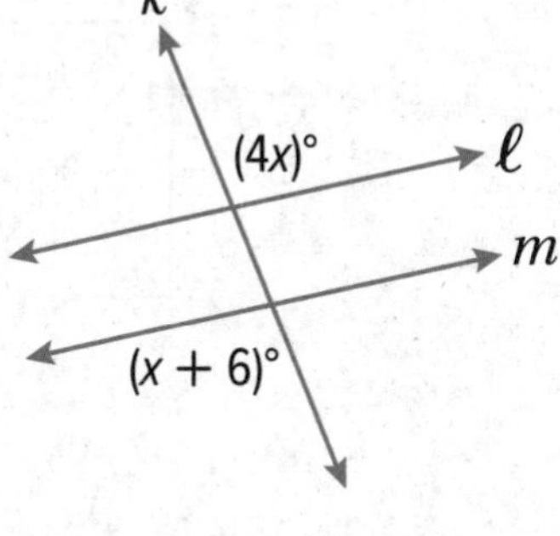

11.

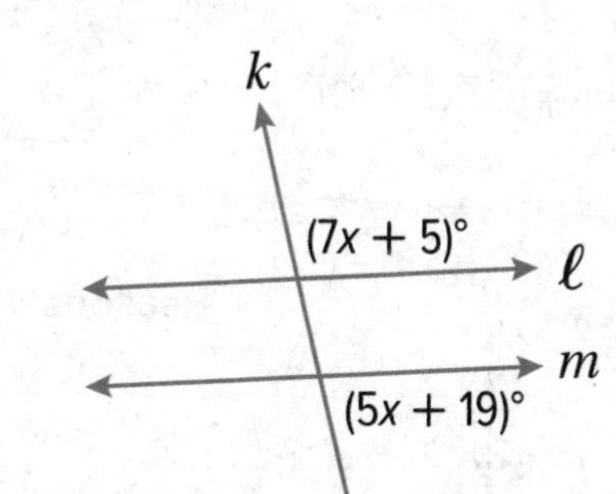

12.

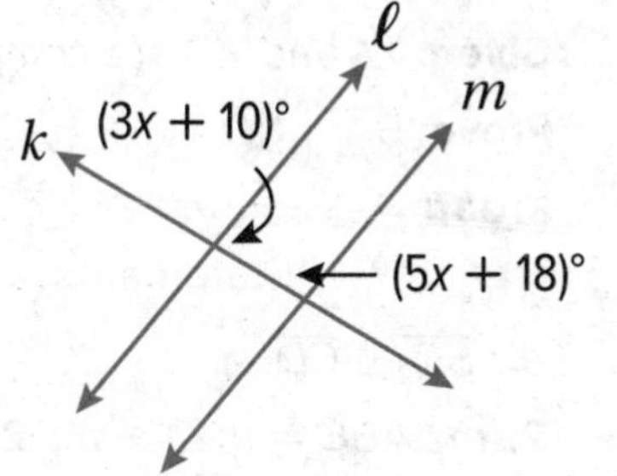

13. Find the value of x so that $\ell \parallel m$.

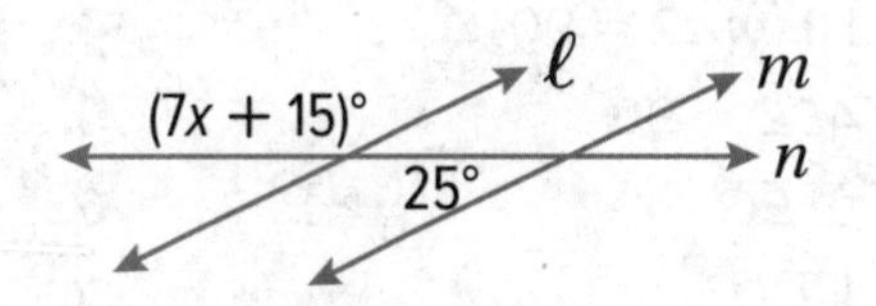

Example 3

14. BOOKS Each orange book on the bookshelf makes a 70° angle with the base of the shelf. What more can you say about these two orange books? Explain.

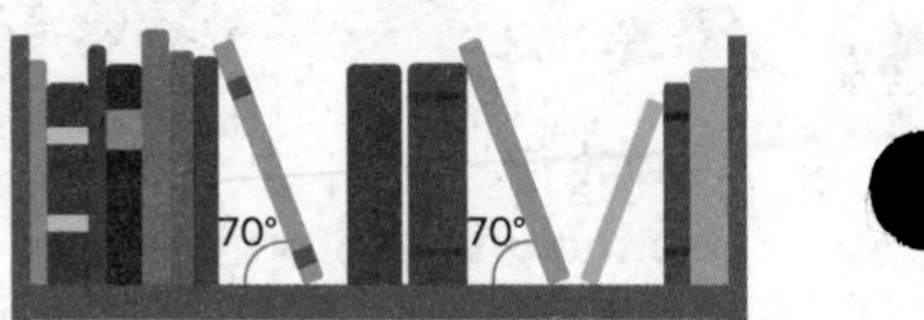

15. PATTERNS A rectangle is cut along the slanted, dashed line shown in the figure. The two pieces are rearranged to form another figure. Describe as precisely as you can the shape of the new figure. Explain.

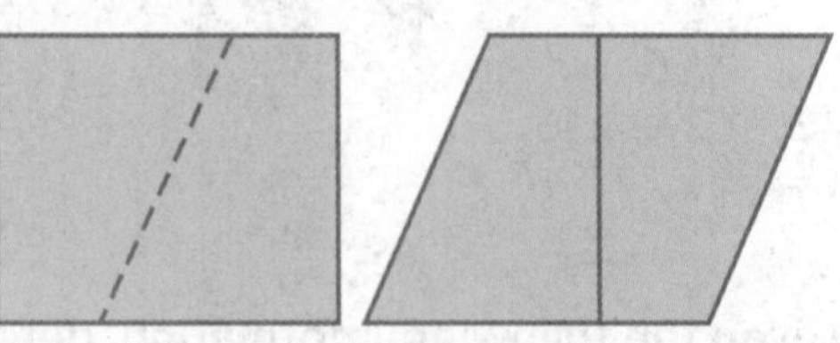

16. FIREWORKS The designers of a fireworks display want to have four fireworks travel along parallel trajectories. They decide to place two launchers on a dock and two launchers on the roof of a building. To make this display work correctly, what should the measure of angle 1 be? Explain.

17. REASONING Chaska is making a giant letter A to put on the rooftop of the A Is for Apple Orchard Store. The figure shows a sketch of the design.

a. What should the measures of angles 1 and 2 be so the horizontal part of the A is truly horizontal? Explain.

b. When building the A, Chaska makes sure that angle 1 is correct, but when he measures angle 2, it is not correct. What does this imply about the A?

Mixed Exercises

18. PROOF Provide a reason for each statement in the proof of the Perpendicular Transversal Converse.

Given: $\angle 1$ and $\angle 2$ are complementary; $\overline{BC} \perp \overline{CD}$

Prove: $\overline{BA} \parallel \overline{CD}$

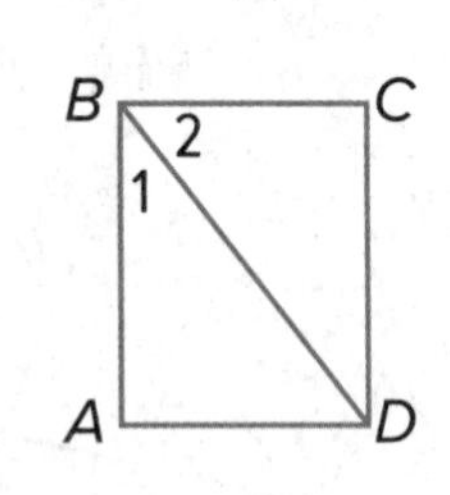

Proof:

Statements	Reasons
1. $\overline{BC} \perp \overline{CD}$	1. ____?____
2. $m\angle ABC = m\angle 1 + m\angle 2$	2. ____?____
3. $\angle 1$ and $\angle 2$ are complementary.	3. ____?____
4. $m\angle 1 + m\angle 2 = 90$	4. ____?____
5. $m\angle ABC = 90$	5. ____?____
6. $\overline{BA} \perp \overline{BC}$	6. ____?____
7. $\overline{BA} \parallel \overline{CD}$	7. ____?____

19. PROOF Write a paragraph proof to prove the Alternate Interior Angles Converse.

Given: $\angle 1 \cong \angle 2$

Prove: $\ell \parallel m$

20. PROOF Write a paragraph proof to prove the Alternate Exterior Angles Converse.

Given: $\angle 1 \cong \angle 2$

Prove: $\ell \parallel m$

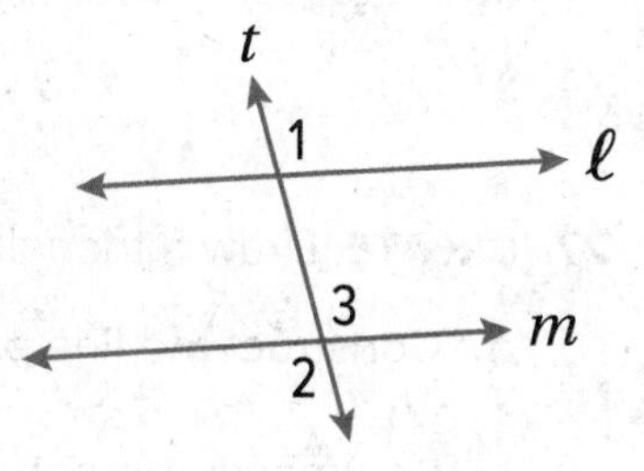

USE TOOLS **Use a compass and straightedge to construct the line through point *P* that is parallel to line *q*.**

21.

22.

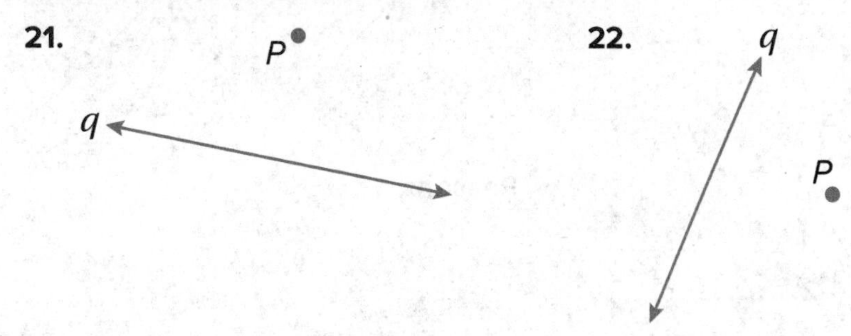

23. PICTURE FRAMES Lindy is making a wooden picture frame. She cuts the top and bottom pieces at a 45° angle. If the corners are right angles, explain how Lindy knows that each pair of opposite sides is parallel.

45°

24. REASONING Jim made a frame for a painting. He wants to check to make sure that opposite sides are parallel by measuring the angles at the corners and seeing whether they are right angles. How many corners must he check to be sure that the opposite sides are parallel?

25. FIND THE ERROR Sean and Daniela are determining which lines are parallel in the figure at the right. Sean says that because $\angle 1 \cong \angle 2$, $\overline{WY} \parallel \overline{XZ}$. Daniela disagrees and says that because $\angle 1 \cong \angle 2$, $\overline{WX} \parallel \overline{YZ}$. Is either of them correct? Explain your reasoning.

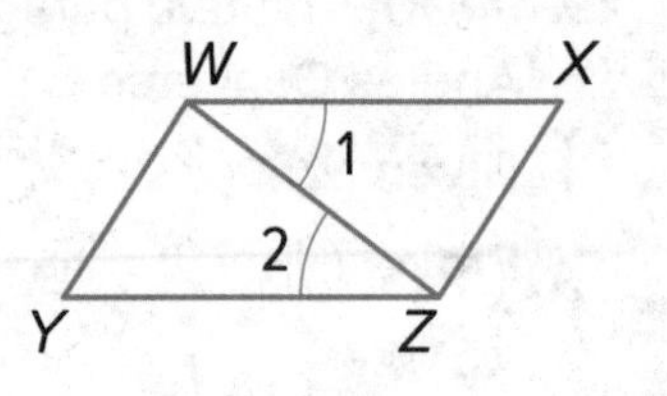

26. ANALYZE Is Theorem 3.23 still true if the two lines are not coplanar? Draw a figure to justify your argument.

27. CREATE Draw a triangle *ABC*.

a. Construct the line parallel to $\overline{BC}$ through point *A*.

b. Use measurements to justify that the line you constructed is parallel to $\overline{BC}$.

c. Justify the construction.

28. PROOF Use the figure at the right to complete the two-column proof to prove that two lines parallel to a third line are parallel to each other.

Given: $a \parallel b$ and $b \parallel c$

Prove: $a \parallel c$

Proof:

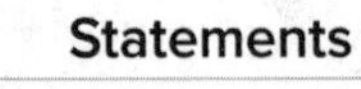

Statements	Reasons
1. $a \parallel b$ and $b \parallel c$	**1.** ___?___
2. $\angle 1 \cong \angle 3$	**2.** ___?___
3.	**3.** ___?___
4.	**4.** ___?___
5.	**5.** ___?___
6. $a \parallel c$	**6.** ___?___

29. WRITE Can a pair of angles be supplementary and congruent? Explain your reasoning.

30. PROOF Refer to the figure at the right.

a. If $m\angle 1 + m\angle 2 = 180°$, prove that $a \parallel c$.

b. Given that $a \parallel c$, if $m\angle 1 + m\angle 3 = 180°$, prove that $t \perp c$.

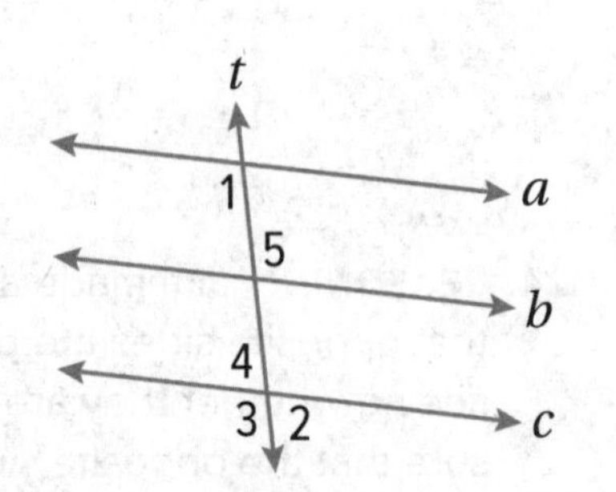

Perpendiculars and Distance

Explore Distance from a Point to a Line

Online Activity Use dynamic geometry software to complete the Explore.

INQUIRY How do you measure the distance between a point and a line?

Learn Distance Between a Point and a Line

There is an infinite number of lines that intersect a line and pass through a given point not on the line. However, when determining the distance between the line and the point, you must find the shortest distance between the two. This distance is the length of the segment that is perpendicular to the line through the point.

Key Concept • Distance Between a Point and a Line

The distance between a line and a point not on the line is the length of the segment perpendicular to the line from the point.

Given $\overleftrightarrow{AB}$ and point C not on the line, there are an infinite number of lines that pass through the point and intersect the line. The shortest distance between the point and the line is the length of the segment that is perpendicular to the line through the point. So, the distance between C and $\overleftrightarrow{AB}$ is CD.

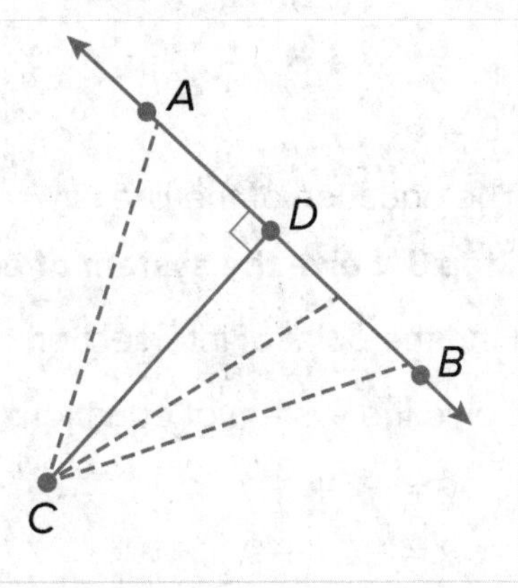

Just as there is one shortest distance from C to $\overleftrightarrow{AB}$, there is exactly one line that passes through C and is perpendicular to $\overleftrightarrow{AB}$.

Postulate 3.14: Perpendicular Postulate

If given a line and a point not on the line, then there exists exactly one line through the point that is perpendicular to the given line.

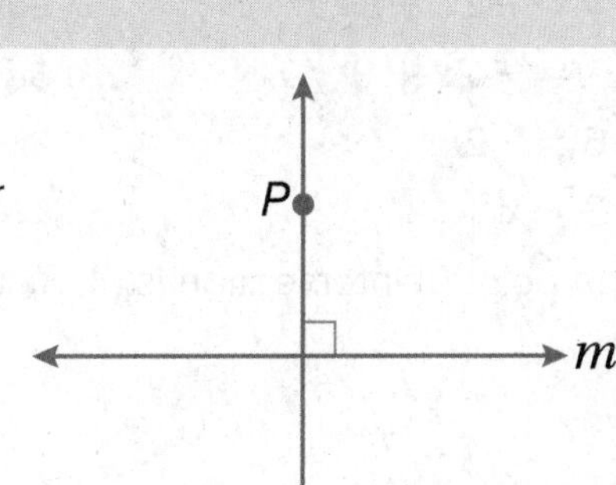

Today's Goals

- Use perpendicular lines to find the distance between a point and a line.
- Find the distance between parallel lines by using perpendicular distance.

Today's Vocabulary

equidistant

Go Online
You can watch a video to see how to find the shortest distance between a point and a line on the coordinate plane.

Think About It!
If line m and point P are on the coordinate plane and P is not on line m, how do you find the distance between P and line m?

Your Notes

Example 1 Distance from a Point to a Line on the Coordinate Plane

Line ℓ contains points (1, 2) and (5, 4). Find the distance between line ℓ and the point $P(1, 7)$.

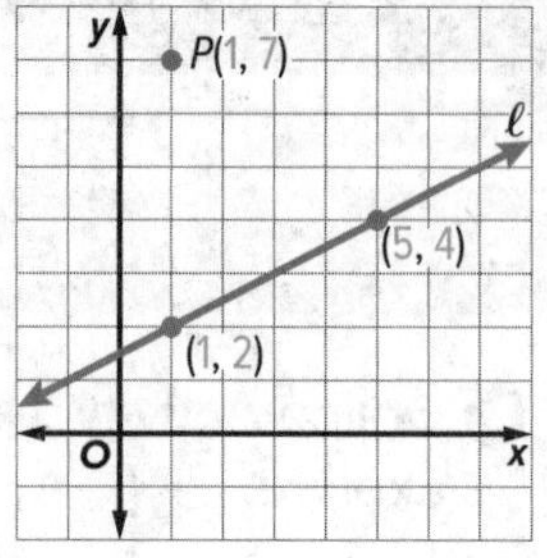

Step 1 Find the equation of line ℓ.

Begin by finding the slope of the line through points (1, 2) and (5, 4).

$m = \frac{y_2 - y_1}{x_2 - x_1} = \frac{4 - 2}{5 - 1} = \frac{2}{4}$ or $\frac{1}{2}$

Then write the equation of the line using the point (1, 2).

$y = mx + b$	Slope-intercept form
$2 = \frac{1}{2}(\_\_) + b$	$m = \frac{1}{2}$ and $(x, y) = (1, 2)$
$\_\_ = \frac{1}{2} + b$	Simplify.
$\_\_ = b$	Subtract $\frac{1}{2}$ from each side.

The equation of line ℓ is $y = \frac{1}{2}x + \frac{3}{2}$.

Step 2 Find the equation of the line perpendicular to line ℓ.

Write the equation of line w that is perpendicular to line ℓ and contains $P(1, 7)$. Because the slope of line ℓ is $\frac{1}{2}$, the slope of line w is -2. Write the equation of line w through $P(1, 7)$ with slope -2.

$y = mx + b$	Slope-intercept form
$\_\_ = \_\_(\_\_) + b$	$m = \_\_$; $(x, y) = (\_\_, \_\_)$
$\_\_ = \_\_ + b$	Simplify.
$9 = b$	Add __ to each side.

The equation of the line is $y =$ ________.

Step 3 Solve the system of equations.

Find the point of intersection of lines ℓ and w.

Solve the system of equations to determine the point of intersection.

$y = \frac{1}{2}x + \frac{3}{2}$	Equation of line ℓ
$y = -2x + 9$	Equation of line w
$y = \_\_$	Solve for y.

Solve for x.

$y = -2x + 9$	Equation of line w
$\_\_ = -2x + 9$	Substitute 3 for y.
$-6 = -2x$	Subtract 9 from each side.
$3 = x$	Divide each side by -2.

The point of intersection is (3, 3). Let this be point Q.

Talk About It!

Why do you use the perpendicular distance to find the distance between a line and a point not on the line?

Study Tip

Solving Systems of Equations Systems of equations can be solved by graphing, substitution, or elimination. Keep this in mind when you are finding the intersection point of perpendicular lines.

Go Online You can complete an Extra Example online.

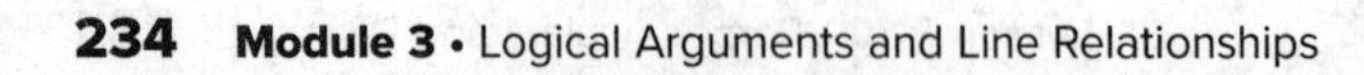

Step 4 Calculate the distance between *P* and *Q*.

Use the Distance Formula to determine the distance between $P(1, 7)$ and $Q(3, 3)$.

$$d = \sqrt{(x_2 - x_1)^2 + (y_2 - y_1)^2} \quad \text{Distance Formula}$$
$$= \sqrt{(3 - 1)^2 + (3 - 7)^2} \quad x_2 = 3, x_1 = 1 \text{ and } y_2 = 3, \text{ and } y_1 = 7$$
$$= \sqrt{20} \quad \text{Simplify.}$$

The distance between point P and line ℓ is $\sqrt{20}$ or about 4.47 units.

Check

Line n contains points $(-5, 3)$ and $(4, -6)$. Find the distance between line n and point $Q(2, 4)$. Round to the nearest tenth, if necessary.

_____ units

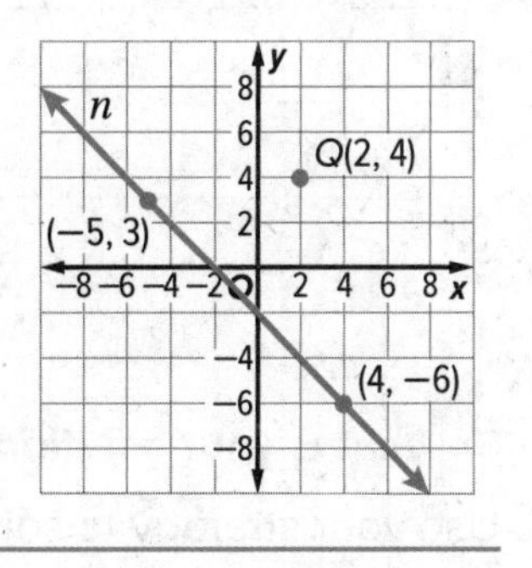

Study Tip

Units of Measure When finding the distance between a point and a line on the coordinate plane, your final measurement should be labeled with *units* unless the problem is set in a real-world context.

Apply Example 2 Solve a Design Problem by Using Distance

AMUSEMENT PARK **The developers of an amusement park want to build a new attraction. According to park regulations, the entrance to each attraction must be at least 10 yards from the center of Main Street. In the design plans, the entrance to the new attraction is located at $A(-6, -10)$, and Main Street contains the points $(-1, 3)$ and $(11, -9)$. If each unit represents 1 yard, will the new attraction comply with park regulations?**

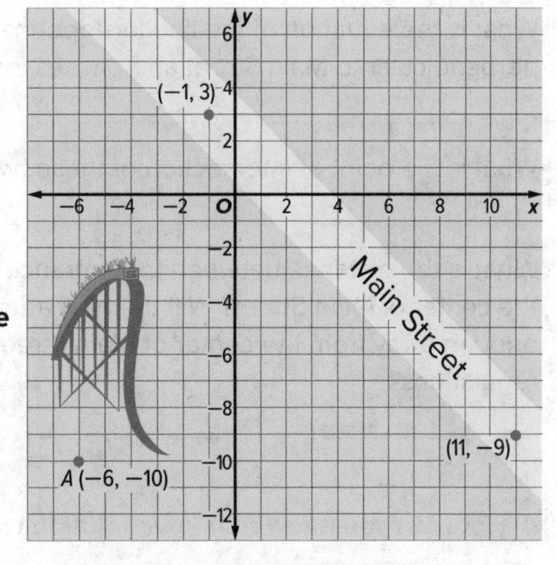

1 What is the task?

Describe the task in your own words. Then list any questions that you may have. How can you find answers to your questions?

____

____

____

____

(*continued on the next page*)

Study Tip

Estimation You can also use the horizontal distance between a line and a point not on the line to estimate the distance between the point and the line. When you graph the horizontal and perpendicular lines that contain the point and intersect the given line, a right triangle is created. The horizontal distance between the given point and line is the same as the length of the hypotenuse of the right triangle. So, you know the perpendicular distance, or the length of the right triangle's leg, must be less than the horizontal distance between the point and the given line.

2 How will you approach the task? What have you learned that you can use to help you complete the task?

3 What is your solution?

Use your strategy to solve the problem.

What is the equation of the line in slope-intercept form that represents Main Street?

What is the equation of the line in slope-intercept form that is perpendicular to Main Street and passes through point A?

What is the point of intersection of these two lines?

What is the distance between the entrance to the new attraction and the center of Main Street? Will the new attraction be located far enough away from the center of Main Street to comply with park regulations?

4 How can you know that your solution is reasonable?

Write About It! Write an argument that can be used to defend your solution.

Go Online You can complete an Extra Example online.

Check

ZONING Javier wants to build a shed on his property. According to zoning laws, the shed must be at least 20 feet from his property line. Javier knows that points A and B fall on his property line. If Javier plans to build the shed behind his house at point C, will he satisfy the zoning laws? If yes, how far away will the shed be from Javier's property line? Each unit on the coordinate plane represents 1 foot.

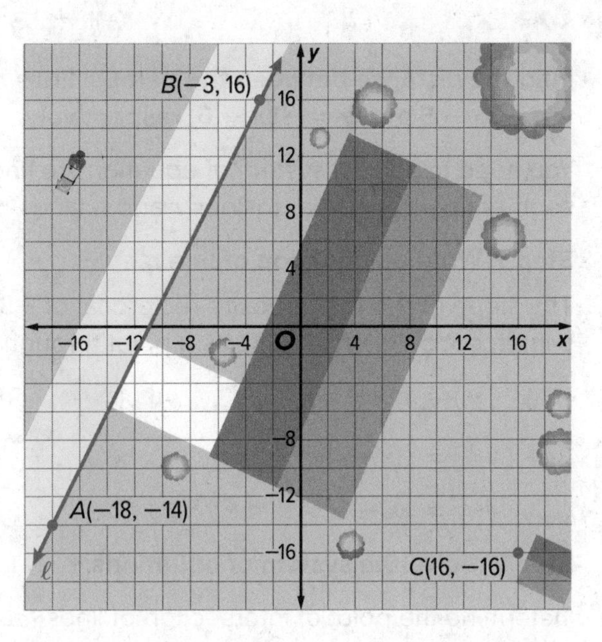

A. no B. yes; 31.3 ft C. yes; 34.1 ft D. yes; 37.2 ft

Learn Distance Between Parallel Lines

By definition, parallel lines do not intersect. An alternate definition states that two lines in a plane are parallel if they are always equidistant. Two lines are **equidistant** from each other if the distance between the two lines, measured along a perpendicular line or segment to the two lines, is always the same.

Think About It! How would you find the distance between two parallel planes?

Key Concept • Distance Between Parallel Lines

The distance between two parallel lines is the perpendicular distance between one of the lines and any point on the other line.

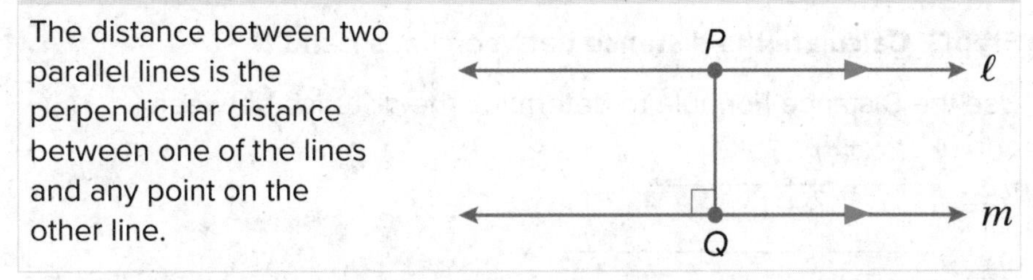

Theorem 3.24: Two Lines Equidistant from a Third

In a plane, if two lines are each equidistant from a third line, then the two lines are parallel to each other.

Example If line w and line v are equidistant from line x, then line w and line v are parallel.

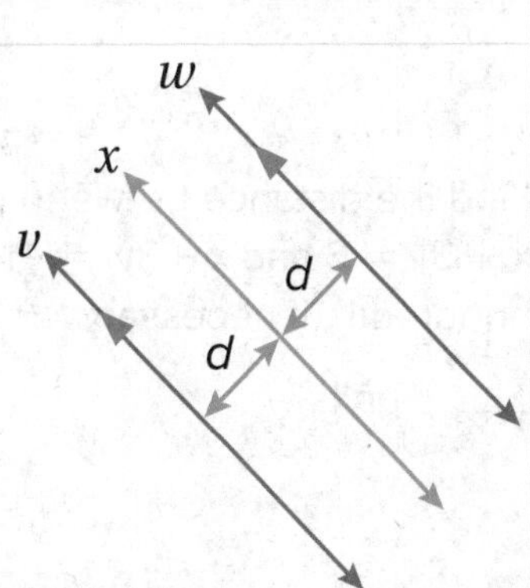

You will prove Theorem 3.24 in Exercise 28.

Example 3 Distance Between Parallel Lines

Find the distance between the parallel lines *r* and *t* with equations $y = -3x - 5$ and $y = -3x + 6$, respectively.

You need to solve a system of equations to find the endpoints of a segment perpendicular to lines *r* and *t*. Lines *r* and *t* have slope −3.

Step 1 Write an equation of line *q*.

The slope of *q* is the opposite reciprocal of −3, or $\frac{1}{3}$. Use the *y*-intercept of line *r*, (0, −5), as a point through which line *q* will pass.

$(y - y_1) = m(x - x_1)$ Point-slope form

$[y - (-5)] = \frac{1}{3}(x - 0)$ $x_1 = 0$, $y_1 = -5$, and $m = \frac{1}{3}$

$y = \frac{1}{3}x - 5$ Solve.

Step 2 Solve the system of equations.

Determine the point of intersection of lines *t* and *q*.

$t: y = -3x + 6$ $q: y = \frac{1}{3}x - 5$

$-3x + 6 = \frac{1}{3}x - 5$ Substitute.

$6 + 5 = \frac{1}{3}x + 3x$ Group like terms.

$\frac{33}{10} = x$ Solve.

Solve for *y* when $x = \frac{33}{10}$.

$y = \frac{1}{3}\left(\frac{33}{10}\right) - 5$ Substitute $\frac{33}{10}$ for *x* in the equation for *q*.

$y = -$ _____ Simplify.

The point of intersection is $\left(\frac{33}{10}, -\frac{39}{10}\right)$ or (_____, _____).

Step 3 Calculate the distance between lines *r* and *t*.

Use the Distance Formula to determine the distance between (0, −5) and (3.3, −3.9).

$d = \sqrt{(x_2 - x_1)^2 + (y_2 - y_1)^2}$ Distance Formula

$= \sqrt{(3.3 - 0)^2 + [-3.9 - (-5)]^2}$ $x_2 = 3.3$, $x_1 = 0$, $y_2 = -3.9$, and $y_1 = -5$

$\approx$ _____ Use a calculator.

The distance between the lines is about 3.5 units.

Check

Find the distance between parallel lines *a* and *b* with equations $x + 3y = 6$ and $x + 3y = -14$, respectively. Round to the nearest hundredth, if necessary.

_____ units

Go Online You can complete an Extra Example online.

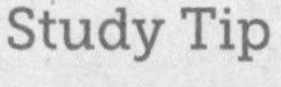

Study Tip

Perpendicular Lines When you are finding the equation of a line perpendicular to a pair of parallel lines, using the *y*-intercept of one of the parallel lines as a point contained by the perpendicular line will allow for easier calculations. The *y*-intercept is also used because it can easily be determined when the equation of a line is given in slope-intercept form.

Think About It!

Compare and contrast the processes for finding the distance between a point and a line and for finding the distance between parallel lines.

Name ______________________ Period ______________ Date ______________

Practice

Go Online You can complete your homework online.

Examples 1 and 2

Find the distance between point *P* and line ℓ.

1. Line ℓ contains points (0, −3) and (7, 4). Point *P* has coordinates (4, 3).

2. Line ℓ contains points (11, −1) and (−3, −11). Point *P* has coordinates (−1, 1).

3. Line ℓ contains points (−2, 1) and (4, 1). Point *P* has coordinates (5, 7).

4. Line ℓ contains points (4, −1) and (4, 9). Point *P* has coordinates (1, 6).

5. Line ℓ contains points (1, 5) and (4, −4). Point *P* has coordinates (−1, 1).

6. Line ℓ contains points (−8, 1) and (3, 1). Point *P* has coordinates (−2, 4).

7. DESIGN Dante is designing a poster for prom using a design program with a coordinate grid. He starts by creating a geometric border. Dante wants the text on the poster to be at least 3 inches away from the top left-hand corner of the border. The border contains the points (0, 7) and (7, 14). If Dante places the text at (7, 8), is the text at least 3 inches away from the border? If yes, how far away is the text from the border? Let every unit represent an inch. Round your answer to the nearest hundredth, if needed.

8. PHYSICS Mrs. Holmes's physics class is using 3D-printing software to create miniature bridges that can hold at least 5 pounds. Teams will print multiple parts of the bridges and then assemble the parts. One team wants there to be at least 6 inches between the upper rail and the lower rail of the bridge. The lower rail of the bridge contains the points (2, 10) and (16, 3). If the upper rail contains the point (5, 1), will the bridge meet the team's specifications? If yes, how far apart are the rails? Let every unit represent an inch. Round your answer to the nearest hundredth, if needed.

Example 3

Find the distance between each pair of parallel lines with the given equations.

9. $y = 7$
$y = -1$

10. $x = -6$
$x = 5$

11. $y = 3x$
$y = 3x + 10$

12. $y = -5x$
$y = -5x + 26$

13. $y = x + 9$
$y = x + 3$

14. $y = -2x + 5$
$y = -2x - 5$

15. $y = \frac{1}{4}x + 2$
$4y - x = -60$

16. $3x + y = 3$
$y + 17 = -3x$

17. $y = -\frac{5}{4}x + 3.5$
$4y + 10.6 = -5x$

Mixed Exercises

Find the distance from the line to the given point.

18. $y = -3$; $(5, 2)$

19. $y = \frac{1}{6}x + 6$; $(-6, 5)$

20. $x = 4$; $(-2, 5)$

21. TELEPHONE WIRES Isaiah works for a telephone company. He rewired some telephone wires on a pole. How can Isaiah use perpendicular distances to confirm that the wires are parallel?

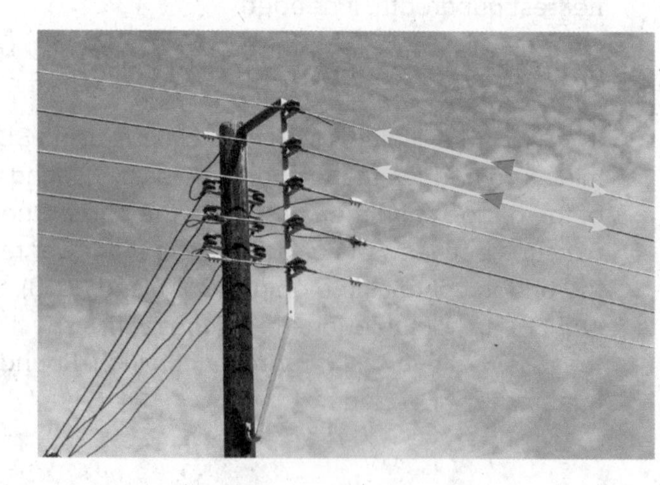

Name ____________________ Period __________ Date __________

22. STATE YOUR ASSUMPTION A city planner is designing a new park using a map of the city on a coordinate plane. The planner wants the entrance of the park to be at least 4 meters away from Washington Avenue. On the map, Washington Avenue contains the points (2, −4) and (11, −1).

a. If the city planner wants to build the entrance of the park at (3, 3), will the entrance be at least 4 meters away from Washington Avenue? If yes, how far away will the entrance be from the street? Let every unit represent 1 meter. Round your answer to the nearest hundredth, if needed.

b. What assumption did you make while solving this problem?

23. Construct the line through *G* perpendicular to $\overleftrightarrow{EF}$.

E G F

24. Construct the line through *P* perpendicular to *m*.

P m

25. REASONING The diagram at the right shows the path that Mark walked from the tee box to where his ball landed on the green. Is the path the shortest possible one from the tee box to the golf ball? Explain why or why not.

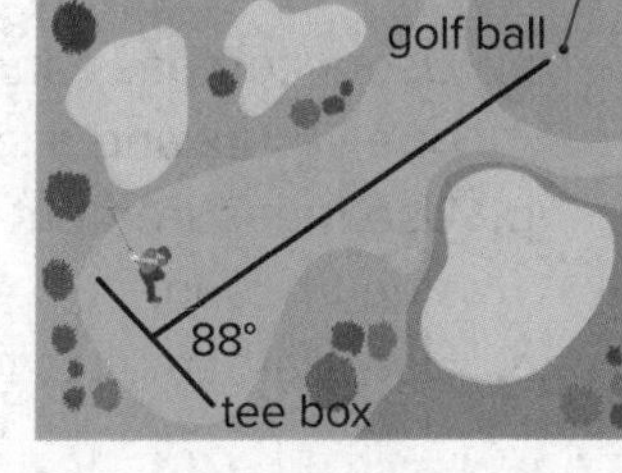

26. $\overline{AB}$ has a slope of 2 and midpoint *M*(3, 2). A segment perpendicular to $\overline{AB}$ has midpoint *P*(4, −1) and shares endpoint *B* with $\overline{AB}$.

a. Graph the segments.

b. Find the coordinates of *A* and *B*.

27. What does it mean if the distance between a point *P* and a line ℓ is zero? If the distance between two lines is zero?

28. PROOF Complete the two-column proof of Theorem 3.24.

Given: ℓ is equidistant to m, and n is equidistant to m.

Prove: $\ell \parallel n$

Proof:

Statements	Reasons
1. ℓ is equidistant to m, and n is equidistant to m.	**1.** ____?____
2.	**2.** Definition of equidistant
3.	**3.** Definition of parallel lines
4.	**4.** Transitive Property of Equality
5. $\ell \parallel n$	**5.** ____?____

29. WRITE Summarize the steps that are necessary to find the distance between a pair of parallel lines given the equations of the two lines.

30. PERSEVERE Suppose a line perpendicular to a pair of parallel lines intersects the lines at the points $(a, 4)$ and $(0, 6)$. If the distance between the parallel lines is $\sqrt{5}$, find the value of a and the equations of the parallel lines.

31. ANALYZE Determine whether the following statement is *sometimes, always,* or *never* true. Justify your argument.

The distance between a line and a plane can be found.

32. CREATE Draw an irregular convex pentagon using a straightedge.

a. Use a compass and straightedge to construct a perpendicular line between one vertex and a side opposite the vertex.

b. Use measurement to justify that the constructed line is perpendicular to the side chosen.

c. Use mathematics to justify this conclusion.

33. WRITE Rewrite Theorem 3.24 in terms of two planes that are equidistant from a third plane. Sketch an example.

34. FIND THE ERROR Harold draws the segments *AB* and *CD* shown below using a straightedge. He claims that these two lines, if extended in both directions, will never intersect. Olga claims that the lines will eventually intersect. Who is correct? Explain your reasoning.

A *B*

C *D*

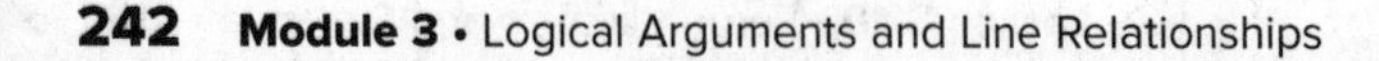

Review

Essential Question

What makes a logical argument, and how are logical arguments used in geometry?

Module Summary

Lessons 3-1 and 3-2

Conjectures and Logical Statements

- To show that a conjecture is not true for all cases, find a counterexample.
- An if-then statement is a compound statement of the form "if p, then q," where p and q are statements.
- The converse of a conditional statement is formed by exchanging the hypothesis and conclusion of the conditional statement. The inverse is formed by negating the hypothesis and the conclusion of the conditional statement. The contrapositive is formed by negating the hypothesis and the conclusion of the converse of the conditional statement.

Lessons 3-3 and 3-4

Reasoning and Proof

- If $p \rightarrow q$ is a true statement and p is true, then q is true.
- If $p \rightarrow q$ and $q \rightarrow r$ are true statements, then $p \rightarrow r$ is a true statement.
- A postulate or axiom is a statement that is accepted as true without proof.
- A proof contains statements and reasons that are organized to show progression from given information to a conclusion. Proofs can be in a two-column format, a flow format (using boxes and arrows), or in a paragraph format.

Lessons 3-5 and 3-6

Proving Segment and Angle Relationships

- The Angle Addition Postulate can be used with other angle relationships to prove theorems about supplementary and complementary angles.
- The properties of algebra that apply to the congruence of segments and the equality of their measures also hold true for the congruence of angles and the equality of their measures.

Lessons 3-7 through 3-10

Relationships Among Angles and Lines

- When two parallel lines are cut by a transversal, there are relationships between specific pairs of angles.
- If two lines are cut by a transversal so corresponding angles are congruent, then the lines are parallel.
- The distance between a line and a point not on the line is the length of the segment perpendicular to the line from the point.

Study Organizer

Foldables

Use your Foldable to review this module. Working with a partner can be helpful. Ask for clarification of concepts as needed.

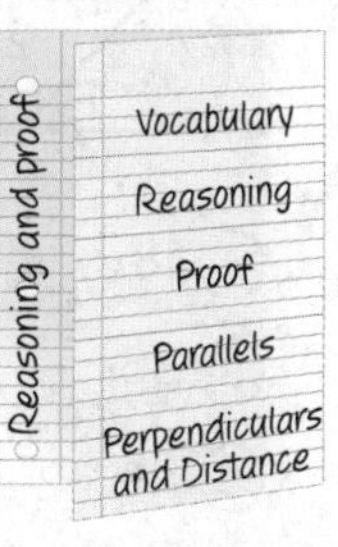

Name ______________ Period ______ Date ______

Test Practice

1. OPEN RESPONSE Point B is the midpoint of $\overline{AC}$, and point C is the midpoint of $\overline{AD}$. If $CD = 12$, what is AB? (Lesson 3-5)

2. OPEN RESPONSE Points X, Y, and Z are collinear, and Y is the midpoint of $\overline{XZ}$. Find the value of b. (Lesson 3-5)

X $2b + 7$ Y $3b - 4$ Z

3. MULTIPLE CHOICE Point B is the midpoint of $\overline{AC}$. $AB = 2x + 5$ and $BC = 5x - 1$. What is the length of $\overline{AB}$? (Lesson 3-5)

Ⓐ 2 units
Ⓑ 9 units
Ⓒ 18 units
Ⓓ 21 units

4. MULTIPLE CHOICE If $m\angle 1 = (2x)°$ and $m\angle 3 = (3x)°$, what is $m\angle 1$ in degrees? (Lesson 3-6)

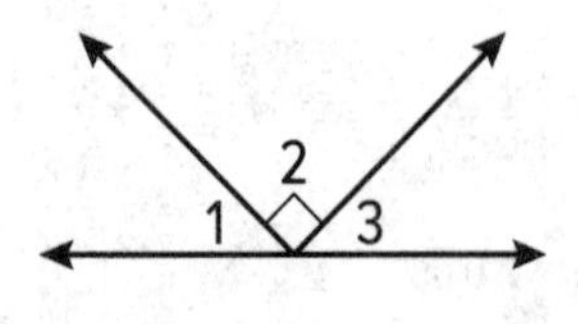

Ⓐ 18
Ⓑ 36
Ⓒ 54
Ⓓ 72

5. OPEN RESPONSE Use the figure below to find x, $m\angle ABD$, and $m\angle DBC$. (Lesson 3-6)

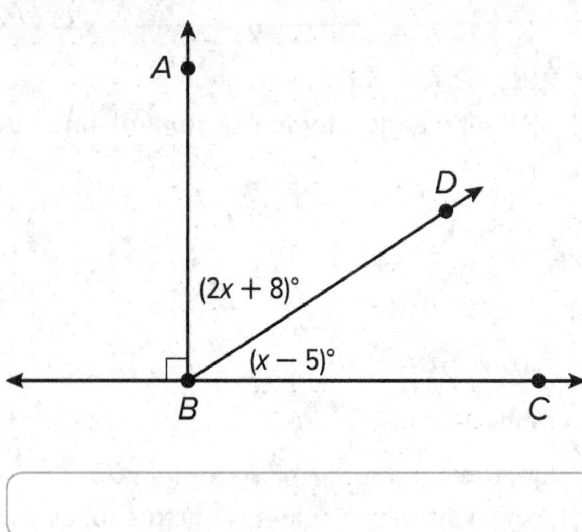

6. OPEN RESPONSE If Pine Street is parallel to Locust Street, find the values of a and b. (Lesson 3-7)

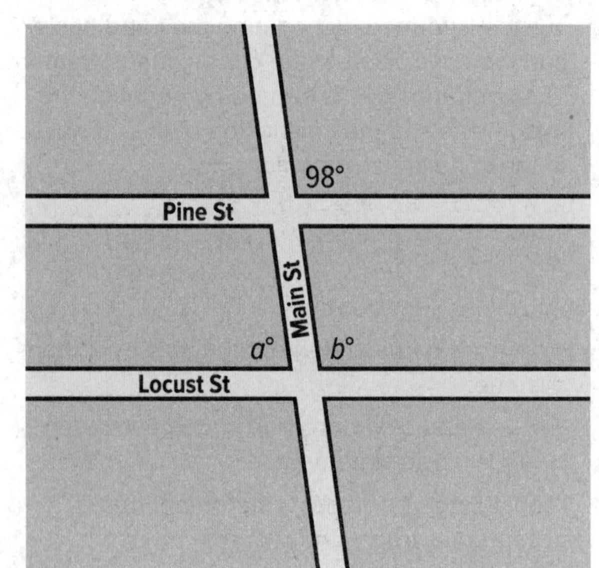

Name ______________________ Period __________ Date __________

7. MULTI-SELECT Select all the statements that describe parallel lines. (Lesson 3-7)

Ⓐ If lines are parallel, then they are coplanar.

Ⓑ If lines are parallel, then they are not coplanar.

Ⓒ If lines are parallel, then they intersect.

Ⓓ If lines are parallel, then they do not intersect.

Ⓔ If lines are parallel, then they are not skew.

8. OPEN RESPONSE A line passes through points at (9, 5) and (4, 3). What is the slope of the line perpendicular to this line? (Lesson 3-8)

9. TABLE ITEM Three lines have these equations:

Line m: $y = \frac{2}{3}x - 7$

Line n: $y = -\frac{2}{3}(x + 1)$

Line p: $y = -\frac{3}{2}x + 4$

Identify the relationship between each pair of lines. (Lesson 3-8)

Lines	m and n	m and p	n and p
parallel			
perpendicular			
neither			

10. OPEN RESPONSE Write an equation in slope-intercept form for the line that passes through (−3, 2), perpendicular to $y = \frac{1}{2}x + 9$. (Lesson 3-8)

11. OPEN RESPONSE If $\overleftrightarrow{AB} \parallel \overleftrightarrow{CD}$, what is $m\angle ACD$? (Lesson 3-9)

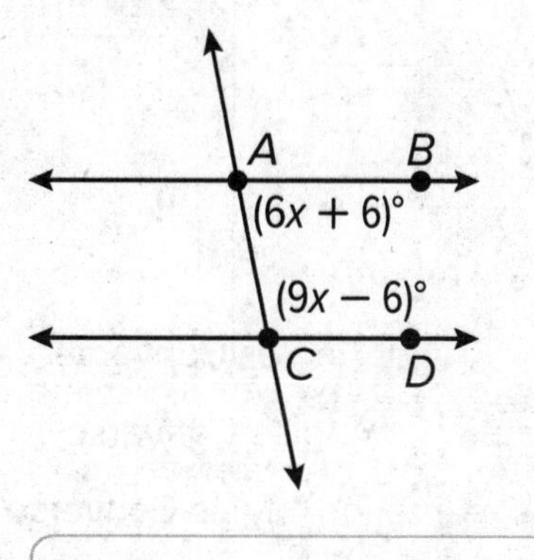

12. MULTIPLE CHOICE In the diagram, $\angle GDE$ and $\angle DEF$ are supplementary, but $\angle GDE$ is not congruent to $\angle EFG$.

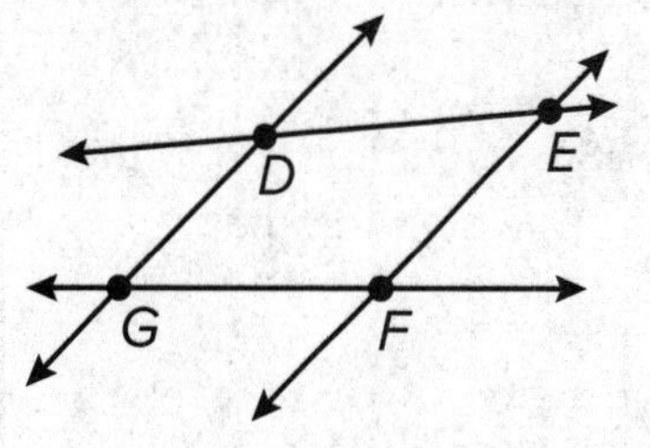

Which lines are parallel? (Lesson 3-9)

Ⓐ $\overleftrightarrow{DE} \parallel \overleftrightarrow{GF}$

Ⓑ $\overleftrightarrow{DG} \parallel \overleftrightarrow{EF}$

Ⓒ $\overleftrightarrow{DE} \parallel \overleftrightarrow{GF}$ and $\overleftrightarrow{DG} \parallel \overleftrightarrow{EF}$

Ⓓ Neither pair of lines is parallel.

13. **MULTI-SELECT** Using the given figure, which theorem(s) could be used to prove the lines are parallel? Select all that apply. (Lesson 3-9)

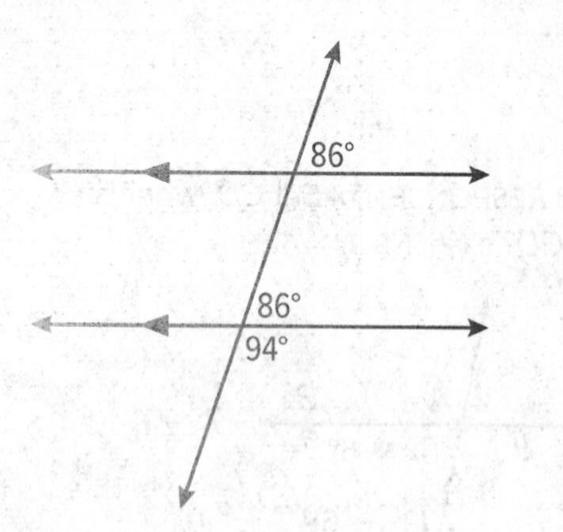

Ⓐ Alternate Exterior Angles Converse

Ⓑ Alternate Interior Angles Converse

Ⓒ Consecutive Interior Angles Converse

Ⓓ Corresponding Angles Converse

Ⓔ Perpendicular Transversal Converse

Ⓕ None of the above

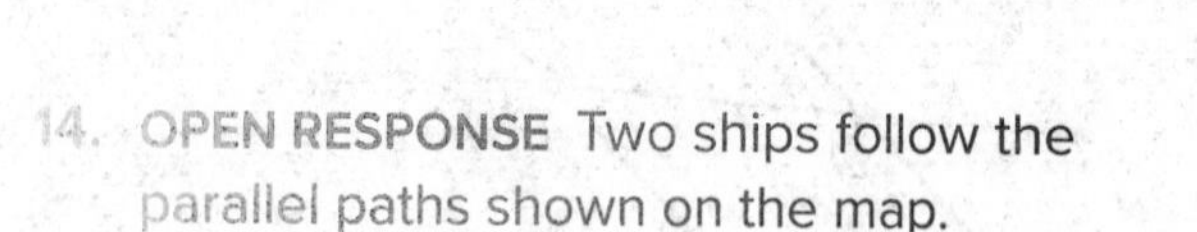

14. **OPEN RESPONSE** Two ships follow the parallel paths shown on the map.

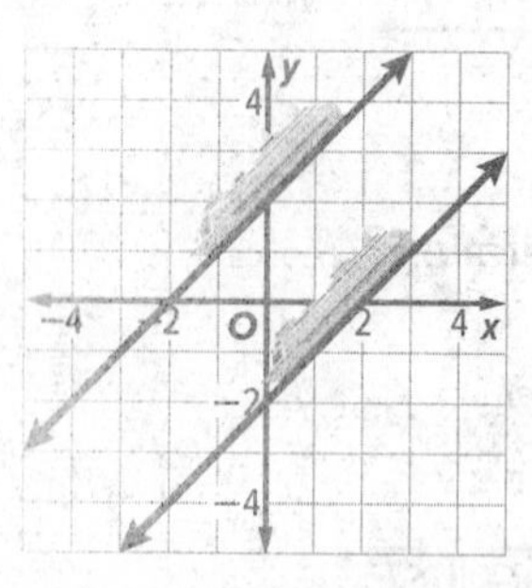

If 1 unit is 1 nautical mile, what is the shortest distance between the two paths? Round your answer to the nearest tenth. (Lesson 3-10)

15. **MULTIPLE CHOICE** Which indicates the correct order of steps for the construction of a perpendicular line through a point on the line using dynamic software? (Lesson 3-10)

W

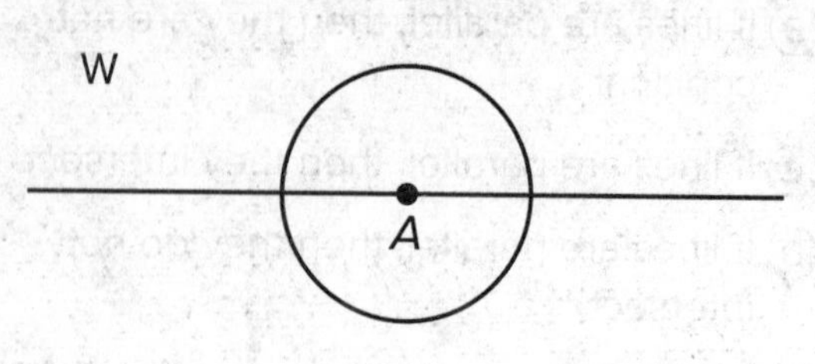

X

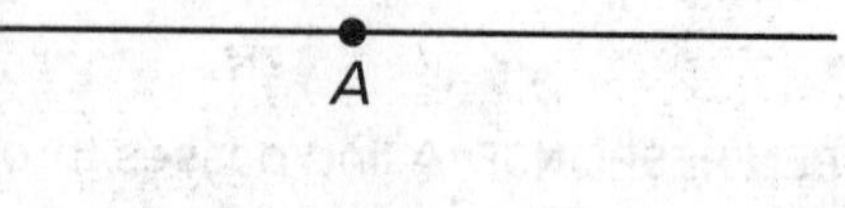

Y

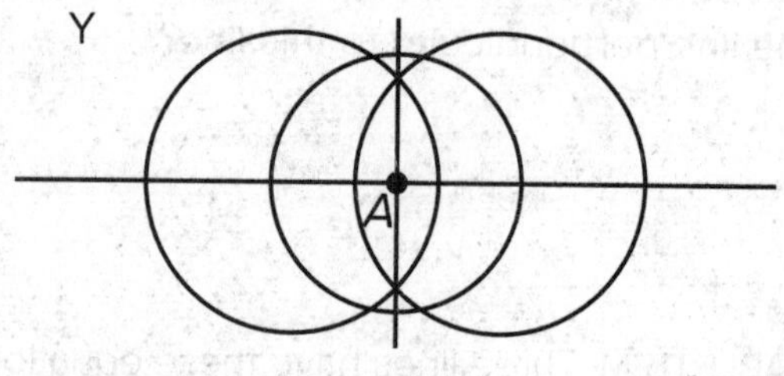

Z

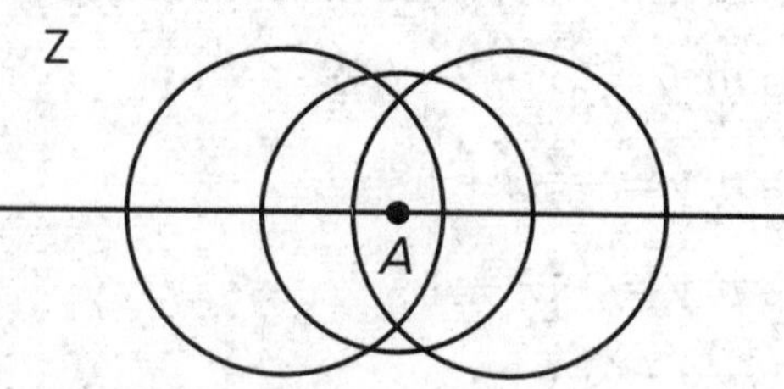

Ⓐ X, W, Y, Z

Ⓑ X, W, Z, Y

Ⓒ W, X, Y, Z

Ⓓ W, X, Z, Y

Module 4

Transformations and Symmetry

Essential Question

How are rigid motions used to show geometric relationships?

What Will You Learn?

Place a check mark (✓) in each row that corresponds with how much you already know about each topic **before** starting this module.

KEY

👎 — I don't know. 👉 — I've heard of it. 👍 — I know it!

	Before			After		
	👎	👉	👍	👎	👉	👍
define congruence in terms of rigid motions						
reflect figures						
draw and analyze reflected figures						
translate figures						
draw and analyze translated figures						
rotate figures						
draw and analyze rotated figures						
draw and analyze figures under multiple transformations						
identify tessellations						
identify line symmetries in two-dimensional figures						
identify rotational symmetries in two-dimensional figures						

Foldables Make this Foldable to help you organize your notes about transformations and symmetry. Begin with two sheets of paper.

1. **Fold** each sheet of paper in half.
2. **Open** the folded papers and fold each paper lengthwise two inches, to form a pocket.
3. **Glue** the sheets side-by-side to create a booklet.
4. **Label** each of the pockets as shown.

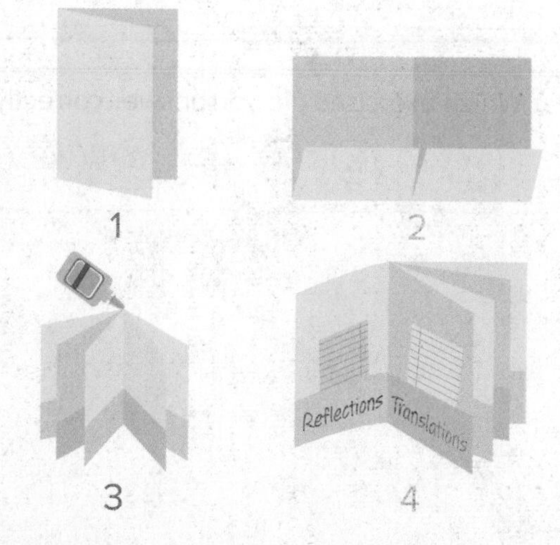

What Vocabulary Will You Learn?

Check the box next to each vocabulary term that you may already know.

- ☐ center of symmetry
- ☐ composition of transformations
- ☐ glide reflection
- ☐ line of symmetry
- ☐ line symmetry
- ☐ magnitude
- ☐ magnitude of symmetry
- ☐ order of symmetry
- ☐ point of symmetry
- ☐ point symmetry
- ☐ regular tessellation
- ☐ rotational symmetry
- ☐ semiregular tessellation
- ☐ symmetry
- ☐ tessellation
- ☐ uniform tessellation

Are You Ready?

Complete the Quick Review to see if you are ready to start this module.
Then complete the Quick Check.

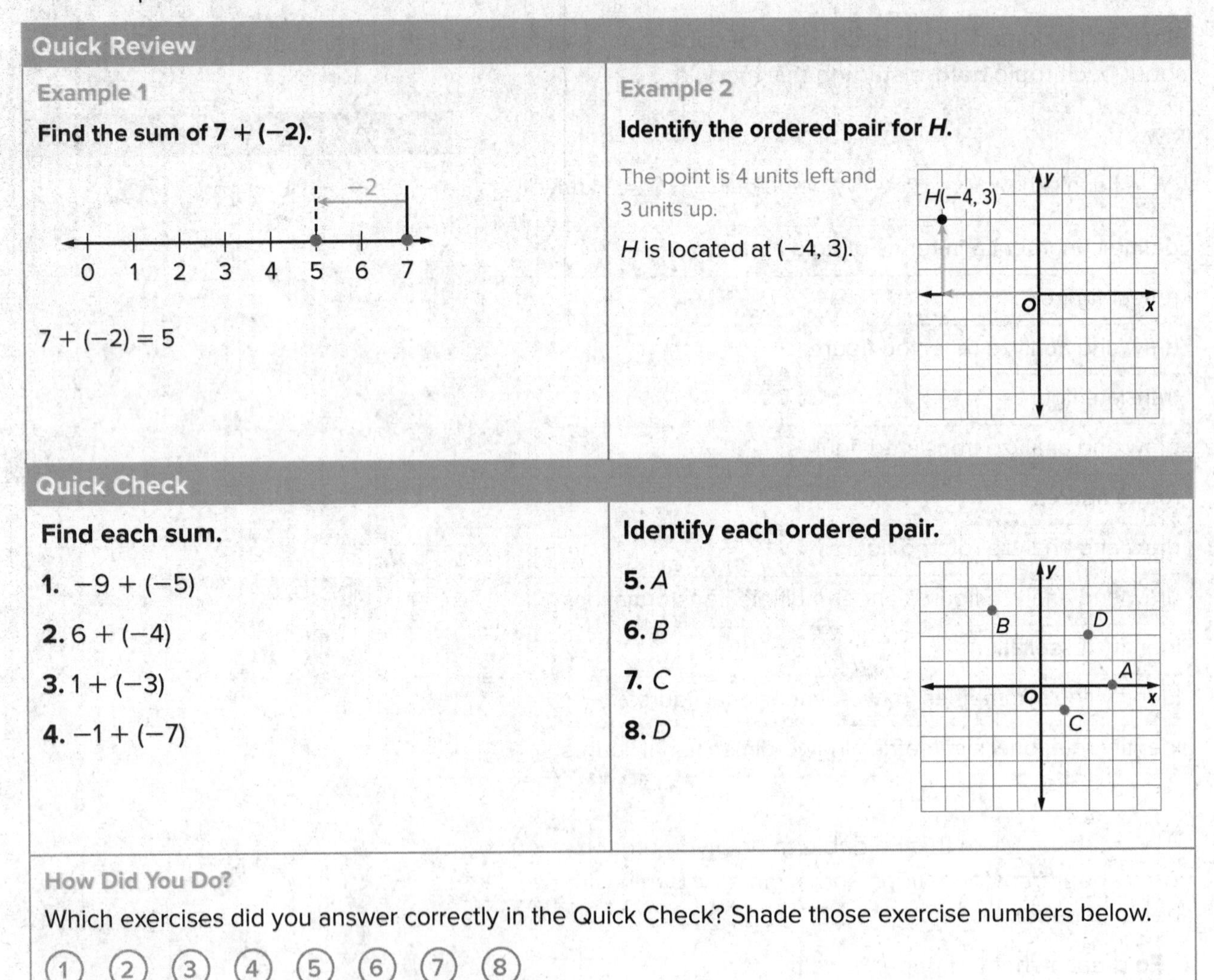

Quick Review

Example 1

Find the sum of $7 + (-2)$.

$7 + (-2) = 5$

Example 2

Identify the ordered pair for *H*.

The point is 4 units left and 3 units up.

H is located at $(-4, 3)$.

Quick Check

Find each sum.

1. $-9 + (-5)$
2. $6 + (-4)$
3. $1 + (-3)$
4. $-1 + (-7)$

Identify each ordered pair.

5. *A*
6. *B*
7. *C*
8. *D*

How Did You Do?

Which exercises did you answer correctly in the Quick Check? Shade those exercise numbers below.

① ② ③ ④ ⑤ ⑥ ⑦ ⑧

Reflections

Today's Goals
- Use rigid motions to reflect figures on the coordinate plane and describe the effects of the reflections.

Explore Developing the Definition of a Reflection

Online Activity Use dynamic geometry software to complete the Explore.

INQUIRY How can you define a reflection?

Learn Reflections

You've learned that when a figure is reflected in a line, each point of the preimage and its corresponding point on the image are the same distance from the line of reflection.

Key Concept • Reflection
Reflection in a Vertical Line
When a figure is reflected in a vertical line that is not the y-axis, the y-coordinates of the image remain the same as the preimage. The distance from a point in the preimage to the line of reflection is the same as the distance from the corresponding point in the image to the line of reflection.
Reflection in a Horizontal Line
When a figure is reflected in a horizontal line that is not the x-axis, the x-coordinates of the image remain the same as the preimage. The distance from a point in the preimage to the line of reflection is the same as the distance from the corresponding point in the image to the line of reflection.
Reflection in $y = x$
To reflect a point in the line $y = x$ interchange the x- and y-coordinates; $(x, y) \rightarrow (y, x)$.

Example 1 Reflection in a Horizontal or Vertical Line

Consider quadrilateral *RSTV* with vertices at *R*(2, 1), *S*(2, 4), *T*(5, 4), and *V*(5, 3). Graph the image of quadrilateral *RSTV* under each reflection. Determine the coordinates of the image.

a. in the line $y = -1$

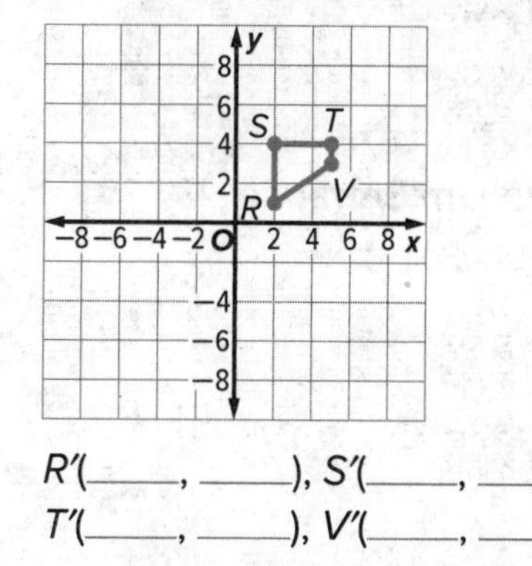

R'(____, ____), S'(____, ____),
T'(____, ____), V'(____, ____)

b. in the line $x = -2$

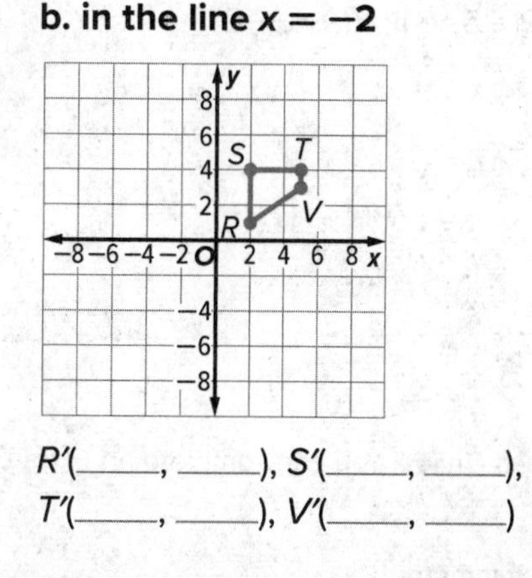

R'(____, ____), S'(____, ____),
T'(____, ____), V'(____, ____)

Talk About It!
Describe the result of the reflection.

Go Online
You can complete an Extra Example online.

Your Notes

Check

Triangle BCD has coordinates $B(-3, 3)$, $C(1, 4)$, and $D(-2, -4)$.

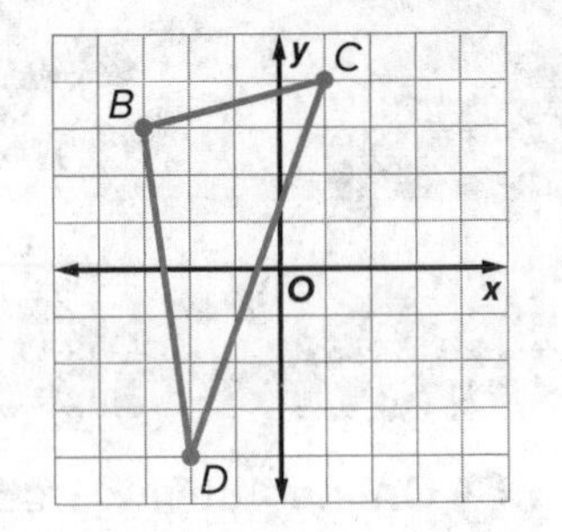

Select the coordinates of the vertices of the image after a reflection in the line $x = 3$.

A $B'(3, 3)$, $C'(-1, 4)$, $D'(2, -4)$

B $B'(-3, -3)$, $C'(1, -4)$, $D'(-2, 4)$

C $B'(9, 3)$, $C'(5, 4)$, $D'(8, -4)$

D $B'(-3, -3)$, $C'(1, 2)$, $D'(-2, 10)$

Example 2 Reflection in the Line $y = x$

DESIGN **Winona is designing a logo for her blog header. She graphs a figure on the coordinate plane and wants to reflect it in the line $y = x$ to complete the basic shape for her logo design. What are the coordinates of the vertices of the image after the reflection?**

$(x, y) \rightarrow (y, x)$

$A(-2, 1) \rightarrow A'(\_\_\_\_, \_\_\_\_)$

$B(1, 8) \rightarrow B'(8, 1)$

$C(1, 4) \rightarrow C'(\_\_\_\_, \_\_\_\_)$

$D(4, 4) \rightarrow D'(\_\_\_\_, \_\_\_\_)$

$E(2, 2) \rightarrow E'(2, 2)$

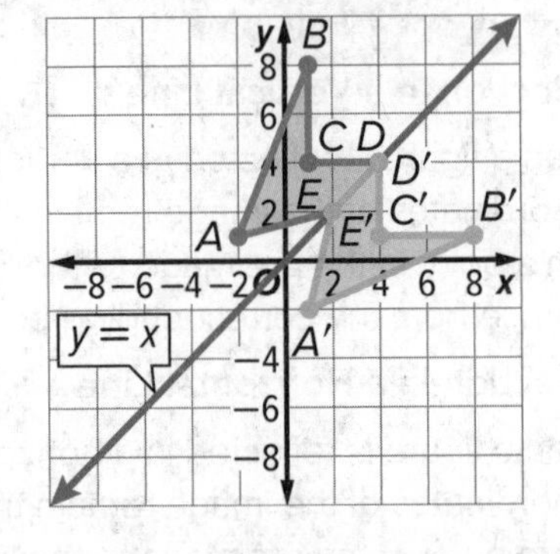

Check

LANDSCAPE Tomas is designing a sculpture garden for an art museum. There is a sidewalk connecting the center of the museum entrance to the edge of the lawn. Tomas has a set of 4 sculptures in a series that he wants to be equidistant from this sidewalk. He plotted positions for Q and X on the graph. If pieces Q and R are a pair and X and Y are a pair, where will pieces R and Y be placed in the garden?

A $R(3, 1)$, $Y(4, 2)$

B $R(3, -1)$, $Y(4, -2)$

C $R(-3, 1)$, $Y(-4, 2)$

D $R(-3, -1)$, $Y(-4, -2)$

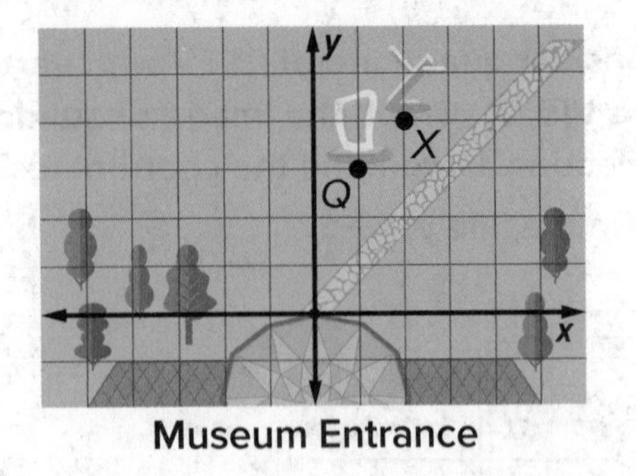

Museum Entrance

Go Online You may want to complete the construction activities for this lesson.

Go Online You can complete an Extra Example online.

Translations

Today's Goals
- Determine the translation vector.

Today's Vocabulary
magnitude

Explore Developing the Definition of a Translation

Online Activity Use dynamic geometry software to complete the Explore.

INQUIRY How can you define a translation?

Learn Translations

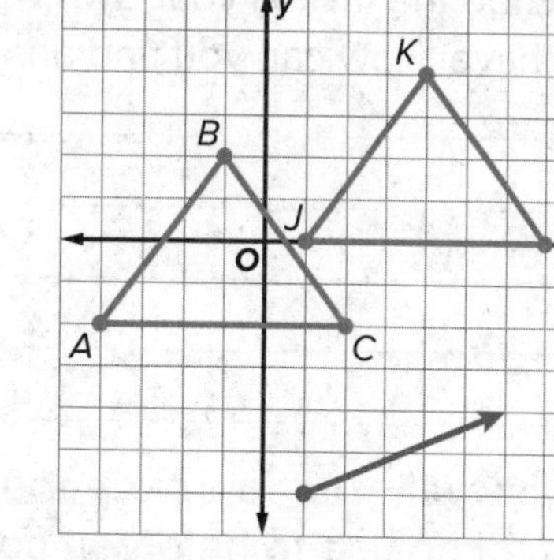

You've learned that a translation is a function in which all of the points of a figure move the same distance in the same direction as described by a translation vector.

When a translation has been applied to a figure:

1. The distance between each pair of corresponding vertices is the same.
2. The segments that connect each pair of corresponding vertices are parallel.

Recall that a translation vector describes the magnitude and direction of the translation. The **magnitude** of a vector is its length from the initial point to the terminal point.

Example 1 Determine a Translation Vector

Determine whether a translation maps $\triangle JKL$ onto $\triangle J'K'L'$. If so, find the translation vector. If not, explain why.

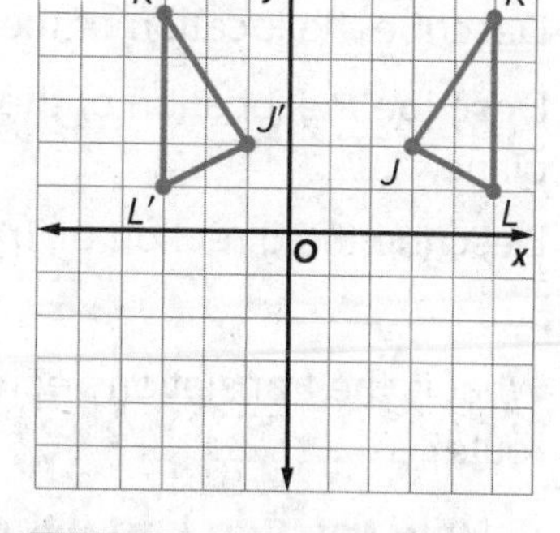

$\triangle J'K'L'$ is not a translation of $\triangle JKL$. The distances between corresponding vertices are not equal.

Study Tip

Translations If the distance between each pair of corresponding vertices is not the same, then the figure was not translated. You do not have to check that the slopes are the same.

Check

Determine whether a translation maps $\triangle JKL$ onto $\triangle J'K'L'$. If so, find the translation vector. If not, explain why.

A yes; $\langle 5, 3\rangle$
B yes; $\langle 3, 5\rangle$
C No; this is a reflection.
D No; the triangles are not congruent.

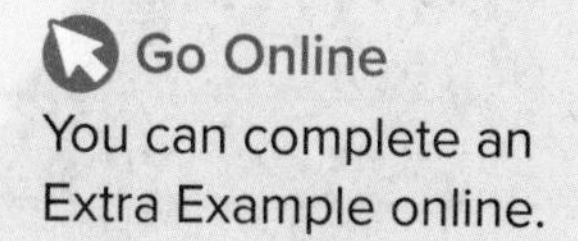

Your Notes

Talk About It!

Is the shortest distance between the buoy and the boat the same length as the translation vector? Justify your argument.

Apply Example 2 Translations on the Coordinate Plane

SCAVENGER HUNT **Travis is on a scavenger hunt at the lake. He needs to swim to the nearest buoy, pick up a card, and then swim until he reaches a fisherman who will give him the next clue in exchange for the card. Describe the translation from the buoy to the fisherman's boat by using a translation vector.**

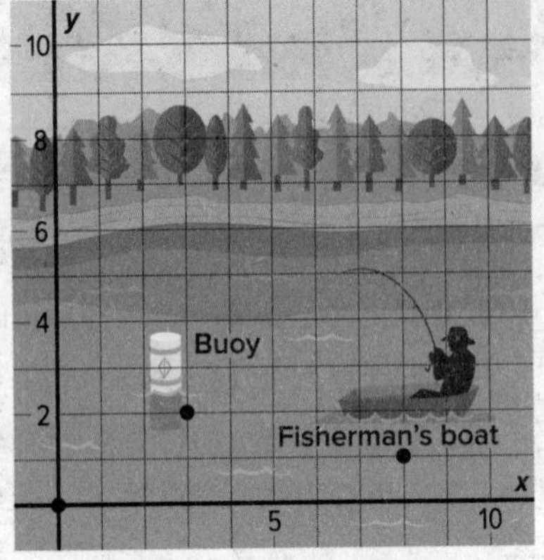

1 What is the task?

Describe the task in your own words. Then list any questions that you may have. How can you find answers to your questions?

2 How will you approach the task? What have you learned that you can use to help you complete the task?

3 What is your solution?

Use your strategy to solve the problem.

Describe the location of the buoy on the coordinate plane. ______

Describe the location of the fisherman's boat on the coordinate plane. ______

Describe the direction of Travis's route. ______

What is the translation vector that can be used to describe Travis's route? ______

4 How can you know that your solution is reasonable?

Write About It! Write an argument that can be used to defend your solution.

n complete an Extra Example online.

Rotations

Explore Developing the Definition of a Rotation

Online Activity Use dynamic geometry software to complete the Explore.

INQUIRY How can you define a rotation?

Learn Rotations About Points that Are Not the Origin

Key Concept • Rotation

A rotation about a fixed point P through an angle of $a°$ is a function that maps point M to point M' such that:

- point P does not move,
- $m\angle MPM'$ is $a°$, and
- $MP = M'P$

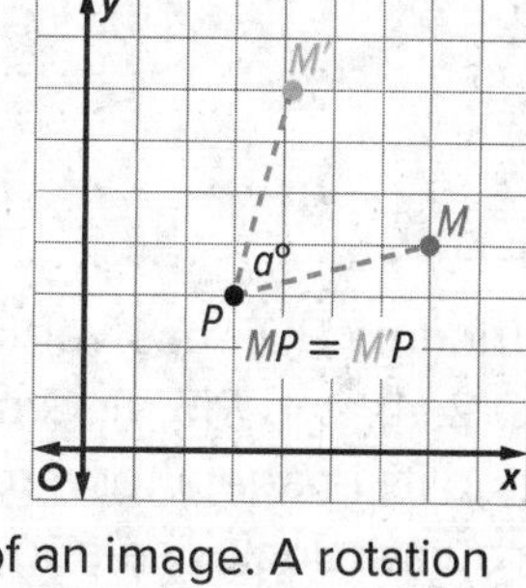

When a point is rotated 90°, 180°, or 270° counterclockwise about the origin, you can use the following rules to determine the coordinates of an image. A rotation of 360° will map an image onto the preimage.

Rotations on the Coordinate Plane (About the Origin)

90° Rotation $(x, y) \rightarrow (-y, x)$

180° Rotation $(x, y) \rightarrow (-x, -y)$

270° Rotation $(x, y) \rightarrow (y, -x)$

When combined with translations, these rules can also be used to rotate figures about points that are not the origin.

Example 1 Rotation About a Point That Is Not the Origin

Triangle ABC has vertices $A(-8, 5)$, $B(-6, 9)$, and $C(-3, 6)$. Graph $\triangle ABC$ and its image after a rotation of 180° about $(-5, 3)$.

Step 1 Graph $\triangle ABC$.

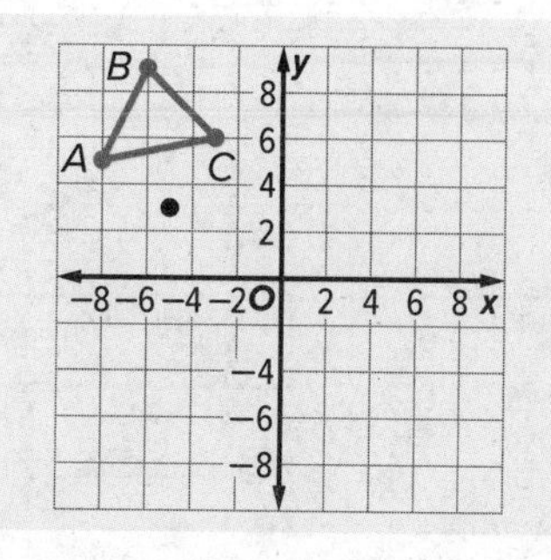

Step 2 Map the center of rotation to the origin.

To map the center of rotation to the origin, translate the center of rotation along the vector $\langle 5, -3 \rangle$. Then translate the vertices of $\triangle ABC$ along the same vector.

$(x, y) \rightarrow (x + 5, y - 3)$

$A(-8, 5) \rightarrow (-3, 2)$ $B(-6, 9) \rightarrow (-1, 6)$ $C(-3, 6) \rightarrow (2, 3)$

(continued on the next page)

Today's Goals

- Use rigid motions to rotate figures about points that are not the origin and describe the effects of the rotations.

Think About It!

Why does MP have to be equal to $M'P$ for the rotation to occur?

Go Online

You can complete an Extra Example online.

Your Notes

Step 3 Rotate 180° about the origin.

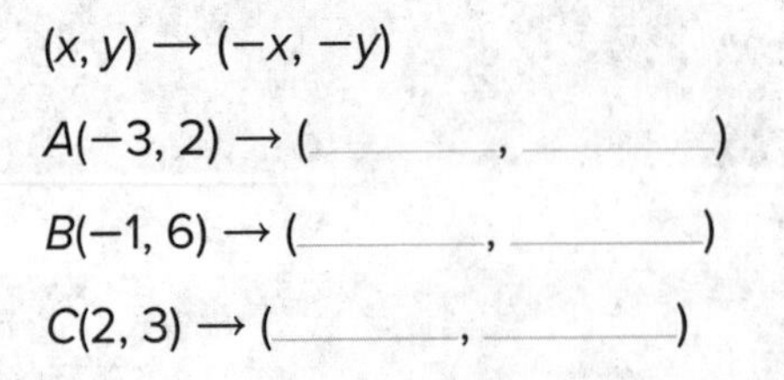

$(x, y) \rightarrow (-x, -y)$

$A(-3, 2) \rightarrow (\_\_\_\_, \_\_\_\_)$

$B(-1, 6) \rightarrow (\_\_\_\_, \_\_\_\_)$

$C(2, 3) \rightarrow (\_\_\_\_, \_\_\_\_)$

Step 4 Map the center of rotation to its original position.

To map the center of rotation to its original position, translate the center of rotation along the vector $\langle -5, 3 \rangle$. Then translate the vertices of the rotated triangle along the same vector.

$(x, y) \rightarrow (x - 5, y + 3)$

$A(3, -2) \rightarrow A'(\_\_\_\_, \_\_\_\_)$

$B(1, -6) \rightarrow B'(\_\_\_\_, \_\_\_\_)$

$C(-2, -3) \rightarrow C'(\_\_\_\_, \_\_\_\_)$

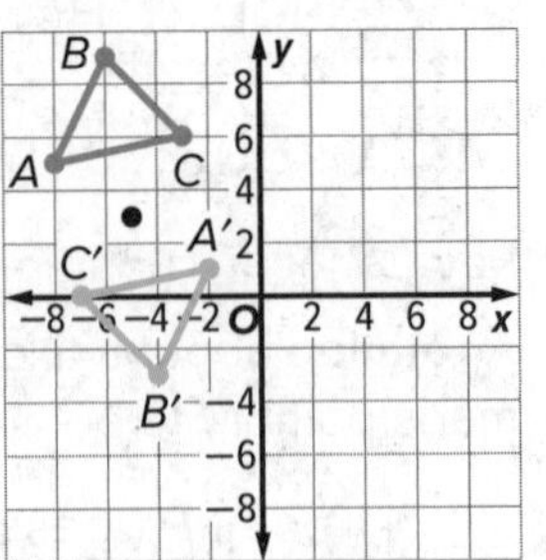

Check

Triangle PQR has vertices $P(2, 1)$, $Q(2, 4)$, and $R(5, 1)$. Graph $\triangle PQR$ and its image after a rotation 270° counterclockwise about (7, 5).

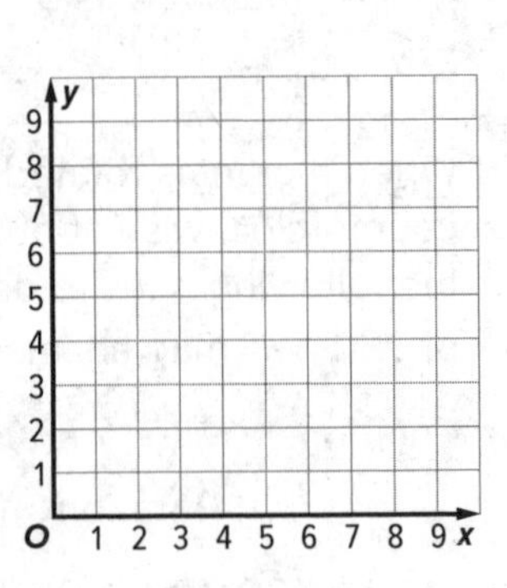

Study Tip

Approximations When you are describing the effects of a rotation, you can approximate the location of the rotated figure without making any calculations. You will know the shape and size of the image because angle measures and lengths are preserved.

Use a Source

Find an example of a flag that has elements that can be created using a rotation. Describe the center and angle of rotation.

Example 2 Describe the Effect of a Rotation

FLAGS **Kendrick is working with a team in his social studies class to create a new country and its government. Kendrick is responsible for creating the country's flag. He is using geometry software to design the flag on the coordinate plane. Describe how the two yellow stars would be affected if they were rotated 90° counterclockwise about the center of the white star.**

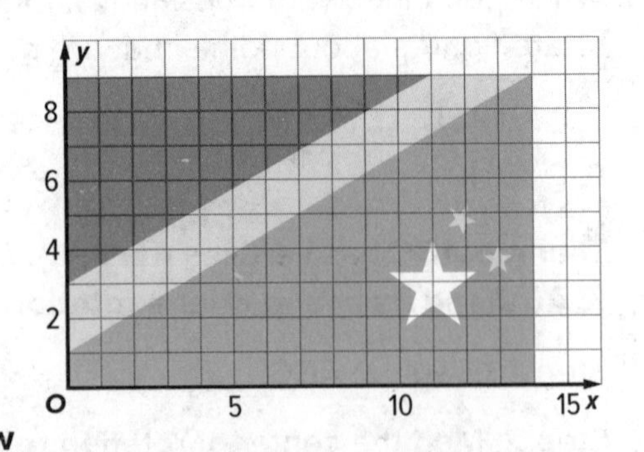

If the stars were rotated, they would curve around the top-left sides of the white star. Together the preimage and the image would create a semicircle of yellow stars above the white star.

Go Online You can complete an Extra Example online.

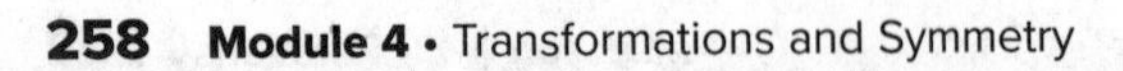

Name ______________________ Period ________ Date ________

Practice

Go Online You can complete your homework online.

Examples 1 and 2

1. Triangle XYZ has vertices $X(0, 2)$, $Y(4, 4)$, and $Z(3, -1)$. Graph $\triangle XYZ$ and its image after a rotation of 180° about $(2, -3)$.

2. Triangle ABC has vertices $A(1, 7)$, $B(3, 2)$, and $C(-2, -2)$. Graph $\triangle ABC$ and its image after a rotation of 270° counterclockwise about $(-4, 2)$.

3. Triangle FGH has vertices $F(-3, 4)$, $G(2, 0)$, and $H(-1, -2)$. Graph $\triangle FGH$ and its image after a rotation of 180° about $(-3, -6)$.

4. Quadrilateral $ABCD$ has vertices $A(-2, 4)$, $B(1, 3)$, $C(2, -3)$, and $D(-3, -1)$. Graph quadrilateral $ABCD$ and its image after a rotation of 90° counterclockwise about $(-1, 2)$.

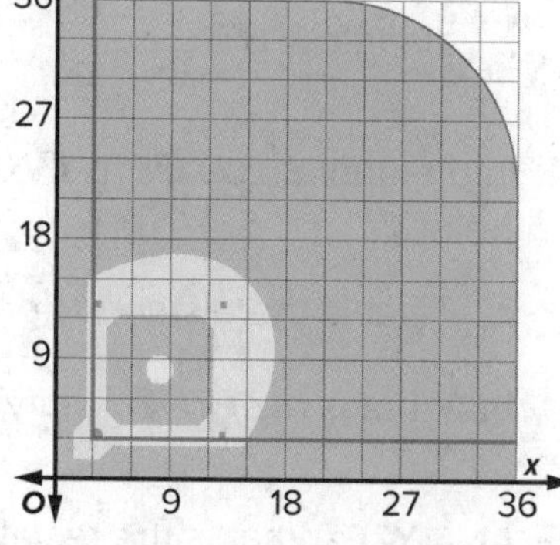

5. **BASEBALL** A scale drawing of a baseball field is shown on the coordinate plane, where home plate is at (3, 3), first base is at (13, 3), second base is at (13, 13), and third base is at (3, 13). Suppose the baseball field is rotated 270° counterclockwise about second base, what are the coordinates of each base?

Mixed Exercises

6. Point Q with coordinates $(4, -7)$ is rotated 270° clockwise about $(5, 1)$. What are the coordinates of its image?

7. Parallelogram $JKLM$ has vertices $J(2, 1)$, $K(7, 1)$, $L(6, -3)$, and $M(1, -3)$. What are the coordinates of the image of K if the parallelogram is rotated 270° counterclockwise about $(-2, -1)$?

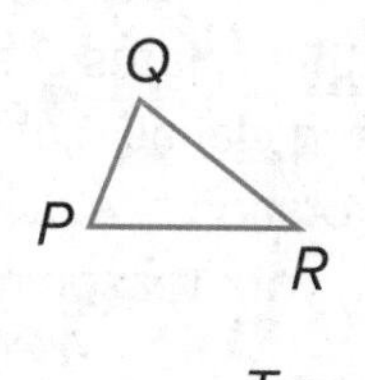

8. **USE TOOLS** Use a protractor and ruler to draw a rotation of $\triangle PQR$ 210° about T.

9. The line segment XY with endpoints $X(3, 1)$ and $Y(2, -2)$ is rotated 90° counterclockwise about $(-6, 4)$. What are the endpoints of $\overline{X'Y'}$?

10. **HIKING** A damaged compass points northwest instead of north. If you travel west by the compass, what is your angle of rotation to true north?

11. A circular dial with the digits 0 through 9 evenly spaced around its edge is rotated clockwise 36°. How many times would you have to perform this rotation to bring the dial back to its original position?

12. Under a rotation about the origin, the point $A(5, -1)$ is mapped to the point $A'(1, 5)$. What is the image of the point $B(-4, 6)$ under this rotation? Explain.

13. **CREATE** Draw a right triangle ABC and point P not on the triangle.

    a. Rotate triangle ABC about point P 90° counterclockwise.

    b. Name a clockwise rotation that would map triangle ABC onto triangle $A'B'C'$.

14. In the figure, $\triangle D'E'F'$ is the image of $\triangle DEF$ after a rotation about point Z.

    a. What is the distance from E' to Z? Justify your reasoning.

    b. What is $m\angle FZF'$? Justify your reasoning.

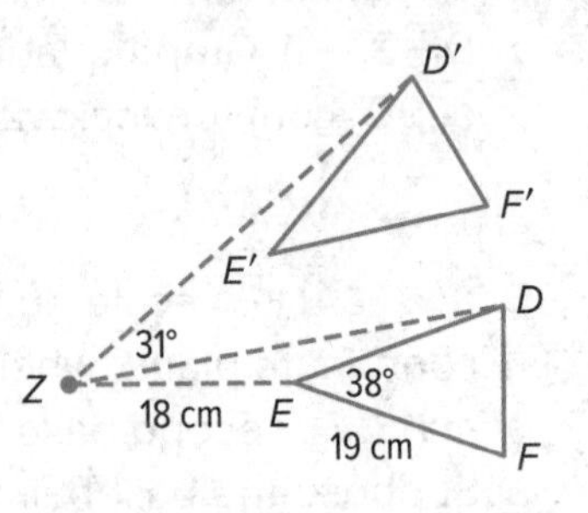

15. **ANALYZE** What is the result of a rotation followed by another rotation about the same point? Give an example.

16. **FIND THE ERROR** Thomas claims that a reflection in the x-axis followed by a reflection in the y-axis is the same thing as a rotation. Is Thomas correct? Explain your reasoning.

17. **WRITE** Which properties of a figure are preserved under a rotation from the preimage to the image? Explain.

18. **FIND THE ERROR** Shanice is looking at the figure shown, which shows two congruent triangles. She measures the angle that rotates A to A' about O and finds it to be 30°. She measures the angle that rotates B to B' about O and also finds it to be 30°. She then claims that because the two triangles are congruent, a 30° rotation has occurred about point O. Is Shanice correct? Explain your reasoning.

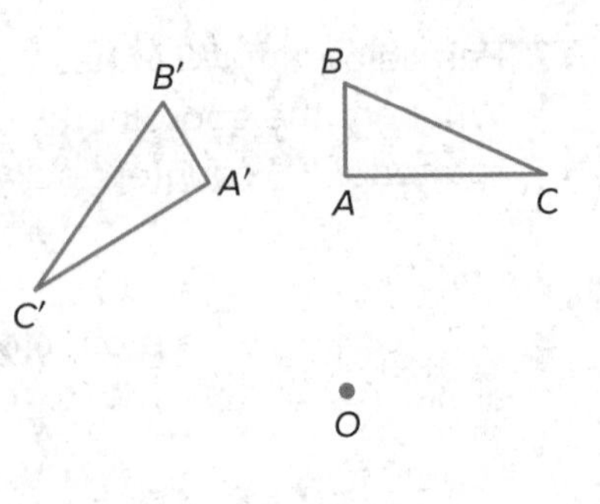

19. **WRITE** Are collinearity and betweenness of points maintained under rotations? Explain.

Compositions of Transformations

Explore Reflections in Two Lines

Online Activity Use dynamic geometry software to complete the Explore.

INQUIRY How is a figure affected by reflections in two lines?

Learn Compositions of Transformations

When a transformation is applied to a figure and then another transformation is applied to its image, the result is called a **composition of transformations**. A glide reflection is one type of composition of transformations.

A **glide reflection** is the composition of a translation followed by a reflection in a line parallel to the translation vector.

Theorem 4.1: Composition of Isometries

The composition of two (or more) isometries is an isometry.

You will prove one case of Theorem 4.1 in Exercise 29.

So, the composition of two or more isometries—reflections, translations, or rotations—results in an image that is congruent to its preimage.

Example 1 Glide Reflection

Triangle PQR has vertices $P(1, 1)$, $Q(2, 5)$, and $R(4, 2)$. Determine the coordinates of the vertices of the image after a translation along $\langle -4, 0\rangle$ and a reflection in the x-axis.

Step 1 Graph $\triangle PQR$.

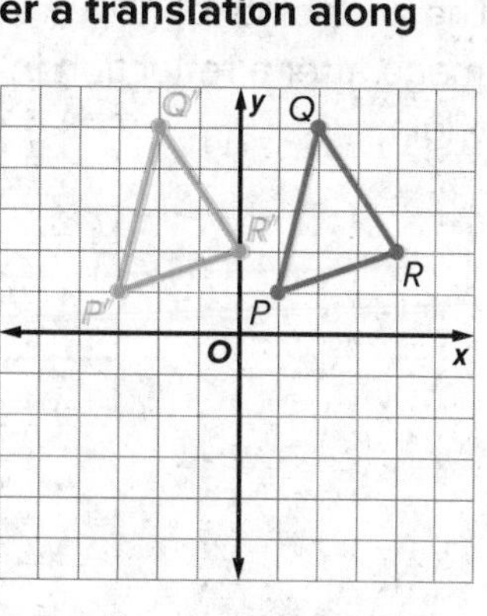

Step 2 Graph the image of $\triangle PQR$ after a translation along the vector $\langle -4, 0\rangle$.

$(x, y) \rightarrow (x - 4, y)$

$P(1, 1) \rightarrow P'(\_\_\_, \_\_\_)$

$Q(2, 5) \rightarrow Q'(\_\_\_, \_\_\_)$

$R(4, 2) \rightarrow R'(\_\_\_, \_\_\_)$

Step 3 Graph the image of $\triangle P'Q'R'$ after a reflection in the x-axis.

$(x, y) \rightarrow (x, -y)$

$P'(-3, 1) \rightarrow P''(\_\_\_, \_\_\_)$

$Q'(-2, 5) \rightarrow Q''(\_\_\_, \_\_\_)$

$R'(0, 2) \rightarrow R''(\_\_\_, \_\_\_)$

Go Online You can complete an Extra Example online.

Today's Goals

- Determine the image of a figure after a composition of transformations.
- Describe the transformation that produces the same image as a reflection in two lines.

Today's Vocabulary

composition of transformations
glide reflection

Think About It!

Compare and contrast glide reflections and compositions of transformations.

Study Tip

Compositions of Transformations Use double primes to indicate the image created by the second transformation in the composition.

Your Notes

Talk About It!

Do any points remain invariant, or unchanged, under glide reflections? under compositions of transformations? Explain.

Check

Triangle JKL has vertices $J(6, -1)$, $K(10, -2)$, and $L(5, -3)$. Determine the coordinates of the vertices of the image after a translation along $\langle 0, 4 \rangle$ and a reflection in the y-axis.

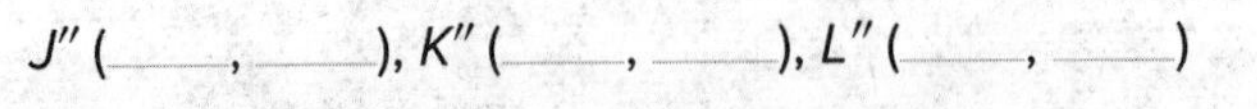

J'' (____, ____), K'' (____, ____), L'' (____, ____)

Example 2 Composition of Isometries

Triangle ABC has vertices $A(-6, -2)$, $B(-5, -5)$, and $C(-2, -1)$. Graph $\triangle ABC$ and its image after a rotation 180° about the origin and a translation along $\langle -2, 4 \rangle$.

Step 1 Graph $\triangle ABC$.

Step 2 Graph the image of $\triangle ABC$ after rotation 180° about the origin.

$(x, y) \rightarrow (-x, -y)$

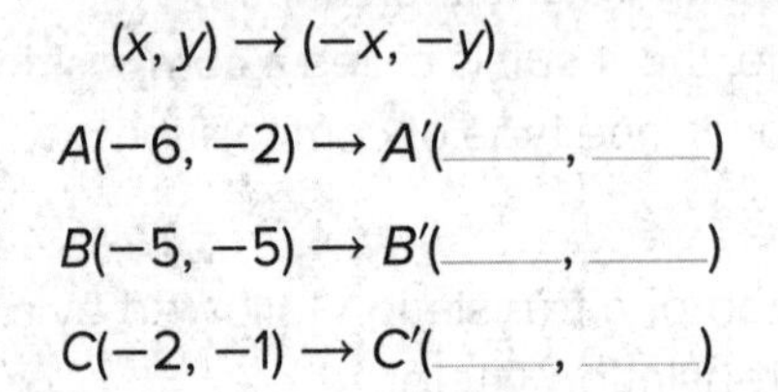

$A(-6, -2) \rightarrow A'($____, ____$)$

$B(-5, -5) \rightarrow B'($____, ____$)$

$C(-2, -1) \rightarrow C'($____, ____$)$

Step 3 Graph the image of $\triangle A'B'C'$ after a translation along $\langle -2, 4 \rangle$.

$(x, y) \rightarrow (x - 2, y + 4)$

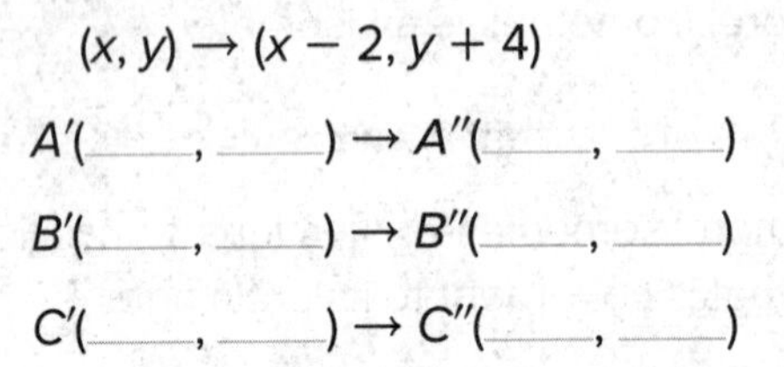

$A'($____, ____$) \rightarrow A''($____, ____$)$

$B'($____, ____$) \rightarrow B''($____, ____$)$

$C'($____, ____$) \rightarrow C''($____, ____$)$

Check

The endpoints of $\overline{CD}$ are $C(-7, 1)$ and $D(-3, 2)$. Graph $\overline{CD}$ and its image after a reflection in the x-axis and a rotation 90° about the origin.

Go Online You can complete an Extra Example online.

Learn Compositions of Two Reflections

The composition of two reflections can result in the same image as a translation or rotation.

Theorem 4.2: Reflections in Parallel Lines

The composition of two reflections in parallel lines can be described by a translation vector that is

- perpendicular to the two lines and
- twice the distance between the two lines.

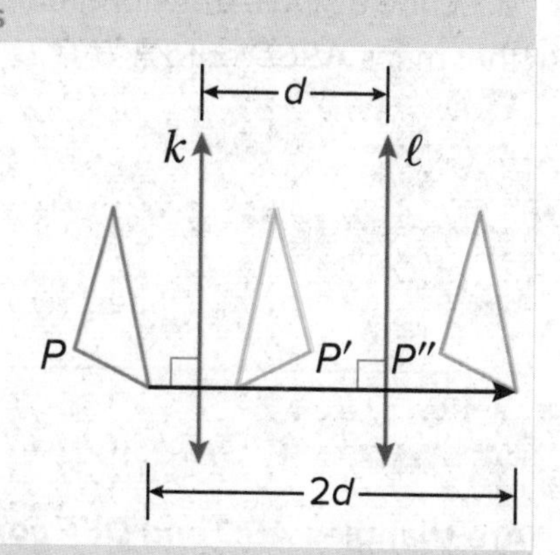

Theorem 4.3: Reflections in Intersecting Lines

The composition of two reflections in intersecting lines can be described by a rotation that is

- about the point where the lines intersect and
- through an angle that is twice the measure of the acute or right angle formed by the lines.

ℓ A′ A″ 2x° k x° P A

You will prove Theorem 4.2 in Exercise 30.

Think About It!

In what type of intersecting lines will a composition of two reflections form a 180° rotation?

Go Online

A proof of Theorem 4.3 is available.

Example 3 Reflect a Figure in Two Lines

Reflect each figure in line n and then line q. Then describe a single transformation that maps the preimage onto the final image.

a.

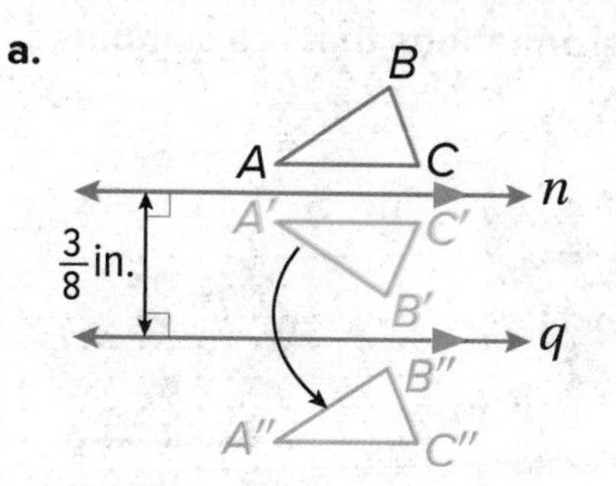

By Theorem 4.2, the composition of two reflections in parallel horizontal lines n and q is equivalent to a vertical translation down $2 \cdot \frac{3}{8}$ or $\frac{3}{4}$ inches.

b.

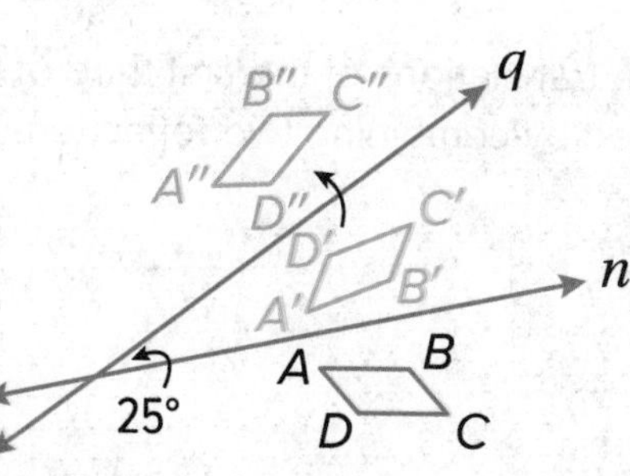

By Theorem 4.3, the composition of two reflections in intersecting lines n and q is equivalent to a $2 \cdot 25°$ or $50°$ counterclockwise rotation about the intersection point of lines n and q.

Go Online You can complete an Extra Example online.

Check

Reflect quadrilateral $ABCD$ in line m and then line n. Then describe a single transformation that maps $ABCD$ onto $A''B''C''D''$.

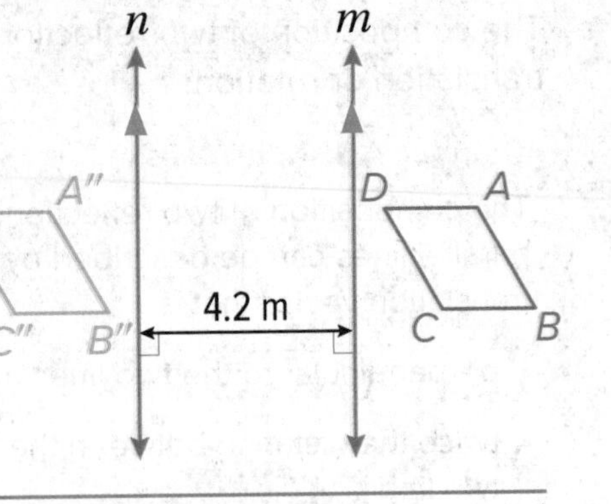

Example 4 Determine Congruence

Are triangles ABC and DEF congruent? If so, what composition of transformations maps $\triangle ABC$ onto $\triangle DEF$?

Because $\triangle ABC$ can be mapped onto $\triangle DEF$ by a ________ about the origin and a ________ along vector ________, $\triangle ABC$ ________ $\triangle DEF$.

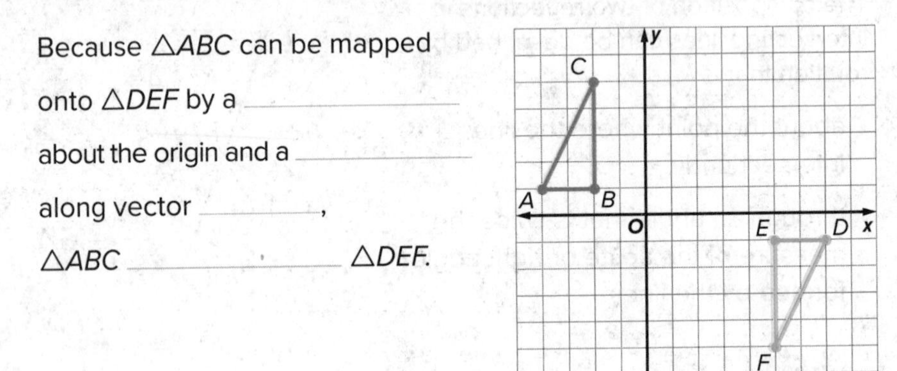

Think About It!

How can you tell that $\triangle ABC$ needs to be rotated 180° before being translated? Use the position of $\triangle DEF$ to justify your argument.

Go Online

An alternate method is available for this example.

Example 5 Describe Transformations

DESIGN PATTERNS **Describe the transformations that are combined to create the pattern shown.**

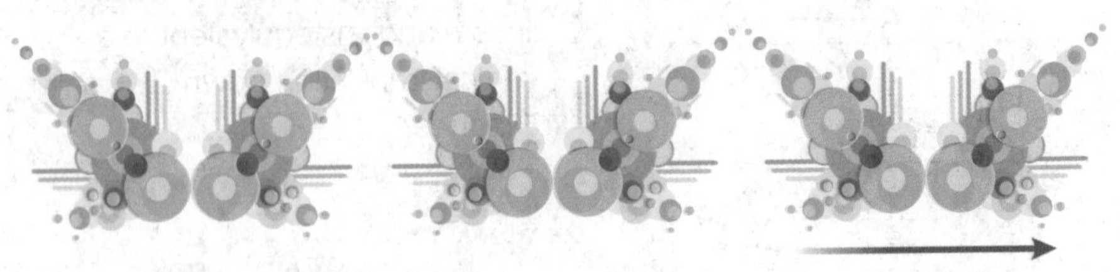

The pattern is created by successive translations of the first third of the design. So this pattern can be created by combining two reflections in a pair of parallel lines.

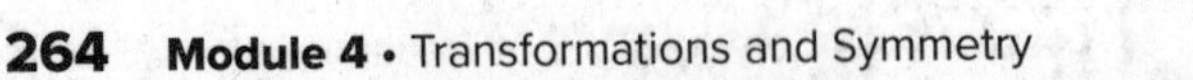

Go Online You can complete an Extra Example online.

Name ______________________ Period ____________ Date ____________

Practice

Go Online You can complete your homework online.

Example 1

Graph each figure with the given vertices and its image after the indicated glide reflection.

1. $\triangle RST$: $R(1, -4)$, $S(6, -4)$, $T(5, -1)$
Translation: along $\langle 2,0 \rangle$
Reflection: in x-axis

2. $\triangle JKL$: $J(1, 3)$, $K(5, 0)$, $L(7, 4)$
Translation: along $\langle -3, 0 \rangle$
Reflection: in x-axis

3. $\triangle DFG$: $D(2, 8)$, $F(1, 2)$, $G(4, 6)$
Translation: along $\langle 3, 3 \rangle$
Reflection: in $y = x$

4. $\triangle MPQ$: $M(-4, 3)$, $P(-5, 8)$, $Q(-1, 6)$
Translation: along $\langle -4, -4 \rangle$
Reflection: in $y = x$

Example 2

Graph each figure with the given vertices and its image after the indicated composition of transformations.

5. $\overline{WX}$: $W(-4, 6)$ and $X(-4, 1)$
Reflection: in x-axis
Rotation: 90° about origin

6. $\overline{AB}$: $A(-3, 2)$ and $B(3, 8)$
Rotation: 90° about origin
Translation: along $\langle 4, 4 \rangle$

7. $\overline{FG}$: $F(1, 1)$ and $G(6, 7)$
Reflection: in x-axis
Rotation: 180° about origin

8. $\overline{RS}$: $R(2, -1)$ and $S(6, -5)$
Translation: along $\langle -2, -2 \rangle$
Reflection: in y-axis

Example 3

Copy and reflect each figure in line u and then line v. Then describe a single transformation that maps the preimage onto the image.

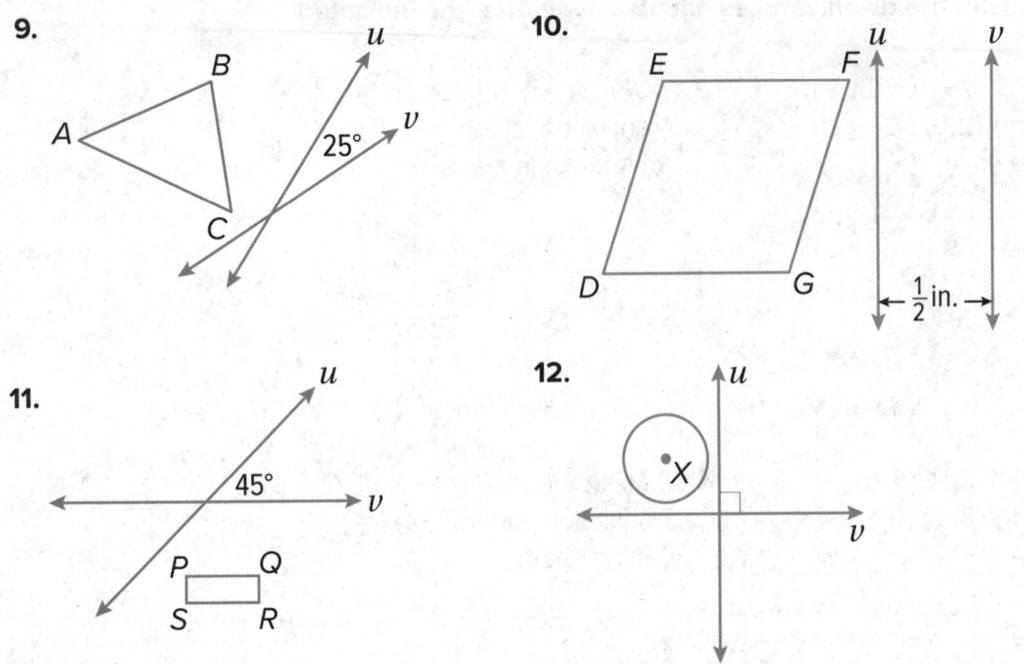

Example 4

Is $\triangle JKL$ congruent to $\triangle MNP$? If so, what composition of transformations maps $\triangle JKL$ onto $\triangle MNP$?

13.

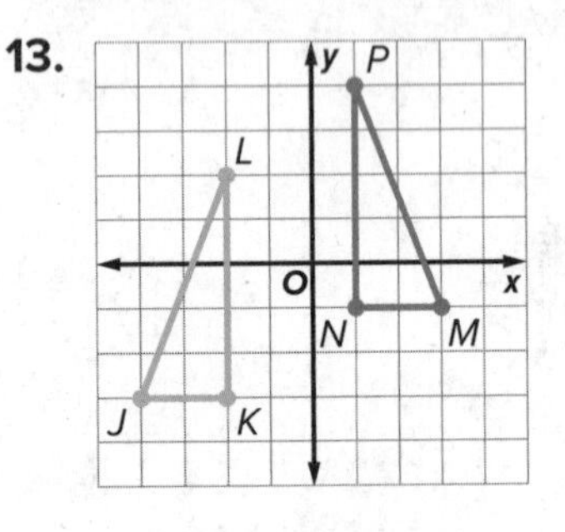

14.

15.

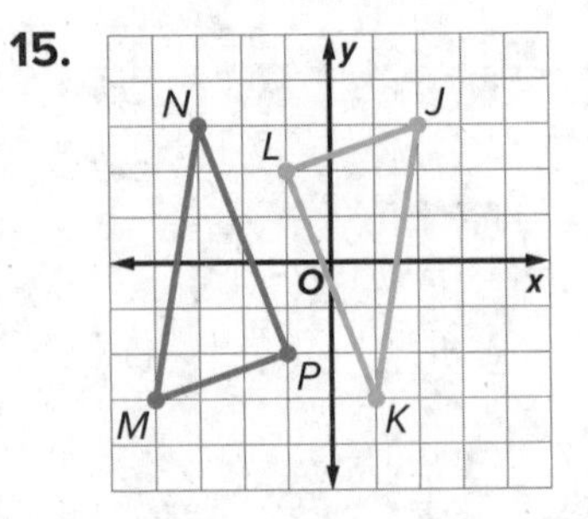

16.

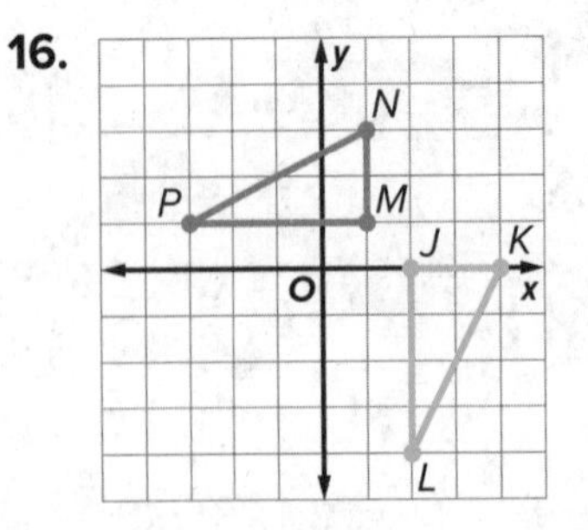

Example 5

17. Describe the transformations that are combined to create the border.

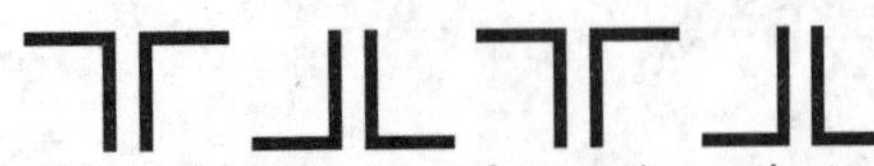

18. Describe the transformations that are combined to create the pattern.

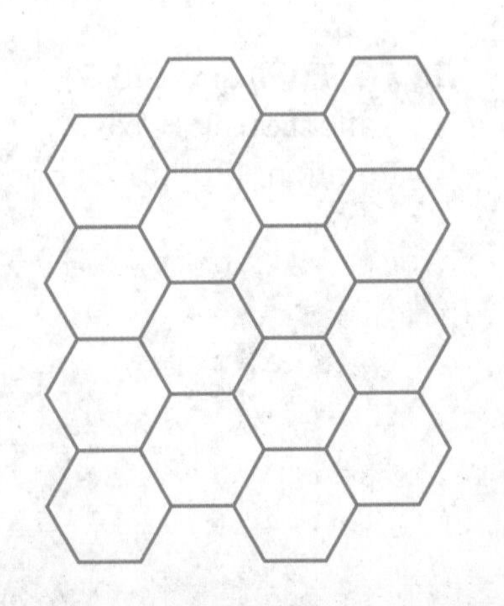

Name ____________ Period ________ Date ________

Mixed Exercises

Draw and label the image of each figure after the given composition of transformations.

19. 270° rotation about the origin followed by translation along ⟨2,−2⟩

20. reflection in the y-axis followed by 180° rotation about the origin

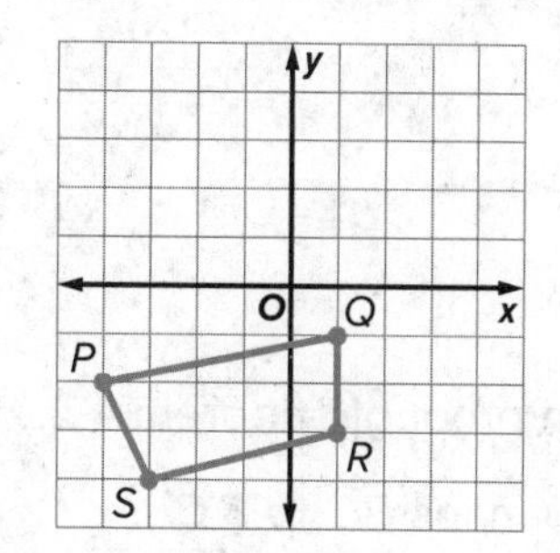

Determine the coordinates of the preimage given the image and composition of transformations.

21. reflection in the x-axis, reflection in the y-axis

22. rotation 180° about origin, translation 3 units up

23. reflection in the y-axis, translation 2 units left

24. Point K is reflected over line p and then over line d. If lines p and d are parallel and 2.8 feet apart, what single translation maps K onto K''?

Determine whether each statement is *always*, *sometimes*, or *never* true. Justify your argument.

25. A composition of two reflections is a rotation.

26. A composition of two translations is a rotation.

27. A reflection in the x-axis followed by a reflection in the y-axis leaves a point in its original location.

28. A translation along ⟨a, b⟩ followed by the translation along ⟨c, d⟩ is the translation along ⟨$a + c$, $b + d$⟩.

29. PROOF Write a paragraph proof for one case of the Composition of Isometries Theorem.

Given: A translation along $\langle a, b\rangle$ maps R to R' and S to S'. A reflection in a maps R' to R'' and S' to S''.

Prove: $\overline{RS} \cong \overline{R''S''}$

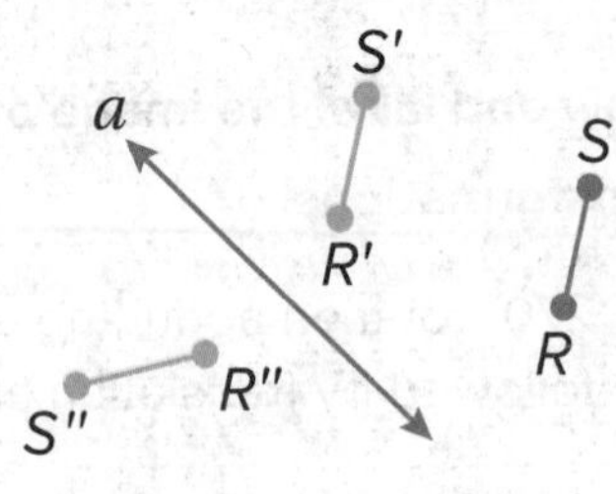

30. PROOF Write a two-column proof of Theorem 4.2.

Given: A reflection in line p maps $\overline{BC}$ to $\overline{B'C'}$. A reflection in line q maps $\overline{B'C'}$ to $\overline{B''C''}$. $p \parallel q$, $AD = x$

Prove: $\overline{BB''} \perp p$, $\overline{BB''} \perp q$; $BB'' = 2x$

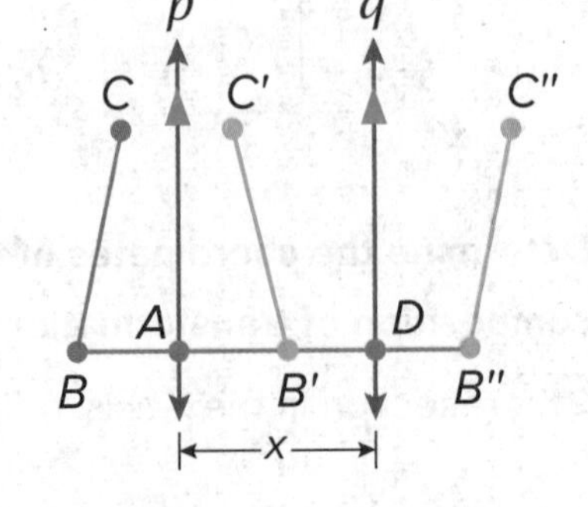

31. ANALYZE When a rotation and a reflection are performed as a composition of transformations on a figure, does the order of the transformations *sometimes*, *always*, or *never* affect the location of the final image? Justify your argument.

32. FIND THE ERROR Daniel and Lolita are translating $\triangle XYZ$ along $\langle 2, 2\rangle$ and reflecting it in the line $y = 2$. Daniel says that the transformation is a glide reflection. Lolita disagrees and says that the transformation is a composition of transformations. Is either of them correct? Explain your reasoning.

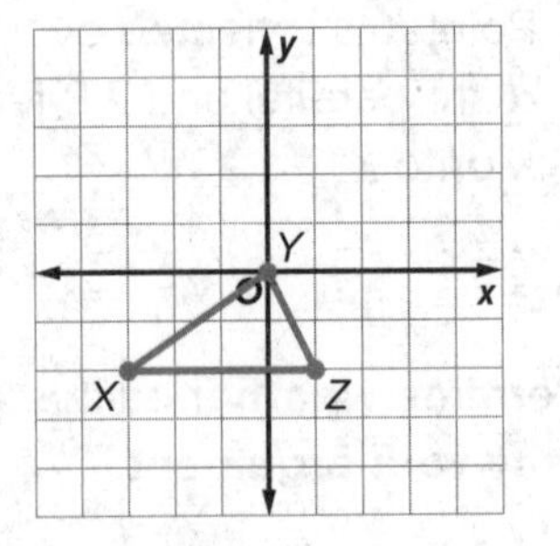

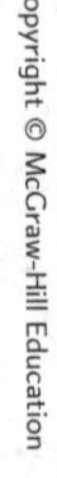

33. PERSEVERE If $PQRS$ is translated along $\langle 3, -2\rangle$, reflected in $y = -1$, and rotated 90° about the origin, what are the coordinates of $P'''Q'''R'''S'''$?

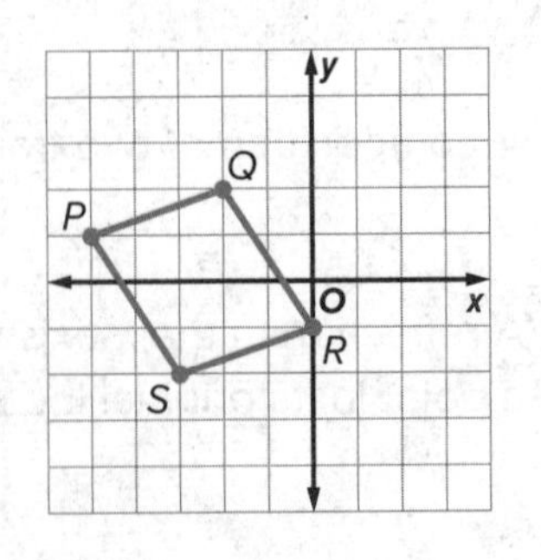

34. ANALYZE If an image will be reflected in the line $y = x$ and the x-axis, does the order of reflections affect the final image? Explain.

Tessellations

Explore Creating Tessellations

Online Activity Use graphing technology to complete the Explore.

INQUIRY When will a regular polygon not tessellate the plane?

Learn Types of Tessellations

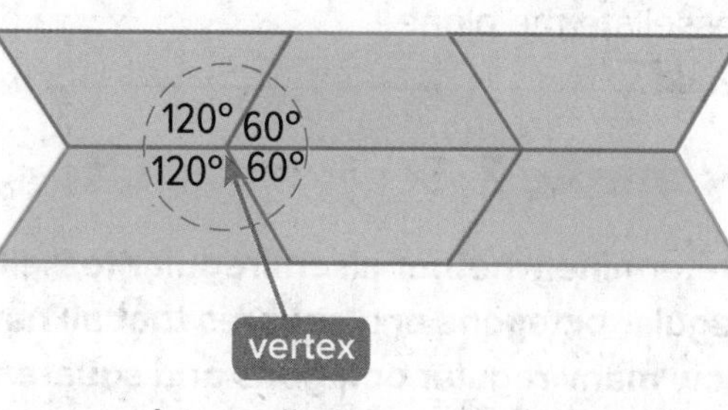
120°
60°
120°
60°
vertex

Compositions of transformations can be used to create patterns from polygons. A **tessellation** is a repeating pattern of one or more figures that covers a plane with no overlapping or empty spaces. A tessellation can be created by transforming the same figure or set of figures in a plane. The sum of the measures of the angles around a vertex of a tessellation is 360°.

A **regular tessellation** is formed by only one type of regular polygon. A regular polygon will tessellate if it has an interior angle measure that is a factor of 360°.

A **semiregular tessellation** is formed by two or more regular polygons. The tessellation shown is made up of only equilateral triangles, so it is a regular tessellation.

A tessellation can contain any type of polygon. A tessellation is a **uniform tessellation** if it contains the same arrangement of shapes and angles at each vertex.

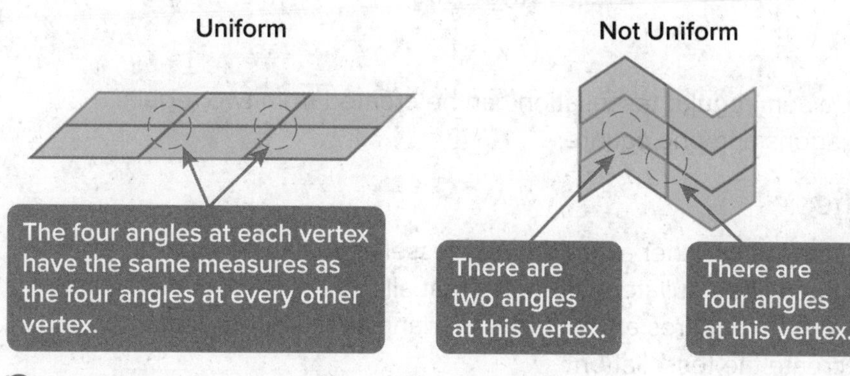
Uniform
Not Uniform
The four angles at each vertex have the same measures as the four angles at every other vertex.
There are two angles at this vertex.
There are four angles at this vertex.

Go Online You can complete an Extra Example online.

Today's Goals

- Use transformations to classify tessellations and identify figures that tessellate the plane.
- Determine whether given polygons tessellate the plane and describe transformations used to create tessellations.

Today's Vocabulary

tessellation
regular tessellation
semiregular tessellation
uniform tessellation

Talk About It!

Can an isosceles trapezoid be used to create a tessellation? a regular tessellation? Justify your arguments.

Your Notes

Math History Minute

Although she had only a high-school education, **Marjorie Rice (1923–2017)** devoted her life to finding ways to tessellate a plane with pentagons. She eventually discovered four new types of tessellating pentagons and more than 60 distinct tessellations by pentagons.

Example 1 Regular Tessellation

Determine whether a regular 16-gon will tessellate the plane. Explain.

Let x represent the measure of an interior angle of a regular 16-gon.

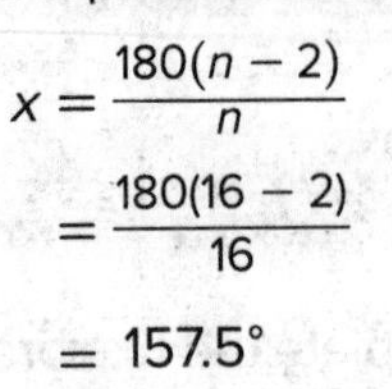

$$x = \frac{180(n-2)}{n}$$

$$= \frac{180(16-2)}{16}$$

$$= 157.5°$$

Because 157.5° is not a factor of 360°, a regular 16-gon will not tessellate the plane.

Check

Determine whether a regular decagon will tessellate the plane. Explain.

Because _______ a factor of 360°, a regular decagon _______ tessellate the plane.

Example 2 Semiregular Tessellation

Determine whether a semiregular tessellation can be created from regular octagons and squares that all have sides 1 unit long. If so, how many regular octagons and squares are needed at each vertex to create the tessellation.

Try to draw a pattern that has no empty spaces using only regular octagons and squares. In the pattern, the vertices are formed by two regular octagons and one square.

Each interior angle of a regular octagon measures $\frac{180(8-2)}{8}$ or 135°. Each interior angle of a square measures 90°.

The sum of the measures of the angles around a vertex of a tessellation is 360°. If there are x regular octagons and y squares at a vertex, then the equation $135x + 90y = 360$ can be used to verify that if there are two regular octagons at a vertex, then there is also a square at the vertex.

Let $x = 2$.

$135x + 90y = 360$	Original equation
$135(2) + 90y = 360$	Substitution
$270 + 90y = 360$	Simplify.
$y = 1$	Solve for y.

So, a semiregular tessellation can be created from two regular octagons and one square.

Check

Determine whether a semiregular tessellation can be created from squares and equilateral triangles that all have sides 1 unit long. If so, how many squares and equilateral triangles are needed at each vertex to create the tessellation? _______________

Go Online

You can complete an Extra Example online.

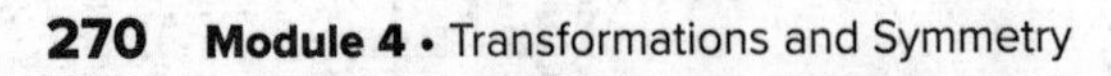

Example 3 Classify a Tessellation

TILES **Tiles for kitchen backsplashes come in many shapes that can create unique patterns. The pattern shown is created with rectangular tiles. Determine whether the pattern is a tessellation. If so, describe it as *uniform, not uniform, regular, not regular,* or *semiregular.***

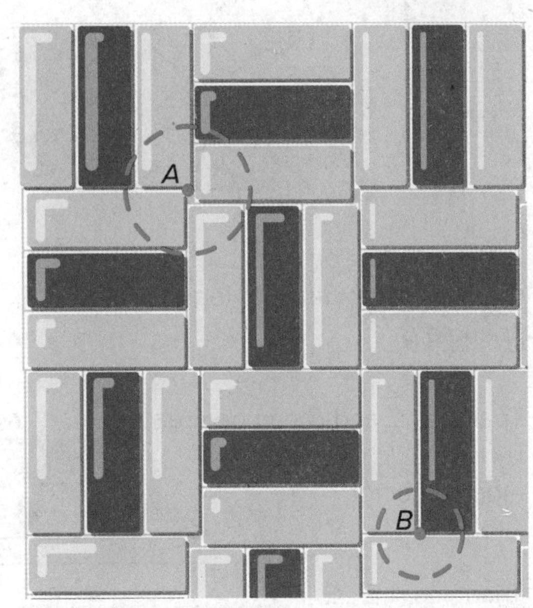

The pattern is a tessellation because there are no empty spaces and the sum of the angles at the different vertices is 360°.

The tessellation is ______________ because at vertex A there are four angles and at vertex B there are three angles.

The tessellation is ______________ because a rectangular tile is used to create the pattern and a rectangle ______________ a regular polygon.

Check

WEAVING Basket weaving is one of the oldest art forms of human civilization, dating back to 5000 B.C. Throughout the years, different cultures have created hundreds of basket patterns. Which terms describe the pattern shown?

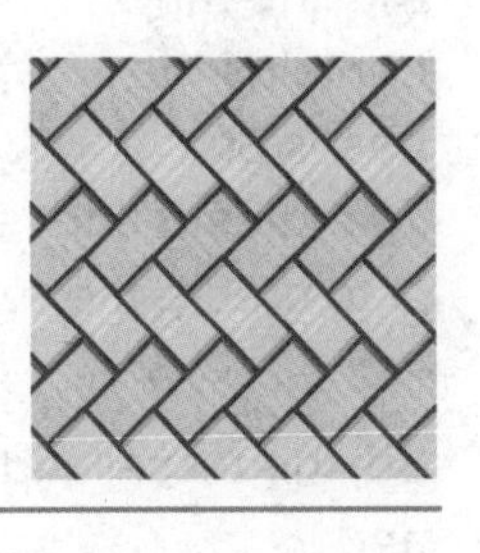

Learn Transformations in Tessellations

Not all polygons have to be regular to tessellate the plane. Any triangle is capable of tessellating the plane because the sum of the measures of its interior angles is 180°.

Any quadrilateral is capable of tessellating the plane. Because a quadrilateral can be formed by two triangles, the sum of the interior angles of a quadrilateral is $2 \cdot 180°$ or 360°.

Even though all triangles and quadrilaterals can tessellate the plane, not all polygons can. Only fifteen known types of convex pentagons and three types of convex hexagons can tessellate the plane. If a convex polygon has seven or more sides, then it cannot tessellate the plane.

Go Online You can complete an Extra Example online.

Go Online
You may want to complete the Concept Check to check your understanding.

> **Think About It!**
> Can the same tessellation be created using only two types of transformations? If so, describe the transformations.

Example 4 Identify Transformations in a Tessellation

Will an isosceles triangle *sometimes*, *always*, or *never* tessellate the plane? Describe the transformation(s) that can be used to create the tessellation shown below.

Because all ______________ tessellate the plane, an isosceles triangle will ______________ tessellate the plane.

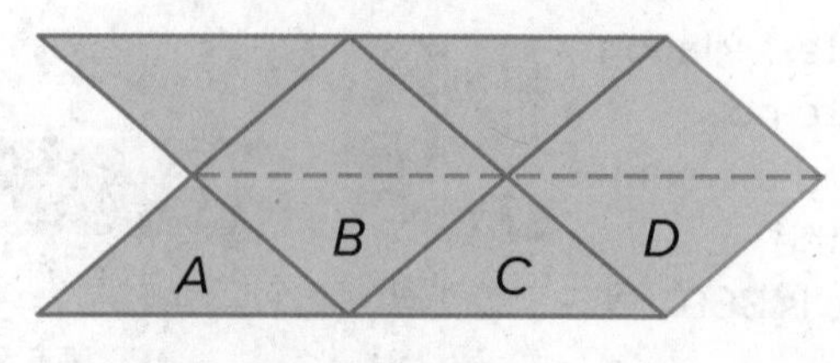

Triangle A can be ______________ about the midpoint of its right leg to create Triangle B.

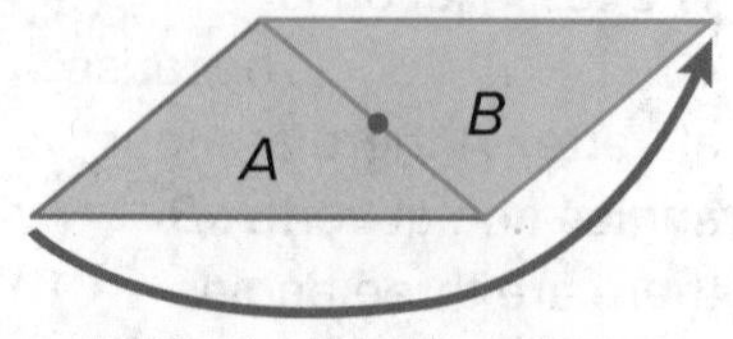

Triangles C and D can be created by ______________ A and B along a vector.

Triangles A, B, C, and D can be ______________ in the line that contains the bases of Triangles B and D to create the tessellation.

So, the tessellation can be created using ______________, ______________, and ______________.

Check

Will a kite *sometimes, always,* or *never* tessellate the plane? ______________

Describe the transformation(s) that can be used to create the tessellation shown. Select all that apply.

A. rotation and translation

B. rotation and reflection

C. reflection and translation

D. translation and translation

E. reflection and rotation

Go Online You can complete an Extra Example online.

Name ______________________ Period __________ Date __________

Practice

Go Online You can complete your homework online.

Example 1

Determine whether each regular polygon will tessellate the plane. Explain.

1. pentagon

2. hexagon

3. 9-gon

Example 2

Determine whether a semiregular uniform tessellation can be created from the given shapes, assuming that all sides are 1 unit long. If so, determine the number of each shape needed at each vertex to create the tessellation.

4. regular pentagons and squares

5. regular hexagons and equilateral triangles

Example 3

Determine whether the pattern is a tessellation. If so, describe it as *uniform, not uniform, regular, not regular, or semiregular.*

6.

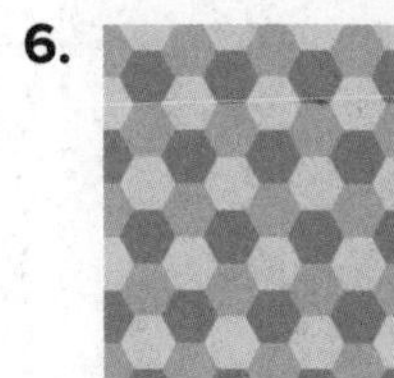

7.

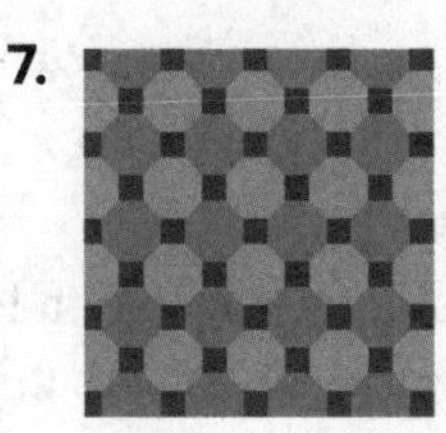

8.

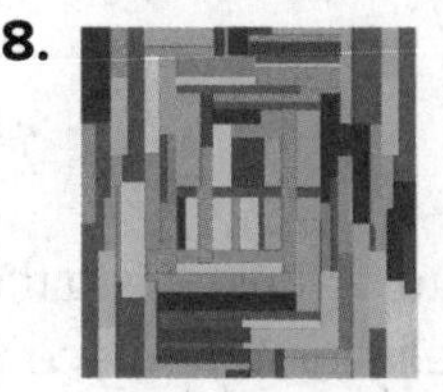

Example 4

Determine whether a tessellation can be created from each figure. If so, describe the transformation(s) that can be used to create the tessellation and draw a picture to support your reasoning.

9. scalene triangle

10. rhombus

11. Determine whether a tessellation can be created from a regular dodecagon. If so, describe the transformation(s) that can be used to create the tessellation. Will a regular dodecagon *sometimes, always,* or *never* tessellate the plane? Justify your argument.

12. Sketch a tessellation that can be created from an isosceles trapezoid. Describe the transformation(s) that can be used to create the tessellation.

13. Will a regular 15-gon *sometimes, always,* or *never* tessellate the plane? Justify your argument.

14. Determine whether a tessellation can be created from a parallelogram. If so, describe the transformation(s) that can be used to create the tessellation and draw a picture to support your reasoning.

Mixed Exercises

Determine the transformation(s) used to make each tessellation.

15. 

16. 

17. 

18. HOME IMPROVEMENT A hardware store sells various shapes of regular polygon paving stones. Kiyoko wants a simple design and only wants to buy one shape of stone. To build a solid base floor for her patio, what type of shape should Kiyoko buy?

19. GIFTS Matthew wants to surprise his girlfriend with a homemade gift. He wants to make a puzzle by tessellating one piece with a picture of a heart on it. What types of transformations can Matthew perform to create his puzzle? Explain.

20. FIND THE ERROR Heather says that if an interior angle of a regular n-gon measures 180°, then the n-gon will tessellate because 180° is a factor of 360°. Do you agree? Explain your reasoning.

21. CREATE Draw a tessellation that can be created by translations or rotations.

22. WRITE How would you accurately describe a tessellation to a person who had never heard the term before?

Symmetry

Explore Symmetry in Figures

Online Activity Use dynamic geometry software to complete the Explore.

INQUIRY How can you tell when a figure can be mapped onto itself?

Learn Line Symmetry

A figure has **symmetry** if there exists a rigid motion—reflection, translation, rotation, or glide reflection—that maps the figure onto itself. Figures that have symmetry are self-congruent. One type of symmetry is *line symmetry*.

A figure in the plane has **line symmetry** (or *reflectional symmetry*) if each half of the figure matches the other side exactly. When a figure has line symmetry, the figure can be mapped onto itself by a reflection in a line, called the **line of symmetry** (or *axis of symmetry*).

line of symmetry

Example 1 Identify Line Symmetry

Determine whether each figure has a line of symmetry. If so, draw the lines of symmetry and state how many lines of symmetry it has.

a.

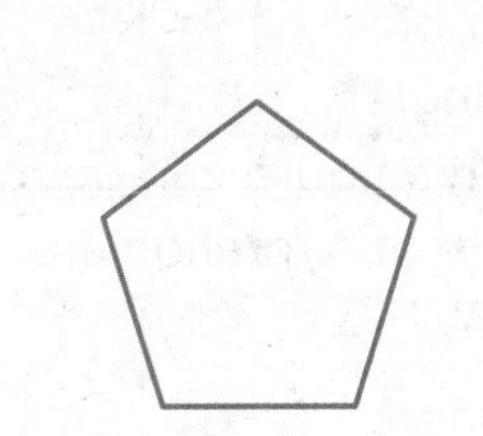

___ lines of symmetry

b.

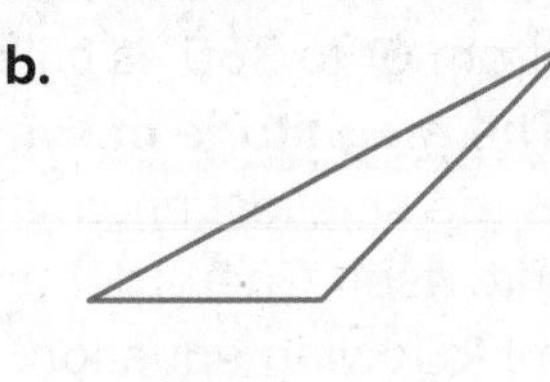

___ lines of symmetry

Go Online You can complete an Extra Example online.

Today's Goal

- Use line symmetry to describe the reflections that carry a figure onto itself.
- Use rotational symmetry to describe the rotations that carry a figure onto itself.

Today's Vocabulary

symmetry
line symmetry
line of symmetry
rotational symmetry
center of symmetry
order of symmetry
magnitude of symmetry
point symmetry
point of symmetry

Talk About It!

Do you think that a figure can have multiple lines of symmetry? Justify your argument.

Your Notes

Think About It!

Josefina argues that you can count the number of lines of symmetry in a circle. Do you agree or disagree? Justify your argument.

Check

Determine whether each figure has a line of symmetry. If so, draw the lines of symmetry and state how many lines of symmetry it has.

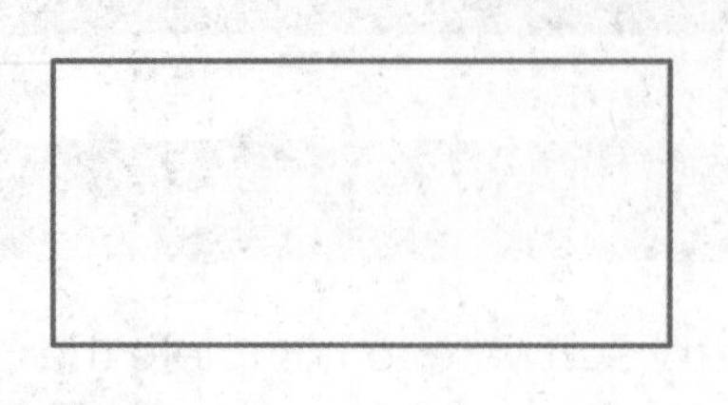

This figure has _____ line(s) of symmetry.

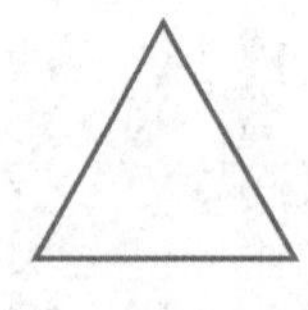

This figure has _____ line(s) of symmetry.

This figure has _____ line(s) of symmetry.

Learn Rotational Symmetry

A figure in the plane has **rotational symmetry** (or *radial symmetry*) if the figure can be mapped onto itself by being rotated less than 360° about the center of the figure so the image and the preimage are indistinguishable. The point in which a figure can be rotated onto itself is called the **center of symmetry** (or *point of symmetry*).

This figure has rotational symmetry because a rotation of 90°, 180°, or 270° maps the figure onto itself.

The number of times that a figure maps onto itself as it rotates from 0° to 360° is called the **order of symmetry.** The **magnitude of symmetry** (or *angle of rotation*) is the smallest angle through which a figure can be rotated so it maps onto itself. The order and magnitude of a rotation are related by the following equation.

$$\text{magnitude} = 360° \div \text{order}$$

This figure has order 4 and magnitude 90°.

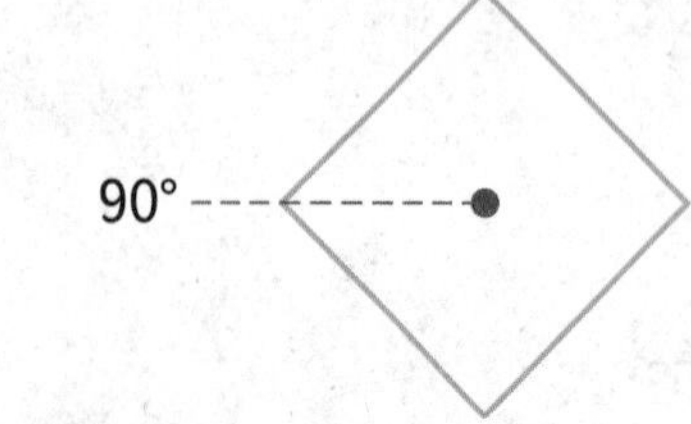

Key Concept • Point Symmetry

A figure has **point symmetry** if it can be mapped onto itself by a rotation of 180°. If a figure has point symmetry, then the center of symmetry in the figure is called the **point of symmetry**.

Example A rhombus has point symmetry because it looks the same right-side up as upside down.

180°

> **Go Online**
> You may want to complete the Concept Check to check your understanding.

Example 2 Identify Rotational Symmetry

NATURE **Objects found in nature often have rotational symmetry.**

Determine whether each figure has rotational symmetry. Explain.

a. **b.** **c.**

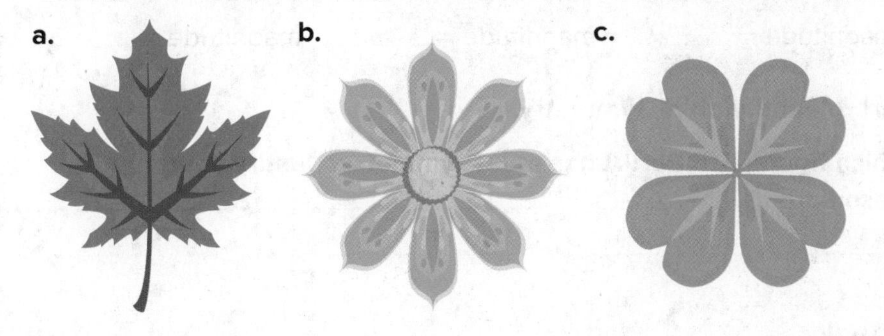

Check

HOUSEHOLD Below are several objects that you might find around your house. Determine whether each figure has rotational symmetry. Explain.

Example 3 Determine Order and Magnitude of Symmetry

Part A State the order and magnitude of symmetry.

Determine whether each figure has rotational symmetry. If so, locate the center of symmetry and state the order and magnitude of symmetry.

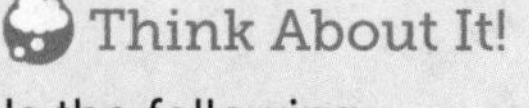

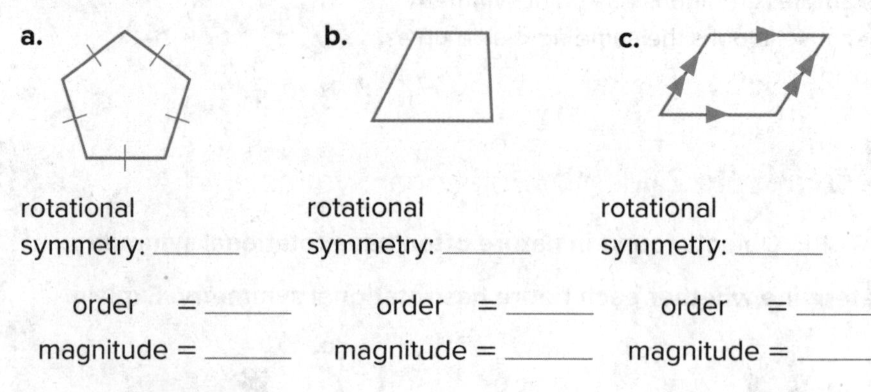

a. rotational symmetry: ______

order = ______

magnitude = ______

b. rotational symmetry: ______

order = ______

magnitude = ______

c. rotational symmetry: ______

order = ______

magnitude = ______

Think About It!

Is the following statement *sometimes*, *always*, or *never* true? Explain. *A polygon with order 4 rotational symmetry has point symmetry.*

Part B Identify point symmetry.

Which figure(s) in Part A has point symmetry? Justify your reasoning. ______________________________

____

Check

Determine whether each figure has rotational symmetry. If so, locate the center of symmetry and state the order and magnitude of symmetry.

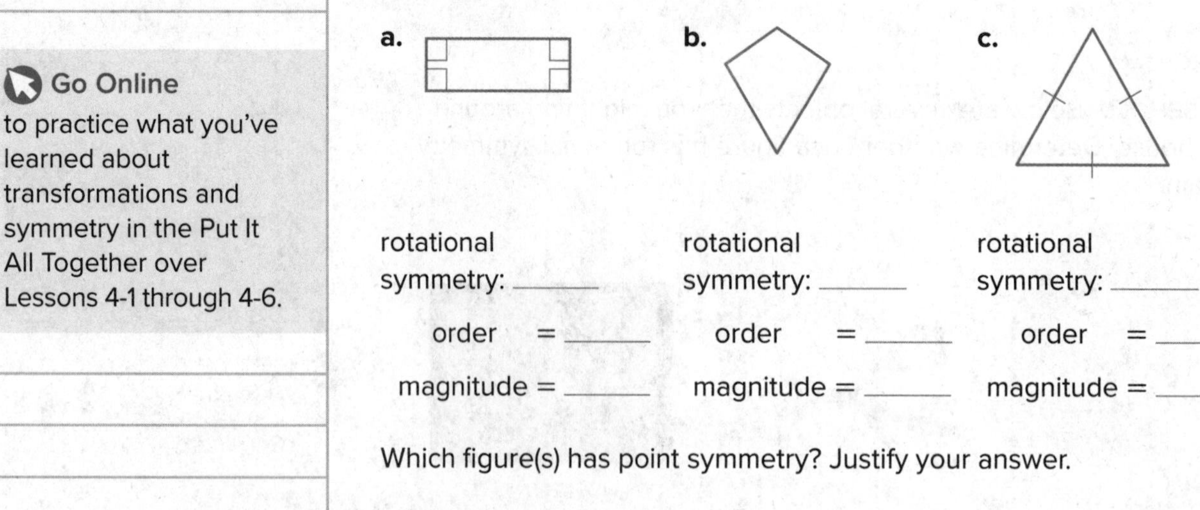

a. rotational symmetry: ______

order = ______

magnitude = ______

b. rotational symmetry: ______

order = ______

magnitude = ______

c. rotational symmetry: ______

order = ______

magnitude = ______

Which figure(s) has point symmetry? Justify your answer.

____

____

Go Online to practice what you've learned about transformations and symmetry in the Put It All Together over Lessons 4-1 through 4-6.

Go Online You can complete an Extra Example online.

Name ______________________ Period ____________ Date ____________

Practice

Go Online You can complete your homework online.

Example 1

Determine whether each figure has a line of symmetry. If so, draw the lines of symmetry and state how many lines of symmetry it has.

1.

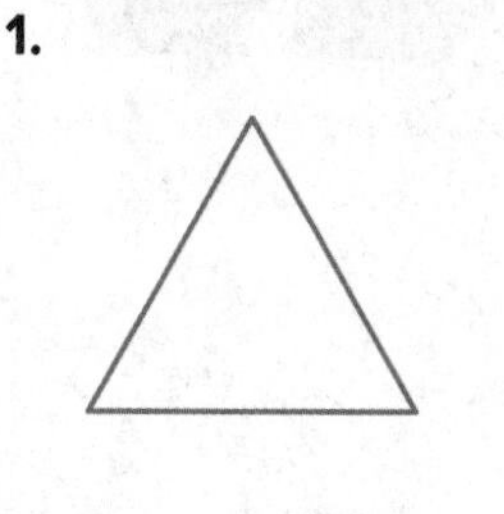

2.

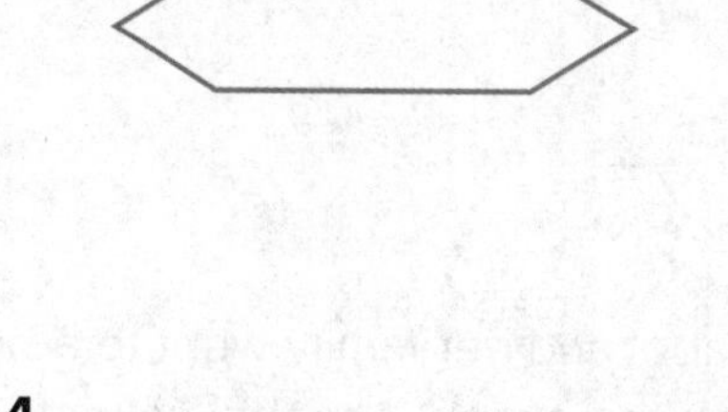

3.

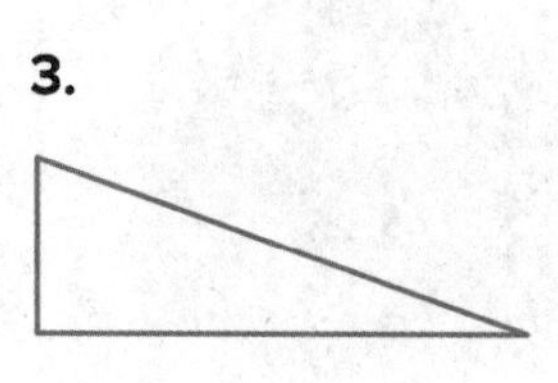

4.

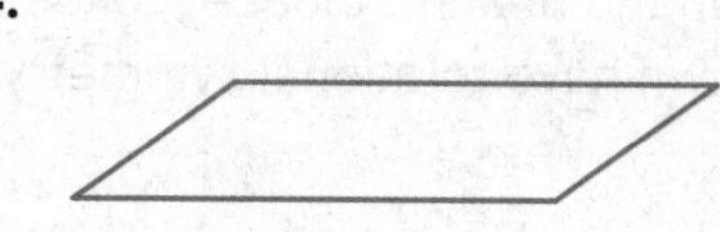

5.

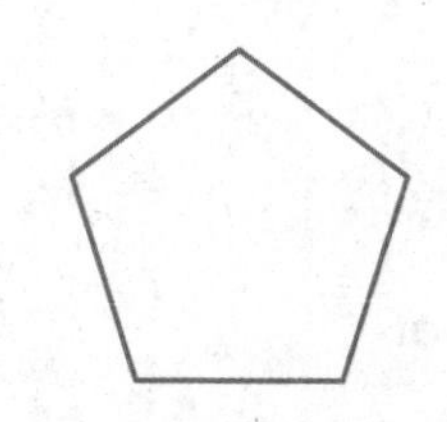

6.

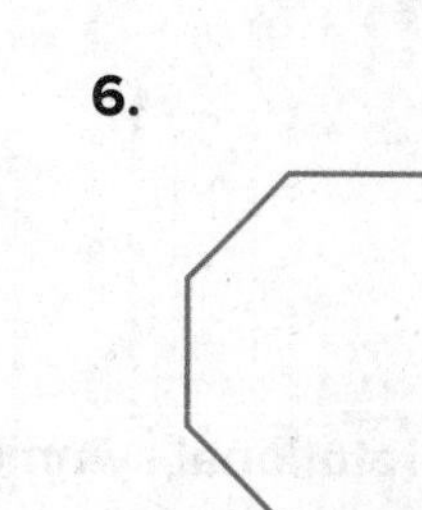

Example 2

7. CARS Steve found the hubcaps shown below at his local junkyard. Determine whether each hubcap has rotational symmetry. Explain.

a.

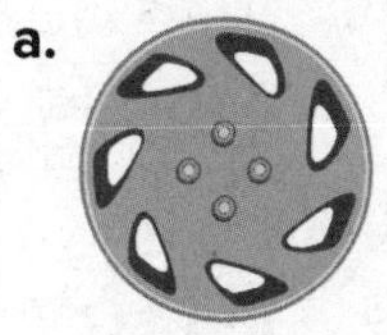

b.

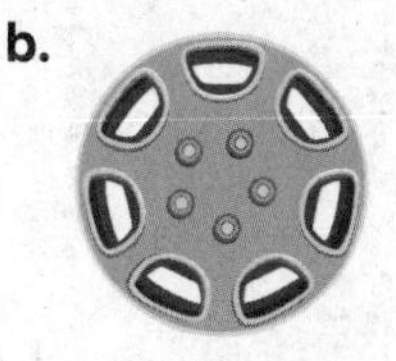

c.

8. FLAGS The figure shows the Union Jack, which is the flag of the United Kingdom. Does the flag have rotational symmetry? Explain.

9. RECYCLING A waste management company offers recycling programs for its clients. Recycling is denoted by the symbol shown. Does the recycling symbol have rotational symmetry? Explain.

10. VACATION Annabel and her family went to a beach for vacation. While she was on the beach, Annabel collected seashells. Does the seashell shown have rotational symmetry? Explain.

Example 3

Determine whether each figure has rotational symmetry. If so, locate the center of symmetry, and state the order and magnitude of symmetry.

11.

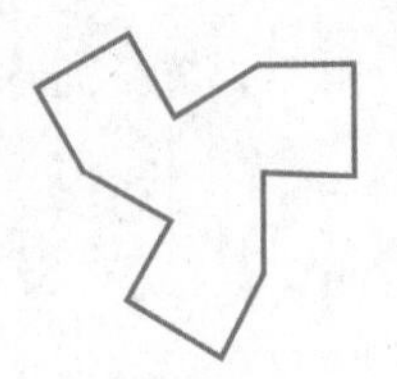

12.

13.

14.

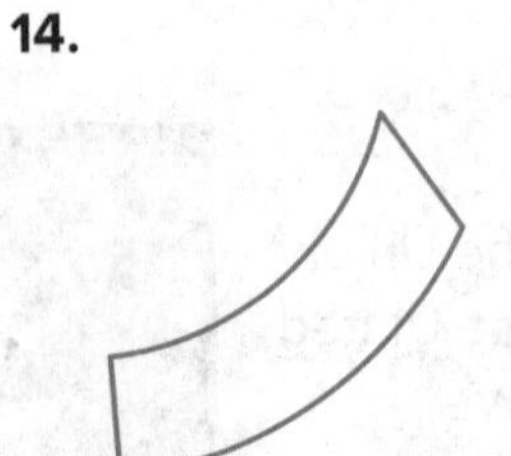

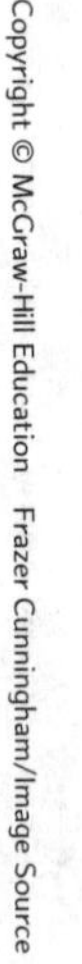

Name ______________________ Period __________ Date __________

Mixed Exercises

Refer to the figure at the right.

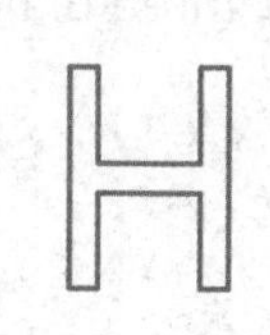

15. Draw the line(s) of symmetry in the figure.

16. Locate the center of symmetry for the figure.

17. What is the order and magnitude of symmetry for the figure?

18. LETTERS Examine each capital letter in the alphabet. Determine which letters have 180° rotational symmetry about a point in the center of the letter.

19. STRUCTURE A regular polygon has rotational symmetry with an order of 5 and a magnitude of 72°. What is the figure?

20. CONSTRUCT ARGUMENTS Consider the symmetry of a circle.

a. How many lines of symmetry does a circle have? Justify your argument.

b. What is the order of rotation for a circle? Justify your argument.

State whether each figure has rotational symmetry. If so, describe the rotations that map the figure onto itself by giving the order of symmetry and magnitude of symmetry.

21. equilateral triangle

22. scalene triangle

23. regular hexagon

24. **PERSEVERE** Draw a three-dimensional object that has a base with line symmetry.

25. **CREATE** Draw an object that has at least one line of symmetry. Describe the lines of symmetry in this object.

26. **ANALYZE** The figure shows the floor plan for a new gallery in an art museum. Describe every reflection or rotation that maps the gallery onto itself.

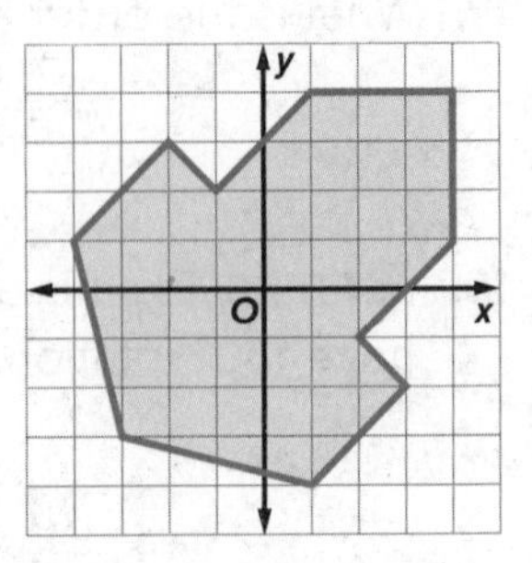

27. **WRITE** A regular polygon has magnitude of symmetry 15°. How many sides does the polygon have? Explain.

28. **FIND THE ERROR** Jaime says that Figure A has only line symmetry, and Jewel says that Figure A has only rotational symmetry. Is either of them correct? Explain your reasoning.

Figure A

29. **PERSEVERE** A quadrilateral in the coordinate plane has exactly two lines of symmetry, $y = x - 1$ and $y = -x + 2$. Find possible vertices for the figure. Graph the figure and the lines of symmetry.

30. **CREATE** Draw a figure that has line symmetry but not rotational symmetry. Explain.

31. **WRITE** How are line symmetry and rotational symmetry related?

Review

Essential Question

How are rigid motions used to show geometric relationships?

Module Summary

Lessons 4-1 through 4-3

Reflections, Translations, and Rotations

- When a figure is reflected in a line, each point of the preimage and its corresponding point on the image are the same distance from the line of reflection.
- A translation is a function in which all the points of a figure move the same distance in the same direction as described by a translation vector.
- A translation vector describes the magnitude and direction of the translation. The magnitude of a vector is its length from the initial point to the terminal point.
- A rotation about a fixed point P through an angle of $a°$ is a function that maps point M to point M' such that point P does not move, $m\angle MPM'$ is $a°$, and $MP = M'P$.

Lesson 4-4

Compositions of Transformations

- When a transformation is applied to a figure and then another transformation is applied to its image, the result is called a composition of transformations.
- A glide reflection is the composition of a translation followed by a reflection in a line parallel to the translation vector.
- The composition of two reflections can result in the same image as a translation or rotation.

Lessons 4-5 and 4-6

Tessellations and Symmetry

- A regular polygon will tessellate if it has an interior angle measure that is a factor of 360°.
- A semiregular tessellation is formed by two or more regular polygons.
- A figure has symmetry if there exists a rigid motion—reflection, translation, rotation, or glide reflection—that maps the figure onto itself.
- A figure in the plane has line symmetry (or reflectional symmetry) if each half of the figure matches the other side exactly.
- A figure in the plane has rotational symmetry (or radial symmetry) if the figure can be mapped onto itself by being rotated less than 360° about the center of the figure so the image and the preimage are indistinguishable.

Study Organizer

Foldables

Use your Foldable to review this module. Working with a partner can be helpful. Ask for clarification of concepts as needed.

Name ______________________ Period __________ Date __________

1. **GRAPH** Graph the image of $\triangle ABC$ with vertices at $A(-5, 0)$, $B(-3, 5)$, $C(-1, 2)$ after a reflection in the line $x = -2$. (Lesson 4-1)

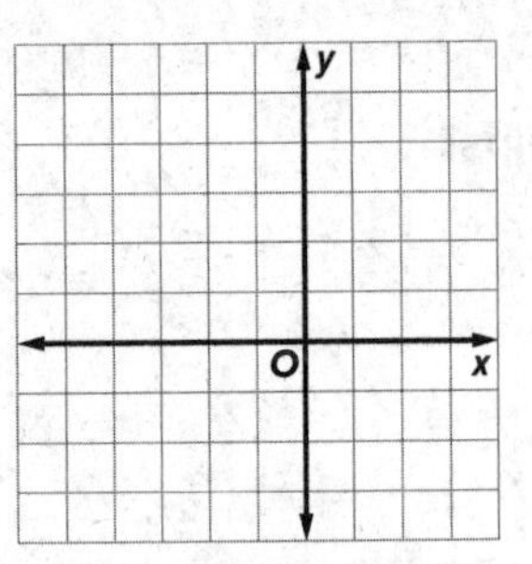

2. **MULTIPLE CHOICE** Which **best** describes a possible step that is used to determine the location of the image of point B when it is reflected in the line $y = x$? (Lesson 4-1)

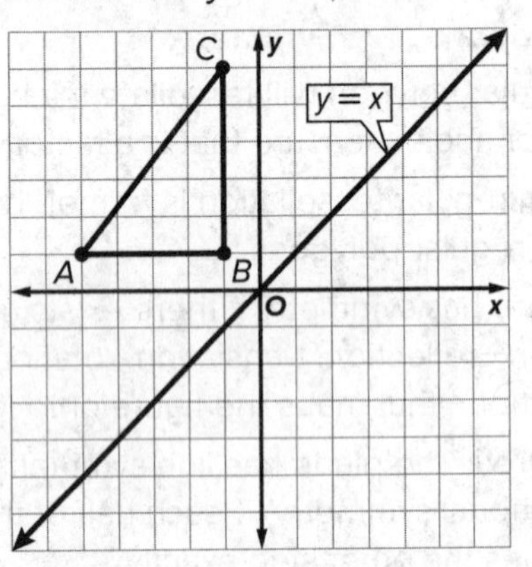

Ⓐ Move down one and right one from (0, 0).

Ⓑ Move down one and right one from (−1, 1).

Ⓒ Move right two from (1, 1).

Ⓓ Move down two from (−1, −1).

3. **OPEN RESPONSE** When point F is reflected in the line $y = x$, the image is located at $F'(6, -9)$. Find the coordinates of point F. (Lesson 4-1)

4. **MULTIPLE CHOICE** Find the vector that translates $A(-2, 7)$ to $A'(6, 4)$. (Lesson 4-2)

Ⓐ $\langle -8, 3 \rangle$

Ⓑ $\langle -3, 8 \rangle$

Ⓒ $\langle 3, 8 \rangle$

Ⓓ $\langle 8, -3 \rangle$

5. **OPEN RESPONSE** Refer to the graph.

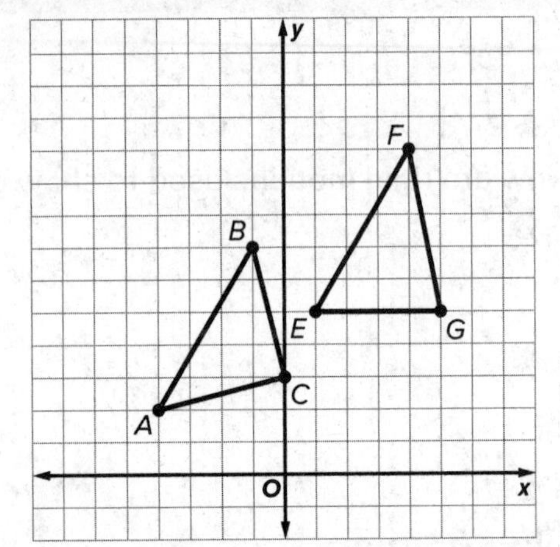

Explain why a translation does not map $\triangle ABC$ to $\triangle EFG$. (Lesson 4-2)

6. **MULTIPLE CHOICE** Which is the image of $P(-5, 11)$ along the vector $\langle 3, -8 \rangle$? (Lesson 4-2)

Ⓐ $P'(-8, 19)$

Ⓑ $P'(-8, 3)$

Ⓒ $P'(-2, -3)$

Ⓓ $P'(-2, 3)$

7. **MULTIPLE CHOICE** Juan is designing a new playground for the elementary school. He needs to determine the shortest distance from the monkey bars to the slide to create a path. Which statement **best** describes the translation from the monkey bars to the slide? (Lesson 4-2)

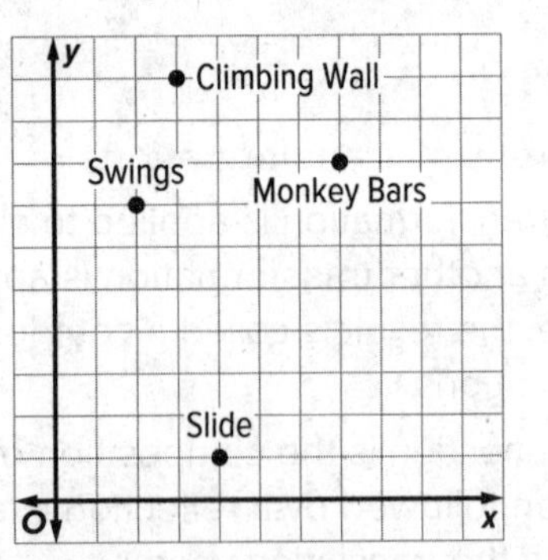

Ⓐ a translation right 11 units and up 9 units

Ⓑ a translation right 3 units and up 7 units

Ⓒ a translation left 3 units and down 7 units

Ⓓ a translation left 11 units and down 9 units

Name ________________ Period ________ Date ________

8. **MULTIPLE CHOICE** Which is the image of $P(3, 0)$ after a counterclockwise rotation of 90° about (2, 4)? (Lesson 4-3)

Ⓐ $P'(-2, 5)$

Ⓑ $P'(6, 5)$

Ⓒ $P'(6, 3)$

Ⓓ $P'(-4, 2)$

9. **OPEN RESPONSE** Refer to the graph.

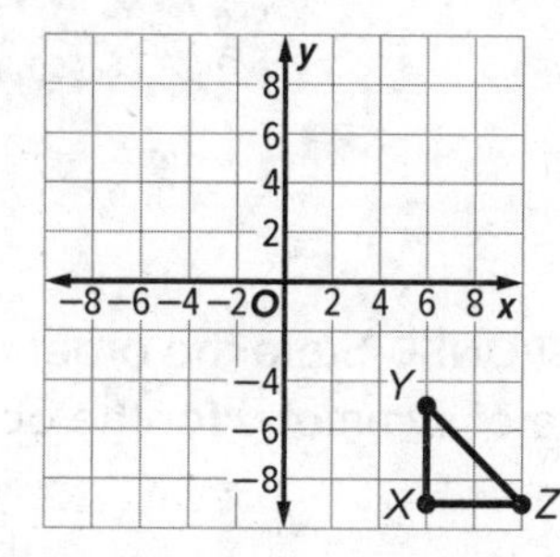

In which quadrant will the image be after a rotation of 180° about (1, −2)? (Lesson 4-3)

10. **MULTIPLE CHOICE** Which is the image of $F(-2, -7)$ after a counterclockwise rotation of 180° about (−1, 5)? (Lesson 4-3)

Ⓐ $F'(-1, 5)$

Ⓑ $F'(0, 17)$

Ⓒ $F'(1, 12)$

Ⓓ $F'(2, 7)$

11. **OPEN RESPONSE** *True or false:* Rotating $M(-5, 1)$ 180° about the origin and then translating along $\langle -3, 4\rangle$ will give the same result as translating along $\langle -3, 4\rangle$ and then rotating 180°. (Lesson 4-4)

12. **MULTPLE CHOICE** Triangle *ABC* is shown. (Lesson 4-4)

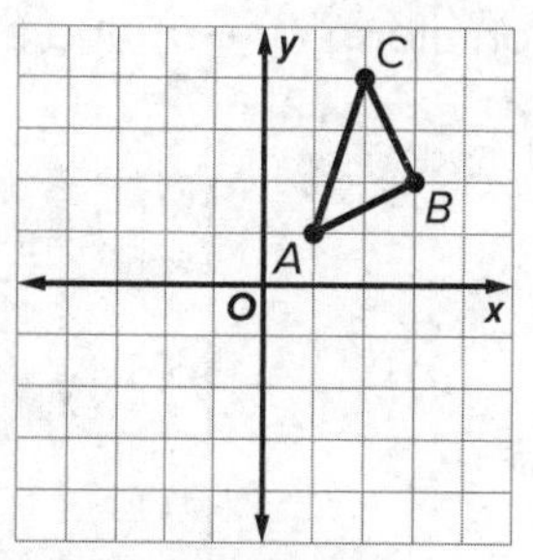

Triangle *ABC* is rotated 90° counterclockwise about the origin and then translated along $\langle -2, 3\rangle$. What is the location of the image of point *B*?

Ⓐ (0, 0)

Ⓑ (1, 1)

Ⓒ (−4, 6)

Ⓓ (−5, 1)

13. **MULTI-SELECT** Select all transformations or compositions of transformations that would map $\triangle ABC$ to $\triangle A'B'C'$. (Lesson 4-4)

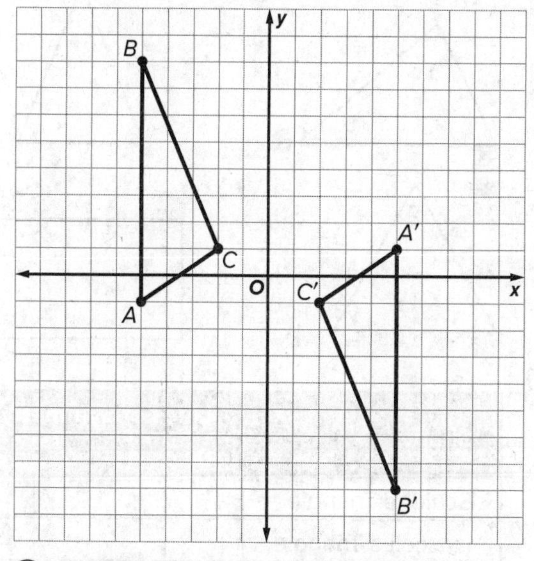

Ⓐ Reflection in the *x*-axis followed by a reflection in the *y*-axis

Ⓑ Reflection in the *x*-axis followed by a rotation of 90° counterclockwise about the origin

Ⓒ Reflection in $y = -x$

Ⓓ Rotation of 180° about the origin

Ⓔ Reflection in $y = x$

14. GRAPH Graph the image of $\triangle PQR$ with vertices at $P(4, 7)$, $Q(7, 3)$, and $R(2, 2)$ after a translation along $\langle 1, -9 \rangle$ and a reflection in $x = 2$. (Lesson 4-4)

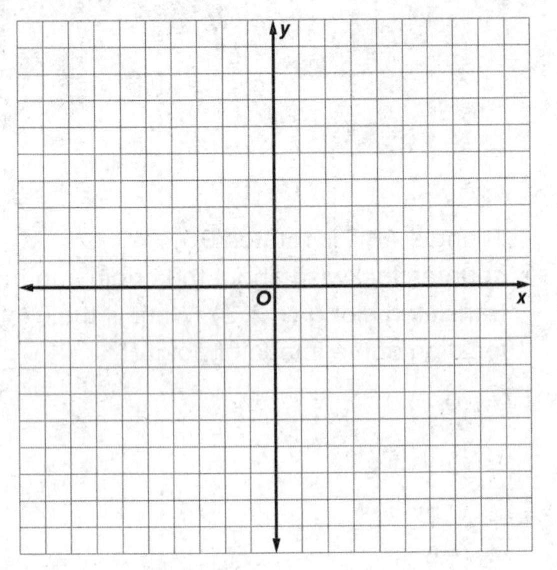

15. TABLE ITEM Match each description to its correct image. (Lesson 4-6)

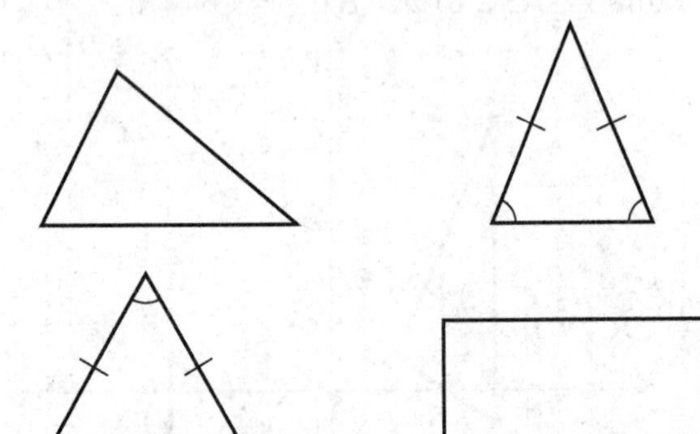

Shape	Lines of Symmetry			
	0	1	2	3
Scalene Triangle				
Isosceles Triangle				
Equilateral Triangle				
Rectangle				

16. OPEN RESPONSE How many lines of symmetry does this figure have? Describe the reflections, if any, that map the figure onto itself. (Lesson 4-6)

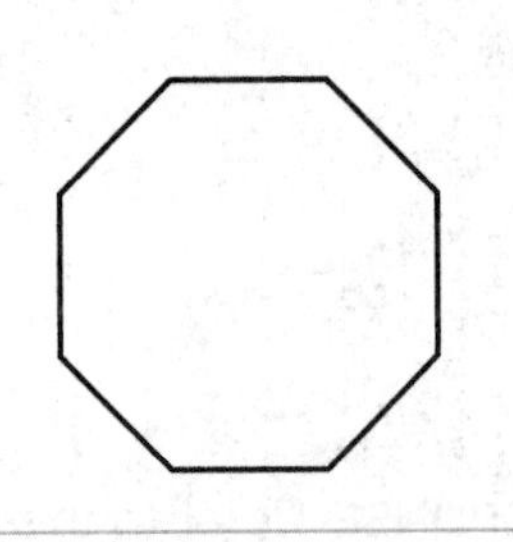

17. OPEN RESPONSE State the order and magnitude of symmetry for the object below. (Lesson 4-6)

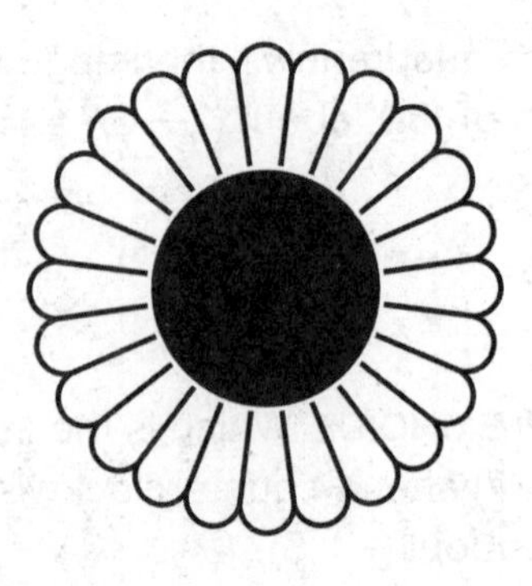

Module 5

Triangles and Congruence

Essential Question

How can you prove congruence and use congruent figures in real-world situations?

What Will You Learn?

Place a check mark (✓) in each row that corresponds with how much you already know about each topic **before** starting this module.

KEY

— I don't know. — I've heard of it. — I know it!

	Before			After		
solve problems using the Triangle Angle-Sum Theorem						
solve problems using the Exterior Angle Theorem						
show that triangles are congruent						
identify corresponding parts of congruent triangles						
solve problems using the SSS Congruence Postulate						
solve problems using the SAS Congruence Postulate						
solve problems using the ASA Congruence Postulate						
solve problems using the AAS Congruence Theorem						
construct congruent triangles						
solve problems using the LL, HA, LA, and HL Theorems						
solve problems involving isosceles and equilateral triangles						
write coordinate proofs						

Foldables Make this Foldable to help you organize your notes about triangles and congruence. Begin with one sheet of paper.

1. **Fold** a sheet of paper as shown, cutting off the excess paper strip to form a taco.
2. **Open** the fold and refold the square the opposite way to form another taco and an X-fold pattern.
3. **Open** and fold the corners toward the center point of the X, forming a small square.
4. **Label** the flaps as shown.

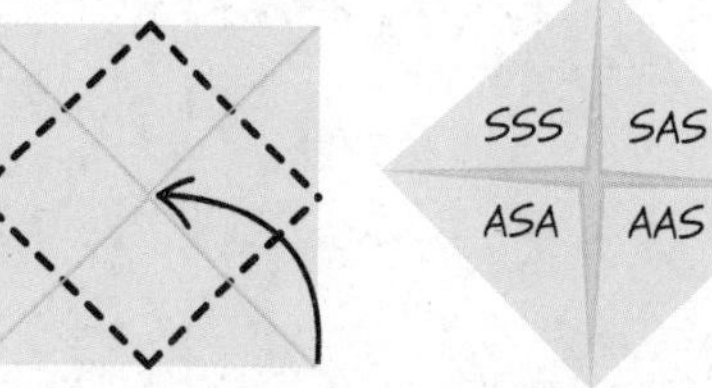

What Vocabulary Will You Learn?

Check the box next to each vocabulary term that you may already know.

- ☐ auxiliary line
- ☐ base angles of an isosceles triangle
- ☐ congruent polygons
- ☐ coordinate proofs
- ☐ corollary
- ☐ corresponding parts
- ☐ exterior angle of a triangle
- ☐ included angle
- ☐ included side
- ☐ interior angle of a triangle
- ☐ isosceles triangle
- ☐ legs of an isosceles triangle
- ☐ principle of superposition
- ☐ remote interior angles
- ☐ vertex angle of an isosceles triangle

Are You Ready?

Complete the Quick Review to see if you are ready to start this module. Then complete the Quick Check.

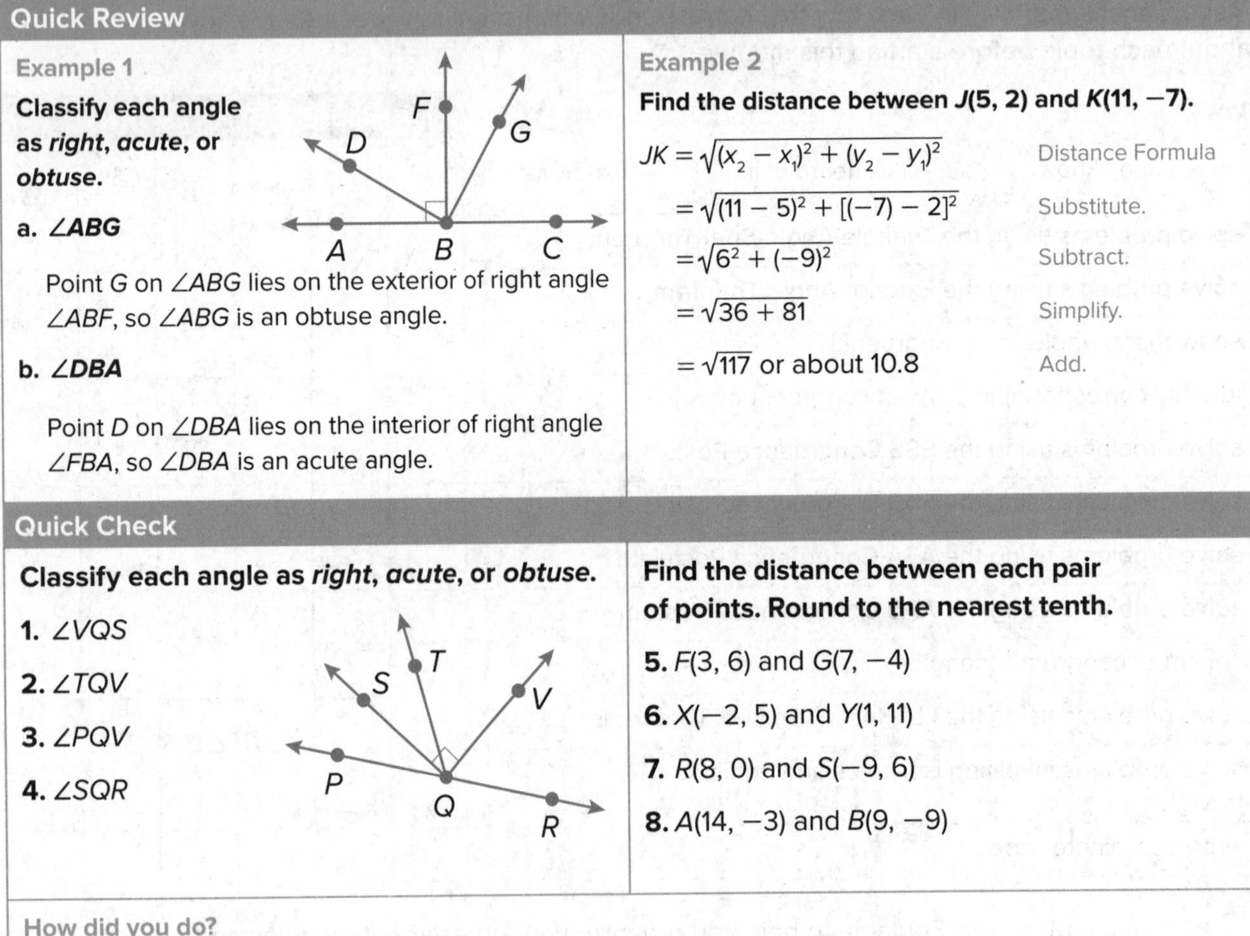

Quick Review

Example 1

Classify each angle as *right*, *acute*, or *obtuse*.

a. $\angle ABG$

Point G on $\angle ABG$ lies on the exterior of right angle $\angle ABF$, so $\angle ABG$ is an obtuse angle.

b. $\angle DBA$

Point D on $\angle DBA$ lies on the interior of right angle $\angle FBA$, so $\angle DBA$ is an acute angle.

Example 2

Find the distance between $J(5, 2)$ and $K(11, -7)$.

$JK = \sqrt{(x_2 - x_1)^2 + (y_2 - y_1)^2}$	Distance Formula
$= \sqrt{(11 - 5)^2 + [(-7) - 2]^2}$	Substitute.
$= \sqrt{6^2 + (-9)^2}$	Subtract.
$= \sqrt{36 + 81}$	Simplify.
$= \sqrt{117}$ or about 10.8	Add.

Quick Check

Classify each angle as *right*, *acute*, or *obtuse*.

1. $\angle VQS$

2. $\angle TQV$

3. $\angle PQV$

4. $\angle SQR$

Find the distance between each pair of points. Round to the nearest tenth.

5. $F(3, 6)$ and $G(7, -4)$

6. $X(-2, 5)$ and $Y(1, 11)$

7. $R(8, 0)$ and $S(-9, 6)$

8. $A(14, -3)$ and $B(9, -9)$

How did you do?

Which exercises did you answer correctly in the Quick Check? Shade those exercise numbers below.

① ② ③ ④ ⑤ ⑥ ⑦ ⑧

Angles of Triangles

Explore Triangle Angle Sums

Online Activity Use dynamic geometry software to complete the Explore.

INQUIRY Is there a relationship associated with the interior angles of a triangle? If so, how do we prove that this relationship is always true?

Learn Interior Angles of Triangles

An **interior angle of a triangle** is the angle at a vertex of a triangle. Because a triangle has three vertices, it also has three interior angles. The Triangle Angle-Sum Theorem describes the relationships among the interior angle measures of any triangle.

Theorem 5.1: Triangle Angle-Sum Theorem

The sum of the measures of the interior angles of a triangle is 180°.

Go Online A proof of Theorem 5.1 is available.

Apply Example 1 Use the Triangle Angle-Sum Theorem

Find the measure of each numbered angle.

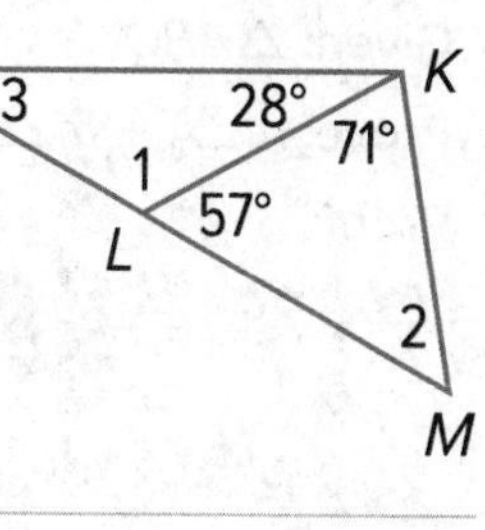

1 What is the task?

Describe the task in your own words. Then list any questions that you may have. How can you find answers to your questions?

2 How will you approach the task? What have you learned that you can use to help you complete the task?

3 What is your solution?

Use your strategy to solve the problem.

$m\angle 1 =$ ____ $m\angle 2 =$ ____ $m\angle 3 =$ ____

4 How can you know that your solution is reasonable?

Write About It! Write an argument that can be used to defend your solution.

Go Online You can complete an Extra Example online.

Today's Goals

- Prove the Triangle Angle-Sum Theorem and apply the theorem to solve problems.
- Prove the Exterior Angle Theorem and apply the theorem to solve problems.
- Prove the corollaries to the Triangle Angle-Sum Theorem and apply the corollaries to solve problems.

Today's Vocabulary

interior angle of a triangle
exterior angle of a triangle
remote interior angles
corollary

Watch Out!

Triangle Angle-Sum Theorem When you are finding missing angle measures of a triangle, check the solution by seeing whether the sum of the measures of the angles of the triangle is 180°.

Talk About It!

Ellie believes that she can solve for $m\angle 3$ before solving for $m\angle 1$. What useful questions can you ask to understand her approach?

Your Notes

Check

Find the measure of each numbered angle.

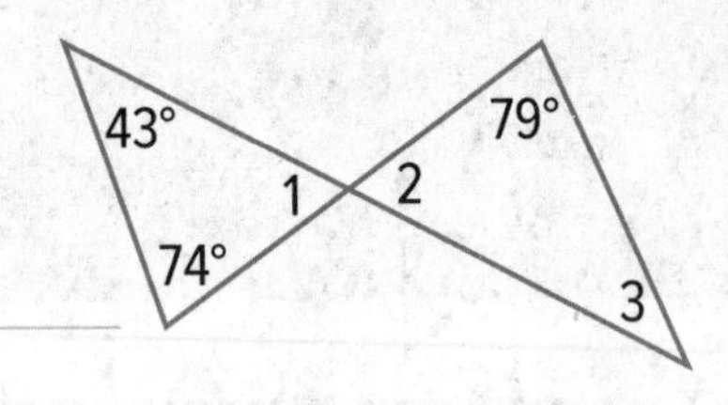

Learn Exterior Angles of Triangles

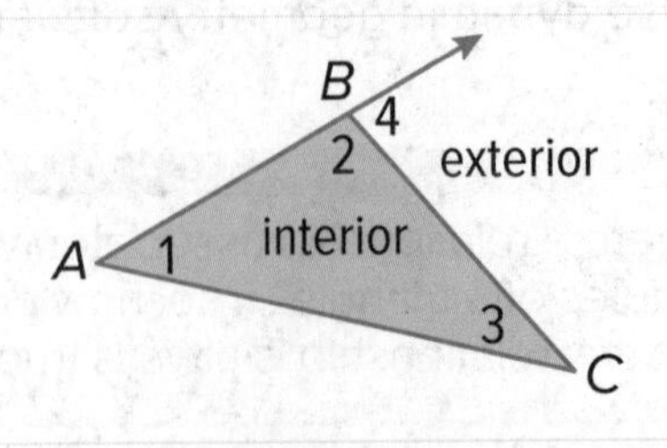

interior angles	The sum of the measures of the interior angles of a triangle is 180°.
exterior angles	An **exterior angle of a triangle** is formed by one side of the triangle and the extension of an adjacent side. A triangle has three exterior angles.
remote interior angles	Each exterior angle of a triangle has two **remote interior angles** that are not adjacent to the exterior angle.

Theorem 5.2 Exterior Angle Theorem

The measure of an exterior angle of a triangle is equal to the sum of the measures of the two remote interior angles.

Given: $\triangle ABC$

Prove: $m\angle A + m\angle B = m\angle 1$

Proof:

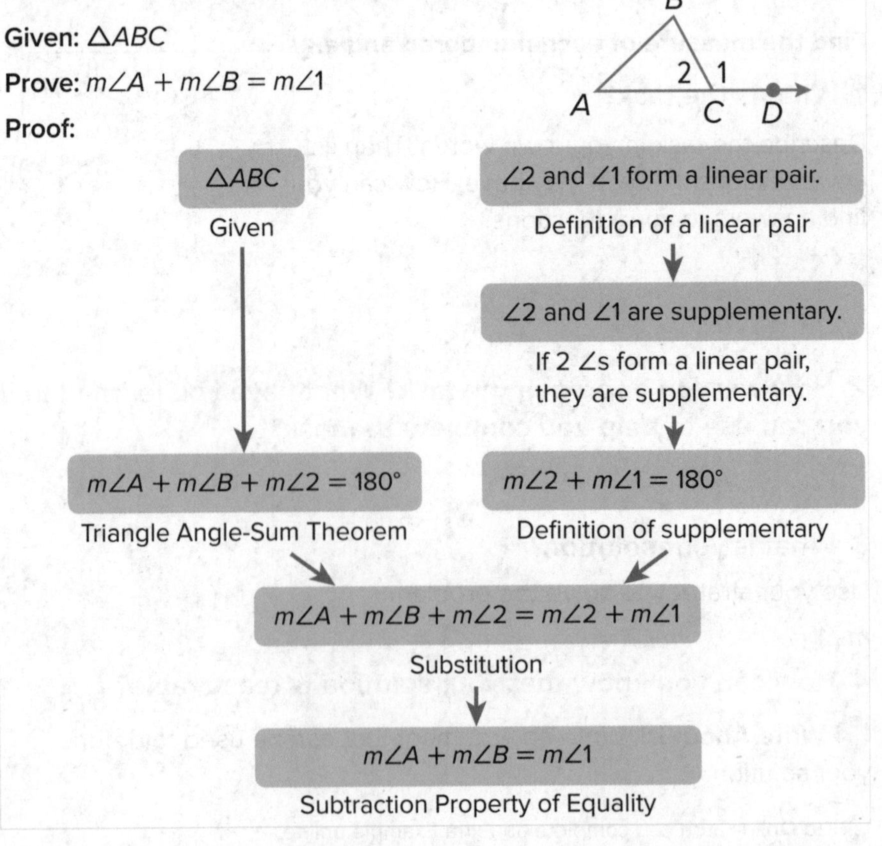

Example 2 Use the Exterior Angle Theorem

ARCHITECTURE **Find the measure of $\angle DAB$ in the front face of the building.**

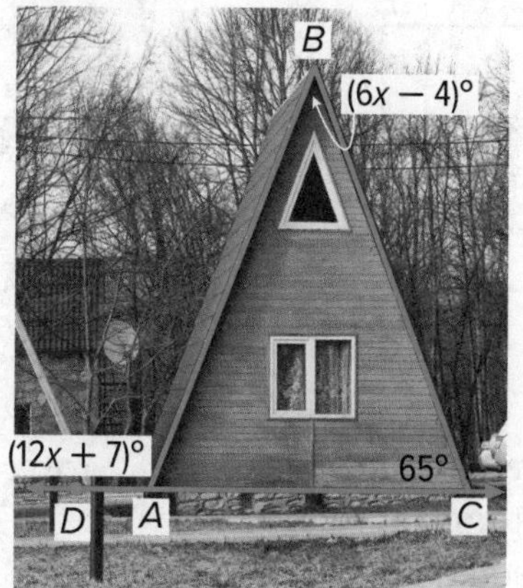

$m\angle DAB = m\angle ABC + m\angle BCA$	Exterior Angle Theorem
$12x + 7 = 6x - 4 +$ ____	Substitution
$x =$ ____	Solve.
$m\angle DAB = 12(9) + 7$ or ____°	

Check

PUZZLES Find the measure of $\angle XYZ$ created by the triangle.

____

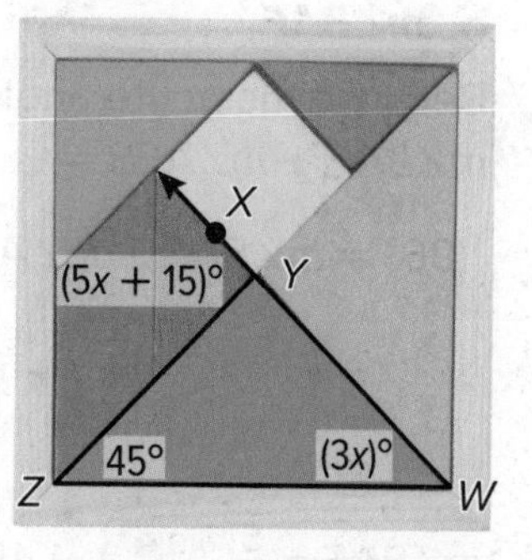

Go Online You can complete an Extra Example online.

Think About It!
What theorems and definitions can you use to check your answer for reasonableness?

What assumption did you make when you were modeling the front face of the building as a triangle?

Learn Triangle Angle-Sum Corollaries

A **corollary** is a theorem with a proof that follows as a direct result of another theorem. As with a theorem, a corollary can be used as a reason in a proof. The corollaries below follow directly from the Triangle Angle-Sum Theorem.

Corollary 5.1
The acute angles of a right triangle are complementary.
Corollary 5.2
There can be at most one right or obtuse angle in a triangle.

You will prove Corollary 5.1 and 5.2 in Exercises 19 and 20, respectively.

Example 3 Find Angle Measures in Right Triangles

Find each measure.

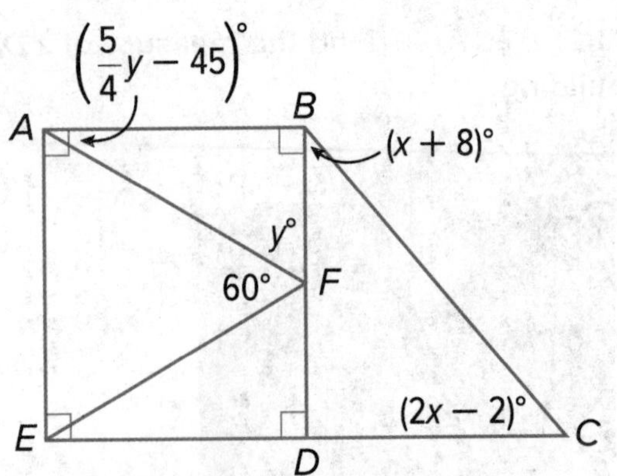

Study Tip

Check for Reasonableness When you are solving for the measure of one or more angles of a triangle, always check to make sure that the sum of the angle measures is 180°.

Think About It!

Do you have enough information to solve for $m\angle FED$? If so, explain your solution process.

a. $m\angle BCD$

Because $\angle BDC$ and $\angle EDF$ form a linear pair and $m\angle EDF =$ ____, $m\angle BDC =$ ____ by the Supplement Theorem. Therefore, $m\angle BCD + m\angle DBC =$ ____ because the acute angles of a right triangle are ________.

$m\angle BCD + m\angle DBC = 90^\circ$	Corollary 5.1
$(x + 8) + (2x - 2) = 90^\circ$	Substitution
$x =$ ____	Solve.

$m\angle BCD = (2(28) - 2)^\circ$ or ____

b. $m\angle BAF$

Because the acute angles of a right triangle are complementary, $m\angle BAF + m\angle AFB =$ ____.

$90^\circ = m\angle BAF + m\angle AFB$	Corollary 5.1
$= ($ ____ $) + y$	Substitution
____ $= y$	Solve.

$m\angle BAF = \frac{5}{4}($ ____ $) - 45$ or ____

Check

Find each measure.

$m\angle BAC =$ ____

$m\angle BCA =$ ____

$m\angle DCF =$ ____

$m\angle CDF =$ ____

$m\angle CFD =$ ____

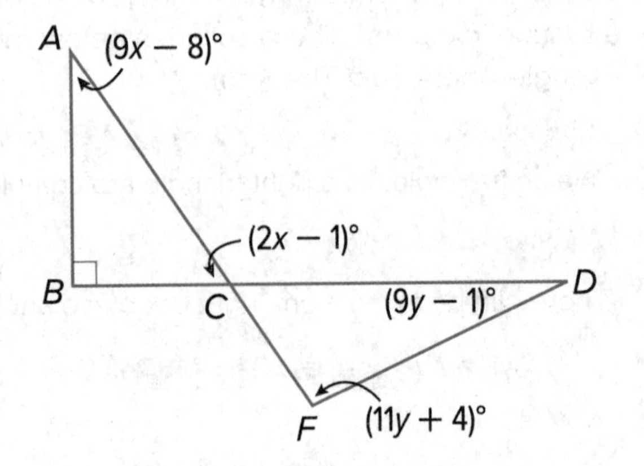

Go Online You can complete an Extra Example online.

Name ______________________ Period __________ Date __________

Practice

Go Online You can complete your homework online.

Example 1

Find the measure of each numbered angle.

1.

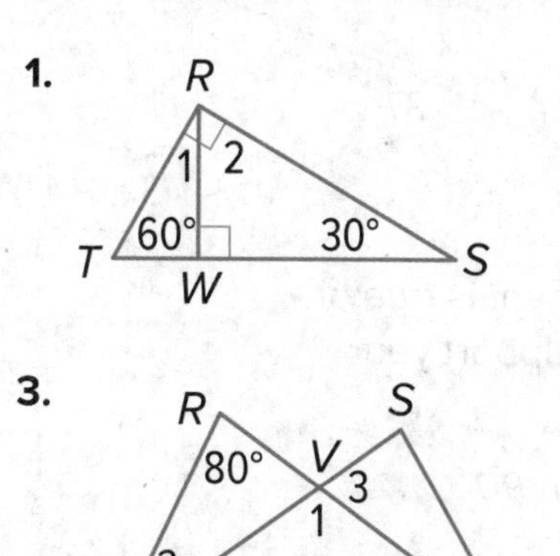

2.

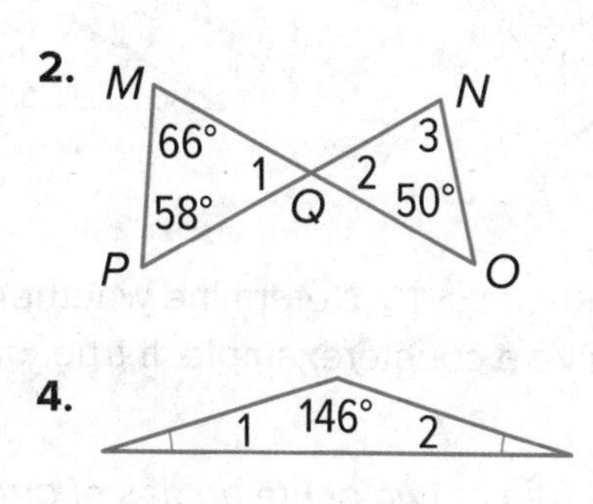

3.

4.

Example 2

Find each measure.

5. $m\angle ABC$

6. $m\angle F$

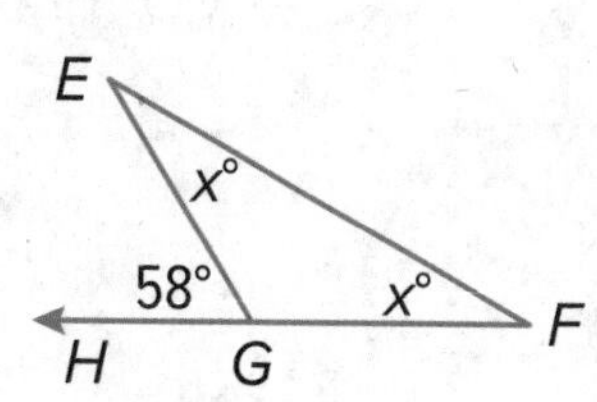

7. TOWERS A lookout tower sits on a network of struts and posts. Leslie measured three angles on the tower. If $m\angle 1 = (7x - 7)°$, $m\angle 2 = (4x + 2)°$, and $m\angle 3 = (2x + 6)°$, what is $m\angle 1$?

8. GARDENING A gardener uses a grow light to grow vegetables indoors. If $m\angle 1 = 8x°$ and $m\angle 2 = (7x - 4)°$, what is $m\angle 1$?

Example 3

Find each measure.

9. $m\angle 1$

10. $m\angle 2$

11. $m\angle 3$

12. $m\angle 4$

13. $m\angle 5$

14. $m\angle 6$

Mixed Exercises

Find the value of x. Then find the measure of each angle.

15.

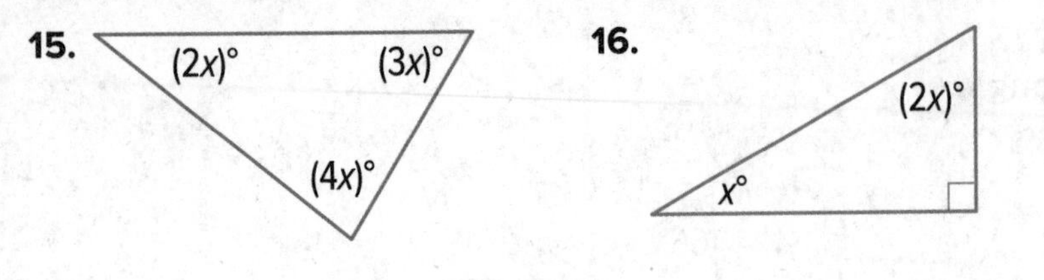

17.

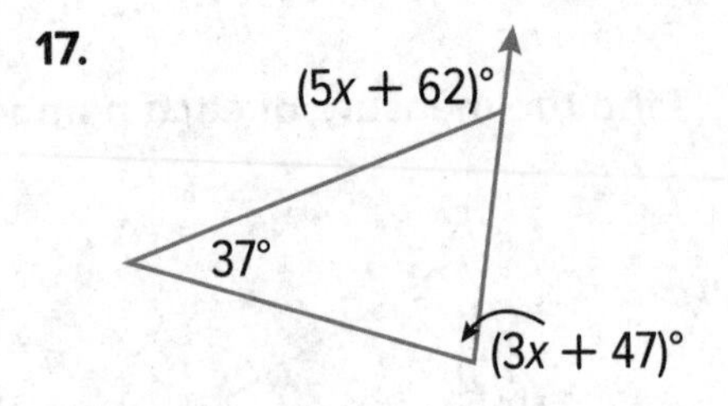

18. CONSTRUCT ARGUMENTS Determine whether the following statement is *true* or *false*. If false, give a counterexample. If true, give an argument to support your conclusion.

If the sum of two acute angles of a triangle is greater than 90°, then the triangle is acute.

PROOF **Write the specified type of proof for each corollary.**

19. flow proof of Corollary 5.1

Given: $\angle R$ is a right angle.

Prove: $\angle S$ and $\angle T$ are complementary.

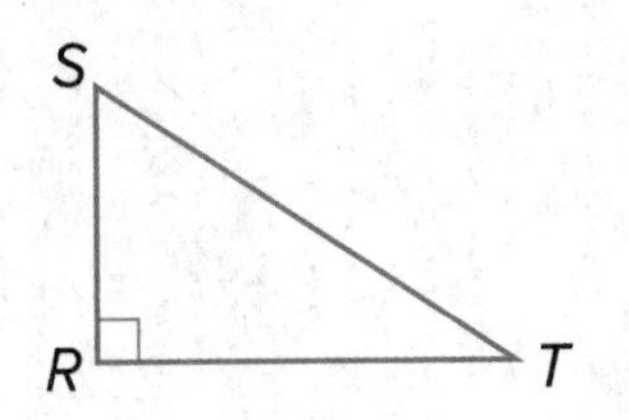

20. paragraph proof of Corollary 5.2

a. Case 1

Given: $\triangle MNO$; $\angle M$ is a right angle.

Prove: There can be at most one right angle in a triangle.

b. Case 2

Given: $\triangle PQR$; $\angle P$ is an obtuse angle.

Prove: There can be at most one obtuse angle in a triangle.

REASONING **Solve each problem.**

21. In triangle DEF, $m\angle E$ is three times $m\angle D$, and $m\angle F$ is 9° less than $m\angle E$. What is the measure of each angle?

22. In triangle RST, $m\angle T$ is 5° more than $m\angle R$, and $m\angle S$ is 10° less than $m\angle T$. What is the measure of each angle?

23. In triangle JKL, $m\angle K$ is four times $m\angle J$, and $m\angle L$ is five times $m\angle J$. What is the measure of each angle?

Name ______ Period ______ Date ______

24. Classify the triangle shown by its angles. Justify your reasoning.

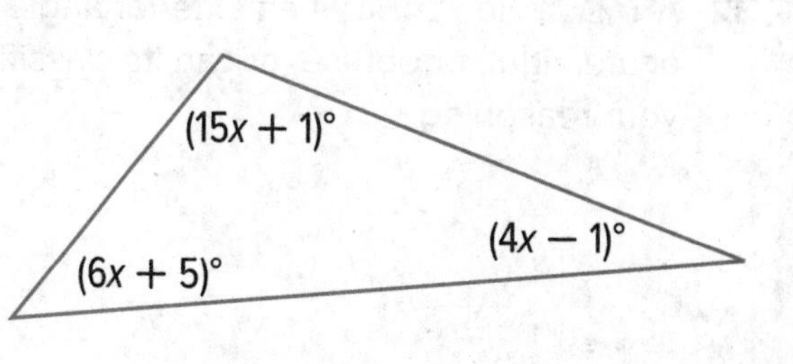

25. In $\triangle XYZ$, $m\angle X = 157°$, $m\angle Y = y°$, and $m\angle Z = z°$. Write an inequality to describe the possible measures of $\angle Z$. Justify your reasoning.

26. AUTOMOBILES Refer to the image at the right.

a. Find $m\angle 1$ and $m\angle 2$.

b. If the hood prop rod were shorter than the one shown, how would $m\angle 1$ change? Explain.

c. If the hood prop rod were shorter than the one shown, how would $m\angle 2$ change? Explain.

27. BASKETBALL Sam, Kendra, and Tony are passing a basketball. If Sam is looking at Kendra, then he needs to turn 40° to pass to Tony. If Tony is looking at Sam, then he needs to turn 50° to pass to Kendra. How many degrees would Kendra have to turn her head to look at Tony if she is looking at Sam?

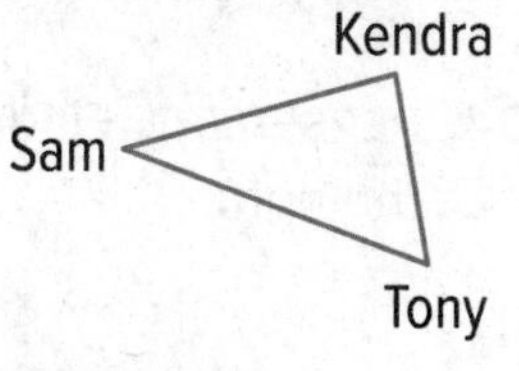

28. CONSTRUCTION The diagram shows an example of the Pratt Truss used in bridge construction. Find $m\angle 1$.

Find the measure of each numbered angle.

29.

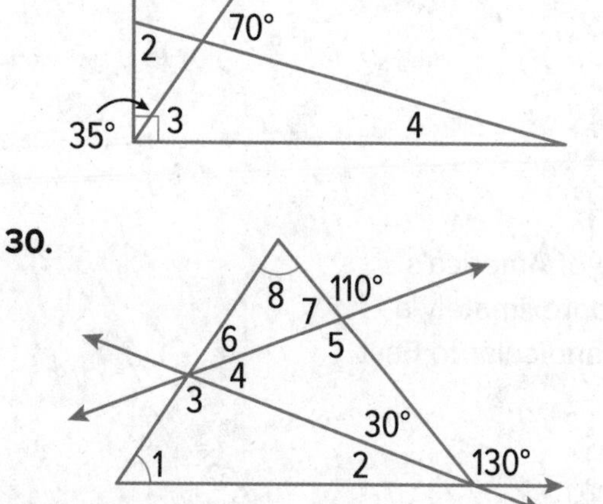

30.

31. USE TOOLS Use tracing paper to verify the Triangle Angle-Sum Theorem. Describe your method and include a sketch.

32. **ANALYZE** In $\triangle ABC$, if an exterior angle adjacent to $\angle A$ is acute, is the triangle acute, right, or obtuse, or can its classification not be determined? Explain your reasoning.

33. **WRITE** Explain why a triangle cannot have an obtuse, acute, and a right exterior angle.

34. **PERSEVERE** Find the values of y and z in the figure at the right.

35. **CREATE** Construct a right triangle and measure one of the acute angles. Calculate the measure of the second acute angle and explain your method. Confirm your result using a protractor.

36. **PERSEVERE** The Flatiron Building in New York City is one of America's oldest skyscrapers, completed in 1902. Its floor plan is approximately a right triangle. As shown in the figure, 5th Avenue is perpendicular to East 22nd Street, and $m\angle B$ is 10 less than 3 times $m\angle C$.

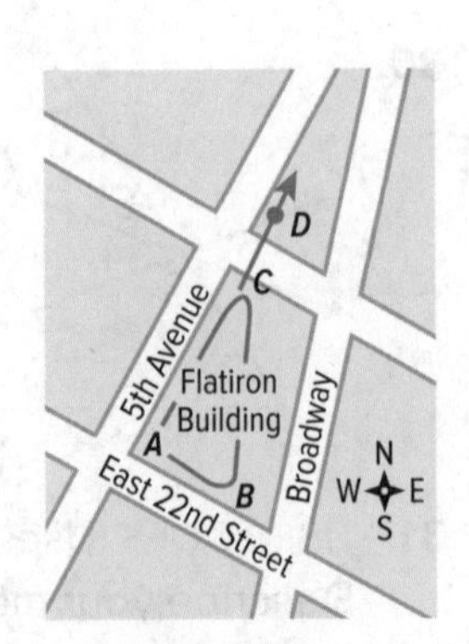

a. Find the angle measures in the floor plan.

b. Find $m\angle BCD$ in two ways. Explain each method.

Congruent Triangles

Explore Relationships in Congruent Triangles

Online Activity Use dynamic geometry software to complete the Explore.

INQUIRY If two triangles are congruent, what is the relationship between their corresponding parts?

Learn Congruent Triangles

The **principle of superposition** states that two figures are congruent if and only if there is a rigid motion or series of rigid motions that maps one figure exactly onto the other. Recall that congruent figures have exactly the same shape and size.

In two **congruent polygons**, all the parts of one polygon are congruent to the **corresponding parts,** or matching parts, of the other polygon. These corresponding parts include *corresponding angles* and *corresponding sides*.

Key Concept • Congruent Triangles

Two triangles are congruent if and only if their corresponding parts are congruent.

For triangles, we say *Corresponding parts of congruent triangles are congruent*, or CPCTC.

Example 1 Identify Corresponding Congruent Parts

Show that the polygons are congruent by identifying all the congruent corresponding parts. Then write a congruence statement.

B C A D Z Y X W

Angles: $\angle A \cong \angle W$; $\angle B \cong \angle X$; $\angle C \cong \angle Y$; $\angle D \cong \angle Z$

Sides: $\overline{BC} \cong \overline{XY}$; $\overline{AB} \cong \overline{WX}$; $\overline{DA} \cong \overline{ZW}$; $\overline{CD} \cong \overline{YZ}$

All corresponding parts of the two polygons are congruent. Therefore, polygon $ABCD \cong$ polygon $WXYZ$.

Go Online You can complete an Extra Example online.

Today's Goals

- Use congruence criterion of corresponding congruent parts of triangles to solve problems.
- Use the Third Angles Theorem and the properties of triangle congruence to solve problems and to prove relationships in geometric figures.

Today's Vocabulary

principle of superposition

congruent polygons

corresponding parts

Go Online

You can watch a video to see how to use transformations to determine whether two triangles are congruent.

Problem-Solving Tip

Get a New Perspective

When comparing two figures, it may be helpful to redraw the figures so they have the same orientation. This would make it easier to compare the corresponding sides and angles.

Your Notes

Check

Show that the polygons are congruent by identifying all the congruent corresponding parts. Then write a congruence statement.

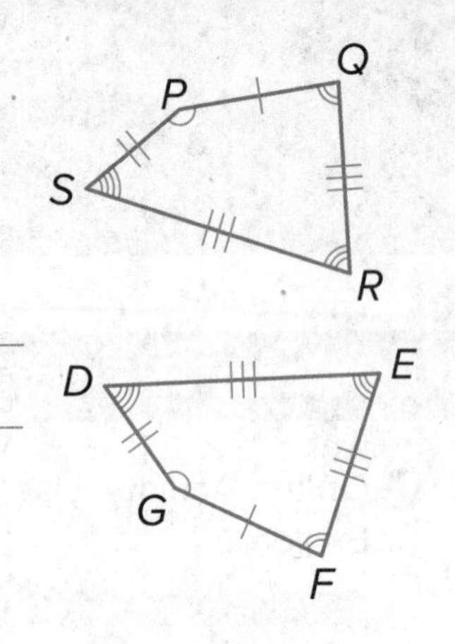

$\angle P \cong$ ____; $\angle Q \cong$ ____; $\angle R \cong$ ____; $\angle S \cong$ ____

$\overline{PQ} \cong$ ____; $\overline{QR} \cong$ ____; $\overline{RS} \cong$ ____; $\overline{SP} \cong$ ____

Complete the congruence statement.

Polygon *PQRS* $\cong$ *Polygon* ______

Example 2 Use Corresponding Parts of Congruent Triangles

In the diagram, $\triangle RSV \cong \triangle TVS$. Find the values of *x* and *y*.

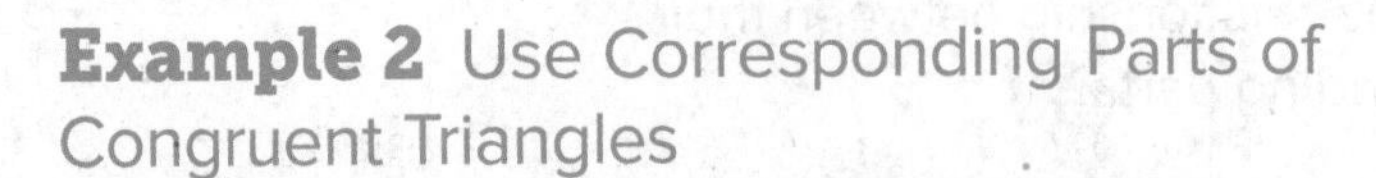

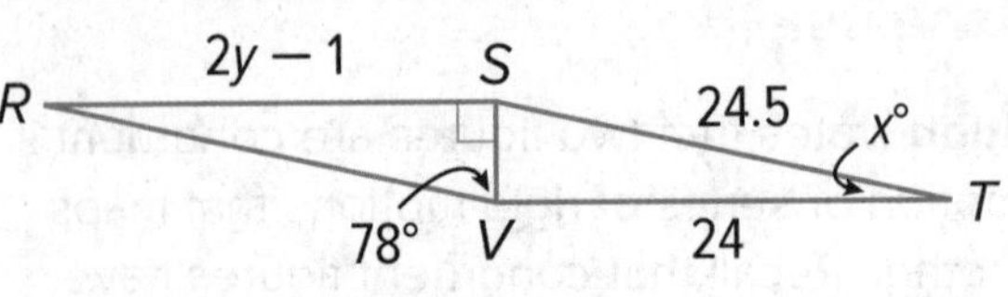

Part A Find the value of *x*.

$\angle T \cong \angle R$	CPCTC
$m\angle T = m\angle R$	Definition of congruence
$= 180° - 90° - 78°$	Triangle Angle-Sum Theorem
$=$ ____	Solve.

The value of *x* is 12.

Part B Find the value of *y*.

$\overline{RS} \cong \overline{TV}$	CPCTC
$RS = TV$	Definition of congruence
$2y - 1 =$ ____	Substitution
$y =$ ____	Solve.

The value of *y* is 12.5.

Study Tip

Use a Congruence Statement You can use a congruence statement to help you correctly identify corresponding sides.
$\triangle RSV \cong \triangle TVS$
$\overline{RS} \cong \overline{TV}$

Talk About It!

Suppose the congruence statement $\triangle RSV \cong \triangle TVS$ was *not* given. Would you be able to solve this problem? Explain.

Check

In the diagram, $\triangle ABC \cong \triangle EDC$. Find the values of *x* and *y*.

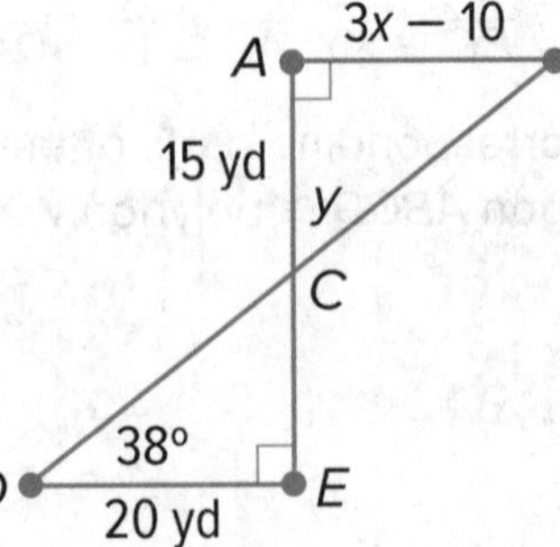

$x =$ ____; $y =$ ____

Go Online You can complete an Extra Example online.

Learn Third Angles Theorem and Triangle Congruence

Theorem 5.3: Third Angles Theorem

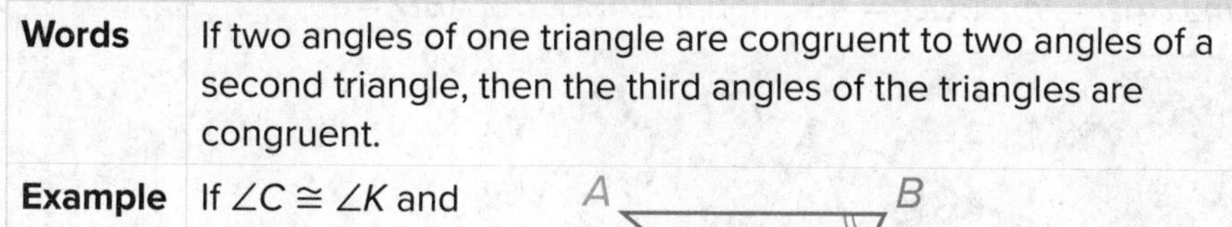

Words	If two angles of one triangle are congruent to two angles of a second triangle, then the third angles of the triangles are congruent.
Example	If $\angle C \cong \angle K$ and $\angle B \cong \angle J$, then $\angle A \cong \angle L$.

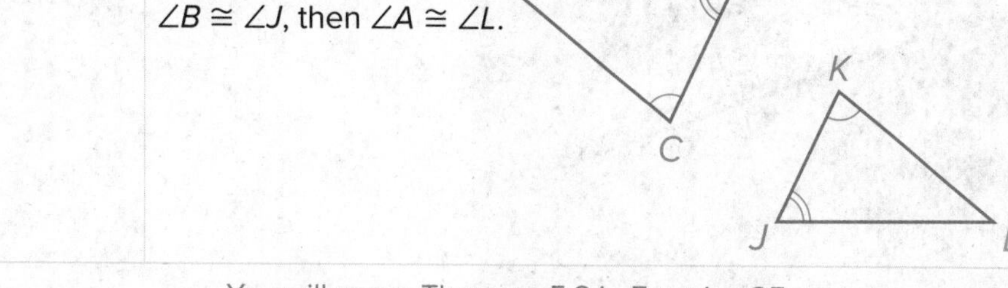

You will prove Theorem 5.3 in Exercise 25.

Like congruence of segments and angles, congruence of triangles is reflexive, symmetric, and transitive.

Theorem 5.4: Properties of Triangle Congruence
Reflexive Property of Triangle Congruence $\triangle ABC \cong \triangle ABC$
Symmetric Property of Triangle Congruence If $\triangle ABC \cong \triangle EFG$, then $\triangle EFG \cong \triangle ABC$.
Transitive Property of Triangle Congruence If $\triangle ABC \cong \triangle EFG$ and $\triangle EFG \cong \triangle JKL$, then $\triangle ABC \cong \triangle JKL$.

Go Online Proofs of Theorem 5.4 are available.

Example 3 Use the Third Angles Theorem

ORIGAMI **Aika is folding origami dragons for a party she is hosting. If $\angle ABD \cong \angle CBD$ and $m\angle BAD = 58°$, find $m\angle CBD$.**

What Do You Know?	How Do You Know It?
$\angle ABD \cong \angle CBD$, $m\angle BAD = 58°$	Given
$\angle BDC \cong \angle BDA$	All rt. $\angle$s are $\cong$.
$\angle BCD \cong \angle BAD$	Third Angles Theorem
$m\angle BCD = m\angle BAD$	Def. of congruence
$m\angle CBD + m\angle BCD = 90°$	The acute $\angle$s of a rt. $\triangle$ are compl.

$m\angle BCD =$ ______ Substitute.

$m\angle CBD +$ ______ $= 90°$ Substitute.

$m\angle CBD =$ ______ Solve.

The measure of $\angle CBD$ is 32°.

Think About It! How could you find the $m\angle CBD$ in a different way?

Go Online You can complete an Extra Example online.

Check

KITES The kite shown is made of two congruent triangles. If $m\angle BAD = m\angle BCD = 45°$, find $m\angle ABD$. $m\angle ABD =$ _______

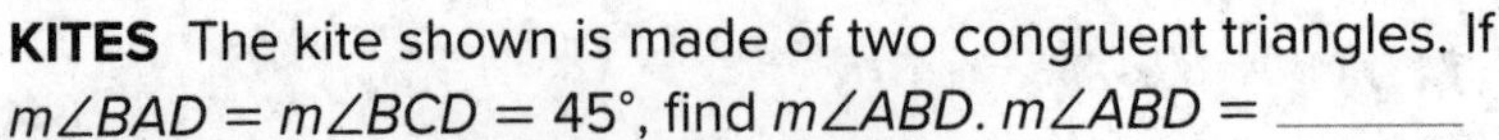

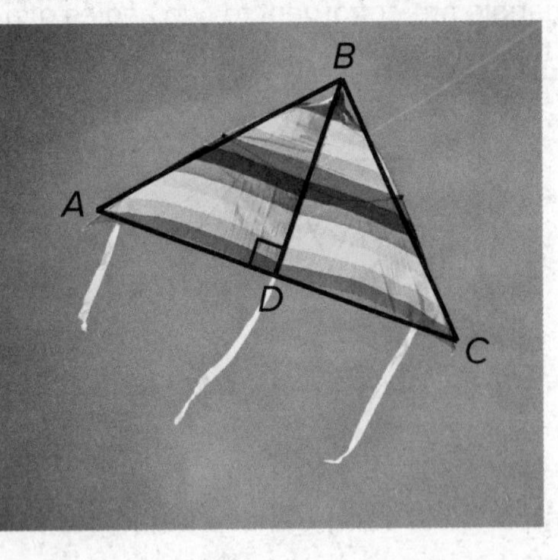

Example 4 Prove that Two Triangles Are Congruent

Write a two-column proof.

Given: $\angle J \cong \angle P$, $\overline{JK} \cong \overline{PM}$, $\overline{JL} \cong \overline{PL}$, and L bisects $\overline{KM}$.

Prove: $\triangle JLK \cong \triangle PLM$

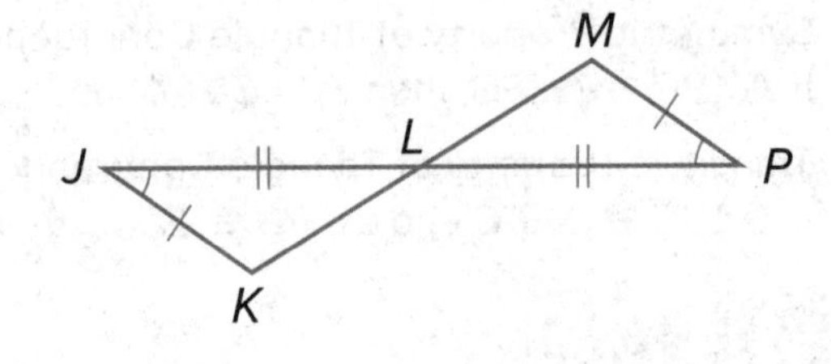

Statements	Reasons
1. $\angle J \cong \angle P$, $\overline{JK} \cong \overline{PM}$, $\overline{JL} \cong \overline{PL}$ and L bisects $\overline{KM}$.	1. Given
2. $\angle JLK \cong \angle PLM$	2. Vertical angles are congruent.
3. $\overline{LK} \cong \overline{LM}$	3. Definition of segment bisector
4. $\angle K \cong \angle M$	4. Third Angles Theorem
5. $\triangle JLK \cong \triangle PLM$	5. Definition of congruent triangles

Study Tip

Symbols To indicate that two triangles are not congruent, write $\triangle ABC \ncong \triangle EFG$. $\triangle ABC \ncong \triangle EFG$ is read as *triangle ABC is not congruent to triangle EFG*.

Check

Write a paragraph proof.

Given: $\angle WXZ \cong \angle YXZ$, $\angle XZW \cong \angle XZY$, $\overline{WX} \cong \overline{YX}$, $\overline{WZ} \cong \overline{YZ}$

Prove: $\triangle WXZ \cong \triangle YXZ$

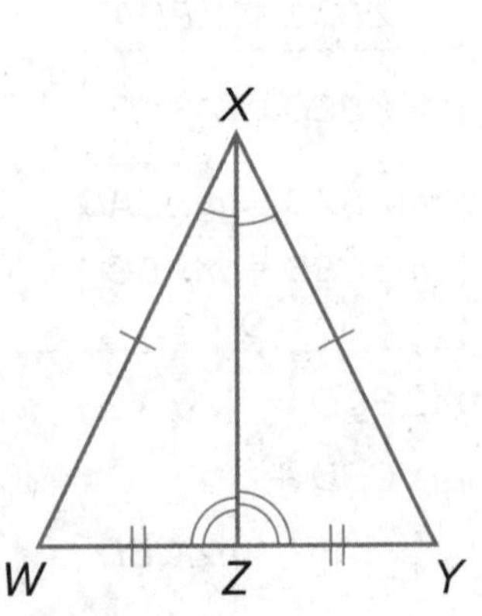

It is given that $\overline{WX} \cong \overline{YX}$ and $\overline{WZ} \cong \overline{YZ}$. By the _______ Property, $\overline{XZ} \cong \overline{XZ}$. It is also given that $\angle WXZ \cong \angle YXZ$ and $\angle XZW \cong \angle XZY$. So, by the _______ Theorem, $\angle W \cong \angle Y$. By the definition of congruent triangles, $\triangle WXZ \cong \triangle YXZ$.

Go Online You can complete an Extra Example online.

Name ______________________ Period ________ Date ________

Practice

Go Online You can complete your homework online.

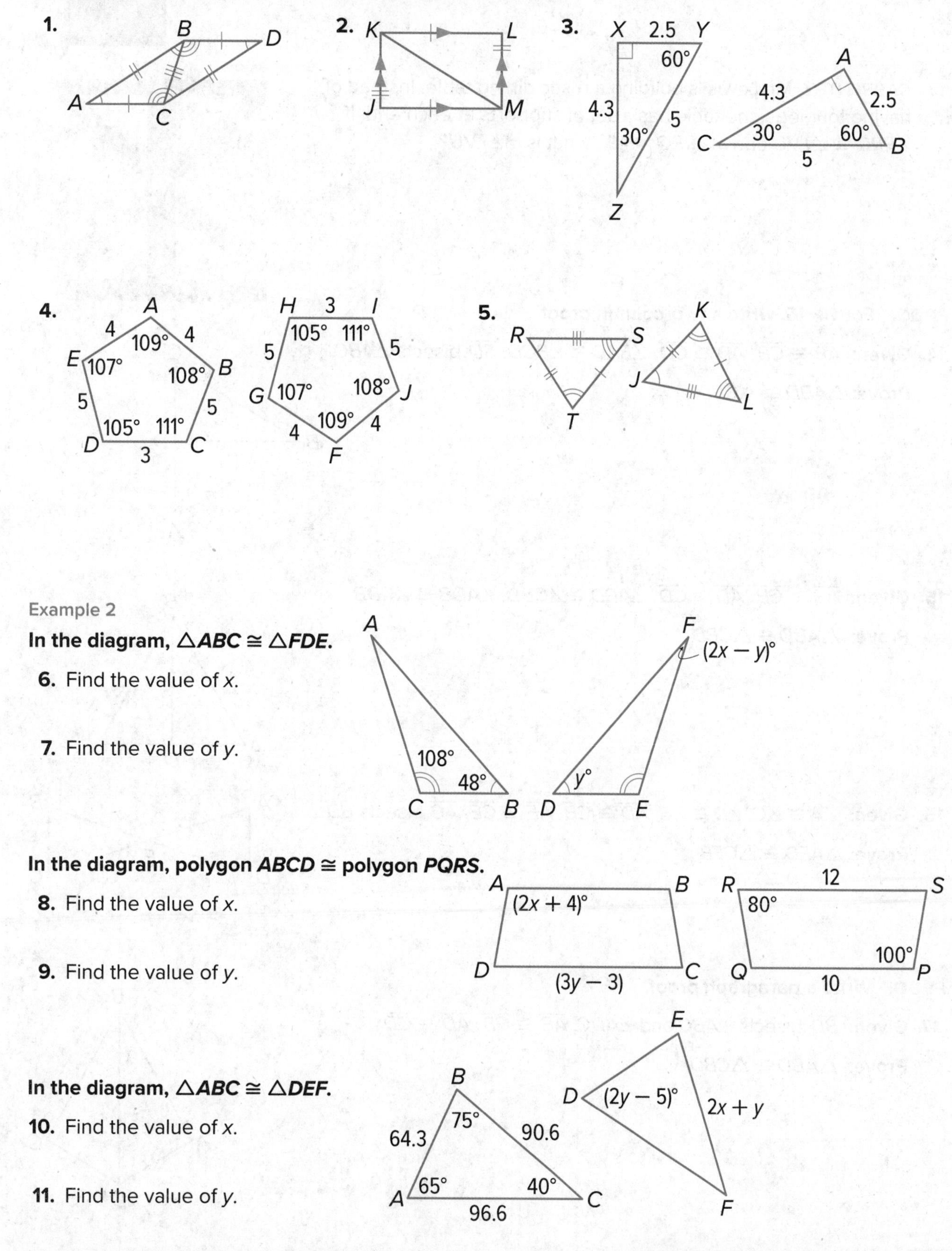

Example 1

Show that the polygons are congruent by identifying all the congruent corresponding parts. Then write a congruence statement.

1.

2.

3.

4.

5.

Example 2

In the diagram, $\triangle ABC \cong \triangle FDE$.

6. Find the value of x.

7. Find the value of y.

In the diagram, polygon $ABCD \cong$ polygon $PQRS$.

8. Find the value of x.

9. Find the value of y.

In the diagram, $\triangle ABC \cong \triangle DEF$.

10. Find the value of x.

11. Find the value of y.

Example 3

12. DESIGN Camila is designing a new image for her cell phone case. If $m\angle ABC = 35°$, $m\angle BAC = 29°$, and $\angle ACB \cong \angle DEB$, what is $m\angle DEB$?

13. CARPENTRY Mr. Lewis is building a rustic dining table. Instead of having four legs, the table has a set of supports at each end. If $\angle PRQ \cong \angle TVU$ and $m\angle RPQ = 49°$, what is $m\angle TVU$?

Example 4

PROOF **For 14–16, write a two-column proof.**

14. Given: $\overline{AB} \cong \overline{CB}$, $\overline{AD} \cong \overline{CD}$, $\angle BAD \cong \angle BCD$, $\overline{BD}$ bisects $\angle ABC$.

Prove: $\triangle ABD \cong \triangle CBD$

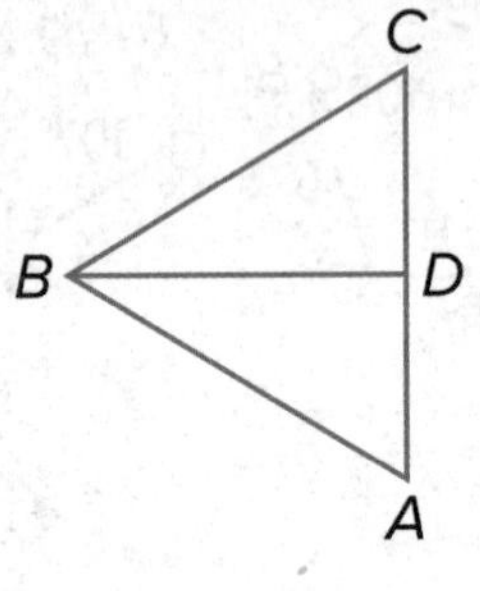

15. Given: $\overline{AB} \cong \overline{CB}$, $\overline{AD} \cong \overline{CD}$, $\angle ABD \cong \angle CBD$, $\angle ADB \cong \angle CDB$

Prove: $\triangle ABD \cong \triangle CBD$

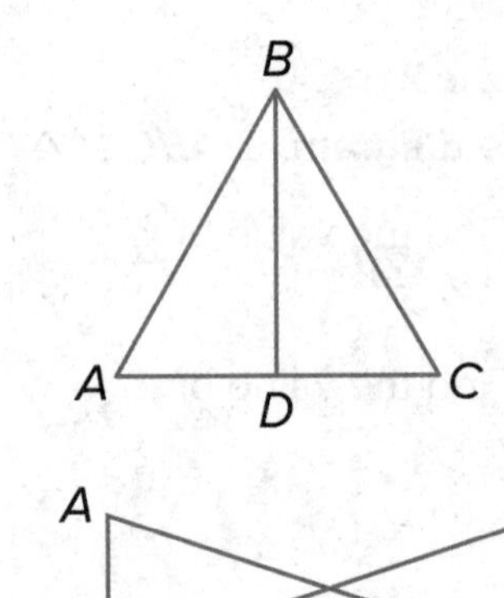

16. Given: $\angle A \cong \angle C$, $\angle D \cong \angle B$, $\overline{AD} \cong \overline{CB}$, $\overline{AE} \cong \overline{CE}$, $\overline{AC}$ bisects $\overline{BD}$.

Prove: $\triangle AED \cong \triangle CEB$

PROOF **Write a paragraph proof.**

17. Given: $\overline{BD}$ bisects $\angle ABC$ and $\angle ADC$, $\overline{AB} \cong \overline{CB}$, $\overline{AD} \cong \overline{CD}$

Prove: $\triangle ABD \cong \triangle CBD$

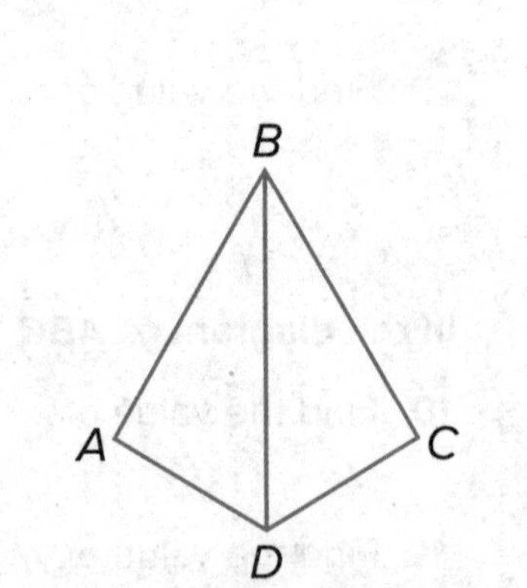

Mixed Exercises

18. PRECISION Beverly is using loyalty cards at her coffee shop. When a customer purchases a cup of coffee, he or she can present a loyalty card to be stamped with a star-shaped stamp that Beverly purchased specifically for this use. When the customer collects nine stamps, they receive their tenth cup of coffee for free. What property guarantees that the stamped designs are congruent?

SECOND STREET COFFEE
Purchase 9 cups, get the 10th free!
FREE

Draw and label a figure to represent the congruent triangles. Then find the values of *x* and *y*.

19. $\triangle ABC \cong \triangle DEF$, $AB = 7$, $BC = 9$, $AC = 11 + x$, $DF = 3x - 13$, and $DE = 2y - 5$

20. $\triangle LMN \cong \triangle RST$, $m\angle L = 49°$, $m\angle M = 10y°$, $m\angle S = 70°$, and $m\angle T = (4x + 9)°$

21. $\triangle JKL \cong \triangle MNP$, $JK = 12$, $LJ = 5$, $PM = 2x - 3$, $m\angle L = 67°$, $m\angle K = (y + 4)°$ and $m\angle N = (2y - 15)°$

22. SIERPINSKI TRIANGLE The figure shown is a portion of the Sierpinski triangle. The triangle has the property that any triangle made from any combination of edges is equilateral. How many triangles in this portion are congruent to the black triangle at the bottom?

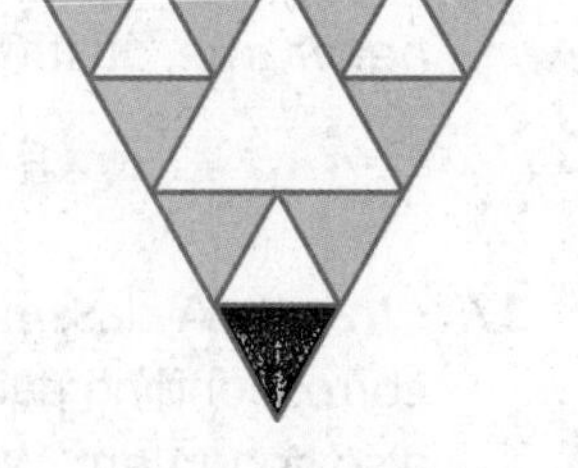

23. LOGO DESIGNS Refer to the design shown.

a. Indicate the triangles that appear to be congruent.

b. Name the congruent angles and congruent sides of a pair of congruent triangles.

24. REASONING Igor noticed on a map that the triangle with vertices that are at the supermarket, the library, and the post office ($\triangle SLP$) is congruent to the triangle with vertices that are at Igor's home, Jasen's home, and Daran's home ($\triangle IJD$). That is, $\triangle SLP \cong \triangle IJD$.

a. The distance between the supermarket and the post office is 1 mile. Which path along the triangle $\triangle IJD$ is congruent to this?

b. The measure of $\angle LPS$ is 40°. Identify the angle that is congruent to this angle in $\triangle IJD$.

25. PROOF Complete the two-column proof of the Third Angles Theorem by providing the reason for each statement.

Given: $\angle P \cong \angle X$ and $\angle Q \cong \angle Y$

Prove: $\angle R \cong \angle Z$

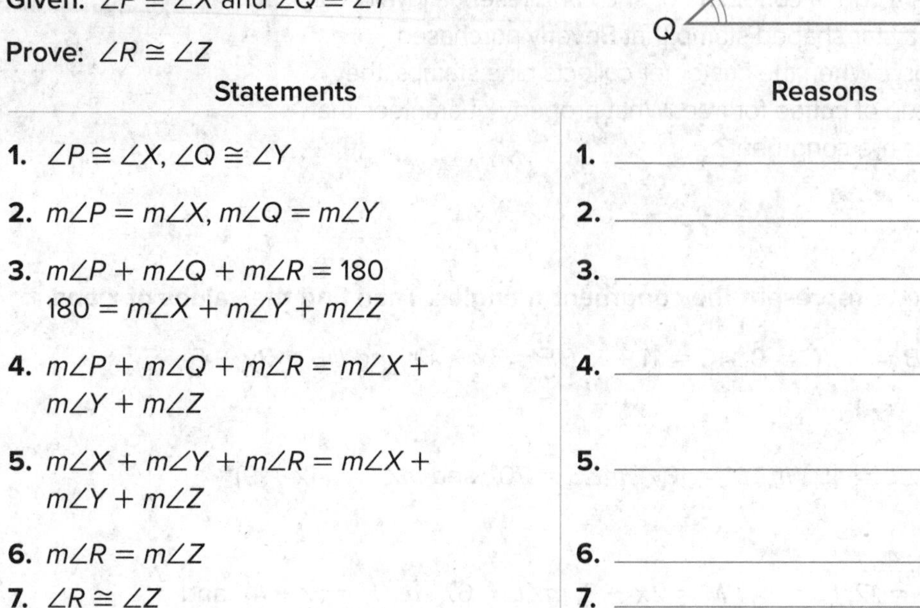

Statements	Reasons
1. $\angle P \cong \angle X$, $\angle Q \cong \angle Y$	1. ________
2. $m\angle P = m\angle X$, $m\angle Q = m\angle Y$	2. ________
3. $m\angle P + m\angle Q + m\angle R = 180$ $180 = m\angle X + m\angle Y + m\angle Z$	3. ________
4. $m\angle P + m\angle Q + m\angle R = m\angle X + m\angle Y + m\angle Z$	4. ________
5. $m\angle X + m\angle Y + m\angle R = m\angle X + m\angle Y + m\angle Z$	5. ________
6. $m\angle R = m\angle Z$	6. ________
7. $\angle R \cong \angle Z$	7. ________

26. ANALYZE Determine whether the following statement is *sometimes*, *always*, or *never* true. Justify your argument.

Equilateral triangles are congruent.

27. CREATE A classmate is using the Third Angles Theorem to show that if two corresponding pairs of the angles of two triangles are congruent, then the third pair is also congruent. Write a question to help him decide whether he can use the same strategy for quadrilaterals.

28. FIND THE ERROR Jasmine and West are evaluating the congruent figures at right. Jasmine says that $\triangle CAB \cong \triangle ZYX$, and West says that $\triangle ABC \cong \triangle YXZ$. Is either of them correct? Explain your reasoning.

29. WRITE Justify why the order of the vertices is important when naming congruent triangles. Give an example to support your argument.

30. PERSEVERE Find the values of x and y if $\triangle PQS \cong \triangle RQS$.

Proving Triangles Congruent: SSS, SAS

Explore Conditions That Prove Triangles Congruent

Online Activity Use dynamic geometry software to complete the Explore.

INQUIRY What conditions can be used to identify whether two triangles are congruent?

Today's Goals

- Use the SSS Congruence criterion for triangles to solve problems and prove relationships in geometric figures.
- Use the SAS Congruence criterion for triangles to solve problems and prove relationships in geometric figures.

Today's Vocabulary

included angle

Learn Proving Triangles Congruent: SSS

You can prove two triangles congruent by showing that all six pairs of corresponding parts are congruent. However, it is possible to prove two triangles congruent using fewer pairs of corresponding parts.

If two triangles have the same three side lengths, then there is a series of rigid motions that will show the two triangles congruent. This leads to the postulate below.

Postulate 5.1: Side-Side-Side (SSS) Congruence

If three sides of one triangle are congruent to three sides of a second triangle, then the triangles are congruent.

Talk About It!

Will two equilateral triangles always be congruent by SSS? Justify your argument.

Example 1 Use SSS to Prove Triangles Congruent

Write a flow proof to show that $\triangle QRT \cong \triangle SRT$.

Given: $\triangle QRS$ is isosceles with $\overline{QR} \cong \overline{SR}$. $\overrightarrow{RT}$ bisects $\overline{QS}$ at point T.

Prove: $\triangle QRT \cong \triangle SRT$

Proof:

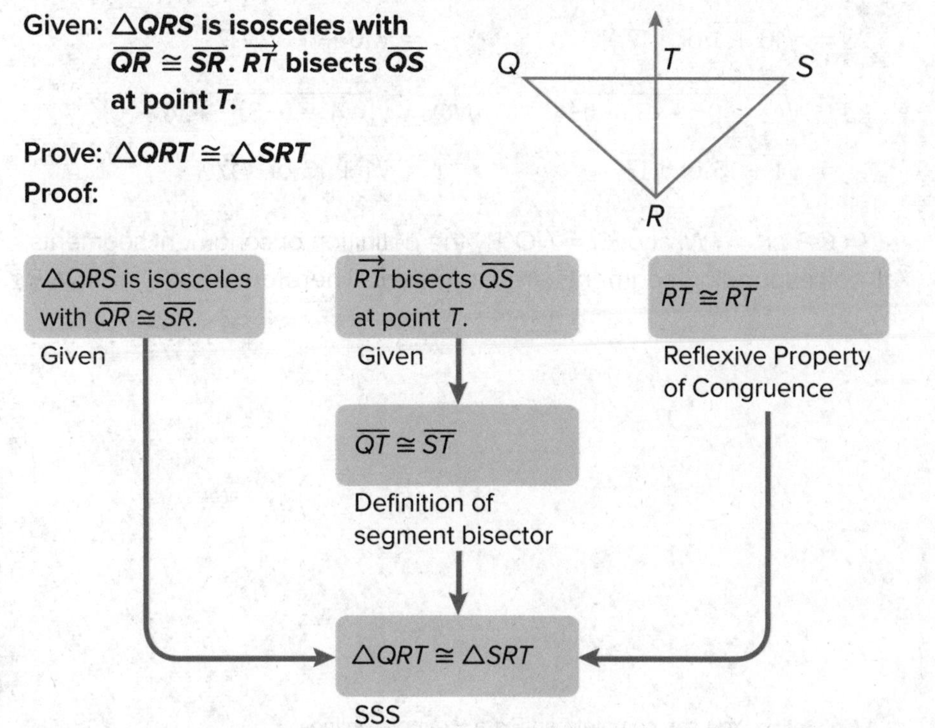

Q
T
S
R
$\triangle QRS$ is isosceles with $\overline{QR} \cong \overline{SR}$.
Given
$\overrightarrow{RT}$ bisects $\overline{QS}$ at point T.
Given
$\overline{RT} \cong \overline{RT}$
Reflexive Property of Congruence
$\overline{QT} \cong \overline{ST}$
Definition of segment bisector
$\triangle QRT \cong \triangle SRT$
SSS

Go Online You can complete an Extra Example online.

Go Online

An alternate method is available for this example.

Your Notes

Think About It!

Is the following statement *true* or *false*? Justify your argument.

If the congruent sides in one isosceles triangle have the same measure as the congruent sides in another isosceles triangle, then the triangles are congruent.

Example 2 Use SSS on the Coordinate Plane

Triangle *JKL* has vertices *J*(2, 5), *K*(1, 1), and *L*(5, 2). Triangle *QNP* has vertices *Q*(−4, 4), *N*(−3, 0), and *P*(−7, 1). Is $\triangle JKL \cong \triangle QNP$?

Part A Graph the triangles.

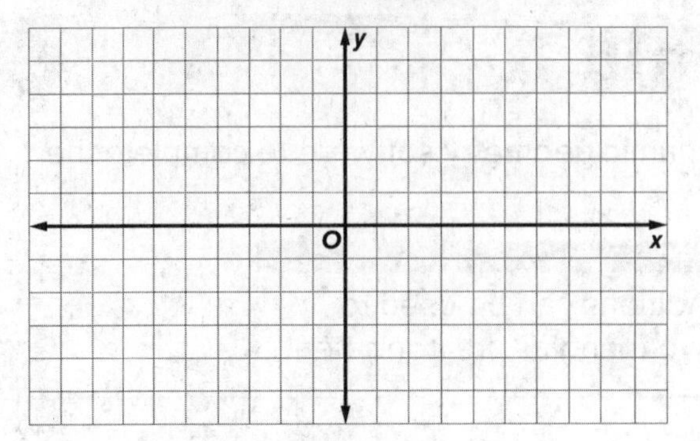

Part B Make a conjecture.

Use your graph to make a conjecture as to whether the triangles are congruent. Explain your reasoning.

From the graph, it appears that the triangles have the same shape and size, so we can conjecture that they are congruent.

Part C Support your conjecture.

Use the Distance Formula to show that all corresponding sides have the same measure.

$JL = \sqrt{(5-2)^2 + (2-5)^2} = \sqrt{9+9}$ or $3\sqrt{2}$

$QP = \sqrt{[-7-(-4)]^2 + (1-4)^2} = \sqrt{9+9}$ or $3\sqrt{2}$

$LK = \sqrt{(1-5)^2 + (1-2)^2} = \sqrt{16+1}$ or $\sqrt{17}$

$PN = \sqrt{[-3-(-7)]^2 + (0-1)^2} = \sqrt{16+1}$ or $\sqrt{17}$

$KJ = \sqrt{(2-1)^2 + (5-1)^2} = \sqrt{1+16}$ or $\sqrt{17}$

$NQ = \sqrt{[-4-(-3)]^2 + (4-0)^2} = \sqrt{1+16}$ or $\sqrt{17}$

$JL = QP$, $LK = PN$, and $KJ = NQ$. By the definition of congruent segments, all corresponding segments are congruent. Therefore, $\triangle JKL \cong \triangle QNP$ by ____.

Go Online You can complete an Extra Example online.

Check

Triangle ABC has vertices $A(1, 1)$, $B(0, 3)$, and $C(2, 5)$. Triangle EFG has vertices $E(1, -1)$, $F(2, -5)$, and $G(4, -4)$. Is $\triangle ABC \cong \triangle EFG$?

Part A

Graph $\triangle ABC$ and $\triangle EFG$ on the same coordinate plane.

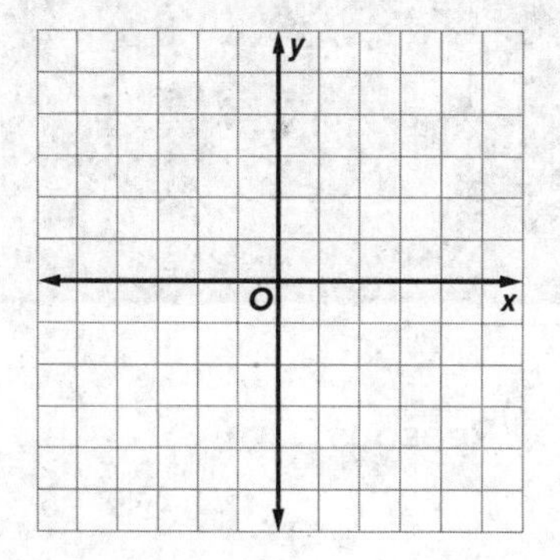

Part B

Find the side lengths of each triangle.

$AB =$ ____; $BC =$ ____; $AC =$ ____; $EF =$ ____; $FG =$ ____; $EG =$ ____

Part C

Is triangle ABC congruent to triangle EFG? Justify your argument.

A. No; $AC \neq FG$, so SSS congruence is not met.

B. No; $BC \neq FG$, so SSS congruence is not met.

C. Yes; all corresponding sides have the same measure, so SSS congruence is met.

D. Yes; all corresponding sides have the same measure, so by the definition of congruent figures, $\triangle ABC \cong \triangle EFG$.

Think About It!

Both legs of one right triangle are congruent to the legs of another right triangle. Are the triangles congruent? Justify your argument.

Learn Proving Triangles Congruent: SAS

The interior angle formed by two adjacent sides of a triangle is called an **included angle**.

If two triangles are formed using the same side lengths and included angle measure, then there is a series of rigid motions that will show that the two triangles are congruent. This leads to the postulate below.

Postulate 5.2: Side-Angle-Side (SAS) Congruence

If two sides and the included angle of one triangle are congruent to two sides and the included angle of a second triangle, then the triangles are congruent.

Go Online You may want to complete the construction activities for this lesson.

Go Online You can complete an Extra Example online.

Study Tip

Assumptions Even though objects in the real world are not two-dimensional, you can still make some assumptions about their geometric relationships. For example, in three-dimensional objects you can apply the Reflexive Property of Congruence and the Vertical Angles Theorem.

Example 3 Use SAS to Prove Triangles Congruent

PLAYGROUND **The playground equipment shown appears to be made of congruent triangles. If $\overline{KL} \cong \overline{LM}$ and $\angle JLK \cong \angle JLM$, write a two-column proof to prove that $\triangle JLK \cong \triangle JLM$. Complete the two-column proof by selecting the correct statements and reasons.**

Statements	Reasons
1. $\overline{KL} \cong \overline{LM}$	1. Given
2. $\angle JLK \cong \angle JLM$	2. ______
3. ______	3. ______
4. $\triangle JLK \cong \triangle JLM$	4. ______

Statements/Reasons:

Given

Definition of congruent segments

SAS

SSS

Definition of angle bisector

Reflexive Property of Congruence

$\overline{JL} \cong \overline{JL}$

$\overline{JK} \cong \overline{JM}$

$\angle KJL \cong \angle MJL$

Check

KITES The kite shown appears to be made up of congruent triangles. If $\overline{DE} \cong \overline{FE}$ and $\overline{EG}$ bisects $\angle DEF$, prove that $\triangle DEG \cong \triangle FEG$. Complete the two-column proof by selecting the correct statements and reasons.

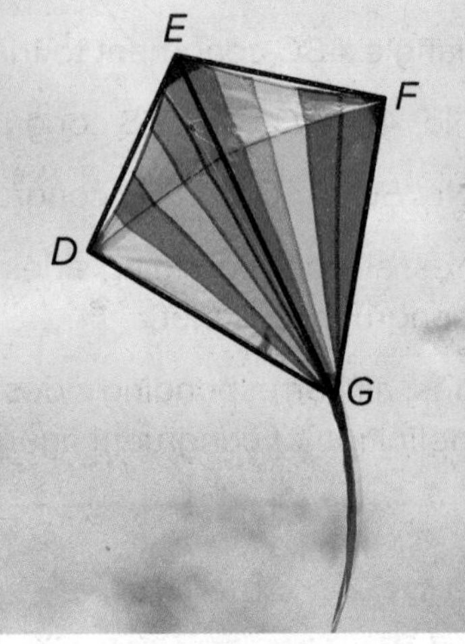
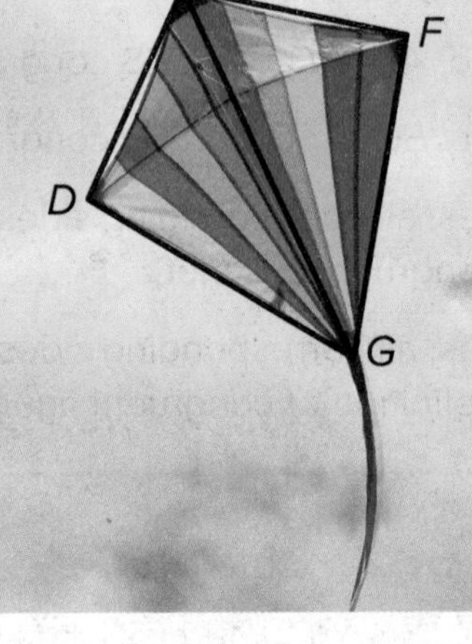

Given: $\overline{DE} \cong \overline{FE}$, $\overline{EG}$ bisects $\angle DEF$.

Prove: $\triangle DEG \cong \angle FEG$

Proof:

Statements	Reasons
1. ______	1. Given
2. $\overline{EG}$ bisects $\angle DEF$.	2. ______
3. ______	3. Definition of angle bisector
4. ______	4. ______
5. $\triangle DEG \cong \triangle FEG$	5. ______

Statements/Reasons:

Given

Definition of congruent segments

SAS

SSS

Definition of angle bisector

Reflexive Property of Congruence

$\overline{DE} \cong \overline{FE}$

$\overline{EG} \cong \overline{EG}$

$\overline{DG} \cong \overline{FG}$

$\angle DEG \cong \angle FEG$

Go Online You can complete an Extra Example online.

Name ______ Period ______ Date ______

Practice

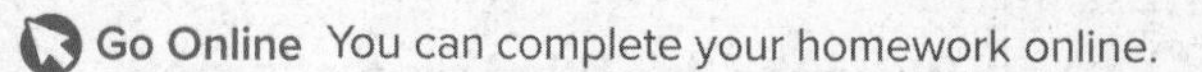

Go Online You can complete your homework online.

Example 1

PROOF **Write the specified type of proof.**

1. two-column proof
Given: $\overline{AB} \cong \overline{XY}$, $\overline{AC} \cong \overline{XZ}$, $\overline{BC} \cong \overline{YZ}$
Prove: $\triangle ABC \cong \triangle XYZ$

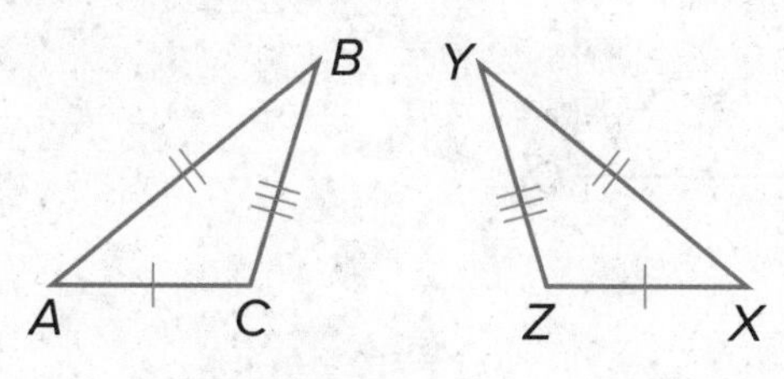

2. flow proof
Given: $\overline{RS} \cong \overline{UT}$, $\overline{RT} \cong \overline{US}$
Prove: $\triangle RST \cong \triangle UTS$

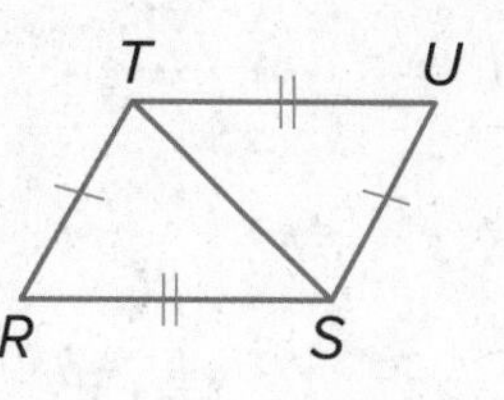

3. two-column proof
Given: $\overline{AB} \cong \overline{CB}$, D is the midpoint of $\overline{AC}$.
Prove: $\triangle ABD \cong \triangle CBD$

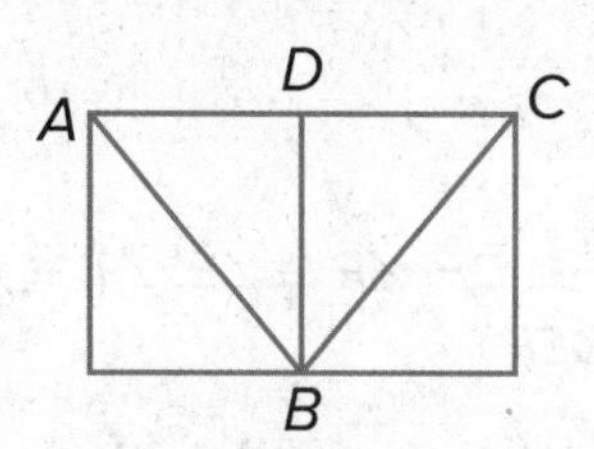

4. flow proof
Given: $\overline{RS} \cong \overline{TS}$, V is the midpoint of $\overline{RT}$.
Prove: $\triangle RSV \cong \triangle TSV$

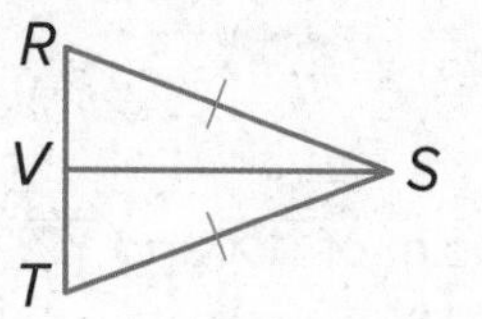

5. paragraph proof
Given: $\overline{QR} \cong \overline{SR}$, $\overline{ST} \cong \overline{QT}$
Prove: $\triangle QRT \cong \triangle SRT$

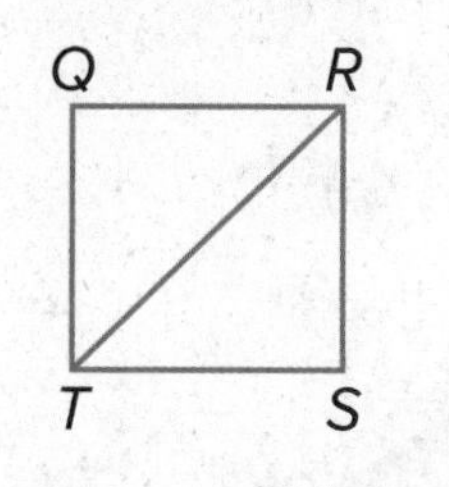

6. two-column proof
Given: $\overline{AB} \cong \overline{ED}$, $\overline{CA} \cong \overline{CE}$, $\overline{AC}$ bisects $\overline{BD}$
Prove: $\triangle ABC \cong \triangle EDC$

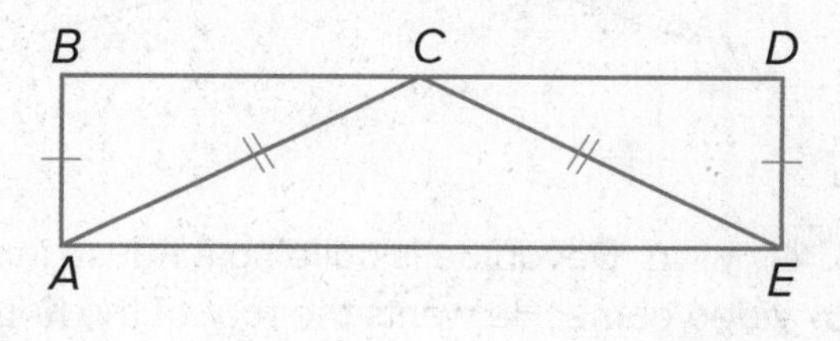

Example 2

REGULARITY **Determine whether $\triangle DEF \cong \triangle PQR$. Explain.**

7. $D(-6, 1)$, $E(1, 2)$, $F(-1, -4)$, $P(0, 5)$, $Q(7, 6)$, $R(5, 0)$

8. $D(-7, -3)$, $E(-4, -1)$, $F(-2, -5)$, $P(2, -2)$, $Q(5, -4)$, $R(0, -5)$

Determine whether $\triangle ABC \cong \triangle KLM$. Explain.

9. $A(-3, 3)$, $B(-1, 3)$, $C(-3, 1)$, $K(1, 4)$, $L(3, 4)$, $M(1, 6)$

10. $A(-4, -2)$, $B(-4, 1)$, $C(-1, -1)$, $K(0, -2)$, $L(0, 1)$, $M(4, 1)$

Example 3

PROOF **Write the specified type of proof.**

11. two-column proof
Given: $NP = PM$, $\overline{NP} \perp \overline{PL}$
Prove: $\triangle NPL \cong \triangle MPL$

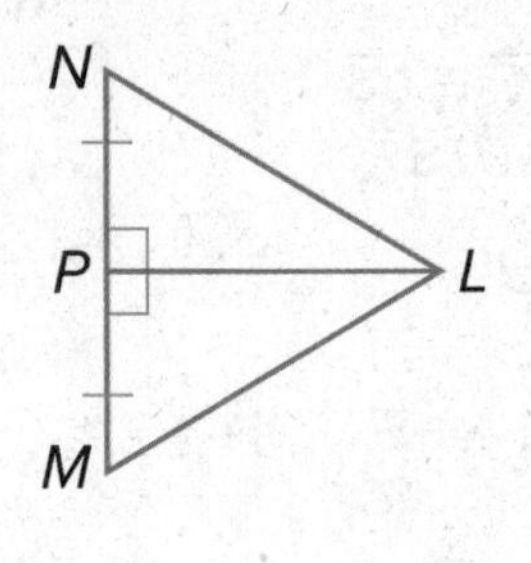

12. two-column proof
Given: $AB = CD$, $\overline{AB} \parallel \overline{CD}$
Prove: $\triangle ACD \cong \triangle DBA$

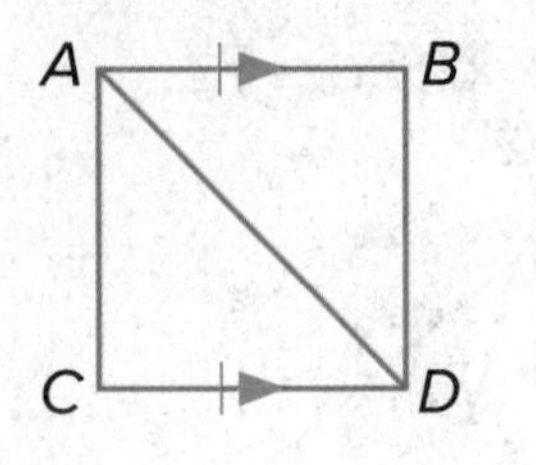

13. paragraph proof
Given: V is the midpoint of $\overline{WX}$ and $\overline{YZ}$.
Prove: $\triangle XVZ \cong \triangle WVY$

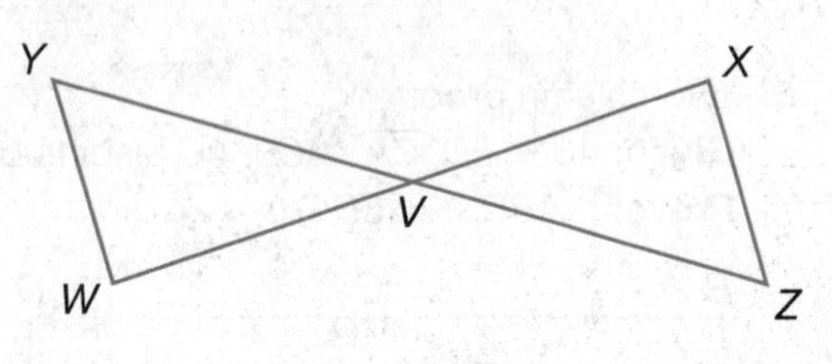

14. flow proof
Given: $\overline{PR} \cong \overline{DE}$, $\overline{PT} \cong \overline{DF}$, $\angle R \cong \angle E$, $\angle T \cong \angle F$
Prove: $\triangle PRT \cong \triangle DEF$

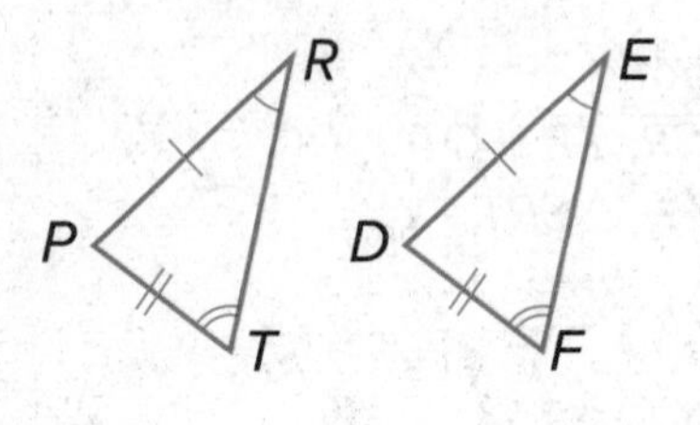

15. GAMING Devontae is building a house in a simulation video game. He wants the roof of the house and the main support beam to create congruent triangles. If $\overline{BD} \perp \overline{AC}$ and $\overline{BD}$ bisects $\overline{AC}$, write a two-column proof to prove $\triangle ABD \cong \triangle CBD$.

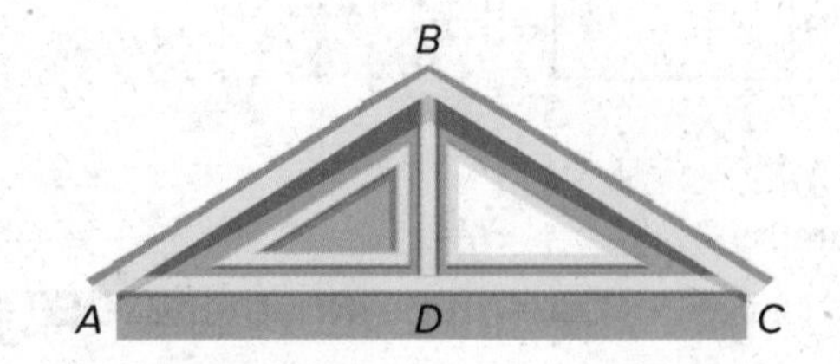

16. TECHNOLOGY Nevaeh has developed a new timer app. The icon for the app contains an hourglass that can be modeled by two triangles. If R is the midpoint of $\overline{QS}$ and $\overline{PT}$, write a paragraph proof to prove $\triangle PRQ \cong \triangle TRS$.

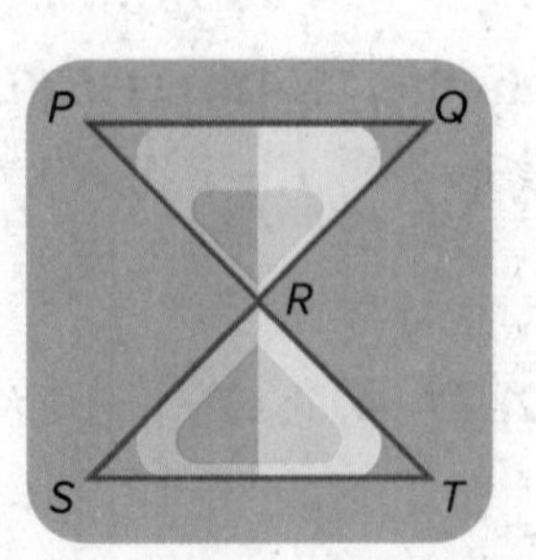

Name ______________________ Period __________ Date __________

Mixed Exercises

Explain whether there is enough information given in each figure to prove that the triangles are congruent using SSS or SAS.

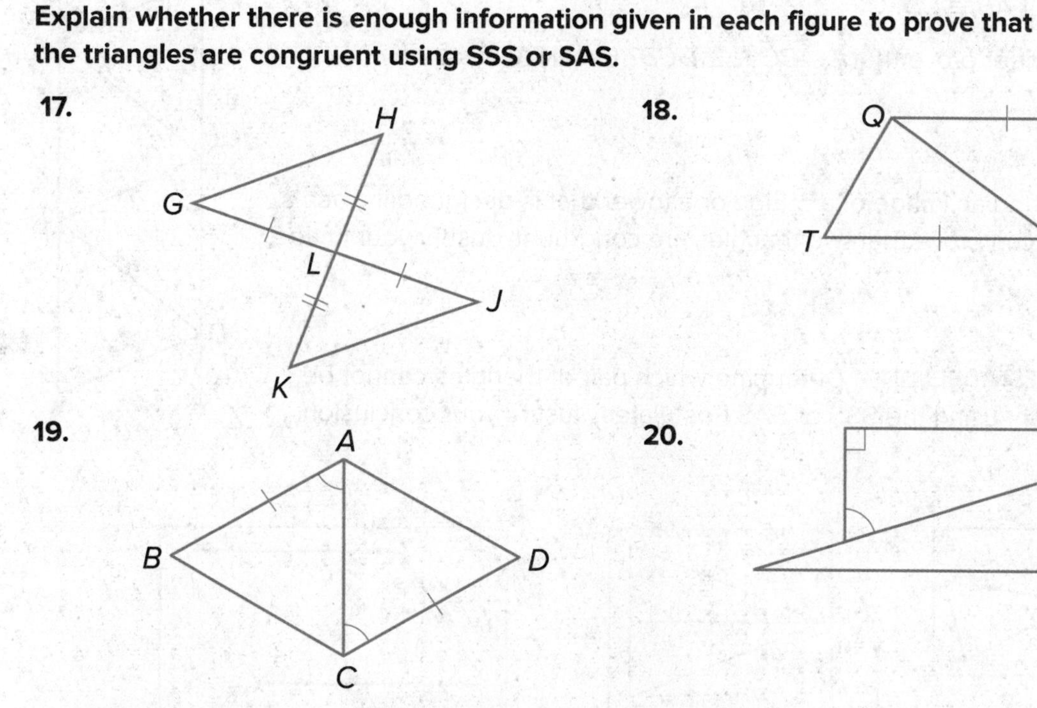

21. **REASONING** Tyson had three sticks of lengths 24 inches, 28 inches, and 30 inches. Is it possible to make two non-congruent triangles using the same three sticks? Explain.

22. **BAKERY** Sonia made a sheet of baklava. She has markings on her pan so that she can cut them into large squares. After she cuts the pastry in squares, she cuts them diagonally to form two congruent triangles, as shown. Which postulate could you use to prove the two triangles congruent?

23. **TILES** Tammy installs bathroom tiles. Her current job requires tiles that are equilateral triangles, and the tiles have to be congruent to each other. She has a sack of tiles that are in the shape of equilateral triangles. She knows that all the tiles are equilateral, but she is not sure whether they are the same size. What must she measure on each tile to be sure that they are congruent?

24. **CAKE** Carl had a piece of cake in the shape of an isosceles triangle with angles measuring 26°, 77°, and 77°. He wanted to divide it into two equal parts, so he cut it through the middle of the 26°-angle to the midpoint of the opposite side. He claims that the two pieces are congruent. Do you agree? Explain.

25. In the figure, $\overline{AC} \cong \overline{AD}$. Suppose you know $\angle C \cong \angle D$. Can you prove that $\triangle ABC \cong \triangle ABD$? Why or why not?

26. **USE A SOURCE** An engineer is designing a new cell phone tower. Part of the tower is shown in the figure. The engineer makes sure that line m is parallel to line n and that $\overline{AB} \cong \overline{CD}$.

   a. Can the engineer prove that $\triangle ABC \cong \triangle DCB$? Explain why or why not.

   b. Go online to find an image of a bridge or a tower that is designed in such a way that you can prove that two triangles are congruent. Justify your image.

27. **WHICH ONE DOESN'T BELONG?** Determine which pair of triangles cannot be proved congruent using the SSS or SAS Postulates. Justify your conclusion.

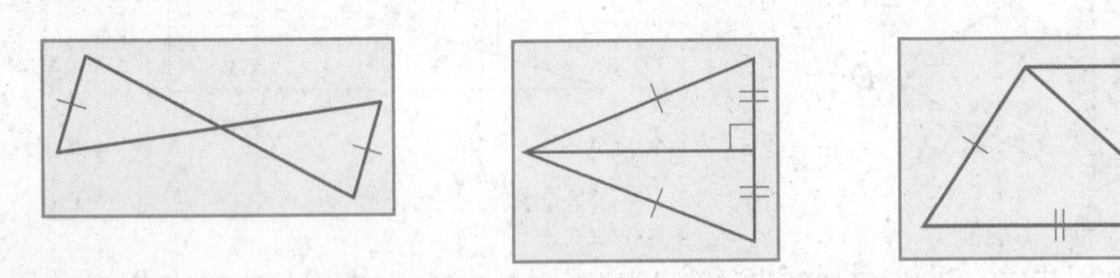

28. **ANALYZE** Determine whether the following statement is *true* or *false*. If true, justify your reasoning. If false, provide a counterexample.

   If the congruent sides in one isosceles triangle have the same measure as the congruent sides in another isosceles triangle, then the triangles are congruent.

29. **WRITE** Two pairs of corresponding sides of two right triangles are congruent. Are the triangles congruent? Explain your reasoning.

30. **CREATE** Use a straightedge to draw obtuse triangle ABC. Then construct $\triangle XYZ$ so it is congruent to $\triangle ABC$ using SSS or SAS. Justify your construction mathematically and verify it using measurement.

31. **FIND THE ERROR** Bonnie says that $\triangle PQR \cong \triangle XYZ$ by SAS. Shada disagrees. She says that there is not enough information to prove that the two triangles are congruent. Is either of them correct? Explain your reasoning.

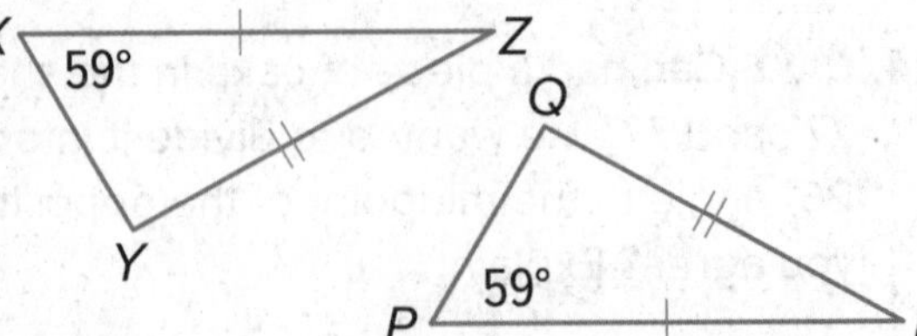

32. **PERSEVERE** **Refer to the graph shown.**

   a. Describe two methods you could use to prove $\triangle WYZ \cong \triangle WYX$. You may not use a ruler or protractor. Which method do you think is more efficient? Explain.

   b. Are $\triangle WYZ$ and $\triangle WYX$ congruent? Explain your reasoning.

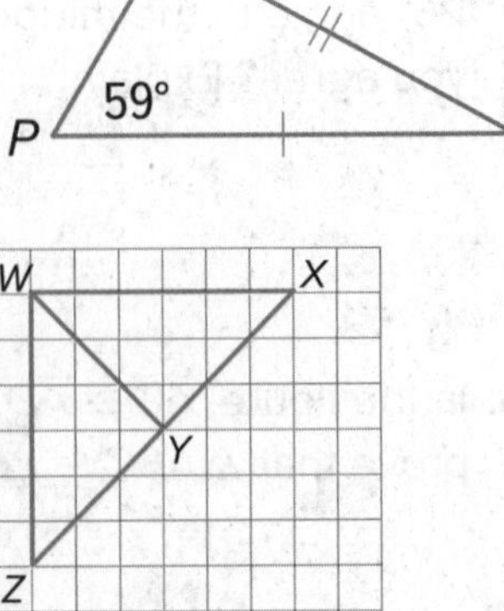

Proving Triangles Congruent: ASA, AAS

Learn Proving Triangles Congruent: ASA

An **included side** is the side of a triangle between two angles.

If two triangles are formed using the same two angle measures and included side length, then there is a series of rigid motions that will show the two triangles congruent. This leads to the postulate below.

Postulate 5.3: Angle-Side-Angle (ASA) Congruence

If two angles and the included side of one triangle are congruent to two angles and the included side of another triangle, then the triangles are congruent.

Today's Goals

- Use the ASA congruence criterion for triangles to solve problems and prove relationships in geometric figures.
- Use the AAS congruence criterion for triangles to prove relationships in geometric figures.

Today's Vocabulary

included side

Example 1 Use ASA to Prove Triangles Congruent

Complete the two-column proof.

Given: $\angle BAC \cong \angle DEC$; $\overline{BD}$ bisects $\overline{AE}$.

Prove: $\triangle ACB \cong \triangle ECD$

Statements	Reasons
1. $\angle BAC \cong \angle DEC$	1. Given
2. $\overline{BD}$ bisects $\overline{AE}$	2. Given
3. $\overline{AC} \cong \overline{EC}$	3. Definition of segment bisector
4. $\angle ACB \cong \angle ECD$	4. Vertical Angles Theorem
5. $\triangle ACB \cong \triangle ECD$	5. ASA

Think About It!

Can the Vertical Angles Theorem always be used to prove angles congruent in any two triangles? Justify your argument.

Check

Complete the two-column proof.

Given: $\overline{WX} \parallel \overline{YZ}$ and $\overline{WZ} \parallel \overline{YX}$

Prove: $\triangle WXZ \cong \triangle YZX$

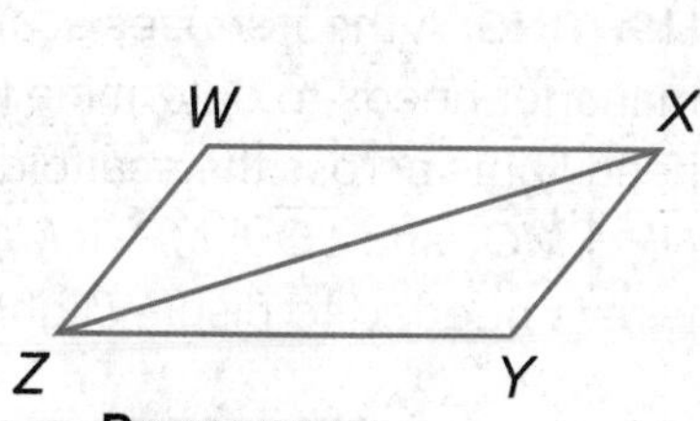

Statements	Reasons
1. ______	1. Given
2. $\overline{WZ} \parallel \overline{YX}$	2. ______
3. $\angle WXZ \cong \angle YZX$	3. ______
4. ______	4. Alternate Interior Angles Theorem
5. ______	5. Reflexive Property of Congruence
6. $\triangle WXZ \cong \triangle YZX$	6. ______

Go Online You can complete an Extra Example online.

Your Notes

Use a Source

Navajo Native Americans often live in octagonal homes known as *hogans*. Use available resources to research the design of octagonal hogans. How can you use ASA and congruent triangles to find the area of the floor of a regular octagonal hogan?

Example 2 Apply ASA Congruence

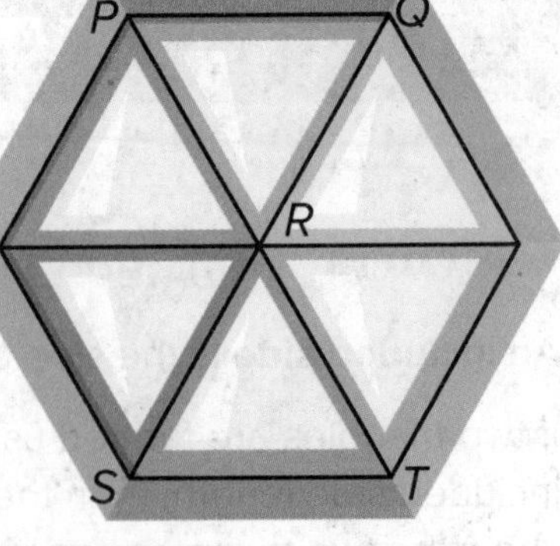

PRODUCTION A company that manufactures windows needs to determine the amount of glass required to make the hexagonal window shown. $\overline{PQ} \parallel \overline{TS}$, R is the midpoint of $\overline{PT}$, and ST is 12 inches.

Part A Determine whether $\triangle PRQ$ is congruent to $\triangle TRS$.

Because $\overline{PQ}$ is parallel to $\overline{TS}$, $\angle RPQ \cong$ ________ by the Alternate Interior Angles Theorem.

Because point R is the midpoint of $\overline{PT}$, $\overline{TR} \cong$ ________ by the Midpoint Theorem.

$\angle TRS$ and ________ are vertical angles, so they are congruent by the Vertical Angles Theorem.

Therefore, by ASA, $\triangle PRQ \cong \triangle TRS$.

Part B Find the area of the window.

If the six triangles that form the window are congruent and the height of $\triangle TRS$ is about 10.39 inches, how much glass is required to manufacture the window?

$A = \frac{1}{2}bh$	Area of a triangle
$\approx \frac{1}{2}(12)(10.39)$	$b = 12$ and $h = 10.39$
≈ 62.34	Simplify.

The area of the window is approximately 62.34(6) or about 374.04 square inches.

Study Tip

Units of Measure Remember that when you are finding the area of a polygon, you are multiplying two dimensions. So, you should use *square units*.

Check

LIGHTING A theater uses scaffolding to hang stage lighting. The stage manager needs to determine how much electrical wire is needed to hang lights across the scaffolding from point L to N. $\overline{LN} \parallel \overline{QP}$, $\overline{NP} \parallel \overline{MQ}$, and $\overline{LQ} \parallel \overline{MP}$. If MN is 4 feet, how many feet of electrical wire is needed to display lights across the scaffolding? ________

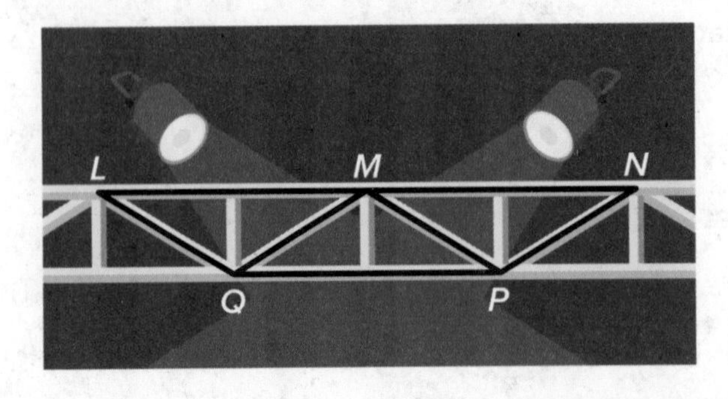

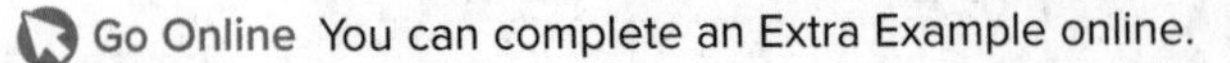
Go Online You can complete an Extra Example online.

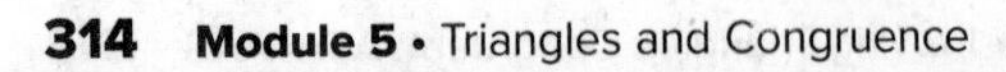

Learn Proving Triangles Congruent: AAS

The congruence of two angles and a nonincluded side is also sufficient to prove two triangles congruent. This congruence relationship is a theorem because it can be proved using the Third Angles Theorem.

Theorem 5.5: Angle-Angle-Side (AAS) Congruence	
Words	If two angles and the nonincluded side of one triangle are congruent to the corresponding two angles and nonincluded side of a second triangle, then the two triangles are congruent.
Example	If $\angle A \cong \angle D$, $\angle B \cong \angle E$, and $\overline{BC} \cong \overline{EF}$, then $\triangle ABC \cong \triangle DEF$.

The proof of the AAS Congruence Theorem is below.

Given: $\angle L \cong \angle Q$
$\angle M \cong \angle R$
$\overline{MN} \cong \overline{RS}$

Prove: $\triangle LMN \cong \triangle QRS$

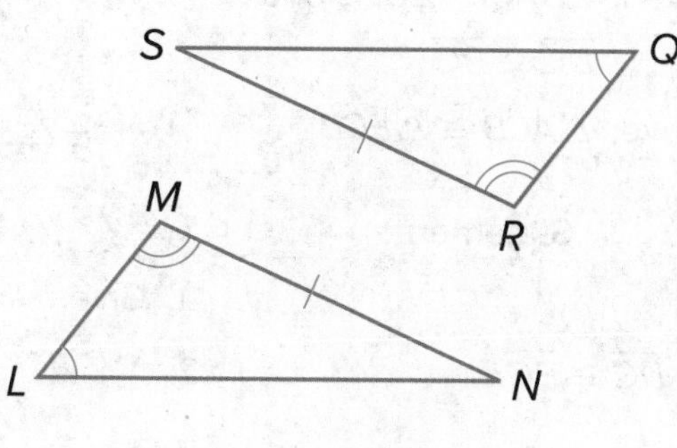

Proof:

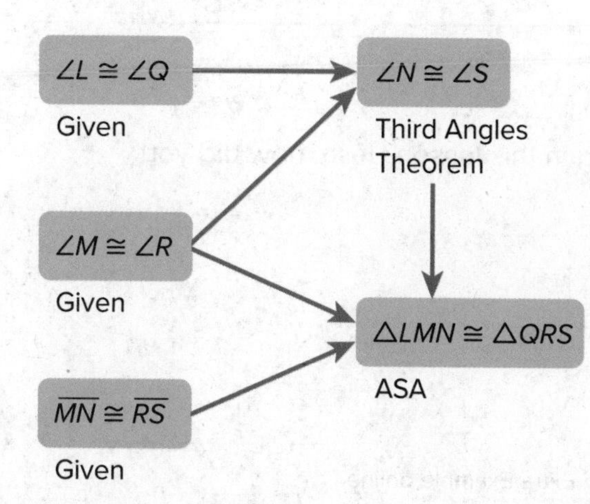

Talk About It!

Do you think angle-angle-angle, or AAA, could be used to prove triangles congruent? Provide an example to justify your reasoning.

Think About It!

Is hexagon *ABCDEF* congruent to hexagon *QRSTUV*? Use triangle congruence to justify your reasoning.

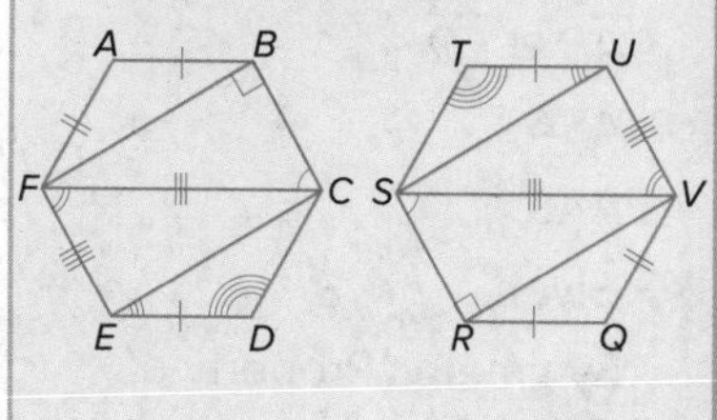

Go Online

You may want to complete the construction activities for this lesson.

Example 3 Use AAS to Prove Triangles Congruent

Choose the correct statements and reasons to complete the flow proof.

Given: $\overline{RQ} \cong \overline{ST}$ and $\overline{RQ} \parallel \overline{ST}$

Prove: $\triangle RUQ \cong \triangle TUS$

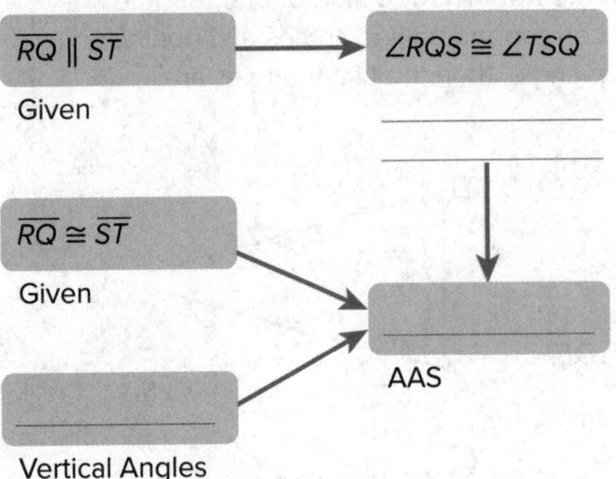

Statements/Reasons:

- Alternate Exterior Angles Theorem
- Alternate Interior Angles Theorem
- Vertical Angles Theorem
- $\angle RUQ \cong \angle TUS$
- $\angle R \cong \angle T$
- $\angle Q \cong \angle S$
- $\triangle RUQ \cong \triangle SUT$
- $\triangle RUS \cong \triangle QUT$
- $\triangle RUQ \cong \triangle TUS$

Check

Choose the correct statements and reasons to complete the two-column proof.

Given: $\angle DAC \cong \angle BEC$ and $\overline{DC} \cong \overline{BC}$

Prove: $\triangle ACD \cong \triangle ECB$

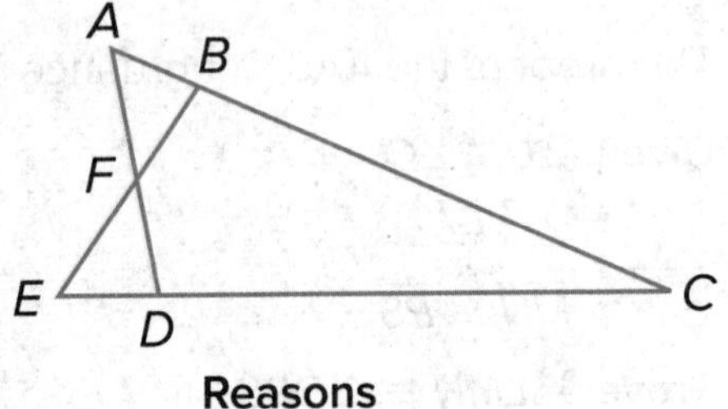

Statements	Reasons
1. ______	1. Given
2. $\overline{DC} \cong \overline{BC}$	2. ______
3. ______	3. ______
4. $\triangle ACD \cong \triangle ECB$	4. ______

Statements/Reasons:

- $\angle DAC \cong \angle BEC$
- $\angle C \cong \angle C$
- $\overline{AB} \cong \overline{ED}$
- ASA
- AAS
- Given
- Symmetric Property of Congruence
- Reflexive Property of Congruence

Pause and Reflect

Did you struggle with anything in this lesson? If so, how did you deal with it?

Record your observations here.

Go Online to practice what you've learned about proving triangles congruent in the Put It All Together over Lessons 5-3 and 5-4.

Go Online You can complete an Extra Example online.

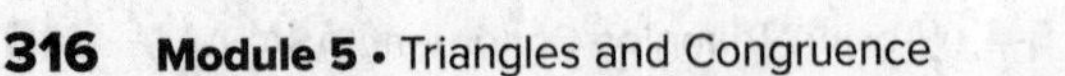

Name ______________________ Period ________ Date ________

Practice

Go Online You can complete your homework online.

Example 1

PROOF **Write the specified type of proof.**

1. two-column proof

Given: $\overline{AB} \parallel \overline{CD}$, $\angle CBD \cong \angle ADB$

Prove: $\triangle ABD \cong \triangle CDB$

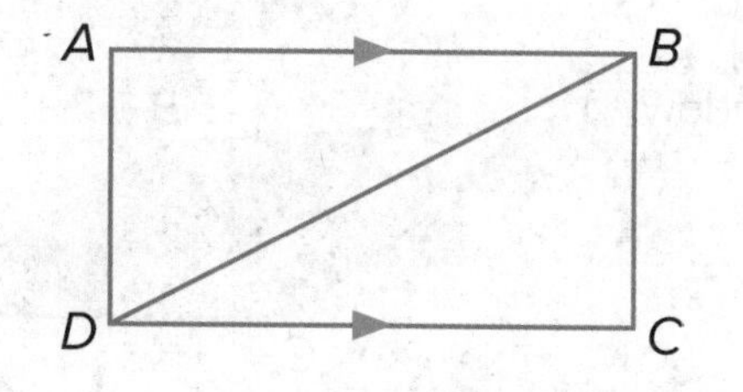

2. two-column proof

Given: $\angle S \cong \angle V$, and T is the midpoint of $\overline{SV}$.

Prove: $\triangle RTS \cong \triangle UTV$

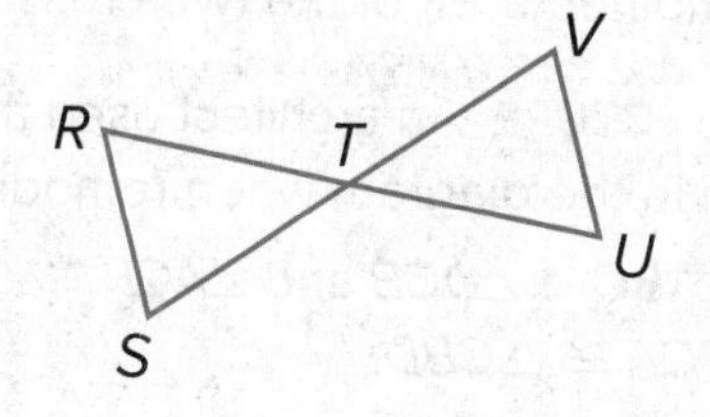

3. flow proof

Given: $\overline{AB} \cong \overline{CB}$, $\angle A \cong \angle C$, and $\overline{DB}$ bisects $\angle ABC$.

Prove: $\overline{AD} \cong \overline{CD}$

B
D
A
C

4. paragraph proof

Given: $\overline{CD}$ bisects $\overline{AE}$, $\overline{AB} \parallel \overline{CD}$, and $\angle E \cong \angle BCA$.

Prove: $\triangle ABC \cong \triangle CDE$

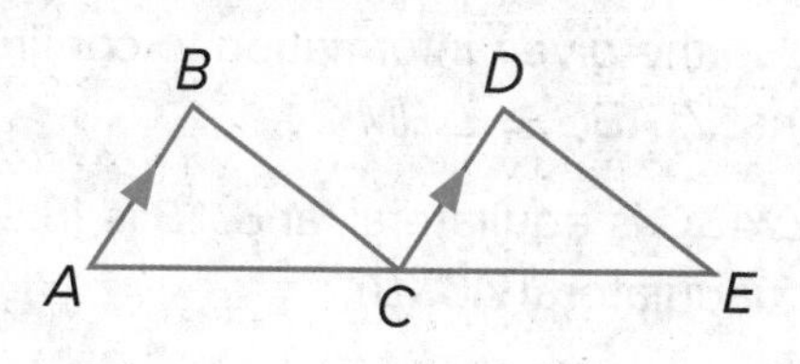

5. paragraph proof

Given: $\overline{CE}$ bisects $\angle BED$; $\angle BCE$ and $\angle ECD$ are right angles.

Prove: $\triangle ECB \cong \triangle ECD$

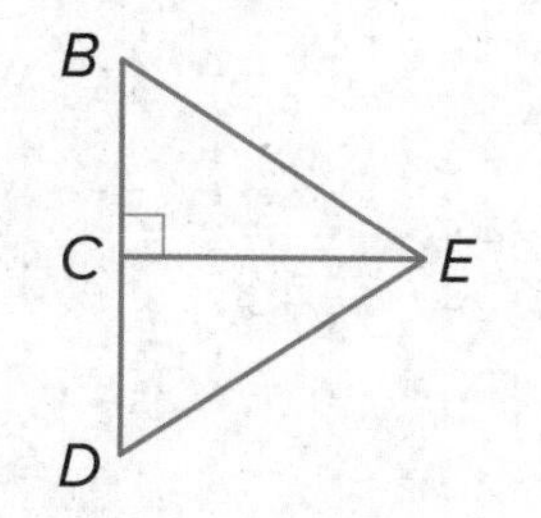

6. paragraph proof

Given: $\angle W \cong \angle Y$, $\overline{WZ} \cong \overline{YZ}$, and $\overline{XZ}$ bisects $\angle WZY$.

Prove: $\triangle XWZ \cong \triangle XYZ$

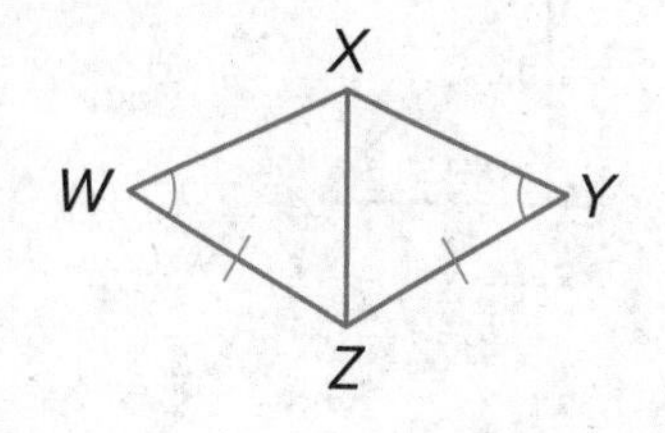

Example 2

7. REASONING Two doorstops have cross sections that are right triangles. Both have a 20° angle, and the lengths of the sides between the 90° and 20° angles are equal.

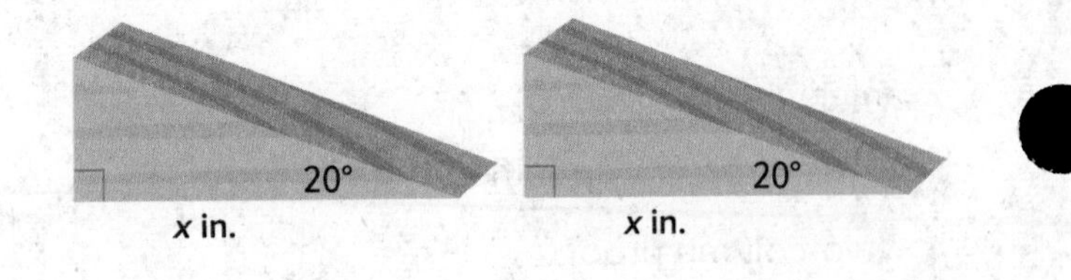

a. Are the cross sections congruent? Explain.

b. If each cross section has a height of 2 inches and $x = 5$, what is the combined area of the two cross sections?

8. ARCHITECTURE An architect used the stained-glass window design in the diagram when remodeling an art studio.

a. If $\angle ABC \cong \angle DCB$ and $\angle ACB \cong \angle DBC$, prove that $\triangle BCA \cong \triangle CBD$.

b. With $\overline{CD}$ as the base, if the height of $\triangle CBD$ is 1.4 meters and CD is 3.5 meters, how much glass is needed to make the entire window?

9. BRIDGES An engineering company that restores bridges needs to determine the amount of steel required to replace some trusses. $\overline{AC} \parallel \overline{BK}$, $\overline{CB} \parallel \overline{KM}$, and B is the midpoint of $\overline{AM}$.

a. Use the given information to confirm that $\triangle ABC \cong \triangle BMK$.

b. $\triangle ABC$ is equilateral, and AB is 18.5 feet. What is the perimeter of quadrilateral $ACKB$?

PROOF **Write the specified type of proof.**

10. two-column proof

Given: $\overline{BC} \parallel \overline{EF}$, $\overline{AB} \cong \overline{DE}$, $\angle C \cong \angle F$

Prove: $\triangle ABC \cong \triangle DEF$

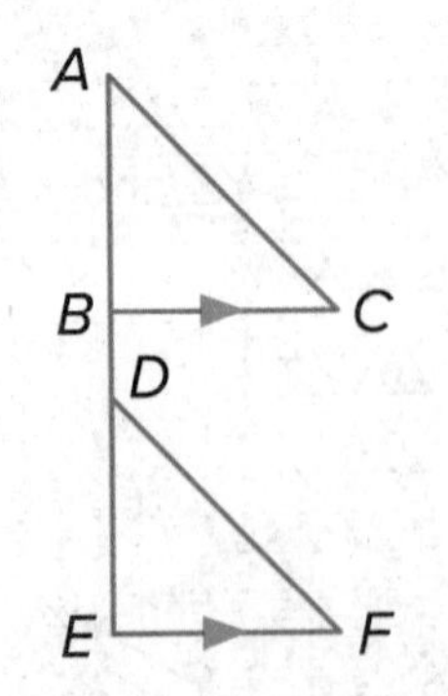

11. flow proof

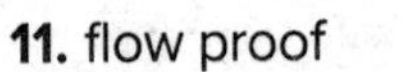

Given: $\angle S \cong \angle U$, and $\overline{TR}$ bisects $\angle STU$.

Prove: $\triangle SRT \cong \triangle URT$

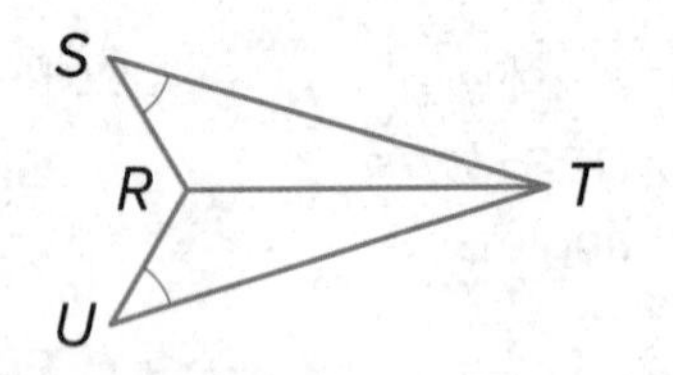

Name ______________________ Period __________ Date __________

Example 3

PROOF Write the specified type of proof.

12. flow proof

Given: $\overline{JK} \cong \overline{MK}$, $\angle N \cong \angle L$

Prove: $\triangle JKN \cong \triangle MKL$

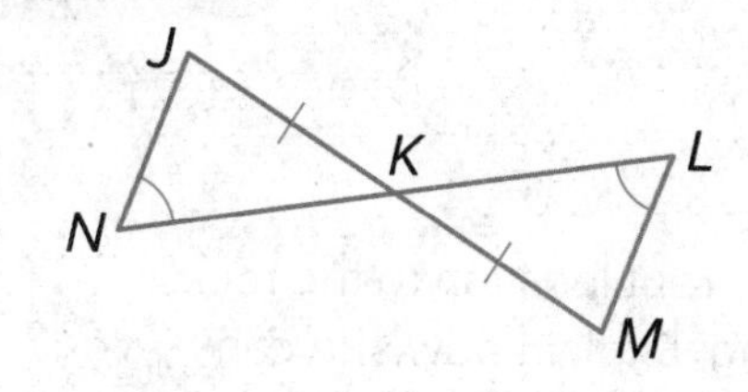

13. paragraph proof

Given: $\overline{DE} \parallel \overline{FG}$, $\angle E \cong \angle G$

Prove: $\triangle DFG \cong \triangle FDE$

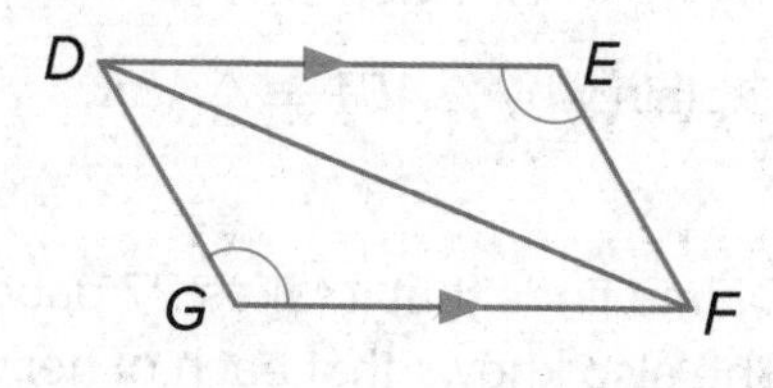

14. two-column proof

Given: V is the midpoint of $\overline{YW}$; $\overline{UY} \parallel \overline{XW}$,

Prove: $\triangle UVY \cong \triangle XVW$

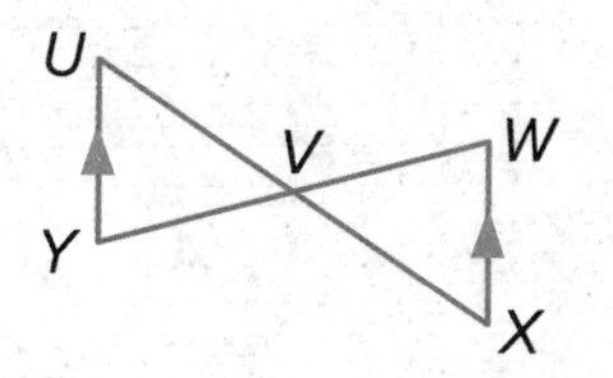

15. two-column proof

Given: $\overline{MS} \cong \overline{RQ}$,

$\overline{MS} \parallel \overline{RQ}$,

Prove: $\triangle MSP \cong \triangle RQP$

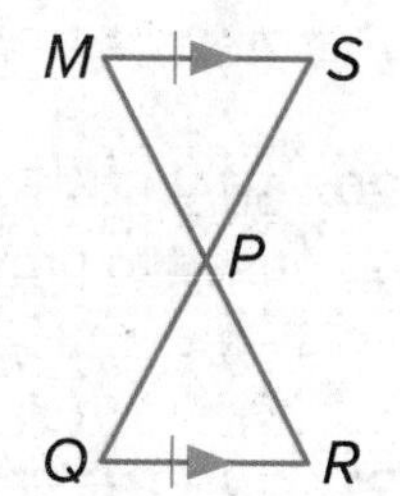

Mixed Exercises

16. USE TOOLS Use a compass and straightedge and the ASA Congruence Postulate to construct a triangle congruent to $\triangle PQR$.

P

Q R

17. PRECISION Two people decide to take a walk. One person is in Bombay, and the other is in Milwaukee. They start by walking straight for 1 kilometer. Then both turn right at an angle of 110° and continue to walk straight again. After a while, both turn right again, but this time at an angle of 120°. Both walk straight for a while in this new direction until they end up where they started. Each person walked in a triangular path at their location. Are the two triangles they formed congruent? Explain.

18. USE ESTIMATION Delma came to a river during a hike, and she wanted to estimate the distance across it. She held her walking stick $\overline{AB}$ vertically on the ground at the edge of the river and sighted along the top of the stick across the river to the base of a tree T. Then she turned without changing the angle of her head and sighted along the top of the stick to a rock R, located on her side of the river.

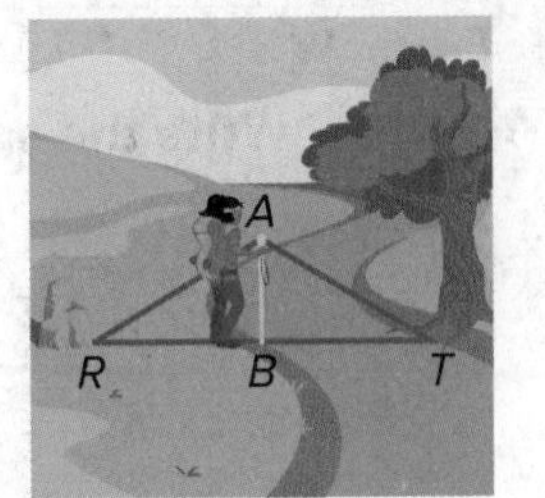

a. Explain why $\triangle ABT \cong \triangle ABR$.

b. Delma finds that it takes 27 paces to walk from her current location to the rock. She also knows that each of her paces is 14 inches long. Explain how she can use this information to estimate the distance across the river.

19. PROOF Write a paragraph proof.

Given: $\angle D \cong \angle F$

$\overline{GE}$ bisects $\angle DEF$.

Prove: $\overline{DG} \cong \overline{FG}$

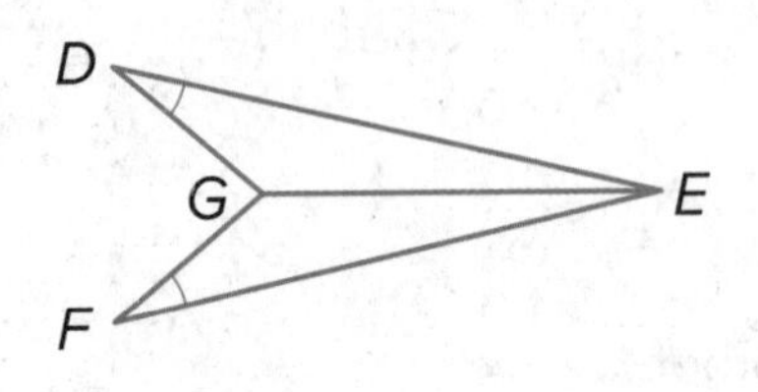

20. ANALYZE Find a counterexample to show why SSA (Side-Side-Angle) cannot be used to prove the congruence of two triangles.

21. FIND THE ERROR Tyrone says that it is not possible to show that $\triangle ADE \cong \triangle ACB$. Lorenzo disagrees, explaining that because $\angle ADE \cong \angle ACB$, $\angle AED \cong \angle ABC$, and $\angle A \cong \angle A$ by the Reflexive Property, $\triangle ADE \cong \triangle ACB$. Who is correct? Explain your reasoning.

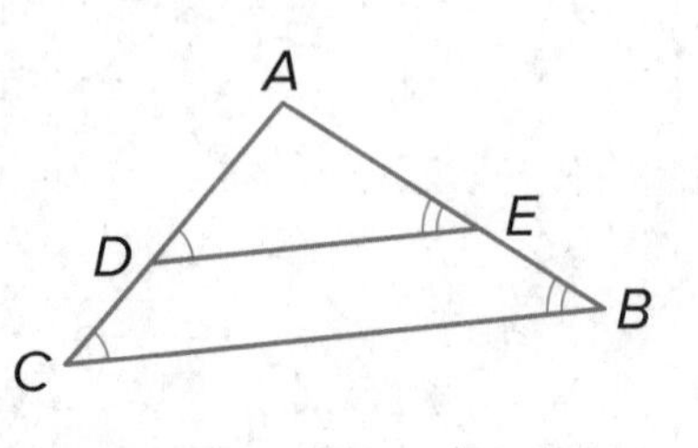

22. CREATE Draw and label two triangles that could be proved congruent by ASA.

23. WRITE How do you know which method (SSS, SAS, and so on) to use when you are proving triangle congruence? Use a table to explain your reasoning.

24. PROOF Using the information given in the diagram, write a flow proof to show that $\triangle PVQ \cong \triangle SVT$.

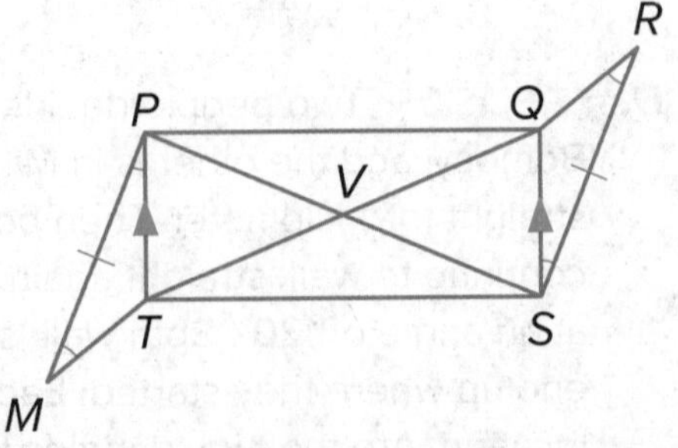

Proving Right Triangles Congruent

Explore Congruence Theorems and Right Triangles

Online Activity Use graphing technology to complete the Explore.

INQUIRY What criteria can be used to prove right triangles congruent?

If two right triangles are formed using the criteria for leg-leg congruence, hypotenuse-angle congruence, leg-angle congruence, or hypotenuse-leg congruence, then there is a series of rigid motions that will show the two triangles congruent. This leads to the theorems below.

Today's Goals

- Use the right triangle congruence theorems to prove relationships in geometric figures.

Learn Right Triangle Congruence

Theorem 5.6: Leg-Leg (LL) Congruence

Words	If the legs of one right triangle are congruent to the corresponding legs of another right triangle, then the triangles are congruent.
Example	Given right $\triangle ABC$ and right $\triangle DEF$, $\overline{AB} \cong \overline{DE}$ and $\overline{CB} \cong \overline{FE}$. So, $\triangle ABC \cong \triangle DEF$ by LL.

Theorem 5.7: Hypotenuse-Angle (HA) Congruence

Words	If the hypotenuse and an acute angle of one right triangle are congruent to the hypotenuse and the corresponding acute angle of another right triangle, then the triangles are congruent.
Example	Given right $\triangle ABC$ and right $\triangle DEF$, $\overline{AC} \cong \overline{DF}$ and $\angle C \cong \angle F$. So, $\triangle ABC \cong \triangle DEF$ by HA.

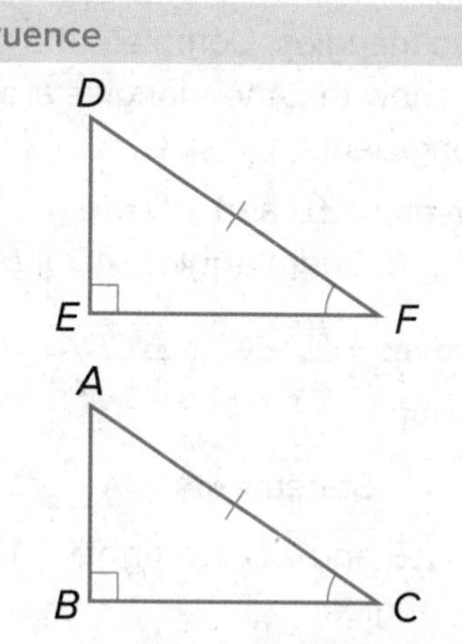

Theorem 5.8: Leg-Angle (LA) Congruence

If one leg and an acute angle of one right triangle are congruent to the corresponding leg and acute angle of another right triangle, then the triangles are congruent.

Theorem 5.9: Hypotenuse-Leg (HL) Congruence

If the hypotenuse and a leg of one right triangle are congruent to the hypotenuse and the corresponding leg of another right triangle, then the triangles are congruent.

Talk About It!

Can you declare that the given triangles are congruent by HA? Justify your argument.

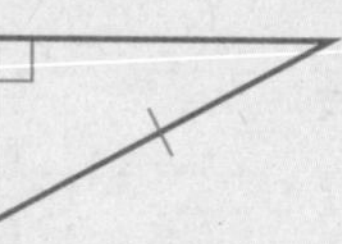

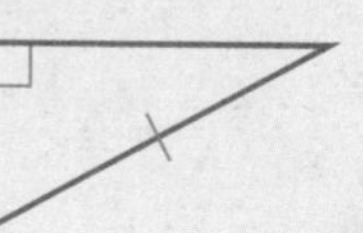

Go Online

Proofs of Theorems 5.6 through 5.9 are available.

Your Notes

Example 1 Problem Solving with Right Triangles

HOME IMPROVEMENT **Craig and his brother are painting a house. The brothers use ladders that are the same length. If they place their ladders an equal distance from the house, will each ladder reach the same height on the house? Construct a logical argument.**

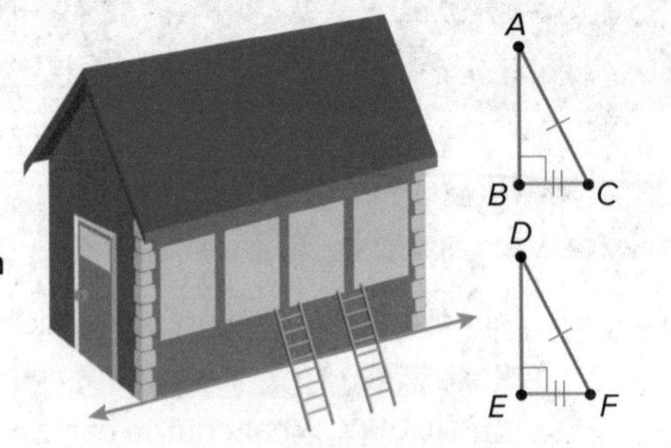

Draw a diagram to model this situation. It is given that the length of the ladders is the same and that they are placed the same distance from the house.

Because the wall of the house is perpendicular to the ground, the triangles formed by the house, the ground, and the ladders are __________ triangles. The __________ are congruent because the ladders are the same length. The corresponding __________ along the ground are congruent because the ladders are placed the same distance from the house. So the triangles are congruent by the __________ Congruence Theorem or HL. Thus, $\overline{AB}$ and $\overline{DE}$ are congruent by CPCTC. You can conclude that the ladders reach to the same height on the house.

Check

FENCES The fence has parallel supports and a crossbar that forms two triangles. Complete the proof to show that the triangles are congruent.

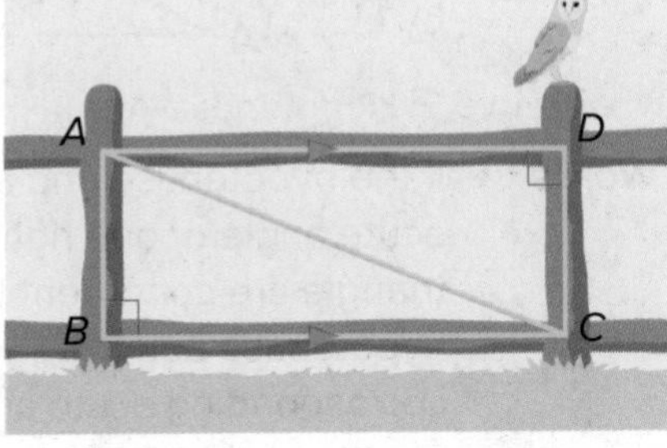

Given: $\angle B$ and $\angle D$ are right angles. $\overline{AD} \parallel \overline{BC}$.

Prove: $\triangle ABC \cong \triangle CDA$

Proof:

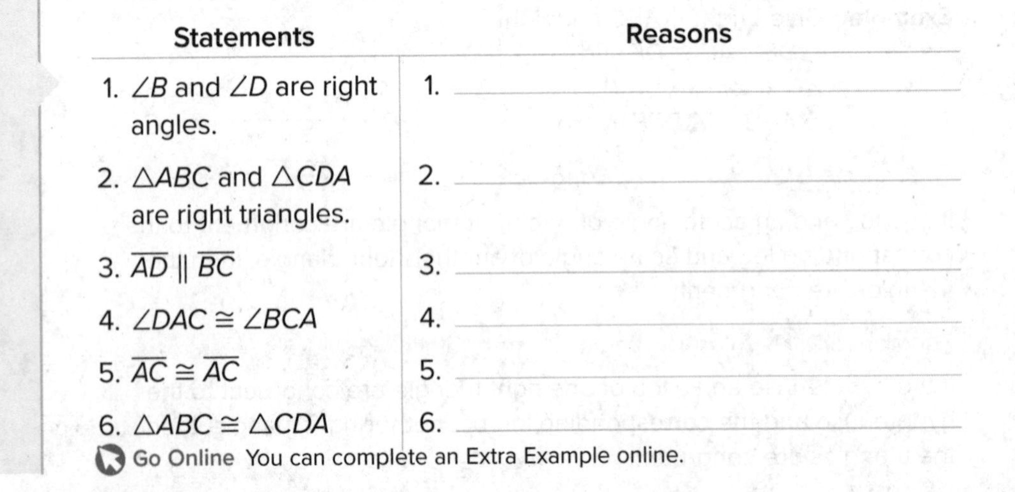

Statements	Reasons
1. $\angle B$ and $\angle D$ are right angles.	1. __________
2. $\triangle ABC$ and $\triangle CDA$ are right triangles.	2. __________
3. $\overline{AD} \parallel \overline{BC}$	3. __________
4. $\angle DAC \cong \angle BCA$	4. __________
5. $\overline{AC} \cong \overline{AC}$	5. __________
6. $\triangle ABC \cong \triangle CDA$	6. __________

Reasons:
- Alternate exterior angles are congruent.
- Alternate interior angles are congruent.
- Consecutive interior angles are congruent.
- Definition of right triangle
- Given
- HA
- HL
- LA
- Reflexive Property
- Symmetric Property

Go Online You can complete an Extra Example online.

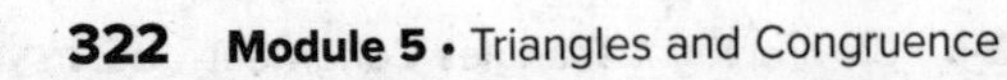

Name _______________ Period _______ Date _______

Practice

Go Online You can complete your homework online.

Example 1

1. CAMPING In the diagram of the pup tent, the support pole is perpendicular to the ground. The base of the support pole is located at the midpoint of the segment connecting the bottom of the sides of the tent. Write a two-column proof to show that the triangles formed by the support pole are congruent.

Given: $\overline{XZ} \perp \overline{WY}$; Z is the midpoint of $\overline{WY}$.

Prove: $\triangle WXZ \cong \triangle YXZ$

2. TOWERS The cell phone tower has parallel poles and diagonal support beams that form two triangles. Write a two-column proof to show that the triangles are congruent.

Given: $\angle H$ and $\angle K$ are right angles.

$\overline{GH} \parallel \overline{KJ}$

Prove: $\triangle GKJ \cong \triangle JHG$

3. BRIDGES In the diagram, the vertical support beam, $\overline{BX}$, is perpendicular to the deck of the bridge. The two diagonals, $\overline{AB}$ and $\overline{CB}$, are equal in length. Write a two-column proof to show that the triangles formed by the vertical support beam are congruent.

Given: $\overline{BX} \perp \overline{AC}$; $AB = CB$

Prove: $\triangle AXB \cong \triangle CXB$

Mixed Exercises

Determine whether each pair of triangles is congruent. If yes, include the theorem that applies.

4\. 5. 6. 7. 8. 9.

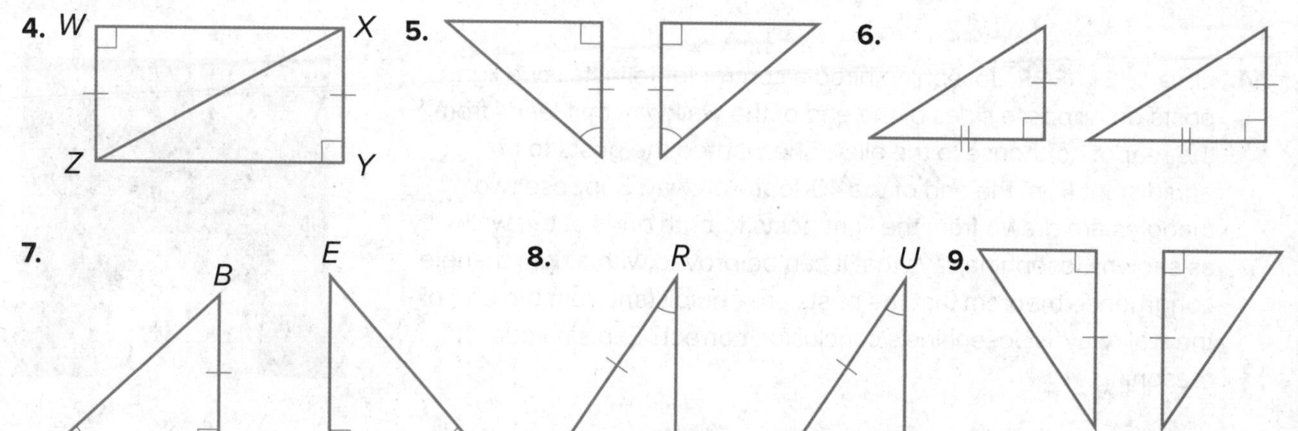

10. Which pairs of corresponding parts need to be congruent to prove that $\triangle ABC \cong \triangle XYZ$ using the indicated theorem?

a. HA

b. LL

11. PROOF Write a two-column proof.

Given: $\overline{BX} \perp \overline{XA}$, $\overline{BY} \perp \overline{YA}$, and $\overline{XA} \cong \overline{YA}$

Prove: $\triangle BXA \cong \triangle BYA$

12. WRITE The sketch shows the side view of a sculpture that is being designed by an artist. Determine whether $\triangle ABC \cong \triangle DCA$. If yes, then provide a paragraph proof. If no, then explain your reasoning.

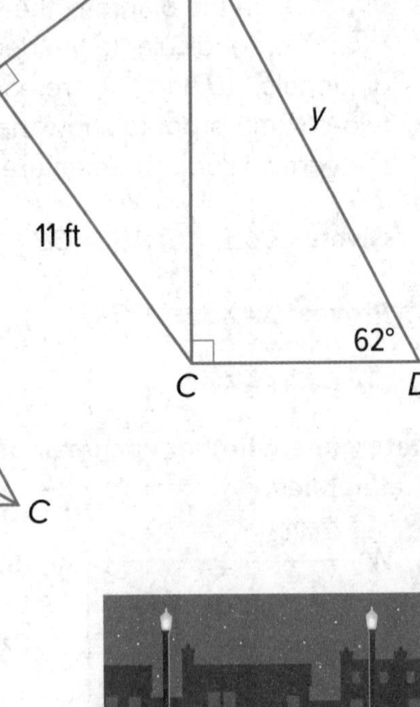

13. PROOF Write a paragraph proof.

Given: $\overline{BY} \perp \overline{AC}$; $\overline{CX} \perp \overline{AB}$; $AX = AY$

Prove: $\triangle ABY \cong \triangle ACX$

14. FIND THE ERROR Josephine hired a contractor to install two light posts on opposite sides of the end of the walkway that leads from the rear of her house to the alley. She wanted the posts to be equidistant from the end of the 40-foot walkway. Suppose two triangles are drawn from the light posts to both ends of the walkway as shown. Josephine says that it can be proved with a right triangle congruence theorem that the posts are equidistant from the end of the walkway. Is Josephine's conclusion correct? Explain your reasoning.

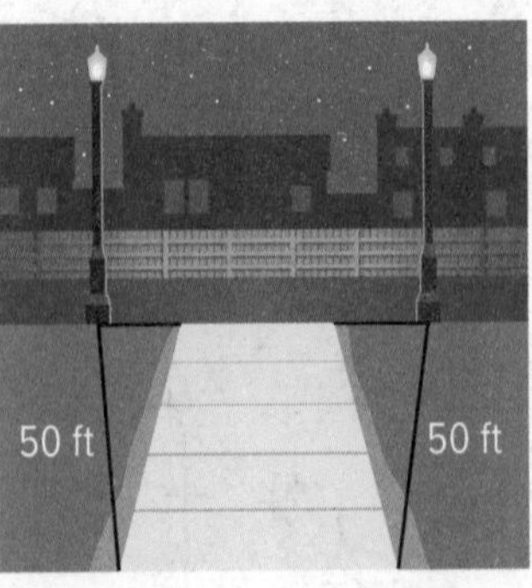

Isosceles and Equilateral Triangles

Explore Properties of Equilateral, Isosceles, and Scalene Triangles

Online Activity Use dynamic geometry software to complete the Explore.

INQUIRY What are the differences between equilateral, isosceles, and scalene triangles?

Explore Isosceles and Equilateral Triangles

Online Activity Use dynamic geometry software to complete the Explore.

INQUIRY What conjecture can you make about the relationship between the parts of isosceles and equilateral triangles?

Learn Isosceles Triangles

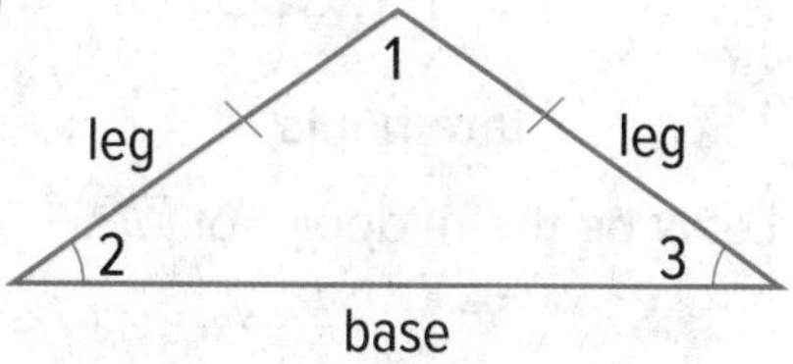

An **isosceles triangle** is a triangle with at least two sides congruent. The two congruent sides are called the **legs of an isosceles triangle**. The angle between the sides that are the legs is called the **vertex angle of an isosceles triangle**. $\angle 1$ is the vertex angle of the triangle. The side of the triangle opposite the vertex angle is called the *base*. The two angles formed by the base and the congruent sides are called the **base angles of an isosceles triangle**. $\angle 2$ and $\angle 3$ are the base angles.

Theorem 5.10: Isosceles Triangle Theorem	
Words	If two sides of a triangle are congruent, then the angles opposite those sides are congruent.
Example	If $\overline{AC} \cong \overline{BC}$, then $\angle 2 \cong \angle 1$. 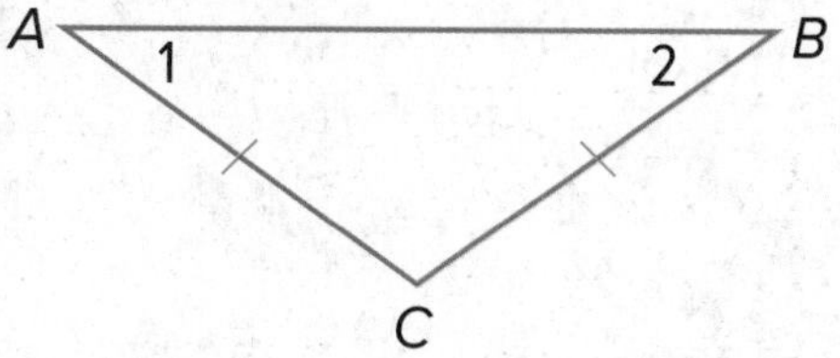

Today's Goals
- Solve problems involving isosceles triangles.
- Solve problems involving equilateral triangles.

Today's Vocabulary
isosceles triangle

legs of an isosceles triangle

vertex angle of an isosceles triangle

base angles of an isosceles triangle

auxiliary line

Math History Minute

Henry Dudeney (1857–1930) was a British government employee who enjoyed creating logic puzzles and mathematical games. One of Dudeney's greatest accomplishments was his success at solving a particular puzzle, the Haberdasher's Puzzle, that requires a person to cut an equilateral triangle into four pieces that can be rearranged to make a square.

Your Notes

Theorem 5.11: Converse of the Isosceles Triangle Theorem	
Words	If two angles of a triangle are congruent, then the sides opposite those angles are congruent.
Example	If $\angle 1 \cong \angle 2$, then $\overline{FE} \cong \overline{DE}$. 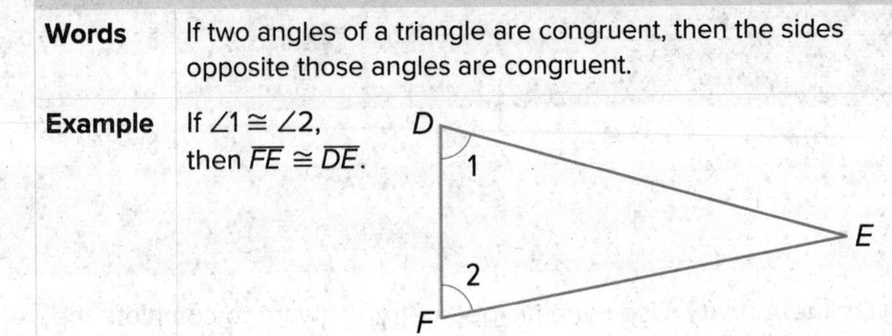

Go Online A proof of Theorem 5.11 is available.

Example 1 Prove Theorems About Isosceles Triangles

Prove the Isosceles Triangle Theorem.

To prove the Isosceles Triangle Theorem, draw an *auxiliary line* and use the two triangles that are formed. An **auxiliary line** is an extra line or segment drawn in a figure to help analyze geometric relationships.

Given: $\triangle LMP$, $\overline{LM} \cong \overline{LP}$

Prove: $\angle M \cong \angle P$

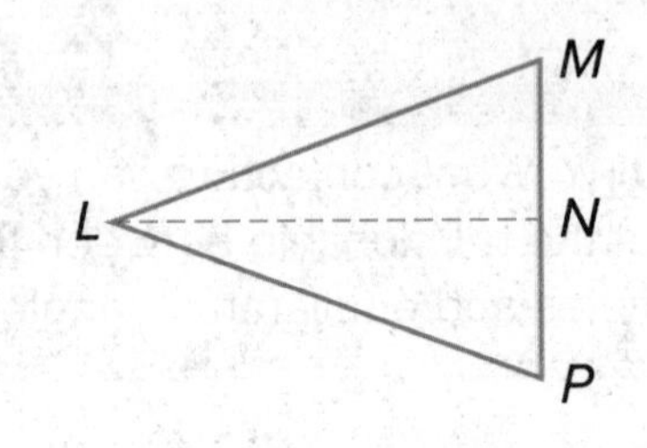

Statements/Reasons:
Definition of congruence
Given
Midpoint Theorem
$\overline{LN} \cong \overline{LN}$
$\overline{MP} \cong \overline{LP}$

Proof:

Statements	Reasons
1. Let N be the midpoint of $\overline{MP}$.	1. Every segment has exactly one midpoint.
2. Draw an auxiliary segment $\overline{LN}$.	2. Two points determine a line.
3. $\overline{MN} \cong \overline{PN}$	3. ______
4. ______	4. Reflexive Property of Congruence
5. $\overline{LM} \cong \overline{LP}$	5. ______
6. $\triangle LMN \cong \triangle LPN$	6. SSS
7. $\angle M \cong \angle P$	7. CPCTC

Go Online You can complete an Extra Example online.

Example 2 Find Missing Measures in Isosceles Triangles

Find $m\angle B$ and $m\angle C$.

Part A Determine side relationships.

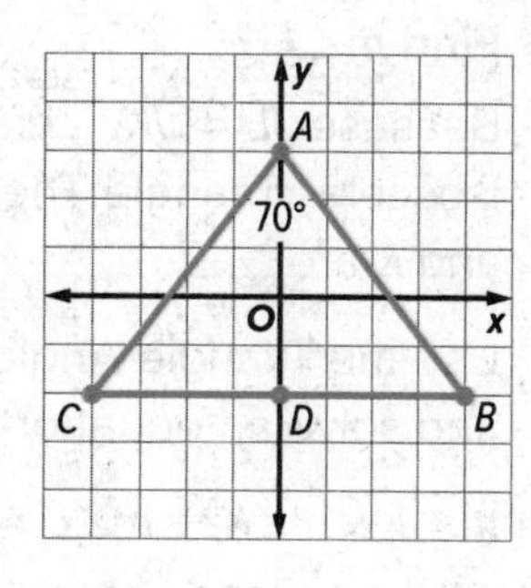

Use the Distance Formula to determine the measures of the sides of $\triangle ABC$. The coordinates of $\triangle ABC$ are $A(0, 3)$, $B(4, -2)$, and $C(-4, -2)$.

$AB = \sqrt{(0-4)^2 + [3-(-2)]^2}$ or $\sqrt{41}$ units

$AC = \sqrt{[0-(-4)]^2 + [3-(-2)]^2}$ or $\sqrt{41}$ units

$BC = \sqrt{[4-(-4)]^2 + [-2-(-2)]^2}$ or 8 units

So, $\triangle ABC$ is an isosceles triangle with $\overline{AB} \cong \overline{AC}$.

Part B Determine the angle measures.

Because $\overline{AB} \cong \overline{AC}$, we know that $\angle C \cong \angle$___ by the Isosceles Triangle Theorem.

$m\angle A + m\angle B + m\angle C = 180°$	Triangle Angle-Sum Theorem
$m\angle A + 2m\angle B = 180°$	Definition of congruent
$70° + 2m\angle B = 180°$	Substitute.
$m\angle B = m\angle C = 55°$	Solve.

Check

Find $m\angle XYZ$ and $m\angle YZX$.

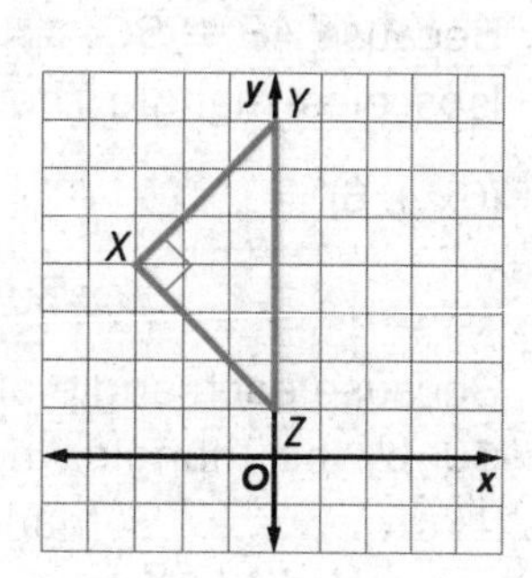

$m\angle XYZ =$ ___

$m\angle YZX =$ ___

Watch Out!

Triangle Relationships We cannot use the Isosceles Triangle Theorem until we show that two sides of $\triangle ABC$ are congruent.

Go Online An alternate method is available for this example.

Learn Equilateral Triangles

The Isosceles Triangle Theorem leads to two corollaries about the angles of an equilateral triangle.

Corollary 5.3
A triangle is equilateral if and only if it is equiangular.
Corollary 5.4
Each angle of an equilateral triangle measures 60°.

You will prove Corollaries 5.3 and 5.4 in Exercises 18 and 19, respectively.

Example 3 Find Missing Measures in Equilateral Triangles

Find $m\angle J$.

Because $JL = JK$, ____________. By the Isosceles Triangle Theorem, base angles L and K are ____________, so $m\angle L = m\angle K$.

Use the Triangle Angle-Sum Theorem to write and solve an equation to find $m\angle J$.

$m\angle J + m\angle K + m\angle L = 180°$	Triangle Angle-Sum Theorem
$m\angle J + 60° + 60° = 180°$	Isosceles Triangle Theorem
$m\angle J = 60°$	Solve.

Check

Find $m\angle R$ and PR.

$m\angle R =$ ____

$PR =$ ____ cm

Study Tip

Isosceles Triangles Any isosceles triangle that has one 60° angle must be an equilateral triangle.

Example 4 Find Missing Values

BILLIARDS **Find the value of each variable.**

Because $\overline{AB} \cong \overline{BC}$, $\angle ACB \cong \angle$____ by the Isosceles Triangle Theorem.

$(6x + 6)° =$ ____	Isosceles Triangle Theorem
$x =$ ____	Solve.

Because each angle of the triangle measures 60° by the Triangle Angle-Sum Theorem, the triangle is an ____________ triangle by Corollary 5.3.

$4y - 2 = 2y + 2$	Corollary 5.3; definition of equilateral $\triangle$
$y = 2$	Solve.

Check

ARCHITECTURE The main entrance to the Louvre Museum is a unique metal and glass pyramid. Find the value of each variable.

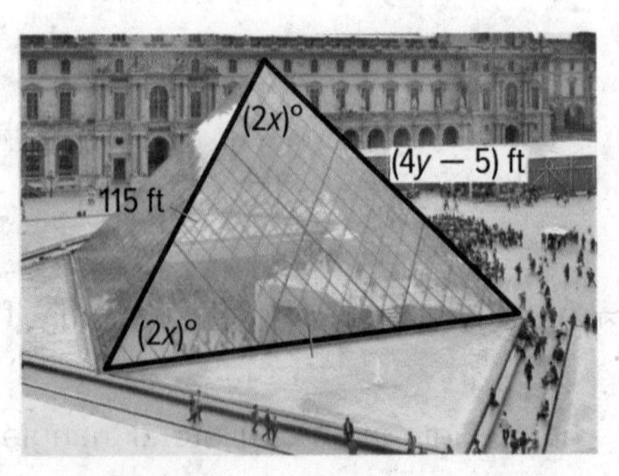

$x =$ ____ and $y =$ ____

Go Online You can complete an Extra Example online.

Talk About It!

Arturo argues that you do not have to use the properties of equilateral triangles to solve for y. Do you agree? Explain your reasoning.

Go Online

You may want to complete the construction activities for this lesson.

Name ______________________ Period ____________ Date ____________

Practice

Go Online You can complete your homework online.

Example 1

1. PROOF Write a two-column proof.

Given: $\angle 1 \cong \angle 2$

Prove: $\overline{AB} \cong \overline{CB}$

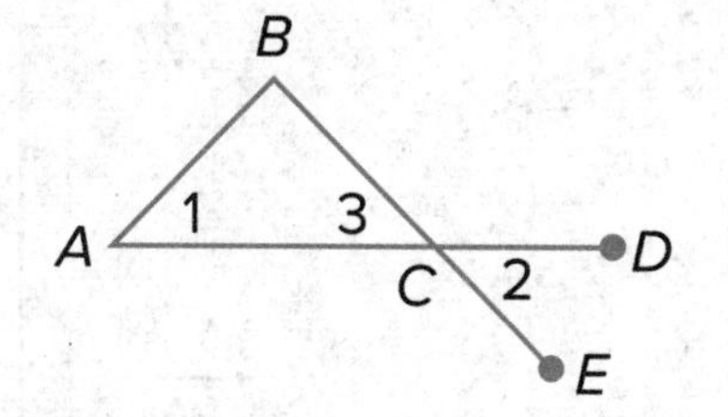

2. PROOF Write a two-column proof.

Given: $\overline{CD} \cong \overline{CG}$

$\overline{DE} \cong \overline{GF}$

Prove: $\overline{CE} \cong \overline{CF}$

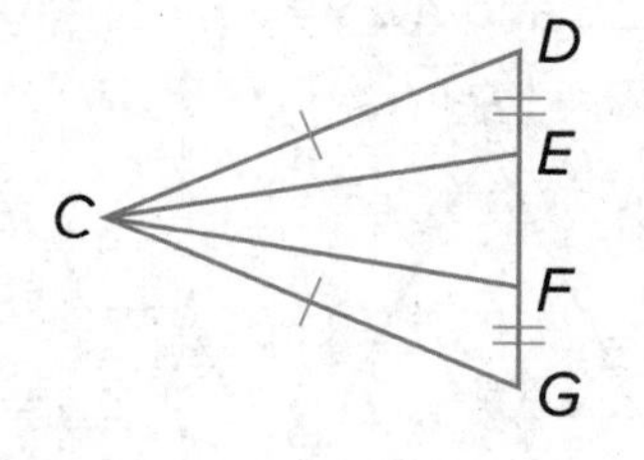

3. PROOF Write a two-column proof.

Given: $\overline{DE} \parallel \overline{BC}$

$\angle 1 \cong \angle 2$

Prove: $\overline{AB} \cong \overline{AC}$

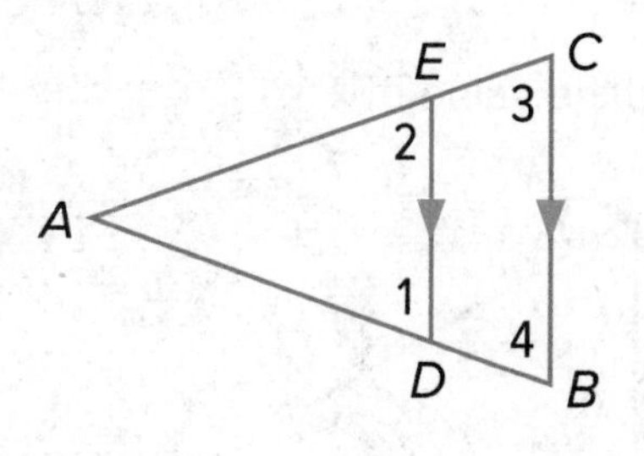

4. ROOFS In the picture, $\overline{BD} \perp \overline{AC}$ and $\triangle ABC$ is an isosceles triangle with base $\overline{AC}$. Write a two-column proof to prove that $\overline{BD}$ bisects the angle formed by the sloped sides of the roof, $\angle ABC$.

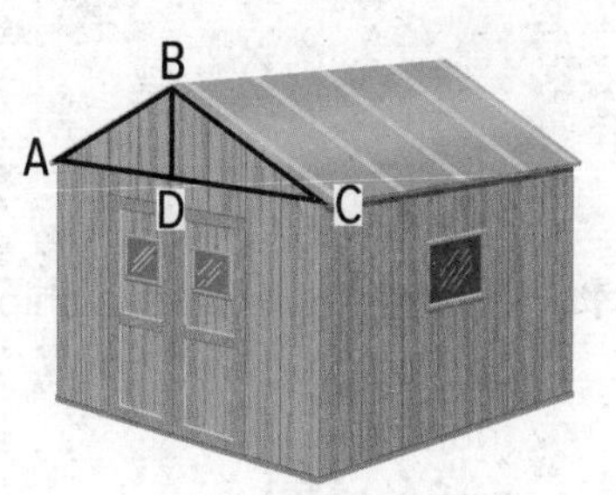

Example 2

5. Refer to the figure.

a. Find the measures of the sides of $\triangle ABC$. Show your work.

b. Find $m\angle A$. Show your work.

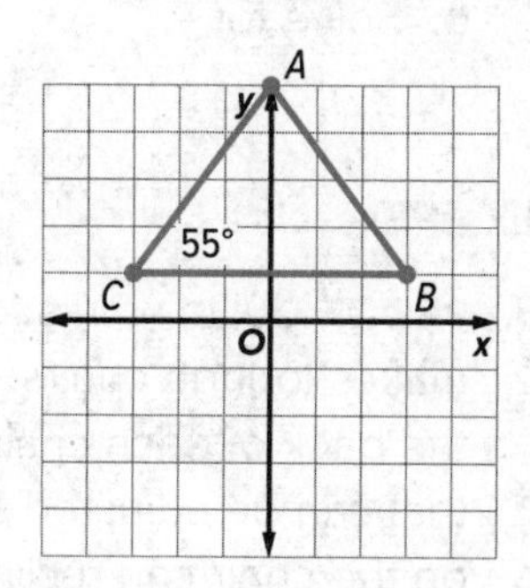

6. Find SR, ST, RT, $m\angle TRS$, and $m\angle RST$. Round to the nearest tenth, if necessary.

7. Find the measures of $\angle DEF$ and $\angle EFD$. Round to the nearest tenth, if necessary.

Examples 3 and 4

8. Find the value of x.

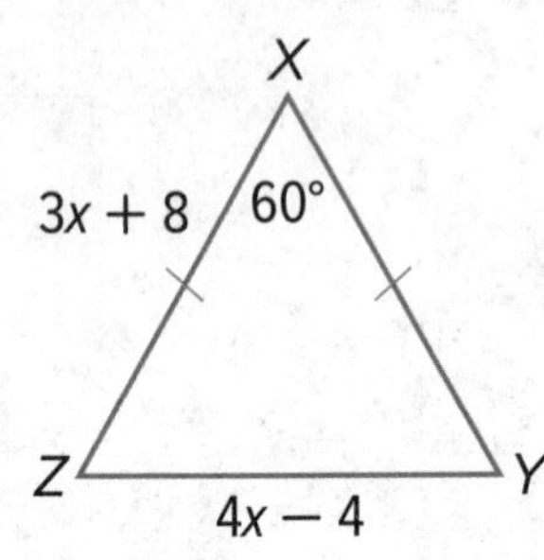

9. Find $m\angle B$ and AC.

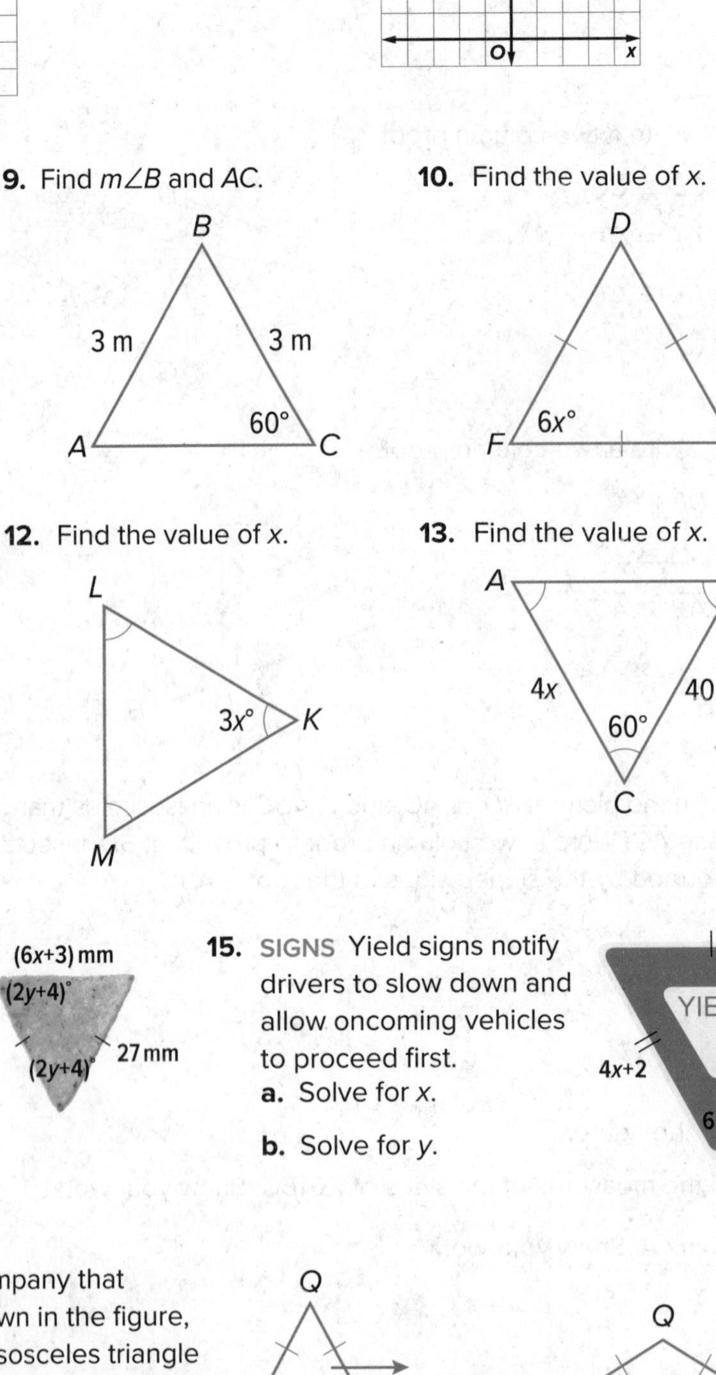

10. Find the value of x.

11. Find $m\angle Y$ and WY.

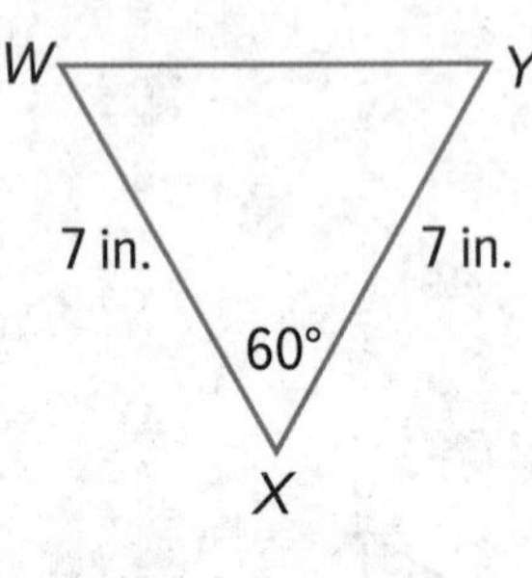

12. Find the value of x.

13. Find the value of x.

14. **CHIPS** Some tortilla chips can be modeled by a triangle.

a. Solve for x.

b. Solve for y.

15. **SIGNS** Yield signs notify drivers to slow down and allow oncoming vehicles to proceed first.

a. Solve for x.

b. Solve for y.

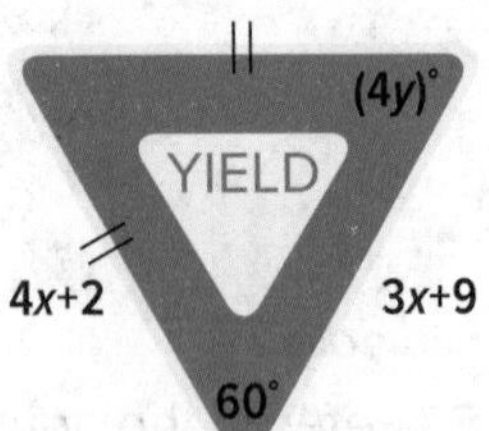

Mixed Exercises

16. **PROOF** Julita works for a company that makes lounge chairs. As shown in the figure, the back of each chair is an isosceles triangle that can be adjusted so the person sitting on the chair can recline.

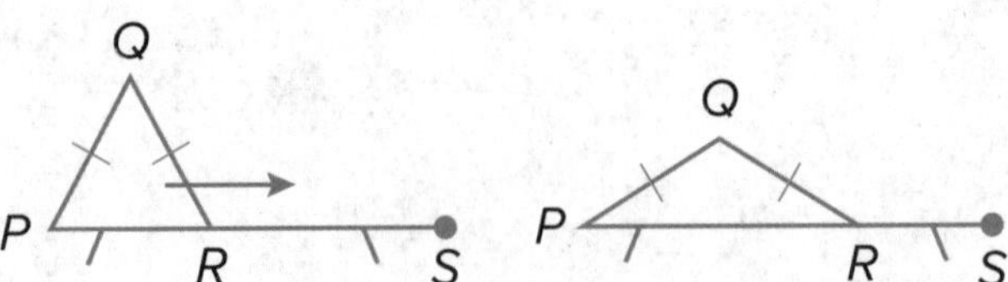

Suppose the chair is adjusted so $m\angle Q = 50°$. What is $m\angle QRS$? Write a paragraph proof to justify your argument.

Name ______________________ Period ____________ Date ____________

17. STRUCTURE Each of the triangles shown is isosceles.

a. Use a ruler to find the midpoint of each side of each triangle. Then draw a triangle formed by connecting the midpoints of each side.

b. Look for patterns in your drawings. Make a conjecture about what you notice.

18. PROOF Write a two-column proof to prove each case of Corollary 5.3.

a. Case 1

Given: $\triangle DEF$ is an equilateral triangle.

Prove: $\triangle DEF$ is an equiangular triangle.

b. Case 2

Given: $\triangle DEF$ is an equiangular triangle.

Prove: $\triangle DEF$ is an equilateral triangle.

19. PROOF Write a two-column proof to prove Corollary 5.4.

Given: $\triangle PQR$ is an equilateral triangle.

Prove: $m\angle P = m\angle Q = m\angle R = 60°$

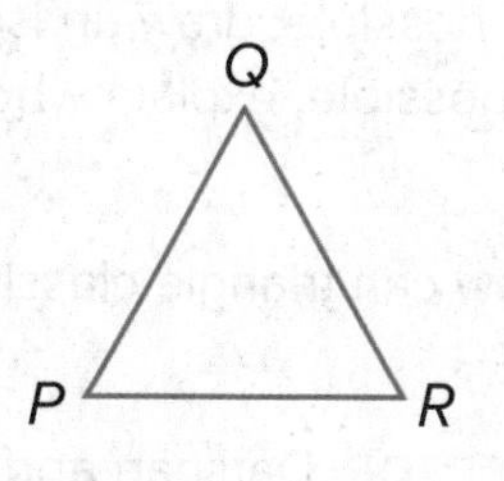

REGULARITY **Find each measure.**

20. $m\angle CAD$

21. $m\angle ACD$

22. $m\angle ACB$

23. $m\angle ABC$

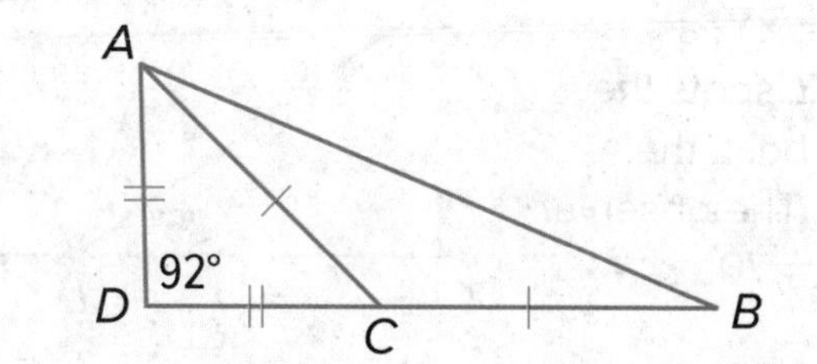

24. PATHS A marble path, as shown at the right, is constructed out of several congruent isosceles triangles. All the vertex angles measure 20°. What is the measure of angle 1 in the figure?

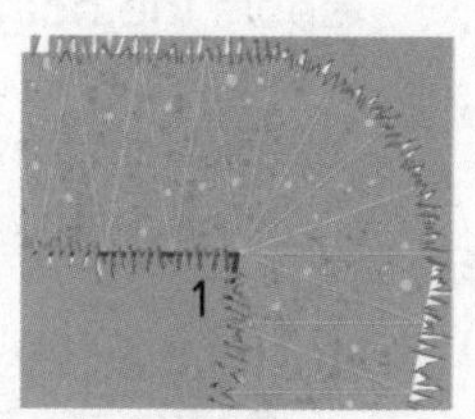

25. PRECISION Construct three different isosceles right triangles. Explain your method. Then verify your constructions using measurement and mathematics.

26. STATE YOUR ASSUMPTIONS Every day, cars drive through approximate isosceles triangles when they go over the Leonard Zakim Bridge in Boston. The ten-lane roadway forms the bases of the triangles.

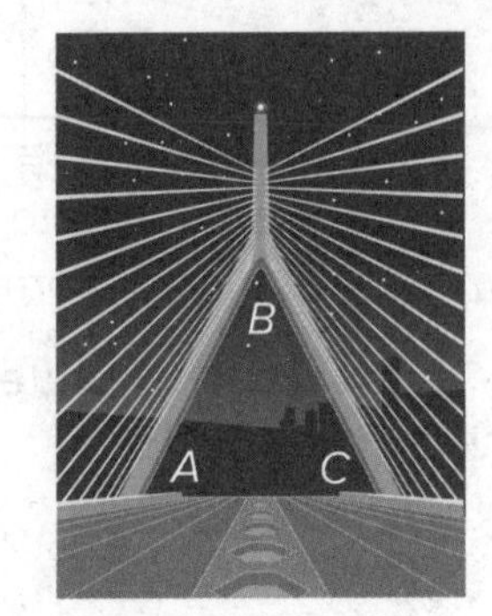

a. If $m\angle A = 67°$, find $m\angle B$.

b. Find $m\angle C$.

c. What assumption is made when approximating that the bridge forms isosceles triangles?

ANALYZE **Determine whether the following statements are *sometimes, always,* or *never* true. Justify your argument.**

27. If the measure of the vertex angle of an isosceles triangle is an integer, then the measure of each base angle is an integer.

28. If the measures of the base angles of an isosceles triangle are integers, then the measure of its vertex angle is odd.

29. CREATE If possible, draw an isosceles triangle with base angles that are obtuse. If it is not possible, explain why not.

30. WRITE How can triangle classifications help you prove triangle congruence?

31. FIND THE ERROR Darshan and Miguela are finding $m\angle G$ in the figure shown. Darshan says that $m\angle G = 35°$, and Miguela says that $m\angle G = 60°$. Is either of them correct? Explain your reasoning.

G
H
70°
F

32. PERSEVERE A boat is traveling at 25 mi/h parallel to a straight section of the shoreline, $\overline{XY}$, as shown. An observer in a lighthouse L spots the boat when the angle formed by the boat, the lighthouse, and the shoreline is 35°. The observer spots the boat again when $m\angle CLX = 70°$.

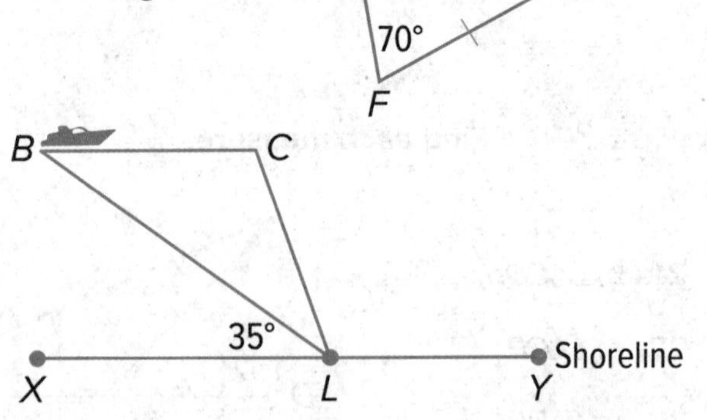

a. Explain how you can prove that $\triangle BCL$ is isosceles.

b. It takes the boat about 15 minutes to travel from point B to point C. When the boat is at point C, what is the distance to the lighthouse?

Triangles and Coordinate Proof

Learn Position and Label Triangles

Coordinate proofs use figures in the coordinate plane and algebra to prove geometric concepts. The first step in a coordinate proof is placing the figure on the coordinate plane.

Key Concept • Placing Triangles on the Coordinate Plane

Steps
1. Use the origin as a vertex, or the center of the triangle.
2. Place at least one side of the triangle on an axis.
3. Keep the triangle within the first quadrant if possible.
4. Use coordinates that make computations as simple as possible.

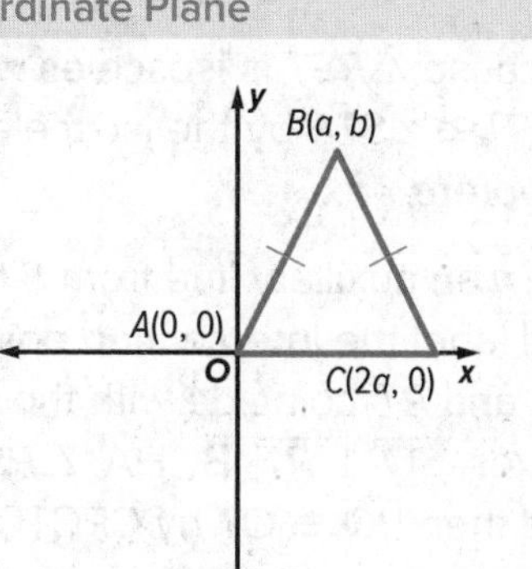

Example 1 Position and Label a Triangle

Position and label right $\triangle ABC$ with legs $\overline{AC}$ and $\overline{AB}$ so $\overline{AC}$ is $2a$ units long and $\overline{AB}$ is $2b$ units long.

Step 1 Position the triangle.

- Position the triangle in the first quadrant.
- Placing the right angle of the triangle, $\angle A$, at the origin will allow the two legs to be along the x- and y-axes.

Step 2 Determine the coordinates.

- Because C is on the y-axis, its x-coordinate is 0. Its y-coordinate is $2a$ because the leg is $2a$ units long.
- Because B is on the x-axis, its y-coordinate is 0. Its x-coordinate is $2b$ because the leg is $2b$ units long.

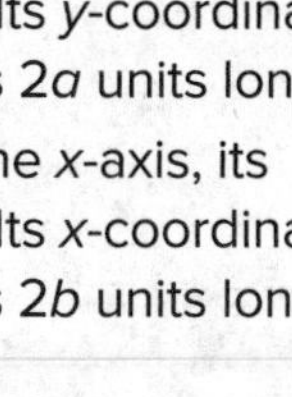

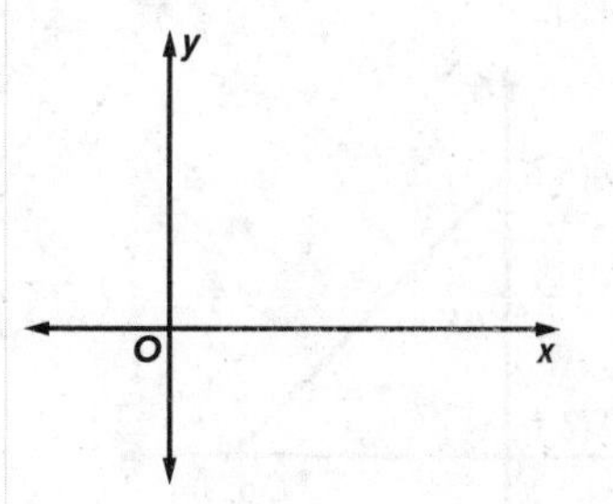

Check

Position and label isosceles triangle JKL on the coordinate plane such that the base $\overline{JL}$ is $2a$ units long, the vertex K is on the y-axis, and the height of the triangle is b units.

y
O
x

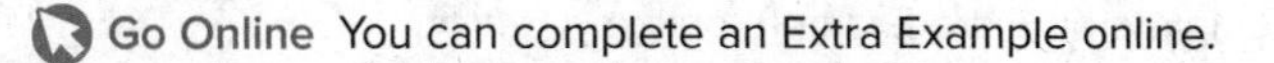

Go Online You can complete an Extra Example online.

Today's Goals

- Position a triangle on the coordinate plane and label the vertices.
- Write coordinate proofs to verify properties and to prove theorems about triangles.

Today's Vocabulary

coordinate proofs

Go Online

You can watch a video to see how to place figures on the plane for coordinate proofs.

Think About It!

The coordinates of two vertices of an equilateral triangle are (0, 0) and ($2a$, 0). The height of the triangle is b units. The coordinates of the third vertex are in terms of a and b. What are the coordinates of the third vertex?

Your Notes

Example 2 Identify Missing Coordinates

Name the missing coordinates of isosceles $\triangle RST$.

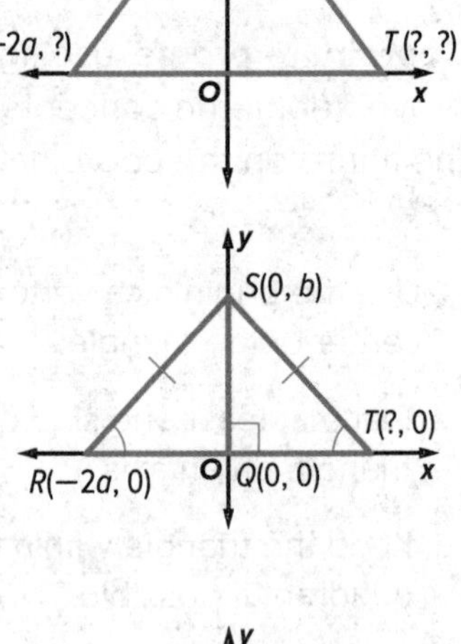

Step 1 Find the *y*-coordinates of *R* and *T*.

The base of the triangle is positioned on the *x*-axis. So, the *y*-coordinate of *R* is ____, and the *y*-coordinate of *T* is ____.

Step 2 Use the properties of $\triangle RST$.

Because $\triangle RST$ is isosceles with $\overline{RS} \cong \overline{TS}$, $\angle SRT \cong \angle STR$ by the Isosceles Triangle Theorem.

Draw an auxiliary line from *S* to the origin and label the intersection point *Q*. Because $\overline{SQ}$ and $\overline{RT}$ coincide with the *x*- and *y*-axes, $\overline{SQ} \perp \overline{RT}$. By HA, $\triangle RSQ \cong \triangle TSQ$, and then $\overline{RQ} \cong QT$ by CPCTC.

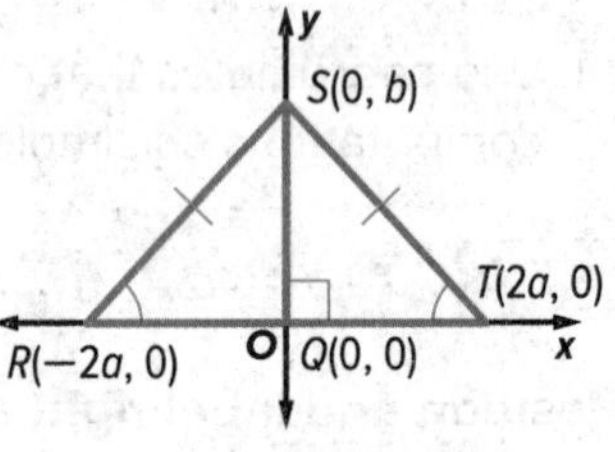

Step 3 Find the *x*-coordinate of *T*.

So, because the *x*-coordinate of *R* is $-2a$, the *x*-coordinate of *T* must be $2a$.

y
S(0, b)
T(2a, 0)
R(−2a, 0)
O
Q(0, 0)
x

Study Tip

Isosceles Triangles You may want to place isosceles and equilateral triangles on the coordinate plane so that the uppermost vertex lies on the *y*-axis. Then you can use the properties of isosceles triangles to label the coordinates of the vertices, making the computations in coordinate proofs easier.

Check

Name the coordinates of isosceles right triangle *ABC* with $\overline{BC}$ *a* units long.

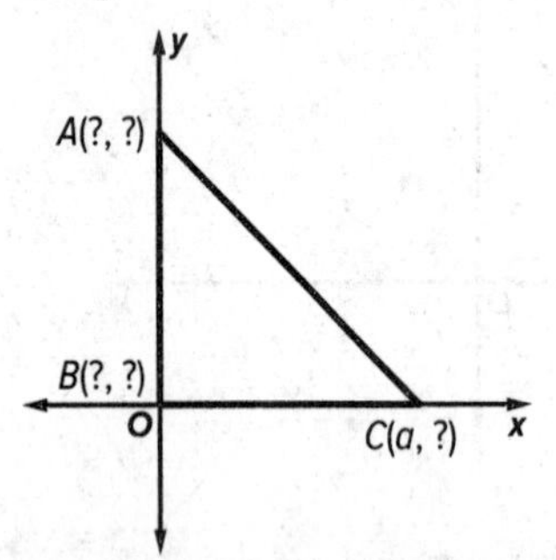

Explore Triangles and Coordinate Proofs

Online Activity Use the guiding exercises to complete the Explore.

INQUIRY How can you assign coordinates to vertices of a triangle if the lengths of the sides are unknown?

Go Online You can complete an Extra Example online.

Learn Triangles and Coordinate Proof

Coordinate proofs use figures on the coordinate plane to prove geometric concepts and theorems.

Key Concept • Writing a Coordinate Proof
Step 1 Place the figure on the coordinate plane.
Step 2 Label the coordinates of the vertices of the figure.
Step 3 Use algebra to prove properties or theorems.

Study Tip

Coordinate Proofs These guidelines apply to all polygons, not just triangles.

Example 3 Write a Coordinate Proof

Write a coordinate proof to show that $\triangle FGH \cong \triangle FDC$.

Use the Distance Formula to find the length of each side of each triangle. If the sides of the triangles are congruent, then the triangles are congruent by SSS.

$DC = \sqrt{[-a-(\underline{\quad})]^2 + (b - \underline{\quad})^2}$ or ____

$GH = \sqrt{(a - \underline{\quad})^2 + (b - \underline{\quad})^2}$ or ____

Because $DC = GH$, $\overline{DC} \cong \overline{GH}$ by the definition of ____.

$DF = \sqrt{[0 - (\underline{\quad})]^2 + \left(\frac{b}{2} - \underline{\quad}\right)^2}$ or $\sqrt{\underline{\qquad}}$

$GF = \sqrt{(a - \underline{\quad})^2 + \left(b - \underline{\quad}\right)^2}$ or $\sqrt{\underline{\qquad}}$

$CF = \sqrt{[0 - (\underline{\quad})]^2 + \left(\frac{b}{2} - \underline{\quad}\right)^2}$ or $\sqrt{\underline{\qquad}}$

$HF = \sqrt{(a - \underline{\quad})^2 + \left(0 - \underline{\quad}\right)^2}$ or $\sqrt{\underline{\qquad}}$

Because $DF =$ ____ $= CF = HF$, $\overline{DF}$ ____ $\overline{GF}$ ____ $\overline{CF}$ ____ $\overline{HF}$, $\triangle FGH \cong \triangle FDC$ by ____.

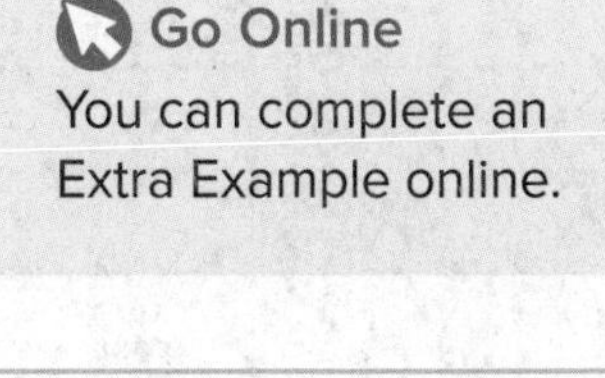

Check

Write a coordinate proof to show that $\triangle ABX \cong \triangle CDX$.

Proof:

The ____ of $\overline{AC}$ is $\left(\frac{0+a+x}{2}, \frac{0+b}{2}\right)$, or $\left(\frac{a+x}{2}, \frac{b}{2}\right)$. The midpoint of $\overline{BD}$ is $\left(\frac{0+x+a}{2}, \frac{b+0}{2}\right)$ or $\left(\frac{a+x}{2}, \frac{b}{2}\right)$. Because x is located at $\left(\frac{a+x}{2}, \frac{b}{2}\right)$, it is the midpoint of $\overline{AC}$ and $\overline{BD}$. By the definition of a segment bisector, $\overline{AC}$ bisects $\overline{BD}$ and $\overline{BD}$ bisects $\overline{AC}$.

Therefore, $\overline{BX} \cong \overline{XD}$ and $\overline{AX} \cong \overline{XC}$. From the ____, $CD = \sqrt{[(a+x)-a]^2 + (b-0)^2}$ or $\sqrt{x^2+b^2}$, and $AB = \sqrt{[(0+x)-0]^2 + (b-0)^2}$ or $\sqrt{x^2+b^2}$. Therefore, $\overline{CD} \cong \overline{AB}$ by the definition of ____, and $\triangle ABX \cong \triangle CDX$ by SSS.

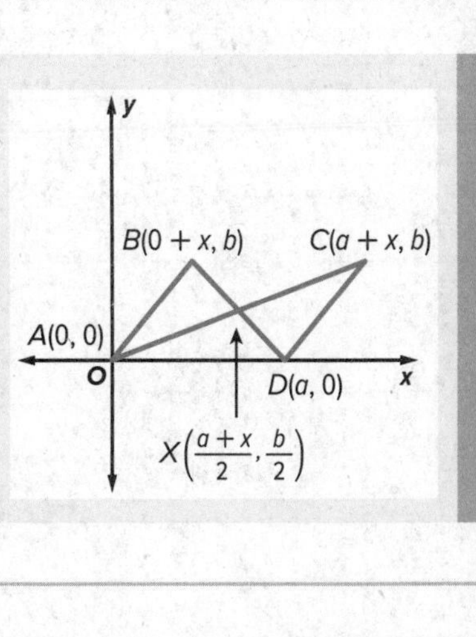

Example 4 Prove a Theorem by Using Coordinate Geometry

Write a coordinate proof to show that if two lines are each equidistant from a third line, then the two lines are parallel to each other.

Given: $\overleftrightarrow{AB}$ **and** $\overleftrightarrow{EF}$ **are equidistant from** $\overleftrightarrow{CD}$**.**

Prove: $\overleftrightarrow{AB} \parallel \overleftrightarrow{EF}$

Proof:

The slope of $\overleftrightarrow{AB} = \frac{b-0}{a-0} =$ ______. The slope of $\overleftrightarrow{EF} = \frac{b-0}{3a-2a} =$ ______.

Because the slopes of $\overleftrightarrow{AB}$ and $\overleftrightarrow{EF}$ are the same, ____________.

Check

Write a coordinate proof to show that the three segments joining the midpoints of the sides of an isosceles triangle form another isosceles triangle.

Given: Isosceles triangle ABC; $\overline{BC} \cong \overline{AC}$; R, S, and T are midpoints of their respective sides.

Prove: $\triangle RST$ is isosceles.

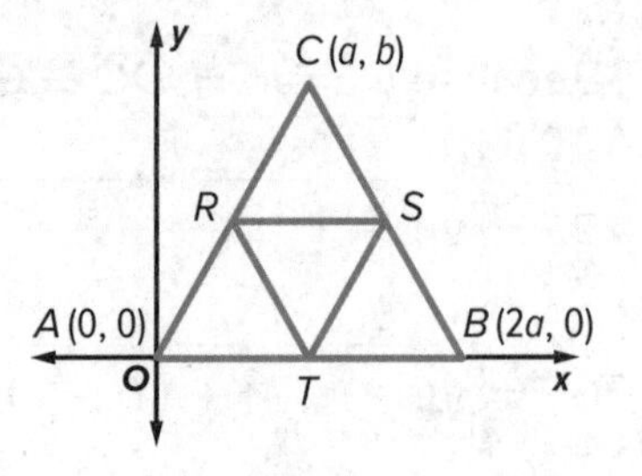

Proof:

Midpoint R is $\left(\frac{a+0}{2}, \frac{b+0}{2}\right)$ or $\left(\frac{a}{2}, \frac{b}{2}\right)$.

Midpoint S is $\left(\frac{a+2a}{2}, \frac{b+0}{2}\right)$ or $\left(\frac{3a}{2}, \frac{b}{2}\right)$.

Midpoint T is $\left(\frac{2a+0}{2}, \frac{0+0}{2}\right)$ or $(a, 0)$.

$RT = \sqrt{\left(\frac{a}{2}-a\right)^2 + \left(\frac{b}{2}-0\right)^2}$ or $\sqrt{\left(\frac{a}{2}\right)^2 + \left(\frac{b}{2}\right)^2}$

$ST = \sqrt{\left(\frac{3a}{2}-a\right)^2 + \left(\frac{b}{2}-0\right)^2}$ or $\sqrt{\left(\frac{a}{2}\right)^2 + \left(\frac{b}{2}\right)^2}$

$RT = ST$ and $\overline{RT} \cong \overline{ST}$ and $\triangle RST$ is isosceles.

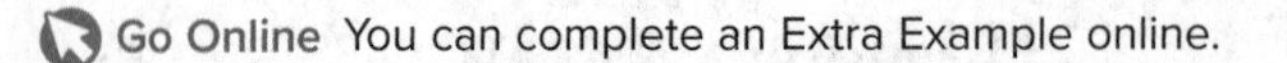
Go Online You can complete an Extra Example online.

Example 5 Classify a Triangle

NAVIGATION **The Polynesian Triangle is a triangle formed between the three Pacific island groups that form the South Pacific region known as Polynesia. The approximate coordinates in latitude and longitude of each vertex are Auckland, New Zealand (−40.9, 174.9), Honolulu, Hawaii (21.3, −157.9), and Easter Island (−27.1, −109.4).**

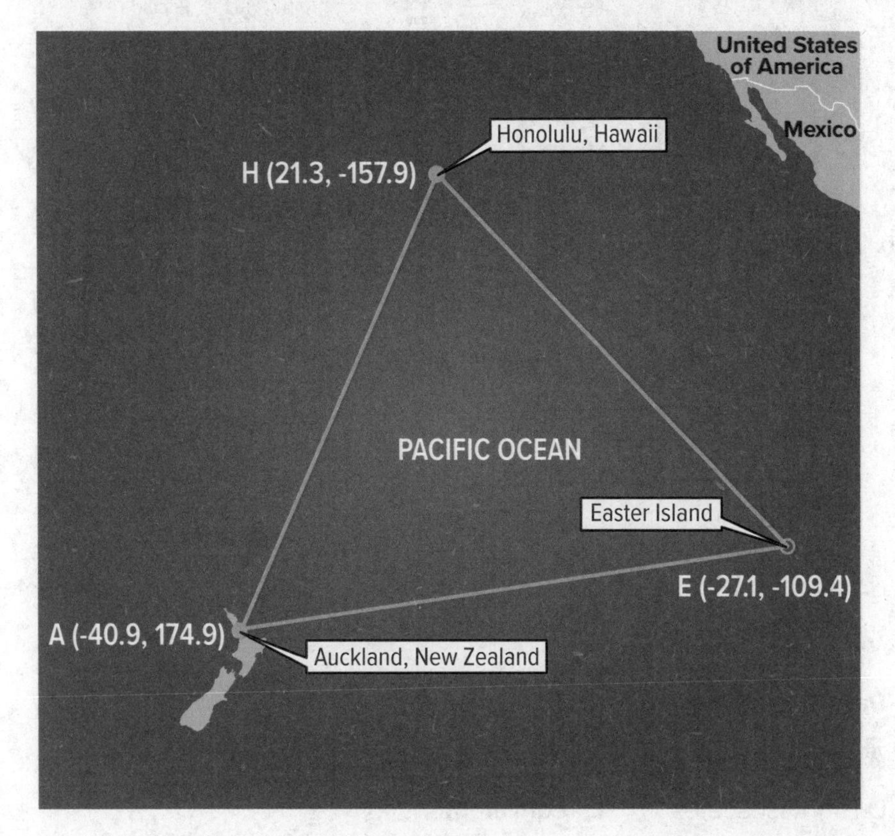

Study Tip

Units of Measure While the distance between cities is usually measured in miles or kilometers, latitude and longitude are measured in degrees relative to the Prime Meridian and the Equator.

Part A Estimate the type of triangle formed by the Polynesian islands.

The triangle appears to be a(n) ______ triangle.

Part B Use coordinate geometry to determine the type of triangle formed.

Use the Distance Formula to determine the length of each side of the triangle.

Round to the nearest tenth.

$AE = \sqrt{[-40.9 - (-27.1)]^2 + [174.9 - (-109.4)]^2}$

$\approx$ ______

$EH = \sqrt{(-27.1 - 21.3)^2 + [-109.4 - (-157.9)]^2}$

$\approx$ ______

$AH = \sqrt{(-40.9 - 21.3)^2 + [174.9 - (-157.9)]^2}$

$\approx$ ______

Because the length of each side is different, the triangle is ______.

Go Online You can complete an Extra Example online.

Check

GEOGRAPHY Eldora's family lives in New Mexico. She lives southwest of Rio Rancho, her uncle lives in Clines Corners, and her grandparents live in Rociada. Eldora has assigned coordinates to each location.

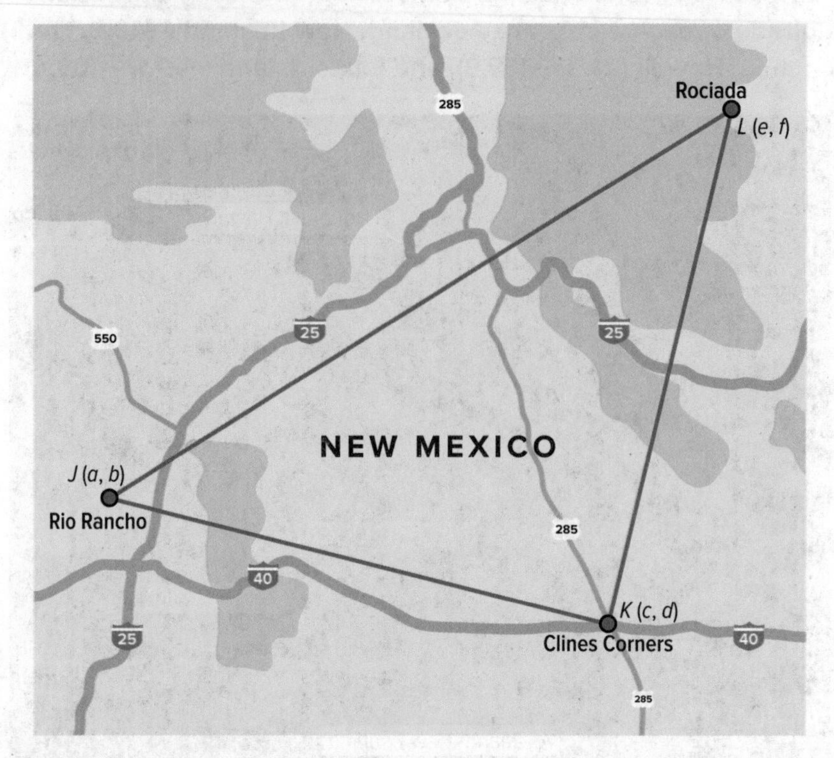

Part A Estimate the type of triangle formed.

A. acute scalene B. obtuse scalene C. right scalene

D. right isosceles E. equilateral

Part B Which of the following can be used in a coordinate proof to show that the estimate chosen above is correct?

A. Use the Distance Formula to find the lengths of $\overline{JL}$, $\overline{JK}$, and $\overline{KL}$. If they are all equal, then the triangle is equilateral.

B. Use the Distance Formula to find the lengths of $\overline{JL}$, $\overline{JK}$, and $\overline{KL}$. If they are different, then the triangle is scalene.

C. Compare the slopes of $\overline{JK}$ and $\overline{KL}$. If the product of the slopes is -1, then the lines are perpendicular. Use the Distance Formula to find the lengths of $\overline{JK}$ and $\overline{KL}$. If the lengths are equal, then the triangle is a right isosceles triangle.

D. Compare the slopes of $\overline{JK}$ and $\overline{KL}$. If the product of the slopes is -1, then the lines are perpendicular. Use the Distance Formula to find the lengths of $\overline{JL}$, $\overline{JK}$, and $\overline{KL}$. If the lengths are different, then the triangle is a right scalene triangle.

Name ______________________ Period ________ Date ________

Practice

Go Online You can complete your homework online.

Example 1

REGULARITY **Position and label each triangle on the coordinate plane.**

1. isosceles $\triangle ABC$ with base $\overline{AB}$ that is a units long and height that is b units

2. right $\triangle XYZ$ with hypotenuse $\overline{YZ}$, leg $\overline{XY}$ that is b units long, and leg $\overline{XZ}$ that is three times the length of $\overline{XY}$

3. isosceles right $\triangle RST$ with hypotenuse $\overline{RS}$ and legs $3a$ units long

4. right $\triangle JKL$ with legs $\overline{JK}$ and $\overline{KL}$ such that $\overline{JK}$ is a units long and leg $\overline{KL}$ is $4b$ units long

Example 2

Name the missing coordinate(s) of each triangle.

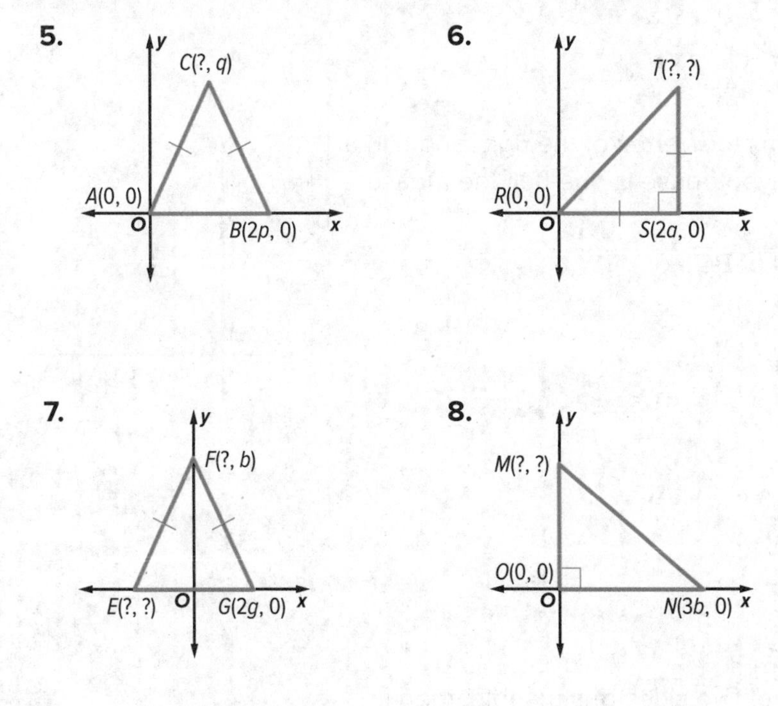

Examples 3 and 4

PROOF **For Exercises 9–13, write a coordinate proof for each statement.**

9. The segments joining the midpoints of the sides of a right triangle form a right triangle.

Given: Point R is the midpoint of $\overline{AB}$.
Point P is the midpoint of $\overline{BC}$.
Point Q is the midpoint of $\overline{AC}$.

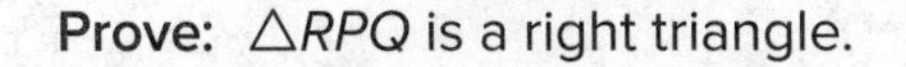

Prove: $\triangle RPQ$ is a right triangle.

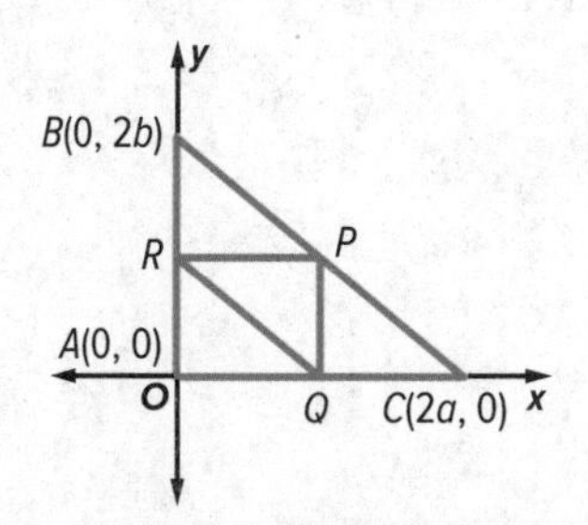

10. A segment from the vertex angle of an isosceles triangle to the midpoint of the base is perpendicular to the base.

Given: Isosceles $\triangle RST$; U is the midpoint of base $\overline{RT}$.

Prove: $\overline{SU} \perp \overline{RT}$

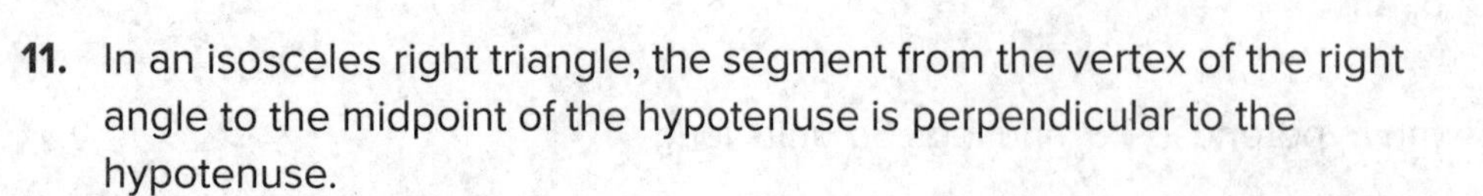

11. In an isosceles right triangle, the segment from the vertex of the right angle to the midpoint of the hypotenuse is perpendicular to the hypotenuse.

Given: isosceles right $\triangle ABC$ with right angle $\angle ABC$; M is the midpoint of $\overline{AC}$.

Prove: $\overline{BM} \perp \overline{AC}$

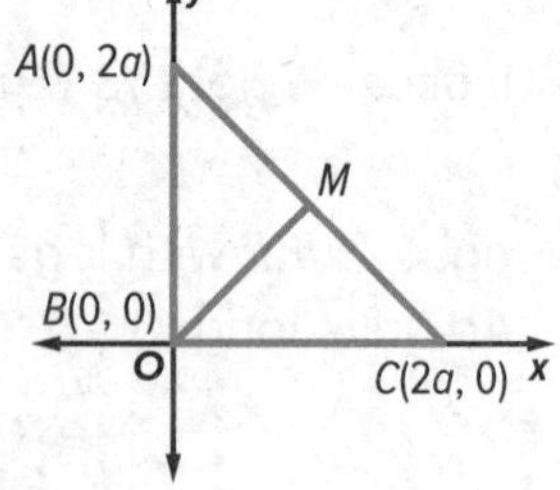

12. The measure of the segment that joins the vertex of the right angle in a right triangle to the midpoint of the hypotenuse is one-half the measure of the hypotenuse.

Given: right $\triangle ABC$; P is the midpoint of $\overline{BC}$.

Prove: $AP = \frac{1}{2}BC$

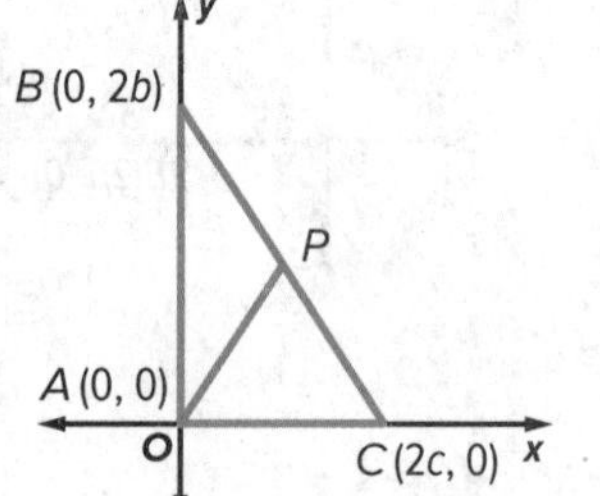

13. If a line segment joins the midpoints of two sides of a triangle, then its length is equal to one-half the length of the third side.

Given: S is the midpoint of $\overline{AC}$.

T is the midpoint of $\overline{BC}$.

Prove: $ST = \frac{1}{2}AB$

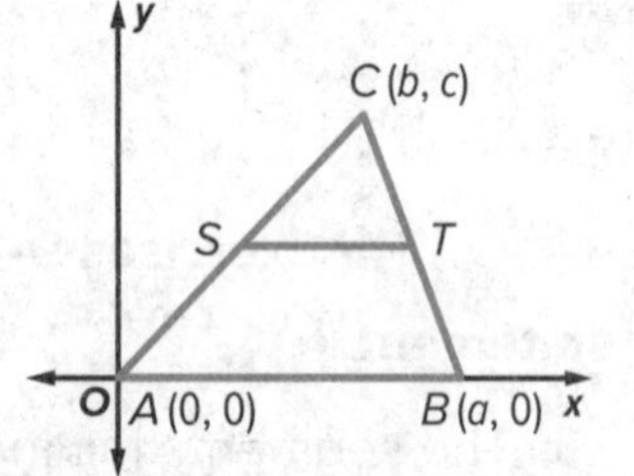

Example 5

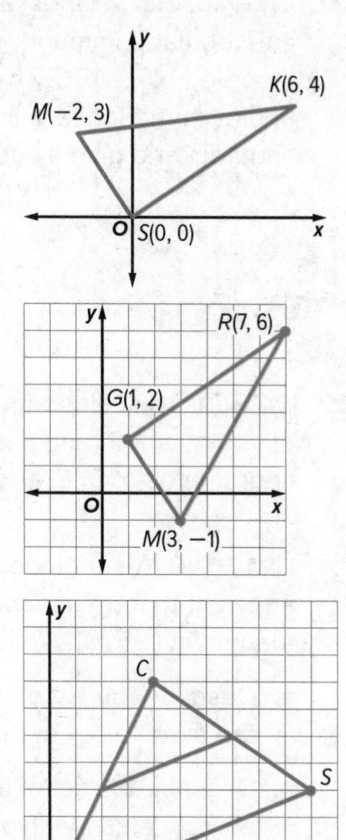

14. NEIGHBORHOODS Kalini lives 6 miles east and 4 miles north of her high school. After school, she works part time at the mall in a music store. The mall is 2 miles west and 3 miles north of the school. Use coordinate geometry to determine the type of triangle formed by Kalini's high school, her home, and the mall.

15. COUNTY FAIR The fair committee wants to print a map to distribute to vendors as they arrive to set up their booths at the fairgrounds. On a coordinate grid, the main gate is located at (3, −1), the grandstand is located at (1, 2), and the rides and games are located at (7, 6). Use coordinate geometry to determine the type of triangle formed by these locations.

16. USE ESTIMATION A town is preparing for a 5K run. The race will start at city hall C. The course will take runners along straight streets to the library L, to the science museum S, and back to city hall for the finish.

a. Estimate the type of triangle formed by the course.

b. Use coordinate geometry to determine the type of triangle formed.

Mixed Exercises

REASONING **For Exercises 17 and 18, determine whether the triangle can be a right triangle. Explain.**

17. $X(0, 0)$, $Y(2h, 2h)$, $Z(4h, 0)$

18. $X(0, 0)$, $Y(1, h)$, $Z(2h, 0)$

19. SHELVES Martha has a shelf bracket shaped like a right isosceles triangle. She wants to know the length of the hypotenuse relative to the sides. She does not have a ruler but remembers the Distance Formula. She places the bracket in Quadrant I of a coordinate grid with the right angle at the origin. The length of each leg is a. What are the coordinates of the vertices that form the two acute angles?

20. FLAGS A flag is shaped like an isosceles triangle. A designer would like to make a drawing of the flag on a coordinate plane. She positions it so the base of the triangle is on the y-axis with one endpoint located at (0, 0). She locates the tip of the flag at $\left(a, \frac{b}{2}\right)$. What are the coordinates of the third vertex?

21. DESIGN Andrew is using a coordinate plane to design a quilt. Two of the triangular patches for the quilt are shown in the figure. Andrew wants to be sure that $\angle A$ and $\angle D$ have the same measure. Describe the main steps you can use to prove that $\angle A \cong \angle D$.

22. **COMMUNITY** A landscape architect is using a coordinate plane to design a triangular community garden. The fence that will surround the garden is modeled by $\triangle ABC$. The architect wants to know whether any of the three angles in the fence will be congruent. Determine the answer for the architect and give a coordinate proof to justify your response.

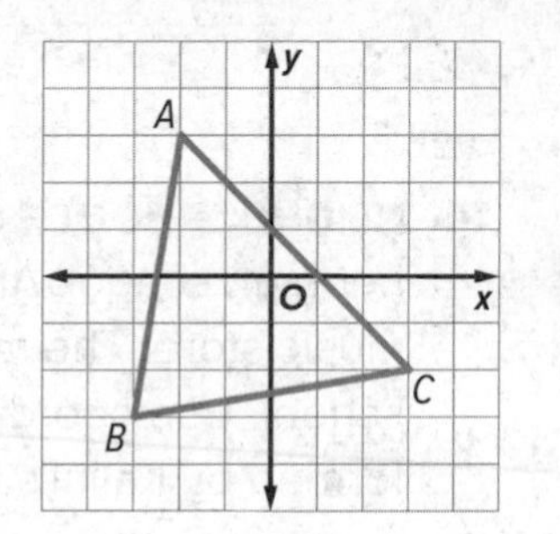

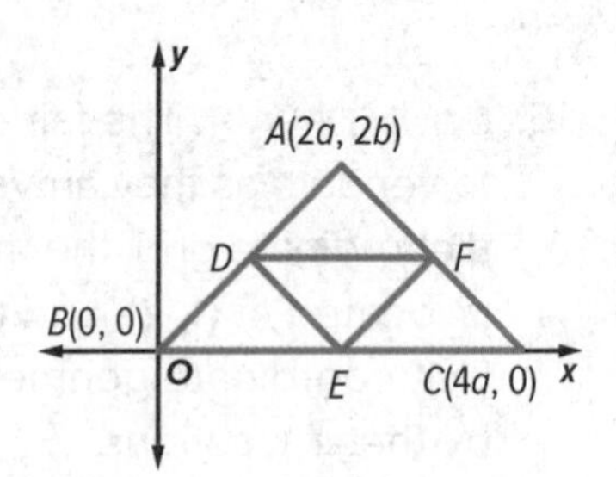

23. $\triangle ABC$ is isosceles with $\overline{AB} \cong \overline{AC}$. D is the midpoint of $\overline{AB}$, E is the midpoint of $\overline{BC}$, and F is the midpoint of $\overline{AC}$. What are the coordinates of D, E, and F?

24. **DRAFTING** An engineer is designing a roadway. Three roads intersect to form a triangle. The engineer marks two vertices of the triangle at $(-5, 0)$ and $(5, 0)$ on a coordinate plane.

a. Describe the set of points in the coordinate plane that could not be used as the third vertex of the triangle.

b. Describe the set of points in the coordinate plane that could be the vertex of an isosceles triangle.

c. Describe the set of points in the coordinate plane that would make a right triangle with the other two points if the right angle is located at $(-5, 0)$.

25. **CREATE** Draw an isosceles right triangle on the coordinate plane so the midpoint of its hypotenuse is the origin. Label the coordinates of the vertex.

26. **WRITE** Explain why following each guideline for placing a triangle on the coordinate plane is helpful in proving coordinate proofs.

a. Use the origin as a vertex of the triangle.

b. Place at least one side of the triangle on the x- or y-axis.

c. Keep the triangle within the first quadrant if possible.

PERSEVERE **Find the coordinates of point L so $\triangle JKL$ is the indicated type of triangle. Point J has coordinates $(0, 0)$, and point K has coordinates $(2a, 2b)$.**

27. scalene triangle

28. right triangle

29. isosceles triangle

30. **ANALYZE** The midpoints of the sides of a triangle are located at $(a, 0)$, $(2a, b)$ and (a, b). If one vertex is located at the origin, what are the coordinates of the other vertices? Explain your reasoning.

Review

Essential Question

How can you prove congruence and use congruent figures in real-world situations?

Module Summary

Lesson 5-1 through 5-2

Angles and Sides

- The sum of the measures of the interior angles of a triangle is 180°.
- Two figures are congruent if and only if there is a rigid motion or series of rigid motions that maps one figure exactly onto the other
- In two congruent polygons, all the parts of one polygon are congruent to the corresponding parts of the other polygon.

Lesson 5-3 through 5-5

Ways to Prove Triangles Congruent

- Side-Side-Side (SSS) Congruence three sides of one triangle congruent to three sides of a second triangle
- Side-Angle-Side (SAS) Congruence two sides and the included angle of one triangle congruent to two sides and the included angle of a second triangle
- Angle-Side-Angle (ASA) Congruence two angles and the included side of one triangle congruent to two angles and the included side of a second triangle
- Angle-Angle-Side (AAS) Congruence two angles and the nonincluded side of one triangle congruent to two angles and the nonincluded side of a second triangle
- For right triangles, use the following ways to prove congruence.

  Leg-Leg Congruence (LL)

  Hypotenuse-Angle Congruence (HA)

  Leg-Angle Congruence (LA)

  Hypotenuse-Leg Congruence (HL)

Lesson 5-6

Isosceles and Equilateral Triangles

- If two sides of a triangle are congruent, then the angles opposite those sides are congruent.
- If two angles of a triangle are congruent, then the sides opposite those angles are congruent.
- Each angle of an equilateral triangle measures 60°.

Lesson 5-7

Coordinate Proof

- To write a coordinate proof: Place the figure on the coordinate plane. Label the vertices. Use algebra to prove properties or theorems.

Study Organizer

Foldables

Use your Foldable to review this module. Working with a partner can be helpful. Ask for clarification of concepts as needed.

SSS SAS
ASA AAS

Name ______________________ Period __________ Date __________

Test Practice

1. **OPEN RESPONSE** Find the measure of $\angle BCD$ in degrees. (Lesson 5-1)

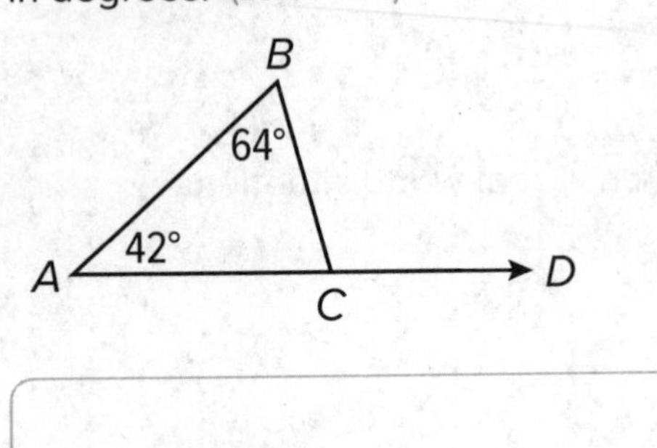

2. **MULTIPLE CHOICE** Find the value of x given the triangle below. (Lesson 5-1)

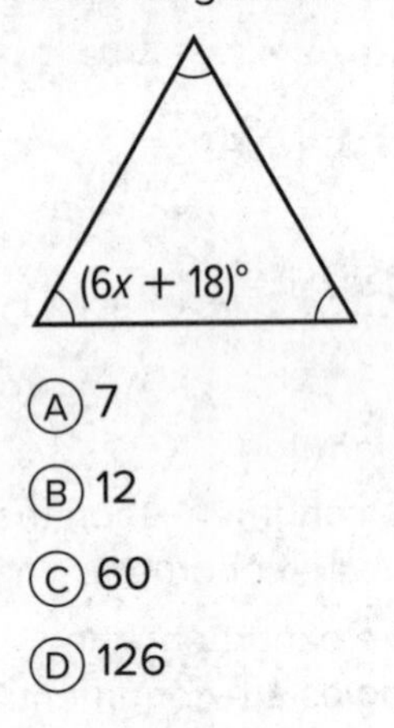

Ⓐ 7
Ⓑ 12
Ⓒ 60
Ⓓ 126

3. **MULTI-SELECT** In $\triangle PQR$, $\angle Q$ is a right angle.

Select all the statements about $\angle P$ and $\angle R$ that must be true. (Lesson 5-1)

Ⓐ $\angle P$ and $\angle R$ are complementary.
Ⓑ $\angle P$ and $\angle R$ are supplementary.
Ⓒ $\angle P$ and $\angle R$ are congruent.
Ⓓ $\angle P$ and $\angle R$ are acute.
Ⓔ $\angle P$ or $\angle R$ is obtuse.

4. **OPEN RESPONSE** $\triangle PRQ$ has side lengths $PR = 6$, $QR = 8$, and $PQ = 5$.

If $\triangle PRQ \cong \triangle CBA$, then put the side lengths of $\triangle CBA$ in order from shortest to longest. (Lesson 5-2)

5. **MULTI-SELECT** Given $\triangle DEF$ and $\triangle JLK$ where $\overline{DE} \cong \overline{JL}$, $\overline{FD} \cong \overline{KJ}$, $\overline{LK} \cong \overline{EF}$, $\angle D \cong \angle J$, $\angle E \cong \angle L$, and $\angle F \cong \angle K$, which of the following conclusions can be made? Select all that apply. (Lesson 5-2)

Ⓐ $\triangle DEF$ and $\triangle JLK$ are congruent.
Ⓑ $\triangle DEF$ and $\triangle JLK$ are not congruent.
Ⓒ A series of rigid motions will map $\triangle DEF$ onto $\triangle JLK$.
Ⓓ A series of rigid motions will not map $\triangle DEF$ onto $\triangle JLK$.
Ⓔ $\triangle FDE$ and $\triangle KJL$ are congruent.
Ⓕ $\triangle FDE$ and $\triangle KJL$ are not congruent.

6. **MULTIPLE CHOICE** Which postulate shows $\triangle ABC \cong \triangle DEF$? (Lesson 5-3)

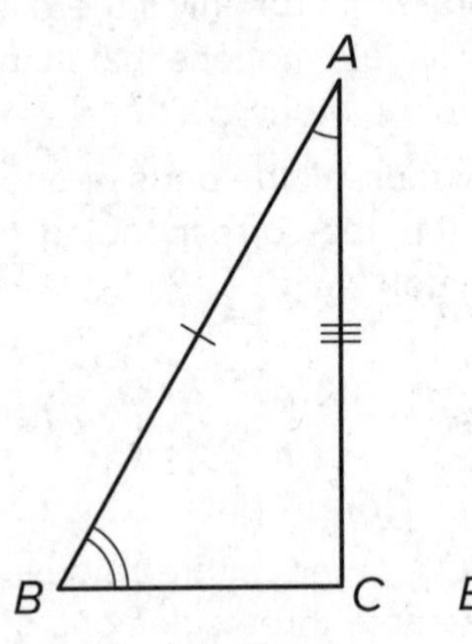

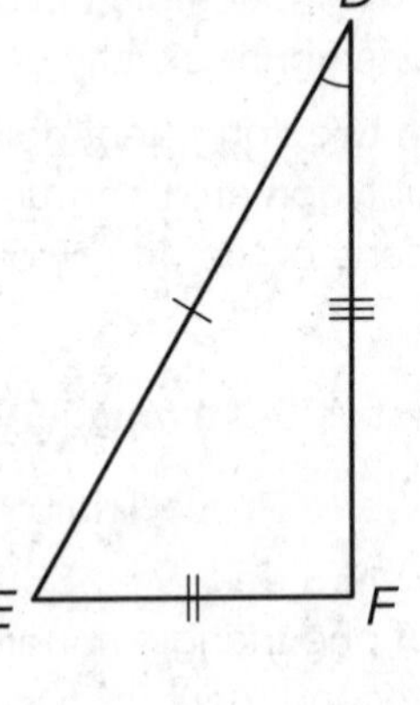

Ⓐ Side-Side-Side
Ⓑ Angle-Side-Angle
Ⓒ Side-Angle-Side
Ⓓ Angle-Angle-Side

7. MULTIPLE CHOICE In $\triangle JKL$ and $\triangle PQR$, $\overline{JK} \cong \overline{PQ}$ and $\overline{JL} \cong \overline{PR}$. Which additional statement would prove that $\triangle JKL \cong \triangle PQR$? (Lesson 5-3)

Ⓐ $\angle J \cong \angle P$

Ⓑ $\angle L \cong \angle R$

Ⓒ $\overline{JK} \cong \overline{PR}$

Ⓓ $\overline{JL} \cong \overline{KL}$

8. OPEN RESPONSE Stephanie and Fernando are building triangular prism birdhouses that have the same dimensions.

- Stephanie says that they should measure the lengths of two pairs of corresponding sides of the triangular bases and use a protractor to measure the included angles to be sure the bases are congruent.
- Fernando says that they can be sure the triangular bases are congruent if they measure the lengths of all three corresponding sides.

Which student is correct? (Lesson 5-3)

9. MULTI-SELECT In $\triangle ABC$ and $\triangle MNP$, $\angle A \cong \angle M$ and $\overline{BC} \cong \overline{NP}$. What additional piece(s) of information could be used to prove $\triangle ABC \cong \triangle MNP$ by AAS? Select all that apply. (Lesson 5-4)

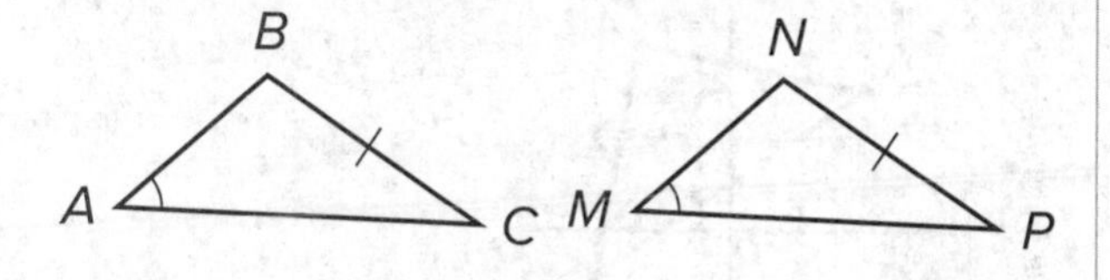

Ⓐ $\angle B \cong \angle N$

Ⓑ $\angle C \cong \angle P$

Ⓒ $\overline{AB} \cong \overline{MN}$

Ⓓ $\overline{AC} \cong \overline{MP}$

Ⓔ $\angle A \cong \angle N$

10. OPEN RESPONSE A technician is assembling parts for a radio antenna. He attaches two metal bars to 3-foot-long crosspieces so a triangle is formed, with each bar meeting the crosspiece at a 40° angle. Which postulate proves that all triangles formed this way are congruent? (Lesson 5-4)

11. MULTIPLE CHOICE In $\triangle RST$, $m\angle R = 85°$, $m\angle S = 33°$, and $RT = 17$.

Which set of measurements would make $\triangle RST \cong \triangle MNP$ by the AAS Theorem? (Lesson 5-4)

Ⓐ $m\angle M = 85°$, $m\angle N = 33°$, and $MP = 17$

Ⓑ $m\angle M = 85°$, $m\angle N = 33°$, and $MN = 17$

Ⓒ $m\angle M = 33°$, $m\angle N = 85°$, and $MP = 17$

Ⓓ $m\angle M = 33°$, $m\angle N = 85°$, and $MN = 17$

12. TABLE ITEM Select the pairs of triangles that must be congruent to each other. (Lesson 5-5)

A. C F A B D E

B. C F A B D E

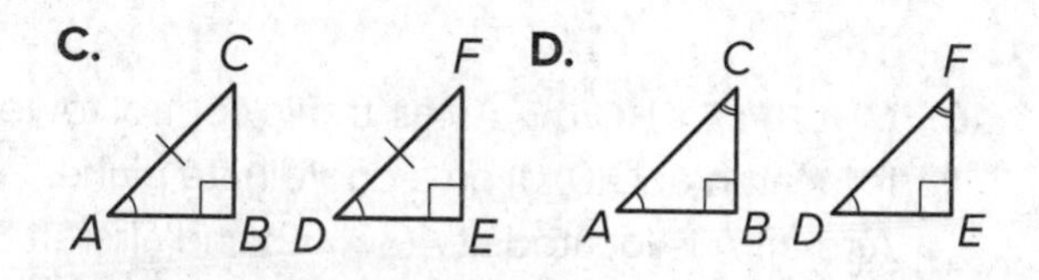

Pair	Congruent? yes	no
A.		
B.		
C.		
D.		

Name ______________________ Period ________ Date ________

13. **OPEN RESPONSE** If the vertex angle of an isosceles triangle measures 86°, what is the angle measure in degrees of one of the base angles? (Lesson 5-6)

14. **MULTIPLE CHOICE** Find the value of x. (Lesson 5-6)

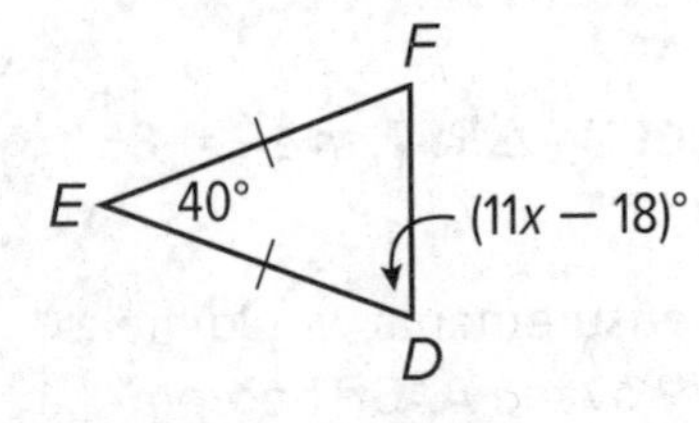

Ⓐ 5.3
Ⓑ 8
Ⓒ 14.4
Ⓓ 70

15. **OPEN RESPONSE** What is the length of segment QR? (Lesson 5-6)

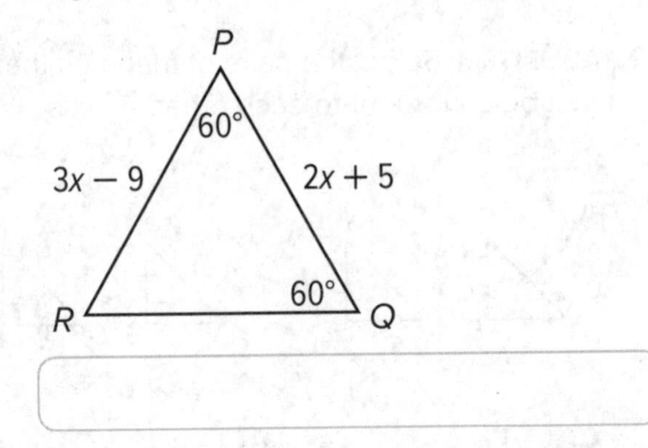

16. **MULTIPLE CHOICE** An air traffic control tower is located at $O(0, 0)$ on a coordinate plane. Aircraft A is located at $A(39, 52)$ and aircraft B is located at $B(25, 60)$.

What statement is true about this situation? (Lesson 5-7)

Ⓐ Aircraft A is closer to the control tower.
Ⓑ Aircraft B is closer to the control tower.
Ⓒ Both aircraft are the same distance from the control tower.
Ⓓ $\triangle OAB$ is an equilateral triangle.

17. **MULTIPLE CHOICE** A triangle drawn on a coordinate plane has vertices $A(0, 0)$, $B(0, 2b)$, and $C(2c, 0)$. Which expression represents the slope of $\overline{BC}$? (Lesson 5-7)

Ⓐ $-\frac{b}{c}$
Ⓑ $\frac{b}{c}$
Ⓒ $\frac{c}{b}$
Ⓓ $-\frac{c}{b}$

18. **MULTIPLE CHOICE** The given triangle will be used in a coordinate proof.

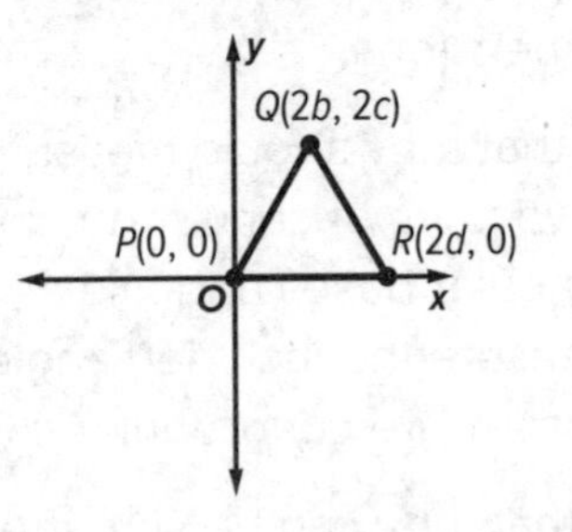

What are the coordinates of the midpoint of $\overline{QR}$? (Lesson 5-7)

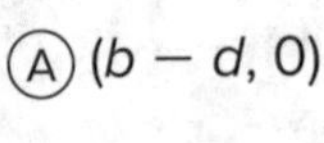

Ⓐ $(b - d, 0)$
Ⓑ $(b + d, 0)$
Ⓒ $(b - d, c)$
Ⓓ $(b + d, c)$

19. **MULTI-SELECT** Use coordinate geometry to determine the type of triangle formed below. Select all that apply. (Lesson 5-7)

C(−3, 3)
A(2, 4)
B(3, −1)

Ⓐ equilateral
Ⓑ isosceles
Ⓒ right
Ⓓ scalene
Ⓔ acute

Module 6

Relationships in Triangles

Essential Question

How can relationships in triangles be used in real-world situations?

What Will You Learn?

Place a check mark (✓) in each row that corresponds with how much you already know about each topic **before** starting this module.

KEY

— I don't know. — I've heard of it. — I know it!

	Before			After		
solve problems using perpendicular bisectors in triangles						
solve problems using angle bisectors						
solve problems using medians in triangles						
solve problems using altitudes in triangles						
solve problems using inequalities in the angles in a triangle						
solve problems using inequalities in the angles and sides in a triangle						
prove algebraic and geometric relationships by using indirect proof						
apply the Triangle Inequality Theorem						
apply the Hinge Theorem and its converse						

Foldables Make this Foldable to help you organize your notes about relationships in triangles. Begin with seven sheets of grid paper.

1. **Stack** the sheets. Fold the top right corner to the bottom edge to form an isosceles triangle.
2. **Fold** the rectangular part in half.
3. **Staple** the sheets along the rectangular fold in four places.
4. **Label** each sheet with a lesson number and the rectangular tab with the module title.

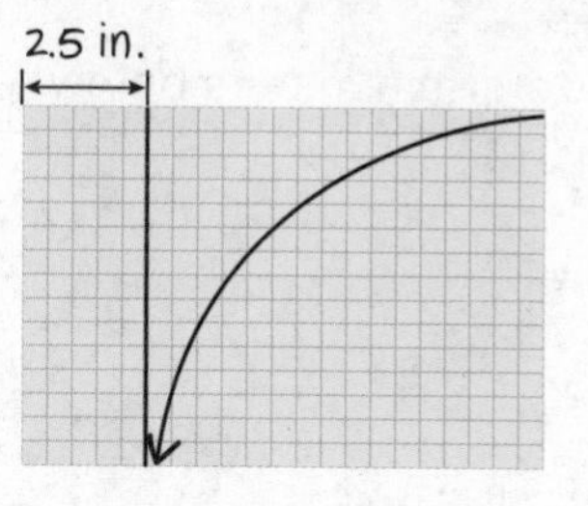

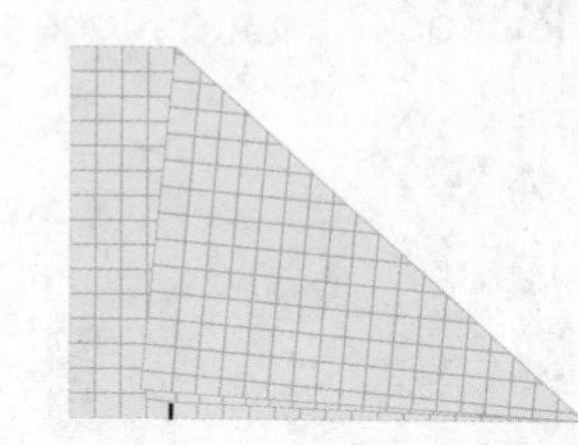

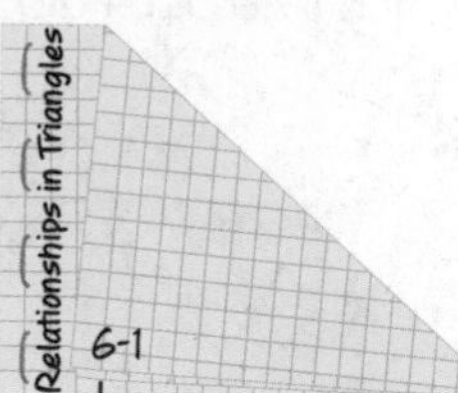

What Vocabulary Will You Learn?

Check the box next to each vocabulary term that you may already know.

☐ altitude of a triangle	☐ incenter	☐ orthocenter
☐ centroid	☐ indirect proof	☐ perpendicular bisector
☐ circumcenter	☐ indirect reasoning	☐ point of concurrency
☐ concurrent lines	☐ median	☐ proof by contradiction

Are You Ready?

Complete the Quick Review to see if you are ready to start this module. Then complete the Quick Check.

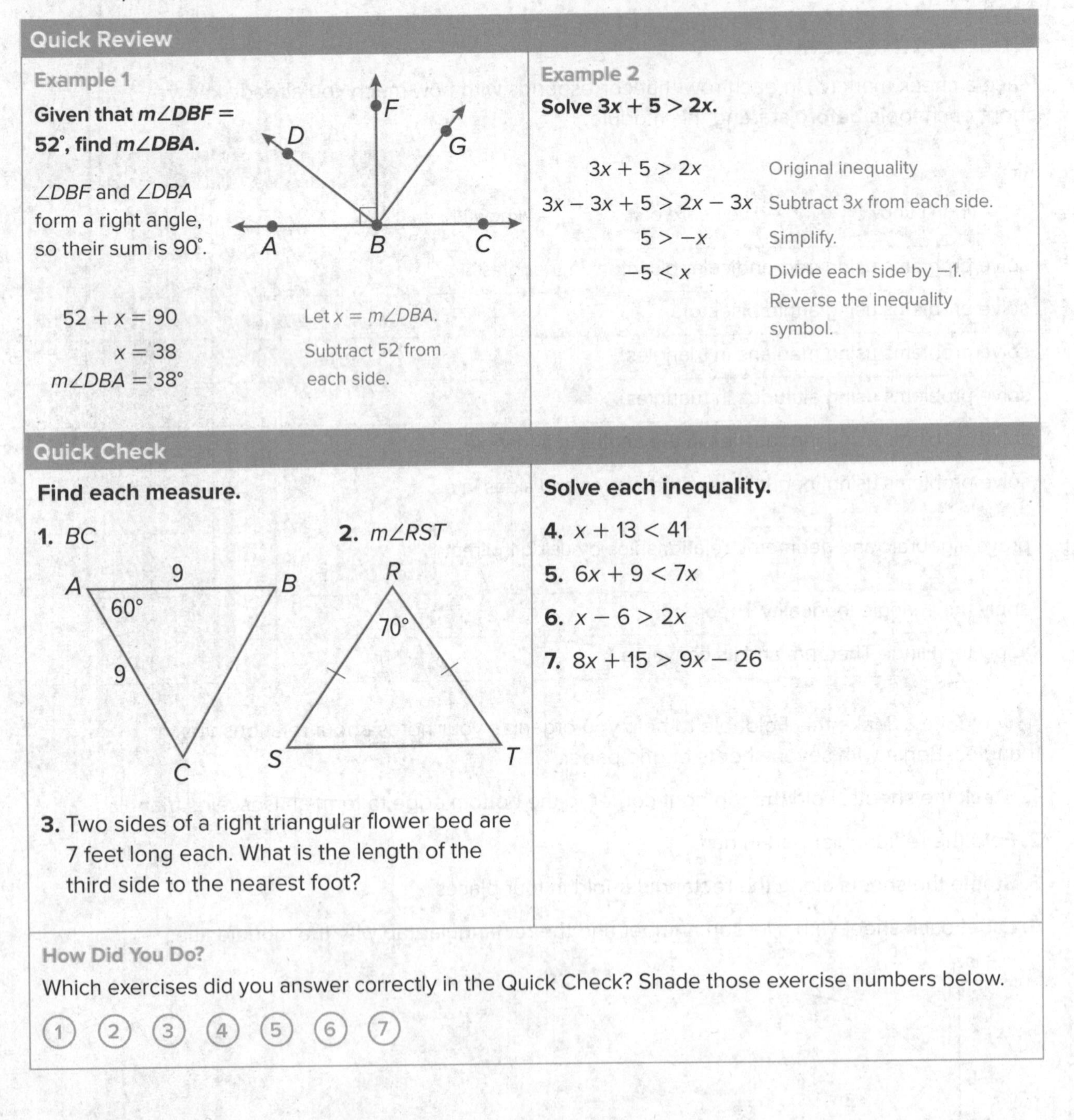

Quick Review

Example 1

Given that $m\angle DBF = 52^\circ$, find $m\angle DBA$.

$\angle DBF$ and $\angle DBA$ form a right angle, so their sum is 90°.

$52 + x = 90$ Let $x = m\angle DBA$.

$x = 38$ Subtract 52 from each side.

$m\angle DBA = 38^\circ$

Example 2

Solve $3x + 5 > 2x$.

$3x + 5 > 2x$	Original inequality
$3x - 3x + 5 > 2x - 3x$	Subtract 3x from each side.
$5 > -x$	Simplify.
$-5 < x$	Divide each side by −1. Reverse the inequality symbol.

Quick Check

Find each measure.

1. BC

2. $m\angle RST$

3. Two sides of a right triangular flower bed are 7 feet long each. What is the length of the third side to the nearest foot?

Solve each inequality.

4. $x + 13 < 41$

5. $6x + 9 < 7x$

6. $x - 6 > 2x$

7. $8x + 15 > 9x - 26$

How Did You Do?

Which exercises did you answer correctly in the Quick Check? Shade those exercise numbers below.

① ② ③ ④ ⑤ ⑥ ⑦

Perpendicular Bisectors

Learn Perpendicular Bisectors of Segments

A **perpendicular bisector** is a line, segment, or ray that passes through the midpoint of a segment and is perpendicular to that segment.

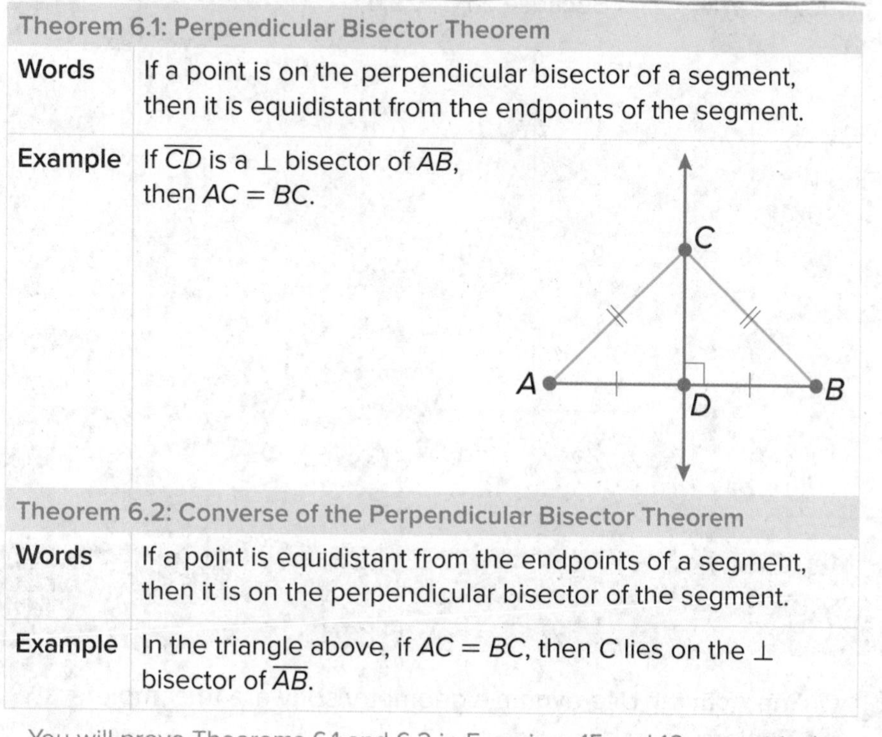

Theorem 6.1: Perpendicular Bisector Theorem	
Words	If a point is on the perpendicular bisector of a segment, then it is equidistant from the endpoints of the segment.
Example	If $\overline{CD}$ is a $\perp$ bisector of $\overline{AB}$, then $AC = BC$.

Theorem 6.2: Converse of the Perpendicular Bisector Theorem	
Words	If a point is equidistant from the endpoints of a segment, then it is on the perpendicular bisector of the segment.
Example	In the triangle above, if $AC = BC$, then C lies on the $\perp$ bisector of $\overline{AB}$.

You will prove Theorems 6.1 and 6.2 in Exercises 15 and 16, respectively.

Example 1 Use the Perpendicular Bisector Theorem

Find *EF*.

$\overleftrightarrow{FG}$ is the perpendicular bisector of $\overline{EH}$.

$EF = HF$ Perpendicular Bisector Theorem

$4a - 15 =$ ____ Substitution

$a =$ ____ Solve.

So, $EF = 4($__$) - 15$ or ____.

E, G, F, H, $4a - 15$, $a + 12$

Check

Find RT.

$RT =$ ____

Q, S, R, T, $2x + 3$, $4x - 7$

Today's Goals

- Prove theorems and solve problems about perpendicular bisectors of line segments.
- Prove theorems and apply geometric methods to solve design problems using the perpendicular bisectors of triangles.

Today's Vocabulary

perpendicular bisector

concurrent lines

point of concurrency

circumcenter

Think About It!

Is your answer reasonable? Explain.

Go Online

You can complete an Extra Example online.

Your Notes

Example 2 Use the Converse of the Perpendicular Bisector Theorem

Find *XY*.

Because $WX = WZ$ and $\overleftrightarrow{WY} \perp \overline{XZ}$, $\overleftrightarrow{WY}$ is the perpendicular bisector of $\overline{XZ}$ by the Converse of the Perpendicular Bisector Theorem. By the definition of segment bisector, $XY =$ ______. Because $ZY = 22.4$, $XY =$ ______.

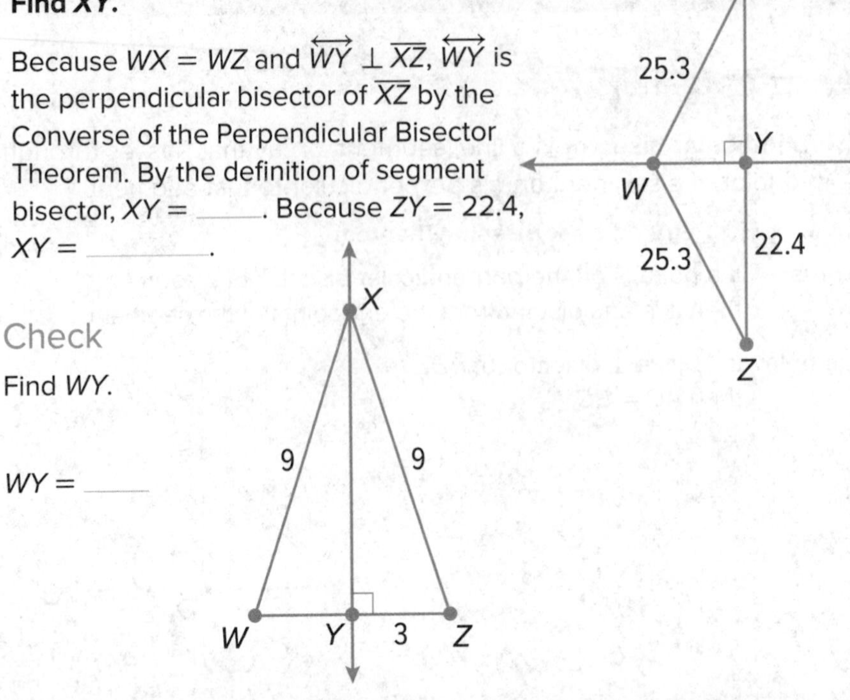

Check

Find *WY*.

$WY =$ ______

Explore Relationships Formed by Perpendicular Bisectors

Online Activity Use dynamic geometry software to complete the Explore.

INQUIRY What relationships exist among the perpendicular bisectors in triangles?

Go Online
You may want to complete the Concept Check to check your understanding.

Learn Perpendicular Bisectors of Triangles

When three or more lines intersect at a common point, the lines are called **concurrent lines**. The point of intersection of concurrent lines is called the **point of concurrency**.

A triangle has three sides, so it also has three perpendicular bisectors. These bisectors are concurrent lines. The point of concurrency of the perpendicular bisectors of the sides of a triangle is called the **circumcenter** of the triangle.

Theorem 6.3: Circumcenter Theorem

The perpendicular bisectors of a triangle intersect at a point called the *circumcenter* that is equidistant from the vertices of the triangle.

Go Online You can complete an Extra Example online.

Study Tip

Circum – The prefix *circum* – means about or around. The circumcenter is the center of a circle around a triangle that contains the vertices of the triangle.

Go Online
A proof of Theorem 6.3 is available.

Example 3 Use the Circumcenter Theorem

Find BF if D is the circumcenter of $\triangle ABC$, $AC = 9$, $DE = 1.83$, and $DF = 1.53$.

D is the circumcenter of $\triangle ABC$, so $\overline{DE}$, $\overline{DF}$, and $\overline{DG}$ are the perpendicular bisectors of the triangle. Because $\overline{DE}$ bisects $\overline{AC}$, $EC = \frac{1}{2}AC$. So, $EC = \frac{1}{2}(\_\_)$ or _____.

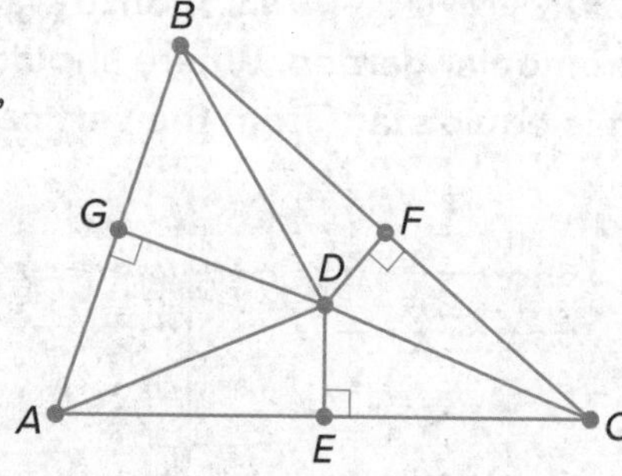

Use the Pythagorean Theorem to find DC.

$c^2 = a^2 + b^2$	Pythagorean Theorem
$DC^2 = 1.83^2 + 4.5^2$	Substitution
$DC^2 = 3.35 + 20.25$	Simplify.
$DC^2 =$ _____	Add.
$DC \approx \pm$ _____	Take the square root of each side.

Because length cannot be negative, use the positive square root, _____.

By the Circumcenter Theorem, $DC = DB$. So, $DB \approx 4.86$.

Because $DB = 4.86$ and $DF = 1.53$, we can use the Pythagorean Theorem to find BF.

$a^2 + b^2 = c^2$	Pythagorean Theorem
$1.53^2 + BF^2 = 4.86^2$	Substitution
_____ $+ BF^2 =$ _____	Simplify.
$BF^2 =$ _____	Subtract.
$BF \approx \pm$ _____	Take the square root of each side.

The length of the segment must be positive, so $BF =$ _____.

Check

Find SV if S is the circumcenter of ΔQPR, $QR = 7.79$, and $PS = 4.25$. If necessary, round your answer to the nearest tenth.

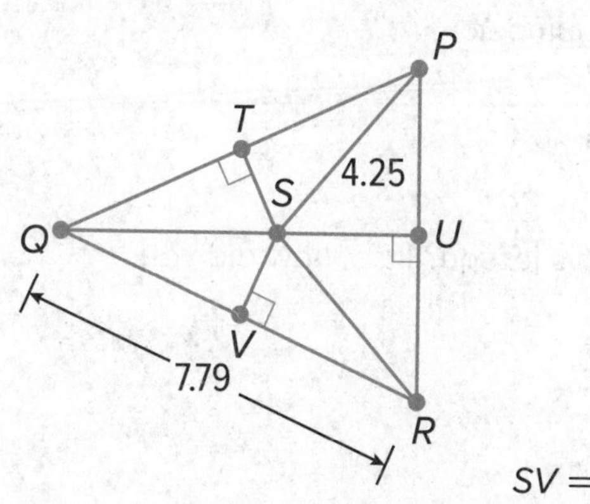

$SV =$ _____

Go Online You can complete an Extra Example online.

Problem-Solving Tip

Make a Plan Before solving for unknown measures, analyze the information you are given, develop a plan, and determine the theorems you will need to apply to find a specific measure.

Talk About It!

Determine whether the statement is *sometimes*, *always*, or *never* true. Justify your argument.

The perpendicular bisectors of a triangle intersect at a point that is equidistant from the sides of the triangle.

Think About It!

If Alonzo is installing the fountain in his backyard garden, what unit of measure is most appropriate for the dimensions of the garden?

Example 4 Use Perpendicular Bisector in Design Problems

GARDEN FOUNTAINS **Alonzo wants to install a fountain in h triangular garden. Where should Alonzo place the fountain so it is equidistant from the vertices of the garden?**

(7, 5)

0)

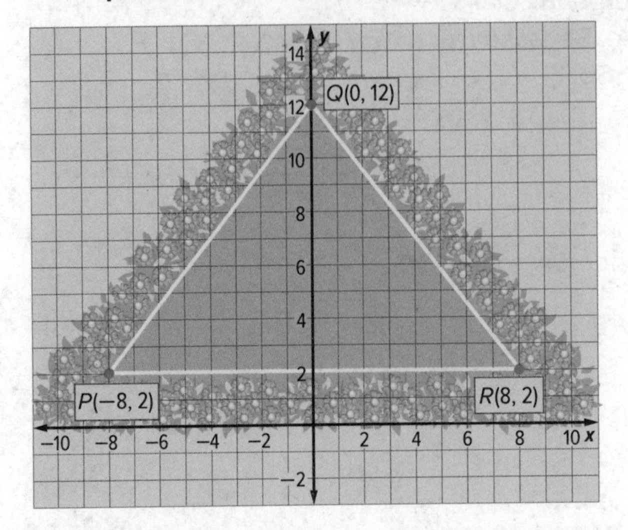

The circumcenter of a triangle is equidistant from the vertices of the triangle. Use a compass and straightedge or dynamic geometry software to construct the perpendicular bisectors of $\triangle PQR$. Find the intersection point of the perpendicular bisectors to determine where Alonzo should install the fountain.

The intersection of the perpendicular bisectors appears to occur at (0, 3.8). So, Alonzo should install the fountain at (0, 3.8).

Check

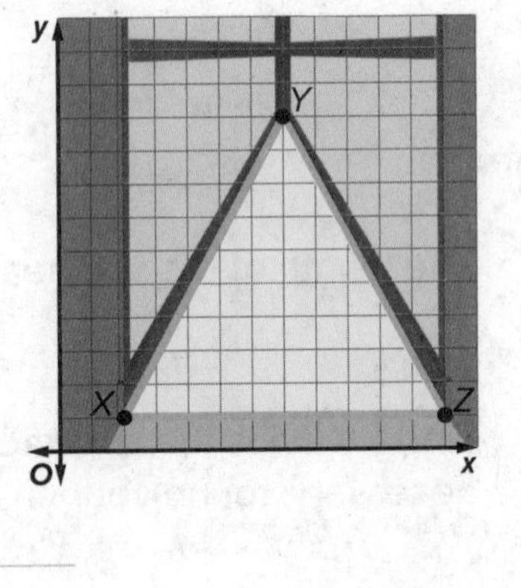

HOME IMPROVEMENT Lana wants to install a skylight in her bedroom. The section of roof where she plans to install the skylight has an incline and is triangular. The vertices of the roof are at $X(2, 1)$, $Y(7, 10)$, and $Z(12, 1)$. Where should the center of the skylight be located so that it is equidistant from the vertices of the roof? If necessary, round your answer to the nearest whole number. ______

Pause and Reflect

Did you struggle with anything in this lesson? If so, how did you deal with it?

Record your observations here.

Go Online

You may want to complete the construction activities for this lesson.

Name ______________________ Period ________ Date ________

PROOF **Use the figure to complete the following proofs.**

15. Write a paragraph proof of the Perpendicular Bisector Theorem (Theorem 6.1).

Given: $\overleftrightarrow{CD}$ is the perpendicular bisector of $\overline{AB}$.

Prove: C is equidistant from A and B.

16. Write a two-column proof of the Converse of the Perpendicular Bisector Theorem (Theorem 6.2).

Given: $\overline{CA} \cong \overline{CB}$, $\overline{AD} \cong \overline{BD}$

Prove: C and D are on the perpendicular bisector of $\overline{AB}$.

Find the coordinates of the circumcenter of the triangle with the given vertices. Explain.

17. $A(0, 0)$, $B(0, 6)$, $C(10, 0)$

18. $J(5, 0)$, $K(5, -8)$, $L(0, 0)$

19. Consider $\overline{CD}$. Describe the set of all points in space that are equidistant from C and D.

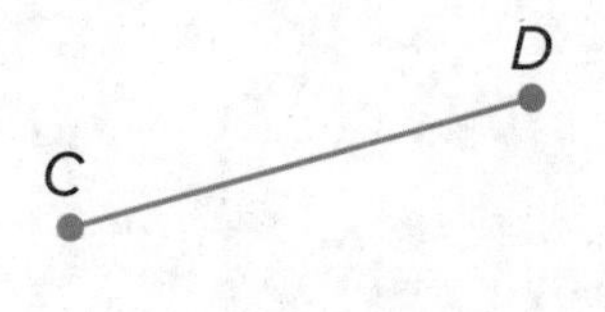

20. YARDWORK Martina has a front yard with three trees. The figure shows the locations of the trees on a coordinate plane. Martina would like to place an inground sprinkler at a location that is the same distance from all three trees. At what point on the coordinate grid should Martina place the sprinkler?

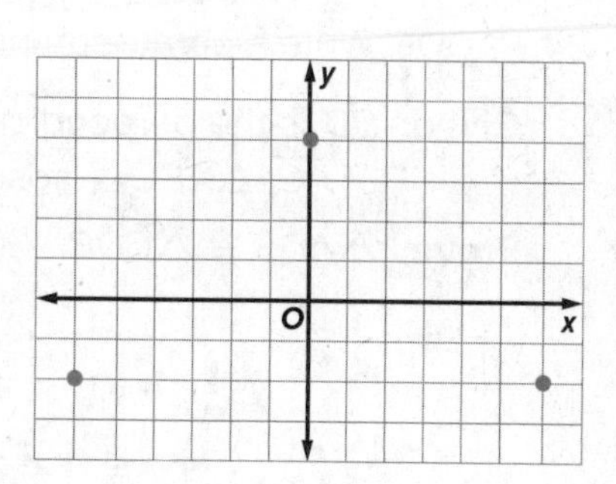

21. CREATE On a baseball diamond, home plate and second base lie on the perpendicular bisector of the line segment that joins first and third base. First base is 90 feet from home plate. How far is it from third base to home plate? Sketch a baseball diamond, labeling home plate as point A, first base as B, second base as C, and third base as D. Label the intersection of $\overline{AC}$ and $\overline{BD}$ as E. Using the Perpendicular Bisector Theorem, determine how far it is from third base to home plate. Describe your conclusion in the context of the situation.

22. FIND THE ERROR Thiago says that from the information supplied in the diagram, he can conclude that K is on the perpendicular bisector of $\overline{LM}$. Caitlyn disagrees. Is either of them correct? Explain your reasoning.

L K J M

23. PROOF Write a two-column proof.

Given: Plane Y is a perpendicular bisector of $\overline{DC}$.

Prove: $\angle ADB \cong \angle ACB$

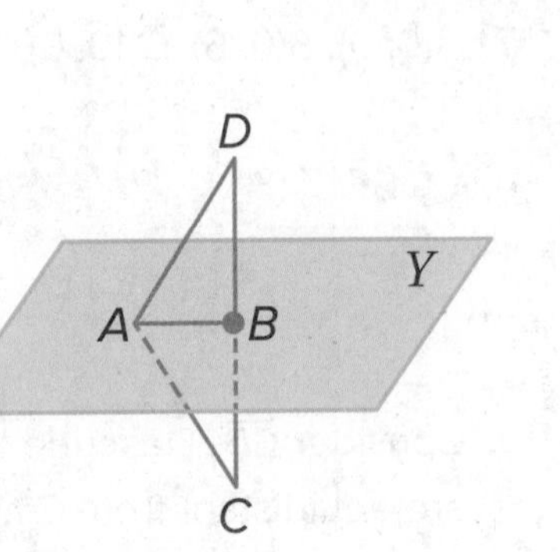

24. PROOF Write a paragraph proof.

Given: $\overline{BD}$ is the perpendicular bisector of $\overline{AC}$. $\triangle ABC$ is isosceles with base $\overline{AC}$.

Prove: $\triangle ADB \cong \triangle CDB$

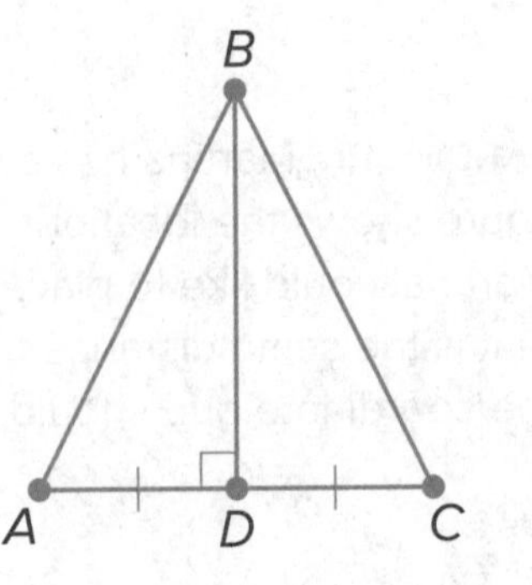

Angle Bisectors

Learn Angle Bisectors

You may recall that an angle bisector divides an angle into two congruent angles. The angle bisector can be a line, segment, or ray.

Theorem 6.4: Angle Bisector Theorem

Words	If a point is on the bisector of an angle, then it is equidistant from the sides of the angle.
Example	If $\overrightarrow{BF}$ bisects $\angle DBE$, $\overline{FD} \perp \overrightarrow{BD}$, and $\overline{FE} \perp \overrightarrow{BE}$, then $DF = FE$. 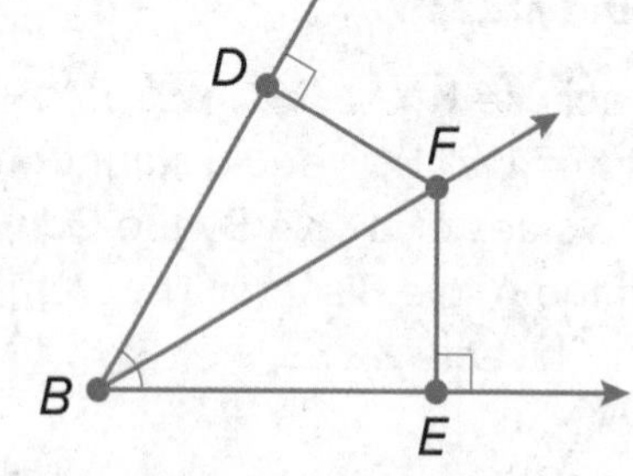

Theorem 6.5: Converse of the Angle Bisector Theorem

Words	If a point in the interior of an angle is equidistant from the sides of the angle, then it is on the bisector of the angle.
Example	If $\overline{FD} \perp \overrightarrow{BD}$, $\overline{FE} \perp \overrightarrow{BE}$, and $DF = FE$, then $\overrightarrow{BF}$ bisects $\angle DBE$. 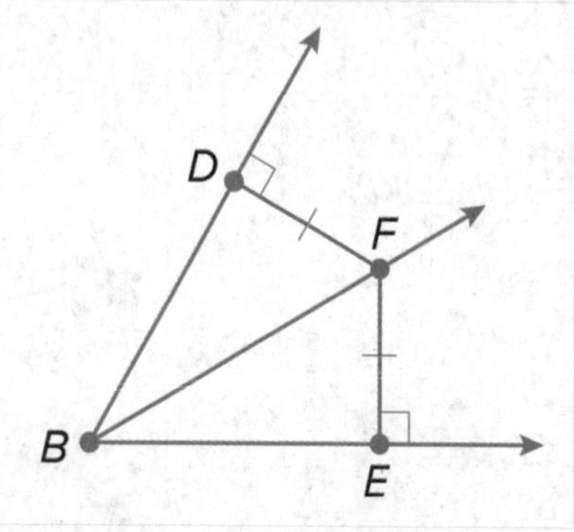

You will prove Theorems 6.4 and 6.5 in Exercises 17 and 18, respectively.

Example 1 Use the Angle Bisector Theorem

Find *QT*.

$\overrightarrow{RT}$ is the angle bisector of $\angle QRS$.

_____	$= ST$	Angle Bisector Theorem
_____	= _____	Substitution
$-5x$	= _____	Simplify.
x	= _____	Simplify.

So, $QT = 4($_____$) + 8$ or _____.

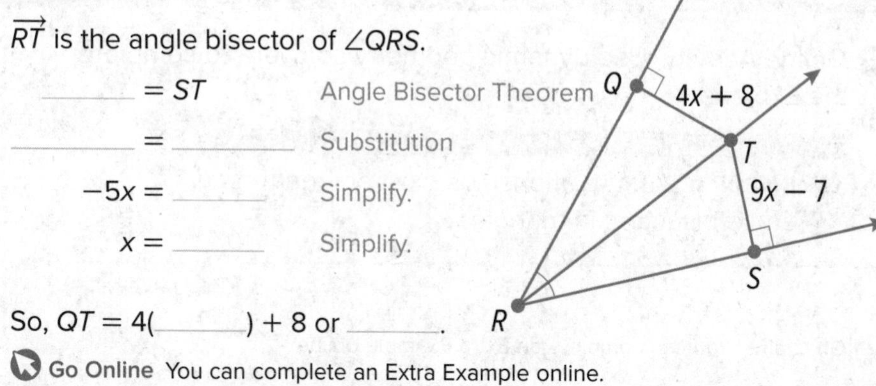

Go Online You can complete an Extra Example online.

Today's Goals

- Prove theorems and solve problems about angle bisectors.
- Prove theorems and apply geometric methods to solve design problems using the angle bisectors of triangles.

Today's Vocabulary

incenter

Study Tip

Angle Bisector In the figure below, there is not enough information to conclude that $\overrightarrow{BD}$ bisects $\angle ABC$. You must also know that $\overline{AD} \perp \overrightarrow{BA}$ and $\overline{CD} \perp \overrightarrow{BC}$.

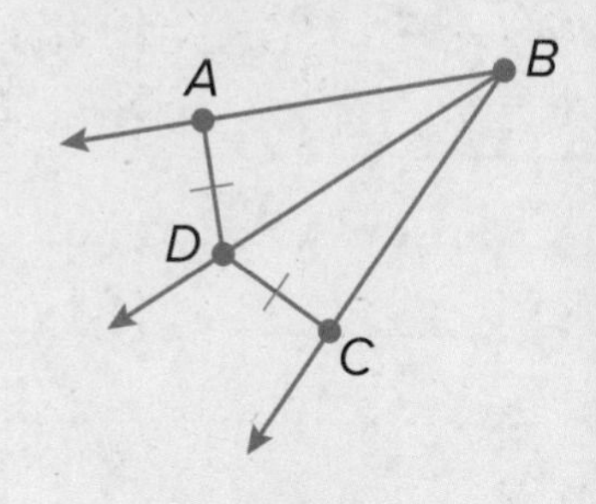

Your Notes

Check

Find SP.

$SP =$ ______

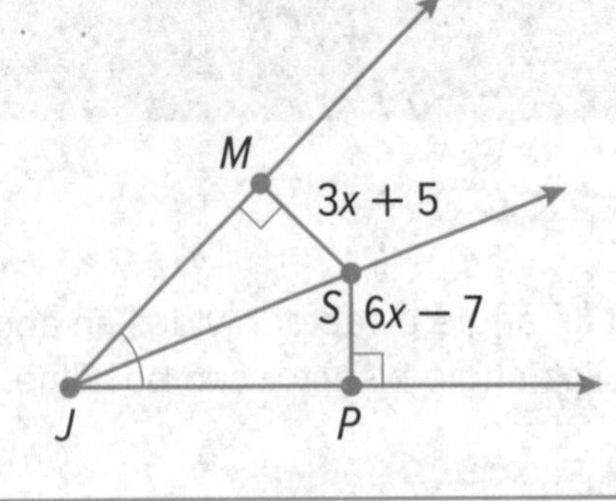

Example 2 Use the Converse of the Angle Bisector Theorem

Find $m\angle ZYW$.

Because $\overline{WX} \perp \overrightarrow{YX}$, $\overline{WZ} \perp \overrightarrow{YZ}$, and $WX = WZ$, W is equidistant from the sides of $\angle XYZ$. By the Converse of the Angle Bisector Theorem, $\overrightarrow{YW}$ bisects $\angle XYZ$.

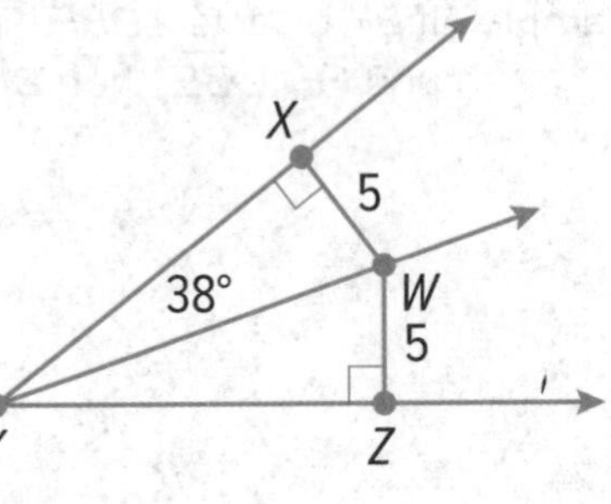

$\angle$ ______ $\cong \angle XYW$ — Definition of angle bisector

$m\angle$ ______ $= m\angle XYW$ — Definition of congruent angles

$m\angle ZYW =$ ______ — Substitution

Check

Find $m\angle JKL$.

$m\angle JKL =$ ______

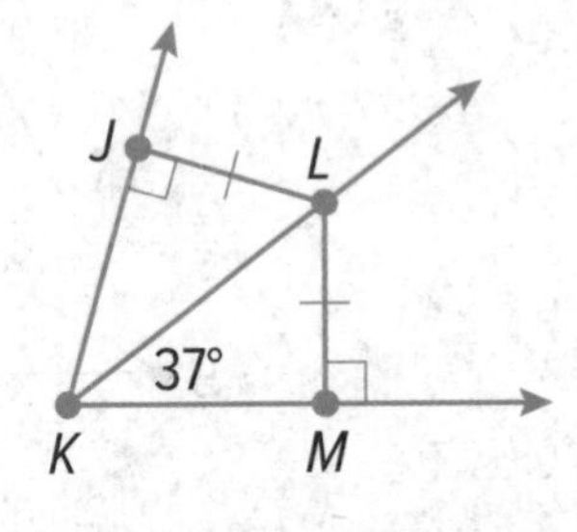

Explore Relationships Formed by Angle Bisectors

Online Activity Use dynamic geometry software to complete the Explore.

INQUIRY What relationships exist among the angle bisectors in triangles?

Go Online You can complete an Extra Example online.

Learn Angle Bisectors of Triangles

Because a triangle has three angles, it also has three angle bisectors. The angle bisectors of a triangle are concurrent, and their point of concurrency is called the **incenter** of a triangle.

Theorem 6.6: Incenter Theorem	
Words	The angle bisectors of a triangle intersect at a point called the incenter that is equidistant from the sides of the triangle.
Example	If P is the incenter of $\triangle ABC$, then $PD = PE = PF$. 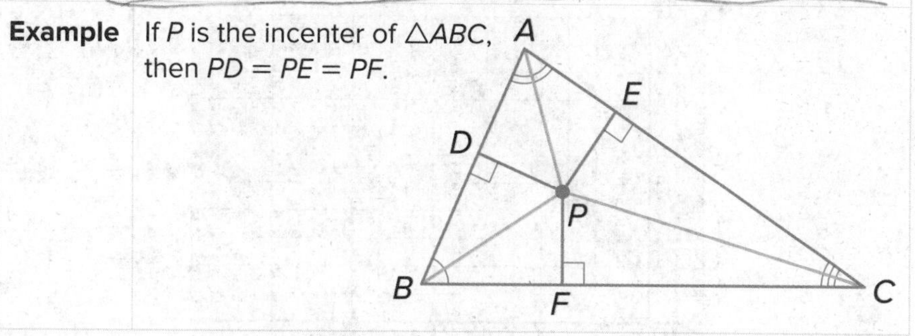

You will prove Theorem 6.6 in Exercise 19.

Think About It!
Determine whether the statement is *sometimes*, *always*, or *never* true. Justify your argument.
The angle bisectors of a triangle intersect at a point that is equidistant from the vertices of the triangle.

Example 3 Use the Incenter Theorem

P is the incenter of $\triangle XYZ$. Find $m\angle LZP$.

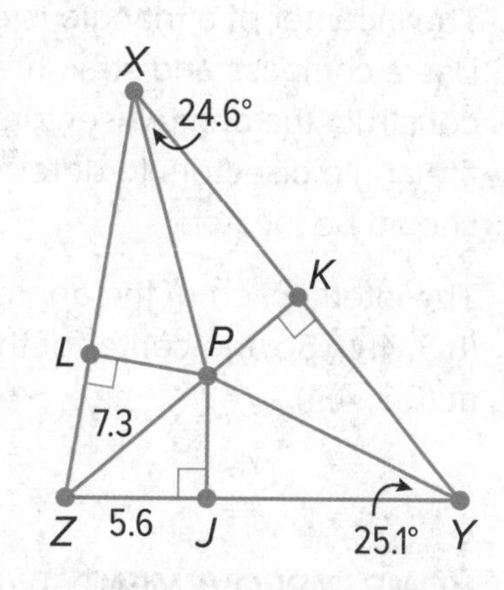

P is the incenter of $\triangle XYZ$, so $\overline{XP}$ bisects $\angle LXK$, $\overline{YP}$ bisects $\angle KYJ$, and $\overline{ZP}$ bisects $\angle LZJ$. Thus, $m\angle LXK = 2m\angle PXK$. So, $m\angle LXK = 2(24.6)$ or $49.2°$.

Likewise, $m\angle KYJ = 2m\angle PYJ$, so $m\angle KYJ = 2(25.1)$ or $50.2°$.

Find $m\angle LZJ$ using the Triangle Angle-Sum Theorem.

$m\angle LXK + m\angle KYJ + m\angle LZJ = 180°$ Triangle Angle-Sum Theorem

____ + ____ + $m\angle LZJ = 180°$ Substitution

$m\angle LZJ =$ ____ Simplify.

Because $\overline{ZP}$ bisects $\angle LZJ$, $2m\angle LZP = m\angle LZJ$. This means that

$m\angle LZP = \frac{1}{2}m\angle LZJ$, so

$m\angle LZP = \frac{1}{2}($____$)$ or ____.

Check

Find each measure if J is the incenter of $\triangle ABC$.

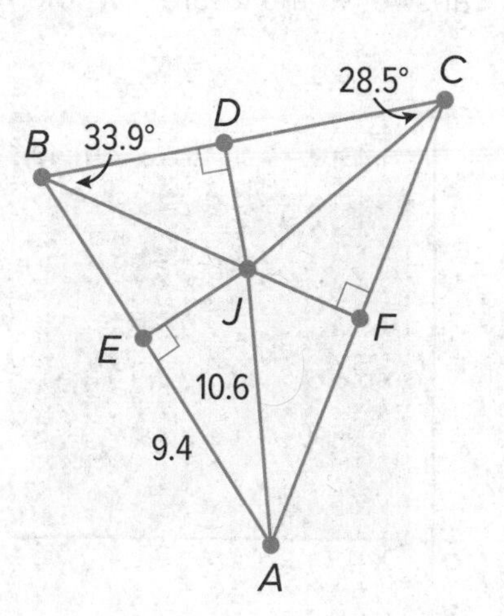

$JF =$ ____

$m\angle JAC =$ ____

Go Online You can complete an Extra Example online.

Example 4 Using Angle Bisectors in Design Problems

COURTYARD A new art sculpture will be placed in the courtyard of a high school. The entrances to the courtyard are located at points *A*, *B*, and *C*. Find the location of the center of the sculpture so that it is equidistant from the sides of the courtyard.

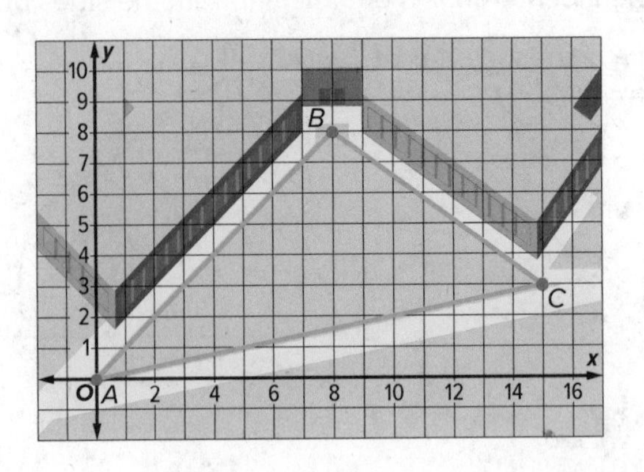

The incenter of a triangle is equidistant from the sides of the triangle. Use a compass and straightedge or dynamic geometry software to construct the angle bisectors of $\triangle ABC$. Find the intersection point of the angle bisectors to determine where the center of the sculpture should be located.

The intersection of the angle bisectors appears to be at about (8.3, 4.4). So, the center of the sculpture should be located at (8.3, 4.4).

Check

HOME IMPROVEMENT Tyrice's parents want to install a hot tub on the back deck of their house. The vertices of the deck are located at points *X*, *Y*, and *Z*. Find the location of the center of the hot tub so it is equidistant from the edges of the deck. If necessary, round your answer to the nearest whole number.

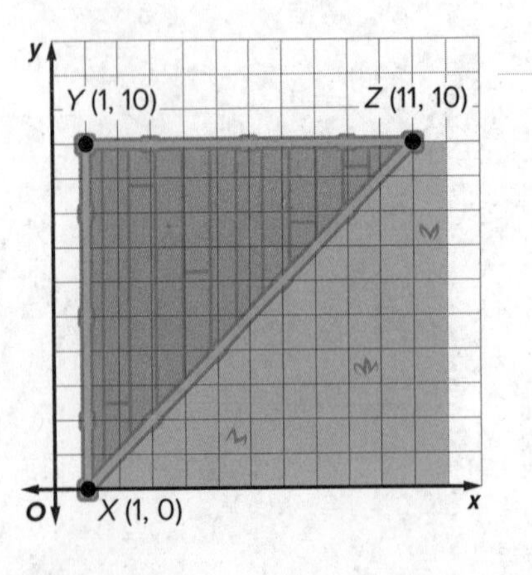

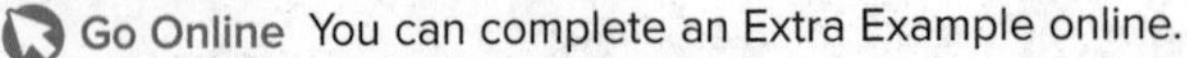

Go Online You can complete an Extra Example online.

Go Online
You may want to complete the construction activities for this lesson.

17. PROOF Write a paragraph proof of the Angle Bisector Theorem.

Given: $\overline{BD}$ is the angle bisector of $\angle ABC$.

Prove: D is equidistant from $\overline{AB}$ and $\overline{BC}$.

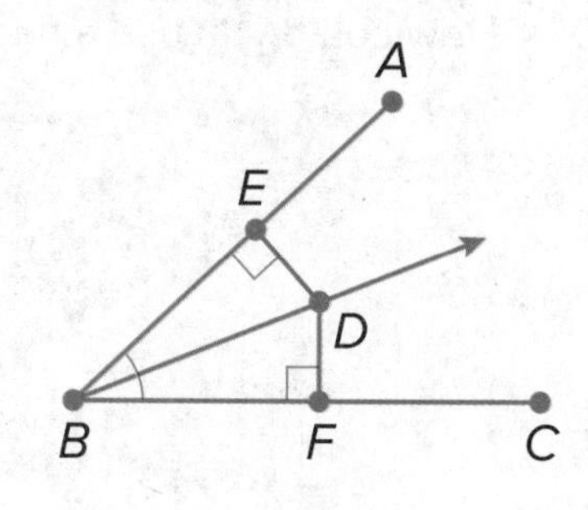

18. PROOF Write a paragraph proof of the Converse of the Angle Bisector Theorem.

Given: P is in the interior of $\angle BAC$, and P is equidistant from $\overline{AB}$ and $\overline{AC}$ at D and E, respectively.

Prove: $\overline{AP}$ is the angle bisector of $\angle BAC$.

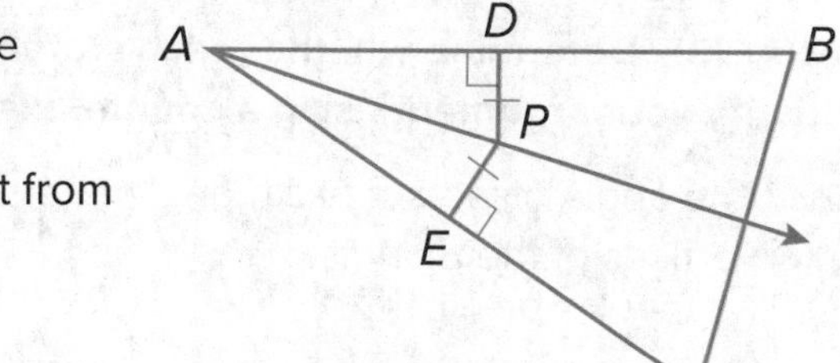

19. PROOF Write a two-column proof of the Incenter Theorem.

Given: $\triangle ABC$ has angle bisectors $\overline{AD}$, $\overline{BE}$, and $\overline{CF}$.
$\overline{KP} \perp \overline{AB}$, $\overline{KQ} \perp \overline{BC}$, $\overline{KR} \perp \overline{AC}$

Prove: $KP = KQ = KR$

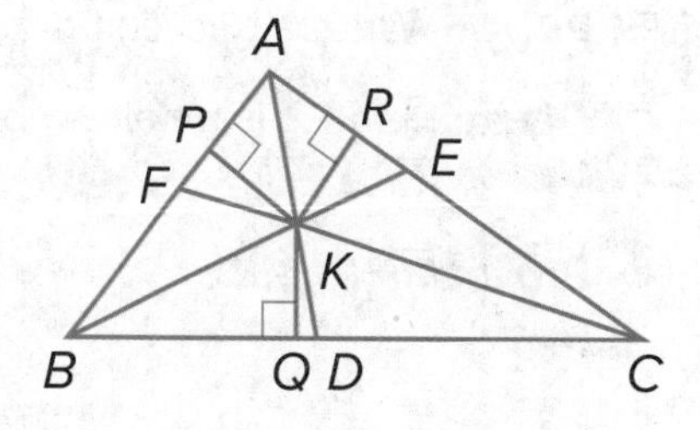

20. CONSTRUCT ARGUMENTS State whether the following sentence is *sometimes, always,* or *never* true. Justify your argument.

The three angle bisectors of a triangle intersect at a point in the exterior of the triangle.

21. USE TOOLS Construct the incenter of the triangle shown. How could you use a ruler to verify your construction?

22. CREATE Draw a triangle with an incenter located inside the triangle and a circumcenter located outside. Justify your drawing using a straightedge and a compass to find both points of concurrency.

ANALYZE **Determine whether each statement is *sometimes*, *always*, or *never* true. Justify your argument using a counterexample or proof.**

23. The angle bisectors of a triangle intersect at a point that is equidistant from the vertices of the triangle.

24. In an isosceles triangle, the perpendicular bisector of the base is also the angle bisector of the opposite vertex.

25. PROOF Write a two-column proof.

Given: Plane Z is an angle bisector of $\angle KJH$.
$\overline{KJ} \cong \overline{HJ}$

Prove: $\overline{MH} \cong \overline{MK}$

K
M
J
Z
H

26. WRITE Compare and contrast the perpendicular bisectors and angle bisectors of a triangle. How are they alike? How are they different? Be sure to compare their points of concurrency.

27. WRITE Write a biconditional statement for Theorem 6.4 and its converse.

Medians and Altitudes of Triangles

Explore Centroid of a Triangle

Online Activity Use dynamic geometry software to complete the Explore.

INQUIRY How is the location of the centroid related to the medians of a triangle?

Learn Medians of Triangles

In a triangle, a **median** is a line segment with endpoints that are a vertex of the triangle and the midpoint of the side opposite the vertex.

Every triangle has three medians that are concurrent. The point of concurrency of the medians of a triangle is called the **centroid**, and it is always inside the triangle.

Theorem 6.7: Centroid Theorem

The medians of a triangle intersect at a point called the centroid that is two-thirds of the distance from each vertex to the midpoint of the opposite side.

You will prove Theorem 6.7 in Exercise 22.

All polygons have a balancing point or *center of gravity*. This is the point at which the weight of a region is evenly dispersed and all sides of the region are balanced. The centroid is the center of gravity for a triangular region.

Example 1 Use the Centroid Theorem

In $\triangle ABC$, P is the centroid and $BL = 6$. Find BP and PL.

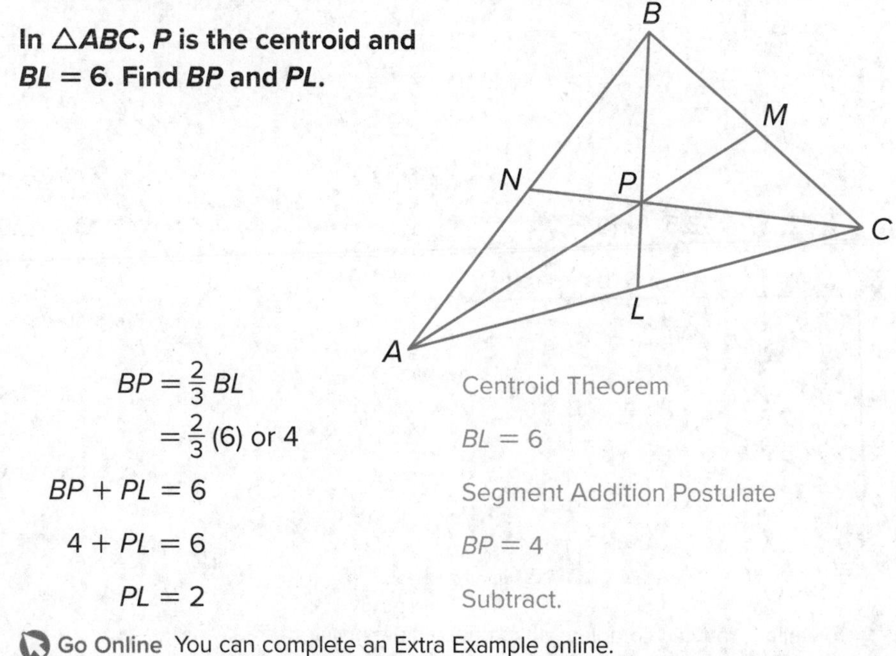

$BP = \frac{2}{3}BL$ Centroid Theorem

$= \frac{2}{3}(6)$ or 4 $BL = 6$

$BP + PL = 6$ Segment Addition Postulate

$4 + PL = 6$ $BP = 4$

$PL = 2$ Subtract.

Go Online You can complete an Extra Example online.

Today's Goals

- Solve problems by applying the Centroid Theorem.
- Use altitudes and the slope criteria for perpendicular lines to determine the coordinates of the orthocenters of triangles on the coordinate plane.

Today's Vocabulary

median

centroid

altitude of a triangle

orthocenter

Think About It!

How could you find the coordinates of the centroid of $\triangle PQR$?

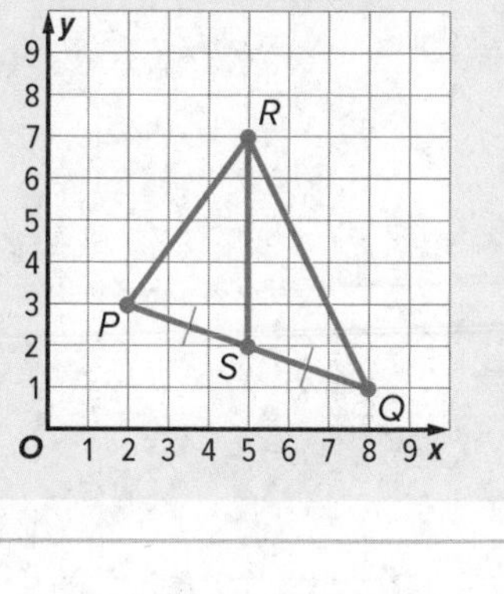

Your Notes

Check

In $\triangle ABC$, Q is the centroid and $BE = 9$. Find BQ and QE.

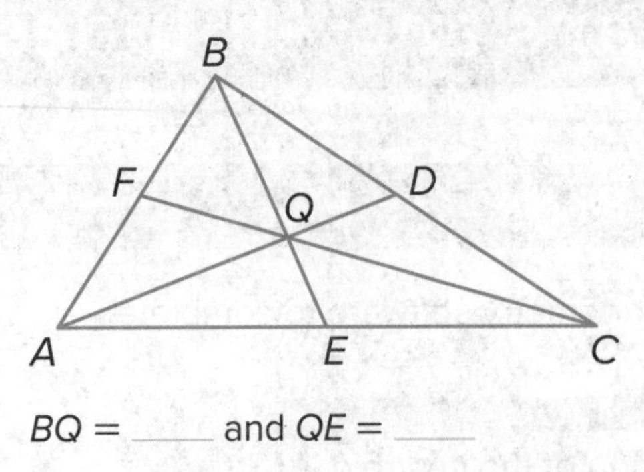

$BQ =$ ____ and $QE =$ ____

Example 2 Apply the Centroid Theorem

In $\triangle LMN$, $PY = 7$. Find LP.

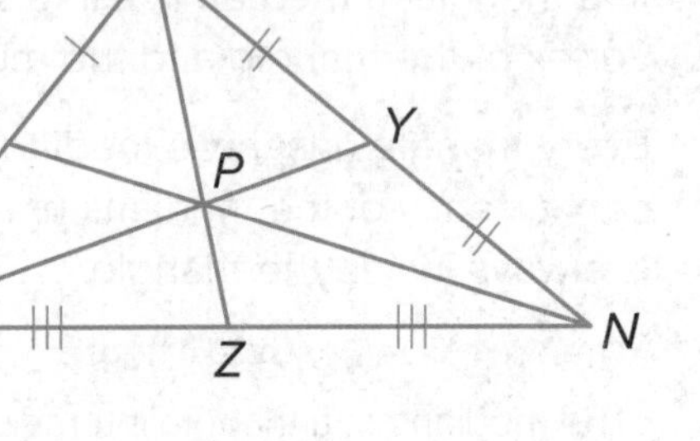

Because $\overline{LX} \cong \overline{XM}$, X is the midpoint of $\overline{LM}$, and $\overline{NX}$ is a median of $\triangle LMN$. Likewise, Y and Z are the midpoints of $\overline{MN}$ and $\overline{LN}$ respectively, so $\overline{LY}$ and $\overline{MZ}$ are also medians of $\triangle LMN$. Therefore, point P is the centroid of $\triangle LMN$.

$LP = \frac{2}{3}LY$	Centroid Theorem
$LP = \frac{2}{3}(LP + PY)$	Segment Addition and Substitution
$LP = \frac{2}{3}(LP + 7)$	$PY = 7$
$LP = \frac{2}{3}LP + \frac{14}{3}$	Distributive Property
$\frac{1}{3}LP = \frac{14}{3}$	Subtract $\frac{2}{3}LP$ from each side.
$LP = 14$	Multiply each side by 3.

Talk About It!

How can you find LP without solving an equation? Justify your argument.

Check

In $\triangle XYZ$, $SP = 3.5$. Find PZ.

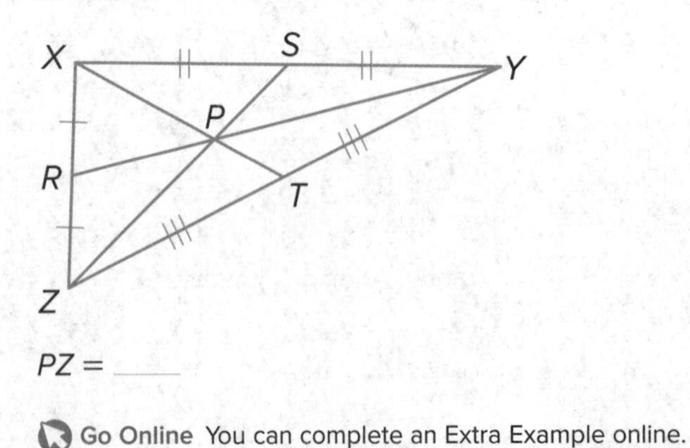

$PZ =$ ____

Go Online You can complete an Extra Example online.

Apply Example 3 Find a Centroid on the Coordinate Plane

CHIMES **Lashaya needs to hang a wind chime with a single piece of cord. The pipes of the wind chime are attached to a triangular platform. When the platform is placed on a coordinate plane, the vertices of the triangle are located at (1, 1), (11, 5), and (7, 10). What are the coordinates of the point where the cord should be attached to the platform so the wind chime stays balanced?**

1 What is the task?

Describe the task in your own words. Then list any questions that you may have. How can you find answers to your questions?

2 How will you approach the task? What have you learned that you can use to help you complete the task?

3 What is your solution?

Use your strategy to solve the problem.

Graph the triangular platform and the medians of $\triangle ABC$.

The midpoint of $\overline{AB}$ ______.

The centroid of $\triangle ABC$ is ______.

y

O x

4 How can you know that your solution is reasonable?

Write About It! Write an argument that can be used to defend your solution.

Go Online You can complete an Extra Example online.

Think About It!

What assumption did you make while solving this problem?

Go Online

You may want to complete the construction activities for this lesson.

Go Online to practice what you've learned about points of concurrency in triangles in the Put It All Together over Lessons 6-1 through 6-3.

Learn Altitudes of Triangles

An **altitude of a triangle** is a segment from a vertex of the triangle to the line containing the opposite side and perpendicular to that side. An altitude can lie in the interior, in the exterior, or on the side of the triangle.

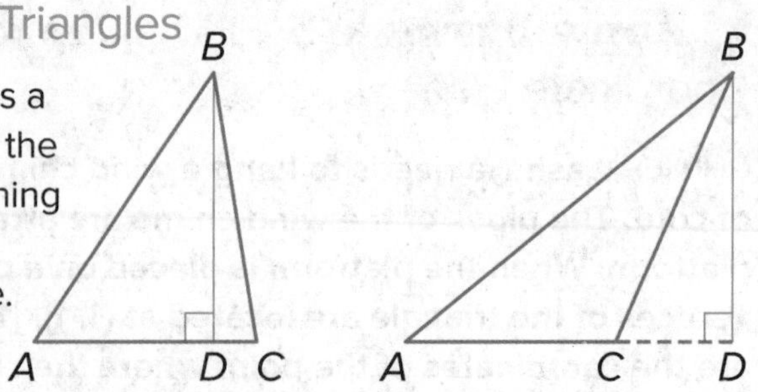

$\overline{BD}$ is an altitude from B to the line containing $\overline{AC}$.

Every triangle has three altitudes. If extended, the altitudes of a triangle are concurrent. The point of concurrency is called the **orthocenter**.

The lines containing altitudes $\overline{AF}$, $\overline{CD}$, and $\overline{BG}$ intersect at P, the orthocenter of $\triangle ABC$.

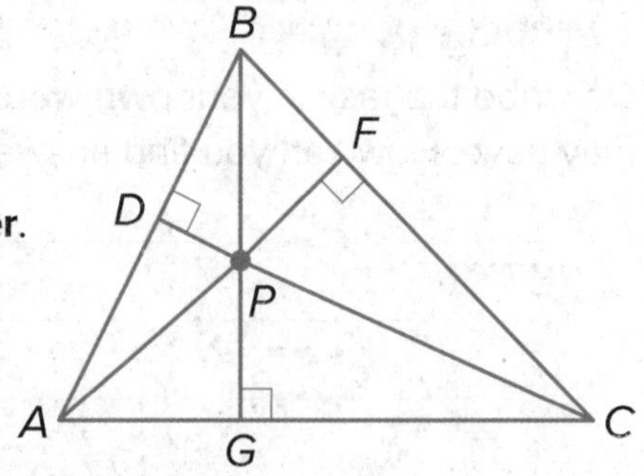

Think About It!

Where do you think the orthocenter will be located in an obtuse triangle and a right triangle?

Example 4 Find an Orthocenter on the Coordinate Plane

The vertices of $\triangle ABC$ are $A(4, 0)$, $B(-2, 4)$, and $C(0, 6)$. Find the coordinates of the orthocenter of $\triangle ABC$.

Step 1 Graph $\triangle ABC$.

To find the orthocenter of $\triangle ABC$, find the point where two of the three altitudes intersect.

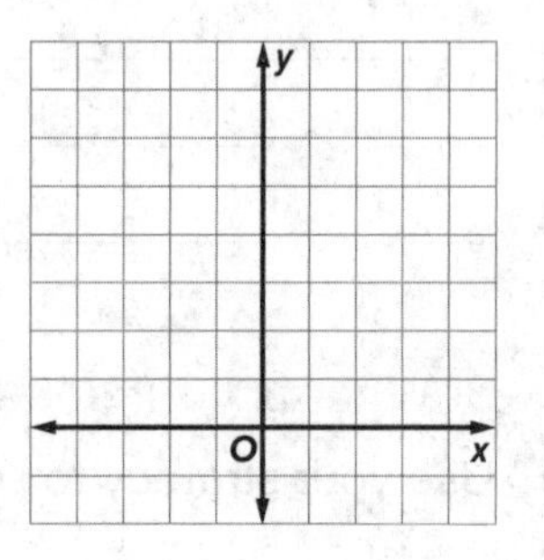

Step 2 Find equations of the altitudes.

Find an equation of the altitude from B to $\overline{AC}$. The slope of $\overline{AC}$ is ____, so the slope of the altitude, which is perpendicular to $\overline{AC}$, is ____. Use the slope and point B on the altitude to find the equation of the line.

$y =$ ________

Find an equation of the altitude from A to $\overline{BC}$. The slope of $\overline{BC}$ is ____, so the slope of the altitude is ____. Use the slope and point A on the altitude to find the equation of the line.

$y =$ ________

Step 3 Solve the system of equations.

$x =$ ____ and $y =$ ____

The coordinates of the orthocenter of $\triangle ABC$ are $\left(-\frac{4}{5}, \frac{24}{5}\right)$.

Check

The vertices of $\triangle FGH$ are $F(-2, 4)$, $G(4, 4)$, and $H(1, -2)$. What are the coordinates of the orthocenter of $\triangle FGH$? If necessary, round your answer to the nearest tenth. ________

Go Online You can complete an Extra Example online.

Math History Minute

Pierre de Fermat (1607–1665) was a French lawyer and mathematician who is considered to be one of the greatest mathematicians of the 17th century. He is best known for his Last Theorem, which wasn't proven until 1995. Fermat is also known for the Fermat point, which minimizes the sum of the distances from the three vertices of a triangle to that point.

Name ______________________ Period __________ Date __________

Practice

Go Online You can complete your homework online.

Examples 1 and 2

In $\triangle CDE$, U is the centroid, $UK = 12$, $EM = 21$, and $UD = 9$. Find each measure.

1. CU

2. MU

3. EU

4. JD

In $\triangle ABC$, $AU = 16$, $BU = 12$, and $CF = 18$. Find each measure.

5. CU

6. AD

7. UF

8. BE

Example 3

Find the coordinates of the centroid of each triangle with the given vertices.

9. $X(-3, 15)$ $Y(1, 5)$, $Z(5, 10)$

10. $S(2, 5)$, $T(6, 5)$, $R(10, 0)$

11. DECORATING Camilla made a collage with pictures of her trip to Europe. She wants to hang the collage from the ceiling in her room so that it is parallel to the ceiling. She draws a model of the collage on a coordinate plane. At what point should she place the string?

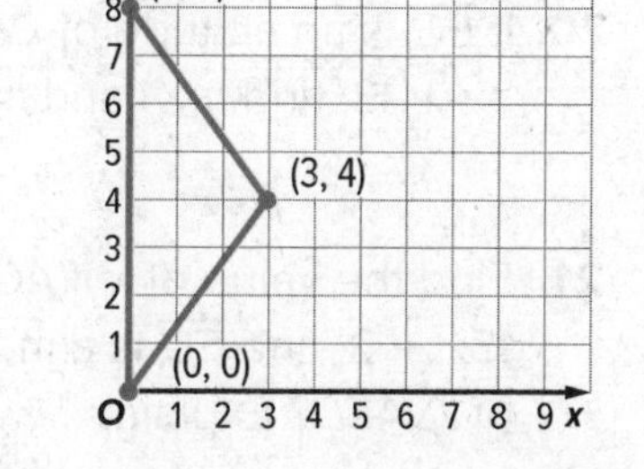

Example 4

Find the coordinates of the orthocenter of the triangle with the given vertices.

12. $J(1, 0)$, $H(6, 0)$, $I(3, 6)$

13. $S(1, 0)$, $T(4, 7)$, $U(8, -3)$

14. $L(8, 0)$, $M(10, 8)$, $N(14, 0)$

15. $D(-9, 9)$, $E(-6, 6)$, $F(0, 6)$

Mixed Exercises

16. REASONING In the figure at the right, if J, P, and L are the midpoints of $\overline{KH}$, $\overline{HM}$, and $\overline{MK}$, respectively, find x, y, and z.

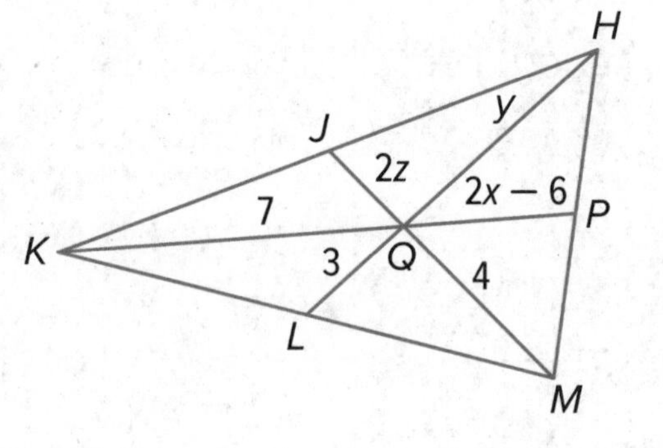

Given $\triangle RST$ with medians $\overline{RM}$, $\overline{SL}$, and $\overline{TK}$, and centroid J, find each value of x.

17. $SL = x(JL)$

18. $JT = x(TK)$

19. $JM = x(RJ)$

For Exercises 20 and 21, refer to the figure at the right.

20. If $\overline{EC}$ is an altitude of $\triangle AED$, $m\angle 1 = 2x + 7$, and $m\angle 2 = 3x + 13$, find $m\angle 1$ and $m\angle 2$.

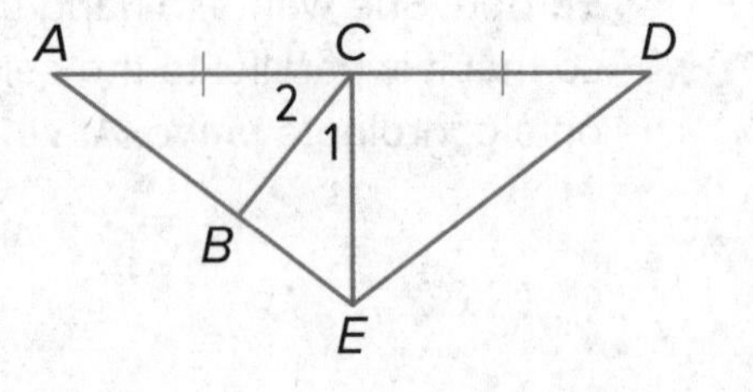

21. Find the value of x if $AC = 4x - 3$, $DC = 2x + 9$, $m\angle ECA = 15x + 2$, and $\overline{EC}$ is a median of $\triangle AED$. Is $\overline{EC}$ also an altitude of $\triangle AED$? Explain.

Name ______________________ Period ____________ Date ______________

22. PROOF Write a coordinate proof to prove the Centroid Theorem.

Given: $\triangle ABC$, medians $\overline{AR}$, $\overline{BS}$, and $\overline{CQ}$

Prove: The medians intersect at point P, and P is two thirds of the distance from each vertex to the midpoint of the opposite side.

(*Hint*: First, find the equations of the lines containing the medians. Then find the coordinates of point P and show that all three medians intersect at point P. Next, use the Distance Formula and multiplication to show $AP = \frac{2}{3}AR$, $BP = \frac{2}{3}BS$, and $CP = \frac{2}{3}CQ$.)

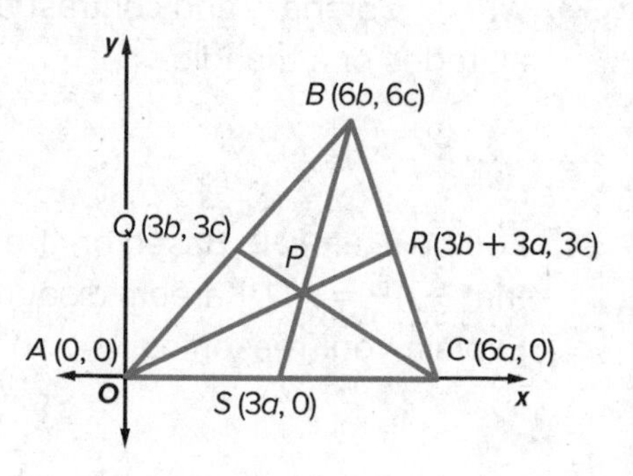

CONSTRUCT ARGUMENTS **Use the given information to determine whether $\overline{LM}$ is a *perpendicular bisector*, *median*, and/or *altitude* of $\triangle JKL$.**

23. $\overline{LM} \perp \overline{JK}$

24. $\triangle JLM \cong \triangle KLM$

25. $\overline{JM} \cong \overline{KM}$

26. $\overline{LM} \perp \overline{JK}$ and $\overline{JL} \cong \overline{KL}$

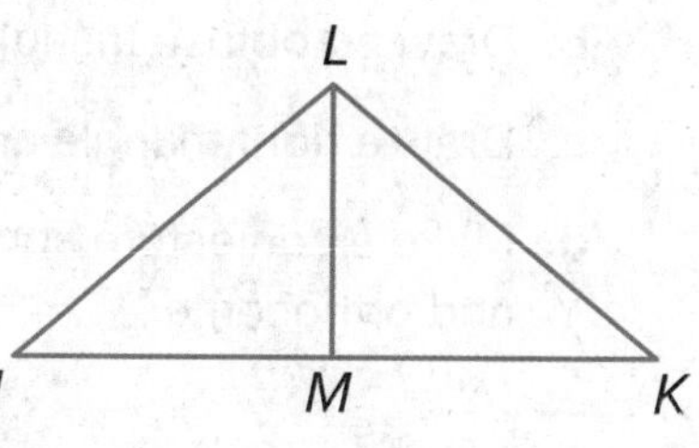

REASONING **In $\triangle JLP$, $m\angle JMP = 3x - 6$, $JK = 3y - 2$, and $LK = 5y - 8$.**

27. If $\overline{JM}$ is an altitude of $\triangle JLP$, find the value of x.

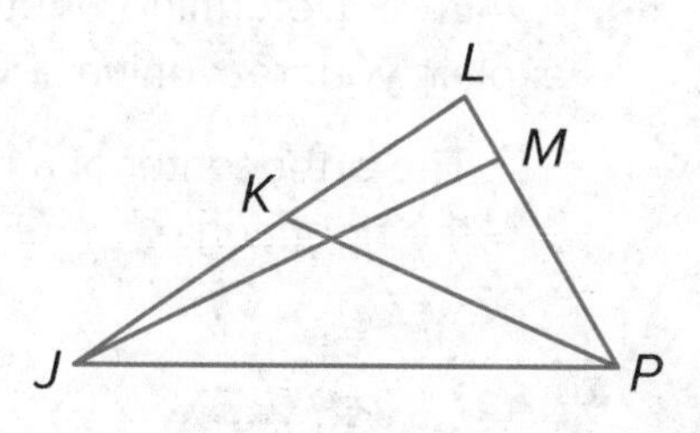

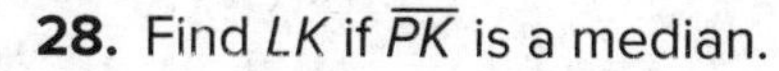

28. Find LK if $\overline{PK}$ is a median.

29. PROOF Write a paragraph proof.

Given: $\triangle XYZ$ Is isosceles.

$\overline{WY}$ bisects $\angle Y$.

Prove: $\overline{WY}$ is a median.

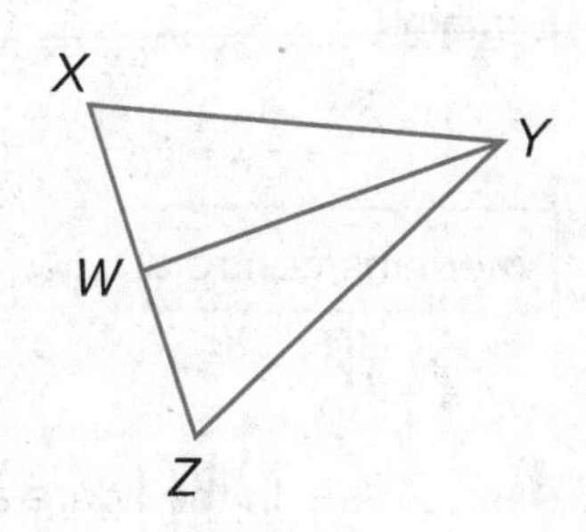

30. WRITE Compare and contrast the perpendicular bisectors, medians, and altitudes of a triangle.

31. FIND THE ERROR Based on the figure at the right, Laura says that $\frac{2}{3}AP = AD$. Kareem disagrees. Is either of them correct? Explain your reasoning.

32. PERSEVERE $\triangle ABC$ has vertices $A(-3, 3)$, $B(2, 5)$, and $C(4, -3)$. What are the coordinates of the centroid of $\triangle ABC$? Explain the process you used to reach your conclusion.

33. CREATE In this problem, you will investigate the relationships among three points of concurrency in a triangle.

a. Draw an acute triangle and find the circumcenter, centroid, and orthocenter.

b. Draw an obtuse triangle and find the circumcenter, centroid, and orthocenter.

c. Draw a right triangle and find the circumcenter, centroid, and orthocenter.

d. Make a conjecture about the relationships among the circumcenter, centroid, and orthocenter.

34. ANALYZE Determine whether the following statement is *true* or *false*. If true, explain your reasoning. If false, provide a counterexample.

The orthocenter of a right triangle is always located at the vertex of the right angle.

35. WHICH ONE DOESN'T BELONG? Choose the term pairing that is not correct. Explain.

medians/centroid	perpendicular bisectors/incenter	altitudes/orthocenter

36. PERSEVERE In the figure at the right, segments $\overline{AD}$ and $\overline{CE}$ are medians of $\triangle ACB$, $\overline{AD} \perp \overline{CE}$, $AB = 10$, and $CE = 9$. Find CA.

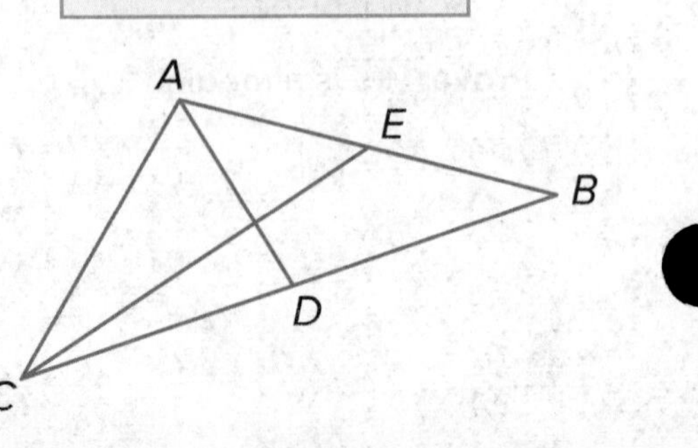

Inequalities in One Triangle

Explore Angle and Angle-Side Inequalities in Triangles

Online Activity Use dynamic geometry software to complete the Explore.

INQUIRY What relationship exists between the sides and angles of a triangle?

Today's Goals

- Solve problems by applying the Exterior Angle Inequality Theorem.
- Prove and apply theorems about inequalities in one triangle.

Learn Angle Inequalities in One Triangle

The inequality relationship between two real numbers is often applied in proofs. Recall that for any real numbers a and b, $a > b$ if and only if there is a positive number c such that $a = b + c$.

Key Concept • Properties of Inequality	
Comparison Property of Inequality	
Words	The value of a must be less than, greater than, or equal to the value of b.
Symbols	For all real numbers a, b, and c, the following is true: $a < b$, $a > b$, or $a = b$.
Transitive Property of Inequality	
Words	If a is less than b and b is less than c, then a is less than c. If a is greater than b and b is greater than c, then a is greater than c.
Symbols	For all real numbers a, b, and c, the following are true. If $a < b$ and $b < c$, then $a < c$. If $a > b$ and $b > c$, then $a > c$.
Addition Property of Inequality	
Words	If the same number is added to each side of a true inequality, then the resulting inequality is also true.
Symbols	For all real numbers a, b, and c, the following are true. If $a > b$, then $a + c > b + c$. If $a < b$, then $a + c < b + c$.
Subtraction Property of Inequality	
Words	If the same number is subtracted from each side of a true inequality, then the resulting inequality is also true.
Symbols	For all real numbers a, b, and c, the following are true. If $a > b$, then $a - c > b - c$. If $a < b$, then $a - c < b - c$.

The definition of inequality and the properties of inequalities can be applied to the measures of angles and segments, because these are real numbers.

Your Notes

Theorem 6.8: Exterior Angle Inequality Theorem

The measure of an exterior angle of a triangle is greater than the measure of either of its corresponding remote interior angles.

Go Online A proof of Theorem 6.8 is available.

Remember, each exterior angle of a triangle has two *remote interior angles* that are not adjacent to the exterior angle.

Go Online
An alternate method is available for this example.

Think About It!
Why must the markings be incorrect in the given diagram? Justify your argument.

Example 1 Use the Exterior Angle Inequality Theorem

List the angles that satisfy the stated condition. Justify your reasoning using the Exterior Angle Inequality Theorem.

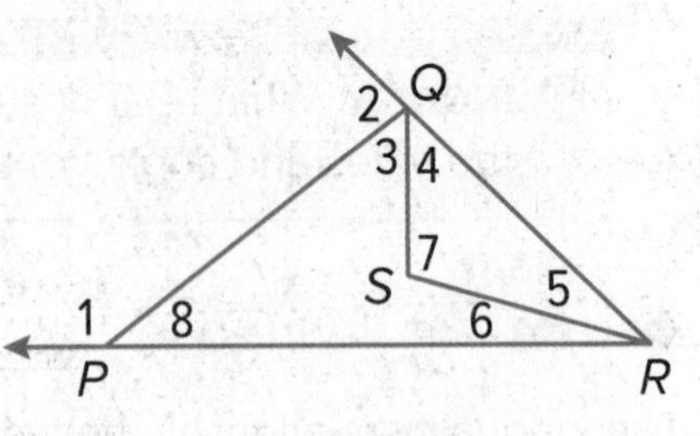

a. measures $< m\angle 1$

$\angle 1$ is an exterior angle of $\triangle PQR$, with $\angle PQR$ and $\angle PRQ$ as corresponding remote interior angles. By the Exterior Angle Inequality Theorem, $m\angle 1 > m\angle PQR$ and $m\angle 1 > m\angle PRQ$. Because $m\angle PQR = m\angle 3 + m\angle 4$ and $m\angle PRQ = m\angle 5 + m\angle 6$, $m\angle 1 > m\angle 3 + m\angle 4$ and $m\angle 1 > m\angle 5 + m\angle 6$ by substitution. So, the angles with measures less than $m\angle 1$ are $\angle 3$, $\angle 4$, $\angle 5$, $\angle 6$, $\angle PQR$, and $\angle PRQ$.

b. measures $> m\angle 8$

$\angle 2$ is an exterior angle of $\triangle PQR$. So, by the Exterior Angle Inequality Theorem, $m\angle 2 > m\angle 8$.

Check

List the angles that satisfy the stated condition. Justify your reasoning using the Exterior Angle Inequality Theorem.

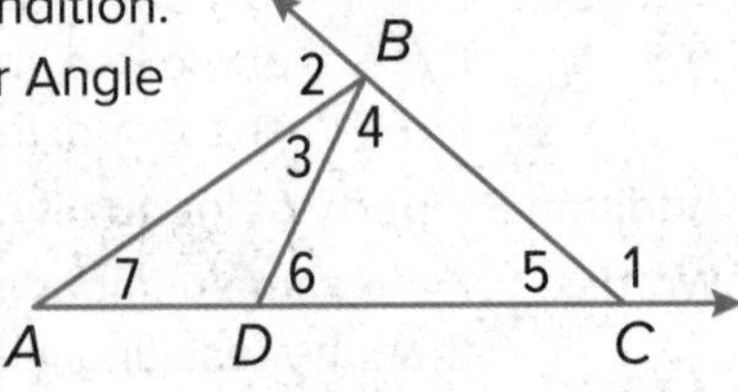

a. measures $< m\angle 1$

b. measures $> m\angle 7$

Learn Angle-Side Inequalities in One Triangle

We know that if two sides of a triangle are congruent, or the triangle is isosceles, then the angles opposite those sides are congruent.

Angle-Side Relationships in Triangles

Theorem 6.9

If one side of a triangle is longer than another side, then the angle opposite the longer side has a greater measure than the angle opposite the shorter side.

Theorem 6.10

If one angle of a triangle has a greater measure than another angle, then the side opposite the angle with the greater measure is longer than the side opposite the angle with the lesser measure.

You will prove Theorem 6.10 in Lesson 6-5, Exercise 19.

Go Online A proof of Theorem 6.9 is available.

Example 2 Order Triangle Angle Measures

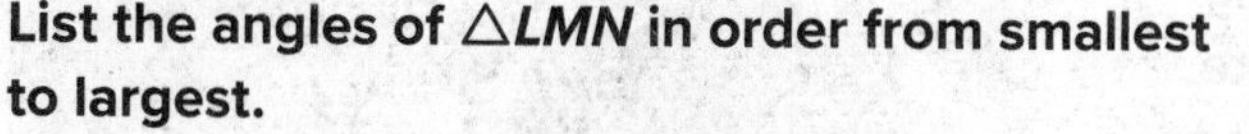

List the angles of $\triangle LMN$ in order from smallest to largest.

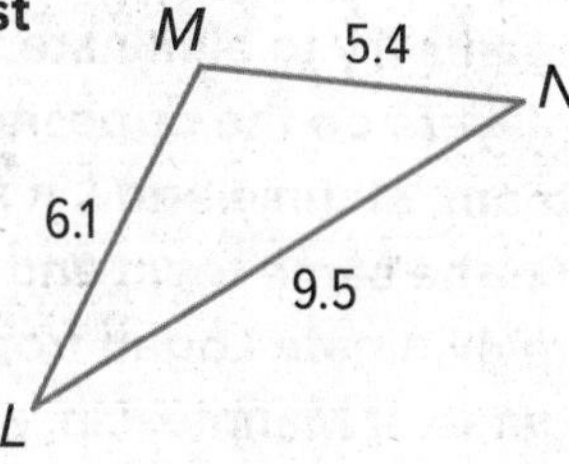

The sides from shortest to longest are $\overline{MN}$, $\overline{LM}$, and $\overline{LN}$.

So, the angles from smallest to largest are

____________.

Check

List the angles of $\triangle ABC$ in order from smallest to largest.

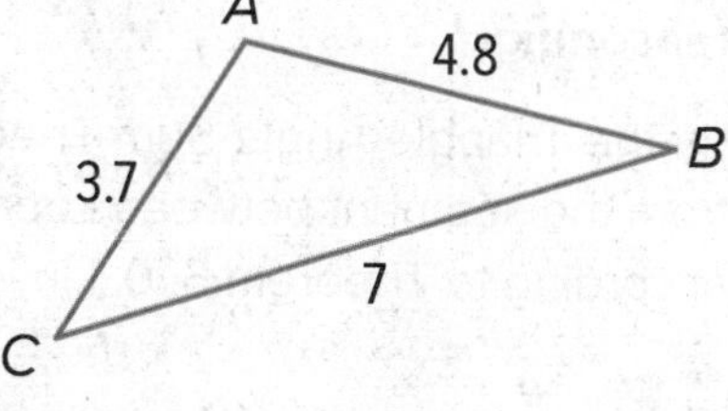

The angles from smallest to largest are

____________.

Example 3 Order Triangle Side Lengths

List the sides of $\triangle WXY$ in order from shortest to longest.

First find the missing angle measure using the Triangle Angle-Sum Theorem.
$m\angle X = 180 - (51 + 90)$ or $39°$

So, the angles from smallest to largest are $\angle X$, $\angle W$, and $\angle Y$. So, the sides from shortest to longest are

____________.

Go Online You can complete an Extra Example online.

Watch Out!

Identifying Side Opposite Be careful to correctly identify the side opposite the angle. The sides that form the angle cannot be the sides opposite the angle.

Talk About It!

Why is the hypotenuse always the longest side of a triangle?

Check

List the sides of $\triangle FGH$ in order from shortest to longest.

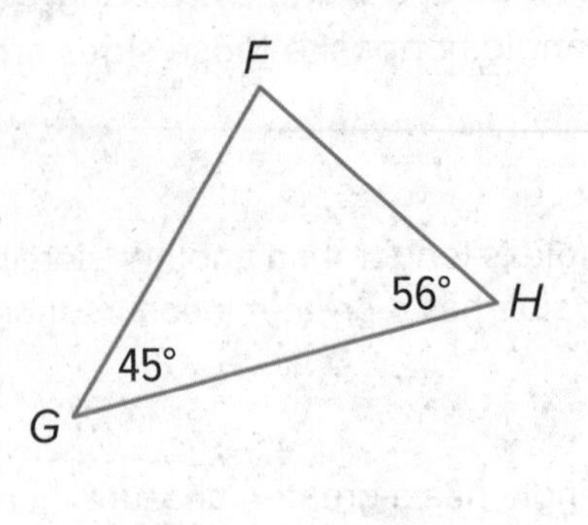

______ ______ ______

shortest longest

Example 4 Use Angle-Side Relationships

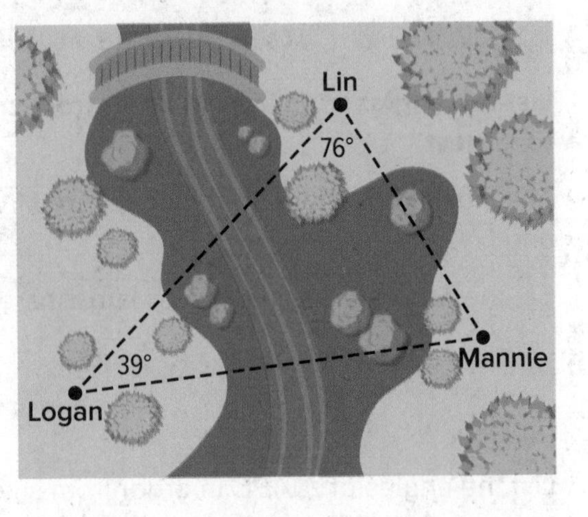

PAINTBALL **During a game of paintball, opposing teams try to eliminate players on the opposite team. Mannie and Lin are on the same team and want to eliminate Logan from the game. If Mannie, Lin, and Logan are located at the positions shown on the diagram, who is closer to Logan? Explain your reasoning.**

By the Triangle Angle-Sum Theorem, the measure of the angle across from the segment between Logan and Lin is 65°. Because $65 < 76$, according to Theorem 6.10, Lin is closer to Logan.

Think About It!

What assumption did you make while solving this problem?

Check

SPORTS Gabrielle, Diego, and Lucy are passing a football. Lucy wants to practice throwing the ball long distances. Which player should she throw the ball to next if she wants to pass the football the farthest distance?

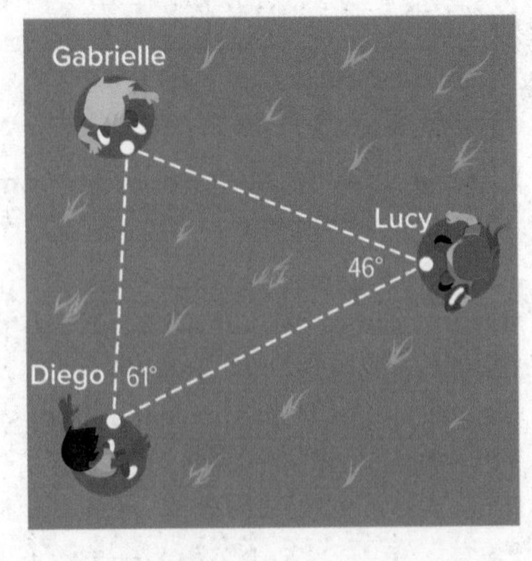

______

Go Online You can complete an Extra Example online.

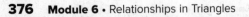

Name ____________________ Period __________ Date __________

Practice

Go Online You can complete your homework online.

Example 1

List the angles that satisfy the stated condition.

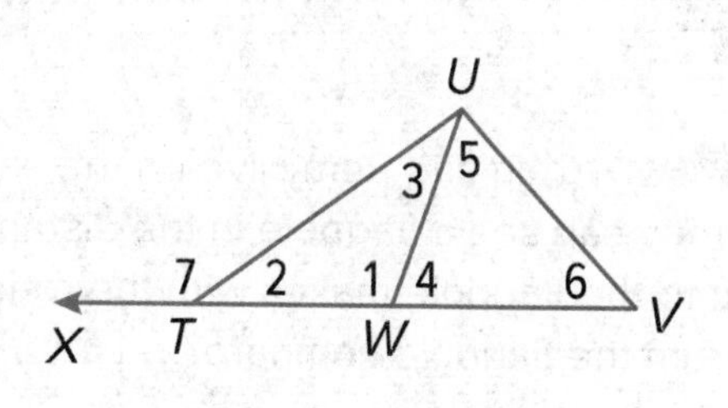

1. measures are greater than $m\angle 3$

2. measures are less than $m\angle 1$

3. measures are greater than $m\angle 1$

4. measures are less than $m\angle 7$

5. measures are greater than $m\angle 2$

Examples 2 and 3

List the angles and sides of each triangle in order from smallest to largest.

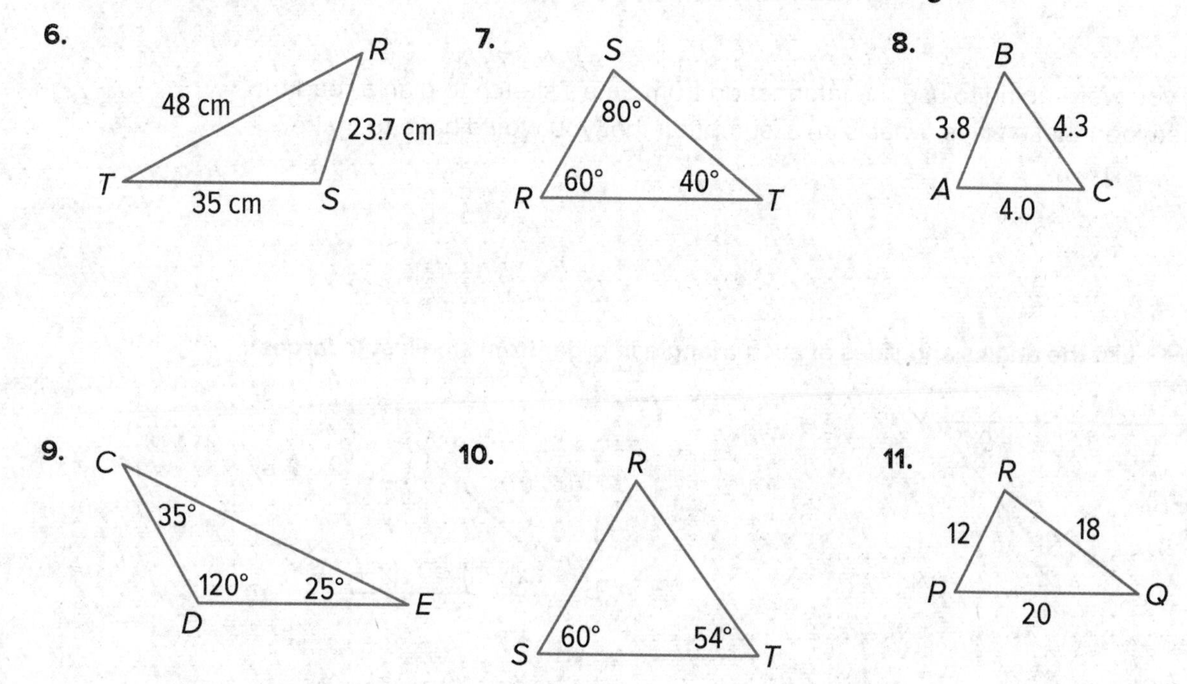

Example 4

12. SPORTS The figure shows the position of three trees on one part of a disc golf course. At which tree position is the angle between the trees the greatest?

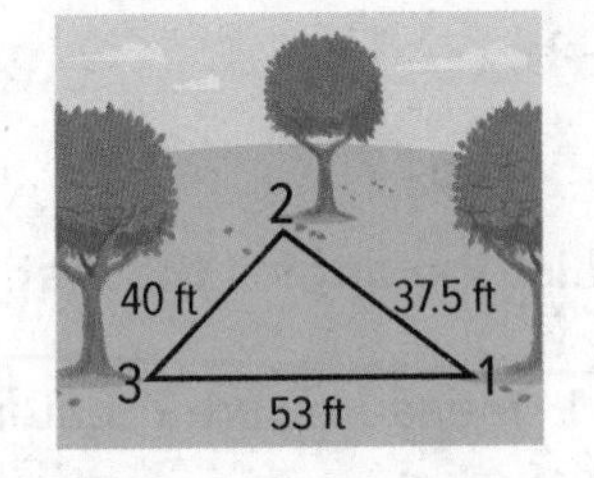

13. NEIGHBORHOOD Cain and Remy live on the same straight road. From their balconies, they can see a flagpole in the distance. The angle that each person's line of sight to the flagpole makes with the road is the same. How do their distances from the flagpole compare?

Mixed Exercises

14. MAPS Sata is going to Texas to visit a friend. As she looked at a map to see where she might want to go, she noticed that Austin, Dallas, and Abilene form a triangle. She wanted to determine how the distances between the cities were related, so she used a protractor to measure two angles.

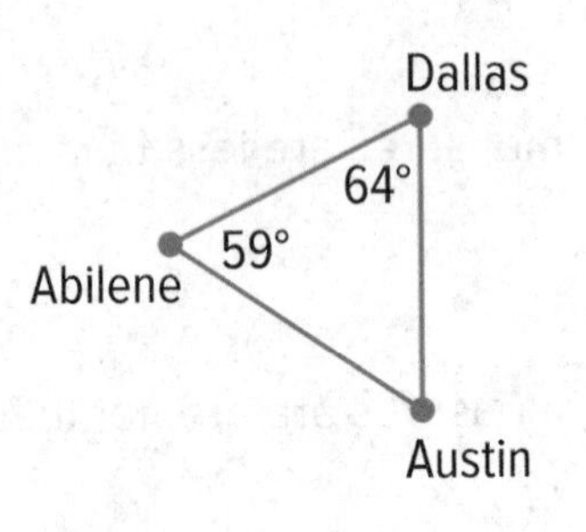

a. Based on the information in the figure, which of the two cities are nearest to each other?

b. Based on the information in the figure, which of the two cities are farthest apart from each other?

c. If you were going to use the information from Sata's sketch to plan a road trip between these cities, what is an assumption that you would have to make?

REASONING **List the angles and sides of each triangle in order from smallest to largest.**

15. **16.**

Name ______________________ Period ____________ Date ____________

In $\triangle ABC$, $\overline{AY}$ bisects $\angle A$, $\overline{BX}$ bisects $\angle B$, and Q is the intersection of $\overline{AY}$ and $\overline{BX}$. Use the figure for Exercises 17–19.

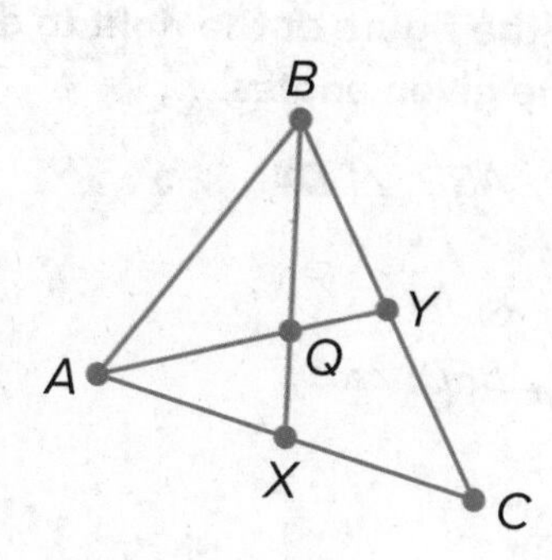

17. Suppose $BC > AC$. Compare $m\angle BAY$ and $m\angle ABX$. Justify your reasoning.

18. Cynda claims that $\overline{XC} \cong \overline{YC}$ if $BC > AC$. Can Cynda make this conclusion? Justify your argument.

19. If $m\angle AQB > m\angle BQY$, which side of $\triangle ABQ$ has the greatest length?

20. PROOF Write a paragraph proof.

Given: $WY > YX$

Prove: $m\angle ZWY > m\angle YWX$

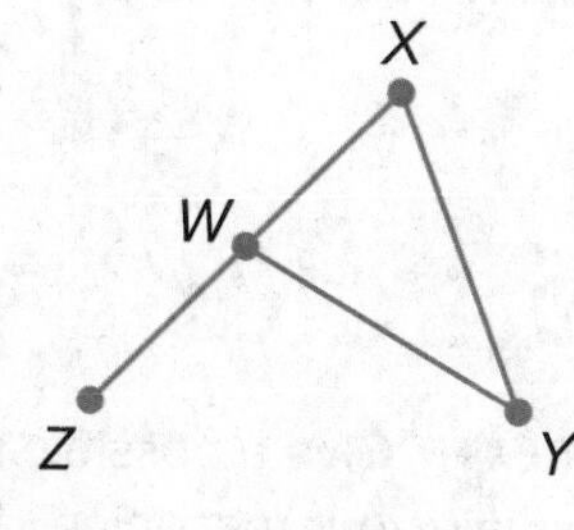

21. SQUARES Mahlik has three different squares. He arranges the squares to form a triangle as shown. Based on the information, list the squares in order from the one with the least perimeter to the one with the greatest perimeter.

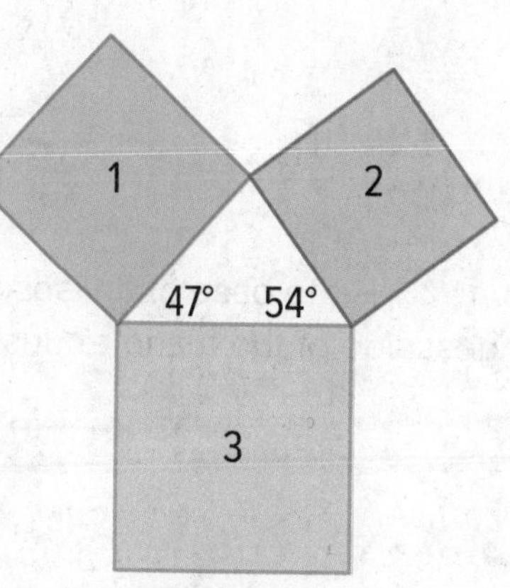

Use the figure at the right to determine which angle has the greatest measure.

22. $\angle 1, \angle 5, \angle 6$

23. $\angle 2, \angle 4, \angle 6$

24. $\angle 7, \angle 4, \angle 5$

25. $\angle 3, \angle 11, \angle 12$

26. $\angle 3, \angle 9, \angle 14$

27. $\angle 8, \angle 10, \angle 11$

Use the figure at the right to determine the relationship between the measures of the given angles.

28. $\angle ABD$, $\angle BDA$

29. $\angle BCF$, $\angle CFB$

30. $\angle BFD$, $\angle BDF$

31. $\angle DBF$, $\angle BFD$

Use the figure at the right to determine the relationship between the given lengths.

32. *SM*, *MR*

33. *RP*, *MP*

34. *RQ*, *PQ*

35. *RM*, *RQ*

36. **PERSEVERE** Using only a ruler, draw $\triangle ABC$ such that $m\angle A > m\angle B > m\angle C$. Justify your drawing.

37. **CREATE** Give a possible measure for $\overline{AB}$ in $\triangle ABC$ shown. Explain your reasoning.

38. **ANALYZE** Is the base of an isosceles triangle *sometimes*, *always*, or *never* the longest side of the triangle? Justify your argument.

39. **PERSEVERE** Use the side lengths in the figure to list the numbered angles in order from smallest to largest given that $m\angle 2 = m\angle 5$. Explain your reasoning.

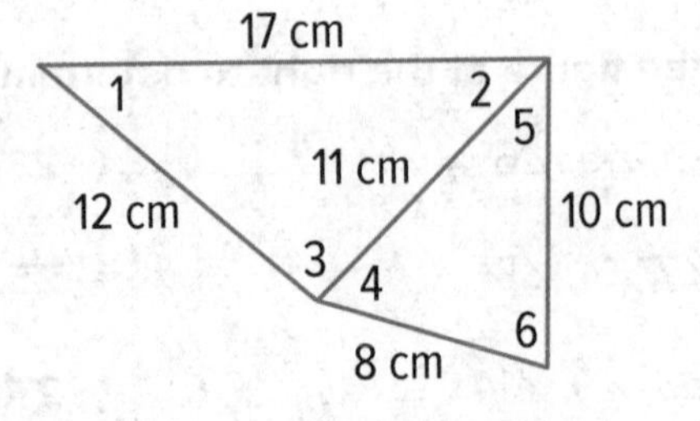

Indirect Proof

Explore Applying Indirect Reasoning

Online Activity Use the video to complete the Explore.

INQUIRY How can you use a contradiction to prove a conclusion?

Learn Indirect Proof

A direct proof is one that starts with a true hypothesis, and the conclusion is proved to be true. **Indirect reasoning** eliminates all possible conclusions but one, so the one remaining conclusion must be true. In an **indirect proof**, or **proof by contradiction**, one assumes that the statement to be proved is false and then uses logical reasoning to deduce that a statement contradicts a postulate, theorem, or one of the assumptions. Once a contradiction is obtained, one concludes that the statement assumed false must in fact be true.

Key Concept • How to Write an Indirect Proof

Step 1 Identify the conclusion that you are asked to prove. Make the assumption that this conclusion is false by assuming that the negation is true.

Step 2 Use logical reasoning to show that this assumption leads to a contradiction of the hypothesis or some other fact such as a definition, postulate, theorem, or corollary.

Step 3 State that because the assumption leads to a contradiction, the original conclusion, what you were asked to prove, must be true.

In indirect proofs, you should assume that the conclusion you are trying to prove is false. If, in the proof, you prove that the hypothesis is then false, this is a *proof by contrapositive*. If, in the proof, you assume that the hypothesis is true and prove that some other known fact is false, this is a *proof by contradiction*.

Example 1 Write an Indirect Algebraic Proof

Write an indirect proof to show that if $-4x - 4 < 12$, then $x > -4$.

Given: $-4x - 4 < 12$

Prove: $x > -4$

Indirect proof:

Step 1 Make an assumption.

The negation of $x > -4$ is $x \leq -4$. So, assume that $x < -4$ or $x = -4$ is true.

(continued on the next page)

Today's Goals

- Prove theorems about triangles by using indirect proof.

Today's Vocabulary

indirect reasoning

indirect proof

proof by contradiction

Talk About It!

Consider the statement: *If 4 is a factor of x, then 2 is a factor of x.* The conclusion of the conditional statement is that *2 is a factor of x.* What assumption is necessary to start an indirect proof?

Think About It!

Why do we have to consider both cases: that $x < -4$ or $x = -4$?

Go Online

You can complete an Extra Example online.

Your Notes

Study Tip

Contradiction A contradiction is a principle of logic stating that an assumption cannot be A and the opposite of A at the same time.

Use a Source

Lateral thinking puzzles are a type of riddle or puzzle that uses indirect reasoning to solve. One popular example is: "A man lives on the tenth floor of a building. Every day he takes the elevator down to the ground floor to go to work. When he returns, he takes the elevator to the seventh floor and walks up the stairs to reach his apartment on the tenth floor. He hates walking, so why does he do it?" Use available resources to find a lateral thinking puzzle with its answer, or find the answer to the example, and describe how indirect reasoning is used in the solution.

Step 2 Contradict the hypothesis.

Case 1: $x = -4$

When $x = -4$, $-4x - 4 = 12$. Because $12 \nless 12$, the assumption contradicts the given information for $x = -4$.

Case 2: $x < -4$

For all values of $x < -4$, $-4x - 4 > 12$, so the assumption contradicts the given information for $x < -4$.

Step 3 Reason indirectly.

In both cases, the assumption leads to a contradiction of the given information that $-4x - 4 < 12$. Therefore, the assumption that $x \leq -4$ must be false, so the original conclusion that $x > -4$ must be true.

Example 2 Apply Indirect Reasoning

HIKING **Marco hiked more than 10.5 miles on a path, making just two stops along the way. Use indirect reasoning to prove that he hiked more than 3.5 miles on at least one leg of his hike.**

Let x equal the distance traveled on the first leg of the hike, y equal the distance traveled on the second leg of the hike, and z equal the distance traveled on the third leg of the hike.

Given: $x + y + z > 10.5$

Prove: $x > 3.5$ or $y > 3.5$ or $z > 3.5$

Indirect proof:

Step 1 Make an assumption.

Assume that no leg of the hike is more than 3.5 miles. That is, x ____ 3.5, y ____ 3.5, and z ____ 3.5.

Step 2 Contradict the hypothesis.

If $x \leq 3.5$, $y \leq 3.5$, and $z \leq 3.5$, then $x + y + z \leq 3.5 + 3.5 + 3.5$ or $x + y + z \leq$ ____.

Step 3 Reason indirectly.

This is a contradiction of the given statement. Therefore, the assumption is ____ and $x > 3.5$ or $y > 3.5$ or $z > 3.5$. Marco traveled more than 3.5 miles on at least one leg of the hike.

Check

FUNDRAISING The senior class is holding a dinner to raise funds for the school's music and arts program. The cost of a nonstudent ticket is \$7, and the cost of a student ticket is \$3.50. If 256 total tickets were sold and the revenue was more than \$1246, prove that at least 100 nonstudent tickets were sold.

Let x equal the number of nonstudent tickets sold and let y equal the number of student tickets sold.

Given: $x + y = 256$; $7x + 3.5y > 1246$

Prove: $x \geq 100$

Indirect Proof:

Assume that ________. If $0 \leq x < 100$, then using the given $x + y = 256$, we know that $157 \leq y \leq 256$. For all values of x, the revenue earned from the sale of nonstudent tickets would be $0 \leq 7x < 700$. For all values of y, the revenue earned from the sale of student tickets would be ________ $\leq 3.5y \leq$ ________. Thus, the total revenue would be ________ $\leq 7x + 3.5y <$ ________, which contradicts the given that the revenue was more than \$1246. So, the assumption that $0 \leq x < 100$ is ________. So, $x \geq 100$ must be ________.

Example 3 Indirect Proofs in Number Theory

Write an indirect proof to show that if x^2 is an odd integer, then x is an odd integer.

Given: x^2 is an odd integer.

Prove: x is an odd integer.

Indirect proof: Assume that x is an even integer. To assume that x is an even integer means that $x = 2k$ for some integer k. Rewrite x^2 in terms of k.

$x^2 = (2k)^2$	Substitution
$= 4k^2$	Simplify.
$= (2 \cdot 2)k^2$	Multiplication Property of Equality
$= 2(2k^2)$	Associative Property

Because k is an integer, $2k^2$ is also an integer. Let n represent the integer $2k^2$. So, x^2 can be represented by $2n$, where n is an integer. This means that x^2 is an even integer, which contradicts the given statement that x^2 is an odd integer. Because the assumption that x is even leads to a contradiction of the given, the original conclusion that x is odd must be true.

Check

Write an indirect proof to show that if xy is an even integer, then either x or y is an even integer.

Given: xy is an even integer.

Prove: x or y is an even integer.

Indirect proof: Assume that x and y are odd integers. Let $x = 2n + 1$ and $y = 2k + 1$, for some integers n and k.

$xy = (2n + 1)(2k + 1)$	Substitution
$= 4nk + 2n + 2k + 1$	Distributive Property
$= 2(2nk + n + k) + 1$	Distributive Property

Because k and n are integers, ________ is also an integer. Let p represent the integer $2nk + n + k$. So, xy can be represented by ________, where p is an integer. This means that xy is an ________ integer, but this contradicts the given that xy is an ________ integer. Because the assumption that x and y are ________ integers leads to a contradiction of the given, the original conclusion that x or y is an ________ integer must be true.

Study Tip

Algebraic Proofs When working with algebraic proofs, it is helpful to remember that you can represent an even number with the expression $2k$ and an odd number with the expression $2k + 1$ for any integer k.

Watch Out!

Counterexamples Proof by contradiction and using a counterexample are not the same. A counterexample helps you disprove a conjecture. It cannot be used to prove a conjecture.

Example 4 Write an Indirect Geometric Proof

If an angle is an exterior angle of a triangle, prove that its measure is greater than the measure of either of its corresponding remote interior angles.

Given: $\angle 1$ is an exterior angle of $\triangle MNO$.

Prove: $m\angle 1 > m\angle 4$ and $m\angle 1 > m\angle 3$

Indirect proof: Assume that $m\angle 1 \leq m\angle 4$ or $m\angle 1 \leq m\angle 3$. $m\angle 1 \leq m\angle 4$ means that either $m\angle 1 = m\angle 4$ or $m\angle 1 < m\angle 4$.

Case 1: $m\angle 1 = m\angle 4$

$m\angle 1 = m\angle 4 + m\angle 3$	Exterior Angle Theorem
$m\angle 1 = m\angle 1 + m\angle 3$	Substitution
$0 = m\angle 3$	Subtract $m\angle 1$ from each side.

This contradicts the fact that the measure of any angle of a triangle is greater than 0. So, $m\angle 1 \neq m\angle 4$.

Case 2: $m\angle 1 < m\angle 4$

By the Exterior Angle Theorem, $m\angle 1 = m\angle 4 + m\angle 3$. Because angle measures in triangles are positive, the definition of inequality implies that $m\angle 1 > m\angle 4$. This contradicts the assumption that $m\angle 1 < m\angle 4$.

The argument for $m\angle 1 \leq m\angle 3$ follows the same reasoning. In both cases, the assumption leads to a contradiction. Therefore, the original conclusion that $m\angle 1 > m\angle 4$ and $m\angle 1 > m\angle 3$ must be true.

Watch Out!

Recognizing Contradictions
Remember that the contradiction in an indirect proof does not always contradict the given information or the assumption. It can contradict a known fact or definition, such as in **Case 1**, that the measure of an angle must be greater than 0.

Check

If a triangle is equilateral, prove that it is also equiangular.

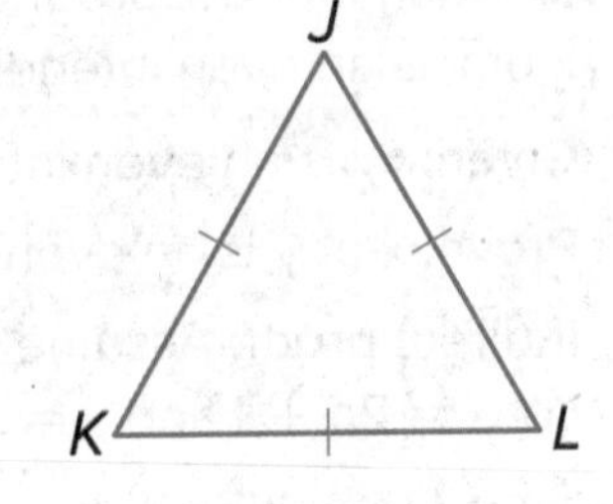

Given: $\triangle JKL$ is equilateral.

Prove: $\triangle JKL$ is equiangular.

Indirect proof: Assume that $\triangle JKL$ is ________________. If $\triangle JKL$ is ________________, then the measure of one angle is greater than the measure of another. Assume that $m\angle K > m\angle L$. Then, by Theorem 6.10, $JL > JK$. This contradicts the given information that $\triangle JKL$ is ________________. Therefore, the assumption that $\triangle JKL$ is ________________ must be false, so the original conclusion that $\triangle JKL$ is ________________ must be true.

Name ______________________ Period __________ Date __________

Practice

Example 1

PROOF **Write an indirect proof of each statement.**

1. If $x^2 + 8 \leq 12$, then $x \leq 2$.

2. If $-4x + 2 < -10$, then $x > 3$.

3. If $2x - 7 > -11$, then $x > -2$.

4. If $5x + 12 < -33$, then $x < -9$.

5. If $-3x + 4 < 7$, then $x > -1$.

6. If $-2x - 6 > 12$, then $x < -9$.

Example 2

7. SHOPPING Desiree buys two bracelets for a little more than $40 before tax. When she gets home, her mother asks her how much each bracelet costs. Desiree cannot remember the individual prices. Use indirect reasoning to show that at least one of the bracelets cost more than $20.

8. CONCESSIONS Kala worked at the concession stand during his school's basketball game. The cost of a small soft drink is $2, and the cost of a large soft drink is $3. If 120 soft drinks were sold and the total soft drink sales was more than $320, prove that at least 81 large soft drinks were sold.

Examples 3 and 4

PROOF **For 9–16, write an indirect proof of each statement. Assume that *x* and *y* are integers.**

9. **Given:** xy is an odd integer.
Prove: x and y are both odd integers.

10. **Given:** x^2 is even.
Prove: x^2 is divisible by 4.

11. **Given:** x is an odd number.
Prove: x is not divisible by 4.

12. **Given:** xy is an even integer.
Prove: x or y is an even integer.

13. In an isosceles triangle, neither of the base angles can be a right angle.

14. A triangle can have only one right angle.

15. Given: $m\angle C = 100°$
Prove: $\angle A$ is not a right angle.

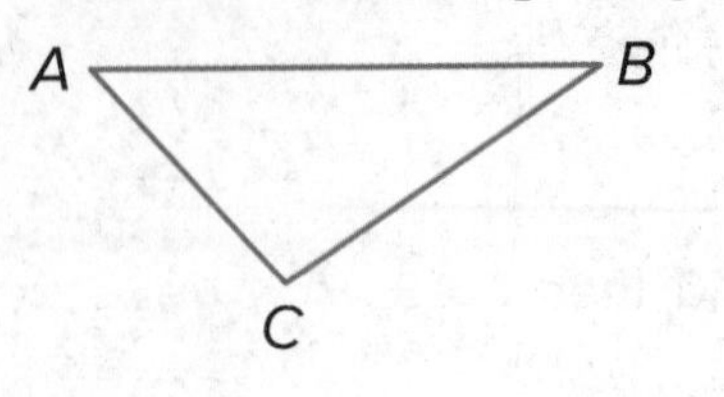

16. Given: $\angle D \not\cong \angle F$
Prove: $DE \neq EF$

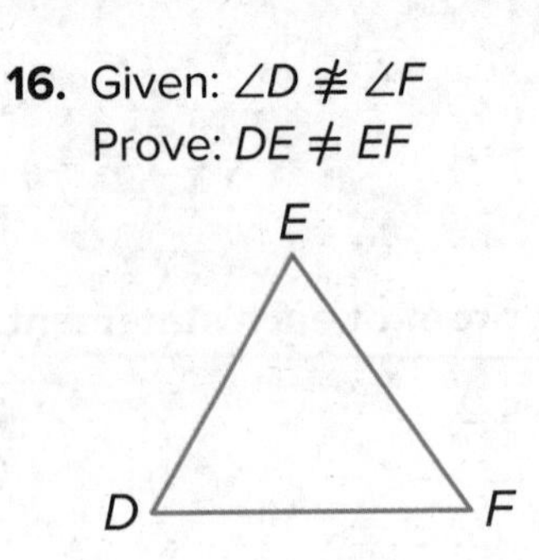

Mixed Exercises

17. **PHYSICS** Sound travels through air at about 344 meters per second when the temperature is 20°C. If Enrique lives 2 kilometers from the fire station and it takes 5 seconds for the sound of the fire station siren to reach him, how can you prove indirectly that it is not 20°C when Enrique hears the siren?

18. **PROOF** Write an indirect proof to show that if $\frac{1}{b} < 0$, then b is negative.

19. **PROOF** Write an indirect proof. (Theorem 6.10)

If one angle of a triangle has a greater measure than another angle, then the side opposite the angle with the greater measure is longer than the side opposite the angle with the lesser measure.

20. **WORDS** The words *accomplishment, counterexample,* and *extemporaneous* all have 14 letters. Use an indirect proof to show that any word with 14 letters must use a repeated letter or have two letters that are consecutive in the alphabet.

21. **WRITE** Explain the procedure for writing an indirect proof.

22. **CREATE** Write a statement that can be proved using an indirect proof. Include the indirect proof of your statement.

23. **PERSEVERE** If x is a rational number, then it can be represented by a quotient $\frac{a}{b}$ for some integers a and b, if $b \neq 0$. An irrational number cannot be represented by the quotient of two integers. Write an indirect proof to show that the product of a nonzero rational number and an irrational number is an irrational number.

The Triangle Inequality

Explore Relationships Among Triangle Side Lengths

Online Activity Use dynamic geometry software to complete the Explore.

INQUIRY How are the three side lengths of a triangle related?

Today's Goals
- Prove and apply the Triangle Inequality Theorem.

Learn The Triangle Inequality

In order for three segments to form a triangle, a special relationship must exist among their lengths.

Theorem 6.11: Triangle Inequality Theorem

The sum of the lengths of any two sides of a triangle must be greater than the length of the third side.

You will prove Theorem 6.11 in Exercise 22.

Study Tip

Make a Model You can use interactive geometry software or a ruler and paper to model a triangle for any given side lengths. The scale of the ruler will not matter as long as the side lengths are given in the same unit of measure and you establish a 1-to-1 conversion factor, for example 1 in. = 1 cm.

Example 1 Identify Possible Triangles Given Side Lengths

Is it possible to form a triangle with the given side lengths? If not, explain why not.

a. 9 cm, 12 cm, 18 cm

Check each inequality.

9 + 12 ____ 18 12 + 18 ____ 9 9 + 18 ____ 12

Because the sum of each pair of side lengths is greater than the third side length, sides with lengths 9, 12, and 18 centimeters ____ form a triangle.

b. 3 in., 5 in., 8 in.

3 + 5 ____ 8

Because the sum of one pair of side lengths is not greater than the third side length, sides with lengths 3, 5, and 8 inches ____ form a triangle.

Think About It!

How can you eliminate triangle side length possibilities by looking at only one inequality?

Check

Is it possible to form a triangle with the given side lengths? If not, explain why not.

a. 2 mm, 5 mm, 6 mm ____

b. 3 yd, 4 yd, 8 yd ____

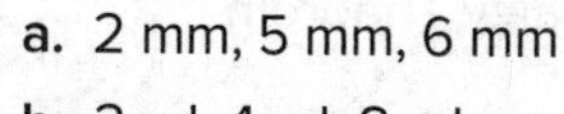 You can complete an Extra Example online.

Your Notes

Study Tip

Multiple Inequality Symbols The compound inequality $4 < x < 12$ is read *x is between 4 and 12.*

Think About It!

What is the greatest possible whole-number measure for the third side?

When the lengths of two sides of a triangle are known, the third side can be any length in a range of values.

Example 2 Find Possible Side Lengths

DRONES **A delivery company uses drones to make speedy deliveries around the city. A drone leaves the home office and flies 8 miles east to its first delivery and then 4 more miles southwest to a second delivery. What is the *least* possible whole-number distance the drone will fly to return to the home office?**

Let x represent the length of the third side. Set up and solve each of the three triangle inequalities.

$4 + 8 > x$ $4 + x > 8$ $x + 8 > 4$

____ $> x$ $x >$ ____ $x >$ ____

The least whole-number value between ____ and 12 is 5. The drone has to fly at least ____ miles to the home office after the two deliveries.

Check

HOME IMPROVEMENT To install a smart thermostat, Kelvin is cutting a triangular hole in the wall. He marks side lengths of $2\frac{1}{8}$ inches and $2\frac{3}{16}$ inches.

Part A

Which assumption do you need to make to determine the range for the measure of the third side of the hole?

A The triangular hole is not equiangular.

B The smart thermostat requires a triangular hole.

C Kelvin measured the side lengths accurately.

D Two sides of the hole are longer than the third side.

Part B

What is the possible range for the third side length?

Example 3 Use the Triangle Inequality Theorem

$\triangle HAB$ and $\triangle HCB$ share side HB, and $HC = HB$. Prove that $HA + AB > HC$.

Given: $HC = HB$

Prove: $HA + AB > HC$

H C B A

Proof:

Statements	Reasons
1. $HC = HB$	1. Given
2. ____	2. Triangle Inequality Theorem
3. $HA + AB > HC$	3. ____

Go Online You can complete an Extra Example online.

Name _______________ Period _______ Date _______

Practice

Go Online You can complete your homework online.

Example 1

REASONING Is it possible to form a triangle with the given side lengths? If not, explain why not.

1. 9, 12, 18

2. 8, 9, 17

3. 14, 14, 19

4. 23, 26, 50

5. 32, 41, 63

6. 2.7, 3.1, 4.3

7. 0.7, 1.4, 2.1

8. 12.3, 13.9, 25.2

Example 2

Find the range for the measure of the third side of a triangle given the measures of two sides.

9. 6 ft and 19 ft

10. 7 km and 29 km

11. 13 in. and 27 in.

12. 18 ft and 23 ft

13. CITIES The distance between New York City, New York, and Boston, Massachusetts, is 187 miles, and the distance between New York City and Hartford, Connecticut, is 97 miles. Hartford, Boston, and New York City form a triangle on a map. What must the distance between Boston and Hartford be greater than?

14. FENCING Capria is planning to fence a triangular plot of land. Two of the sides of the plot measure 230 yards and 490 yards.

a. Find the range for the measure of the third side of the triangular plot of land.

b. What are the maximum and minimum lengths of fencing Capria will need?

Example 3

PROOF Write a two-column proof.

15. Given: $\overline{PL} \parallel \overline{MT}$; K is the midpoint of $\overline{PT}$.

Prove: $PK + KM > PL$

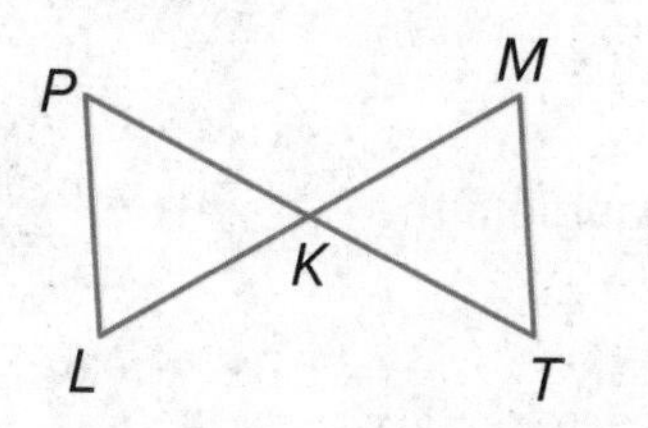

16. Given: $\triangle ABC \cong \triangle DEC$

Prove: $AB + DE > AD - BE$

D
C
B
E
A

Mixed Exercises

Find the range of possible measures of x if each set of expressions represents measures of the sides of a triangle.

17. x, 4, 6

18. $x - 2$, 10, 12

REGULARITY **Determine whether the given coordinates are the vertices of a triangle. Explain.**

19. $X(1, -3)$, $Y(6, 1)$, $Z(2, 2)$

20. $J(-7, -1)$, $K(9, -5)$, $L(21, -8)$

21. **GARDENING** Ennis has 4 lengths of wood from which he plans to make a border for a triangular-shaped herb garden. The lengths of the wood borders are 8 inches, 10 inches, 12 inches, and 18 inches. How many different triangular borders can Ennis make?

22. **PROOF** Write a two-column proof. (Theorem 6.11)

Given: $\triangle ABC$

Prove: $AC + BC > AB$

(*Hint:* Draw an auxiliary segment $\overline{CD}$, so that C is between B and D and $\overline{CD} \cong \overline{AC}$.)

A

C B

23. **PERSEVERE** The sides of an isosceles triangle are whole numbers, and its perimeter is 30 units. What is the probability that the triangle is equilateral?

24. **CREATE** The length of one side of a triangle is 2 inches. Draw a triangle in which the 2-inch side is the shortest side and one in which the 2-inch side is the longest side. Include side and angle measures on your drawing.

25. **WRITE** What can you tell about a triangle when given three side lengths? Include at least two items.

26. **ANALYZE** What is the range of lengths of each leg of an isosceles triangle if the measure of the base is 6 inches? Explain.

Inequalities in Two Triangles

Explore Analyzing Inequalities in Two Triangles

Online Activity Use dynamic geometry software to complete the Explore.

INQUIRY How do the included angle measures of two triangles with two pairs of congruent sides compare?

Today's Goals

- Prove and apply the Hinge Theorem.
- Prove and apply the Converse of the Hinge Theorem.

Learn Hinge Theorem

Theorem 6.12: Hinge Theorem

If two sides of a triangle are congruent to two sides of another triangle, and the included angle of the first is larger than the included angle of the second triangle, then the third side of the first triangle is longer than the third side of the second triangle.

You will prove Theorem 6.12 in Exercise 18.

Think About It!

In $\triangle ADB$ and $\triangle CDB$, side $\overline{BD}$ is a common side, $\overline{AB} \cong \overline{BC}$, and $m\angle ABD > m\angle CBD$. Because $m\angle ABD > m\angle CBD$, what is the relationship between AD and DC? Explain your reasoning.

Example 1 Use the Hinge Theorem

BOATING **Two families set sail on their boats from the same dock. The Nguyens sail 3.5 nautical miles north, turn 85° east of north, and then sail 2 nautical miles. The Griffins sail 3.5 nautical miles south, turn 95° east of south, and then sail 2 nautical miles. At this point, which boat is farther from the dock? Explain your reasoning.**

Step 1 Draw a diagram of the situation.

The courses of each boat and the straight-line distance from each stopping point back to the boat dock form two triangles. Each boat sails _____ nautical miles, turns, and then sails another _____ nautical miles.

N
2 n.m.
85°
SANDY HARBOR
3.5 n.m.
L
G
3.5 n.m.
2 n.m.
95°

Step 2 Determine the interior angle measures.

Use linear pairs to find the measures of the included angles. The measure of the included angle for the Nguyens is 180 – 85 or 95°. The measure of the included angle for the Griffins is 180 – 95 or 85°.

Step 3 Compare the distance each boat is from the boat launch.

Use the Hinge Theorem to compare the distance each boat is from the boat launch.

Because 95 > 85, $NL > GL$ by the Hinge Theorem. So, the _____ are farther from the boat launch.

Go Online You can complete an Extra Example online.

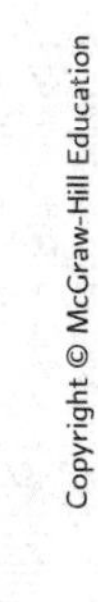

Your Notes

Check

DRONES Esperanza and Landon each fly a drone from the same place in a park and at the same altitude. Esperanza flies her drone 440 feet east, then turns it 122° south of east and flies 300 more feet. Landon flies his drone 440 feet east, then turns it 140° north of east and flies 300 more feet. Whose drone is closer to its original position? ______

You can use the Hinge Theorem to prove relationships in two triangles.

Example 2 Prove Triangle Relationships by Using the Hinge Theorem

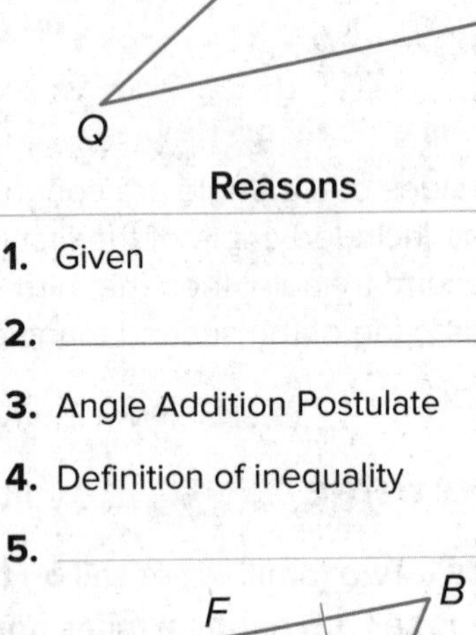

Complete the two-column proof.

Given: $\overline{PQ} \cong \overline{PM}$

Prove: $NQ > NM$

Proof:

Statements	Reasons
1. $\overline{PQ} \cong \overline{PM}$	1. Given
2. $\overline{PN} \cong \overline{PN}$	2. ______
3. ______	3. Angle Addition Postulate
4. ______	4. Definition of inequality
5. $NQ > NM$	5. ______

Study Tip

Inequalities in Triangles Because we know that $m\angle NPQ$ is the sum of $m\angle NPM$ and $m\angle MPQ$ and neither angle has a degree measure of 0, $m\angle NPQ$ has to be greater than both $m\angle NPM$ and $m\angle MPQ$.

Statements/Reasons:

Hinge Theorem

Reflexive Property

$m\angle NPQ = m\angle NPM + m\angle MPQ$

$m\angle NPQ > m\angle NPM$

Check

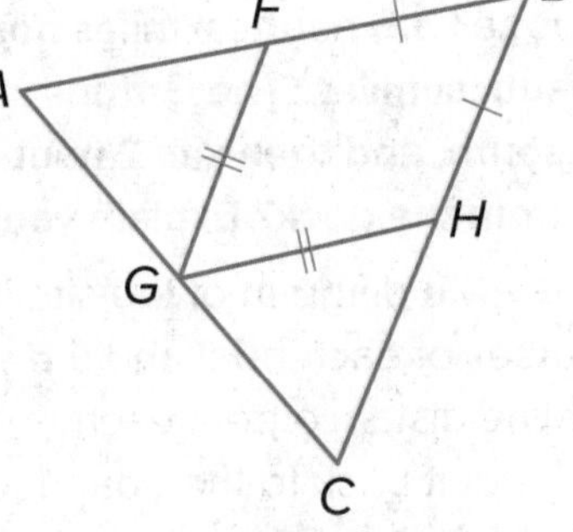

Complete the two-column proof.

Given: $\overline{HG} \cong \overline{FG}$, $\overline{BH} \cong \overline{BF}$
G is the midpoint of $\overline{AC}$.
$m\angle CGH > m\angle AGF$

Prove: $BC > AB$

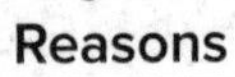

Statements	Reasons
1. $\overline{HG} \cong \overline{FG}$, $\overline{BH} \cong \overline{BF}$	1. Given
2. G is the midpoint of $\overline{AC}$.	2. Given
3. $m\angle CGH > m\angle AGF$	3. Given
4. $CG = AG$	4. ______
5. $CH > AF$	5. ______
6. $BH = BF$	6. ______
7. $CH + BH > AF + BH$	7. Addition Property
8. $CH + BH > AF + BF$	8. Substitution
9. $BC = CH + BH$, $AB = AF + BF$	9. ______
10. $BC > AB$	10. ______

Statements/Reasons:

Definition of midpoint

Hinge Theorem

Substitution

Segment Addition Postulate

Definition of congruent segments

Go Online You can complete an Extra Example online.

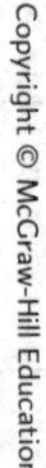

Learn Converse of the Hinge Theorem

Theorem 6.13: Converse of the Hinge Theorem

If two sides of a triangle are congruent to two sides of another triangle and the third side of the first triangle is longer than the third side of the second triangle, then the included angle measure of the first triangle is greater than the included angle measure of the second triangle.

You will prove Theorem 6.13 in Exercise 19.

Example 3 Apply Algebra to Relationships in Triangles

Find the range of possible values for x.

$(3x + 45)°$, H, J, 68°, G, 10, 7.3, E

Step 1 Use the Converse of the Hinge Theorem.

From the diagram, we know that $\overline{GH} \cong \overline{JH}$, $\overline{EH} \cong \overline{EH}$, and $JE > GE$.

_____ > _____ Converse of the Hinge Theorem

$3x + 45 > 68$ __________

_____ Solve for x.

Step 2 Use your knowledge of the interior angles of a triangle.

Use the fact that the measure of any angle in a triangle is less than 180° to write a second inequality.

_____ < _____

$3x + 45 < 180°$ __________

_____ Solve for x.

Step 3 Complete the compound inequality.

____ $< x <$ ____

Check

Find the range of possible values for x.

____

19, 15, 55°, $(3x + 10)°$

Go Online You can complete an Extra Example online.

Think About It!

In $\triangle XYW$ and $\triangle ZWY$, $\overline{YW}$ is a common side, $\overline{WZ} \cong \overline{YX}$, and $XW > ZY$. Because $XW > ZY$, what is the relationship between $m\angle XYW$ and $m\angle ZWY$? Explain your reasoning.

Think About It!

Do you think that the value of x is likely to be closer to $\frac{23}{3}$ or 45? Justify your argument.

Example 4 Prove Relationships by Using the Converse of the Hinge Theorem

Complete the flow proof.

Given: T is the midpoint of $\overline{ZX}$.

$\overline{ST} \cong \overline{WT}$, $SZ > WX$

Prove: $m\angle XTR > m\angle ZTY$

Proof:

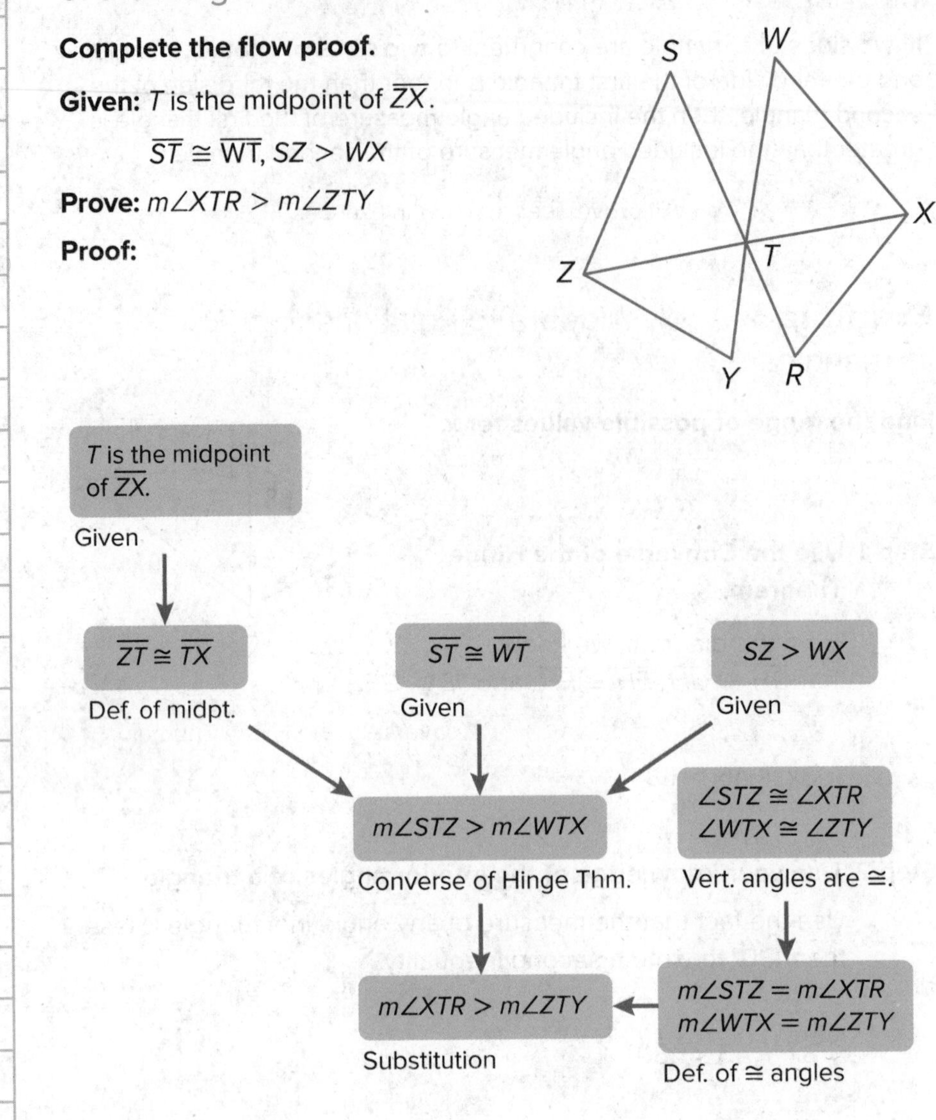

Pause and Reflect

Did you struggle with anything in this lesson? If so, how did you deal with it?

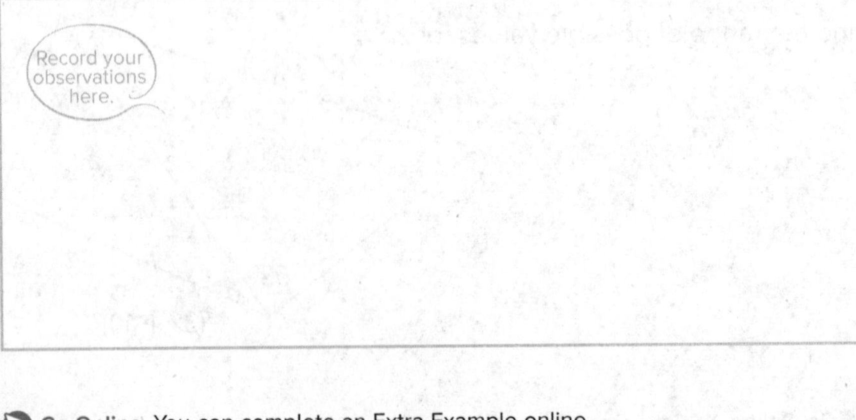

Go Online You can complete an Extra Example online.

Name ______________________ Period __________ Date __________

Practice

Go Online You can complete your homework online.

Example 1

1. FLIGHT Two planes take off from the same airstrip. The first plane flies west for 150 miles and then flies 30° south of west for 220 miles. The second plane flies east for 220 miles and then flies $x°$ south of east for 150 miles. If $x < 30$, which plane is farther from the airstrip after the second leg? Justify your answer.

2. HIKING Gen and Ari start hiking from the same point. Gen hikes 5 miles due east and turns to hike 4.5 miles 30° south of east. Ari hikes 5 miles due west and turns to hike 4.5 miles 15° north of west.

   a. Draw a model to represent the situation.

   b. Who is closer to the starting point? Explain your reasoning.

Example 2

PROOF **Write a two-column proof.**

3. **Given:** $RX = XS$; $m\angle SXT = 97$

   Prove: $ST > RT$

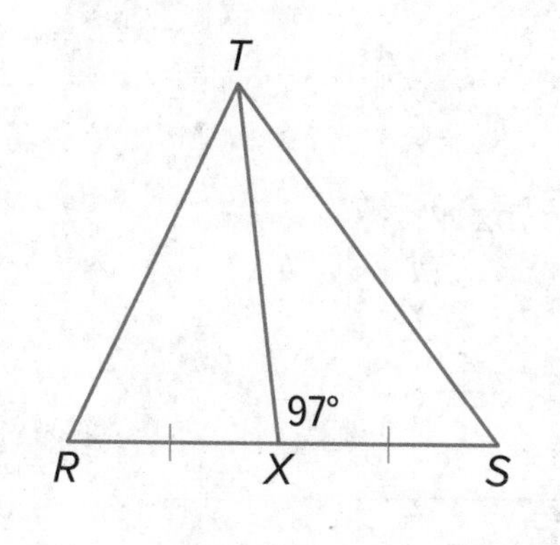

4. **Given:** $\overline{LK} \cong \overline{JK}$, $\overline{RL} \cong \overline{RJ}$, K is the midpoint of $\overline{QS}$, and $m\angle SKL > m\angle QKJ$.

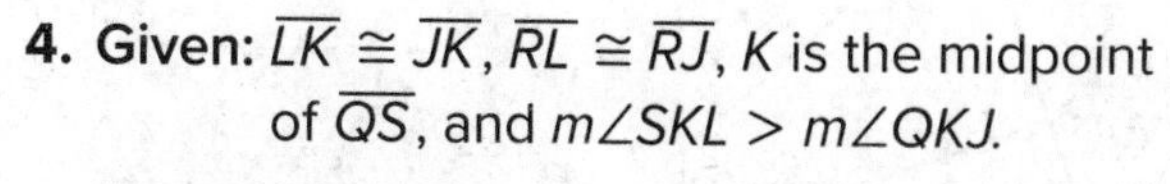

   Prove: $RS > QR$

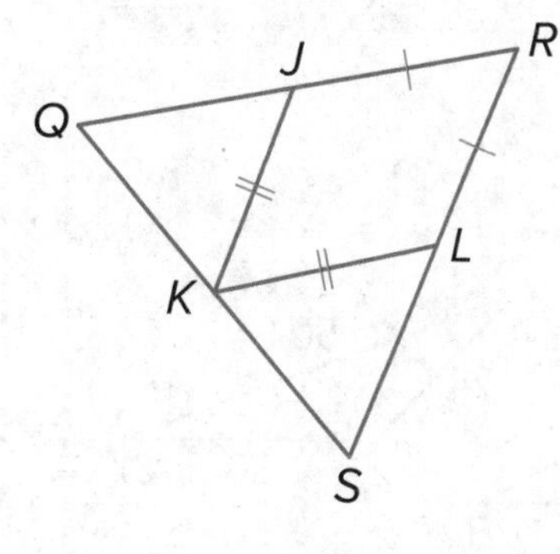

Example 3

Find the range of possible values for x.

5.

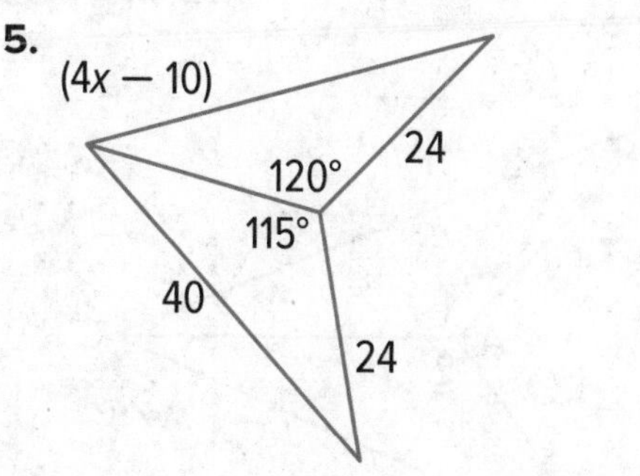

6.

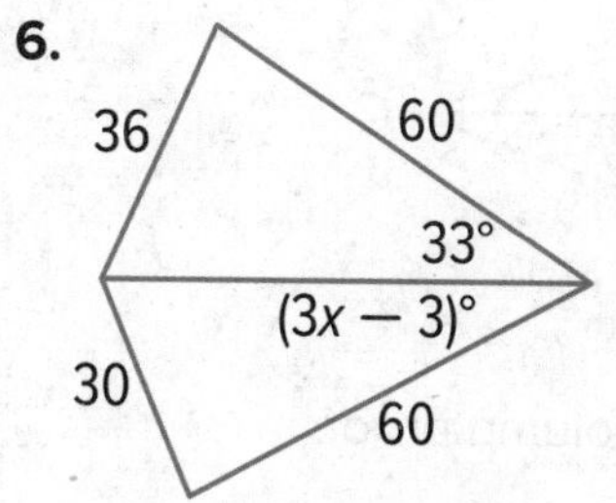

Example 4

PROOF **Write a two-column proof.**

7. **Given:** $\overline{XU} \cong \overline{VW}$, $VW > XW$
$\overline{XU} \parallel \overline{VW}$

Prove: $m\angle XZU > m\angle UZV$

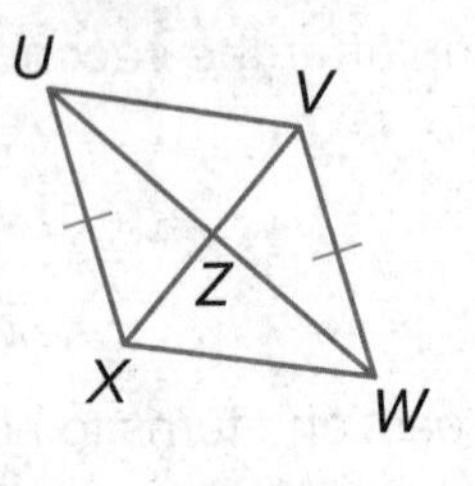

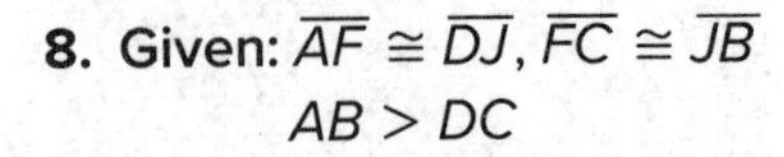

8. **Given:** $\overline{AF} \cong \overline{DJ}$, $\overline{FC} \cong \overline{JB}$
$AB > DC$

Prove: $m\angle AFC > m\angle DJB$

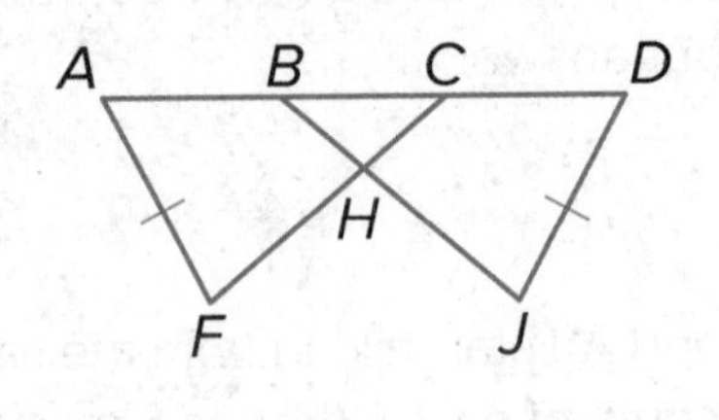

Mixed Exercises

Compare the given measures.

9. MR and RP

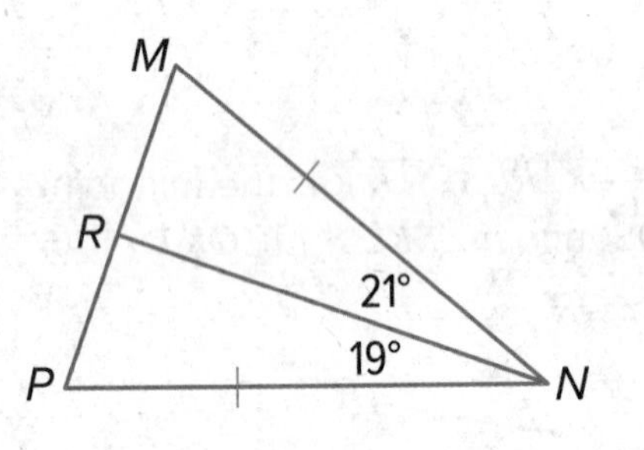

10. AD and CD

C
D
22°
B
38°
A

11. $m\angle C$ and $m\angle Z$

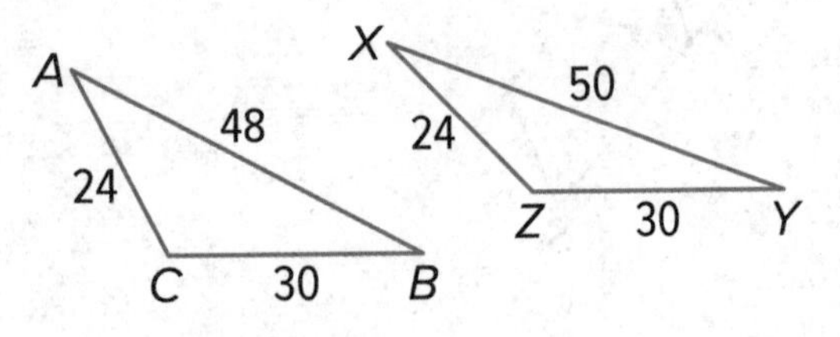

12. $m\angle XYW$ and $m\angle WYZ$

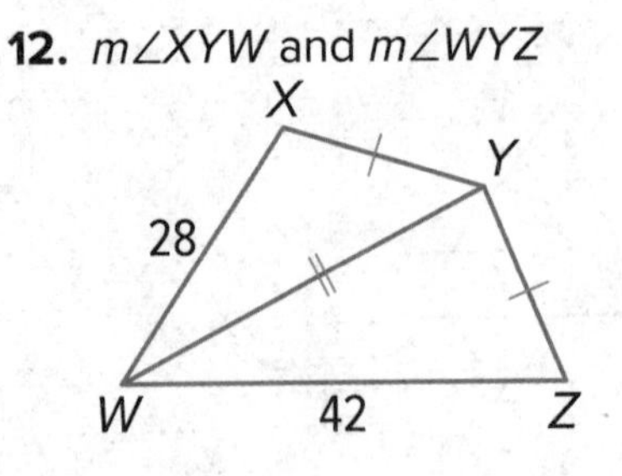

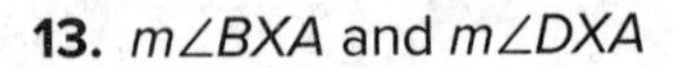

13. $m\angle BXA$ and $m\angle DXA$

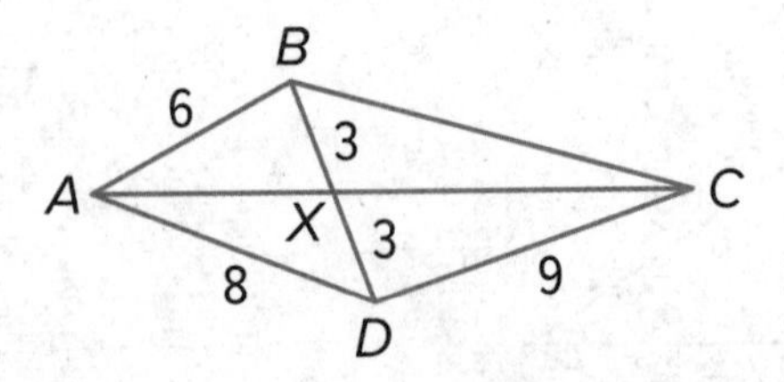

14. PROOF Write a two-column proof.

Given: $\overline{BA} \cong \overline{DA}$, $BC > DC$

Prove: $m\angle 1 > m\angle 2$

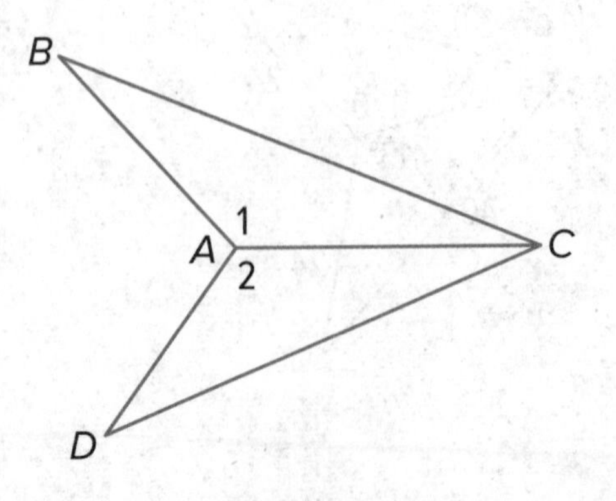

Name ____________________ Period ________ Date ________

15. PROOF Write a paragraph proof.

Given: $\overline{EF} \cong \overline{GH}$, $m\angle F > m\angle G$.

Prove: $EG > FH$

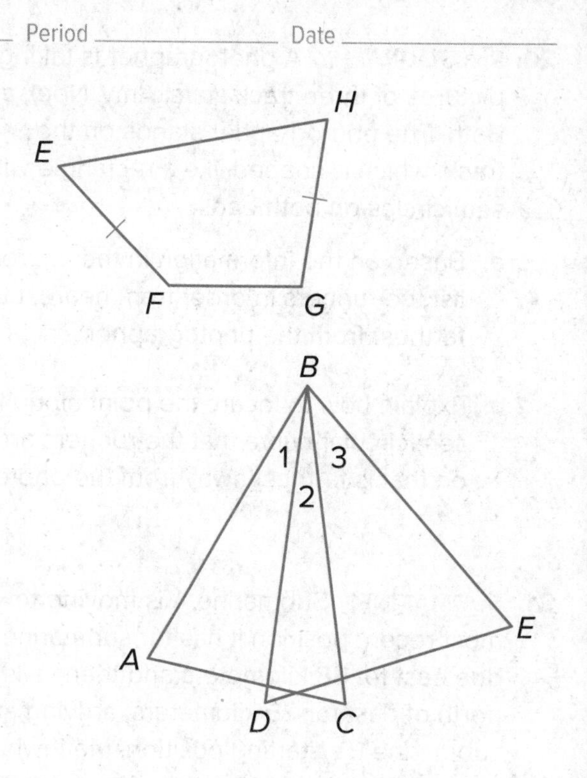

16. REASONING In the figure, $\overline{BA}$, $\overline{BD}$, $\overline{BC}$, and $\overline{BE}$ are congruent and $AC < DE$. How does $m\angle 1$ compare with $m\angle 3$? Explain your thinking.

17. CLOCKS The minute hand of a grandfather clock is 3 feet long, and the hour hand is 2 feet long. Is the distance between their ends greater at 3:00 or at 8:00?

18. PROOF Write a paragraph proof to prove the Hinge Theorem. (Theorem 6.12)

Given: $\triangle ABC$ and $\triangle DEF$
$\overline{AC} \cong \overline{DF}$, $\overline{BC} \cong \overline{EF}$
$m\angle F > m\angle C$

Prove: $DE > AB$

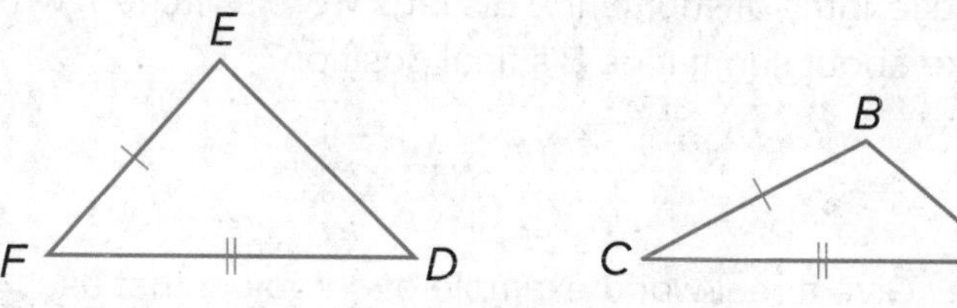

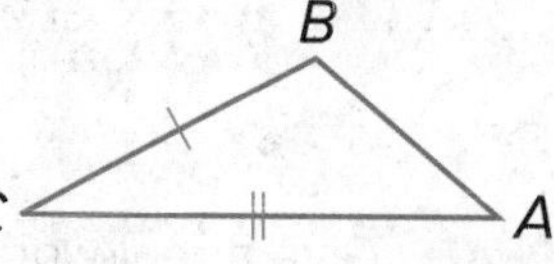

19. PROOF Use an indirect proof to prove the Converse of the Hinge Theorem. (Theorem 6.13)

Given: $\overline{RS} \cong \overline{UW}$
$\overline{ST} \cong \overline{WV}$
$RT > UV$

Prove: $m\angle S > m\angle W$

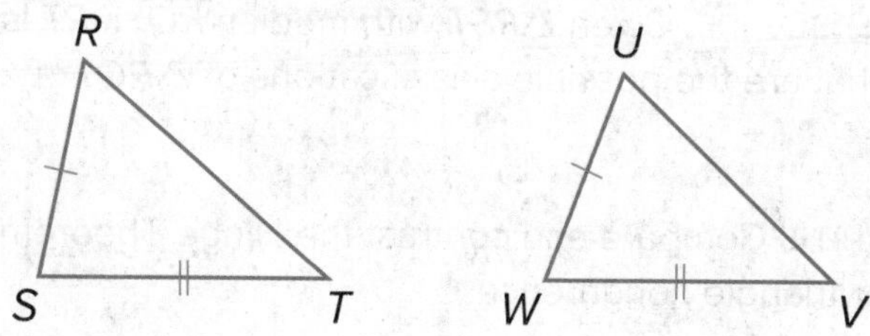

20. PHOTOGRAPHY A photographer is taking pictures of three track stars, Amy, Noel, and Beth. The photographer stands on the track, which is shaped like a rectangle with semicircles on both ends.

a. Based on the information in the figure, list the runners in order from nearest to farthest from the photographer.

b. Explain how to locate the point along the semicircular curve that the runners are on that is farthest away from the photographer.

21. SUBMARINES Submarine A is moving toward the most recent position it has for submarine B. It sails due east for 38 kilometers and then sails 52° north of east for 25 kilometers, arriving at submarine B's starting position. Meanwhile, submarine B sails 38° south of west for 25 kilometers and then due west for 38 kilometers.

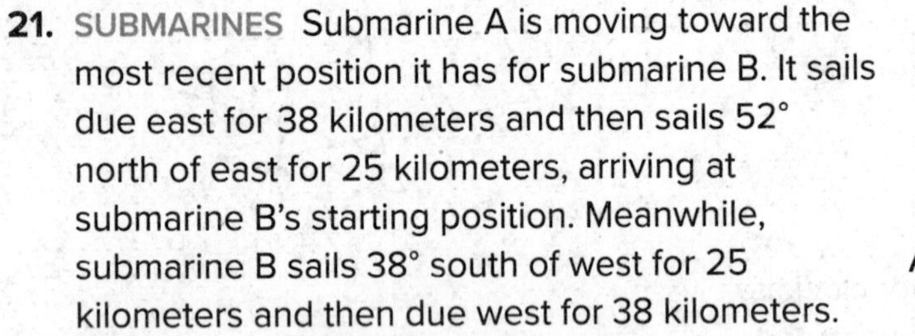

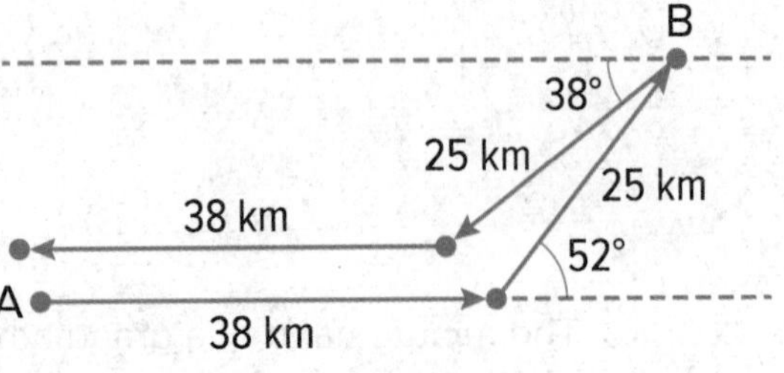

a. Does submarine B end up northeast, northwest, southeast, or southwest of submarine A's starting point?

b. If the lengths of submarine B's legs were switched, what conclusions could you make about submarine B's final position?

22. CREATE Give a real-world example of an object that uses a hinge. Draw two sketches in which the hinge on your object is adjusted to two different positions. Use your sketches to explain why Theorem 6.12 is called the Hinge Theorem.

23. PERSEVERE Given $\triangle RST$ with median $\overline{RQ}$, if RT is greater than or equal to RS, what are the possible classifications of $\triangle RQT$? Explain your reasoning.

24. WRITE Compare and contrast the Hinge Theorem to the SAS Postulate for triangle congruence.

25. ANALYZE If $\overline{BD}$ is a median and $AB < BC$, then is $\angle BDC$ *sometimes, always,* or *never* an acute angle? Justify your argument.

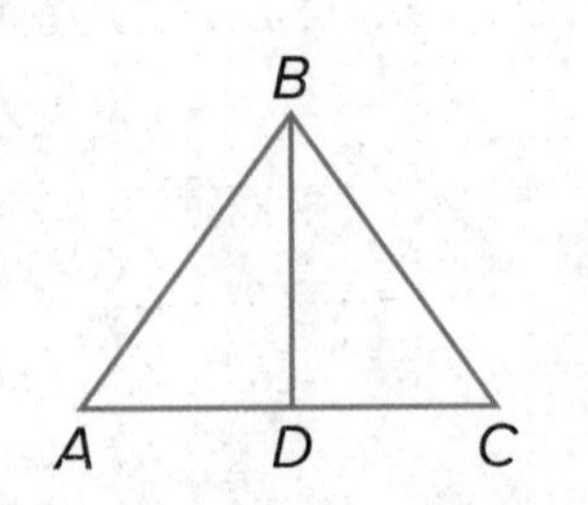

Review

Essential Question

How can relationships in triangles be used in real-world situations?

Module Summary

Lessons 6-1 and 6-2

Perpendicular Bisectors and Angle Bisectors

- If a point is on the perpendicular bisector of a segment, then it is equidistant from the endpoints of the segment.
- The perpendicular bisectors of a triangle intersect at the circumcenter that is equidistant from the vertices of the triangle.
- If a point is on the bisector of an angle, then it is equidistant from the sides of the angle.
- The angle bisectors of a triangle intersect at the incenter, which is equidistant from the sides of the triangle.

Lesson 6-3

Medians and Altitudes

- A median of a triangle is a line segment with endpoints that are a vertex of the triangle and the midpoint of the side opposite the vertex.
- The medians of a triangle intersect at the centroid, which is two-thirds of the distance from each vertex to the midpoint of the opposite side.
- An altitude of a triangle is a segment from a vertex of the triangle to the line that contains the opposite side and is perpendicular to that side.
- The altitudes of a triangle intersect at a point called the orthocenter.

Lessons 6-4, 6-6, and 6-7

Inequalities in Triangles

- If one side of a triangle is longer than another side, then the angle that is opposite the longer side has a greater measure than the angle that is opposite the shorter side.
- The sum of the lengths of any two sides of a triangle must be greater than the length of the third side.
- If two sides of a triangle are congruent to two sides of another triangle, and the included angle of the first is larger than the included angle of the second triangle, then the third side of the first triangle is longer than the third side of the second triangle.

Lesson 6-5

Indirect Proof

- To write an indirect proof:

  Identify the conclusion you are asked to prove. Make the assumption that this conclusion is false by assuming that the opposite is true.

  Use logical reasoning to show that this assumption leads to a contradiction of the hypothesis or some other fact such as a definition, postulate, theorem, or corollary.

  State that because the assumption leads to a contradiction, the original conclusion, what you were asked to prove, must be true.

Study Organizer

Foldables

Use your Foldable to review this module. Working with a partner can be helpful. Ask for clarification of concepts as needed.

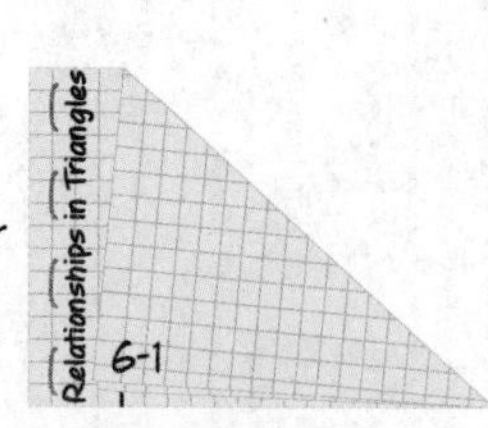

Name ______________________ Period __________ Date __________

Test Practice

1. OPEN RESPONSE In triangle WXZ, what is the length of segment ZY? (Lesson 6-1)

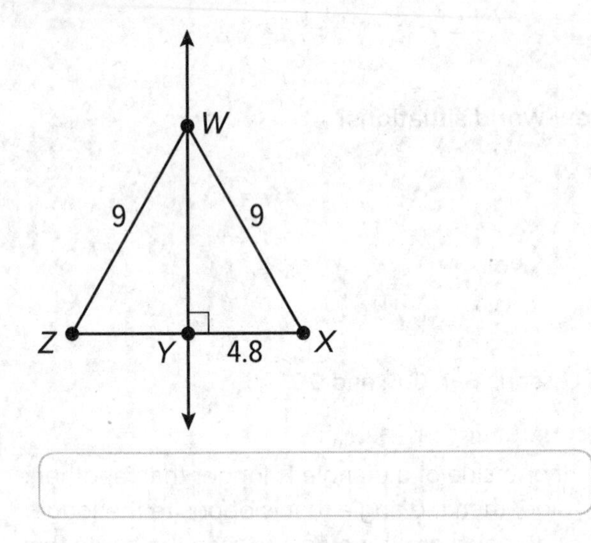

2. MULTIPLE CHOICE Macha is a day care provider and has three tables with children eating lunch, as shown on the grid.

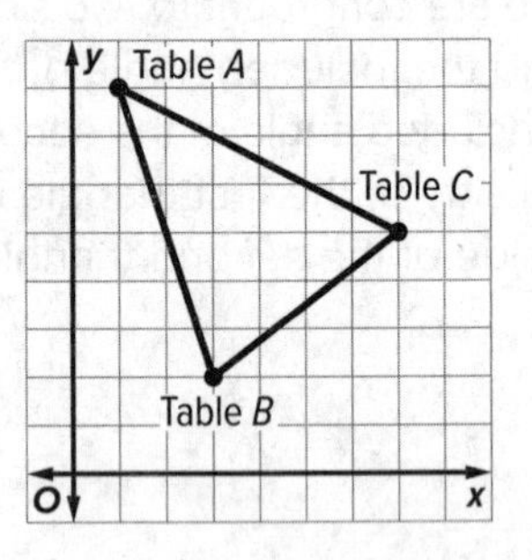

Where should Macha stand to be equidistant from all three tables? (Lesson 6-1)

Ⓐ (3.5, 5.5)

Ⓑ $\left(\frac{11}{3}, 5\right)$

Ⓒ (3.85, 4.72)

Ⓓ (4, 4)

3. MULTIPLE CHOICE What is the length of side AB? (Lesson 6-1)

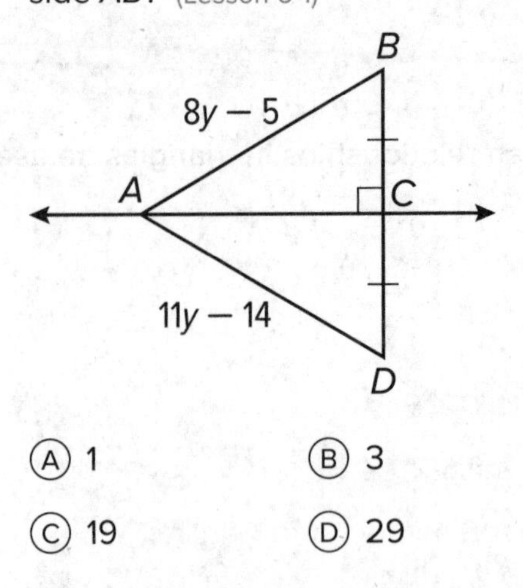

Ⓐ 1 Ⓑ 3

Ⓒ 19 Ⓓ 29

4. OPEN RESPONSE Given the following, where D is the incenter of $\triangle ABC$, $DF = 6$, and $DB = 10$, find the measure of EB. (Lesson 6-2)

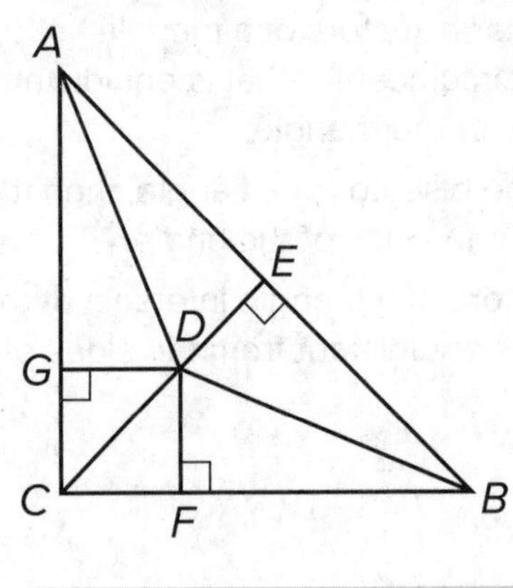

5. MULTIPLE CHOICE Use the figure to find the length of segment RS. (Lesson 6-2)

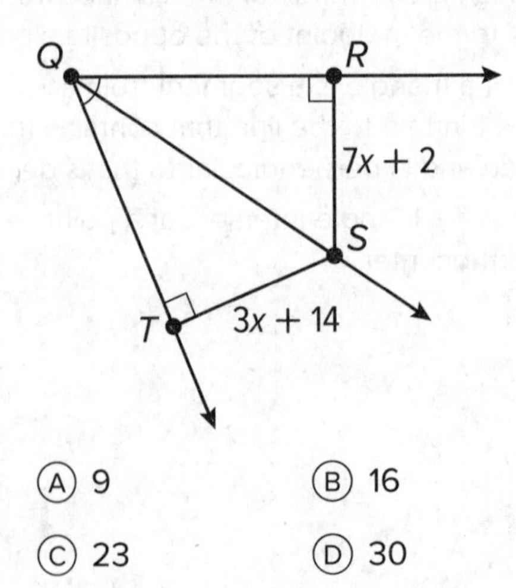

Ⓐ 9 Ⓑ 16

Ⓒ 23 Ⓓ 30

6. OPEN RESPONSE Find $m\angle EFH$. (Lesson 6-2)

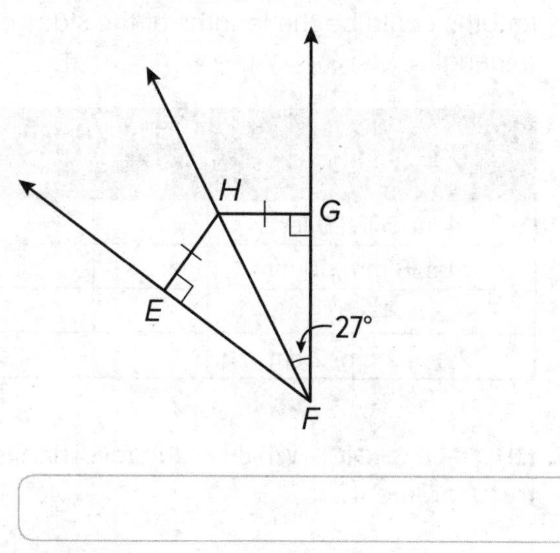

7. GRAPH Graph the orthocenter of triangle *ABC*. (Lesson 6-3)

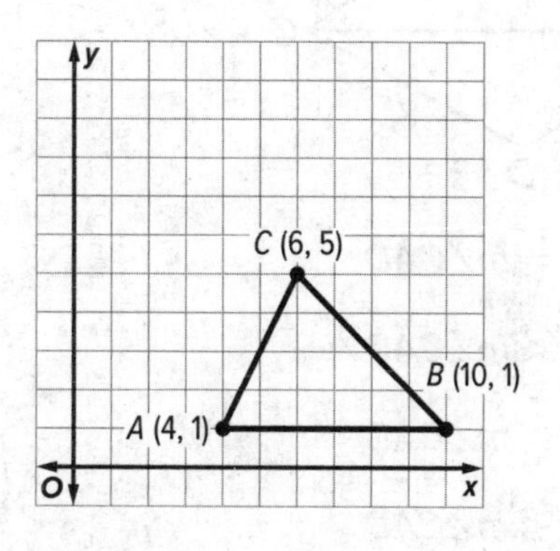

8. OPEN RESPONSE In $\triangle ABC$, G is the centroid and $GE = 4$. Find AG. (Lesson 6-3)

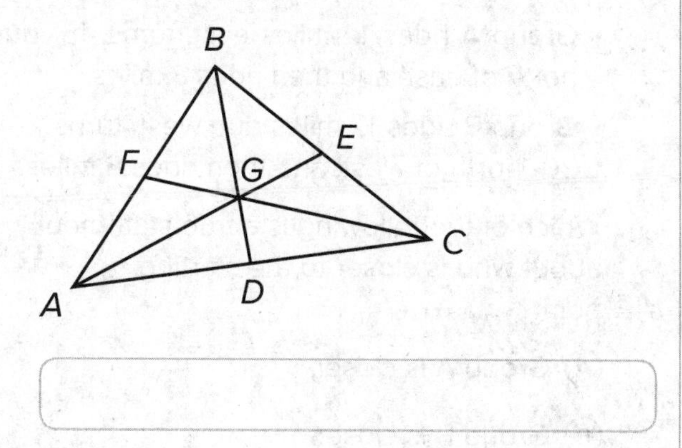

9. OPEN RESPONSE In $\triangle ABC$, G is the centroid and $DC = 36$. Find DG and GC. (Lesson 6-3)

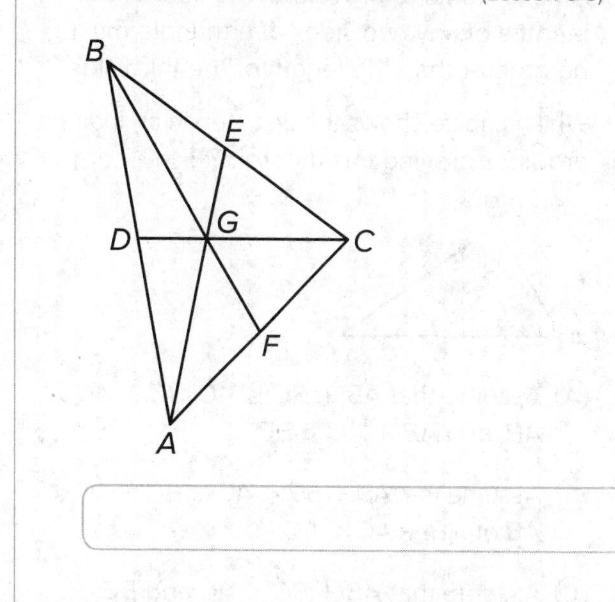

10. MULTIPLE CHOICE Order the angles of the triangle from smallest to largest. (Lesson 6-4)

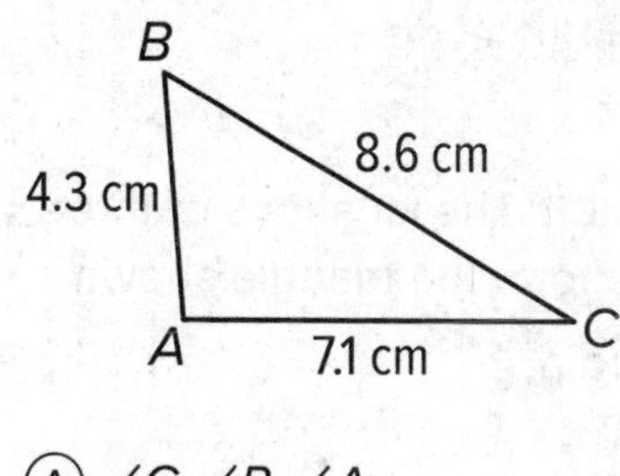

Ⓐ $\angle C$, $\angle B$, $\angle A$

Ⓑ $\angle A$, $\angle B$, $\angle C$

Ⓒ $\angle C$, $\angle A$, $\angle B$

Ⓓ $\angle A$, $\angle C$, $\angle B$

11. OPEN RESPONSE Using an indirect proof, Antonia must prove that if a triangle is a right triangle, then none of the angles have measures greater than 90°. What assumption must she start with? (Lesson 6-5)

Name ______________________ Period __________ Date __________

12. MULTIPLE CHOICE Using an indirect proof, Amelia wants to prove that the sum of the lengths of any two sides of a triangle must be greater than the length of the third side.

Which choice shows the first step in an indirect proof for proving this theorem? (Lesson 6-5)

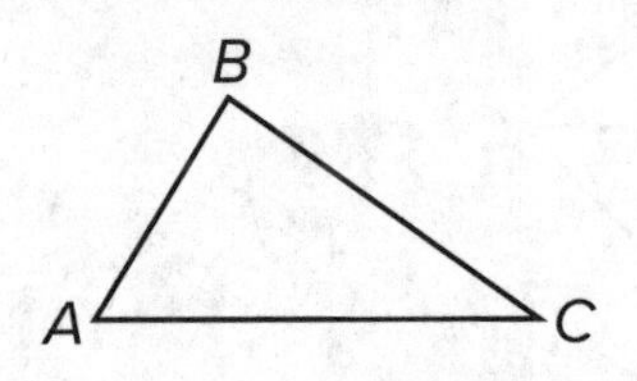

Ⓐ Assume that $AB + BC \leq AC$, $BC + AC > AB$, and $AB + AC > BC$.

Ⓑ Assume that $AB + BC < AC$ or $BC + AC < AB$ or $AB + AC < BC$.

Ⓒ Assume that $AB + BC < AC$ and $BC + AC < AB$ and $AB + AC < BC$.

Ⓓ Assume that $AB + BC \leq AC$ or $BC + AC \leq AB$ or $AB + AC \leq BC$.

13. MULTIPLE CHOICE The locations of three friends' homes form the triangle shown.

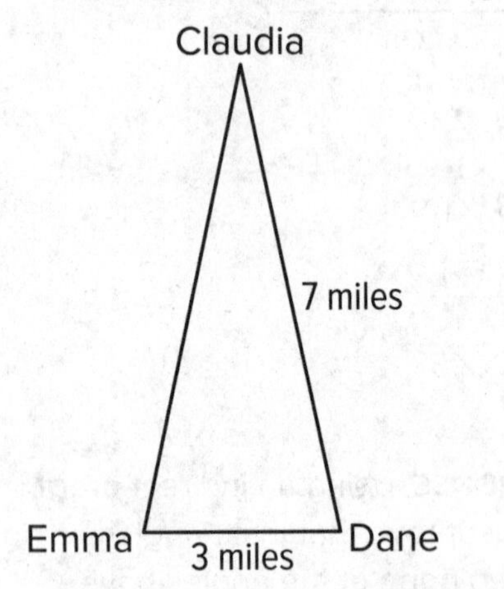

What is the possible range for the distances between Emma's house and Claudia's house? (Lesson 6-6)

Ⓐ $x > 4$

Ⓑ $x < 10$

Ⓒ $4 < x < 21$

Ⓓ $4 < x < 10$

14. TABLE ITEM State whether each set of lengths could be the lengths of the sides of a triangle. (Lesson 6-6)

Side Lengths	Make a triangle?	
	Yes	No
4 in., 5 in., 9 in.		
7 mm, 8 mm, 10 mm		
4.2 ft, 4.2 ft, 9.1 ft		
2 cm, 2 cm, 2 cm		

15. MULTIPLE CHOICE Which of the following is a true statement? (Lesson 6-7)

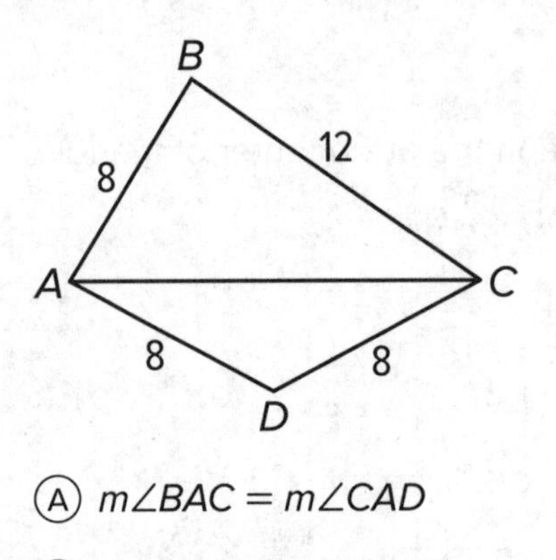

Ⓐ $m\angle BAC = m\angle CAD$

Ⓑ $m\angle BAC < m\angle CAD$

Ⓒ $m\angle BAC > m\angle CAD$

Ⓓ $m\angle BAC \leq m\angle CAD$

16. MULTIPLE CHOICE Two groups of cyclists leave from the same starting point.

- Group A rides 12 miles east, turns 45° due north of east, and then rides 8 miles.
- Group B rides 12 miles due west, turns 30° north of west, and then rides 8 miles.

Which of the following is a true statement about who is closer to the starting point? (Lesson 6-7)

Ⓐ Group A is closer.

Ⓑ Group B is closer.

Ⓒ The groups are the same distance.

Ⓓ Not enough information to tell

Module 1

Quick Check

1.

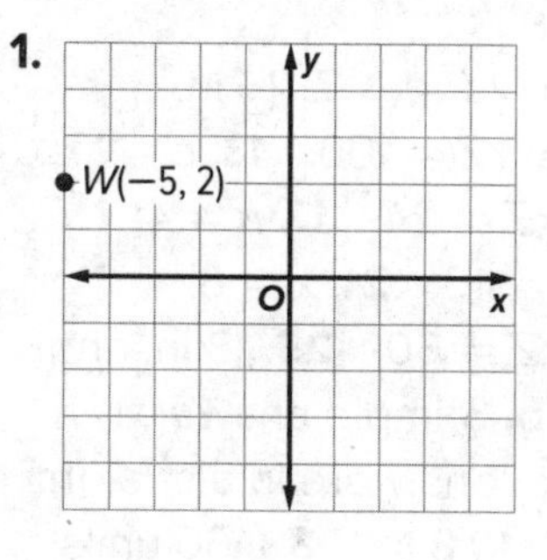

3.

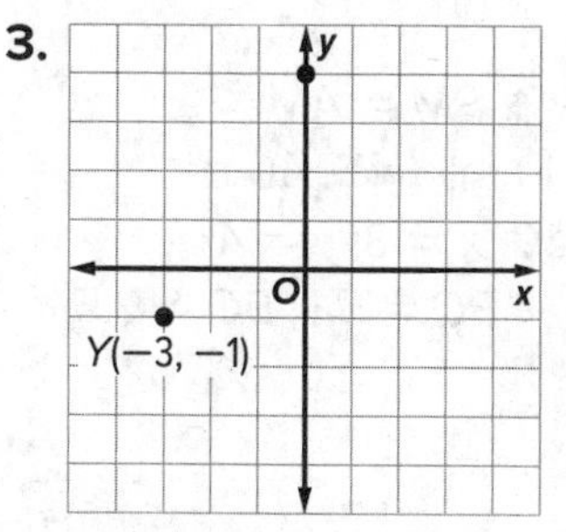

5. 20 **7.** 68

Lesson 1-1

1. Sample answer: Kelsey's jersey number is greater than 5 and less than 11. Marie and Kelsey are on the same team. Kylie's team scored 26 points. **3.** Sample answer: Tom mulched the yard of all three clients this week. Ms. Martinez paid Tom $115 this week. Mr. Hansen paid Tom to mow his lawn and mulch his yard. Mrs. Johnson used all of Tom's services this week. **5.** analytic geometry **7.** synthetic geometry **9.** synthetic geometry **11.** Sample answer: Pedro and Rafael ate the same type of salad. **13.** Sample answer: Theo is likely doing analytic geometry, because he is using a graph with points. **15.** Sample answer: Because Sydney's plan is on a grid, she is likely using analytic geometry; that is, assuming the grid is used as a coordinate system. **17.** Sample answer: The three routes are not a model for the axiomatic system. Axioms 1, 2, and 4 are satisfied. Axiom 3 is not satisfied because Route 3 visits Stadium District twice and it is not the first/last stop. Axiom 5 is not satisfied because all three routes visit Stadium District. **19.** Sample answer: The rules of a game are like the axioms of an axiomatic system. They establish what can happen within the game. Plays are like theorems. They are tested against the rules or axioms to see whether they are legal in the game. In basketball, it is a rule that during playing time 5 players from each team shall be on the playing court. A play in which 6 players are on the court is a violation because the rules allow exactly 5 players. **21.** Sample answer: The second figure does not satisfy all the axioms. The axioms do not specify that the line segments connecting the points need to be straight, so the first and third figures would work. **23.** Sample answer: The triangle on the coordinate grid does not belong because it illustrates analytic geometry, while the other two figures illustrate synthetic geometry.

Lesson 1-2

1. Sample answer: n and q **3.** plane R **5.** Sample answer: point P **7.** Yes; sample answer: Line n intersects line q when the lines are extended. **9.** plane **11.** plane **13.** point on a line **15.** line **17.** plane **19.** plane **21.** Sample answer:

23. Sample answer:

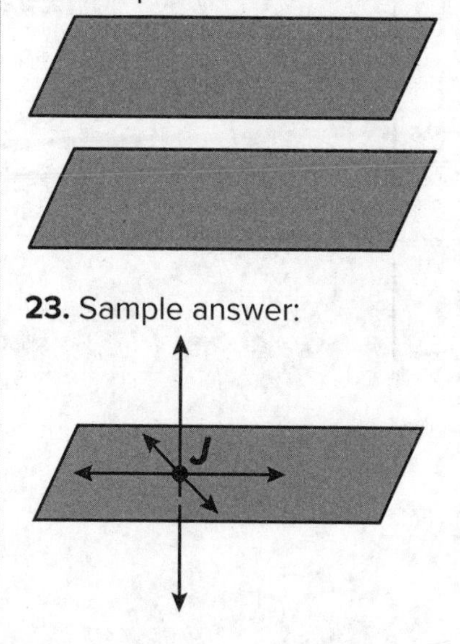

25. 5 **27.** *A, B, E, F* or *B, C, D, E* or *A, C, D, F* **29.** $\overleftrightarrow{AB}$, $\overleftrightarrow{AG}$, $\overleftrightarrow{AH}$, $\overleftrightarrow{BC}$, $\overleftrightarrow{BH}$, $\overleftrightarrow{CD}$, $\overleftrightarrow{DE}$, $\overleftrightarrow{DH}$, $\overleftrightarrow{EF}$, $\overleftrightarrow{FG}$, $\overleftrightarrow{FH}$ **31a.** The lines intersect at the vanishing points. **31b.** Sample answer: The walls of the building and the ground form planes.
33. Sample answer:

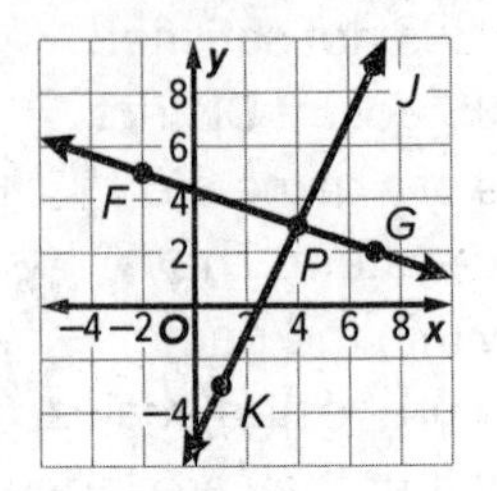

35. Sample answer: line *q* **37.** $\overleftrightarrow{CD}$ or $\overleftrightarrow{DC}$
39. Sample answer:

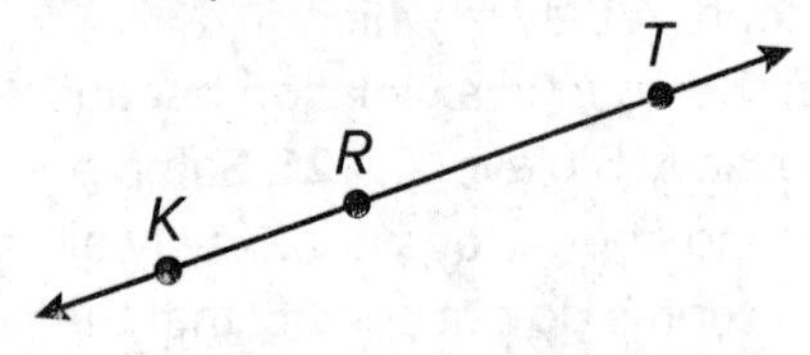

41. Sample answer:

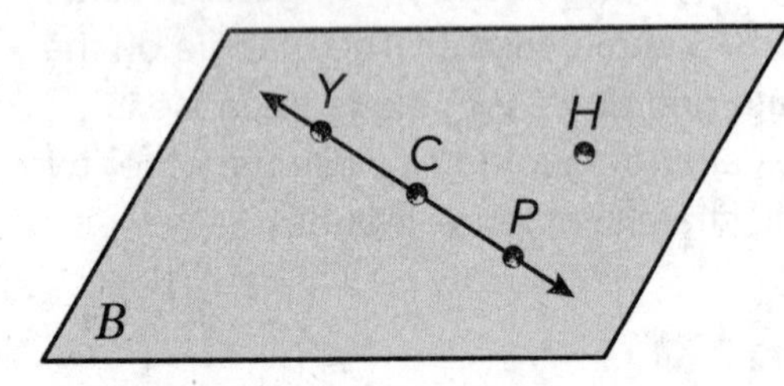

43. lines perpendicular to a plane
45. Sample answer:

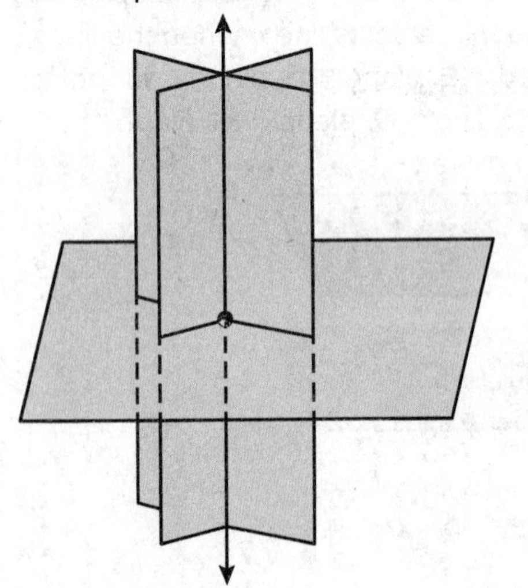

47. Sample answer: Hiroshi is correct. After you draw 3 lines from the first point to the other three points, one of the lines from each of the other three points is already drawn. **49.** Sample answer: A table is a finite plane. It is not possible to have a real-life object that is an infinite plane because all real-life objects have boundaries.

Lesson 1-3

1. 2.1 mm **3.** 1.1 cm **5.** 2.0 m **7.** $2\frac{1}{4}$ in. **9.** 5.3 mm **11.** $b = 12.5$; $YZ = 100$ **13.** $c = 1.7$; $YZ = 3.4$ **15.** $w = 4$; $YZ = 24$ **17.** $n = 4$; $YZ = 1$ **19.** $k = 6$; $YZ = 46$ **21.** $x = 6$; $YZ = 18$ **23.** $x = 10$; $YZ = 60$ **25.** 13 in. and 65 in. **27a.** 6 mi **27b.** Sample answer: I assumed the three locations were in a straight line. **29.** 4.4 mm **31.** 10.8 in. **33.** 66 units
35. 3 **37.** 4
39. Sample answer: $AP + PM = AM$
41. 5184 ft **43.** $x = 3$; 13 mi **45.** 40 ft
47. $JK = 12$, $KL = 16$ **49.** $x = 3$; $y = 4$
51. Sample answer: $2.8 + BC = 5.3$; $BC = 2.5$ in.

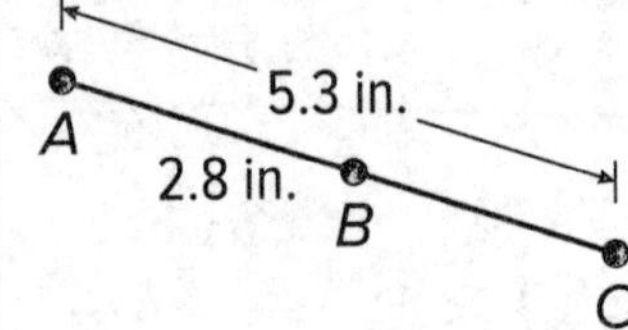

Lesson 1-4

1. 5 **3.** 9 **5.** 12 **7.** 3 **9.** 3 **11.** 9 **13.** 6
15. yes **17.** no **19.** no **21.** 10 units
23. $\sqrt{89}$ or about 9.4 units
25. $\sqrt{20}$ or about 4.5 units **27.** $\sqrt{20}$ or approximately 4.5 units **29.** Yes; sample answer: The distance between Mariah's house and the library is $\sqrt{74}$ or about 8.6 miles. Because $\frac{2}{3}$ of 12 miles is 8 miles, Mariah's bike ride is more than $\frac{2}{3}$ of the cycling portion of the triathlon. **31.** $\sqrt{37}$ or about 6.1 units **33.** $\sqrt{29}$ or about 5.4 units **35.** $\sqrt{13}$ or about 3.6 units **37.** $\sqrt{89}$ or about 9.4 units **39.** $3\sqrt{37}$ or about 18.2 units **41.** $\sqrt{52}$ or about 7.2 units **43.** no **45.** (0, −4), (0, 10) **47.** 10 in. **49.** No; sample answer: We know that $QU + UR = QR = 4$ and $QU = UR$, so $QU = 2$. Further, we know that $RV + VS = RS = 2$, and $RV = VS$, so $RV = 1$. Because QU is not equal to RV, we know that $\overline{QU}$ is not congruent to $\overline{RV}$. **51.** (5, 10)

53. Sample answer: Substitute 10 for d, (1, 3) for (x_1, y_1), and (9, y) for (x_2, y_2) in the Distance Formula: $10 = \sqrt{(9-1)^2 + (y-3)^2}$. Solve for y:

$$100 = (9-1)^2 + (y-3)^2$$
$$= 8^2 + (y-3)^2$$
$$= 64 + (y-3)^2$$
$$36 = (y-3)^2$$
$$6 = y - 3 \text{ or } -6 = y - 3$$
$$9 = y \text{ or } -3 = y$$

So, the y-coordinate of point B is 9 or −3.

Lesson 1-5

1. 6 **3.** 9 **5.** 8.4 **7.** −1 **9.** −5.5 **11.** −4 **13.** −1 **15.** −2 **17.** Y **19.** −4 **21.** −3 **23a.** 4.36 mi **23b.** 10 mi **25.** 720 mi **27.** $2\frac{2}{5}$ in. **29.** Sample answer: Draw $\overline{AB}$. Next, draw a construction line and place point C on it. From C, strike 6 arcs in succession of length AB. On the sixth segment of length AB, perform a segment bisector two times to create a $\frac{1}{4}AB$ length. Label the endpoint D. **31.** Sometimes; sample answer: If the coordinate of X is 0 and the coordinate of Y is negative, then the coordinate of W will be negative and less than the coordinate of X. If the coordinate of X is positive and the coordinate of Y is greater than the coordinate of X, then the coordinate of W will be greater than the coordinate of X.

Lesson 1-6

1. (−3.6, −2.2) **3.** $\left(1, 1\frac{2}{3}\right)$ **5.** $\left(\frac{14}{3}, 1\right)$ **7.** $\left(-\frac{7}{5}, 4\right)$ **9.** $\left(\frac{16}{7}, -3\right)$ **11.** $\left(1, \frac{5}{4}\right)$ **13.** $\left(\frac{20}{7}, -1\right)$ **15.** (−7, 11) and (−1, 1) **17.** (−3, −2) **19.** $\frac{3}{4}$ **21a.** Julianne substituted the wrong values for (x_1, y_1) and (x_2, y_2). **21b.** (1.6, 4.2) **23.** Sample answer: Because the distance from A to P is twice the distance from P to D, the distance from A to P could be 2 and the distance from P to D could be 1. Therefore, the fractional distance that P is from A to D is $\frac{2}{2+1}$ or $\frac{2}{3}$. The coordinates of point P are (5, 10).

25. Sample answer:

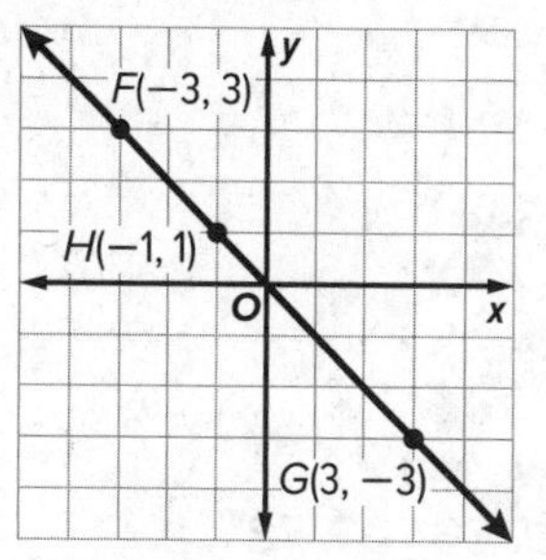

H is $\frac{1}{3}$ of the distance from F to G.

Lesson 1-7

1. −2 **3.** 0.5 **5.** 1.5 **7.** 3 **9.** −4.5 **11.** 8.5 **13.** 9 **15.** 3 **17.** 9 yd **19.** (4, 6) **21.** (−3, −3) **23.** (−1.5, 3.5) **25.** (5, 4) **27.** (18.5, 5.5) **29.** (−6.5, −3) **31.** (4.2, 10.4) **33.** $A(1, 6)$ **35.** $C(16, -4)$ **37.** $C(-12, 13.25)$ **39.** 16 **41.** 2.5 **43.** 8 **45.** 7 **47.** 0.5 **49.** $Q(8, 4)$ **51.** $N(-5, 1)$ **53.** (1, 1) **55.** (6, 7.5) **57.** 48 m **59.** Sample answer: The midpoint of a segment is the average of the coordinates of the endpoints. Divide each coordinate of the endpoint that is not located at the origin by 2. For example, if the segment has coordinates (0, 0) and (−10, 6), the midpoint is located at $\left(-\frac{10}{2}, \frac{6}{2}\right)$ or (−5, 3). Using the Midpoint Formula, if the endpoints of the segment are (0, 0) and (a, b), the midpoint is $\left(\frac{a+0}{2}, \frac{b+0}{2}\right)$ or $\left(\frac{a}{2}, \frac{b}{2}\right)$. **61.** Sample answer:

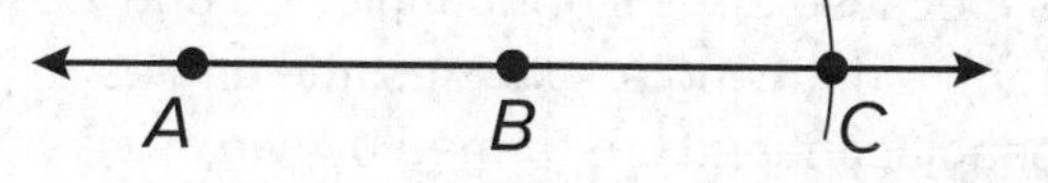

Module 1 Review

1.

Real-World Object	Point	Line	Plane
Electronic Tablet			X
Pool Stick		X	
Scoop of Ice Cream	X		
Light Pole		X	
Emoji	X		

3. intersecting planes **5.** 11 **7.** A **9.** B **11.** D **13.** (3, 6) **15.** B **17.** C

Module 2

Quick Check

1. 63 **3.** 3 **5.** 1980 **7.** 344

Lesson 2-1

1. *A* **3.** $\angle ADC$, $\angle CDA$ **5.** 66° **7.** 78° **9.** 61° **11.** 56° **13.** Sample answer: $\angle SRQ$ and $\angle TRP$ **15.** 65 **17.** $x = 35$, $y = 85$ **19.** *R* **21.** *P* **23.** $\overrightarrow{NM}$ and $\overrightarrow{NV}$ **25.** $\overrightarrow{RP}$ and $\overrightarrow{RQ}$ or $\overrightarrow{RT}$ and $\overrightarrow{RQ}$ **27.** $\angle TPQ$ **29.** $\angle TPN$, $\angle NPT$, $\angle TPM$, $\angle MPT$ **31.** *S*, *Q* **33.** Sample answer: $\angle MPR$, $\angle PRQ$ **35.** $x = 48$, $y = 21$ **37.** Sample answer: $\angle HGE$, $\angle DGE$ **39.** Sample answer: $\angle BFC$, $\angle BFD$ **41.** Sample answer: $\angle STV$, $\angle VTW$, $\angle UTW$ **43a.** yes **43b.** no **43c.** no **43d.** no **45.** 45° **47a.** 45° **47b.** $\overleftrightarrow{EF}$ **49.** 168; sample answer: If $m\angle RMP = 21°$ and $\overrightarrow{MR}$ bisects $\angle QMP$, then $m\angle QMP = 2(21)$ or 42°. If $m\angle QMP = 42°$ and $\overrightarrow{MQ}$ bisects $\angle LMP$, then $m\angle LMP = 2(42)$ or 84°. If $m\angle LMP = 84°$ and $\overrightarrow{MP}$ bisects $\angle LMN$, then $m\angle LMN = 2(84) = 168°$.

Lesson 2-2

1. 72.5°, 107.5° **3.** 128°; 52° **5.** 45°; 135° **7.** $m\angle ABD = 47°$; $m\angle DBC = 43°$ **9.** $a = 20$ **11.** Yes; because $\angle 7$ is a right angle, $\angle 6$ and $\angle 8$ must form a right angle. **13.** Yes; the angles are nonadjacent and are formed by two intersecting lines. **15.** 40° **17.** 7 **19.** 92° **21.** No; the measures of the angles are unknown. **23.** No; the angles are not adjacent. **25.** $x = 25$ **27.** $x = 94$, $y = 79$, $z = 26$ **29.** Yes; sample answer: Angles that are right or obtuse do not have complements because their measures are greater than or equal to 90°. **31.** Sample answer: You can determine whether an angle is right if it is marked with a right angle symbol, if the angle is a vertical pair with a right angle, or if the angle forms a linear pair with a right angle.

33. Sample answer:

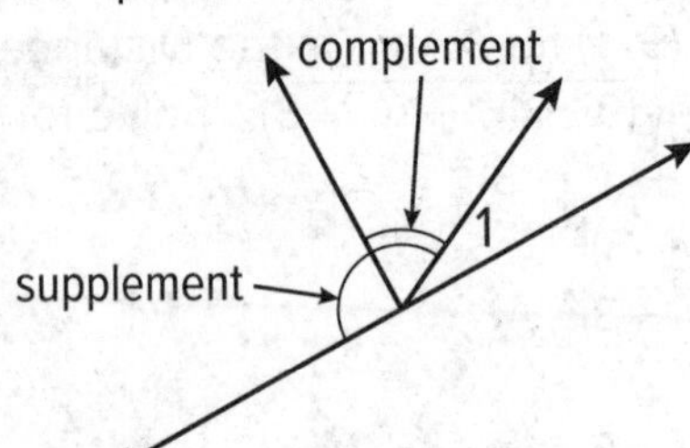

35. No; sample answer: Straight angles or angles that are greater than 180° do not have supplementary angles because their measures are greater than or equal to 180°.

Lesson 2-3

1. 20.9 cm; 16 cm^2 **3.** 17.8 cm; 25.1 cm^2 **5.** 11.3 cm; 8 cm^2 **7.** 15.7 in.; 19.6 in^2 **9.** 42.1 ft; 92.5 ft^2 **11.** 35.1 m^2 **13a.** 398 m **13b.** In part a, I assumed that there was no space between the field and the first lane of the track. I also assumed the athlete's body was centered on the border of the track. **13c.** 402 m **13d.** 30 cm **15.** quadrilateral; 20 units; 24 $units^2$ **17.** quadrilateral; 20 units; 25 $units^2$ **19.** \$650; sample answer: The side of the play area that is adjacent to the house does not need fencing. The remaining three sides of the play area on the grid have lengths 4, 5, and 4 units. The perimeter of the play area to be fenced on the grid is $P = 4 + 5 + 4 = 13$ units. Each unit on the grid represents 5 feet, so Derek will need 13(5 ft) or 65 ft of fencing. The cost of the fencing is \$10 per foot, so the total cost will be 65(\$10) = \$650. **21.** 78.5 square miles **23a.** $s = 4$, so $A = 4^2 = 16$ $units^2$ **23b.** $b = 2$, $h = \sqrt{3}$, so $A = \frac{1}{2}(2)(\sqrt{3}) = \sqrt{3}$ $units^2$ **23c.** $16 - \sqrt{3}$ or about 14.3 $units^2$; sample answer: The area not covered by the triangle is equal to the area of the square minus the area of the triangle.
So, $A = (16 - \sqrt{3})$ $units^2$ or about 14.3 $units^2$.
25. 290.93 $units^2$
27. 39 units; Sample answer: An equilateral triangle has congruent side lengths. Use the Distance Formula to find $KM = \sqrt{(10 - (-2))^2 + (6 - 1)^2} = 13$. So, $P = 3(KM) = 3(13)$ or 39 units.

Lesson 2-4

1. reflection **3.** rotation **5.** translation **7.** rotation **9.** $A'(2, 0)$, $B'(-1, -5)$, and $C'(4, -3)$ **11.** $A'(2, 2)$, $B'(-1, 7)$, and $C'(4, 5)$ **13.** $A'(-2, 0)$, $B'(1, -5)$, and $C'(-4, -3)$ **15.** $D'(4, 1)$, $E'(5, -2)$, and $F'(1, -2)$ **17.** $D'(5, -1)$, $E'(6, 2)$, and $F'(2, 2)$ **19.** $D'(-4,1)$, $E'(-5, -2)$, and $F'(-1,-2)$ **21.** translation along vector $\langle -10, 35 \rangle$ **23a.** rotation **23b.** translation **25.** reflection; translation

27.

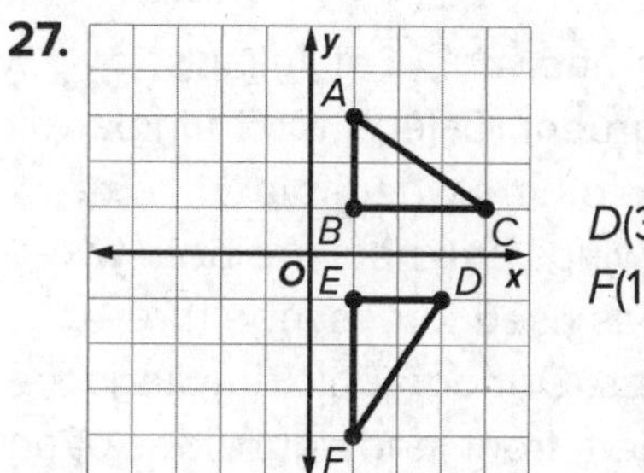

$D(3, -1)$, $E(1, -1)$, $F(1, -4)$

29. reflection in the x-axis **31.** translation 3 units right and 2 units down **33.** Antwan; sample answer: When you reflect a point across the x-axis, the reflected point is in the same place horizontally, but not vertically. When (2, 3) is reflected across the x-axis, the coordinates of the reflected point are (2, −3) because it is in the same location horizontally, but the other side of the x-axis vertically.
35. Sample answer:

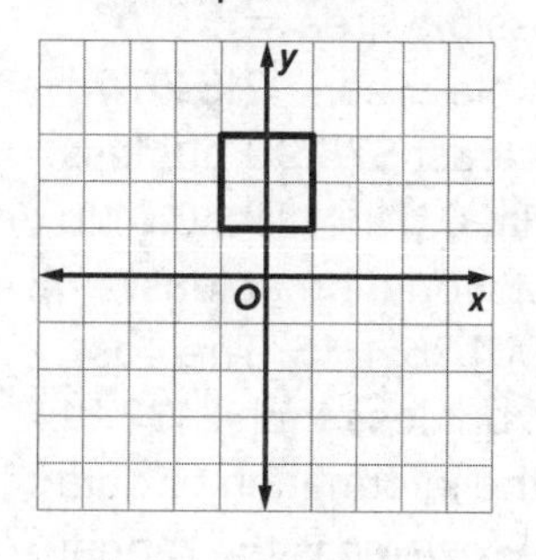

Lesson 2-5

1. not a polyhedron; cone **3.** polyhedron; rectangular pyramid; base $WXYZ$; faces $\square WXYZ$, $\triangle VWX$, $\triangle VXY$, $\triangle VYZ$, $\triangle VZW$; edges $\overline{WX}$, $\overline{XY}$, $\overline{YZ}$, $\overline{ZW}$, $\overline{WV}$, $\overline{XV}$, $\overline{YV}$, $\overline{ZV}$; vertices W, X, Y, Z, V **5.** sphere; not a polyhedron **7.** 26.9 cm²; 7.7 cm³ **9.** 800 ft²; 1280 ft³ **11.** 90π or about 282.7 yd²; 100π or about 314.2 yd³ **13a.** 35 ft² **13b.** 13 bags **15.** 2004 in² **17.** 3.9 cm **19.** 3 in. **21a.** 5027 ft³ **21b.** 402 ft³ **21c.** 5429 ft³ **23a.** 0.012 m³; $V = \frac{1}{3}Bh$, so $V = \frac{1}{3}(900)(40)$ or 12,000 cm³. One cubic meter equals 1 million cubic centimeters, so the volume of one pyramid shaped lawn ornament is 0.012 m³. **23b.** concrete: 28.5 kg; granite: 32.2 kg; marble 32.5 kg **23c.** If the volume of the lawn ornament stays the same, then the weight of the ornament increases as the density of the material used to make it increases. **25.** Neither; sample answer: The surface area is twice the sum of the areas of the top, front, and left side of the prism or $2(5 \cdot 3 + 5 \cdot 4 + 3 \cdot 4)$, which is 94 square inches. **27.** Sample answer: The cone and pyramid have nearly the same volumes. Cone: The area of the base is approximately 154 square centimeters, so $V \approx \frac{1}{3}(154)(28)$ or about 1437 cubic centimeters. The volume of the pyramid is greater by such a small amount that we can say the volumes are approximately equal. **29.** 27 mm³

Lesson 2-6

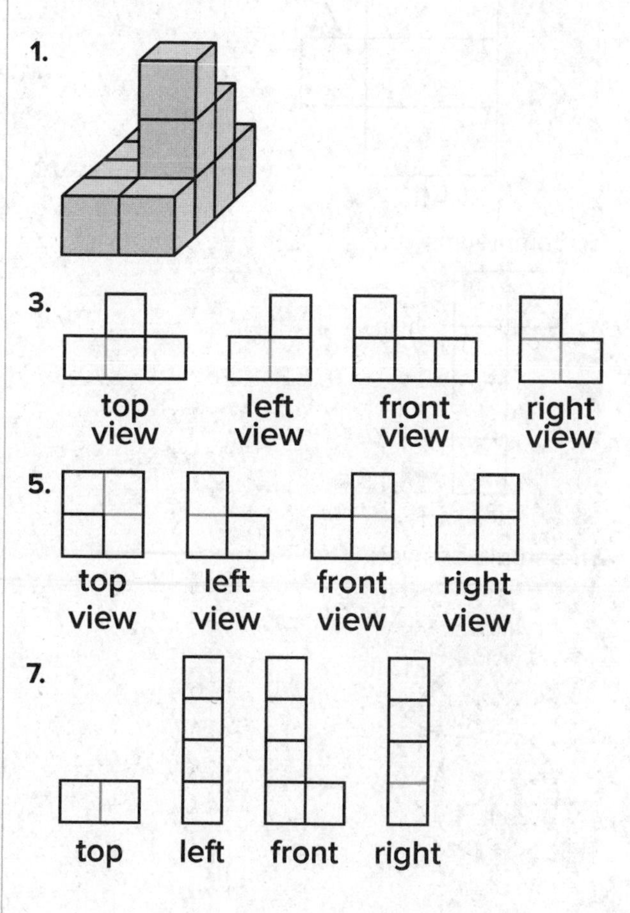

9. square pyramid, 144 cm^2

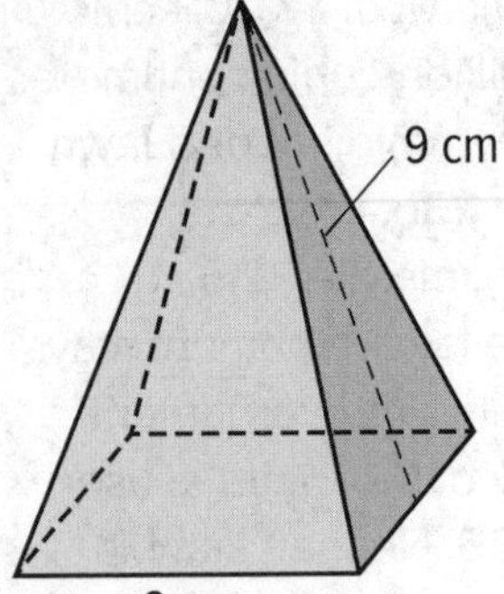

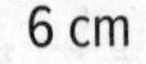

11. cylinder; 3.5π in^2 or 11.0 in^2

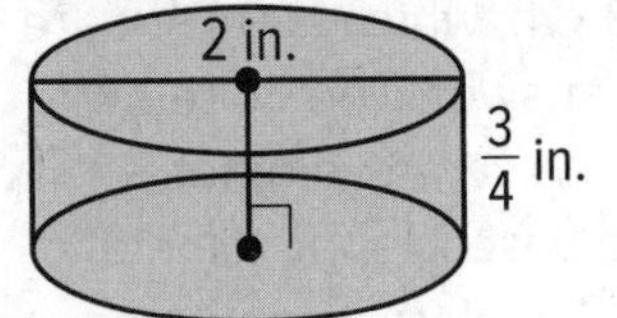

13. tetrahedron

15. octahedron

17. Sample answer:

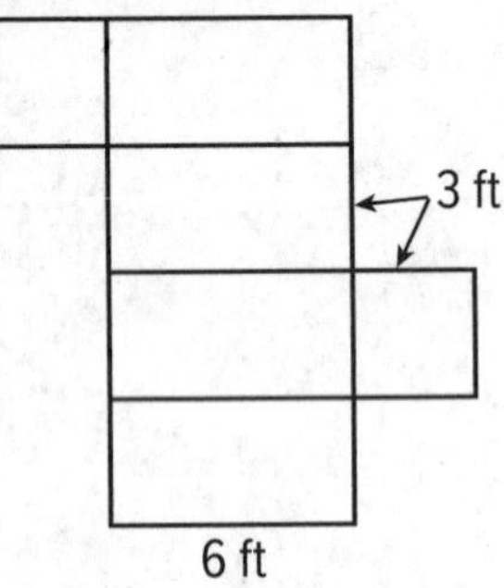

19. Sample answer:

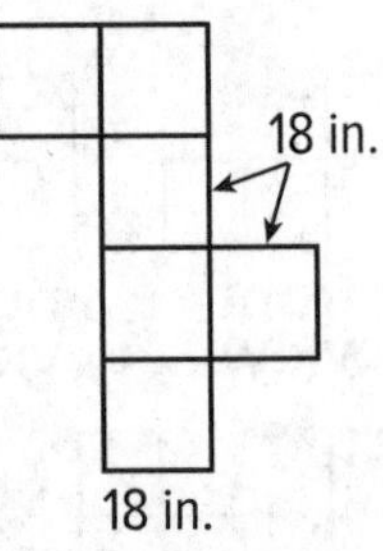

21. Sample answer:

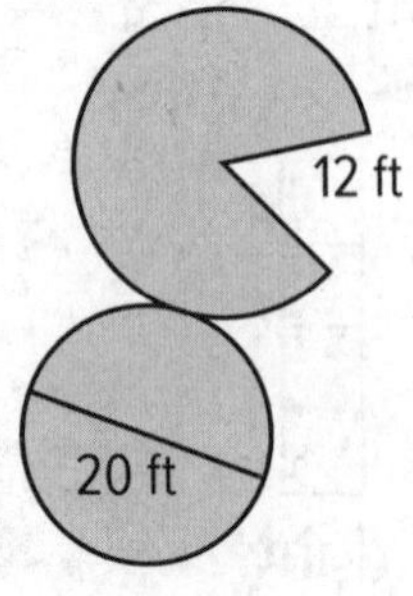

23.

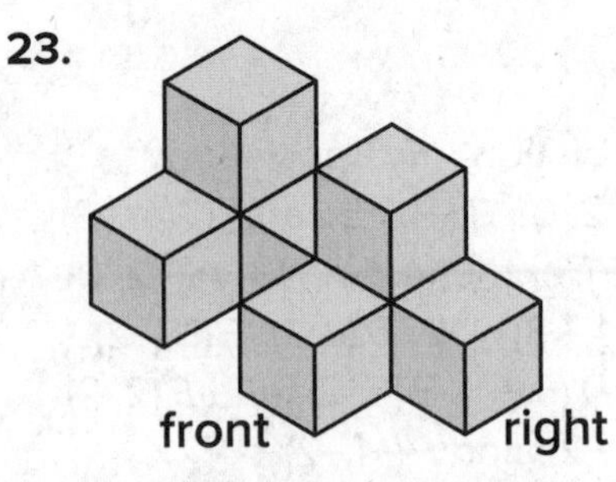

25. Sample answer: ice cream cone

27. No; the area of this net is 48 square centimeters.

29. Julian; sample answer: All of Julian's triangles are congruent. Caleb's top triangle will not match with the others when folded. **31.** Orthographic drawings and nets are both two-dimensional shapes used to describe three-dimensional figures. Orthographic drawings are views of the top, left, front, and right sides of an object, whereas nets can be folded to create a three-dimensional object.

Lesson 2-7

1. The sample is precise because there are consistently 17 or 18 rice cakes in each package. The sample indicates an innacurate claim of 20 rice cakes per package.
3. 0.05 milliamp **5.** 1.1°F **7a.** 16.5 yd **7b.** 12.5 yd **7c.** least possible area = 189.75 yd^2; greatest possible area = 218.75 yd^2 **7d.** The cost would be at least \$75.90 but less than \$87.50. **9.** 1 in. **11a.** 0.11 lb **11b.** 0.01 lb **11c.** 0.64 lb **13.** The cost would be at least \$955.94 but less than \$961.36. **15.** The cost would be at least \$107.81 but less than \$132.81. **17.** Accuracy is how well the information or data matches the true values. Precision is the repeatability of the measurement and level of measurement. Sample answer: 16.1 oz, 16.3 oz, 15.93 oz, 15.8 oz. **19a.** There are two faces that have an area of 111.4 in^2, two faces that have an area of 35.82 in^2, and two faces that have an area of 75.3 in^2. **19b.** 445.0 in^2 **19c.** The calculation of surface area is accurate to the nearest tenth. The true surface area falls between 443.31 in^2 and 446.60 in^2. **19d.** 440.1 in^2

Lesson 2-8

1. 5 **3.** 2 **5.** 1 **7.** Michelle **9.** 4.5 to 5.5 in. **11.** 2.5 to 3.5 cm **13.** 4.375 to 4.625 in. **15.** 2 **17.** 4 **19.** 2 **21.** 7.2 mL **23.** 15.1 cm^2 **25.** The area of the circle is between 31.2 and 33.2 cm^2. **27.** 4 **29.** 5 **31.** 5 **33.** 0.663 gram **35.** Scale 1: 4; Scale 2: 3; Scale 3: 4 **37.** 79.40 yd^2 **39.** 8.3 g **41.** 624 in^3

43. Sample answer: The 0 in each dimension, 1.40 cm and 1.60 cm, is significant. The answer should be given with 3 significant figures as 2.24 square centimeters. **45.** Yes; sample answer: The zeros before and after the decimal are not significant because a nonzero number did not come before them. Therefore, both numbers have two significant figures.
47. Sometimes; sample answer: A zero between two nonzero significant figures is always significant, a leading zero is never significant, and a zero at the end of a number is only significant when a decimal point is given in the number.

Module 2 Review

1. A, C **3.** Sample answer: Use paddy paper or wax paper. Draw an angle on the paper. Fold the paper so that one side of the angle is directly on top of the other side. Draw the angle bisector in the crease of the fold. **5.** B **7.** Perimeter: ≈22.4 units; Area: 30 square units **9.** B **11.** C **13.** 2144.66 in^3 **15.** B

Module 3

Quick Check

1. 5 **3.** 2.3 **5.** $\angle BXD$, $\angle AXE$ **7.** 6

Lesson 3-1

1. Each term in the pattern is four more than the previous term; 24.

3. Each term is one half the previous term; $\frac{1}{16}$.

5. Each arrival time is 2 hours and 30 minutes prior to the previous arrival time; 7:30 A.M.

7. The shaded section in each circle has moved one section counterclockwise from its location in the previous circle.

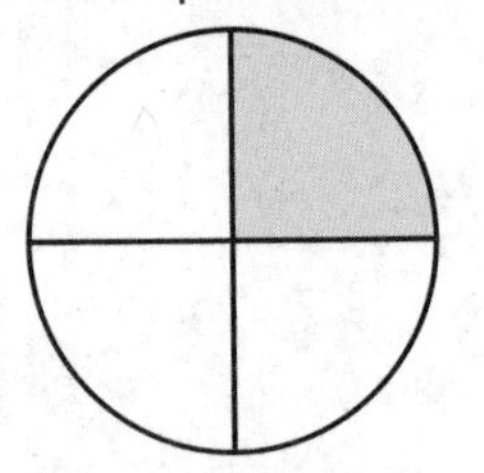

9. The product is an odd number.

11. They are equal. **13.** The lines are perpendicular.

15. $NP = PQ$ **17.** 180 cm

19. False; sample answer: Suppose $x = 2$, then $-x = -2$.

21. false;

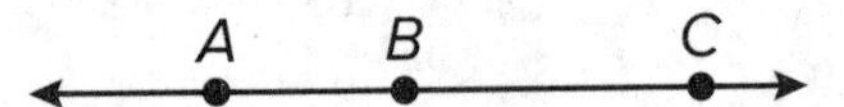

23. False; sample answer: The length could be 4 m, and the width could be 5 m.

25. Yes; sample answer: If no team got more than 5 medals, then the total number of medals could not be more than 5×6 or 30 medals.

27a. 1, 3, 5 **27b.** You get all the odd numbers.

27c. Their sum equals n^2.

29. Sample answer: When $n = 4$, $2^n - 1 = 15$ and $15 = 3 \times 5$.

31a. Sample answer: The number of regions doubles when you add a point on the circle.

31b. For six points, there should be 32 regions; however only 31 regions are formed. The conjecture is false.

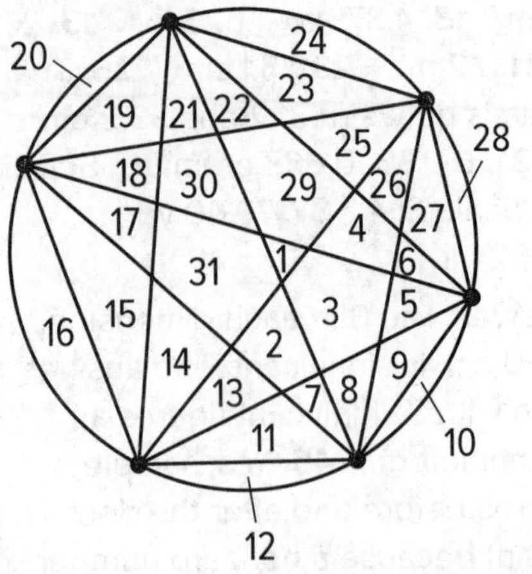

33. False; sample answer: If the two points create a straight angle that includes the third point, then the conjecture is true. If the two points do not create a straight angle with the third point, then the conjecture is false.

35. Sample answer: A postulate states that a line contains at least two points. These two points and all the points between them are line segments, by the definition of a line segment.

Lesson 3-2

1. $-3 - 2 = -5$, and vertical angles are congruent; p is true, and q is true, so p and q is true.

3. Vertical angles are congruent, or $2 + 8 \leq 10$; q is true, and $\sim r$ is true, so $q \vee \sim r$ is true.

5. $-3 - 2 \neq -5$, and not all vertical angles are congruent; $\sim p$ is false, and $\sim q$ is false, so $\sim p \wedge \sim q$ is false. **7.** H: there is no struggle; C: there is no progress **9.** H: you lead; C: I will follow **11.** H: two angles are vertical; C: they are congruent **13.** H: you were at the party; C: you received a gift; If you were at the party, then you received a gift. **15.** H: a figure is a circle; C: the area is πr^2; If a figure is a circle, then the area is πr^2. **17.** H: an angle is right; C: the angle measures 90°; If an angle is right, then the angle measures 90°. **19.** Converse: If you can buy five raffle tickets, then you have five dollars. The converse is true. Inverse: If you do not have five dollars, then you cannot buy five raffle tickets. The inverse is true.

Contrapositive: If you cannot buy five raffle tickets, then you do not have five dollars. The contrapositive is true.

21. Converse: If you do not take this medicine, then you can drive. The converse is true. Inverse: If you are not driving, then you can take this medicine. The inverse is true. Contrapositive: If you take this medicine, then you are not driving. The contrapositive is true.
23. Conditional: If it is raining, then the game will be cancelled. The conditional is true. Converse: If the game is cancelled, then it is raining. Counterexample: The game could be cancelled, and it is not raining. The converse is false. Because the converse is false, a biconditional statement cannot be written.
25. Conditional: If a polygon has four sides, then it is a quadrilateral. Converse: If a polygon is a quadrilateral, then it has four sides. The conditional and the converse are true, so the biconditional is true. **27.** true **29.** false
31. yes **33.** nothing **35.** If a ray does not bisect an angle, then it does not divide the angle into two congruent angles.
37. If two angles are right angles, then they are congruent.
39a. Sample answer: If you are in Houston, then you are in Texas. **39b.** Sample answer: Converse: If you are in Texas, then you are in Houston. Inverse: If you are not in Houston, then you are not in Texas. Contrapositive: If you are not in Texas, then you are not in Houston. **39c.** Converse: false; Inverse: false; Contrapositive: true
41. There exists at least one student at Hammond High school that does not have a locker.
43. For every real number x, $x^2 \neq x$. **45.** There exists a real number that does not have a real square root.
47. Truth table with the following columns:

p	q	$\sim p$	$\sim q$	$p \rightarrow q$	$q \rightarrow p$	$\sim p \rightarrow \sim q$	$\sim q \rightarrow \sim p$
T	T	F	F	T	T	T	T
T	F	F	T	F	T	T	F
F	T	T	F	T	F	F	T
F	F	T	T	T	T	T	T

Because column 5 is the same as column 8, the conditional is equivalent to its contrapositive. Because column 6 is the same as column 7, the converse and the inverse are equivalent.
49. Kiri; sample answer: When the hypothesis of a conditional is false, the conditional is always true.

Lesson 3-3

1. deductive **3.** deductive **5.** inductive
7. valid; Law of Detachment **9.** Invalid; your battery could be dead because it was old.
11. valid; Law of Detachment **13.** If Tina has a grade point average of 3.0 or greater, then she will have her name in the school paper.
15. If the measure of an angle is between 90° and 180°, then it is not acute. **17.** No valid conclusion; the conclusion of statement (1) is not the hypothesis of statement (2). **19.** The sum of the measures of ∠1 and ∠2 is 90°; Law of Detachment **21.** No valid conclusion; the conclusion of statement (1) is not the hypothesis of statement (2).
23. If Terryl completes a course with a grade of C, then he will have to take the course again; Law of Syllogism. **25.** Valid; Theo is inside the small and large circles, so the conclusion is valid.

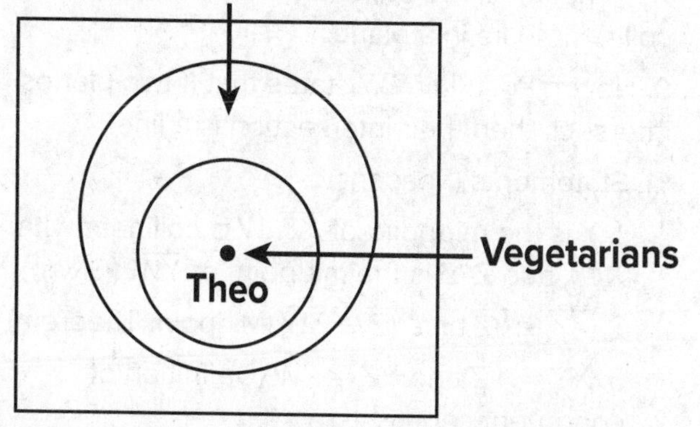

27. then Mozart did not live in Vienna in the early 1800s. **29.** The child is at least 5 years old. **31.** Energy costs will be higher in Florida.
33. Law of Detachment: $[(p \rightarrow q) \wedge p] \rightarrow q$. Law of Syllogism: $[(p \rightarrow q) \wedge (q \rightarrow r)] \rightarrow (p \rightarrow r)$.
35. Sample answer: Jonah's statement can be restated as, "Jonah is in Group B, and Janeka is in Group B." For this compound statement

to be true, both parts of the statement must be true. If Jonah was in Group A, he would not be able to say that he is in Group B, because students in Group A must always tell the truth. Therefore, the statement that Jonah is in Group B is true. For the compound statement to be false, the statement that Janeka is in Group B must be false. Therefore, Jonah is in Group B, and Janeka is in Group A.

37. Sample answer: Given: If you are at the Willis Tower, then you are in Chicago. If you are in Chicago, then you are in Illinois. Conclusion: Therefore, if you are at the Willis Tower, then you are in Illinois.

39. D; Statement D follows logically from statements (1) and (2). Statements A, B, and C do not follow logically from statements (1) and (2).

Lesson 3-4

1. The two planes meet at the edge, which lies on line t. Postulate: If two planes intersect, then their intersection is a line.

3. Postulate 3.1; Through any two points, there is exactly one line.

5. Always; Postulate 3.2 states that through any three noncollinear points, there is exactly one plane.

7. Sometimes; the points do not have to be collinear to lie in a plane.

9. Never; Postulate 3.7 states that if two planes intersect, then their intersection is a line.

11. Statements (Reasons)

1. Y is the midpoint of $\overline{XZ}$. W is collinear with X, Y, and Z. Z is the midpoint of $\overline{YW}$. (Given)
2. $\overline{XY} \cong \overline{YZ}$ and $\overline{YZ} \cong \overline{ZW}$ (Midpoint Theorem)
3. $XY = YZ$ and $YZ = ZW$ (Definition of congruent segments)
4. $XY = ZW$ (Transitive Property of Equality)
5. $\overline{XY} \cong \overline{ZW}$ (Definition of congruent segments)

13. Statements (Reasons)

1. $SR = RT$ (Given)
2. R is the midpoint of $\overline{ST}$. (Definition of midpoint)
3. $SR = UR$ and $RT = RV$ (Given)
4. $SR = RT$, so $UR = RV$. (Substitution)
5. R is the midpoint of $\overline{UV}$. (Definition of midpoint)

15.

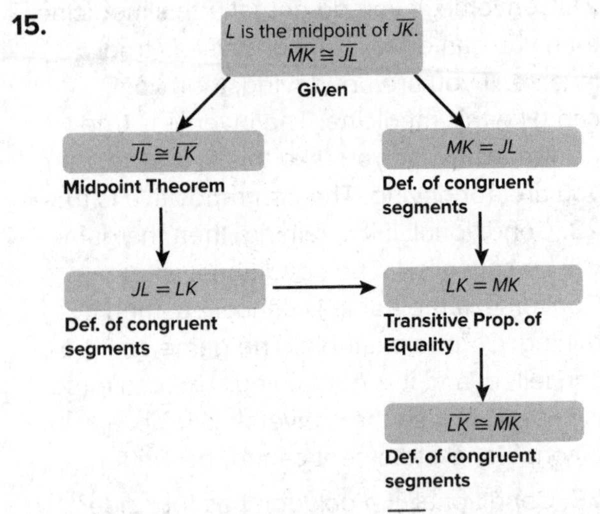

17. Given: B is the midpoint of $\overline{AC}$. C is the midpoint of $\overline{BD}$.
Prove: $AB = CD$
Proof: Because B is the midpoint of $\overline{AC}$ and C is the midpoint of $\overline{BD}$, we know by the Midpoint Theorem that $\overline{AB} \cong \overline{BC}$ and $\overline{BC} \cong \overline{CD}$. Because congruent segments have equal measures, $AB = BC$ and $BC = CD$. Thus, by the Transitive Property of Equality, $AB = CD$.

19. Postulate 3.3: A line contains at least two points. **21.** Iza is because a line can lie in a plane and intersect it in infinite points. **23.** 15 **25.** Ana is correct. Sample answer: The proof should begin with the given, which is that $\overline{AB}$ is congruent to $\overline{BD}$ and A, B, and D are collinear. Therefore, Ana began the proof correctly. **27a.** Plane Q is perpendicular to plane P. **27b.** Line a is perpendicular to plane P. **29.** Sometimes; if the points were noncollinear, then there would be exactly one plane by Postulate 3.2 shown by Figure 1. If the points were collinear, then there would be infinitely many planes. Figure 2 shows what two planes through collinear points would look like. More planes would rotate around the three points.

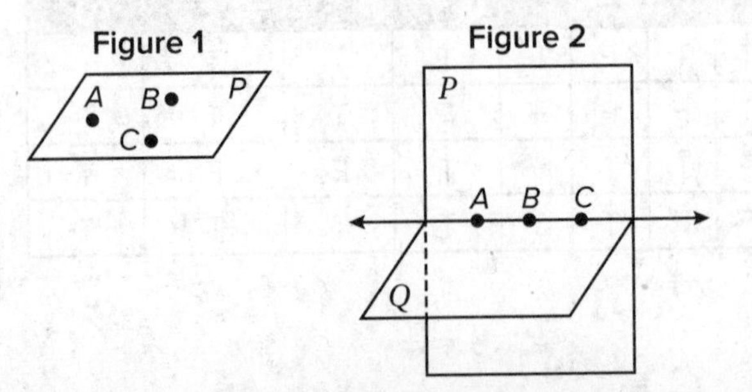

Lesson 3-5

1.

1. C is the midpoint of $\overline{AE}$.
   C is the midpoint of $\overline{BD}$.
   $\overline{AE} \cong \overline{BD}$
2. Definition of midpoint
3. Definition of $\cong$ segments
4. $AE = AC + CE$
   $BD = BC + CD$
5. Substitution Property
6. Substitution Property
7. $2AC = 2CD$
8. $AC = CD$
9. Definition of $\cong$ segments

3. Given: $\overline{VZ} \cong \overline{VY}$ and $\overline{WY} \cong \overline{XZ}$

Prove: $\overline{VW} \cong \overline{VX}$

Proof

Statements (Reasons)

1. $\overline{VZ} \cong \overline{VY}$ and $\overline{WY} \cong \overline{XZ}$ (Given)
2. $VZ = VY$ and $WY = XZ$ (Def. of $\cong$ segments)
3. $VZ = VX + XZ$ and $VY = VW + WY$ (Segment Addition Postulate)
4. $VX + XZ = VW + WY$ (Substitution Prop.)
5. $VX + WY = VW + WY$ (Substitution Prop.)
6. $VX = VW$ (Sub. Prop. of =)
7. $VW = VX$ (Symmetric Property)
8. $\overline{VW} \cong \overline{VX}$ (Def. of $\cong$ segments)

5. Clara is shorter than Chad when Maria is shorter than Luna; Clara and Chad are the same height when Maria is the same height as Luna.

7. No, it's not possible. Lola's house must be less than a mile from each house because she lives between them.

9a. Given: $\overline{AC} \cong \overline{GI}$, $\overline{FE} \cong \overline{LK}$, $AC + CF + FE = GI + IL + LK$

Prove: $\overline{CF} \cong \overline{IL}$

Proof

Statements (Reasons)

1. $\overline{AC} \cong \overline{GI}$, $\overline{FE} \cong \overline{LK}$, $AC + CF + FE = GI + IL + LK$ (Given)
2. $AC = GI$ and $FE = LK$ (Def. of $\cong$ segments)
3. $AC + CF + FE = AC + IL + LK$ (Substitution Property)
4. $AC - AC + CF + FE = AC - AC + IL + LK$ (Subtraction Property of Equality)
5. $CF + FE = IL + LK$ (Substitution Property)
6. $CF + FE = IL + FE$ (Substitution Property)
7. $CF + FE - FE = IL + FE - FE$ (Subtraction Property of Equality)
8. $CF = IL$ (Substitution Property)
9. $\overline{CF} \cong \overline{IL}$ (Def. of $\cong$ segments)

11a. Both segments are half the length of two congruent segments, so the lengths of the shorter segments must be the same.

11b.

1. $\overline{AB} \cong \overline{CD}$; M is the midpoint of $\overline{AB}$ and $\overline{CD}$.
2. Congruent segments have equal lengths.
3. Definition of midpoint
5. Segment Addition Postulate
6. Substitution Property of Equality
8. Substitution Property of Equality
9. $AM = CM$
10. $\overline{AM} \cong \overline{CM}$

13. Neither; because $\overline{AB} \cong \overline{CD}$ and $\overline{CD} \cong \overline{BF}$, then $\overline{AB} \cong \overline{BF}$ by the Transitive Property of Congruence.

15. Sample diagram:

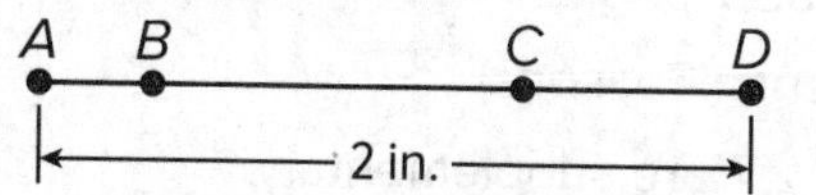

17. No; The Segment Addition Postulate only applies to points that are collinear, but points P, Q, and R are not collinear.

19. Because $\overline{PQ} \cong \overline{RS}$ and congruent segments have equal lengths, $PQ = RS$. Because Q is the midpoint of $\overline{PR}$, $PQ = QR$. By the Substitution Property of Equality, $QR = RS$ so R is the midpoint of $\overline{QS}$.

Lesson 3-6

1. 113° **3.** 74° **5.** 46°

7. Statements (Reasons)

1. $\angle 1$ and $\angle 2$ form a linear pair.
   $\angle 3$ and $\angle 4$ form a linear pair. (Def. of linear pair)
2. $\angle 1$ and $\angle 2$ are supplementary.
   $\angle 3$ and $\angle 4$ are supplementary. (Supp. Thm)
3. $\angle 1 \cong \angle 3$ (Given)
4. $\angle 2 \cong \angle 4$ ($\cong$ Supp. Thm)

9. Statements (Reasons)

1. $m\angle ABC = m\angle DEF$ (Given)
2. $\angle ABC \cong \angle DEF$ (Def. of $\cong$ angles)
3. $\angle ABC$ and $\angle DEF$ are supplementary. (Given)
4. $\angle ABC$ and $\angle DEF$ are rt. angles. (If two $\angle$s are $\cong$ and supp., then each $\angle$ is a rt. $\angle$.)

11. 36°; 84°

13. $m\angle 6 = m\angle 8 = 73°$, $m\angle 7 = 107°$ ($\cong$ Supp. Thm. and Vert. $\angle$s Thm.)

15. $m\angle 3 = 62°$, $m\angle 1 = m\angle 4 = 45°$ ($\cong$ Comp. and Supp. Thm.)

17. $m\angle 9 = 156°$, $m\angle 10 = 24°$ ($\cong$ Supp. Thm.)

19. Statements (Reasons)

1. $\angle PQT$ and $\angle TQR$ form a linear pair. (Given)
2. $\angle PQR$ is a straight angle. (Given from figure)
3. $m\angle PQR = 180°$ (Def. of straight angle)
4. $m\angle PQT + m\angle TQR = m\angle PQR$ (Angle Add. Post.)
5. $m\angle PQT + m\angle TQR = 180°$ (Subs.)
6. $\angle PQT$ and $\angle TQR$ are supplementary. (Def. of supp. angles)

21. Statements (Reasons)

1. $\angle 1$ and $\angle 2$ are supplementary.
   $\angle 2$ and $\angle 3$ are supplementary. (Given)
2. $m\angle 1 + m\angle 2 = 180°$
   $m\angle 2 + m\angle 3 = 180°$ (Def. of supp. angles)
3. $m\angle 1 + m\angle 2 = m\angle 2 + m\angle 3$ (Subs.)
4. $m\angle 1 = m\angle 3$ (Subtraction Prop. of =)
5. $\angle 1 \cong \angle 3$ (Def. of $\cong$ angles)

23. Statements (Reasons)

1. $\ell \perp m$ (Given)
2. $\angle 1$ is a right angle. (Def. of $\perp$)
3. $m\angle 1 = 90°$ (Def. of rt. angles)
4. $\angle 1 \cong \angle 4$ (Vert. Angles Thm)
5. $m\angle 1 = m\angle 4$ (Def. of $\cong$ angles)
6. $m\angle 4 = 90°$ (Subs.)
7. $\angle 1$ and $\angle 2$ form a linear pair.
   $\angle 3$ and $\angle 4$ form a linear pair. (Def. of linear pairs)
8. $m\angle 1 + m\angle 2 = 180°$, $m\angle 4 + m\angle 3 = 180°$ (Linear pairs are supp.)
9. $90° + m\angle 2 = 180°$, $90° + m\angle 3 = 180°$ (Subs.)
10. $m\angle 2 = 90°$, $m\angle 3 = 90°$ (Subtraction)
11. $\angle 2$, $\angle 3$, and $\angle 4$ are rt. angles. (Def. of rt. angles (Steps 6, 10))

25. Statements (Reasons)

1. $\angle 1 \cong \angle 2$, $\angle 1$ and $\angle 2$ are supplementary. (Given)
2. $m\angle 1 + m\angle 2 = 180°$ (Def. of supp. angles)
3. $m\angle 1 = m\angle 2$ (Def. of $\cong$ angles)
4. $m\angle 1 + m\angle 1 = 180°$ (Subs.)
5. $2(m\angle 1) = 180°$ (Subs.)
6. $m\angle 1 = 90°$ (Div. Prop.)
7. $m\angle 2 = 90°$ (Subs. (steps 3, 6))
8. $\angle 1$ and $\angle 2$ are rt. angles. (Def. of rt. angles)

27. By the Transitive Property, if any two angles are equal to the angle of the template, then they must be equal to each other.

29. Sample answer: $m\angle WXY = 90°$

Given: $m\angle WXZ = 45°$, $\angle WXZ \cong \angle YXZ$

Prove: $m\angle WXY = 90°$

Proof:

Statements (Reasons)

1. $m\angle WXZ = 45°$, $\angle WXZ \cong \angle YXZ$ (Given)
2. $m\angle WXZ = m\angle YXZ$ (Def. of $\cong$ $\angle$s)
3. $m\angle YXZ = 45°$ (Substitution)
4. $m\angle WXY = m\angle WXZ + m\angle YXZ$ (Angle Add. Post.)
5. $m\angle WXY = 45° + 45°$ (Substitution)
6. $m\angle WXY = 90°$ (Substitution)

31. Each of these theorems uses the words *or to congruent angles* indicating that this case of the theorem must also be proved true. The first proof of each theorem only addressed the *to the same angle* case of the theorem.

Proof of the Congruent Complements Theorem (Case 2: Congruent Angles)

Given: $\angle ABC \cong \angle DEF$, $\angle GHI$ is complementary to $\angle ABC$, $\angle JKL$ is complementary to $\angle DEF$.

Prove: $\angle GHI \cong \angle JKL$

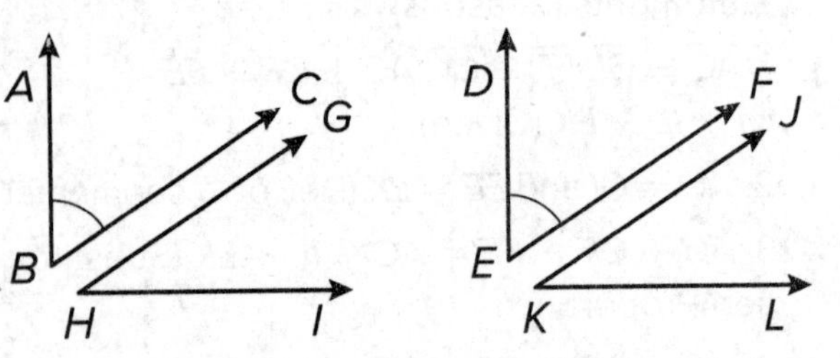

Proof:

Statements (Reasons)

1. $\angle ABC \cong \angle DEF$, $\angle GHI$ is complementary to $\angle ABC$, $\angle JKL$ is complementary to $\angle DEF$. (Given)
2. $m\angle ABC + m\angle GHI = 90°$, $m\angle DEF + m\angle JKL = 90°$ (Def. of compl. angles)
3. $m\angle ABC = m\angle DEF$ (Def. of $\cong$ angles)
4. $m\angle ABC + m\angle JKL = 90°$ (Subs.)
5. $90° = m\angle ABC + m\angle JKL$ (Symm. Prop.)
6. $m\angle ABC + m\angle GHI = m\angle ABC + m\angle JKL$ (Trans. Prop.)
7. $m\angle ABC - m\angle ABC + m\angle GHI = m\angle ABC - m\angle ABC + m\angle JKL$ (Subt. Prop.)
8. $m\angle GHI = m\angle JKL$ (Subs.)
9. $\angle GHI \cong \angle JKL$ (Def. of $\cong$ angles)

Proof of the Congruent Supplements Theorem (Case 2: Congruent Angles)

Given: $\angle ABC \cong \angle DEF$, $\angle GHI$ is supplementary to $\angle ABC$, $\angle JKL$ is supplementary to $\angle DEF$.

Prove: $\angle GHI \cong \angle JKL$

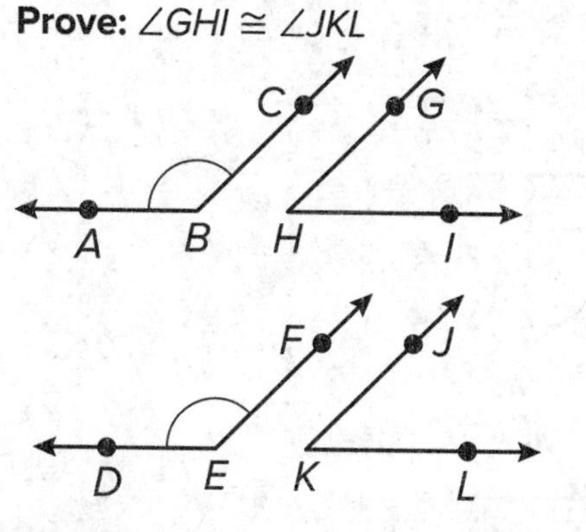

Proof:

Statements (Reasons)

1. $\angle ABC \cong \angle DEF$, $\angle GHI$ is suppl. to $\angle ABC$, $\angle JKL$ is suppl. to $\angle DEF$. (Given)
2. $m\angle ABC + m\angle GHI = 180°$, $m\angle DEF + m\angle JKL = 180°$ (Def. of suppl. angles)
3. $m\angle ABC = m\angle DEF$ (Def. of $\cong$ angles)
4. $m\angle ABC + m\angle JKL = 180°$ (Subs.)
5. $180° = m\angle ABC + m\angle JKL$ (Symm. Property)
6. $m\angle ABC + m\angle GHI = m\angle ABC + m\angle JKL$ (Trans. Prop.)
7. $m\angle ABC - m\angle ABC + m\angle GHI = m\angle ABC - m\angle ABC + m\angle JKL$ (Subt. Prop.)
8. $m\angle GHI = m\angle JKL$ (Subs.)
9. $\angle GHI \cong \angle JKL$ (Def. of $\cong$ angles)

Lesson 3-7

1. $\overline{BF}$, $\overline{CG}$, and $\overline{DH}$ **3.** *ABCD* and *EFGH* or *ABFE* and *CDHG* **5.** $\overline{EF}$, $\overline{AB}$, and $\overline{DC}$
7. Sample answer: *ABCD* and *DCGH* could be characterized as perpendiculars, because *DCGH* contains segment $\overline{CG}$, which is perpendicular to *ABCD*. **9.** line *d*; alternate exterior **11.** line *d*; corresponding **13.** line *a*; consecutive interior **15.** line *c*; alternate interior **17.** line *d*; corresponding **19.** line *p*
21. 100° **23.** 80° **25.** 100° **27.** 170°
29. $x = 28$, $y = 47$; Use supplementary angles to find *x*. Then use alternate exterior angles to find *y*. **31.** $x = 12$, $y = 31$ **33.** 105° **35.** 105° **37.** 75° **39.** 94°

41. 64°; sample answer: Opposite sides of a rectangle are parallel. So, the top and bottom lines on the side panel are parallel and cut by a transversal, which is the dashed line. Therefore, $\angle 1$ and the 116°-angle are consecutive interior angles, so their sum is 180°. $m\angle 1 + 116° = 180°$, so $m\angle 1 = 64°$.

43. Sample answer:

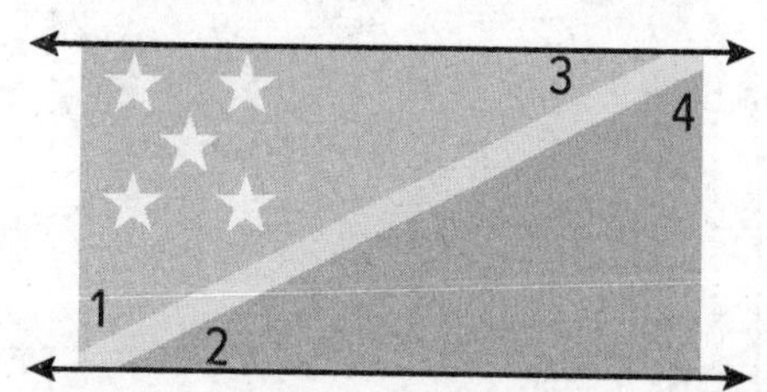

$\angle 1$ and $\angle 4$ are alternate interior angles. $\angle 2$ and $\angle 3$ are alternate interior angles. $\angle 1$ and $\angle 2$ are complementary angles, and $\angle 3$ and $\angle 4$ are complementary angles.

45. By the Corresponding Angles Postulate, $\angle 1 \cong \angle 13$ and $\angle 13 \cong \angle 9$. By the Transitive Property, $\angle 1 \cong \angle 9$. So, $m\angle 1 = m\angle 9$. By the Corresponding Angles Postulate, $\angle 4 \cong \angle 8$ and $\angle 8 \cong \angle 12$. By the Transitive Property, $\angle 4 \cong \angle 12$. So, $m\angle 4 = m\angle 12$. It is given that $m\angle 1 - m\angle 4 = 25°$. By the Substitution Property, $m\angle 9 - m\angle 12 = 25°$.

47. By the Vertical Angles Theorem, $\angle 7 \cong \angle 5$. By the Corresponding Angles Theorem $\angle 5 \cong \angle 1$. By the Transitive Property, $\angle 1 \cong \angle 7$.

49.

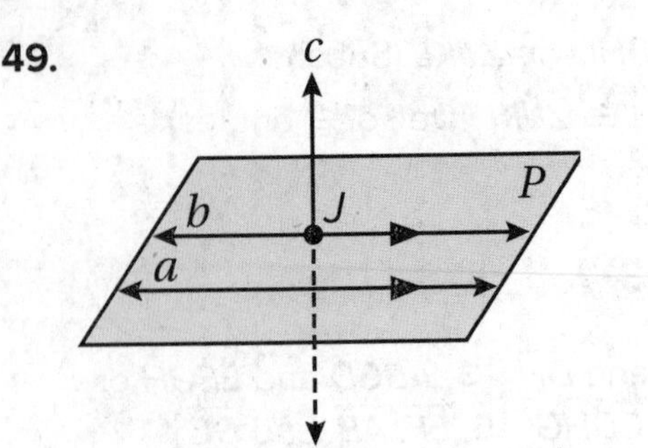

51. Sometimes; sample answer: $\overleftrightarrow{AB}$ intersects $\overleftrightarrow{EF}$ depending on where the planes intersect.

53. $x = 171$ or $x = 155$; $y = 3$ or $y = 5$

55. No; sample answer: From the definition of skew lines, the lines must not intersect and cannot be coplanar. Different planes cannot be coplanar, but they are always parallel or intersecting. Therefore, planes cannot be skew.

Lesson 3-8

1. parallel

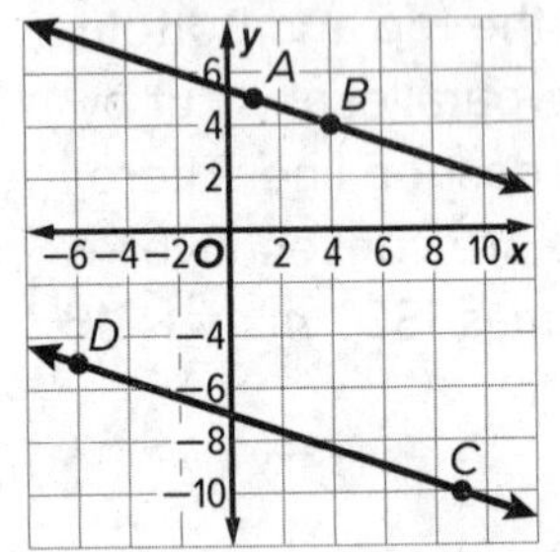

3. neither

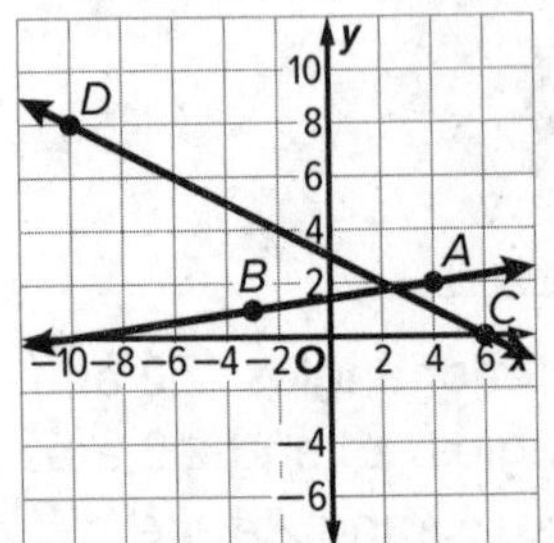

5. perpendicular

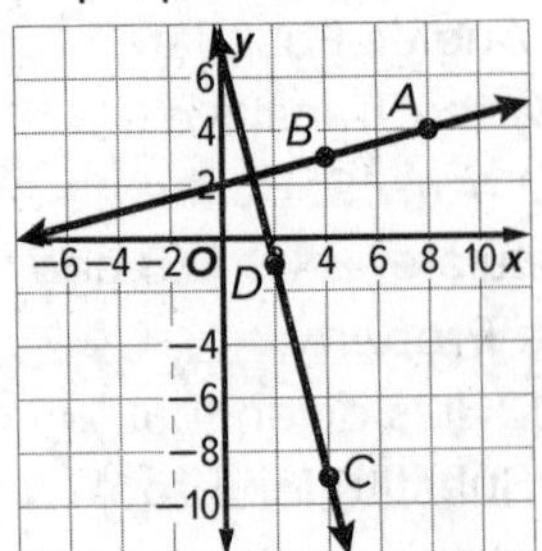

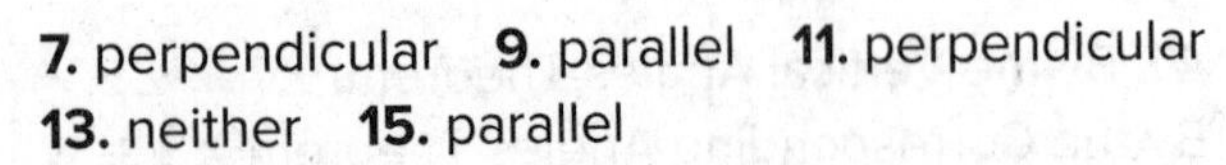

7. perpendicular **9.** parallel **11.** perpendicular **13.** neither **15.** parallel

17.

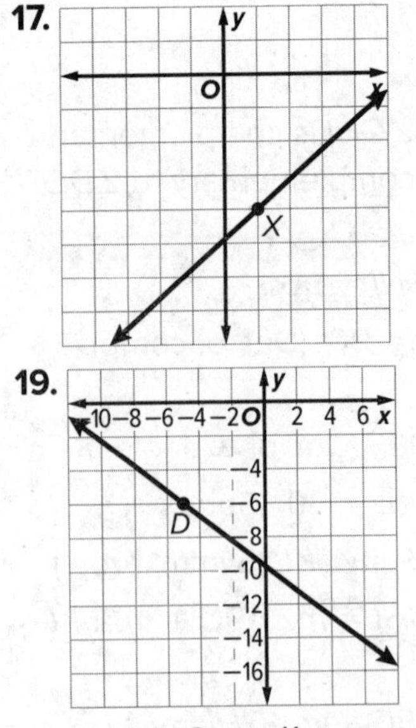

19.

21a. $y = -\frac{2}{3}x + \frac{11}{3}$ **21b.** When the drummer crosses the x-axis, the y-coordinate will be 0, so solve $0 = -\frac{2}{3}x + \frac{11}{3}$ to find the x-coordinate; solving shows that $x = \frac{11}{2}$ or $5\frac{1}{2}$, so the x-coordinate will be greater than 5.

23. $y = -10$ **25.** $y = \frac{1}{5}x + \frac{12}{5}$

27. $y = -8$

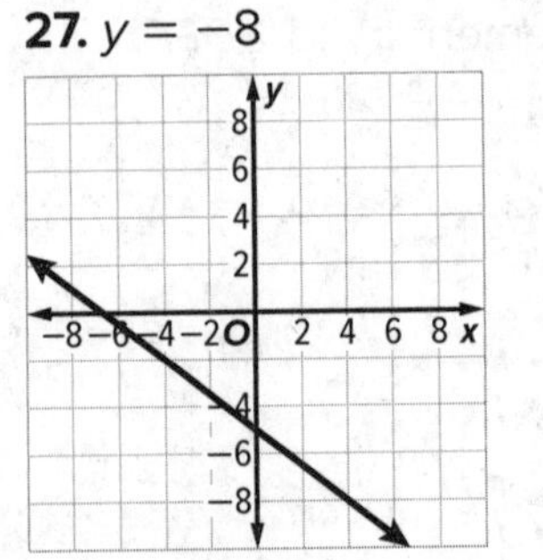

29. $y = 0$

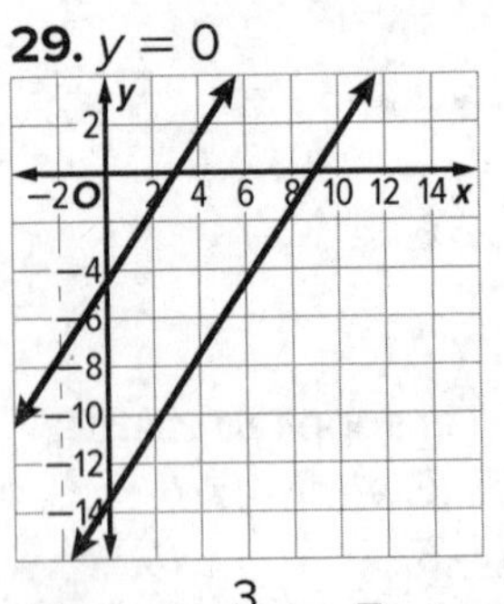

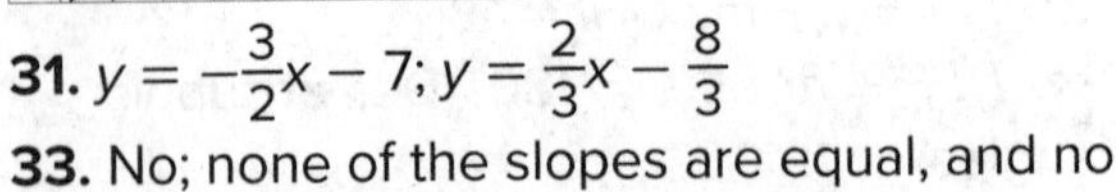

31. $y = -\frac{3}{2}x - 7$; $y = \frac{2}{3}x - \frac{8}{3}$

33. No; none of the slopes are equal, and no two of the slopes have a product of -1.

35a. 450 yd **35b.** -3; Ford Street and 6th Street are parallel, so they have the same slope. **35c.** Both have a slope of $\frac{1}{3}$ because both are perpendicular to Ford and 6th, and the slope of a perpendicular is given by the negative reciprocal. **35d.** 200 yd

37. $S\left(0, -5\frac{1}{2}\right)$; The slope of $\overline{QR}$ is $\frac{2-4}{3-(-2)} = -\frac{2}{5}$, so the slope of $\overline{RS}$ is $\frac{5}{2}$. Let the coordinates of S be $(0, y)$ because S must be on the y-axis. Solve $\frac{5}{2} = \frac{y-2}{0-3}$ for y. $y = -5\frac{1}{2}$, so the coordinates of S are $\left(0, -5\frac{1}{2}\right)$.

39a. $y - 1 = a(x - 5)$; the line must have slope a to be parallel to line p.

39b. $y = -\frac{1}{a}x + \frac{2}{a} + 3$; the line must have slope $-\frac{1}{a}$ to be perpendicular to line p.

41a. $B(2, 4)$ and $D(10, -4)$

41b. Sample answer: The slopes of $\overline{AB}$ and $\overline{DC}$ are undefined, so they are parallel to each other. The slopes of $\overline{AD}$ and $\overline{BC}$ are 0, so they are parallel to each other.

41c. Sample answer: Because the slope of $\overline{AB}$ is undefined and the slope of $\overline{BC}$ is zero, the lines are perpendicular to each other. Therefore, they form a right angle, which measures 90°. The same logic applies to all the sides.

43. Yes; the slope of the line through the points (−2, 2) and (2, 5) is $\frac{3}{4}$. The slope of the line through the points (2, 5) and (6, 8) is $\frac{3}{4}$. Because these lines have the same slope and have a point in common, their equations would be the same. Therefore, all the points are on the same line, and all the points are collinear.

45. Two nonvertical lines are parallel if and only if they have the same slope. Two nonvertical lines are perpendicular if and only if the product of their slopes is −1.

Lesson 3-9

1. $a \parallel b$; Alternate Interior Angles Converse

3. $\ell \parallel m$; Alternate Exterior Angles Converse

5. $g \parallel h$; Converse of Corresponding Angles Thm.

7. 22 **9.** 6 **11.** 13 **13.** 20

15. Parallelogram; sample answer: The top edges are perpendicular to the vertical line, so they are a single line. The bottom edge is also a single line and perpendicular to the same line as the top, so it is parallel to the top. The top edge is a transversal to the left and right slanted edges, and the angles are supplementary. So, the left and right edges are parallel.

17a. 108°; sample answer: To ensure that the horizontal part of the A is truly horizontal, it should be parallel to the dashed line. Therefore, $\angle 2$ and the 108°-angle are alternate interior angles, and $m\angle 2 = 108°$. $\angle 1$ and $\angle 2$ are congruent angles, so $m\angle 1 = 108°$.

17b. Sample answer: One side of the A is longer than the other.

19. Sample answer: It is given that $\angle 1 \cong \angle 2$. Also, $\angle 1 \cong \angle 3$, because these are vertical angles. Therefore, $\angle 2 \cong \angle 3$ by the Transitive Property of Congruence. This shows that $\ell \parallel m$ by the Converse of Corresponding Angles Theorem.

21.

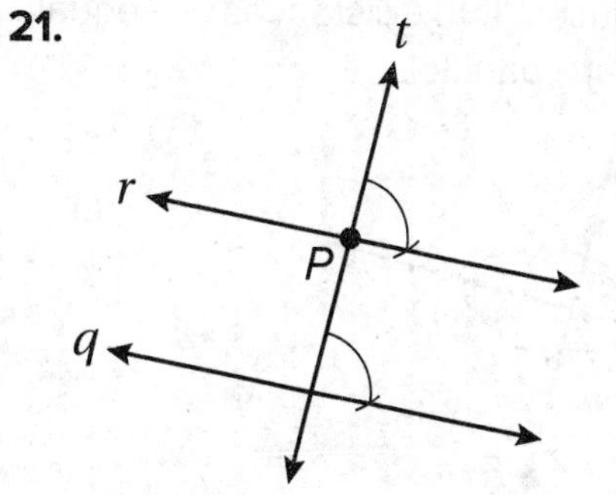

23. Sample answer: Because the corners are right angles, each pair of opposite sides is perpendicular to the same line. Therefore, each pair of opposite sides is parallel.

25. Daniela is correct. $\angle 1$ and $\angle 2$ are alternate interior angles for $\overline{WX}$ and $\overline{YZ}$. So, if alternate interior angles are congruent, then the lines are parallel.

27a.

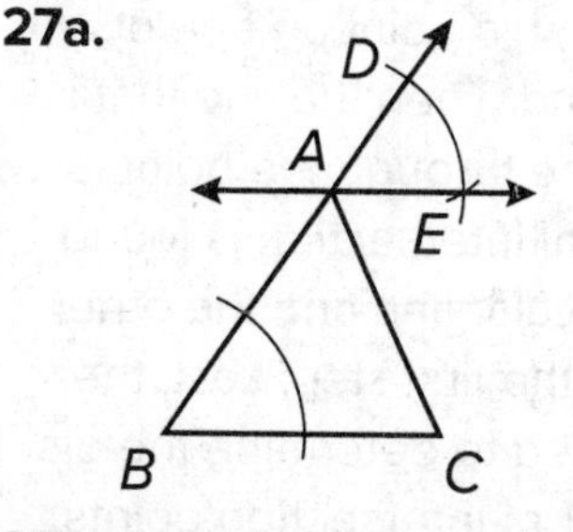

27b. Sample answer: Using a straightedge, the lines are equidistant. So, they are parallel.

27c. Sample answer: $\angle ABC$ was copied to construct $\angle DAE$. So, $\angle ABC \cong \angle DAE$. $\angle ABC$ and $\angle DAE$ are corresponding angles, so by the Converse of the Corresponding Angles Theorem, $\overleftrightarrow{AE} \parallel \overleftrightarrow{BC}$.

29. Yes; sample answer: A pair of angles can be supplementary and congruent if the measure of both angles is 90°, because the sum of the angle measures would be 180°.

Lesson 3-10

1. $\sqrt{2}$ or about 1.41 units **3.** 6 units
5. $\sqrt{10}$ or about 3.16 units **7.** yes; 4.24 in.
9. 8 units **11.** $\sqrt{10}$ or about 3.16 units
13. $3\sqrt{2}$ or about 4.24 units
15. $4\sqrt{17}$ or about 16.49 units
17. $\sqrt{14.76}$ or about 3.84 units **19.** 0 units
21. Sample answer: Isaiah can measure the perpendicular distance between the wires in two different places. If the distances are equal, then the wires are parallel.
23.

25. No; sample answer: A path that is perpendicular to the tee box would be the shortest. The angle that the tee box makes with the path that Mark walked is less than 90°, so it is not the shortest possible path.
27. Sample answer: The point is on the line. The two lines are the same line.
29. Sample answer: First, a point on one of the parallel lines is found. Then the line that is perpendicular to the line through the point is found. Then the point of intersection is found between the perpendicular line and the other line that is not used in the first step. Last, the Distance Formula is used to determine the distance between the pair of intersection points. This value is the distance between the pair of parallel lines.

31. Sometimes; sample answer: The distance can only be found if the line is parallel to the plane.
33. If two planes are each equidistant from a third plane, then the two planes are parallel to each other.

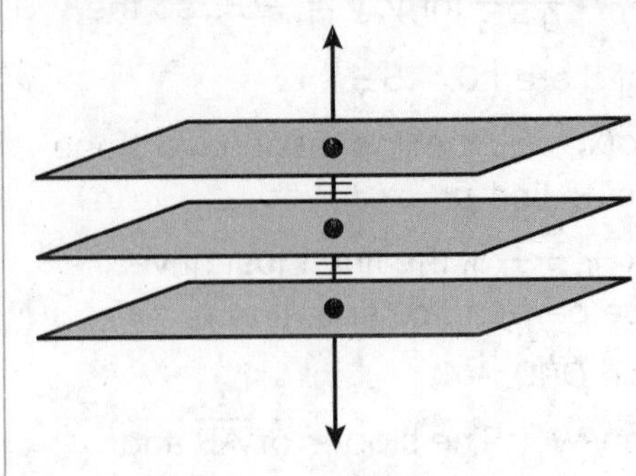

Module 3 Review

1. 6
3. B
5. $x = 29$, $m\angle ABD = 66°$, $m\angle DBC = 24°$
7. A, D, E
9.

Lines	m and n	m and p	n and p
parallel			
perpendicular		X	
neither	X		X

11. 102° **13.** A, B, C, D **15.** B

Module 4

Quick Check

1. −14 **3.** −2 **5.** (3, 0) **7.** (1, −1)

Lesson 4-1

1. $A'(2, -3)$, $B'(1, 0)$, $C'(-3, -2)$

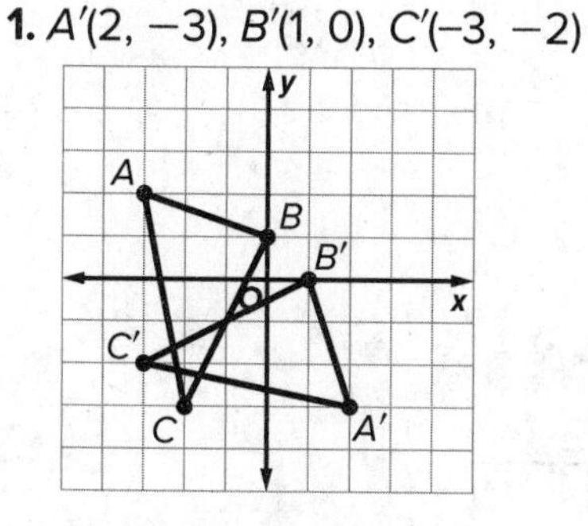

3. $R'(3, -2)$, $S'(4, 2)$, $T'(-3, 2)$, $U'(-4, -2)$

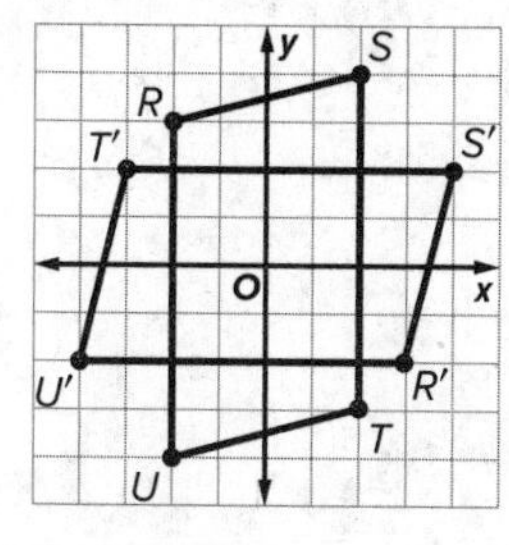

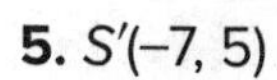
5. $S'(-7, 5)$

7.

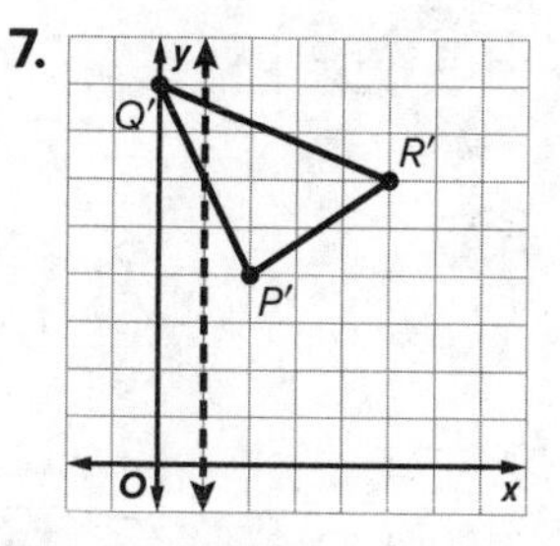

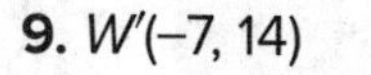
9. $W'(-7, 14)$

11.

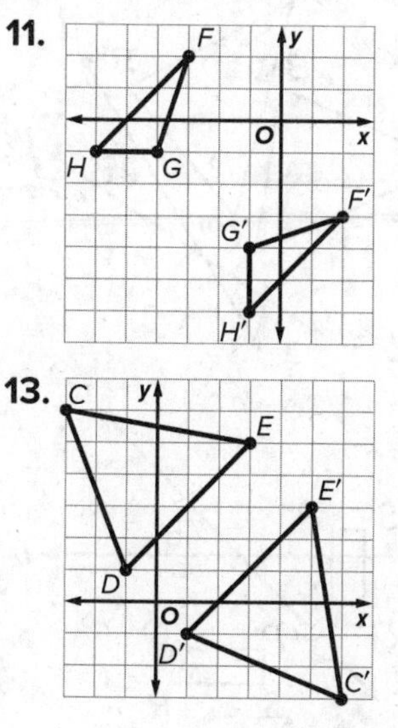

13.

15. reflection in the line $y = -1$

17. Sample answer: The M can be represented with the points (0, 0), (0, 3), (1, 1), (2, 0), and (2, 3). Reflecting in the line $y = x$ gives (0, 0), (3, 0), (1, 1), (0, 2), and (3, 2). **19.** $(-b, a)$

Lesson 4-2

1. $\triangle J'K'L'$ is a translation of $\triangle JKL$. This translation vector can be represented as $\langle 2, 5 \rangle$.

3. (4, 3), (5, 5), (6, 3)

5.

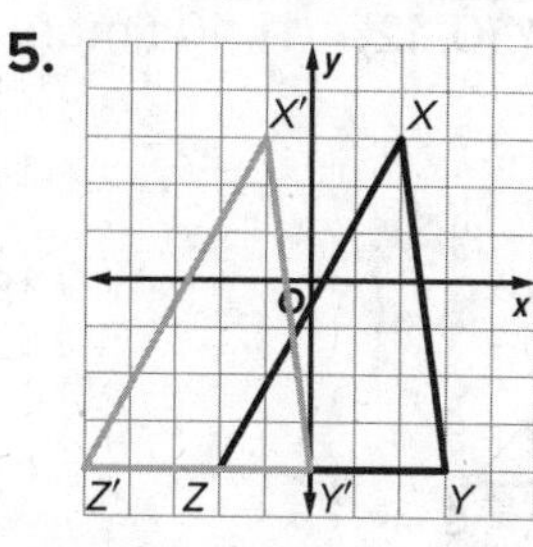

7.

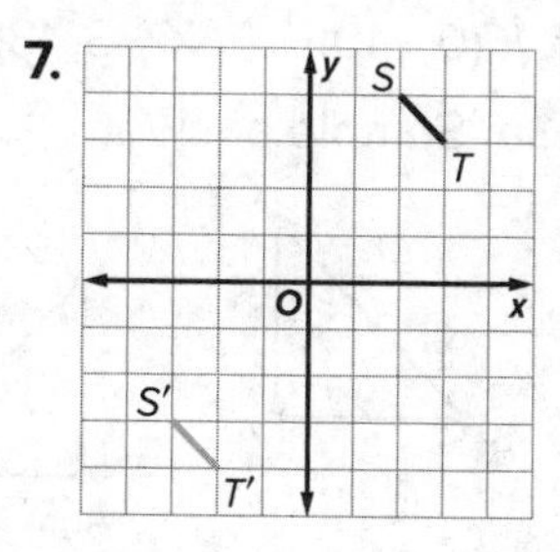

9. $Q'(2, -7)$ **11.** $B'(12, 2)$

13. No; sample answer: The size has been changed.

15. $(x, y) \rightarrow (x, y + 4)$

17. (0, 2), (2, 2), (0, 0), (2, 0); Sample answer: To minimize the length of the vector, I used the vertex closest to the origin, (2, 1), as the preimage for the point translated to the origin.

Lesson 4-3

1.

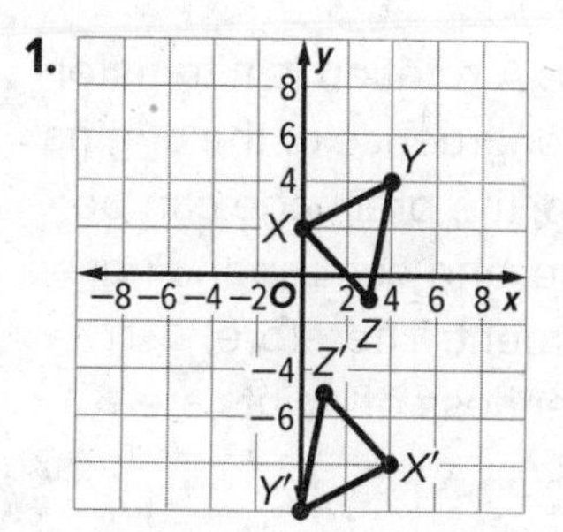

3. 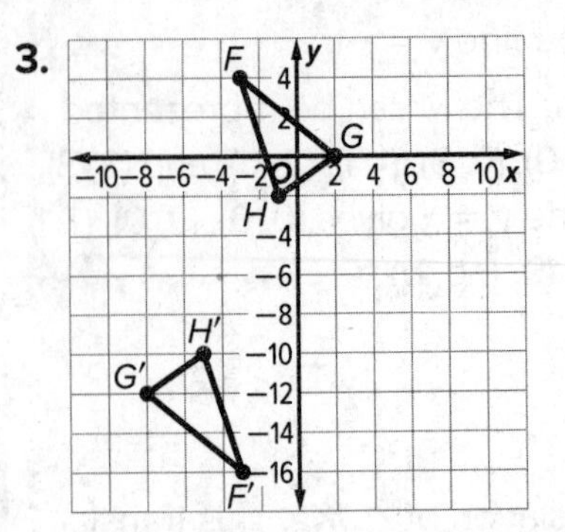

5. home plate: (3, 23), first base: (3, 13), second base: (13, 13), third base: (13 , 23)

7. $K'(0, -10)$ **9.** $X'(-3, 13)$, $Y'(0, 12)$ **11.** 10 times

13a. Sample answer:

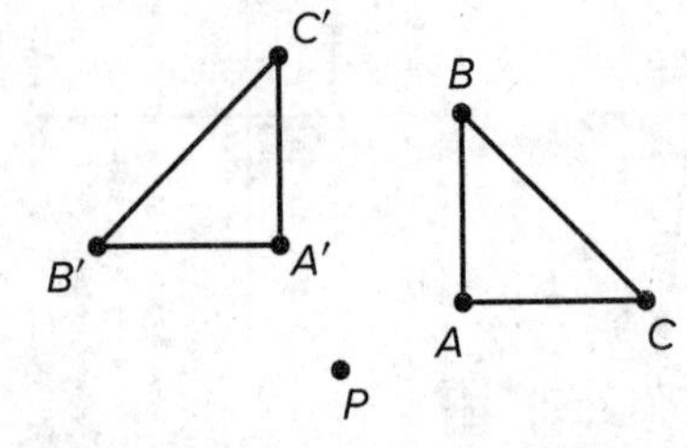

13b. Sample answer: 270° clockwise rotation about point P

15. Sample answer: A rotation followed by another rotation is still a rotation. For example, a rotation of 30° clockwise followed by a rotation of 20° counterclockwise is the same as a rotation of 10° clockwise. A rotation of 30° counterclockwise followed by a rotation of 15° counterclockwise is the same as a rotation of 45° counterclockwise.

17. Sample answer: Distance is preserved because the lengths of segments remain the same measure. Angle measures are preserved because angle measures remain the same measure. Parallelism is preserved because parallel lines remain parallel. Collinearity is preserved because points remain on the same lines.

19. Yes; sample answer: A rotation is a transformation that maintains congruence of the original figure and its image. So, the preimage can be mapped onto the image, and corresponding segments will be congruent. Therefore, collinearity and betweenness of points are maintained in rotations.

Lesson 4-4

1.

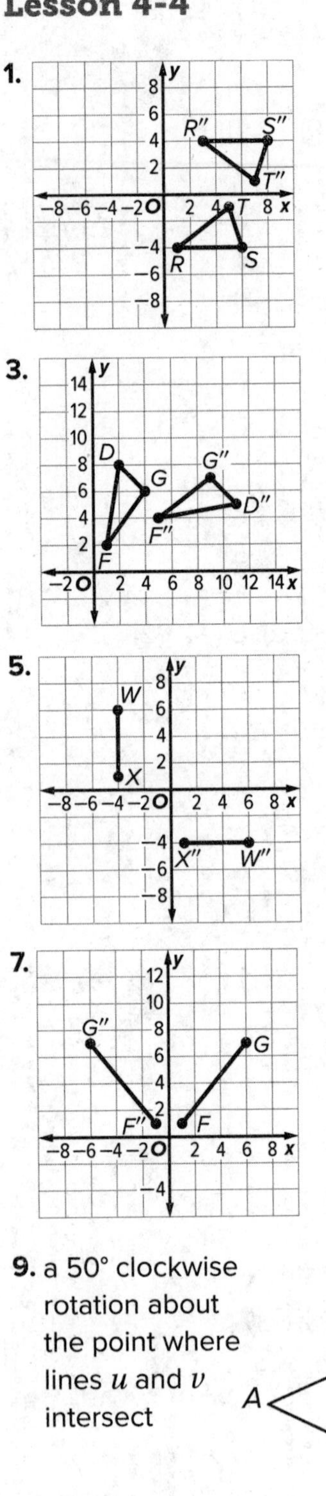

3.

5.

7.

9. a 50° clockwise rotation about the point where lines u and v intersect

11. a 90° clockwise rotation about the point where lines u and v intersect

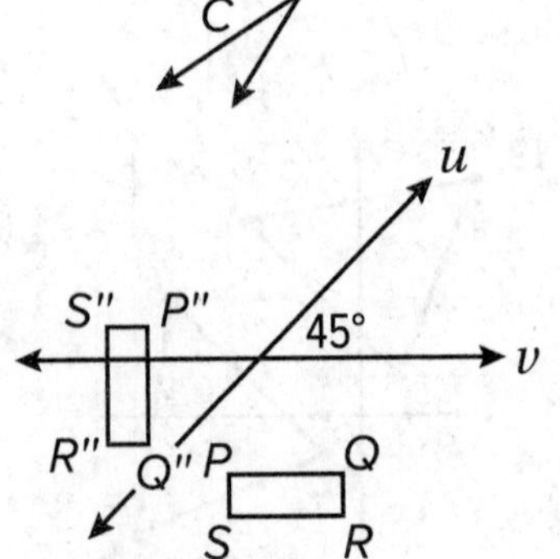

13. $\triangle JKL \cong \triangle MNP$; reflection in y-axis followed by translation along $\langle -1, 2 \rangle$

15. $\triangle JKL \cong \triangle MNP$; translation along $\langle 2, 0 \rangle$ followed by 180° rotation about the origin

17. Sample answer: Reflect the first two shapes in a horizontal line through the midpoints of the vertical segments, and then translate to the right and repeat.

19.

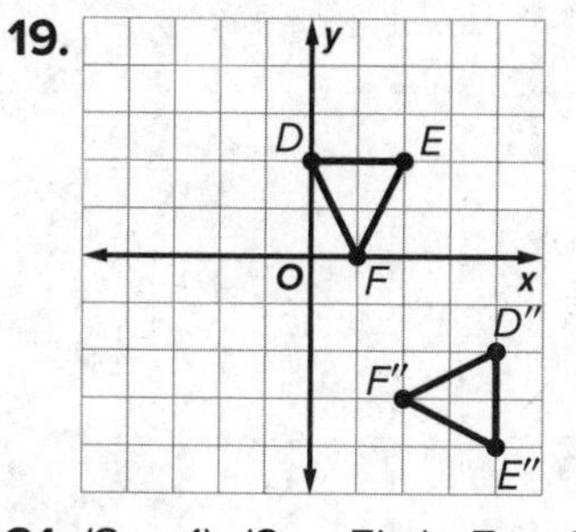

21. (3, −1), (2, −5), (−5, −2)

23. (1, 1), (0, 5), (−7, 2)

25. Sometimes; sample answer: If the lines of reflection intersect, the composition is a rotation.

27. Sometimes; sample answer: This is true if the point is the origin.

29. Sample answer: Proof: It is given that a translation along $\langle a, b \rangle$ maps R to R' and S to S'. Using the definition of a translation, points R and S move the same distance in the same direction, therefore $\overline{RS} \cong \overline{R'S'}$. It is also given that a reflection in a maps R' to R'' and S' to S''. Using the definition of a reflection, points R' and R'' and points S' and S'' are the same distance from line a, so $\overline{R'S'} \cong \overline{R''S''}$. By the Transitive Property of Congruence, $\overline{RS} \cong \overline{R''S''}$.

31. Sometimes; sample answer: The order of rotating by 180° about the origin and reflecting in the line $y = x$ does not change the location of the final image.

33. $P'''(1, -2)$, $Q'''(2, 1)$, $R'''(-1, 3)$, $S'''(-2, 0)$

Lesson 4-5

1. $x = \frac{180(5-2)}{5} = 108°$; Because 108° is not a factor of 360°, a regular pentagon will not tessellate the plane.

3. $x = \frac{180(9-2)}{9} = 140°$; Because 140° is not a factor of 360°, a regular 9-gon will not tessellate the plane.

5. yes; 2 regular hexagons, 2 equilateral triangles

7. tessellation; uniform, semiregular

9. Yes; sample answer: reflection, rotation, translation

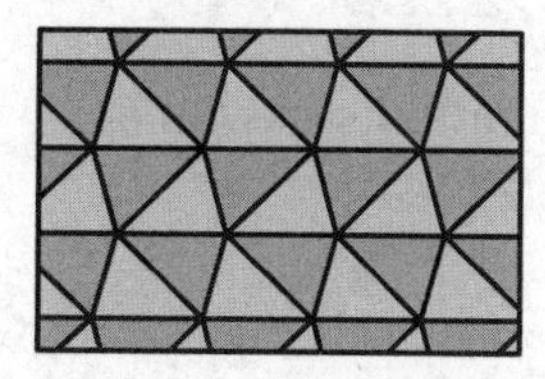

11. Never; sample answer: Each interior angle of a regular dodecagon is $\frac{180°(12-2)}{12} = 150°$. Because 150° is not a factor of 360°, a regular dodecagon will not tessellate the plane.

13. Never; sample answer: Each interior angle of a regular 15-gon is $\frac{180°(15-2)}{15} = 156°$. Because 156° is not a factor of 360°, a regular 15-gon will not tessellate the plane.

15. translation **17.** rotation **19.** Sample answer: Translations can be performed because the pieces slide. Rotations can be performed because each piece can be turned. Reflections cannot be performed because the back of a piece cannot be used to create the puzzle.

21.

Lesson 4-6

1. yes; 3

3. no

5. yes; 5

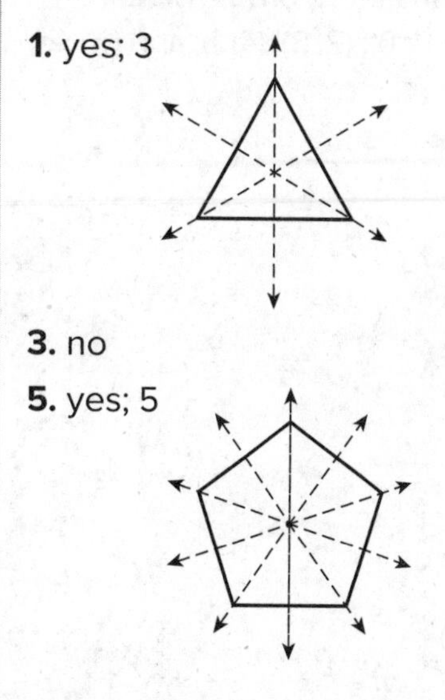

7a. Yes; the hubcap can map onto itself with a rotation that is less than 360°.

7b. Yes; the hubcap can map onto itself with a rotation that is less than 360°.

7c. Yes; the hubcap can map onto itself with a rotation that is less than 360°.

9. Yes; the symbol can map onto itself with a rotation that is less than 360°.

11. yes; 3; 120°

13. yes; 2; 180°

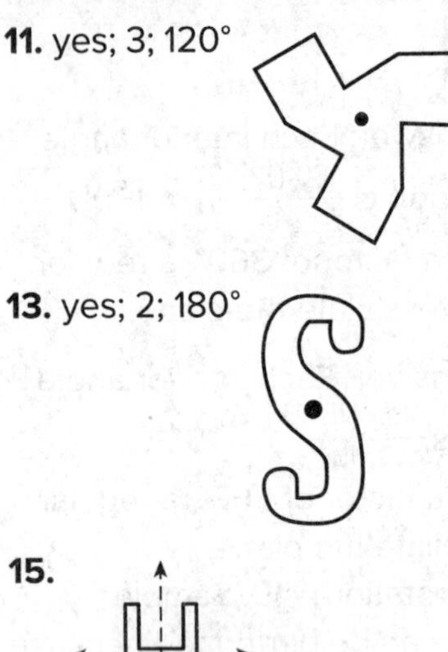

15.

17. 2; 180° **19.** pentagon **21.** yes; order of symmetry: 3; magnitude of symmetry: 120° **23.** yes; order of symmetry: 6; magnitude of symmetry: 60°

25. Sample answer: A rectangular mirror with two lines of symmetry, one vertical and one horizontal, through the middle or a spoon with one line of symmetry down the middle.

27. 24; 360° ÷ 15° = 24, so the order of symmetry is 24. This means there are 24 sides.

29. Sample answer: (–1, 0), (2, 3), (4, 1), and (1, –2);

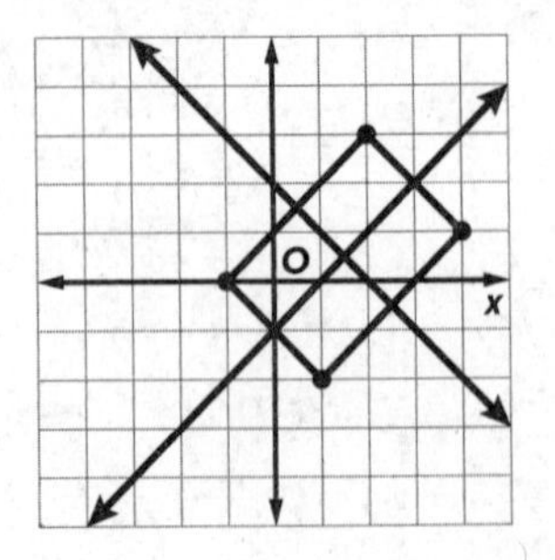

31. Sample answer: In both rotational and line symmetry a figure is mapped onto itself. However, in line symmetry the figure is mapped onto itself by a reflection, and in rotational symmetry a figure is mapped onto itself by a rotation. A figure can have line symmetry and rotational symmetry.

Module 4 Review

1.

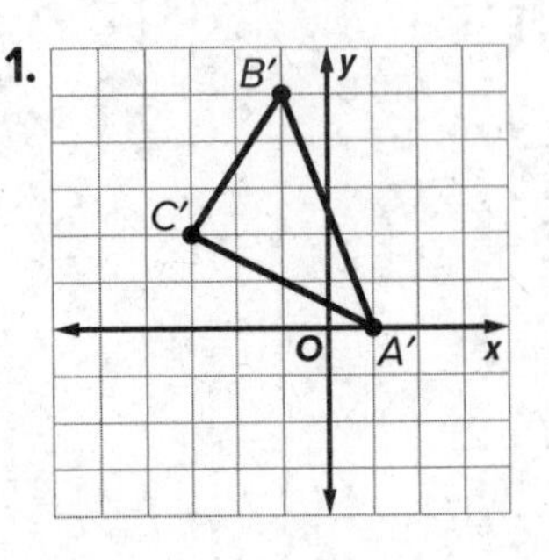

3. (–9, 6)

5. Sample answer: The length of $\overline{AC}$ is not the same as the length of $\overline{EG}$.

7. C **9.** Quadrant II **11.** false **13.** A, D

15.

Shape	Lines of Symmetry			
	0	1	2	3
Scalene Triangle	X			
Isosceles Triangle		X		
Equilateral Triangle				X
Rectangle			X	

17. order = 24; magnitude = 15°

Module 5

Quick Check

1. right **3.** obtuse **5.** 10.8 **7.** 18.0

Lesson 5-1

1. $m\angle 1 = 30°$, $m\angle 2 = 60°$
3. $m\angle 1 = 109°$, $m\angle 2 = 29°$, $m\angle 3 = 71°$
5. 50° **7.** 98° **9.** 62° **11.** 26° **13.** 55°
15. $x = 20$; 40°, 60°, 80° **17.** $x = 11$; 80°, 117°
19.

Proof:

$\angle R$ is a rt. $\angle$.
Given

$m\angle R = 90°$
Def. of rt. $\angle$

$m\angle R + m\angle S + m\angle T = 180°$
Triangle Angle-Sum Thm.

$90 + m\angle S + m\angle T = 180°$
Substitution

$m\angle S + m\angle T = 90°$
Subtraction Prop.

$\angle S$ and $\angle T$ are complementary.
Def. of comp. angles

21. $m\angle D = 27°$, $m\angle E = 81°$, $m\angle F = 72°$
23. $m\angle J = 18°$, $m\angle K = 72°$, $m\angle L = 90°$
25. $m\angle Z < 23°$; Sample answer: Because the sum of the measures of the angles of a triangle is 180° and $m\angle X = 157°$, $157° + m\angle Y + m\angle Z = 180°$, so $m\angle Y + m\angle Z = 23°$. If $m\angle Y$ was 0°, then $m\angle Z$ would equal 23°. But because an angle must have a measure greater than 0°, $m\angle Z$ must be less than 23°, so $m\angle Z < 23°$.
27. 90° **29.** $m\angle 1 = 55°$, $m\angle 2 = 75°$, $m\angle 3 = 55°$, $m\angle 4 = 15°$ **31.** Sample answer: Draw a triangle and then tear the corners off the triangle. Arrange the three corners so the angles are adjacent. The angles now form a straight angle. Because a straight angle measures 180°, the sum of the measures of the angles of a triangle is 180°.

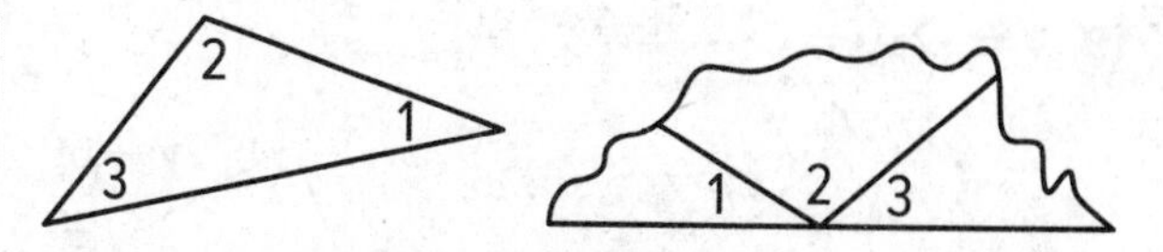

33. Sample answer: Because an exterior angle is acute, the adjacent angle must be obtuse. Because another exterior angle is right, the adjacent angle must be right. A triangle cannot contain a right angle and an obtuse angle because the sum would be greater than 180°. Therefore, a triangle cannot have an obtuse, an acute, and a right exterior angle.
35. Sample answer: I found the measure of the second angle by subtracting the first angle from 90° because the acute angles of a right triangle are complementary.

56°
34°

Lesson 5-2

1. $\angle A \cong \angle D$; $\angle ABC \cong \angle DCB$; $\angle ACB \cong \angle DBC$; $\overline{AC} \cong \overline{DB}$; $\overline{AB} \cong \overline{DC}$; $\triangle ABC \cong \triangle DCB$
3. $\angle X \cong \angle A$; $\angle Y \cong \angle B$; $\angle Z \cong \angle C$; $\overline{XY} \cong \overline{AB}$; $\overline{XZ} \cong \overline{AC}$; $\overline{YZ} \cong \overline{BC}$; $\triangle XYZ \cong \triangle ABC$
5. $\angle R \cong \angle J$; $\angle T \cong \angle K$; $\angle S \cong \angle L$; $\overline{RT} \cong \overline{JK}$; $\overline{TS} \cong \overline{KL}$; $\overline{RS} \cong \overline{JL}$; $\triangle RTS \cong \triangle JKL$
7. 48 **9.** 5 **11.** 35 **13.** 41°
15. Proof:

Statements (Reasons)
1. $\overline{AB} \cong \overline{CB}$, $\overline{AD} \cong \overline{CD}$ (Given)
2. $\overline{BD} \cong \overline{BD}$ (Reflexive Prop. of Congruence)
3. $\angle ABD \cong \angle CBD$, $\angle ADB \cong \angle CDB$ (Given)
4. $\angle A \cong \angle C$ (Third Angles Theorem)
5. $\triangle ABD \cong \triangle CBD$ (Def. of congruent triangles)

17. Proof: It is given that $\overline{BD}$ bisects $\angle ABC$ and $\angle ADC$. Therefore, $\angle ABD \cong \angle CBD$ and $\angle ADB \cong \angle CDB$ by the definition of angle bisector. By the Third Angles Theorem, $\angle A \cong \angle C$. It is given that $\overline{AB} \cong \overline{CB}$ and $\overline{AD} \cong \overline{CD}$. By the Reflexive Property of Congruence, $\overline{BD} \cong \overline{BD}$. Therefore, $\triangle ABD \cong \triangle CBD$ by the definition of congruent triangles.

Selected Answers

19. $x = 12; y = 6$

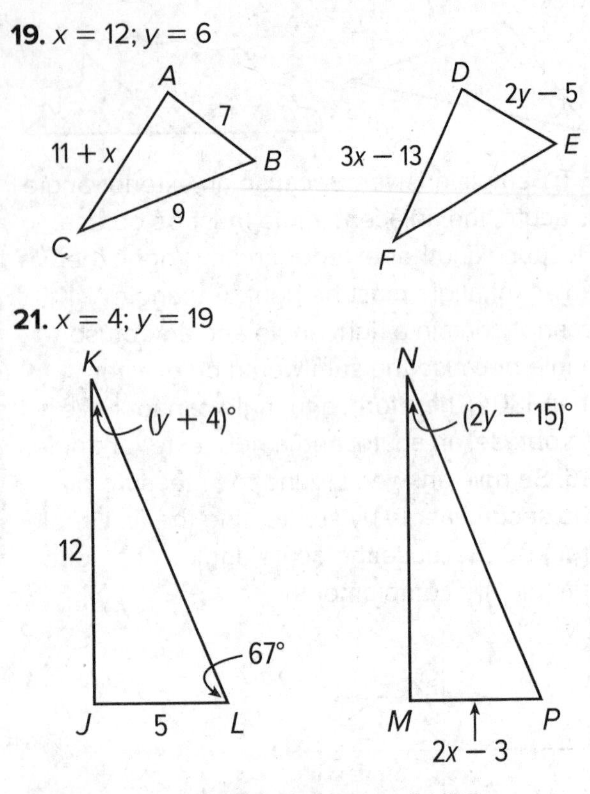

21. $x = 4; y = 19$

23a. $\triangle ABI \cong \triangle EBF$, $\triangle CBD \cong \triangle HBG$

23b. Sample answer: $\angle A \cong \angle E$, $\angle ABI \cong \angle EBF$, $\angle I \cong \angle F$; $\overline{AB} \cong \overline{EB}$, $\overline{BI} \cong \overline{BF}$, $\overline{AI} \cong \overline{EF}$

25. Statements (Reasons)

1. $\angle P \cong \angle X$, $\angle Q \cong \angle Y$ (Given)
2. $m\angle P = m\angle X$, $m\angle Q = m\angle Y$ (Def. of congruent angles)
3. $m\angle P + m\angle Q + m\angle R = 180$
   $180 = m\angle X + m\angle Y + m\angle Z$ (Triangle Angle-Sum Thm.)
4. $m\angle P + m\angle Q + m\angle R = m\angle X + m\angle Y + m\angle Z$ (Transitive Property)
5. $m\angle X + m\angle Y + m\angle R = m\angle X + m\angle Y + m\angle Z$ (Substitution Property)
6. $m\angle R = m\angle Z$ (Subtraction Prop. of Eq.)
7. $\angle R \cong \angle Z$ (Def. of congruent angles)

27. Sample answer: Do you think that the sum of the angles of a quadrilateral is constant? If so, do you think that the final pair of corresponding angles will be congruent if three other pairs of corresponding angles are congruent for a pair of quadrilaterals?

29. Sample answer: When naming congruent triangles, it is important that the corresponding vertices be in the same location for both triangles because the location indicates congruence. For example, if $\triangle ABC$ is congruent to $\triangle DEF$, then $\angle A \cong \angle D$, $\angle B \cong \angle E$, and $\angle C \cong \angle F$.

Lesson 5-3

1. Statements (Reasons)

1. $\overline{AB} \cong \overline{XY}$
   $\overline{AC} \cong \overline{XZ}$
   $\overline{BC} \cong \overline{YZ}$ (Given)
2. $\triangle ABC \cong \triangle XYZ$ (SSS Post.)

3. Proof:

Statements (Reasons)

1. $\overline{AB} \cong \overline{CB}$, D is the midpoint of $\overline{AC}$. (Given)
2. $\overline{AD} \cong \overline{DC}$ (Definition of midpoint)
3. $\overline{BD} \cong \overline{BD}$ (Reflexive Property of Congruence)
4. $\triangle ABD \cong \triangle CBD$ (SSS)

5. Proof: We know that $\overline{QR} \cong \overline{SR}$ and $\overline{ST} \cong \overline{QT}$. $\overline{RT} \cong \overline{RT}$ by the Reflexive Property. Because $\overline{QR} \cong \overline{SR}$, $\overline{ST} \cong \overline{QT}$, and $\overline{RT} \cong \overline{RT}$, $\triangle QRT \cong \triangle SRT$ by SSS.

7. $DE = 5\sqrt{2}$, $PQ = 5\sqrt{2}$, $EF = 2\sqrt{10}$, $QR = 2\sqrt{10}$, $DF = 5\sqrt{2}$, $PR = 5\sqrt{2}$; $\triangle DEF \cong \triangle PQR$ by SSS because corresponding sides have the same measure and are congruent. **9.** $AB = 2$, $KL = 2$, $BC = 2\sqrt{2}$, $LM = 2\sqrt{2}$, $AC = 2$, $KM = 2$; The corresponding sides have the same measure and are congruent, so $\triangle ABC \cong \triangle KLM$ by SSS.

11. Proof:

Statements (Reasons)

1. $NP = PM$, $\overline{NP} \perp \overline{PL}$ (Given)
2. $\overline{NP} \cong \overline{MP}$ (Def. of congruence)
3. $\angle MPL$ and $\angle NPL$ are rt. angles. ($\perp$ lines form rt. angles.)
4. $\angle MPL \cong \angle NPL$ (All right angles are congruent.)
5. $\overline{PL} \cong \overline{PL}$ (Reflexive Property of $\cong$)
6. $\triangle NPL \cong \triangle MPL$ (SAS)

13. Proof: Because V is the midpoint of $\overline{YZ}$ and the midpoint of $\overline{WX}$, by the Midpoint Theorem, $\overline{YV} \cong \overline{VZ}$ and $\overline{WV} \cong \overline{XV}$. Because $\angle YVW$ and $\angle ZVX$ are vertical angles, by the Vertical Angles Theorem, the angles are congruent. Therefore, by SAS, $\triangle XVZ \cong \triangle WVY$.

15. Proof:

Statements (Reasons)

1. $\overline{BD} \perp \overline{AC}$; $\overline{BD}$ bisects $\overline{AC}$. (Given)
2. $\angle BDA$ and $\angle BDC$ are rt. angles. ($\perp$ lines form rt. angles.)

3. $\angle BDA \cong \angle BDC$ (All right angles are congruent.)
4. $\overline{AD} \cong \overline{DC}$ (Def. of segment bisector)
5. $\overline{BD} \cong \overline{BD}$ (Reflexive Property of $\cong$)
6. $\triangle ABD \cong \triangle CBD$ (SAS)

17. Yes; sample answer: $\angle GLH$ and $\angle JLK$ are vertical angles, so they are congruent. Therefore, $\triangle GLH \cong \triangle JLK$ by the SAS Congruence Postulate. **19.** Yes; sample answer: The triangles share the side $\overline{AC}$, so they have two pairs of congruent sides. The given congruent angles are included angles, so $\triangle ABC \cong \triangle CDA$ by SAS. **21.** No; sample answer: The sticks do not change size, so any arrangement will yield a congruent triangle. **23.** Sample answer: She needs to measure one side of each tile because all the tiles are equilateral triangles. **25.** No; sample answer: You cannot use SAS because the angle congruence that we are given is not an included angle between two sides that are known to be congruent, and SSS cannot be used because only 2 sides of each triangle are known to be congruent. **27.** First pair; sample answer: The second pair can be shown congruent by SAS or SSS, and the third pair can be shown congruent by SSS. **29.** Case 1: You know that the hypotenuses are congruent and that one pair of legs are congruent. Then the Pythagorean Theorem says that the other pair of legs are congruent, so the triangles are congruent by SSS. Case 2: You know that the pairs of legs are congruent and that the right angles are congruent, so the triangles are congruent by SAS. **31.** Shada; to use SAS, the angle must be the included angle.

Lesson 5-4

1. Proof:

Statements (Reasons)

1. $\overline{AB} \parallel \overline{CD}$ (Given)
2. $\angle CBD \cong \angle ADB$ (Given)
3. $\angle ABD \cong \angle CDB$ (Alternate Interior Angles Theorem)
4. $\overline{BD} \cong \overline{BD}$ (Reflexive Property of Congruence)
5. $\triangle ABD \cong \triangle CDB$ (ASA)

3. Proof:

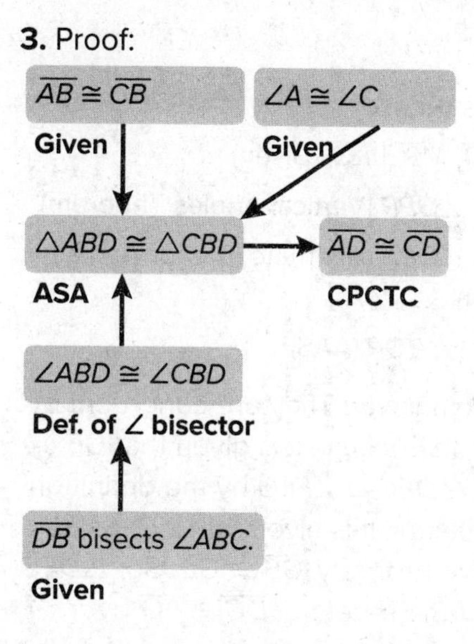

5. Proof: We are given that $\overline{CE}$ bisects $\angle BED$ and that $\angle BCE$ and $\angle ECD$ are right angles. Because all right angles are congruent, $\angle BCE \cong \angle ECD$. By the definition of angle bisector, $\angle BEC \cong \angle DEC$. The Reflexive Property tells us that $\overline{EC} \cong \overline{EC}$. By the Angle-Side-Angle Congruence Postulate, $\triangle ECB \cong \triangle ECD$.
7a. yes; by the ASA Congruence Postulate
7b. 10 in^2 **9a.** Sample answer: Because $\overline{AC} \parallel \overline{BK}$, $\angle CAB \cong \angle KBM$ by the Corresponding Angles Theorem. Because $\overline{CB} \parallel \overline{KM}$, $\angle ABC \cong \angle BMK$ by the Corresponding Angles Theorem. Because B is the midpoint of $\overline{AM}$, $\overline{AB} \cong \overline{BM}$ by the Midpoint Theorem. Therefore, by the ASA Congruence Postulate, $\triangle ABC \cong \triangle BMK$.
9b. 74 ft

11. Proof:

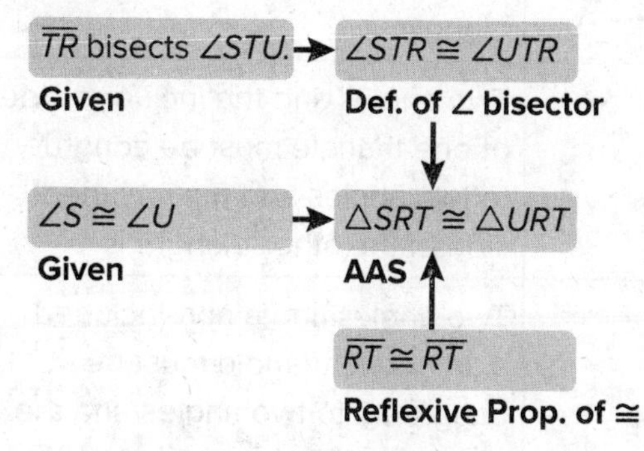

13. Proof: It is given that $\angle E \cong \angle G$ and $\overline{DE} \parallel \overline{FG}$. By the Alternate Interior Angles Theorem, $\angle DFG \cong \angle FDE$. $\overline{DF} \cong \overline{DF}$ by the Reflexive Property of Congruence. Therefore, $\triangle DFG \cong \triangle FDE$ by AAS.

Selected Answers

15. Proof:

Statements (Reasons)

1. $\overline{MS} \cong \overline{RQ}$, $\overline{MS} \parallel \overline{RQ}$ (Given)
2. $\angle SPM \cong \angle QPR$ (Vertical Angles Theorem)
3. $\angle SMP \cong \angle QRP$ (Alternate Interior Angles Theorem)
5. $\triangle MSP \cong \triangle RQP$ (AAS)

17. Yes; sample answer: They are congruent by AAS. **19.** Proof: Because it is given that $\overline{GE}$ bisects $\angle DEF$, $\angle DEG \cong \angle FEG$ by the definition of an angle bisector. It is given that $\angle D \cong \angle F$. By the Reflexive Property, $\overline{GE} \cong \overline{GE}$. So, $\triangle DEG \cong \triangle FEG$ by AAS. Therefore, $\overline{DG} \cong \overline{FG}$ by CPCTC. **21.** Tyrone; Lorenzo showed that all three corresponding angles were congruent, but AAA is not a proof of triangle congruence.

23.

Method	Use when...
Definition of Congruent Triangles	All corresponding parts of one triangle are congruent to the corresponding parts of the other triangle.
SSS	The three sides on one triangle must be congruent to the three sides of the other triangle.
SAS	Two sides and the included angle of one triangle must be congruent to two sides and the included angle of the other triangle.
ASA	Two angles and the included side of one triangle must be congruent to two angles and the included side of the other triangle.
AAS	Two angles and a non-included side of one triangle must be congruent to two angles and the corresponding non-included side of the other triangle.

Lesson 5-5

1. Statements (Reasons)

1. $\overline{XZ} \perp \overline{WY}$ (Given)
2. $\angle XZW$ and $\angle XZY$ are right angles. ($\perp$ lines form right angles.)
3. $\triangle WXZ$ and $\triangle YXZ$ are right triangles. (Definition of right triangle)
4. Z is the midpoint of $\overline{WY}$. (Given)
5. $\overline{WZ} \cong \overline{ZY}$ (Definition of midpoint)
6. $\overline{XZ} \cong \overline{XZ}$ (Reflexive Property of Congruence)
7. $\triangle WXZ \cong \triangle YXZ$ (LL Congruence Theorem)

3. Proof:

Statements (Reasons)

1. $\overline{BX} \perp \overline{AC}$ (Given)
2. $\angle AXB$ and $\angle CXB$ are rt. $\angle$s. (Definition of $\perp$ lines)
3. $\triangle AXB$ and $\triangle CXB$ are rt. $\triangle$s. (Definition of right $\triangle$s)
4. $\overline{XB} \cong \overline{XB}$ (Reflexive Property of Congruence)
5. $AB = CB$ (Given)
6. $\overline{AB} \cong \overline{CB}$ (Definition of congruent)
7. $\triangle AXB \cong \triangle CXB$ (HL Congruence Theorem)

5. Yes; LA **7.** No; not enough information

9. No; not enough information

11. Proof:

Statements (Reasons)

1. $\overline{BX} \perp \overline{XA}$, $\overline{BY} \perp \overline{YA}$ (Given)
2. $\angle BXA$ and $\angle BYA$ are rt. $\angle$s. (Definition of $\perp$ lines)
3. $\triangle BXA$ and $\triangle BYA$ are rt. $\triangle$s. (Definition of right $\triangle$s)
4. $\overline{XA} \cong \overline{YA}$ (Given)
5. $\overline{BA} \cong \overline{BA}$ (Reflexive Property of Congruence)
6. $\triangle BXA \cong \triangle BYA$ (HL Congruence Theorem)

13. Proof: By the definition of $\perp$ segments, $\angle AYB$ and $\angle AXC$ are right angles. By the definition of right triangles, $\triangle AYB$ and $\triangle AXC$ are right triangles. By the definition of congruent segments, $\overline{AX}$ is congruent to $\overline{AY}$. By the Reflexive Property of Congruence, $\angle BAY$ is congruent to $\angle CAX$. Therefore by LA, $\triangle ABY$ is congruent to $\triangle ACX$.

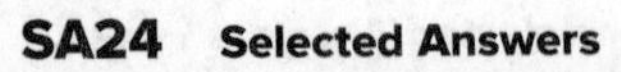

Lesson 5-6

1. Proof:

Statements (Reasons)

1. $\angle 1 \cong \angle 2$ (Given)

2. $\angle 2 \cong \angle 3$ (Vertical Angles Thm.)

3. $\angle 1 \cong \angle 3$ (Transitive Prop. of $\cong$)

4. $\overline{AB} \cong \overline{CB}$ (Conv. of Isos. Triangle Thm.)

3. Proof:

Statements (Reasons)

1. $\overline{DE} \parallel \overline{BC}$ (Given)
2. $\angle 1 \cong \angle 4$, $\angle 2 \cong \angle 3$ (Corresponding angles are $\cong$.)
3. $\angle 1 \cong \angle 2$ (Given)
4. $\angle 1 \cong \angle 3$ (Transitive Property of $\cong$)
5. $\angle 3 \cong \angle 4$ (Substitution)
6. $\overline{AB} \cong \overline{AC}$ (Converse of Isosceles Triangle Thm.)

5a. The coordinates of $\triangle ABC$ are $A(0, 5)$, $B(3, 1)$, and $C(-3, 1)$.

$AC = \sqrt{[0-(-3)]^2 + (5-1)^2}$ or 5 units

$AB = \sqrt{(0-3)^2 + (5-1)^2}$ or 5 units

$BC = 6$ units

So, $\triangle ABC$ is an isosceles triangle with $\overline{AB} \cong \overline{AC}$.

5b. Because $\overline{AB} \cong \overline{AC}$, we know that $\angle C \cong \angle B$ by the Isosceles Triangle Theorem.

$m\angle A + m\angle B + m\angle C = 180°$	Triangle Angle-Sum Theorem
$m\angle A + 2m\angle C = 180°$	Definition of congruent
$m\angle A + 2(55) = 180°$	Substitute.
$m\angle A + 110 = 180°$	Multiply.
$m\angle A = 70°$	Solve.

7. $m\angle DEF = 45°$ and $m\angle EFD = 45°$ **9.** 60°; 3 m **11.** 60°; 7 in. **13.** 10 **15a.** 7 **15b.** 15

17a.

17b. Sample answer: The triangle formed by connecting the midpoints of the sides of an isosceles triangle is an isosceles triangle.

19. Proof:

Statements (Reasons)

1. $\triangle PQR$ is an equilateral triangle. (Given)
2. $\overline{PQ} \cong \overline{QR} \cong \overline{PR}$ (Def. of equilateral triangle)
3. $\angle P \cong \angle Q \cong \angle R$ (Isosceles Triangle Theorem)
4. $m\angle P = m\angle Q = m\angle R$ (Def. of congruence)
5. $m\angle P + m\angle Q + m\angle R = 180°$ (Triangle Angle-Sum Thm.)
6. $3m\angle P = 180°$ (Substitution)
7. $m\angle P = 60°$ (Division Property)
8. $m\angle P = m\angle Q = m\angle R = 60°$ (Substitution)

21. 44° **23.** 22°

25.

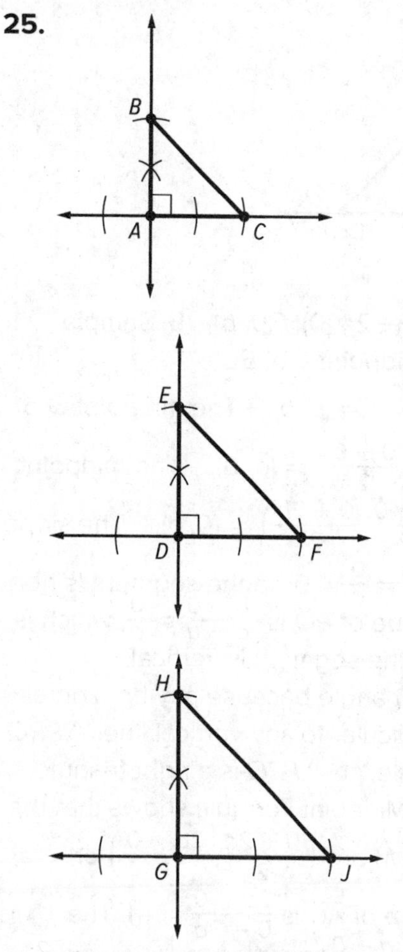

Sample answer: I constructed a pair of perpendicular segments and then used the same compass setting to mark points that are equidistant from their intersection. I measured both legs for each triangle. Because $AB = AC = 1.3$ cm, $DE = DF = 1.9$ cm, and $GH = GJ = 2.3$ cm, the triangles are isosceles. I used a protractor to confirm that $\angle A$, $\angle D$, and $\angle G$ are all right angles.

27. Sometimes; sample answer: Only if the measure of the vertex angle is even.
29. Sample answer: It is not possible because a triangle cannot have more than one obtuse angle. **31.** No; $m\angle G = \frac{180-70}{2}$ or 55°

Lesson 5-7

1.

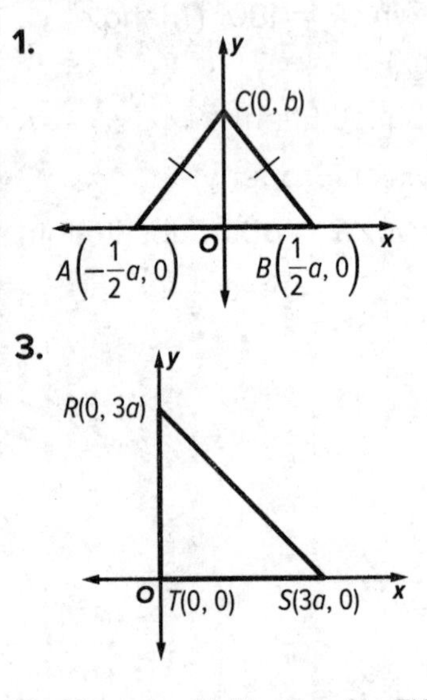

3.

5. $C(p, q)$ **7.** $E(-2g, 0)$; $F(0, b)$ **9.** Sample answer: The midpoint P of $\overline{BC}$ is $\left(\frac{0+2a}{2}, \frac{2b+0}{2}\right) = (a, b)$. The midpoint Q of $\overline{AC}$ is $\left(\frac{0+2a}{2}, \frac{0+0}{2}\right) = (a, 0)$. The midpoint R of $\overline{AB}$ is $\left(\frac{0+0}{2}, \frac{0+2b}{2}\right) = (0, b)$. The slope of $\overline{RP}$ is $\frac{b-b}{a-0} = \frac{0}{a} = 0$, so the segment is horizontal. The slope of $\overline{PQ}$ is $\frac{b-0}{a-a} = \frac{0}{a}$, which is undefined, so the segment is vertical. $\angle RPQ$ is a right angle because any horizontal line is perpendicular to any vertical line. $\triangle PRQ$ has a right angle, so $\triangle PRQ$ is a right triangle.
11. Proof: The Midpoint Formula shows that the coordinates of M are $\left(\frac{0+2a}{2}, \frac{2a+0}{2}\right)$ or (a, a). The slope of $\overline{AC}$ is $\frac{2a-0}{0-2a} = -1$. The slope of $\overline{BM}$ is $\frac{a-0}{a-0} = 1$. The product of the slopes is -1, so $\overline{BM} \perp \overline{AC}$. **13.** Proof: The coordinates of S are $\left(\frac{b}{2}, \frac{c}{2}\right)$, and the coordinates of T are $\left(\frac{a+b}{2}, \frac{c}{2}\right)$.

$ST = \sqrt{\left(\frac{a+b}{2} - \frac{b}{2}\right)^2 + \left(\frac{c}{2} - \frac{c}{2}\right)^2}$ or $\frac{a}{2}$
$AB = \sqrt{(a-0)^2 + (0-0)^2}$ or a
$ST = \frac{1}{2}AB$
15. The slope between the grandstand and the rides and games is $\frac{2}{3}$. The slope between the grandstand and the main gate is $-\frac{3}{2}$. Because $\frac{2}{3} \cdot -\frac{3}{2} = -1$, the triangle formed by these three locations is a right triangle.
17. Slope of $\overline{XY} = 1$, slope of $\overline{YZ} = -1$, slope of $\overline{ZX} = 0$; because $1(-1) = -1$, $\overline{XY} \perp \overline{YZ}$. Therefore, $\triangle XYZ$ is a right triangle.

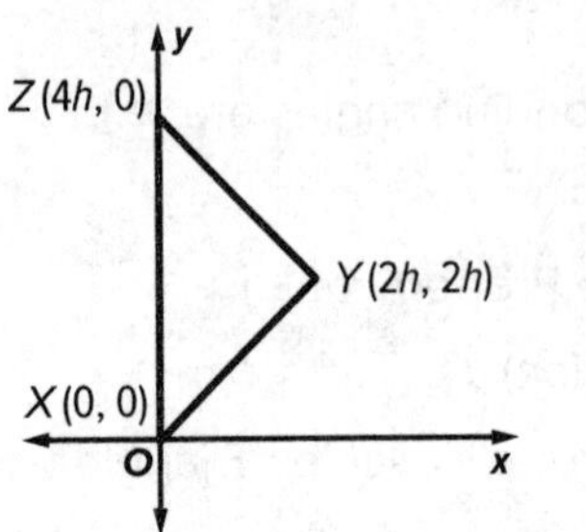

19. $(a, 0)$ and $(0, a)$ **21.** Sample answer: Use the Distance Formula to find the length of each side of each triangle. Show that the triangles are congruent by SSS. Conclude that $\angle A \cong \angle D$ using CPCTC. **23.** $D(a, b)$, $E(2a, 0)$, $F(3a, b)$
25. Sample answer:

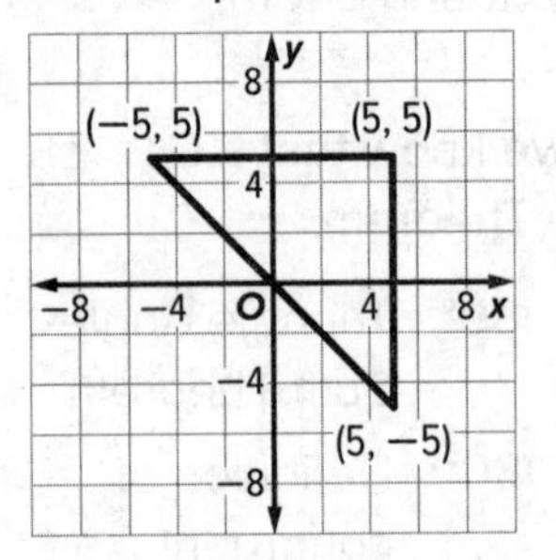

27. Sample answer: $(a, 0)$
29. Sample answer: $(4a, 0)$ or $(0, 4b)$

Module 5 Review

1. 106 **3.** A, D **5.** A, C, E **7.** A **9.** A, B
11. A **13.** 47 **15.** 33 units **17.** A **19.** B

Module 6

Quick Check

1. 9 **3.** 10 ft **5.** $x > 9$ **7.** $x < 41$

Lesson 6-1

1. 2.8 **3.** 4.2 **5.** 7 **7.** 9.8 **9.** (4, 1.8) **11.** 7

13. No; we need to know if the segment bisector is a perpendicular bisector.

15. It is given that $\overleftrightarrow{CD}$ is the perpendicular bisector of $\overline{AB}$. By the definition of bisector, E is the midpoint of $\overline{AB}$. Thus, $\overline{AE} \cong \overline{BE}$ by the Midpoint Theorem. Because two points determine a line, you can draw line segments from A to C and from B to C. $\angle CEA$ and $\angle CEB$ are right angles by the definition of perpendicular. Because all right angles are congruent, $\angle CEA \cong \angle CEB$. By the Reflexive Property of Congruence, $\overline{CE} \cong \overline{CE}$. Thus, $\triangle CEA \cong \triangle CEB$ by SAS. $\overline{CA} \cong \overline{CB}$ by CPCTC, and by the definition of congruence, $CA = CB$. By the definition of equidistant, C is equidistant from A and B.

17. The equation of a line of one of the perpendicular bisectors is $y = 3$. The equation of the line of another perpendicular bisector is $x = 5$. These lines intersect at (5, 3). The circumcenter is located at (5, 3).

19. a plane perpendicular to the plane in which $\overline{CD}$ lies and bisecting $\overline{CD}$ **33.** $\left(\frac{39}{10}, \frac{19}{10}\right)$

21.

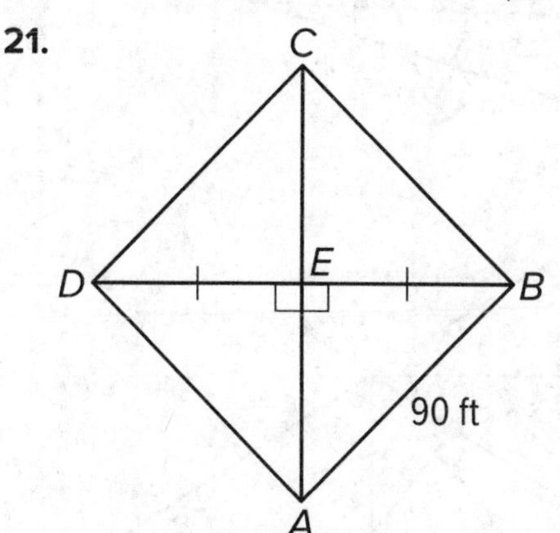

Sample answer: A is equidistant from B and D. Therefore, $AD = AB$. So, $AD = 90$ feet. AD is the distance from home plate to third base. Therefore, it is 90 feet from third base to home plate.

23. Proof:

Statements (Reasons)

1. Plane Y is a perpendicular bisector of $\overline{DC}$. (Given)
2. $\angle DBA$ and $\angle CBA$ are right angles, and $\overline{DB} \cong \overline{CB}$ (Definition of perpendicular bisector)
3. $\angle DBA \cong \angle CBA$ (All right angles are congruent.)
4. $\overline{AB} \cong \overline{AB}$ (Reflexive Property of $\cong$)
5. $\triangle DBA \cong \triangle CBA$ (SAS)
6. $\angle ADB \cong \angle ACB$ (CPCTC)

Lesson 6-2

1. 43° **3.** 19 **5.** 7 **7.** 40° **9.** 28° **11.** 13
13. (5.5, 6) **15.** 6

17. Sample answer: Given that $\overline{BD}$ is the angle bisector of $\angle ABC$, $\angle ABD \cong \angle CBD$ by the definition of angle bisector. Because there is a perpendicular line between any line and a point not on the line by the Perpendicular Postulate, let F be on $\overline{BC}$ such that $\overline{DF} \perp \overline{BC}$, and let E be on $\overline{AB}$ such that $\overline{DE} \perp \overline{AB}$. Therefore, $\angle BFD$ and $\angle BED$ are right angles by the definition of perpendicular, and $m\angle BFD = 90°$ and $m\angle BED = 90°$ by the definition of right angle. Further, $m\angle BFD = m\angle BED$ by substitution, and $\angle BFD \cong \angle BED$ by the definition of congruence. Also, $\overline{BD} \cong \overline{BD}$ by the Reflexive Property of Congruence. Because $\angle EBD \cong \angle DBF$, $\angle BFD \cong \angle BED$, and $\overline{BD} \cong \overline{BD}$, $\triangle BED \cong \triangle BFD$ by AAS. Therefore, $\overline{ED} \cong \overline{FD}$ by CPCTC. $ED = FD$ by the definition of congruence, so D is equidistant from $\overline{AB}$ and $\overline{BC}$ by the definition of equidistant.

19. Proof:

Statements (Reasons)

1. $\overline{AD}$, $\overline{BE}$, and $\overline{CF}$ are all angle bisectors, and $\overline{KP} \perp \overline{AB}$, $\overline{KQ} \perp \overline{BC}$, and $\overline{KR} \perp \overline{AC}$. (Given)
2. $KP = KQ$, $KQ = KR$, $KP = KR$ (Any point on the angle bisector is equidistant from the sides of the angle.)
3. $KP = KQ = KR$ (Transitive Property of Equality)

21. Sample answer: I constructed the angle bisector of each angle. The intersection point is the incenter. To verify the construction, I could construct segments perpendicular to each side through the incenter and then measure each distance with the ruler to verify they are all equal.

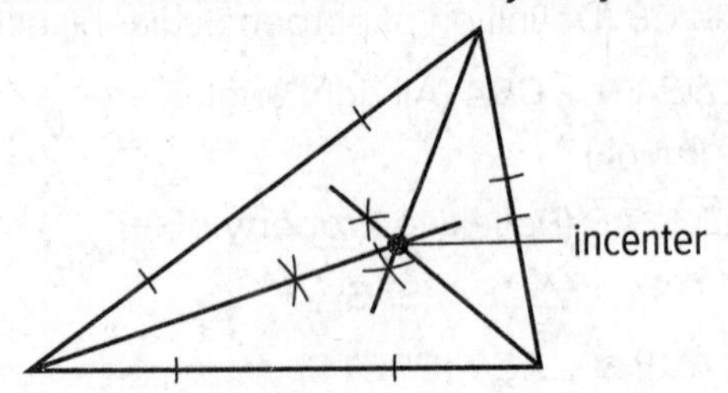

23. Sometimes; sample answer: If the triangle is equilateral, then this is true, but if the triangle is isosceles or scalene, the statement is false.

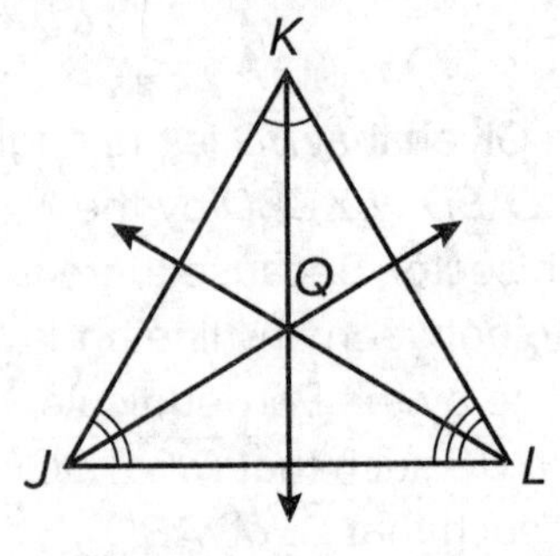

$JQ = KQ = LQ$

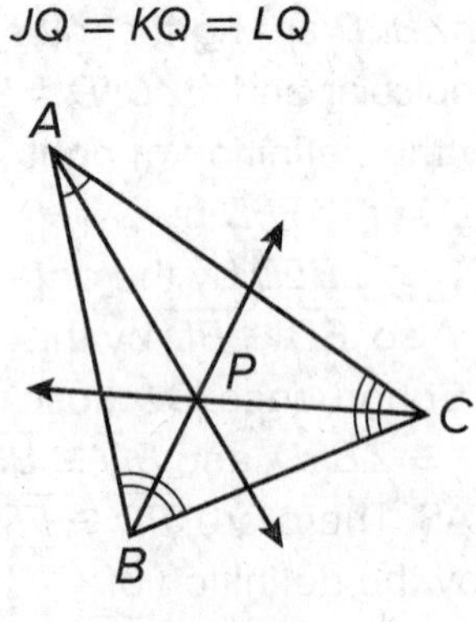

$AP \neq BP \neq CP$

25. Proof: Statements (Reasons)

1. Plane Z is an angle bisector of $\angle KJH$; $\overline{KJ} \cong \overline{HJ}$. (Given)
2. $\angle KJM \cong \angle HJM$ (Definition of angle bisector)
3. $\overline{JM} \cong \overline{JM}$ (Reflexive Property of Congruence)
4. $\triangle KJM \cong \triangle HJM$ (SAS)
5. $\overline{MH} \cong \overline{MK}$ (CPCTC)

27. A point is on the bisector of an angle if and only if it is equidistant from the sides of the angle.

Lesson 6-3

1. 24 **3.** 14 **5.** 12 **7.** 6 **9.** (1, 10) **11.** (1, 4) **13.** (1, 0) **15.** (−9, −3) **17.** 3 **19.** $\frac{1}{2}$

21. 6; no; because $m\angle ECA = 92°$

23. altitude **25.** median **27.** 32

29. Proof: Because $\triangle XYZ$ is isosceles, $\overline{XY} \cong \overline{YZ}$. By the definition of angle bisector, $\angle XYW \cong \angle ZYW$. $\overline{YW} \cong \overline{YW}$ by the Reflexive Property. So, by SAS, $\triangle XYW \cong \triangle ZYW$. By CPCTC, $\overline{XW} \cong \overline{ZW}$. By the definition of midpoint, W is the midpoint of $\overline{XZ}$. By the definition of median, $\overline{WY}$ is a median.

31. Sample answer: Kareem is correct. According to the Centroid Theorem, $AP = \frac{2}{3}AD$. The segment lengths are transposed.

33a.

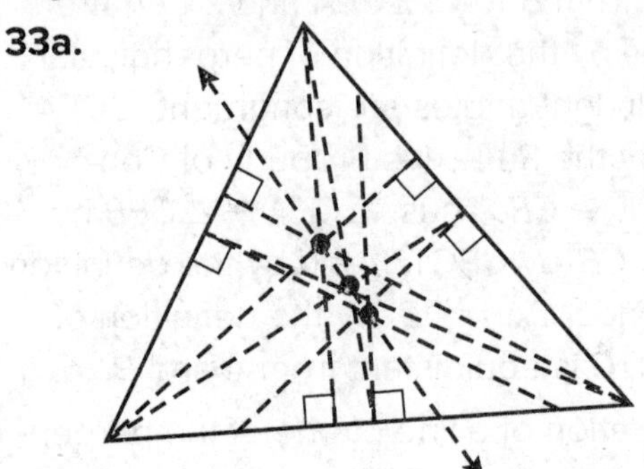

33b.

33c.

33d. Sample answer: The circumcenter, centroid, and orthocenter are all collinear.

35. perpendicular bisectors/incenter; The point of concurrency for perpendicular bisectors is the circumcenter. The incenter is the point of concurrency for the angle bisectors of a triangle.

Lesson 6-4

1. $\angle 4, \angle 7$ **3.** $\angle 7$ **5.** $\angle 4$ **7.** $\angle T, \angle R, \angle S, \overline{RS}, \overline{ST}, \overline{RT}$ **9.** $\angle E, \angle C, \angle D, \overline{CD}, \overline{DE}, \overline{CE}$ **11.** $\angle Q, \angle P, \angle R, \overline{PR}, \overline{RQ}, \overline{QP}$ **13.** They are equal.
15. $\angle X, \angle Y, \angle Z, \overline{YZ}, \overline{XZ}, \overline{XY}$
17. Sample answer: Because $BC > AC$, $m\angle BAC > m\angle ABC$ by Theorem 6.9. $\overline{AY}$ and $\overline{BX}$ are angle bisectors, so $m\angle ABX = \frac{1}{2}m\angle ABC$ and $m\angle BAY = \frac{1}{2}m\angle BAC$. Then $m\angle BAY > m\angle ABX$.

19. Sample answer: $\overline{AB}$ is the longest. $m\angle BQY > m\angle ABQ$ and $m\angle BQY > m\angle BAQ$ by the Exterior Angle Inequality Theorem. Therefore, if $m\angle AQB > m\angle BQY$, then $m\angle AQB > m\angle ABQ$ and $m\angle AQB > m\angle BAQ$ by the Transitive Property of Inequality. Then, by Theorem 6.10, $AB > AQ$ and $AB > BQ$. Therefore, $\overline{AB}$ is the longest side of $\triangle ABQ$.

21. 2, 1, 3 **23.** $\angle 2$ **25.** $\angle 3$ **27.** $\angle 8$
29. $m\angle BCF > m\angle CFB$
31. $m\angle DBF < m\angle BFD$ **33.** $RP > MP$
35. $RM > RQ$ **37.** Sample answer: 10; $m\angle C > m\angle B$, so if $AB > AC$, then Theorem 6.10 is satisfied. Because $10 > 6$, $AB > AC$.
39. $m\angle 1, m\angle 2 = m\angle 5, m\angle 4, m\angle 6, m\angle 3$; Sample answer: The side opposite $\angle 5$ is the smallest side in that triangle, and $m\angle 2 = m\angle 5$; so, we know that $m\angle 4$ and $m\angle 6$ are both greater than $m\angle 2$ and $m\angle 5$. The side opposite $m\angle 6$ is greater than the side opposite $m\angle 4$. Because the side opposite $\angle 2$ is greater than the side opposite $\angle 1$, we know that $m\angle 1 < m\angle 2$ and $m\angle 5$. Because $m\angle 2 = m\angle 5$, $m\angle 1 + m\angle 3 = m\angle 4 + m\angle 6$. Because $m\angle 1 < m\angle 4$, then $m\angle 3 > m\angle 6$.

Lesson 6-5

1. Proof:

Step 1: Assume that $x > 2$.

Step 2: For all values of $x > 2$, $x^2 + 8 > 12$, so the assumption contradicts the given information for $x > 2$.

Step 3: The assumption of $x > 2$ leads to a contradiction of the given information that $x^2 + 8 \leq 12$. Therefore, the assumption of $x > 2$ must be false, so the original conclusion of $x \leq 2$ must be true.

3. Proof:

Step 1: Assume that $x < -2$ or $x = -2$.

Step 2: When $x = -2$, $2x - 7 = -11$. Because $-11 \not> -11$, the assumption contradicts the given information for $x = -2$.

For all values of $x < -2$, $2x - 7 < -11$, so the assumption contradicts the given information for $x < -2$.

Step 3: In both cases, the assumption leads to a contradiction of the given information that $2x - 7 > -11$. Therefore, the assumption of $x \leq -2$ must be false, so the original conclusion of $x > -2$ must be true.

5. Proof:

Step 1: Assume that $x < -1$ or $x = -1$.

Step 2: When $x = -1$, $-3x + 4 = 7$. Because $7 \not< 7$, the assumption contradicts the given information for $x = -1$.

For all values of $x < -1$, $-3x + 4 > 7$, so the assumption contradicts the given information for $x < -1$.

Step 3: In both cases, the assumption leads to a contradiction of the given information that $-3x + 4 < 7$. Therefore, the assumption of $x \leq -1$ must be false, so the original conclusion of $x > -1$ must be true.

7. Let x be the cost of one bracelet and the cost of the other bracelet be y.

Step 1: Given: $x + y > 40$; Prove: $x > 20$ or $y > 20$; Indirect Proof: Assume that $x \leq 20$ and $y \leq 20$.

Step 2: If $x \leq 20$ and $y \leq 20$, then $x + y \leq 20 + 20$ or $x + y \leq 40$. This is a contradiction because we know that $x + y > 40$.

Step 3: Because the assumption that $x \leq 20$ and $y \leq 20$ leads to a contradiction of a known fact, the assumption must be false. Therefore, the conclusion that $x > 20$ or $y > 20$ must be true. Thus, at least one of the bracelets had to cost more than \$20.

9. Step 1: Assume that x and y are not both odd integers. That is, assume that either x or y is an even integer, or that x and y are both even integers.

Step 2: **Case 1:** You only need to show that the assumption that x is an even integer leads to a contradiction, because the argument for y is an even integer follows the same reasoning. So, assume that x is an even integer and that y is an odd integer. This means that $x = 2k$ for some integer k and that $y = 2m + 1$ for some integer m.

$xy = (2k)(2m + 1)$	Statement of assumption
$= 4km + 2k$	Distributive Property
$= 2(2km + k)$	Distributive Property

Because k and m are integers, $2km + k$ is also an integer. Let p represent the integer $2km + k$. So, xy can be represented by $2p$. This means that xy is an even integer, which contradicts the given information that xy is an odd integer.

Case 2: Assume that x and y are both even integers. This means that $x = 2k$ and $y = 2m$ for some integers k and m.

$xy = (2k)(2m)$	Statement of assumption
$= 4km$	Simplify.
$= 2(2km)$	Distributive Property

Because k and m are integers, $2km$ is also an integer. Let p represent the integer $2km$. So, xy can be represented by $2p$. This means that xy is an even integer which contradicts the given information that xy is an odd integer.

Step 3: In both cases, the assumption leads to a contradiction of the given information, so the original conclusion that both x and y are odd integers must be true.

11. Step 1: Assume that x is divisible by 4. That is, assume that 4 is a factor of x.

Step 2: Let $x = 4n$ for some integer n. So, $x = 2(2n)$. So, 2 is a factor of x, which means that x is an even number. This contradicts the given information.

Step 3: Because the assumption that x is divisible by 4 leads to a contradiction of the given information, the original conclusion that x is not divisible by 4 must be true.

13.

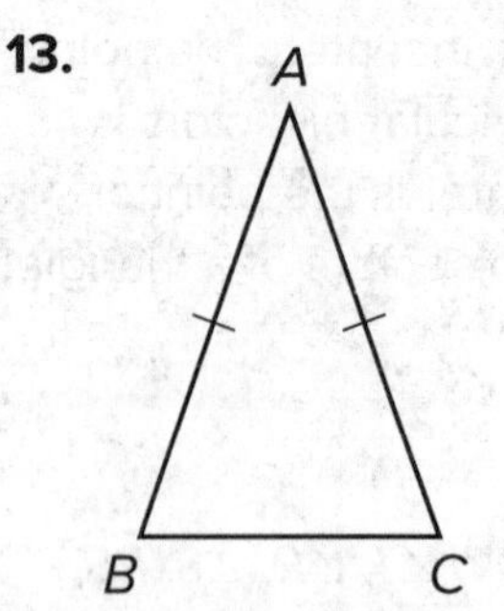

Step 1: Assume that $\angle B$ is a right angle.

Step 2: By the Isosceles Triangle Theorem, $\angle C$ is also a right angle. This contradicts the fact that a triangle can have no more than one right angle.

Step 3: Because the assumption that $\angle B$ is a right angle must be false, the original conclusion that neither of the base angles is a right angle must be true.

15. Step 1: Assume that $\angle A$ is a right angle.

Step 2: Show that this leads to a contradiction. If $\angle A$ is a right angle, then $m\angle A = 90°$ and $m\angle C + m\angle A = 100 + 90 = 190°$. Thus, the sum of the measures of the angles of $\triangle ABC$ is greater than 180°.

Step 3: The conclusion that the sum of the measures of the angles of $\triangle ABC$ is greater than 180° is a contradiction of a known property. The assumption that $\angle A$ is a right angle must be false, which means that the statement $\angle A$ *is not a right angle* must be true.

17. Assume that it is 20°C when Enrique hears the siren; then show that at this temperature it will take more than 5 seconds for the sound of the siren to reach him. Because the assumption is false, it must not be 20°C when Enrique hears the siren.

19. Given: $\triangle ABC$; $m\angle A > m\angle ABC$

Prove: $BC > AC$

Proof: Assume that $BC \not> AC$. By the Comparison Property, $BC = AC$ or $BC < AC$.

Case 1: If $BC = AC$, then $\angle ABC \cong \angle A$ by the Isosceles Triangle Theorem. But, $\angle ABC \cong \angle A$ contradicts the given statement that $m\angle A > m\angle ABC$. So, $BC \neq AC$.

Case 2: If $BC < AC$, then there must be a point D between A and C such that $\overline{DC} \cong \overline{BC}$. Draw the auxiliary segment $\overline{BD}$. Because $DC = BC$, by the Isosceles Triangle

Theorem, $\angle BDC \cong \angle DBC$. Now, $\angle BDC$ is an exterior angle of $\triangle BAD$ and by the Exterior Angles Inequality Theorem $m\angle BDC > m\angle A$. By the Angle Addition Postulate, $m\angle ABC = m\angle ABD + m\angle DBC$. Then, by the definition of inequality, $m\angle ABC > m\angle DBC$. By substitution and the Transitive Property of Inequality, $m\angle ABC > m\angle A$. But this contradicts the given statement that $m\angle A > m\angle ABC$. In both cases, a contradiction was found, and hence our assumption must have been false. Therefore, $BC > AC$.

21. Sample answer: First identify the statement you need to prove and assume that this statement is false by assuming that the opposite of the statement is true. Next, reason logically until you reach a contradiction. Finally, conclude that the statement you wanted to prove must be true because the contradiction proves that the assumption you made was false.

23. Step 1: Let x be a nonzero rational number such that $x = \frac{a}{b}$ for some integers a and b, $b \neq 0$. Let y represent an irrational number. Substituting, $xy = \frac{ay}{b}$. Assume that xy is a rational number such that $xy = \frac{c}{d}$ for some integers c and d, $d \neq 0$.

Step 2:

$xy = \frac{ay}{b}$	x is a rational number.
$\frac{c}{d} = \frac{ay}{b}$	Substitution of assumption
$cb = ayd$	Multiply each side by db. This is possible because $d \neq 0$ and $b \neq 0$.
$\frac{cb}{ad} = y$	Divide each side by ad. $a \neq 0$ because $x = \frac{a}{b}$ and x is nonzero.

Because a, b, c, and d are integers, $a \neq 0$, and $d \neq 0$, $\frac{cb}{ad}$ is the quotient of two integers. Therefore, y is a rational number. This contradicts the given statement that y is an irrational number.

Step 3: Because the assumption that xy is a rational number leads to a contradiction of the given, the original conclusion that xy is irrational must be true.

Lesson 6-6

1. yes **3.** yes **5.** yes **7.** no; $0.7 + 1.4 = 2.1$
9. 13 ft $< n <$ 25 ft **11.** 14 in. $< n <$ 40 in.
13. 90 mi
15. Proof:

Statements (Reasons)

1. $\overline{PL} \parallel \overline{MT}$ (Given)
2. $\angle P \cong \angle T$ (Alternate Interior Angles Theorem)
3. K is the midpoint of $\overline{PT}$ (Given)
4. $PK = KT$ (Definition of midpoint)
5. $\angle PKL \cong \angle MKT$ (Vertical Angles Theorem)
6. $\triangle PKL \cong \triangle TKM$ (ASA)
7. $PK + KL > PL$ (Triangle Inequality Theorem)
8. $KL = KM$ (CPCTC)
9. $PK + KM > PL$ (Substitution)

17. $2 < x < 10$ **19.** yes; $XY + YZ > XZ$, $XY + XZ > YZ$, and $XZ + YZ > XY$ **21.** 3 **23.** $\frac{1}{7}$
25. Sample answer: whether or not the side lengths actually form a triangle, what the smallest and largest angles are, whether the triangle is equilateral. isosceles, or scalene

Lesson 6-7

1. the second plane; Sample answer: The legs are congruent. If $x < 30$, then the measure of the included angle, $(180 - x)°$, is greater for the second plane; so, by the Hinge Theorem, the second plane if farther away from the airstrip.
3. Proof:

Statements (Reasons)

1. $\angle SXT$ and $\angle RXT$ are supplementary. (Def. of linear pair)
2. $m\angle SXT + m\angle RXT = 180°$ (Def. of supplementary)
3. $m\angle SXT = 97°$ (Given)
4. $97° + m\angle RXT = 180°$ (Substitution)
5. $m\angle RXT = 83°$ (Subtraction)
6. $97 > 83$ (Inequality)

7. $m\angle SXT > m\angle RXT$ (Substitution)
8. $RX = XS$ (Given)
9. $TX = TX$ (Reflexive Property of Equality)
10. $ST > RT$ (Hinge Theorem)

5. $x > 12.5$ **7.** Proof:
Statements (Reasons)
1. $\overline{XU} \cong \overline{VW}$, $\overline{XU} \parallel \overline{VW}$ (Given)
2. $\angle UXV \cong \angle XVW$, $\angle XUW \cong \angle UWV$ (Alternate Interior Angles Theorem)
3. $\triangle XZU \cong \triangle VZW$ (ASA)
4. $\overline{XZ} \cong \overline{VZ}$ (CPCTC)
5. $\overline{WZ} \cong \overline{WZ}$ (Reflexive Property)
6. $VW > XW$ (Given)
7. $m\angle VZW > m\angle XZW$ (Converse of Hinge Theorem)
8. $\angle VZW \cong \angle XZU$, $\angle XZW \cong \angle VZU$ (Vertical angles are congruent.)
9. $m\angle VZW = m\angle XZU$, $m\angle XZW = m\angle VZU$ (Definition of congruent angles)
10. $m\angle XZU > m\angle UZV$ (Substitution Property)

9. $MR > RP$ **11.** $m\angle C < m\angle Z$
13. $m\angle BXA < m\angle DXA$ **15.** It is given that $\overline{EF} \cong \overline{GH}$. Also, $\overline{FG} \cong \overline{FG}$ by the Reflexive Property. It is also given that $m\angle F > m\angle G$. Therefore, by the Hinge Theorem, $EG > FH$.
17. 8:00
19. Given: $\overline{RS} \cong \overline{UW}$
$\overline{ST} \cong \overline{WV}$
$\overline{RT} > \overline{UV}$

Prove: $m\angle S > m\angle W$

Indirect Proof

Step 1: Assume that $m\angle S \leq m\angle W$.

Step 2: If $m\angle S \leq m\angle W$, then either $m\angle S < m\angle W$ or $m\angle S = m\angle W$.

Case 1: If $m\angle S < m\angle W$, then $RT < UV$ by SAS Inequality.

Case 2: If $m\angle S = m\angle W$, then $\triangle RST = \triangle UVW$ by SAS, and $\overline{RT} \cong \overline{UV}$ by CPCTC. Thus, $RT = UV$.

Step 3: Both cases contradict the given $RT > UV$. Therefore, the assumption must be false, and the conclusion, $m\angle S < m\angle W$, must be true.

21a. Northwest; sample answer: Submarine B travels farther than submarine A by the Hinge Theorem, and on an overall heading closer to east-west, so it ends up farther west than submarine A's starting position. Also submarine B's southwesterly leg is nearer to the westerly direction, so it does not travel as far south, and finishes to the north of submarine A's starting position. **21b.** Sample answer: Submarine B would have traveled farther than submarine A and would not be to the northeast or southeast of submarine A's starting position.

23. Right or obtuse; sample answer: If $RT = RS$, then the triangle is isosceles, and the median is also perpendicular to $\overline{TS}$. That would mean that both triangles formed by the median, $\triangle RQT$ and $\triangle RQS$, are right. If $RT > RS$, that means that $m\angle RQT > m\angle RQS$. Because they are a linear pair and the sum of the angle measures must be 180°, $m\angle RQT$ must be greater than 90° and $\triangle RQT$ is obtuse.

25. Never; sample answer: From the Converse of the Hinge Theorem, $m\angle ADB < m\angle BDC$. $\angle ADB$ and $\angle BDC$ form a linear pair. So, $m\angle ADB + m\angle BDC = 180°$. Because $m\angle BDC > m\angle ADB$, $m\angle BDC$ must be greater than 90° and $m\angle ADB$ must be less than 90°. So, by the definition of obtuse and acute angles, $\angle BDC$ is always obtuse and $\angle ADB$ is always acute.

Module 6 Review

1. 4.8
3. C
5. C
7.

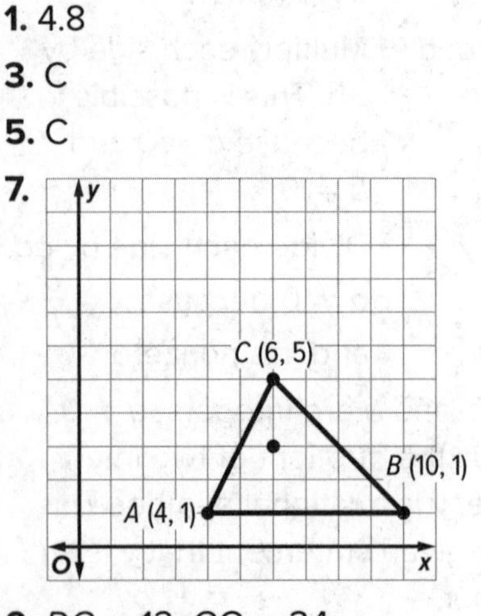

9. $DG = 12$, $GC = 24$
11. Antonia must assume that one of the angles is greater than 90°.
13. D **15.** C

English / Español

A

30°-60°-90° triangle (Lesson 9-4) A right triangle with two acute angles that measure 30° and 60°.

triángulo 30°-60°-90° Un triángulo rectángulo con dos ángulos agudos que miden 30° y 60°.

45°-45°-90° triangle (Lesson 9-4) A right triangle with two acute angles that measure 45°.

triángulo 45°-45°-90° Un triángulo rectángulo con dos ángulos agudos que miden 45°.

accuracy (Lesson 2-7) The nearness of a measurement to the true value of the measure.

exactitud La proximidad de una medida al valor verdadero de la medida.

adjacent angles (Lesson 2-1) Two angles that lie in the same plane and have a common vertex and a common side but have no common interior points.

ángulos adyacentes Dos ángulos que se encuentran en el mismo plano y tienen un vértice común y un lado común, pero no tienen puntos comunes en el interior.

adjacent arcs (Lesson 10-2) Arcs in a circle that have exactly one point in common.

arcos adyacentes Arcos en un circulo que tienen un solo punto en común.

alternate exterior angles (Lesson 3-7) When two lines are cut by a transversal, nonadjacent exterior angles that lie on opposite sides of the transversal.

ángulos alternos externos Cuando dos líneas son cortadas por un ángulo transversal, no adyacente exterior que se encuentran en lados opuestos de la transversal.

alternate interior angles (Lesson 3-7) When two lines are cut by a transversal, nonadjacent interior angles that lie on opposite sides of the transversal.

ángulos alternos internos Cuando dos líneas son cortadas por un ángulo transversal, no adyacente interior que se encuentran en lados opuestos de la transversal.

altitude of a parallelogram (Lesson 11-1) A perpendicular segment between any two parallel bases.

altitud de un paralelogramo Un segmento perpendicular entre dos bases paralelas.

altitude of a prism or cylinder (Lesson 11-4) A segment perpendicular to the bases that joins the planes of the bases.

altitud de un prisma o cilindro Un segmento perpendicular a las bases que une los planos de las bases.

altitude of a pyramid or cone (Lesson 11-4) A segment perpendicular to the base that has the vertex as one endpoint and a point in the plane of the base as the other endpoint.

altitud de una pirámide o cono Un segmento perpendicular a la base que tiene el vértice como un punto final y un punto en el plano de la base como el otro punto final.

altitude of a triangle (Lesson 6-3) A segment from a vertex of the triangle to the line containing the opposite side and perpendicular to that side.

altitud de triángulo Un segmento de un vértice del triángulo a la línea que contiene el lado opuesto y perpendicular a ese lado.

ambiguous case (Lesson 9-7) When two different triangles could be created or described using the given information.

caso ambiguo Cuando dos triángulos diferentes pueden ser creados o descritos usando la información dada.

analytic geometry (Lesson 1-1) The study of geometry that uses the coordinate system.

geometría analítica El estudio de la geometría que utiliza el sistema de coordenadas.

angle (Lesson 2-1) The intersection of two noncollinear rays at a common endpoint.

ángulo La intersección de dos rayos no colineales en un extremo común.

angle bisector (Lesson 2-1) A ray or segment that divides an angle into two congruent angles.

bisectriz de un ángulo Un rayo o segmento que divide un ángulo en dos ángulos congruentes.

angle of depression (Lesson 9-6) The angle formed by a horizontal line and an observer's line of sight to an object below the horizontal line.

ángulo de depresión El ángulo formado por una línea horizontal y la línea de visión de un observador a un objeto por debajo de la línea horizontal.

angle of elevation (Lesson 9-6) The angle formed by a horizontal line and an observer's line of sight to an object above the horizontal line.

ángulo de elevación El ángulo formado por una línea horizontal y la línea de visión de un observador a un objeto por encima de la línea horizontal.

angle of rotation (Lesson 2-4) The angle through which a figure rotates.

ángulo de rotación El ángulo a través del cual gira una figura.

apothem (Lesson 11-2) A perpendicular segment between the center of a regular polygon and a side of the polygon or the length of that line segment.

apotema Un segmento perpendicular entre el centro de un polígono regular y un lado del polígono o la longitud de ese segmento de línea.

approximate error (Lesson 2-7) The positive difference between an actual measurement and an approximate or estimated measurement.

error aproximado La diferencia positiva entre una medida real y una medida aproximada o estimada.

arc (Lesson 10-2) Part of a circle that is defined by two endpoints.

arco Parte de un círculo que se define por dos puntos finales.

arc length (Lesson 10-2) The distance between the endpoints of an arc measured along the arc in linear units.

longitude de arco La distancia entre los extremos de un arco medido a lo largo del arco en unidades lineales.

area (Lesson 2-3) The number of square units needed to cover a surface.

área El número de unidades cuadradas para cubrir una superficie.

auxiliary line (Lesson 5-1) An extra line or segment drawn in a figure to help analyze geometric relationships.

línea auxiliar Una línea o segmento extra dibujado en una figura para ayudar a analizar las relaciones geométricas.

axiom (Lesson 1-1) A statement that is accepted as true without proof.

axioma Una declaración que se acepta como verdadera sin prueba.

axiomatic system (Lesson 1-1) A set of axioms from which theorems can be derived.

sistema axiomático Un conjunto de axiomas de los cuales se pueden derivar teoremas.

axis symmetry (Lesson 11-5) If a figure can be mapped onto itself by a rotation between 0° and 360° in a line.

eje simetría Si una figura puede ser asignada sobre sí misma por una rotación entre 0° y 360° en una línea.

B

base angles of a trapezoid (Lesson 7-6) The two angles formed by the bases and legs of a trapezoid.

ángulos de base de un trapecio Los dos ángulos formados por las bases y patas de un trapecio.

base angles of an isosceles triangle (Lesson 5-6) The two angles formed by the base and the congruent sides of an isosceles triangle.

ángulo de la base de un triángulo isosceles Los dos ángulos formados por la base y los lados congruentes de un triángulo isosceles.

base edge (Lesson 11-4) The intersection of a lateral face and a base in a solid figure.

arista de la base La intersección de una cara lateral y una base en una figura sólida.

base of a parallelogram (Lesson 11-1) Any side of a parallelogram.

base de un paralelogramo Cualquier lado de un paralelogramo.

base of a pyramid or cone (Lesson 2-5) The face of the solid opposite the vertex of the solid.

base de una pirámide o cono La cara del sólido opuesta al vértice del sólido.

bases of a prism or cylinder (Lesson 2-5) The two parallel congruent faces of the solid.

bases de un prisma o cilindro Las dos caras congruentes paralelas de la figura sólida.

bases of a trapezoid (Lesson 7-6) The parallel sides in a trapezoid.

bases de un trapecio Los lados paralelos en un trapecio.

betweenness of points (Lesson 1-3) Point C is between A and B if and only if A, B, and C are collinear and $AC + CB = AB$.

intermediación de puntos El punto C está entre A y B si y sólo si A, B, y C son colineales y $AC + CB = AB$.

biconditional statement (Lesson 3-2) The conjunction of a conditional and its converse.

declaración bicondicional La conjunción de un condicional y su inverso.

bisect (Lesson 1-7) To separate a line segment into two congruent segments.

bisecar Separe un segmento de línea en dos segmentos congruentes.

C

center of a circle (Lesson 10-1) The point from which all points on a circle are the same distance.

centro de un círculo El punto desde el cual todos los puntos de un círculo están a la misma distancia.

center of dilation (Lesson 8-1) The center point from which dilations are performed.

centro de dilatación Punto fijo en torno al cual se realizan las homotecias.

center of a regular polygon (Lesson 11-2) The center of the circle circumscribed about a regular polygon.

centro de un polígono regular El centro del círculo circunscrito alrededor de un polígono regular.

center of rotation (Lesson 2-4) The fixed point about which a figure rotates.

centro de rotación El punto fijo sobre el que gira una figura.

center of symmetry (Lesson 4-6) A point in which a figure can be rotated onto itself.

centro de la simetría Un punto en el que una figura se puede girar sobre sí misma.

central angle of a circle (Lesson 10-2) An angle with a vertex at the center of a circle and sides that are radii.

ángulo central de un círculo Un ángulo con un vértice en el centro de un círculo y los lados que son radios.

central angle of a regular polygon (Lesson 11-2) An angle with its vertex at the center of a regular polygon and sides that pass through consecutive vertices of the polygon.

ángulo central de un polígono regular Un ángulo con su vértice en el centro de un polígono regular y lados que pasan a través de vértices consecutivos del polígono.

centroid (Lesson 6-3) The point of concurrency of the medians of a triangle.

baricentro El punto de intersección de las medianas.

chord of a circle or sphere (Lessons 10-1, 11-4) A segment with endpoints on the circle or sphere.

cuerda de un círculo o esfera Un segmento con extremos en el círculo o esfera.

circle (Lesson 10-1) The set of all points in a plane that are the same distance from a given point called the center.

círculo El conjunto de todos los puntos en un plano que están a la misma distancia de un punto dado llamado centro.

circumcenter (Lesson 6-1) The point of concurrency of the perpendicular bisectors of the sides of a triangle.

circuncentro El punto de concurrencia de las bisectrices perpendiculares de los lados de un triángulo.

circumference (Lesson 2-3) The distance around a circle.

circunferencia La distancia alrededor de un círculo.

circumscribed angle (Lesson 10-5) An angle with sides that are tangent to a circle.

ángulo circunscrito Un ángulo con lados que son tangentes a un círculo.

circumscribed polygon (Lesson 10-5) A polygon with vertices outside the circle and sides that are tangent to the circle.

poligono circunscrito Un polígono con vértices fuera del círculo y lados que son tangentes al círculo.

collinear (Lesson 1-2) Lying on the same line.

colineal Acostado en la misma línea.

combination (Lesson 12-4) A selection of objects in which order is not important.

combinación Una selección de objetos en los que el orden no es importante.

common tangent (Lesson 10-5) A line or segment that is tangent to two circles in the same plane.

tangente común Una línea o segmento que es tangente a dos círculos en el mismo plano.

complement of A (Lesson 12-2) All of the outcomes in the sample space that are not included as outcomes of event A.

complemento de A Todos los resultados en el espacio muestral que no se incluyen como resultados del evento A.

complementary angles (Lesson 2-2) Two angles with measures that have a sum of 90°.

ángulo complementarios Dos ángulos con medidas que tienen una suma de 90°.

component form (Lesson 2-4) A vector written as $\langle x, y\rangle$, which describes the vector in terms of its horizontal component x and vertical component y.

forma de componente Un vector escrito como $\langle x, y\rangle$, que describe el vector en términos de su componente horizontal x y componente vertical y.

composite figure (Lesson 11-2) A figure that can be separated into regions that are basic figures, such as triangles, rectangles, trapezoids, and circles.	**figura compuesta** Una figura que se puede separar en regiones que son figuras básicas, tales como triángulos, rectángulos, trapezoides, y círculos.
composite solid (Lesson 11-4) A three-dimensional figure that is composed of simpler solids.	**solido compuesta** Una figura tridimensional que se compone de figuras más simples.
composition of transformations (Lesson 4-4) When a transformation is applied to a figure and then another transformation is applied to its image.	**composición de transformaciones** Cuando una transformación se aplica a una figura y luego se aplica otra transformación a su imagen.
compound event (Lesson 12-5) Two or more simple events.	**evento compuesto** Dos o más eventos simples.
compound statement (Lesson 3-2) Two or more statements joined by the word *and* or *or*.	**enunciado compuesto** Dos o más declaraciones unidas por la palabra *y* o *o*.
concave polygon (Lesson 2-3) A polygon with one or more interior angles with measures greater than 180°.	**polígono cóncavo** Un polígono con uno o más ángulos interiores con medidas superiores a 180°.
concentric circles (Lesson 10-1) Coplanar circles that have the same center.	**círculos concéntricos** Círculos coplanarios que tienen el mismo centro.
conclusion (Lesson 3-2) The statement that immediately follows the word *then* in a conditional.	**conclusión** La declaración que inmediatamente sigue la palabra *entonces* en un condicional.
concurrent lines (Lesson 6-1) Three or more lines that intersect at a common point.	**líneas concurrentes** Tres o más líneas que se intersecan en un punto común.
conditional probability (Lesson 12-7) The probability that an event will occur given that another event has already occurred.	**probabilidad condicional** La probabilidad de que un evento ocurra dado que otro evento ya ha ocurrido.
conditional statement (Lesson 3-2) A compound statement that consists of a premise, or hypothesis, and a conclusion, which is false only when its premise is true and its conclusion is false.	**enunciado condicional** Una declaración compuesta que consiste en una premisa, o hipótesis, y una conclusión, que es falsa solo cuando su premisa es verdadera y su conclusión es falsa.
cone (Lesson 2-5) A solid figure with a circular base connected by a curved surface to a single vertex.	**cono** Una figura sólida con una base circular conectada por una superficie curvada a un solo vértice.
congruent (Lesson 1-3) Having the same size and shape.	**congruente** Tener el mismo tamaño y forma.
congruent angles (Lesson 2-1) Two angles that have the same measure.	**ángulo congruentes** Dos ángulos que tienen la misma medida.

congruent arcs (Lesson 10-2) Arcs in the same or congruent circles that have the same measure.

arcos congruentes Arcos en los mismos círculos o congruentes que tienen la misma medida.

congruent polygons (Lesson 5-2) All of the parts of one polygon are congruent to the corresponding parts or matching parts of another polygon.

polígonos congruentes Todas las partes de un polígono son congruentes con las partes correspondientes o partes coincidentes de otro polígono.

congruent segments (Lesson 1-3) Line segments that are the same length.

segmentos congruentes Línea segmentos que son la misma longitud.

congruent solids (Lesson 11-8) Solid figures that have exactly the same shape, size, and a scale factor of 1:1.

sólidos congruentes Figuras sólidas que tienen exactamente la misma forma, tamaño y un factor de escala de 1:1.

conic sections (Lesson 11-5) Cross sections of a right circular cone.

secciones cónicas Secciones transversales de un cono circular derecho.

conjecture (Lesson 3-1) An educated guess based on known information and specific examples.

conjetura Una suposición educada basada en información conocida y ejemplos específicos.

conjunction (Lesson 3-2) A compound statement using the word *and*.

conjunción Una declaración compuesta usando la palabra *y*.

consecutive interior angles (Lesson 3-7) When two lines are cut by a transversal, interior angles that lie on the same side of the transversal.

ángulos internos consecutivos Cuando dos líneas se cortan por un ángulo transversal, interior que se encuentran en el mismo lado de la transversal.

constructions (Lesson 1-3) Methods of creating figures without the use of measuring tools.

construcciones Métodos de creación de figuras sin el uso de herramientas de medición.

contrapositive (Lesson 3-2) A statement formed by negating both the hypothesis and the conclusion of the converse of a conditional.

antítesis Una afirmación formada negando tanto la hipótesis como la conclusión del inverso del condicional.

converse (Lesson 3-2) A statement formed by exchanging the hypothesis and conclusion of a conditional statement.

recíproco Una declaración formada por el intercambio de la hipótesis y la conclusión de la declaración condicional.

convex polygon (Lesson 2-3) A polygon with all interior angles measuring less than 180°.

polígono convexo Un polígono con todos los ángulos interiores que miden menos de 180°.

coordinate proofs (Lesson 5-7) Proofs that use figures in the coordinate plane and algebra to prove geometric concepts.

pruebas de coordenadas Pruebas que utilizan figuras en el plano de coordenadas y álgebra para probar conceptos geométricos.

coplanar (Lesson 1-2) Lying in the same plane.

coplanar Acostado en el mismo plano.

corollary (Lesson 5-1) A theorem with a proof that follows as a direct result of another theorem.

corolario Un teorema con una prueba que sigue como un resultado directo de otro teorema.

corresponding angles (Lesson 3-7) When two lines are cut by a transversal, angles that lie on the same side of a transversal and on the same side of the two lines.

ángulos correspondientes Cuando dos líneas se cortan transversalmente, los ángulos que se encuentran en el mismo lado de una transversal y en el mismo lado de las dos líneas.

corresponding parts (Lesson 5-2) Corresponding angles and corresponding sides of two polygons.

partes correspondientes Ángulos correspondientes y lados correspondientes.

cosine (Lesson 9-5) The ratio of the length of the leg adjacent to an angle to the length of the hypotenuse.

coseno Relación entre la longitud de la pierna adyacente a un ángulo y la longitud de la hipotenusa.

counterexample (Lesson 3-1) An example that contradicts the conjecture showing that the conjecture is not always true.

contraejemplo Un ejemplo que contradice la conjetura que muestra que la conjetura no siempre es cierta.

cross section (Lesson 11-5) The intersection of a solid and a plane.

sección transversal Intersección de un sólido con un plano.

cylinder (Lesson 2-5) A solid figure with two congruent and parallel circular bases connected by a curved surface.

cilindro Una figura sólida con dos bases circulares congruentes y paralelas conectadas por una superficie curvada.

D

decomposition (Lesson 11-1) Separating a figure into two or more nonoverlapping parts.

descomposición Separar una figura en dos o más partes que no se solapan.

deductive argument (Lesson 3-4) An argument that guarantees the truth of the conclusion provided that its premises are true.

argumento deductivo Un argumento que garantiza la verdad de la conclusión siempre que sus premisas sean verdaderas.

deductive reasoning (Lesson 3-3) The process of reaching a specific valid conclusion based on general facts, rules, definitions, or properties.

razonamiento deductivo El proceso de alcanzar una conclusión válida específica basada en hechos generales, reglas, definiciones, o propiedades.

defined term (Lesson 1-1) A term that has a definition and can be explained.

término definido Un término que tiene una definición y se puede explicar.

definitions (Lesson 1-1) An explanation that assigns properties to a mathematical object.

definiciones Una explicación que asigna propiedades a un objeto matemático.

degree (Lesson 10-2) $\frac{1}{360}$ of the circular rotation about a point.

grado $\frac{1}{360}$ de la rotación circular alrededor de un punto.

density (Lesson 11-9) A measure of the quantity of some physical property per unit of length, area, or volume.

densidad Una medida de la cantidad de alguna propiedad física por unidad de longitud, área o volumen.

dependent events (Lesson 12-5) Two or more events in which the outcome of one event affects the outcome of the other events.

eventos dependientes Dos o más eventos en que el resultado de un evento afecta el resultado de los otros eventos.

diagonal (Lesson 7-1) A segment that connects any two nonconsecutive vertices within a polygon.

diagonal Un segmento que conecta cualquier dos vértices no consecutivos dentro de un polígono.

diameter of a circle or sphere (Lessons 10-1, 11-4) A chord that passes through the center of a circle or sphere.

diámetro de un círculo o esfera Un acorde que pasa por el centro de un círculo o esfera.

dilation (Lesson 8-1) A nonrigid motion that enlarges or reduces a geometric figure.

dilatación Un movimiento no rígido que agranda o reduce una figura geométrica.

directed line segment (Lesson 1-5) A line segment with an initial endpoint and a terminal endpoint.

segment de línea dirigido Un segmento de línea con un punto final inicial y un punto final terminal.

directrix (Lesson 10-8) An exterior line perpendicular to the line containing the foci of a curve.

directriz Una línea exterior perpendicular a la línea que contiene los focos de una curva.

disjunction (Lesson 3-2) A compound statement using the word *or*.

disyunción Una declaración compuesta usando la palabra *o*.

distance (Lesson 1-4) The length of the line segment between two points.

distancia La longitud del segmento de línea entre dos puntos.

E

edge of a polyhedron (Lesson 2-5) A line segment where the faces of the polyhedron intersect.

arista de un poliedro Un segmento de línea donde las caras del poliedro se cruzan.

enlargement (Lesson 8-1) A dilation with a scale factor greater than 1.

ampliación Una dilatación con un factor de escala mayor que 1.

equiangular polygon (Lesson 2-3) A polygon with all angles congruent.

polígono equiangular Un polígono con todos los ángulos congruentes.

equidistant (Lesson 1-7) A point is equidistant from other points if it is the same distance from them.

equidistante Un punto es equidistante de otros puntos si está a la misma distancia de ellos.

equidistant lines (Lesson 3-10) Two lines for which the distance between the two lines, measured along a perpendicular line or segment to the two lines, is always the same.

líneas equidistantes Dos líneas para las cuales la distancia entre las dos líneas, medida a lo largo de una línea o segmento perpendicular a las dos líneas, es siempre la misma.

equilateral polygon (Lesson 2-3) A polygon with all sides congruent.

polígono equilátero Un polígono con todos los lados congruentes.

event (Lesson 12-1) A subset of the sample space.

evento Un subconjunto del espacio de muestra.

experiment (Lesson 12-1) A situation involving chance.

experimento Una situación de riesgo.

exterior of an angle (Lesson 2-1) The area outside of the two rays of an angle.

exterior de un ángulo El área fuera de los dos rayos de un ángulo.

exterior angle of a triangle (Lesson 5-1) An angle formed by one side of the triangle and the extension of an adjacent side.

ángulo exterior de un triángulo Un ángulo formado por un lado del triángulo y la extensión de un lado adyacente.

exterior angles (Lesson 3-7) When two lines are cut by a transversal, any of the four angles that lie outside the region between the two intersected lines.

ángulos externos Cuando dos líneas son cortadas por una transversal, cualquiera de los cuatro ángulos que se encuentran fuera de la región entre las dos líneas intersectadas.

F

face of a polyhedron (Lesson 2-5) A flat surface of a polyhedron.

cara de un poliedro Superficie plana de un poliedro.

factorial of *n* (Lesson 12-4) The product of the positive integers less than or equal to *n*.

factorial de *n* El producto de los enteros positivos inferiores o iguales a *n*.

finite sample space (Lesson 12-1) A sample space that contains a countable number of outcomes.

espacio de muestra finito Un espacio de muestra que contiene un número contable de resultados.

flow proof (Lesson 3-4) A proof that uses boxes and arrows to show the logical progression of an argument.

demostración de flujo Una prueba que usa cajas y flechas para mostrar la progresión lógica de un argumento.

focus (Lesson 10-8) A point inside a parabola having the property that the distances from any point on the parabola to them and to a fixed line have a constant ratio for any points on the parabola.

foco Un punto dentro de una parábola que tiene la propiedad de que las distancias desde cualquier punto de la parábola a ellos ya una línea fija tienen una relación constante para cualquier punto de la parábola.

fractional distance (Lesson 1-5) An intermediary point some fraction of the length of a line segment.

distancia fraccionaria Un punto intermediario de alguna fracción de la longitud de un segmento de línea.

G

geometric mean (Lesson 9-1) The *n*th root, where *n* is the number of elements in a set of numbers, of the product of the numbers.

media geométrica La enésima raíz, donde *n* es el número de elementos de un conjunto de números, del producto de los números.

geometric model (Lesson 2-3) A geometric figure that represents a real-life object.

modelo geométrico Una figura geométrica que representa un objeto de la vida real.

geometric probability (Lesson 12-3) Probability that involves a geometric measure such as length or area.

probabilidad geométrica Probabilidad que implica una medida geométrica como longitud o área.

glide reflection (Lesson 4-4) The composition of a translation followed by a reflection in a line parallel to the translation vector.

reflexión del deslizamiento La composición de una traducción seguida de una reflexión en una línea paralela al vector de traslación.

H

height of a parallelogram (Lesson 11-1) The length of an altitude of the parallelogram.

altura de un paralelogramo La longitud de la altitud del paralelogramo.

height of a solid (Lesson 11-4) The length of the altitude of a solid figure.

altura de un sólido La longitud de la altitud de una figura sólida.

height of a trapezoid (Lesson 11-1) The perpendicular distance between the bases of a trapezoid.

altura de un trapecio La distancia perpendicular entre las bases de un trapecio.

hypothesis (Lesson 3-2) The statement that immediately follows the word *if* in a conditional.

hipótesis La declaración que sigue inmediatamente a la palabra *si* en un condicional.

I

if-then statement (Lesson 3-2) A compound statement of the form *if p*, *then q*, where *p* and *q* are statements.

enunciado si-entonces Enunciado compuesto de la forma *si p*, *entonces q*, donde *p* y *q* son enunciados.

image (Lesson 2-4) The new figure in a transformation.

imagen La nueva figura en una transformación.

incenter (Lesson 6-2) The point of concurrency of the angle bisectors of a triangle.

incentro El punto de intersección de las bisectrices interiors de un triángulo.

included angle (Lesson 5-3) The interior angle formed by two adjacent sides of a triangle.

ángulo incluido El ángulo interior formado por dos lados adyacentes de un triángulo.

included side (Lesson 5-4) The side of a triangle between two angles.

lado incluido El lado de un triángulo entre dos ángulos.

independent events (Lesson 12-5) Two or more events in which the outcome of one event does not affect the outcome of the other events.

eventos independientes Dos o más eventos en los que el resultado de un evento no afecta el resultado de los otros eventos.

indirect measurement (Lesson 9-6) Using similar figures and proportions to measure an object.

medición indirecta Usando figuras y proporciones similares para medir un objeto.

indirect proof (Lesson 6-5) One assumes that the statement to be proven is false and then uses logical reasoning to deduce that a statement contradicts a postulate, theorem, or one of the assumptions.

demostración indirecta Se supone que la afirmación a ser probada es falsa y luego utiliza el razonamiento lógico para deducir que una afirmación contradice un postulado, teorema o uno de los supuestos.

indirect reasoning (Lesson 6-5) Reasoning that eliminates all possible conclusions but one so that the one remaining conclusion must be true.

razonamiento indirecto Razonamiento que elimina todas las posibles conclusiones, pero una de manera que la conclusión que queda una debe ser verdad.

inductive reasoning (Lesson 3-1) The process of reaching a conclusion based on a pattern of examples.

razonamiento inductive El proceso de llegar a una conclusión basada en un patrón de ejemplos.

infinite sample space (Lesson 12-1) A sample space with outcomes that cannot be counted.

espacio de muestra infinito Un espacio de muestra con resultados que no pueden ser contados.

inscribed angle (Lesson 10-4) An angle with its vertex on a circle and sides that contain chords of the circle.

ángulo inscrito Un ángulo con su vértice en un círculo y lados que contienen acordes del círculo.

inscribed polygon (Lesson 10-4) A polygon inside a circle in which all of the vertices of the polygon lie on the circle.

polígono inscrito Un polígono dentro de un círculo en el que todos los vértices del polígono se encuentran en el círculo.

intercepted arc (Lesson 10-4) The part of a circle that lies between the two lines intersecting it.

arco intersecado La parte de un círculo que se encuentra entre las dos líneas que se cruzan.

interior of an angle (Lesson 2-1) The area between the two rays of an angle.

interior de un ángulo El área entre los dos rayos de un ángulo.

interior angle of a triangle (Lesson 5-1) An angle at the vertex of a triangle.

ángulo interior de un triángulo Un ángulo en el vértice de un triángulo.

interior angles (Lesson 3-7) When two lines are cut by a transversal, any of the four angles that lie inside the region between the two intersected lines.

ángulos interiores Cuando dos líneas son cortadas por una transversal, cualquiera de los cuatro ángulos que se encuentran dentro de la región entre las dos líneas intersectadas.

intersection (Lesson 1-2) A set of points common to two or more geometric figures.

intersección Un conjunto de puntos communes a dos o más figuras geométricas.

intersection of *A* and *B* (Lesson 12-2) The set of all outcomes in the sample space of event *A* that are also in the sample space of event *B*.

intersección de *A* y *B* El conjunto de todos los resultados en el espacio muestral del evento *A* que también se encuentran en el espacio muestral del evento *B*.

inverse (Lesson 3-2) A statement formed by negating both the hypothesis and conclusion of a conditional statement.

inverso Una declaración formada negando tanto la hipótesis como la conclusión de la declaración condicional.

inverse cosine (Lesson 9-5) The ratio of the length of the hypotenuse to the length of the leg adjacent to an angle.

inverso del coseno Relación de la longitud de la hipotenusa con la longitud de la pierna adyacente a un ángulo.

inverse sine (Lesson 9-5) The ratio of the length of the hypotenuse to the length of the leg opposite an angle.

inverso del seno Relación de la longitud de la hipotenusa con la longitud de la pierna opuesta a un ángulo.

inverse tangent (Lesson 9-5) The ratio of the length of the leg adjacent to an angle to the length of the leg opposite the angle.

inverso del tangente Relación de la longitud de la pierna adyacente a un ángulo con la longitud de la pierna opuesta a un ángulo.

isosceles trapezoid (Lesson 7-6) A quadrilateral in which two sides are parallel and the legs are congruent.

trapecio isósceles Un cuadrilátero en el que dos lados son paralelos y las patas son congruentes.

isosceles triangle (Lesson 5-6) A triangle with at least two sides congruent.

triángulo isósceles Un triángulo con al menos dos lados congruentes.

J

joint frequencies (Lesson 12-8) In a two-way frequency table, the frequencies in the interior of the table.

frecuencias articulares En una tabla de frecuencia bidireccional, las frecuencias en el interior de la tabla.

K

kite (Lesson 7-6) A convex quadrilateral with exactly two distinct pairs of adjacent congruent sides.

cometa Un cuadrilátero convexo con exactamente dos pares distintos de lados congruentes adyacentes.

L

lateral area (Lesson 11-4) The sum of the areas of the lateral faces of the figure.

área lateral La suma de las áreas de las caras laterales de la figura.

lateral edges (Lesson 11-4) The intersection of two lateral faces.

aristas laterales La intersección de dos caras laterales.

lateral faces (Lesson 11-4) The faces that join the bases of a solid.

caras laterales Las caras que unen las bases de un sólido.

lateral surface of a cone (Lesson 11-4) The curved surface that joins the base of a cone to the vertex.

superficie lateral de un cono La superficie curvada que une la base de un cono con el vértice.

lateral surface of a cylinder (Lesson 11-4) The curved surface that joins the bases of a cylinder.

superficie lateral de un cilindro La superficie curvada que une las bases de un cilindro.

legs of an isosceles triangle (Lesson 5-6) The two congruent sides of an isosceles triangle.

patas de un triángulo isósceles Los dos lados congruentes de un triángulo isósceles.

legs of a trapezoid (Lesson 7-6) The nonparallel sides in a trapezoid.

patas de un trapecio Los lados no paralelos en un trapezoide.

line (Lesson 1-2) A line is made up of points, has no thickness or width, and extends indefinitely in both directions.

línea Una línea está formada por puntos, no tiene espesor ni anchura, y se extiende indefinidamente en ambas direcciones.

line of reflection (Lesson 2-4) A line midway between a preimage and an image.

línea de reflexión Una línea a medio camino entre una preimagen y una imagen.

line of symmetry (Lesson 4-6) An imaginary line that separates a figure into two congruent parts.

línea de simetría Una línea imaginaria que separa una figura en dos partes congruentes.

line segment (Lesson 1-3) A measurable part of a line that consists of two points, called endpoints, and all of the points between them.

segmento de línea Una parte medible de una línea que consta de dos puntos, llamados extremos, y todos los puntos entre ellos.

line symmetry (Lesson 4-6) Each half of a figure matches the other half exactly.

simetría de línea Cada mitad de una figura coincide exactamente con la otra mitad.

linear pair (Lesson 2-1) A pair of adjacent angles with noncommon sides that are opposite rays.

par lineal Un par de ángulos adyacentes con lados no comunes que son rayos opuestos.

logically equivalent (Lesson 3-2) Statements with the same truth value.

lógicamente equivalentes Declaraciones con el mismo valor de verdad.

M

magnitude (Lesson 4-2) The length of a vector from the initial point to the terminal point.

magnitud La longitud de un vector desde el punto inicial hasta el punto terminal.

magnitude of symmetry (Lesson 4-6) The smallest angle through which a figure can be rotated so that it maps onto itself.

magnitud de la simetria El ángulo más pequeño a través del cual una figura se puede girar para que se cargue sobre sí mismo.

major arc (Lesson 10-2) An arc with measure greater than 180°.

arco mayor Un arco con una medida superior a 180°.

marginal frequencies (Lesson 12-8) In a two-way frequency table, the frequencies in the totals row and column.

frecuencias marginales En una tabla de frecuencias de dos vías, las frecuencias en los totales de fila y columna.

median of a triangle (Lesson 6-3) A line segment with endpoints that are a vertex of the triangle and the midpoint of the side opposite the vertex.

mediana de un triángulo Un segmento de línea con extremos que son un vértice del triángulo y el punto medio del lado opuesto al vértice.

midpoint (Lesson 1-7) The point on a line segment halfway between the endpoints of the segment.

punto medio El punto en un segmento de línea a medio camino entre los extremos del segmento.

midsegment of a trapezoid (Lesson 7-6) The segment that connects the midpoints of the legs of a trapezoid.

segment medio de un trapecio El segmento que conecta los puntos medios de las patas de un trapecio.

midsegment of a triangle (Lesson 8-5) The segment that connects the midpoints of the legs of a triangle.

segment medio de un triángulo El segmento que conecta los puntos medios de las patas de un triángulo.

minor arc (Lesson 10-2) An arc with measure less than 180°.

arco menor Un arco con una medida inferior a 180°.

mutually exclusive (Lesson 12-6) Events that cannot occur at the same time.

mutuamente exclusivos Eventos que no pueden ocurrir al mismo tiempo.

N

negation (Lesson 3-2) A statement that has the opposite meaning, as well as the opposite truth value, of an original statement.

negación Una declaración que tiene el significado opuesto, así como el valor de verdad opuesto, de una declaración original.

net (Lesson 2-6) A two-dimensional figure that forms the surfaces of a three-dimensional object when folded.

red Una figura bidimensional que forma las superficies de un objeto tridimensional cuando se dobla.

nonrigid motion (Lesson 8-1) A transformation that changes the dimensions of a given figure.

movimiento no rígida Una transformación que cambia las dimensiones de una figura dada.

O

octant (Lesson 9-3) One of the eight divisions of three-dimensional space.

octante Una de las ocho divisiones del espacio tridimensional.

opposite rays (Lesson 2-1) Two collinear rays with a common endpoint.

rayos opuestos Dos rayos colineales con un punto final común.

order of symmetry (Lesson 4-6) The number of times a figure maps onto itself.

orden de la simetría El número de veces que una figura se asigna a sí misma.

ordered triple (Lesson 9-3) Three numbers given in a specific order used to locate points in space.

triple ordenado Tres números dados en un orden específico usado para localizar puntos en el espacio.

orthocenter (Lesson 6-3) The point of concurrency of the altitudes of a triangle.

ortocentro El punto de concurrencia de las altitudes de un tri patas de un triángulo.

orthographic drawing (Lesson 2-6) The two-dimensional views of the top, left, front, and right sides of an object.

dibujo ortográfico Las vistas bidimensionales de los lados superior, izquierdo, frontal y derecho de un objeto.

outcome (Lesson 12-1) The result of a single performance or trial of an experiment.

resultado El resultado de un solo rendimiento o ensayo de un experimento.

P

parabola (Lesson 10-8) A curved shape that results when a cone is cut at an angle by a plane that intersects the base.

parábola Forma curvada que resulta cuando un cono es cortado en un ángulo por un plano que interseca la base.

paragraph proof (Lesson 3-4) A paragraph that explains why the conjecture for a given situation is true.

prueba de párrafo Un párrafo que explica por qué la conjetura para una situación dada es verdadera.

parallel lines (Lesson 3-7) Coplanar lines that do not intersect.

líneas paralelas Líneas coplanares que no se intersecan.

parallel planes (Lesson 3-7) Planes that do not intersect.

planos paralelas Planos que no se intersecan.

parallelogram (Lesson 7-2) A quadrilateral with both pairs of opposite sides parallel.

paralelogramo Un cuadrilátero con ambos pares de lados opuestos paralelos.

perimeter (Lesson 2-3) The sum of the lengths of the sides of a polygon.

perimetro La suma de las longitudes de los lados de un polígono.

permutation (Lesson 12-4) An arrangement of objects in which order is important.

permutación Un arreglo de objetos en el que el orden es importante.

perpendicular (Lesson 2-2) Intersecting at right angles.

perpendicular Intersección en ángulo recto.

perpendicular bisector (Lesson 6-1) Any line, segment, or ray that passes through the midpoint of a segment and is perpendicular to that segment.

mediatriz Cualquier línea, segmento o rayo que pasa por el punto medio de un segmento y es perpendicular a ese segmento.

pi (Lesson 10-1) The ratio $\frac{\text{cricumference}}{\text{diameter}}$.

pi Relación $\frac{\text{circunferencia}}{\text{diámetro}}$.

plane (Lesson 1-2) A flat surface made up of points that has no depth and extends indefinitely in all directions.

plano Una superficie plana compuesta de puntos que no tiene profundidad y se extiende indefinidamente en todas las direcciones.

plane symmetry (Lesson 11-5) When a plane intersects a three-dimensional figure so one half is the reflected image of the other half.

simetría plana Cuando un plano cruza una figura tridimensional, una mitad es la imagen reflejada de la otra mitad.

Platonic solid (Lesson 2-5) One of five regular polyhedra.

sólido platónico Uno de cinco poliedros regulares.

point (Lesson 1-2) A location with no size, only position.

punto Una ubicación sin tamaño, solo posición.

point of concurrency (Lesson 6-1) The point of intersection of concurrent lines.

punto de concurrencia El punto de intersección de líneas concurrentes.

point of symmetry (Lesson 4-6) The point about which a figure is rotated.

punto de simetría El punto sobre el que se gira una figura.

point of tangency (Lesson 10-5) For a line that intersects a circle in one point, the point at which they intersect.

punto de tangencia Para una línea que cruza un círculo en un punto, el punto en el que se cruzan.

point symmetry (Lesson 4-6) A figure or graph has this when a figure is rotated 180° about a point and maps exactly onto the other part.

simetría de punto Una figura o gráfica tiene esto cuando una figura se gira 180° alrededor de un punto y se mapea exactamente sobre la otra parte.

polygon (Lesson 2-3) A closed plane figure with at least three straight sides.

polígono Una figura plana cerrada con al menos tres lados rectos.

polyhedron (Lesson 2-5) A closed three-dimensional figure made up of flat polygonal regions.

poliedros Una figura tridimensional cerrada formada por regiones poligonales planas.

postulate (Lesson 1-1) A statement that is accepted as true without proof.

postulado Una declaración que se acepta como verdadera sin prueba.

precision (Lesson 2-7) The repeatability, or reproducibility, of a measurement.

precisión La repetibilidad, o reproducibilidad, de una medida.

preimage (Lesson 2-4) The original figure in a transformation.

preimagen La figura original en una transformación.

principle of superposition (Lesson 5-2) Two figures are congruent if and only if there is a rigid motion or series of rigid motions that maps one figure exactly onto the other.

principio de superposición Dos figuras son congruentes si y sólo si hay un movimiento rígido o una serie de movimientos rígidos que traza una figura exactamente sobre la otra.

---

prism (Lesson 2-5) A polyhedron with two parallel congruent bases connected by parallelogram faces.

prisma Un poliedro con dos bases congruentes paralelas conectadas por caras de paralelogramo.

proof (Lesson 3-4) A logical argument in which each statement is supported by a statement that is accepted as true.

prueba Un argumento lógico en el que cada sentencia está respaldada por una sentencia aceptada como verdadera.

proof by contradiction (Lesson 6-5) One assumes that the statement to be proven is false and then uses logical reasoning to deduce that a statement contradicts a postulate, theorem, or one of the assumptions.

prueba por contradicción Se supone que la afirmación a ser probada es falsa y luego utiliza el razonamiento lógico para deducir que una afirmación contradice un postulado, teorema o uno de los supuestos.

pyramid (Lesson 2-5) A polyhedron with a polygonal base and three or more triangular faces that meet at a common vertex.

pirámide Poliedro con una base poligonal y tres o más caras triangulares que se encuentran en un vértice común.

Pythagorean triple (Lesson 9-2) A set of three nonzero whole numbers that make the Pythagorean Theorem true.

triplete Pitágorico Un conjunto de tres números enteros distintos de cero que hacen que el Teorema de Pitágoras sea verdadero.

R

radian (Lesson 10-2) A unit of angular measurement equal to $\frac{180°}{\pi}$ or about 57.296°.

radián Una unidad de medida angular igual o $\frac{180°}{\pi}$ alrededor de 57.296°.

radius of a circle or sphere (Lessons 10-1, 11-4) A line segment from the center to a point on a circle or sphere.

radio de un círculo o esfera Un segmento de línea desde el centro hasta un punto en un círculo o esfera.

radius of a regular polygon (Lesson 11-2) The radius of the circle circumscribed about a regular polygon.

radio de un polígono regular El radio del círculo circunscrito alrededor de un polígono regular.

ray (Lesson 2-1) Part of a line that starts at a point and extends to infinity.

rayo Parte de una línea que comienza en un punto y se extiende hasta el infinito.

rectangle (Lesson 7-4) A parallelogram with four right angles.

rectángulo Un paralelogramo con cuatro ángulos rectos.

reduction (Lesson 8-1) A dilation with a scale factor between 0 and 1.

reducción Una dilatación con un factor de escala entre 0 y 1.

reflection (Lesson 2-4) A function in which the preimage is reflected in the line of reflection.

reflexión Función en la que la preimagen se refleja en la línea de reflexión.

regular polygon (Lesson 2-3) A convex polygon that is both equilateral and equiangular.

polígono regular Un polígono convexo que es a la vez equilátero y equiangular.

regular polyhedron (Lesson 2-5) A polyhedron in which all of its faces are regular congruent polygons and all of the edges are congruent.

poliedro regular Un poliedro en el que todas sus caras son polígonos congruentes regulares y todos los bordes son congruentes.

regular pyramid (Lesson 11-4) A pyramid with a base that is a regular polygon.

pirámide regular Una pirámide con una base que es un polígono regular.

regular tessellation (Lesson 4-5) A tessellation formed by only one type of regular polygon.

teselado regular Un teselado formado por un solo tipo de polígono regular.

relative frequency (Lesson 12-8) In a two-way frequency table, the ratios of the number of observations in a category to the total number of observations.

frecuencia relativa En una tabla de frecuencia bidireccional, las relaciones entre el número de observaciones en una categoría y el número total de observaciones.

remote interior angles (Lesson 5-1) Interior angles of a triangle that are not adjacent to an exterior angle.

ángulos internos no adyacentes Ángulos interiores de un triángulo que no están adyacentes a un ángulo exterior.

rhombus (Lesson 7-5) A parallelogram with all four sides congruent.

rombo Un paralelogramo con los cuatro lados congruentes.

rigid motion (Lesson 2-4) A transformation that preserves distance and angle measure.

movimiento rígido Una transformación que preserva la distancia y la medida del ángulo.

rotation (Lesson 2-4) A function that moves every point of a preimage through a specified angle and direction about a fixed point.

rotación Función que mueve cada punto de una preimagen a través de un ángulo y una dirección especificados alrededor de un punto fijo.

rotational symmetry (Lesson 4-6) A figure can be rotated less than 360° about a point so that the image and the preimage are indistinguishable.

simetría rotacional Una figura puede girar menos de 360° alrededor de un punto para que la imagen y la preimagen sean indistinguibles.

S

sample space (Lesson 12-1) The set of all possible outcomes.

espacio muestral El conjunto de todos los resultados posibles.

scale factor of a dilation (Lesson 8-1) The ratio of a length on an image to a corresponding length on the preimage.

factor de escala de una dilatación Relación de una longitud en una imagen con una longitud correspondiente en la preimagen.

secant (Lesson 10-6) Any line or ray that intersects a circle in exactly two points.

secante Cualquier línea o rayo que cruce un círculo en exactamente dos puntos.

sector (Lesson 11-3) A region of a circle bounded by a central angle and its intercepted arc.

sector Una región de un círculo delimitada por un ángulo central y su arco interceptado.

segment bisector (Lesson 1-7) Any segment, line, plane, or point that intersects a line segment at its midpoint.

bisectriz del segmento Cualquier segmento, línea, plano o punto que interseca un segmento de línea en su punto medio.

semicircle (Lesson 10-2) An arc that measures exactly 180°.

semicírculo Un arco que mide exactamente 180°.

semiregular tessellation (Lesson 4-5) A tessellation formed by two or more regular polygons.

teselado semiregular Un teselado formado por dos o más polígonos regulares.

sides of an angle (Lesson 2-1) The rays that form an angle.

lados de un ángulo Los rayos que forman un ángulo.

significant figures (Lesson 2-8) The digits of a number that are used to express a measure to an appropriate degree of accuracy.

dígitos significantes Los dígitos de un número que se utilizan para expresar una medida con un grado apropiado de precisión.

similar polygons (Lesson 8-2) Two figures are similar polygons if one can be obtained from the other by a dilation or a dilation with one or more rigid motions.

polígonos similares Dos figuras son polígonos similares si uno puede ser obtenido del otro por una dilatación o una dilatación con uno o más movimientos rígidos.

similar solids (Lesson 11-8) Solid figures with the same shape but not necessarily the same size.

sólidos similares Figuras sólidas con la misma forma pero no necesariamente del mismo tamaño.

similar triangles (Lesson 8-3) Triangles in which all of the corresponding angles are congruent and all of the corresponding sides are proportional.

triángulos similares Triángulos en los cuales todos los ángulos correspondientes son congruentes y todos los lados correspondientes son proporcionales.

similarity ratio (Lesson 8-2) The scale factor between two similar polygons.

relación de similitud El factor de escala entre dos polígonos similares.

similarity transformation (Lesson 8-2) A transformation composed of a dilation or a dilation and one or more rigid motions.

transformación de similitud Una transformación compuesto por una dilatación o una dilatación y uno o más movimientos rígidos.

sine (Lesson 9-5) The ratio of the length of the leg opposite an angle to the length of the hypotenuse.

seno La relación entre la longitud de la pierna opuesta a un ángulo y la longitud de la hipotenusa.

skew lines (Lesson 3-7) Noncoplanar lines that do not intersect.

líneas alabeadas Líneas no coplanares que no se cruzan.

slant height of a pyramid or right cone (Lesson 11-4) The length of a segment with one endpoint on the base edge of the figure and the other at the vertex.

altura inclinada de una pirámide o cono derecho La longitud de un segmento con un punto final en el borde base de la figura y el otro en el vértice.

slope (Lesson 3-8) The ratio of the change in the y-coordinates (rise) to the corresponding change in the x-coordinates (run) as you move from one point to another along a line.

pendiente La relación entre el cambio en las coordenadas y (subida) y el cambio correspondiente en las coordenadas x (ejecución) a medida que se mueve de un punto a otro a lo largo de una línea.

slope criteria (Lesson 3-8) Outlines a method for proving the relationship between lines based on a comparison of the slopes of the lines.

criterios de pendiente Describe un método para probar la relación entre líneas basado en una comparación de las pendientes de las líneas.

solid of revolution (Lesson 11-5) A solid figure obtained by rotating a shape around an axis.

sólido de revolución Una figura sólida obtenida girando una forma alrededor de un eje.

solving a triangle (Lesson 9-5) When you are given measurements to find the unknown angle and side measures of a triangle.

resolver un triángulo Cuando se le dan mediciones para encontrar el ángulo desconocido y las medidas laterales de un triángulo.

space (Lesson 1-2) A boundless three-dimensional set of all points.

espacio Un conjunto tridimensional ilimitado de todos los puntos.

sphere (Lesson 2-5) A set of all points in space equidistant from a given point called the center of the sphere.

esfera Un conjunto de todos los puntos del espacio equidistantes de un punto dado llamado centro de la esfera.

square (Lesson 7-5) A parallelogram with all four sides and all four angles congruent.

cuadrado Un paralelogramo con los cuatro lados y los cuatro ángulos congruentes.

statement (Lesson 3-2) Any sentence that is either true or false, but not both.

enunciado Cualquier oración que sea verdadera o falsa, pero no ambas.

straight angle (Lesson 2-1) An angle that measures 180°.

ángulo recto Un ángulo que mide 180°.

supplementary angles (Lesson 2-2) Two angles with measures that have a sum of 180°.

ángulos suplementarios Dos ángulos con medidas que tienen una suma de 180°.

surface area (Lesson 2-5) The sum of the areas of all faces and side surfaces of a three-dimensional figure.

área de superficie La suma de las áreas de todas las caras y superficies laterales de una figura tridimensional.

symmetry (Lesson 4-6) A figure has this if there exists a rigid motion—reflection, translation, rotation, or glide reflection—that maps the figure onto itself.

simetría Una figura tiene esto si existe un movimiento rígido–reflexión, una traducción, una rotación o una reflexión de deslizamiento rígida–que mapea la figura sobre sí misma.

synthetic geometry (Lesson 1-1) The study of geometric figures without the use of coordinates.

geometría sintética El estudio de figuras geométricas sin el uso de coordenadas.

T

tangent (Lesson 9-5) The ratio of the length of the leg opposite an angle to the length of the leg adjacent to the angle.

tangente La relación entre la longitud de la pata opuesta a un ángulo y la longitud de la pata adyacente al ángulo.

tangent to a circle (Lesson 10-5) A line or segment in the plane of a circle that intersects the circle in exactly one point and does not contain any points in the interior of the circle.

tangente a un círculo Una línea o segmento en el plano de un círculo que interseca el círculo en exactamente un punto y no contiene ningún punto en el interior del círculo.

tangent to a sphere (Lesson 11-4) A line that intersects the sphere in exactly one point.

tangente a una esfera Una línea que interseca la esfera exactamente en un punto.

tessellation (Lesson 4-5) A repeating pattern of one or more figures that covers a plane with no overlapping or empty spaces.

teselado Patrón repetitivo de una o más figuras que cubre un plano sin espacios superpuestos o vacíos.

theorem (Lesson 1-1) A statement that can be proven true using undefined terms, definitions, and postulates.

teorema Una afirmación o conjetura que se puede probar verdad utilizando términos, definiciones y postulados indefinidos.

transformation (Lesson 2-4) A function that takes points in the plane as inputs and gives other points as outputs.

transformación Función que toma puntos en el plano como entradas y da otros puntos como salidas.

translation (Lesson 2-4) A function in which all of the points of a figure move the same distance in the same direction.

traslación Función en la que todos los puntos de una figura se mueven en la misma dirección.

translation vector (Lesson 2-4) A directed line segment that describes both the magnitude and direction of the slide if the magnitude is the length of the vector from its initial point to its terminal point.

vector de traslación Un segmento de línea dirigido que describe tanto la magnitud como la dirección de la diapositiva si la magnitud es la longitud del vector desde su punto inicial hasta su punto terminal.

transversal (Lesson 3-7) A line that intersects two or more lines in a plane at different points.

transversal Una línea que interseca dos o más líneas en un plano en diferentes puntos.

trapezoid (Lesson 7-6) A quadrilateral with exactly one pair of parallel sides.

trapecio Un cuadrilátero con exactamente un par de lados paralelos.

trigonometric ratio (Lesson 9-5) A ratio of the lengths of two sides of a right triangle.

relación trigonométrica Una relación de las longitudes de dos lados de un triángulo rectángulo.

trigonometry (Lesson 9-5) The study of the relationships between the sides and angles of triangles.

trigonometría El estudio de las relaciones entre los lados y los ángulos de los triángulos.

truth value (Lesson 3-2) The truth or falsity of a statement.

valor de verdad La verdad o la falsedad de una declaración.

two-column proof (Lesson 3-4) A proof that contains statements and reasons organized in a two-column format.

prueba de dos columnas Una prueba que contiene declaraciones y razones organizadas en un formato de dos columnas.

two-way frequency table (Lesson 12-8) A table used to show the frequencies of data from a survey or experiment classified according to two variables, with the rows indicating one variable and the columns indicating the other.

tabla de frecuencia bidireccional Una tabla utilizada para mostrar las frecuencias de datos de una encuesta o experimento clasificados de acuerdo a dos variables, con las filas indicando una variable y las columnas que indican la otra.